S0-BPR-640

Estuarine Research

Academic Press Rapid Manuscript Reproduction

produced by the
Estuarine Research Federation
with support from the
Office of the Coastal Zone Management
National Oceanic and Atmospheric Administration
United States Department of Commerce

Estuarine Research

VOLUME I

Chemistry, Biology, and the Estuarine System

Edited by
L. Eugene Cronin
Estuarine Research Federation

Academic Press, Inc. NEW YORK SAN FRANCISCO LONDON 1975
A Subsidiary of Harcourt Brace Jovanovich, Publishers

ACADEMIC PRESS, INC.
111 Fifth Avenue, New York, New York 10003

United Kingdom Edition published by
ACADEMIC PRESS, INC. (LONDON) LTD.
24/28 Oval Road, London NW1

Library of Congress Cataloging in PUblication Data

International Estuarine Research Conference, 2d,
Myrtle Beach, S. C., 1973.
Estuarine research.

Papers presented at a conference held by the
Estuarine Research Federation and cosponsored by the
American Society of Limnology and Oceanography and the
Estuarine and Brackish Water Sciences Association.
Bibliography: p.
Includes index.
CONTENTS: v. 1. Chemistry and biology. —v. 2.
Geology and engineering.
1. Estuaries—Congresses. 2. Estuarine ocean-
ography—Congresses. 3. Estuarine biology—Congresses.
I. Cronin, Lewis Eugene, (date) II. Estuarine
Research Federation. III. American Society of
Limnology and Oceanography. IV. Estuarine and
Brackish-water Sciences Association. V. Title.
GC96.I57 1973 551.4′609 75-29370
ISBN 0-12-197501-0 (v. 1)

PRINTED IN THE UNITED STATES OF AMERICA

Contents

Preface ix
Contents of Volume II xi

Part I
Chemistry: Cycling of Elements in Estuaries

Sediment-Water Exchange in Chesapeake Bay
Owen P. Bricker, III and Bruce N. Troup 3

The Accumulation of Metals in and
Release from Sediments of Long Island Sound
John Thomson, Karl K. Turekian, and Richard J. McCaffrey 28

Role of Juvenile Fish in Cycling of
Mn, Fe, Cu, and Zn in a Coastal-Plain Estuary
F. A. Cross, J. N. Willis, L. H. Hardy, N. Y. Jones, and J. M. Lewis 45

Geochemistry of Mercury in the Estuarine Environment
Steven E. Lindberg, Anders W. Andren, and Robert C. Harriss 64

Phosphorus Flux and Cycling in Estuaries
David L. Correll, Maria A. Faust, and David J. Severn 108

Heavy Metal Fluxes Through Salt-Marsh Estuaries
Herbert L. Windom 137

Processes Controlling the Dissolved Silica Distribution in San Francisco Bay
D. H. Peterson, T. J. Conomos, W. W. Broenkow, and E. P. Scrivani 153

Processes Affecting the Composition of Estuarine Waters
(HCO_3, Fe, Mn, Zn, Cu, Ni, Cr, Co, and Cd).
J. H. Carpenter, W. L. Bradford, and V. Grant 188

Part II
Biology: Dynamics of Food Webs in Estuaries

Detritus Production in Coastal Georgia Salt Marshes
Robert J. Reimold, John L. Gallagher, Rick A. Linthurst, and William J. Pfeiffer 217

Microbial ATP and Organic Carbon in Sediments of the Newport River Estuary, North Carolina
Randolph L. Ferguson and Marianne B. Murdoch 229

Preliminary Studies with a Large Plastic Enclosure
J. M. Davies, J. C. Gamble, and J. H. Steele 251

The Detritus-Based Food Web of an Estuarine Mangrove Community
William E. Odum and Eric J. Heald 265

Sources and Fates of Nutrients of the Pamlico River Estuary, North Carolina
J. E. Hobbie, B. J. Copeland, and W. G. Harrison 287

Nutrient Inputs to the Coastal Zone: The Georgia and South Carolina Shelf
Evelyn Brown Haines 303

Population Dynamics of Zooplankton in the Middle St. Lawrence Estuary
E. L. Bousfield, G. Filteau, M. O'Neill, and P. Gentes 325

The Ecological Significance of the Zooplankton in the Shallow Subtropical Waters of South Florida
Michael R. Reeve 352

Relationship of Larval Dispersal, Gene-flow and Natural Selection to Geographic Variation of Benthic Invertebrates in Estuaries and Along Coastal Regions
Rudolf S. Scheltema 372

Geographical Distribution and Morphological Divergence in American Coastal-zone Planktonic Copepods of the Genus *Labidocera*
Abraham Fleminger 392

Nektonic Food Webs in Estuaries
Donald P. de Sylva 420

Consumption and Utilization of Food by Various Postlarval and Juvenile Fishes of North Carolina Estuaries
D. S. Peters and M. A. Kjelson 448

Some Aspects of Fish Production and Cropping in Estuarine Systems
Saul B. Saila 473

The Effects of Power Plants on Productivity of the Nekton
S. G. O'Connor and A. J. McErlean 494

Structural and Functional Aspects of a Recently Established *Zostera marina* Community
Gordon W. Thayer, S. Marshall Adams, and Michael W. LaCroix 518

Quantitative and Dynamic Aspects of the Ecology of Turtle Grass, *Thalassia testudinum*
Joseph C. Zieman 541

The Role of Resuspended Bottom Mud in Nutrient Cycles of Shallow Embayments
Donald C. Rhoads, Kenneth Tenore, and Mason Browne 563

Part III
The Estuarine System: Estuarine Modeling

A Preliminary Ecosystem Model of Coastal Georgia *Spartina* Marsh
R. G. Wiegert, R. R. Christian, J. L. Gallagher, J. R. Hall, R. D. H. Jones and R. L. Wetzel 583

The *A posteriori* Aspects of Estuarine Modeling
Robert E. Ulanowicz, David A. Flemer, Donald R. Heinle, and Curtis D. Mobley 602

Utility of Systems Models: A Consideration of Some Possible Feedback Loops of the Peruvian Upwelling Ecosystem
John J. Walsh 617

Relationship Between Morphometry and Biological Functioning in Three Coastal Inlets of Nova Scotia
K. H. Mann 634

The Estuarine Ecosystem(s) at Beaufort, North Carolina
Douglas A. Wolfe 645

CONTENTS

An Ecological Simulation Model of Narragansett Bay –
The Plankton Community
James N. Kremer and Scott W. Nixon 672

A Tophic Level Ecosystem Model Analysis of the Plankton Community
in a Shallow-water subtropical Estuarine Embayment
John Caperon 691

Educing and Modeling the Functional Relationships Within
Sublittoral Salt-marsh Aufwuchs Communities – Inside one of the Black Boxes
*John J. Lee, John H. Tietjen, Norman M. Saks, George G. Ross,
Howard Rubin, and William A. Muller* 710

Index 735

Preface

These publications are the first of a biennial series planned by the Estuarine Research Federation to present new information and concepts relating to the estuaries of the world. Volumes I and II contain the papers presented in the Second International Estuarine Research Conference, held by the Federation at Myrtle Beach, South Carolina in October of 1973. The Conference was cosponsored by the American Society of Limnology and Oceanography and by the Estuarine and Brackish Water Sciences Association.

There has been a rapid and recent increase in research on estuaries, their components and processes, and their responses to human activities. The increase has followed recognition of the exceptional value of these coastal systems, awareness of the abuse many of them have received, and expanding scientific interest in these complex and highly dynamic bodies of water which link the fresh water and the seas. As the number of persons engaged in estuarine research, and of those who wish to use the product of such research increased, so, too, did the need for improved communications among and from investigators. A small Atlantic Estuarine Research Society was organized in 1947 to provide frequent, informal exchange. In later years, the New England Estuarine Society, the South Atlantic Estuarine Research Society, and the Gulf Estuarine Research Society have emerged to serve their respective regions. All of these have joined to form the Estuarine Research Federation, an umbrella organization for the constituent societies and their 1200 members, with potential for adding additional, interested organizations. The Federation conducts and publishes biennial symposia on "Recent Advances in Estuarine Research," implements estuarine research, and provides assistance on national and international policies and practices related to estuaries.

A valuable symposium on estuaries was held under multiple sponsorship in 1964 at Jekyll Island, Georgia, and produced the classic volume Estuaries edited by George Lauff and published by AAAS. That volume was comprehensive. The Federation held its First International Conference on Long Island in 1971 but publication of papers was not feasible. The Federation recognizes that total coverage is no longer feasible at any one point in time because of the expanding production of new results of research. The Executive Board has therefore decided to select, for each biennial meeting, those topics in which major recent advances have indeed been achieved, design a symposium for their presentation and discussion, and arrange for publication. These are the first products. Volume I contains papers on *Chemistry*, focused on the Cycling of Elements and Estuaries; *Biology*, including sessions on the Dynamics of Food Webs, Nutrient Cycling, Zooplankton, Nekton, and Benthos; and *The Estuarine System*. Volume II provides publications on *Geology*, with collections on Estuaries with Small Tidal Ranges, Intermediate Tidal Ranges, and Large Tidal Ranges, and an additional section on Wide-Mouthed Estuaries. It also includes new materials on *Engineering*, with emphasis on Use of Vegetation in Coastal Engineering and on Estuarine Dredging Problems and Effects. The Third International Conference will be held by the Federation in October of 1975 at Galveston, Texas. The present publications are somewhat delayed in production, but rapid completion of future volumes is a foremost goal and commitment.

We wish to express exceptional appreciation to the conveners, chairman, and contributors, identified elsewhere, for the innovative and dedicated efforts they put into the creation and conduct of the Conference. Dr. Robert J. Reimold of the University of Georgia gave excellent supervision to the preparation and arrangement of all materials for camera-ready copy.

Quite special acknowledgment is given to the Office of Coastal Zone Management of the U.S. National Oceanic and Atmospheric Administration and its Director, Dr. Robert Knecht, for considerately administered financial support which made possible participation by scientists from distant laboratories and the preparation of final materials for publication.

L. Eugene Cronin
Chairman

Austin B. Williams

Jerome Williams

For the Editorial Committee

CONTENTS OF VOLUME II

Part I
Geology: Coarse Grained Sediment Transport and Accumulation in Estuaries

Morphology of Sand Accumulation in Estuaries: An Introduction to the Symposium
Miles O. Hayes 3

Hurricanes as Geologic Agents on the Texas Coast
Joseph H. McGowen and Alan J. Scott 23

Tide and Fair-Weather Wind Effects in a Bar-Built Louisiana Estuary
Björn Kjerfve 47

Processes of Sediment Transport and Tidal Delta Development in a Stratified Tidal Inlet
L. D. Wright and C. J. Sonu 63

Origin and Processes of Cuspate Spit Shorelines
Peter S. Rosen 77

Moveable-bed Model Study of Galveston Bay Entrance
F. A. Herrmann, Jr. 93

Simulation of Sediment Movement for Masonboro Inlet, North Carolina
William C. Seabergh 111

Sediment Transport Processes in the Vicinity of Inlets
with Special Reference to Sand Trapping
Robert G. Dean and Todd L. Walton 129

A Recent History of Masonboro Inlet, North Carolina
L. Vallianos 151

The Recent History of Wachapreague Inlet, Virginia
J. T. DeAlteris and R. J. Byrne 167

The Influence of Waves on the Origin and Development of the
Offset Coastal Inlets of the Southern Delmarva Peninsula, Virginia
V. Goldsmith, R. J. Byrne, A. H. Sallenger and David M. Drucker 183

Response Chracteristics of a Tidal Inlet: A Case Study
R. J. Byrne, P. Bullock and D. G. Tyler 201

Genesis of Bedforms in Mesotidal Estuaries
J. C. Boothroyd and D. K. Hubbard 217

Bedform Distribution and Migration Patterns on Tidal Deltas
in the Chatham Harbor Estuary, Cape Cod, Massachusetts
Albert C. Hine 235

Morphology and Hydrodynamics of the Merrimack River Ebb–Tidal Delta
D. K. Hubbard 253

Ebb-Tidal Deltas of Georgia Estuaries
George F. Oertel 267

Hydrodynamics and Tidal Deltas of North Inlet, South Carolina
R. J. Finley 277

Intertidal Sand Bars in Cobequid Bay (Bay of Fundy)
R. W. Dalrymple, R. J. Knight and G. V. Middleton 293

Sediment Transport and Deposition in a Macrotidal River
Channel: Ord River, Western Australia
L. D. Wright, J. M. Coleman and B. G. Thom 309

A Study of Hydraulics and Bedforms at the Mouth of the Tay Estuary, Scotland
Christopher D. Green 323

Circulation and Salinity Distribution in the Rio Guayas Estuary, Ecuador
S. Murray, D. Conlon, A. Siripong and J. Santoro 345

Tidal Currents, Sediment Transport and Sand Banks in Chesapeake Bay Entrance, Virginia
John C. Ludwick 365

High-Energy Bedforms in the Non-tidal Great Belt Linking North Sea and Baltic Sea
Friedrich Werner and Robert S. Newton 381

Part II
Engineering: 1) Use of Vegetation in Coastal Engineering

The Influence of Environmental Changes in Heavy Metal Concentrations on *Spartina alterniflora*
William M. Dunstan and Herbert L. Windom 393

Biotic Techniques for Shore Stabilization
Edgar W. Garbisch, Paul B. Woller, William J. Bostian, and Robert J. McCallum 405

Salt-Water Marsh Creation
E. D. Seneca, W. W. Woodhouse and S. W. Broome 427

Submergent Vegetation for Bottom Stabilization
Lionel N. Eleuterius 439

Vegetation for Creation and Stabilization of Foredunes, Texas Coast
B. E. Dahl, Bruce A. Fall and Lee C. Otteni 457

Management of Salt-Marsh and Coastal-Dune Vegetation
D. S. Ranwell 471

Some Estuarine Consequences of Barrier Island Stabilization
Paul J. Godfrey and Melinda M. Godfrey 485

Where Do We Go From Here?
Donald W. Woodard 517

2) Estuarine Dredging Problems and Effects

An Overview of the Technical Aspects of the Corps of
Engineers National Dredged Material Research Program
C. J. Kirby, J. W. Keeley and J. Harrison 523

Aspects of Dredged Material Research in New England
Carl Hard 537

Effects of Suspended and Deposited Sediments on
Estuarine Environments
J. A. Sherk, J. M. O'Connor and D. A. Neumann 541

Water-Quality Aspects of Dredging and Dredge-Spoil
Disposal in Estuarine Environments
Herbert L. Windom 559

Meiobenthos Ecosystems as Indicators of the Effects of Dredging
Willis E. Pequegnat 573

Index 585

PART I

CHEMISTRY: CYCLING OF ELEMENTS IN ESTUARIES

Convened By:
James H. Carpenter
Division of Chemical Oceanography
School of Marine and Atmospheric Science
University of Miami
10 Rickenbacker Causeway
Miami, Florida 33149

SEDIMENT-WATER EXCHANGE IN CHESAPEAKE BAY

Owen P. Bricker III[1]

and

Bruce N. Troup[1]

ABSTRACT

In Chesapeake Bay, diagenetic reactions in the sediment-interstitial water environment result in the enrichment of many dissolved species relative to their concentrations in the overlying water. Rapid exchange of dissolved species across the sediment-water interface in response to physical and chemical processes leads to the establishment of strong gradients in these species in the upper meter of the sediment column. In spite of the non-equilibrium nature of the overall system, the concentrations of species that participate in reactions whose time scales are rapid relative to their transport through the system, can be adequately described in terms of equilibrium models. The concentrations of chemically non-reactive dissolved materials and species that are involved in slow reactions relative to movement through the system, must be modeled on the basis of non-equilibrium diffusional transport.

INTRODUCTION

Estuaries are complex systems which receive chemical inputs from a variety of different sources (Fig. 1). River run-off contributes dissolved species derived from chemical weathering of rocks in the watershed, suspended material from mechanical weathering of terrigenous matter, and dissolved and particulate organic material of biogenic origin. The influx of sea water provides a strong

1. Department of Earth and Planetary Sciences, The Johns Hopkins University, Baltimore, Maryland 21218.

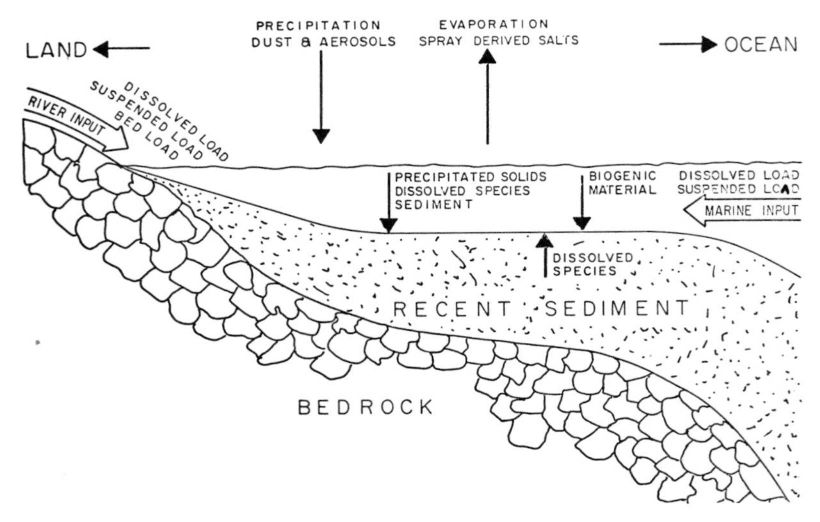

Figure 1. Exchanges of matter in an idealized estuarine system.

electrolyte solution of nearly constant relative composition with respect to the major ions Ca^{++}, Mg^{++}, Na^{+}, K^{+}, Cl^{-}, and $SO_4^{=}$, dissolved and particulate organic material, and suspended sediment. Superimposed on the natural sources are inputs resulting from the activities of man.

The estuary is an open system in which the inputs are balanced by outputs in the form of flow through the system and sinks within the system. Some of the more soluble elements pass through the estuary to the ocean essentially unchanged, whereas others combine and precipitate as solid phases or settle to the bottom sorbed on particulate matter to be stored in the sediment reservoir. Still other materials leave via atmospheric pathways. The biogeochemical characteristics of each element determine the type of behavior exhibited in its passage through the estuarine system. The balance between the inputs and the outputs of the elements, together with the physical flow regimen, determines the overall chemistry of the estuary.

The least understood component of the system is the influence that sediment reservoir has on the chemistry of estuarine waters. The sediment and its contained pore waters may be either a source of dissolved species or a sink for them, depending upon the reactions that take place between pore fluids and sediment and the direction of transfer of material across the sediment-water interface.

We have been investigating the chemistry and mineralogy of the interstitial water-sediment system of the Chesapeake Bay in an attempt to define the reactions that occur in this system, to evaluate their diagenetic significance, and to assess their effect on the overlying waters of the bay in response to transfer of material across the sediment-water interface.

THE CHESAPEAKE BAY ESTUARY

The Chesapeake Bay is one of the world's major estuarine systems. It dominates the mid-Atlantic seaboard and is the receiving basin for rivers draining much of southern New York state, Pennsylvania, Maryland, and Virginia (Figs. 2a and 2b). The Susquehanna River supplies nearly 90% of the fresh-water input to the bay above the mouth of the Potomac River, and approximately 50% of

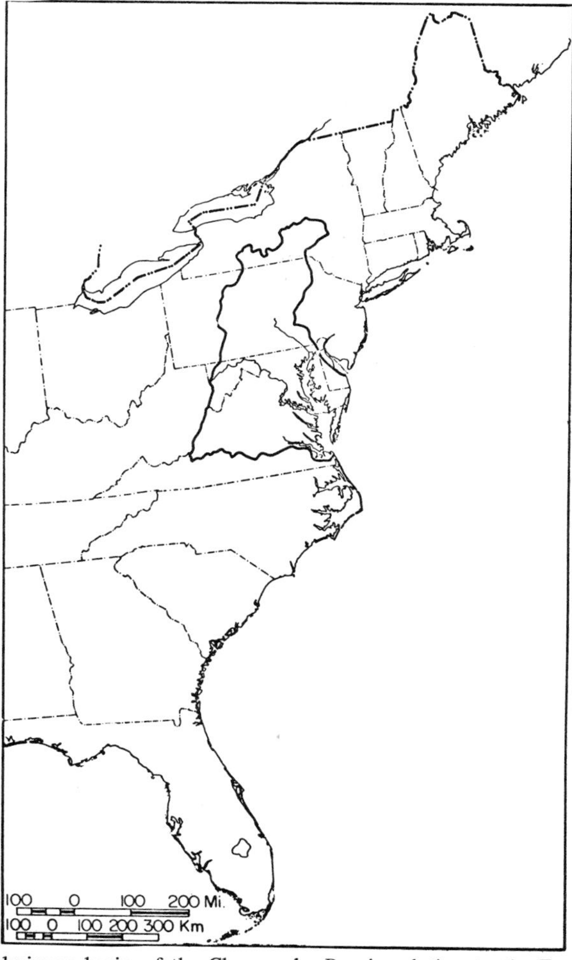

Figure 2a. The drainage basin of the Chesapeake Bay in relation to the East coast of the United States.

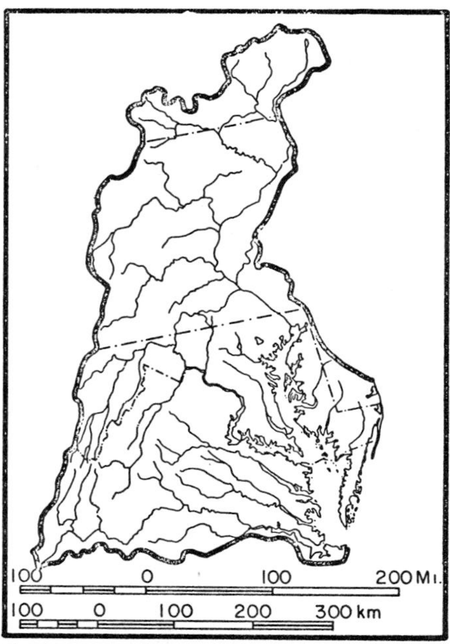

Figure 2b. The drainage basin of the Chesapeake Bay.

the fresh-water input to the entire bay. The Potomac and James rivers supply most of the additional fresh-water input (Table 1). Annual mean discharge into the bay from the entire watershed is approximately 2200 meters^3/sec (23 and 24). The water discharged by the rivers carries dissolved and particulate materials to the bay. Particulate matter is also derived from shoreline erosion (17 and 20).

In addition to the river-borne dissolved constituents, tidal mixing transports sea salts into the estuarine system. The seasonal salinity distribution in the bay, which results from the mixing of river water with sea water, has been well documented by the Chesapeake Bay Institute (18). Salinity varies from less than $1^{o}/oo$ in the vacinity of the mouth of the Susquehanna, to nearly normal sea-water salinity at the mouth of the bay. In early spring, when river-discharge is high, salinities in the bay are at their lowest. As the ratio of river water to sea water decreases throughout the summer and fall, salinities increase up the bay. On an annual basis, the estuarine system is filled approximately half by sea water and half by river water.

Solids carried to the bay by rivers consist of highly weathered materials from the watersheds. The coarser sediments collect near the mouths of the rivers, and the bay proper receives predominantly fine silt and clay-sized particles. Most of the silt-sized fraction is quartz, which is also present in the clay-sized range, but declines in abundance as size decreases. The finer fractions consist primarily

TABLE 1

River discharge to Chesapeake Bay*

River	Discharge m^3/s
Susquehanna	1141
Potomac	388
James	284
All others	300

*Wilson and others, 1967; USGS Estimated Stream Discharge entering Chesapeake Bay, 1950-1972.

of illite and chlorite, with traces of kaolinite and montmorillonite. Small amounts of presumably authigenic phosphates, carbonates, and sulfides are also present. Additionally, rivers transport to the bay large amounts of iron in the form of hydrous oxide coatings on mineral grains.

Schubel(17) has estimated that the mean sedimentation rate in the bay is on the order of 2-3 mm/year. Estimates of sedimentation rates in the bay may be misleading, however, as flood discharge from tropical storm Agnes deposited in one week an amount of sediment equivalent to perhaps 100 years of normal sedimentation. Even though they do not occur often, high-intensity storms such as Agnes are probably much more important factors in the deposition of sediment in the bay than are the normal daily transport and accumulation mechanisms. The bottom of Chesapeake Bay is blanketed by up to 60 meters of recent sediment. Data concerning the distribution of sediment types and the geometry of the sediment cover, however, are limited (2, 15, and 17).

Sediment-Water Exchange

The nature of the sediment-interstitial water environment is determined by a number of factors. As sediment accumulates, bay water is entrained and trapped in the interstices between sediment particles. Initially, then, the interstitial water has the same composition as the bottom water at that locality in the bay. As sedimentation continues, the water and sediment that were at the top of the sediment column become increasingly deeply buried beneath the sediment-water interface. There is no longer free exchange with the overlying water, and chemical and biological processes begin to exert an influence on the composition of both the sediment and its contained water. If the organic content is large, as is usually the case with estuarine sediments, bacterial activity establishes strongly reducing conditions within a short distance below the sediment-water interface.

Hydrous iron and manganese oxide coatings on mineral grains carried to the bay by rivers become unstable and dissolve. Sulfate is reduced to sulfide during bacterial oxidation of organic material with a concomitant production of bicarbonate and release of phosphate. Dissolved iron may react with sulfide, phosphate, or carbonate to form new minerals in the sediment-pore water system. Clays undergo a shift in the relative proportions of cations occupying exchange sites (14) and may undergo exchange of iron for magnesium in their structures (5).

As a result of reactions such as these, the chemical environment of the pore water is very different from that of the overlying water, even at shallow depths beneath the sediment-water interface. If the sediment-pore water system were completely isolated from exchange with the water column upon burial, the bulk composition of the system would remain constant and changes would be limited to recombinations of the species present in the solids and aqueous phase. However, in the high porosity sediments of the upper meter, diffusion of dissolved materials readily occurs both within the sediment column and across the sediment-water interface. In addition to diffusional transport, the activities of burrowing benthic organisms, the movement of sediment by storm-generated waves and currents, and disturbance of the sediment by internally generated gas bubbles resulting from the decay of organic matter, may all contribute locally to the transfer of material across the sediment-water interface. The net result of these processes is that gradients in the concentrations of dissolved species are established in the pore water of the sediment column. Thus, in order to understand the chemistry of the estuarine system, it is necessary to have a knowledge of (a) the reactions occurring in the sediment reservoir that take up material from, or release material to, the interstitial waters, and (b) the mechanisms and rates of transfer of material across the sediment-water interface. Equilibrium models are useful in identifying the reactions occurring in the sediment reservoir, and diffusion models are useful in determining the rates of transfer of materials within the sediment and across the sediment-water interface.

The Sediment-Interstitial Water System

In order to understand the reactions occurring in the system, it is necessary to know something of the chemical composition of the aqueous phase and the nature of the solid phases present. Our sampling program was primarily oriented toward obtaining data on the interstitial-water compositions of the upper meter of sediments along the salinity gradient of the Chesapeake Bay. At each sampling locality the gross mineralogy of the sediments was also determined, and detailed examinations for certain minor phases were made.

Samples were taken along the axis of the bay from Howell Point (Station

922) to Wolf Trap Lighthouse (Station 724) and, in addition, station 856 was sampled monthly for a period of a year (Fig. 3). In all the sampling operations a modification of the Benthos gravity-coring apparatus was used.

Ten samples from each core were examined. The upper ten centimeters of cores were sampled in successive 2-cm layers and the rest of the core was sampled in 5- to 8-cm layers at depths of 15, 20, 40, 70, and 100 cm. These sample intervals were chosen on the basis of observed gradients in dissolved species with depth in our initial cores.

The water content of the upper meter of sediments ranges from nearly 85% (wt%) at the sediment-water interface to approximately 60% at a depth of one meter. Although their water content is very high, these sediments are sufficiently

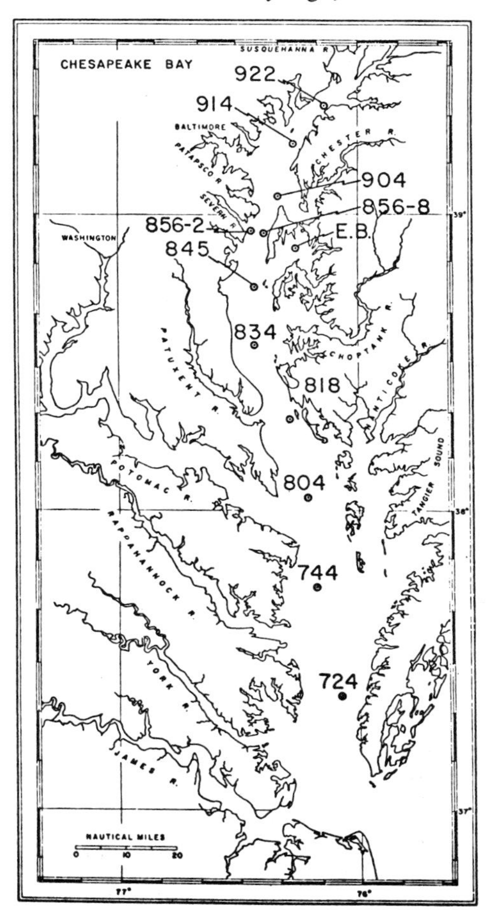

Figure 3. Sampling stations along the salinity gradient of the Chesapeake Bay.

cohesive to allow dissection and sampling without destruction of their structures.

The samples whose interstitial water was to be analyzed were extruded directly from the core into nylon squeezers of the type described by Reeburgh (12). The pressure of nitrogen gas acting against a rubber diaphragm compressed the sediment and expressed interstitial water through a 0.22 μ m membrane filter. In this operation a portion of the water sample was diverted directly from the squeezer into an electrode cell for measurement of pH, $pS^{=}$, platinum electrode potential, and temperature. Another portion of the sample was collected in tubes containing acidified o-phenanthroline for analysis of iron. The reminder of the sample was collected in a polyethylene bottle from which aliquots were removed for shipboard analysis of other dissolved constituents. The samples were then acidified and refrigerated for onshore analysis of the major cations. Our analytical techniques are summarized in Table 2.

TABLE 2

Analytical Techniques

Type	Species
Colorimetric	Phosphate: molybedenum blue Ferrous iron: o-phenanthroline
Titrimetric	Carbonate alkalinity Chlorinity Sulfate
Potentiometric	pH Sulfide: Orion $Ag^{+}/S^{=}$ electrode Eh: Pt electrode
Atomic absorption	Mg^{2+}, Ca^{2+}
Flame emission	Na^{+}, K^{+}

At the start of our interstitial-water sampling, we could not obtain reproducible results for iron and phosphate. We attributed this to the effect of temperature and/or pressure changes existing between the bay bottom and the shipboard laboratory. The squeezing of duplicate sediment samples at different temperatures and pressures failed to produce any differences in the behavior of iron and phosphate. Further investigation of this phenomenon disclosed that ferrous iron was oxidized during the short period the sediment samples were exposed to the atmosphere during the extruding and squeezing operations. The resulting ferric compound removed phosphate and trace metals, which led to erroneously low values for these species and iron.

We subsequently developed a procedure for handling the core and carrying out all sample manipulations under an inert atmosphere (3). This procedure permited reducing waters, rich in ferrous iron, to be sampled and analyzed without losses of dissolved species caused by oxidation effects (Fig. 4).

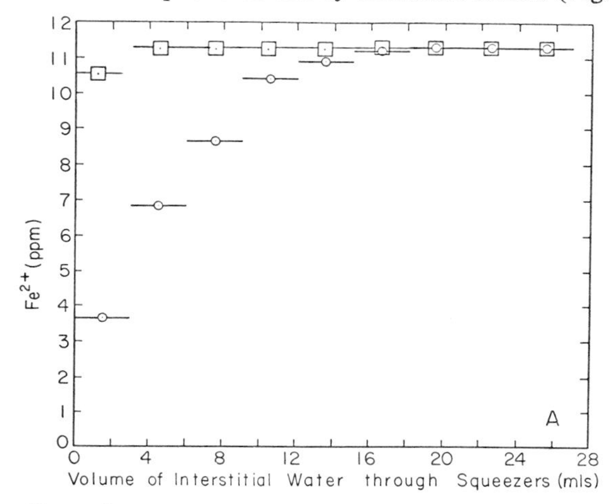

Figure 4. The effect of oxygen on the concentration of ferrous iron in squeezed interstitial-water samples. Each data point represents a concentration of a 3 ml aliquot. A single sediment sample was homogenized and halved. One half (⊙) was squeezed exposed to the atmosphere; the second half (⊡) was squeezed in a glove bag filled with nitrogen gas. The maximum value was 4 ppm.

Interpretation of the reactions between the aqueous phase and solids requires knowledge of the activities of the aqueous species. In order to evaluate the activity of a dissolved species it is necessary to know not only its concentration but also the ionic strength of the solution in which it occurs. In the Chesapeake Bay the major ionic species in the overlying waters and in the sediment pore waters are Na^+, Mg^{++}, Ca^{++}, K^+, and Cl^-. The ionic strength of the waters is largely determined by these ions, with some additional contribution from $SO_4^=$ and HCO_3^-. The waters, therefore, were analyzed for these species to provide the information necessary to permit realistic calculations of chemical equilibria in the system.

We used the iron-phosphate-bicarbonate system as an example to illustrate the application of equilibrium modeling to the Chesapeake Bay system. The analytical data for iron, phosphate, and carbonate alkalinity in cores taken at stations along the axis of the bay are shown in Figures 5 and 6. The most striking features of these data are the steep gradients observed in the upper part of the sediment column. The gradients imply fluxes of material both within the sediment column and across the sediment-water interface, pointing up the potential significance of the bottom sediments as a source for chemical species in

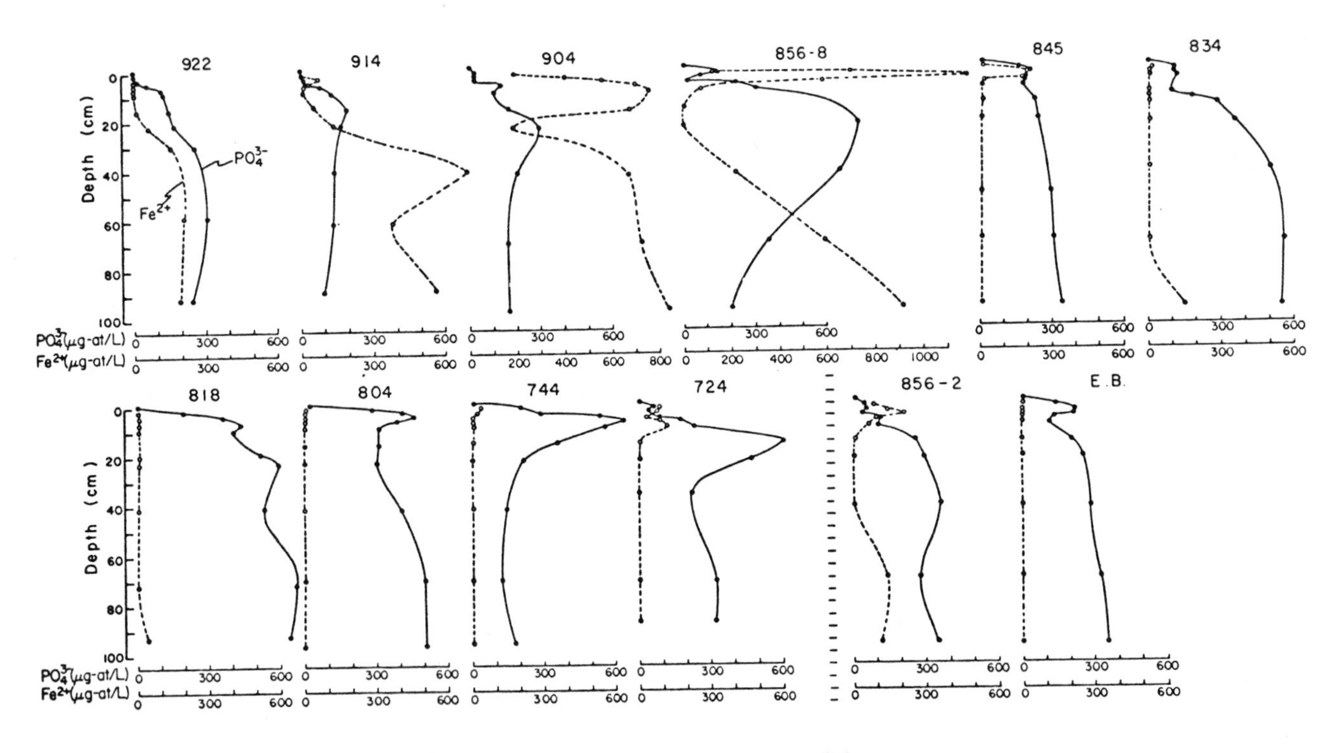

Figure 5. Reactive phosphate and dissolved ferrous iron in the interstitial water of the upper meter of sediments in Chesapeake Bay, August, 1972.

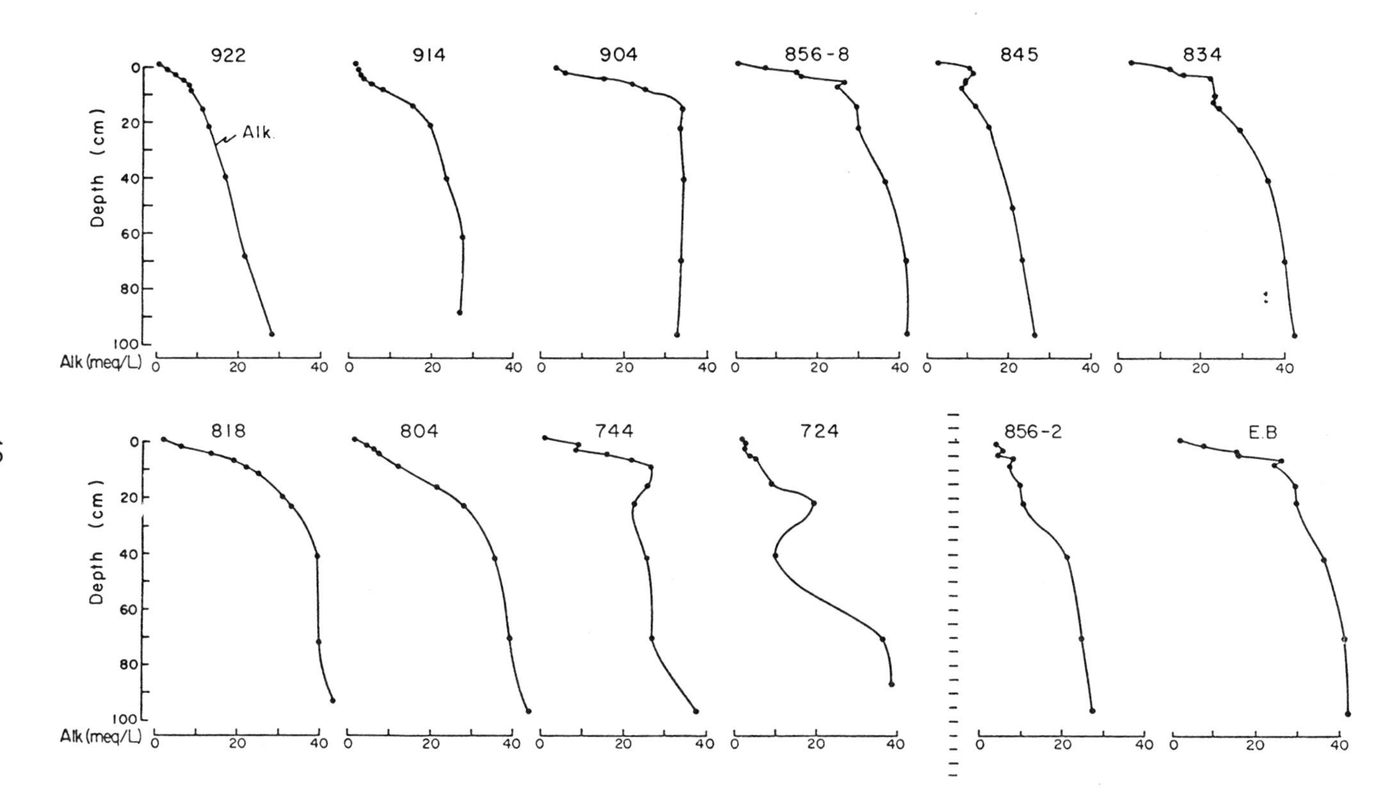

Figure 6. Carbonate alkalinity in the interstitial water of the upper meter of sediments in Chesapeake Bay, August, 1972.

the estuary. The observed changes in concentration suggest that the interstitial-water reservoir behaves as an open system. Even though the steep gradients imply fluxes of material across the sediment-water interface, species that undergo chemical reactions whose rates are rapid with respect to the time scale of transport of these species through the system may approach equilibrium in the system. The concentrations of species that behave in this manner will be controlled at each point in the system by the reaction and can be modeled in terms of the appropriate equilibrium expression.

Equilibrium Modeling

Chemical equilibrium can be closely approached only if the residence times of the reacting species are long enough to allow completion of kinetically limiting reactions. For homogeneous aqueous solutions the kinetics of the reactions are so fast that equilibrium is usually closely approached. A good example is the equilibrium model developed by Garrels and Thomson (7), which accurately predicts the distribution of dissolved species in sea water.

The kinetics of reactions between solid phases and solutions are slower than those of homogeneous reactions; therefore, longer residence times are required if equilibrium is to be approached. It has usually been found that overlying waters are rarely in equilibrium with the sediments below them because of the short time the overlying water is in contact with the sediment. In contrast, interstitial waters are ideal natural environments in which to investigate equilibrium between solids and aqueous phases because of the long residence times of the aqueous phase. Furthermore, heterogeneous chemical reactions may approach equilibrium faster in interstitial water because of the large ratio of solid surface area to water volume in fine-grained sediments.

Simplifying assumptions are necessary if equilibrium thermodynamics are applied to natural-water systems. A model (MODAC) developed to compute the activities of various dissolved species and ion activity products of possible authigenic phases in the interstitial water of the upper meter of sediments of the Chesapeake Bay illustrates the typical assumptions and calculations inherent to the technique.

In MODAC the sediments were assumed to be an isobaric (1 atm), isothermal (25°C) system containing pure, stoichiometric solid phases. The deepest sediment core was sampled in 30 m of water; effect that the resultant 3 atm difference between the *in situ* pressure and the assumed pressure has on the chemical equilibria is so small as to be negligible. The temperature gradient in the upper meter of sediments was small, and the temperature averaged 24°C. Although many of the solid phases of the sediments are probably not pure, X-ray diffraction patterns of our sediment samples indicated that essentially pure, crystalline siderite and vivianite were present.

It is impossible to measure individual ion activity coefficients, but they can be estimated through the use of the Debye-Huckel approximation. There are several forms of the Debye-Huckel equation involving different numbers of fitted coefficients from which activity coefficients can be calculated. The form of the Debye-Huckel equation used in MODAC is

$$\log \gamma_i = -\frac{A z_i^2 \sqrt{I}}{1 + aB\sqrt{I}} \tag{1}$$

where A and B are constants for a given solvent at a specified temperature, a is the "effective diameter" of hydrated ions (a fitted parameter determined by Kielland), z_i the charge of the ion, γ_i the activity coefficient, I the ionic strength equal to $\frac{1}{2} \Sigma c_i z_i^2$, where c_i is the concentration of the ion i. Equation 1 represents the activity coefficients "with very good accuracy" up to an ionic strength of about 0.1 (13). The calculated ionic strengths in the northern bay range from 0.02 to 0.36. The use of the Debye-Huckel equation for ionic strengths up to 0.36 results in a maximum error of 5% in the activity coefficient for monovalent ions and 13% for divalent ions, errors which approximate the accuracy of activity coefficients in a diluted sea-water medium of I = 0.4.

Computation of the ionic strength, which is required for the calculation of activity coefficients, is facilitated in the sediment pore waters of Chesapeake Bay by the constant cation-to-chlorinity ratio in almost all the samples. At all stations, except for the one at the mouth of the Susquehanna River, the concentrations of the major cations reflected those of a diluted sea-water medium and could be predicted from the chlorinities. Thus, except for the station at the mouth of river, the contribution of the major cations to the ionic strength was calculated from the chlorinity. A significant fraction of the total sulfate and bicarbonate concentrations consists of ion pairs which reduce the contribution of sulfate and bicarbonate species to the ionic strength. To compensate for this, it was assumed that only one-half of the analytical concentration of sulfate and one-tenth of the bicarbonate concentration contributed to the ionic strengths. Although this method of calculating ionic strengths is not exact, the difference between those calculated in MODAC and those rigorously calculated did not exceed 2% along the salinity gradient of the bay.

Calculation of the distribution of the ferrous iron, phosphate, and carbonate species involves the manipulation of three mass balance equations. For pH's < 7.7 and $I < .39$, which were the maximum values observed in Chesapeake Bay interstitial waters, the carbonate alkalinity can be expressed solely in terms of the bicarbonate ion.

$$CA = (HCO_3^-)_T = (HCO_3^-)_f + (CaHCO_3^+) + (MgHCO_3^+) + (NaHCO_3^0) \tag{2}$$

The mass-action equations for the ion pairs are

$$K_a = \frac{a_{CaHCO_3^+}}{a_{Ca^{2+}}\, a_{HCO_3^-}} = \frac{\gamma_{CaHCO_3^+}(CaHCO_3^+)}{\gamma_{Ca^{2+}}(Ca^{2+})\,\gamma_{HCO_3^-}(HCO_3^-)_f} \tag{3}$$

$$K_b = \frac{a_{MgHCO_3^+}}{a_{Mg^{2+}}\, a_{HCO_3^-}} = \frac{\gamma_{MgHCO_3^+}(MgHCO_3^+)}{\gamma_{Mg^{2+}}(Mg^{2+})\,\gamma_{HCO_3^-}(HCO_3^-)_f} \tag{4}$$

$$K_c = \frac{a_{NaHCO_3^o}}{a_{Na^+}\, a_{HCO_3^-}} = \frac{\gamma_{NaHCO_3^o}(NaHCO_3^o)}{\gamma_{Na^+}(Na^+)\,\gamma_{HCO_3^-}(HCO_3^-)_f} \tag{5}$$

Substitution of the mass action equations for the ion pairs into equation 2, use of the assumption that the activity coefficients of singly charged 1:1 ion pairs and the bicarbonate ion are equal and those of neutral ion pairs are unity (7), and subsequent rearrangement yields

$$(HCO_3^-)f = \frac{CA}{1 + K_a\gamma_{Ca^{2+}}(Ca^{2+}) + K_b\gamma_{Mg^{2+}}(Mg^{2+}) + K_c\gamma_{Na^+}\gamma_{HCO_3^-}(Na^+)} \tag{6}$$

Which can be solved directly from the carbonate alkalinity, the concentrations of the cations, and the published stability constants of the ion pairs (19).

The mass-balance equations for ferrous iron and phosphate are

$$Fe(II)_T = (Fe^{2+})_f + (FeHPO_4^o) + (FeH_2PO_4^+) + (FeCl_2^o) + (FeCl^+) + (FeOH^+) + (FeSO_4^o) \tag{7}$$

$$P_T = (HPO_4^{2-})_f + (FeHPO_4^o) + (CaHPO_4^o) + (MgHPO_4^o) + (NaHPO_4^-) + (H_2PO_4^-)_f + (FeH_2PO_4^+) + (MgH_2PO_4^+) \tag{8}$$

Equations 7 and 8 can be rearranged and solved similarly to equation 6.

Solution of equations 6, 7, and 8 yields the free concentrations of ferrous iron, phosphate, and bicarbonate. The product of the free concentrations and the appropriate single-ion activity coefficients gives the activity of each of the species.

The equilibrium state between sediment and interstitial waters depends on the stability of various solid phases. Carbonate, phosphate, sulfide, and hydroxyl ion are the ligands with which iron commonly precipitates in natural-water systems. The activities of sulfide in the interstitial waters of northern Chesapeake Bay were so exceedingly small that, in effect, there was no free sulfide in them. Therefore, ferrous sulfide is not an important regulator of the solubility of ferrous iron. At the observed concentrations of bicarbonate and phosphate, it can be shown thermodynamically that ferrous hydroxide is unstable with respect to both siderite and vivianite in the interstitial waters of the bay. Thus, in the sediments of the northern bay, carbonate and phosphate are only two ligands capable of significantly affecting the solubility of ferrous iron.

The calculated activities of ferrous iron, phosphate, and carbonate were used to derive the ion activity products of various solid phases including vivianite and siderite:

$$\log IAP_{viv} = \log a^3_{Fe^{2+}} a^2_{PO_4^{3-}} \tag{9}$$

$$\log IAP_{sid} = \log a_{Fe^{2+}} a_{CO_3^{2-}} \tag{10}$$

For convenience in comparing the saturation states of solids containing more than one cation and one anion, a log-saturation function can be used to weight the deviation of the IAP from the solubility product, K_{so}:

$$\log \text{saturation} = \frac{2}{n + m} \log \frac{a^n_{Fe^{2+}} a^m_{X^{-\frac{Zn}{m}}}}{K_{so}(Fe_nX_m)} \tag{11}$$

Figure 7 is a plot of the log-saturation values of siderite and vivianite computed from the composition of the interstitial-water samples. Each level of several representative cores along the length of the Bay is plotted. (Note: the upper six points are from the upper 14 cm of the cores, the lower four points are from the lower 86 cm of the cores.) Positive log-saturation values indicate supersaturation, and negative values indicate undersaturation: For both siderite and vivianite the interstitial waters of the southern stations are just saturated with respect to the two solid phases. For both minerals the northern stations are slightly supersaturated, indicating that these minerals should be found in the sediments.

To investigate the equilibrium between these two minerals, which possibly control the iron concentration in the interstitial waters, the activity of carbonate was plotted against the activity of phosphate for station 856-8. The theoretical boundary describing the equilibrium between siderite and vivianite is given by:

$$Fe_3(PO_4)_2 \cdot 8H_2O + 3CO_3^{2-} = 3FeCO_3 + 2\,PO_4^{3-} + 8H_2O \tag{12}$$

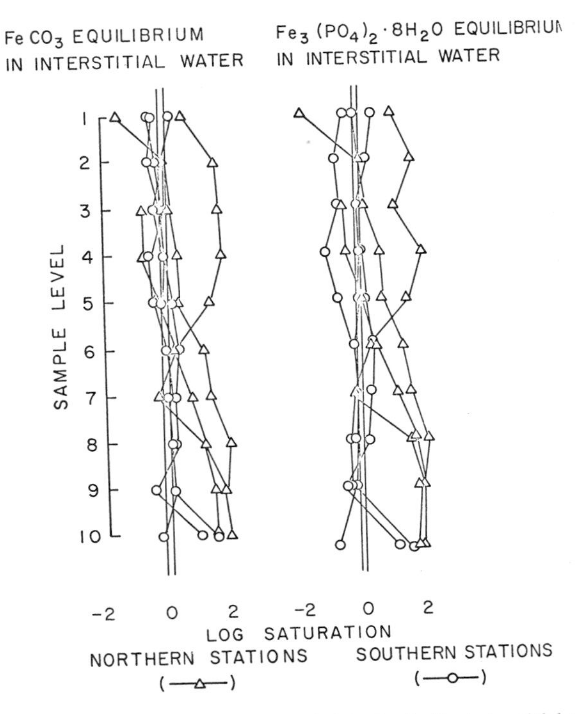

Figure 7. Log-saturation plots of siderite and vivianite in the interstitial water at northern and southern stations in the Chesapeake Bay.

$$K = \frac{a^2{}_{PO_4{}^{3-}}}{a^3{}_{CO_3{}^{2-}}} \tag{13}$$

$$\log a_{CO_3{}^{2-}} = 2/3 \log a_{PO_4{}^{3-}} + 1/3 \log K \tag{14}$$

Equation 14 was constructed on Figure 8, and the activities of the interstitial-water samples from each level of two representative cores at station 856 were plotted on the figure. The close correspondence between the data points and the theoretical equilibrium line suggests that these pore waters are in equilibrium with both siderite and vivianite.

Many investigators have shown various natural-water systems to be supersaturated with respect to certain solid phases, the implication being that the solids do indeed precipitate and influence the chemistry of the aqueous phase. However, it is not reasonable to assume that the compositions of the

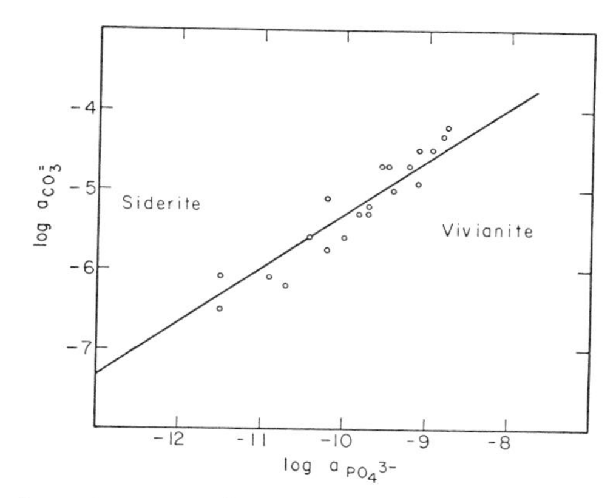

Figure 8. Thermodynamic stability diagram for siderite, vivianite, and solution. The points represent the calculated activities of phosphate and carbonate from two cores at station 856-8. The straight line is a theoretical boundary separating the areas in which the two minerals predominate.

waters are controlled by the predicted solids without verifying the existence of these solids in the sediments.

Using X-ray diffraction techniques, we identified vivianite and siderite in the sediments of the northern Chesapeake Bay. Examination of sediments for easily oxidizable phases such as vivianite requires that the sample have minimum contact with oxygen. The effect of oxygen on reduced sediment is illustrated in Figure 9, which shows three X-ray diffraction traces of the same sample run consecutively over a period of 30 minutes. The trace on the left was run five minutes after the slide was prepared, the second trace, three minutes later, and the final trace, 20 minutes after the second. Within the span of time needed for the three runs, the vivianite peak at $12.8^{\circ}2\theta$ disappeared into the background.

The successful prediction of the presence of vivianite and siderite clearly indicates the advantages of an equilibrium model. Although the sediment-interstitial water system is not strictly a closed system, the equilibrium model does appear to adequately describe the behavior of these phases in the system.

The fact that equilibrium may exist at all levels in a core between the solution and a solid phase does not require that the activities of the species in solution be fixed, but only that the mass-action equation be satisfied at all levels. If concentration differences exist between levels, diffusion will occur.

Diffusion

The simplest model of diffusive transport involves only molecular diffusion, which is described by Fick's first and second laws of diffusion:

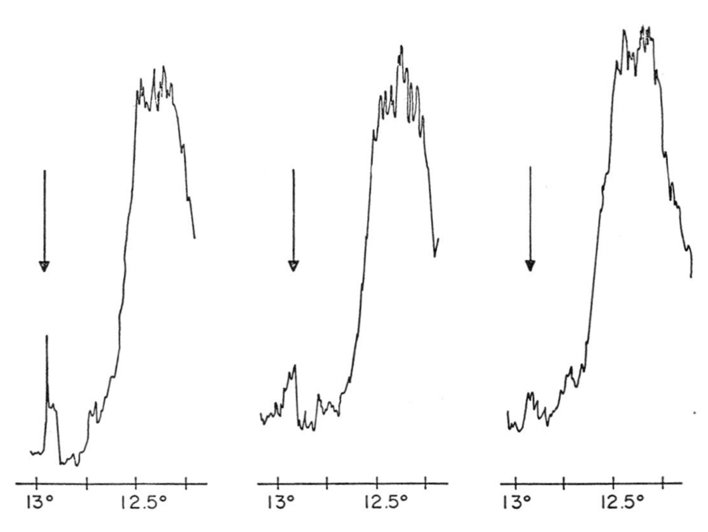

Figure 9. Consecutive X-ray diffraction spectra of a single sample from station 914 for the angular span of 12.0-13.2°2θ. The arrows indicate the major peaks associated with vivianite.

$$J = -D \frac{\partial c_i}{\partial z} \quad (15)$$

$$\frac{\partial c_i}{\partial t} = \frac{\partial^2 c_i}{\partial z} \quad (16)$$

where c_i is the concentration of the species i, z is depth, D is the diffusion coefficient, J is the flux of i through a plane normal to z, and t is time. Since spatial variations x, y are generally negligible compared to vertical variations, the diffusion equations can be expressed açcurately in terms of depth z only.

The total variation of the concentration of i with respect to time caused by all physical processes is given by:

$$\frac{dc_i}{dt} = \left(\frac{\partial c_i}{\partial t}\right)_z + w\left(\frac{\partial c_i}{\partial z}\right)_t \quad (17)$$

where $\left(\frac{\partial c_i}{\partial t}\right)_z$ is the local time rate of change of the concentration of i, $\left(\frac{\partial c_i}{\partial z}\right)_t$ is the concentration gradient of i at time t, and w is the rate of change of depthof a layer with time. During the time sediments are deposited any given layer is progressively buried deeper. Thus, w represents a net rate of deposition, which can be measured empirically (1). The last term in equation 17 becomes important for long-term processes in sediments.

When the species i is involved in a chemical reaction, the total variation of the concentration of i with time is the sum of the effect of the chemical reaction on the concentration of i as well as the effect of diffusion. This can be expressed mathematically as:

$$\frac{dc_i}{dt} = D \frac{\partial^2 c_i}{\partial z^2} + \sum_j r_{ij} \tag{18}$$

where $_j r_{ij}$ is the sum of the rates of change of the concentration of i with respect to time due to the reactions j (1). Substitution of equation 18 into equation 17 and rearrangement yields the local time rate of change of the concentration of i:

$$\left(\frac{\partial c_i}{\partial t}\right)_z = D \frac{\partial^2 c_i}{\partial z^2} + \sum_j r_{ij} - w\left(\frac{\partial c_i}{\partial t}\right)_t \tag{19}$$

The relative importance of diffusion across the sediment-water interface can be determined by studying an ion that does not participate in any chemical reactions in the sediments, in which case the $\Sigma_j r_{ij}$ term in equation 19 disappears. Such an ion is chloride which is not involved in mineral equilibria or ion-exchange reactions in brackish-water sediments.

Many investigators have successfully modeled chloride diffusion across the sediment-water interface in fresh-water lakes or in the sea with constant overlying water chlorinity (9, 22, 6) or with monotonically increasing chlorinity over long periods of time (10). Few have attempted to model an estuarine system in which there are seasonal variations in the overlying water chlorinity such as those we see in the Chesapeake Bay.

Scholl and Johnson (16) developed a model for chloride diffusion across the sediment-water interface in an estuarine system closely analogous to a heat-conduction model described in Carslaw and Jaeger(4). The chlorinity boundary at the sediment-water interface is a sinusoidal function that oscillates seasonally about a mean chlorinity value. For the short time scale relevant to seasonal transport in estuaries the last term in equation 19 can be neglected. The chlorinity-depth distributions generated by the Scholl and Johnson model closely mimic typical chlorinity distributions observed in Chesapeake Bay sediments (Fig. 10).

As can be seen in this yearly cycle of the distribution of chlorinity with depth at station 858-8, which was typical, significant monthly variations extend to 10 cm, while variations for the entire year extend to about 30 cm. These observations agree well with the expected thickness of a layer responding to diffusive transport from the overlying water calculated from equation 20:

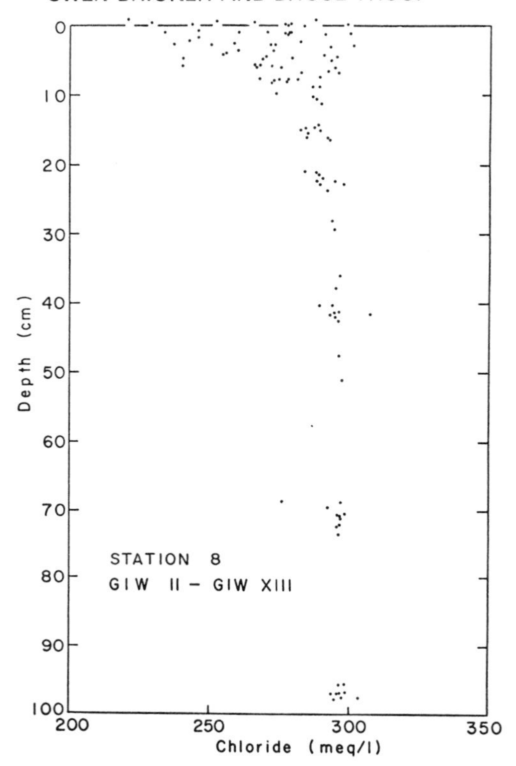

Figure 10. The distribution with depth of chlorinity from 12 cores analyzed at monthly intervals at station 856-8.

$$\text{thickness of "diffusive layer"} = \sqrt{D \cdot t} \tag{20}$$

Using a diffusion coefficient of 3 x $10^{-5}cm^2sec^{-1}$ for the choloride ion in unconsolidated sediments (10) and a time scale of a month, the thickness of the diffusive layer is 9 cm. For a time scale of a year, the thickness is 31 cm.

The rapid exchange of chloride across the sediment-water interface can also be illustrated by data taken immediately after tropical storm Agnes in 1972. After Agnes, record amounts of fresh water were discharged into the Chesapeake Bay. Figure 11 shows the excellent correlation between the minimum and maximum chlorinities in the overlying water and the upper 10 cm of interstitial water, which occurred in July and August of 1972, respectively. The chlorinity changes are most pronounced in the upper level of the core (0-2 cm) and are dampened with depth.

Calculation of the flux of chemically reactive species, such as iron and phosphate, across the sediment-water interface involves the use of reaction terms in the diffusion equation 18. The distributions with depth of iron and phosphate

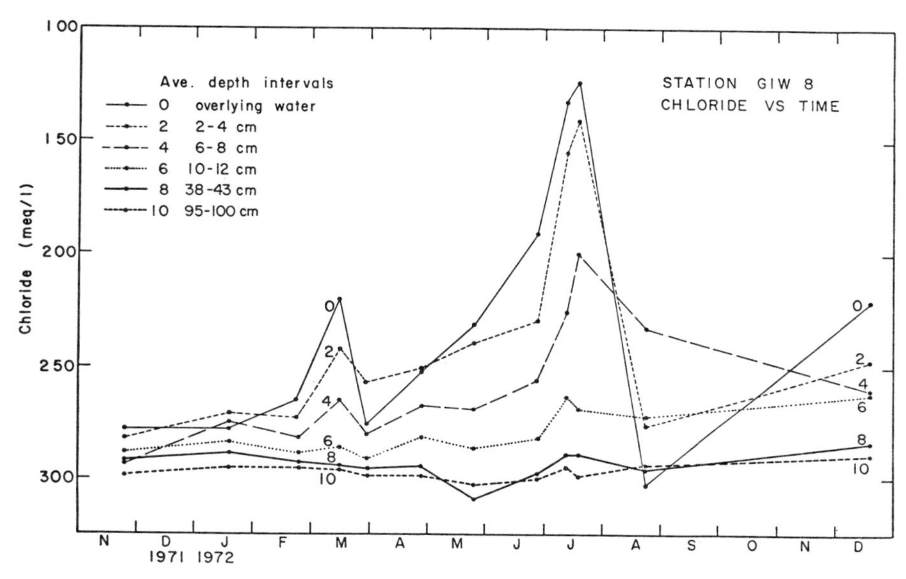

Figure 11. The relation of the chlorinities in the sediment column to the chlorinities of the overlying water at station 856-8 for a one-year period.

are determined primarily by diffusion, the rate of organic decomposition, and the rate of controlling mineral-equilibrium reactions. As discussed above, an important controlling mineral-equilibrium reaction for phosphate is the vivianite equilibrium. Thus, for phosphate the $\sum_j r_{ij}$ term in equation 18 represents the sum of the rates of production of phosphate from organic decomposition and the rate of precipitation or dissolution of vivianite:

$$\frac{dc_P}{dt} = D \frac{\partial^2 c_P}{\partial z^2} + \sum_j r_{P,j} \tag{21}$$

where

$$\sum_j r_{P,j} = \left(\frac{dc_P}{dt}\right)_{\substack{\text{organic}\\ \text{decomposition}}} + \left(\frac{dc_P}{dt}\right)_{\text{vivanite}} \tag{22}$$

If the dissolution and precipitation of vivianite is diffusion-controlled rather than surface-reaction-controlled, the last term in equation 22 can be expressed as a function of the departure from equilibrium:

$$\left(\frac{dc_P}{dt}\right)_{\text{vivianite}} = k(c_P^{eq} - c_P) \tag{23}$$

where c_P^{eq} is the phosphate concentration in equilibrium with vivianite, c_P is the actual concentration in the interstitial waters, and k may be a rate constant or a function of particle size. Equation 23 describes a process in which the larger the departure from equilibrium, the faster equilibrium is approached.

An apparent flux of phosphate across the sediment-water interface in the Chesapeake Bay has been calculated by Bray et al (3) to be 9×10^{-3} μ g-atom/cm^2/wk; in a week, a flux of this magnitude would transport enough phosphate across the sediment-water interface to increase the total phosphate content of the water column by 5%. However, this calculation represents a maximum value since reactions occurring at the sediment-water interface were not considered. Mortimer (11) has pointed out that the existence of an oxidized layer inhibits sediment-water exchange because of coprecipitation reactions. In iron-rich environments the adsorption of phosphate on hydrous ferric oxides in an oxidized layer may greatly reduce the flux of phosphate (3).

The large apparent fluxes of ferrous iron across the sediment-water interface implied by the pronounced concentration gradients found in the upper 10 cm of Chesapeake Bay sediments (Figure 5) are, of course, gross overestimates. Almost all of the ferrous iron diffusing upwards is precipitated upon contact with oxygen. The temporal variation of ferrous iron at any point in the sediment column, therefore, is the sum of several different chemical reactions as well as diffusion:

$$\frac{dc_{Fe}}{dt} = D\frac{\partial^2 c_{Fe}}{\partial z^2} + \sum_j r_{Fe,j} \tag{24}$$

where

$$\sum_j r_{Fe,j} = \left(\frac{dc_{Fe}}{dt}\right)_{reduction} + \left(\frac{dc_{Fe}}{dt}\right)_{vivianite} + \left(\frac{dc_{Fe}}{dt}\right)_{iron\ sulfide} + \left(\frac{dc_{Fe}}{dt}\right)_{siderite} + \left(\frac{dc_{Fe}}{dt}\right)_{oxidation} \tag{25}$$

The vivianite, siderite, and iron sulfide equilibrium terms can be expressed analogously to equation 23, the reduction term represents the rate of dissolution of limonitic ferric oxide coatings from fine-grained sediment particles as a result of oxygen consumption in the sediments, and the oxidation term is the rate expression found by Stumm and Lee (21):

$$\left(\frac{dc_{Fe}}{dt}\right)_{oxidation} = -k[C_{Fe}]\ [O_2]\ [OH^-]^2 \tag{26}$$

The insertion of exponential terms such as those found in equations 23 and 26 makes the overall diffusion equation non-linear. As a consequence of the increased complexity introduced by these terms and the lack of data on the kinetics of many pore water-sediment reactions, rigorous solution of equation (25) is not possible at the present time. The equation does indicate, however, the type of data needed to model the exchange of chemically reactive species across the sediment-water interface and within the sediment column.

SUMMARY

We have attempted to illustrate the general behavior of the sediment-interstitial water environment and its importance to the chemistry of the estuarine system. In the Chesapeake Bay diagenetic reactions in the pore water-sediment environment result in many dissolved species being enriched by one or more orders of magnitude above that of their concentrations in the overlying water. Rapid exchange between overlying water and the upper part of the sediment column leads to the establishment of strong gradients in dissolved species extending to depths of at least 30 cm beneath the sediment-water interface.

Dissolved species are exchanged across the sediment-water interface in response to physical and chemical processes in the estuary. Scouring of sediments by tidal- or storm-induced currents and bioturbation of the bottom by benthic infauna lead to direct mixing and exchange of waters across the interface. Diffusion arising from concentration differences above and below the interface may result in transport through the sediment column and across the interface of such non-reactive species as chloride. Chemically reactive species are diffused in response to concentration differences coupled with the chemical reactions that take place in the system.

Because of their localized nature in time and space, it is difficult to evaluate the relative importance of mechanical and biological transport mechanisms in the transfer of dissolved species across the sediment-water interface. Movement of non-reacting species in estuarine systems can be reasonably well approximated in terms of a simple diffusion model using an oscillating boundary to mimic seasonal variations in the composition of overlying water. In order to model the diffusion of chemically reactive species in the estuarine system, the controlling chemical reactions and their rates must be known.

Equilibrium models can be useful in identifying chemical reactions whose rates are rapid relative to the time scale of exchange and transport of participating dissolved species through the estuarine system. These models allow the concentrations of dissolved species involved in kinetically rapid reactions to be predicted at any point in the estuarine system where such reactions occur. Chemically reactive dissolved species that participate in reactions whose time scales are slower than the time scale of movement of the dissolved species through the system are controlled by non-equilibrium transport processes. The concentration of these species must be described by more complex models involving diffusion and chemical reaction terms. Data of this type for systems as complex as the estuarine pore-water environment are not yet available; consequently the concentrations of species behaving in this manner cannot be accurately modeled.

ACKNOWLEDGEMENT

We thank John Bray, Virginia Grant, Rich Holdren, Mimi Uhlfelder, and the crew of the R/V *Ridgely Warfield* for their help in collecting and analyzing the samples. This work was supported by the Atomic Energy Commission, AEC AT(11-1)-3292.

REFERENCES

1. Berner, R. A.
 1971 **Principles of chemical sedimentology.** McGraw-Hill, New York. 240 p.
2. Biggs, R. B.
 1967 The sediments of Chesapeake Bay. In **Estuaries,** pp. 239-260. (ed. Lauff, G. H.), Amer. Asso. Adv. Sci. Publ. 83. Washington, D. C.
3. Bray, J. T., Bricker, O. P., and Troup, B. N.
 1973 Phosphate in interstitial waters of anoxic sediments: oxidation effects during sampling procedure. **Science,** 180: 1362-1364.
4. Carslaw, H. S. and Jaeger, J. C.
 1959 **Conduction of heat in solids.** 2nd ed. Oxford University Press, Oxford, England, p. 51-67.
5. Drever, J. I.
 1971 Magnesium-iron replacement in clay minerals in anoxic sediments. **Science,** 172: 1334-1336.
6. Duursma, E. K. and Rosch, C. J.
 1970 Theoretical, experimental and field studies concerning diffusion of radioisotopes in sediments and suspended particles of the sea. B. Methods and experiements. **Netherl. Jour. Sea Res.,** 4: 395-469.
7. Garrels, R. M. and Thompson, M. E.
 1962 A chemical model for sea water at 25°C and one atmosphere total pressure. **Amer. Jour. Sci.,** 260: 57-66.
8. Kielland, J.
 1937 Individual activity coefficients of ions in aqueous solutions. **Jour. Am. Chem. Soc.,** 59: 1675-1678.
9. Lerman, A.
 1971 Time to chemical steady-states in lakes and ocean. In **Nonequilibrium Systems in Natural Water Chemistry,** (ed. Hem, J. D.). Advances in Chemistry Series 106, Am. Chem. Soc., Washington, D. C., p. 30-76.
10. Lerman, A. and Weiler, R.
 1970 Diffusion and accumulation of chloride and sodium in Lake Ontario sediment. **Earth Plan. Sci. Letts..,** 10: 150-156.
11. Mortimer, C. H.

1971 Chemical exchanges between sediments and water in the Great Lakes – speculations on probable regulatory mechanisms. **Limnol. Oceanogr.,** 16: 387-404.

12. Reeburgh, W. S.
1967 An improved interstitial water sampler. **Limnol. Oceanogr..,** 12: 163-165.

13. Robinson, R. A. and Stokes, R. H.
1970 **Electrolyte Solutions.** 2nd ed. (rev.). Butterworth & Co., London. 571 p.

14. Russell, K. L.
1970 Geochemistry and halmyrolysis of clay minerals, Rio Ameca, Mexico. **Geochim. Cosmochim. Acta.,** 34: 893-907.

15. Ryan, J. D.
1953 The sediments of the Chesapeake Bay. Dept. of Geol. Mines and Water Resources Bull. 12: 120 p. Md. Board of Natural Resources.

16. Scholl, D. W. and Johnson, W. L.
1967 Effects of diffusion on interstitial water chemistry in deltaic areas. Seventh International Sedimentological Congress.

17. Schubel, J. R.
1968 Suspended sediments of the northern Chesapeake Bay. Chesapeake Bay Inst. Tech. Rept. 35: 264 p. Reference 68-2. The Johns Hopkins University.

18. Seitz, R. C.
1971 Temperature and salinity distributions in vertical sections along the longitudinal axis and across the entrance of the Chesapeake Bay (April 1968 to March 1969). Graphical Summary Rept. No. 5, Ref. 71-7, Chesapeake Bay Inst., The Johns Hopkins University, 99 p.

19. Sillén, L. G. and Martell, A. E.
1964 **Stability constants of metal ion complexes.** Special publ. No. 17, The Chem. Soc. of London. 754 p.

20. Singewald, J. T. and Slaughter, T. H.
1949 Shore erosion in tidewater Maryland. Maryland Dept. of Geol., Mines, and Water Resources, Bull. 6, 140 p.

21. Stumm, W. and Lee G. F.
1961 Oxygenation of ferrous iron. **Ind. Eng. Chem.,** 53: 143.

22. Tzur, Y.
1971 Interstitial diffusion and advection of solute in accumulating sediments. **Jour. Geophys. Res.,** 76: 4208-4211.

23. U. S. Geological Survey
1950-1973 Estimated stream discharge entering Chesapeake Bay, Distributed by District Chief, USGS, Parkville, Maryland 21234.

24. Wilson, A. and others.
1967 River discharge to the sea from the shores of the conterminous United States. Hydrologic Investigations Atlas, HA-282. U. S. Geological Survey.

THE ACCUMULATION OF METALS IN AND RELEASE FROM SEDIMENTS OF LONG ISLAND SOUND

John Thomson, Karl K. Turekian and Richard J. McCaffrey[1]

ABSTRACT

Detailed analyses of a short diver-obtained core and of a long gravity core along a line south of New Haven harbor were used to determine the modes and rates of accumulation of metals over time in sediments of central Long Island Sound and the patterns of their release. The Pb^{210} dating of the long core shows an increase in Zn, Cu, Pb, Hg and Mn (the metals determined) over the last 70 years. This is presumed to be the consequence of human activity around the Sound. The short diver-obtained core was analyzed for Pb^{210}, Ra^{226}, U^{234}, U^{238}, Th^{228}, Th^{230} and Th^{232} as well as Cu, Pb, Zn, Cd and Mn. Interpretation of the sedimentary and chemical data shows that about 16 years ago there was an episodic deposition of approximately 18 cm of sediment with subsequent bioturbation of the upper 10 cm and retention of stratification in the lower portion. It also shows loss of U and Ra^{228} (and, by analogy, probably of Ra^{226}) and gain of Pb^{210}, Th^{228}, and, possibly, some Mn since the sediment load was deposited. The metals Cu, Pb and Zn showed no loss from the sediment.

INTRODUCTION

Marked increases in the concentrations of some metals at the tops of sediment cores relative to deeper parts have been noted from many nearshore marine and lake deposits (2, 3, 4, 10, and 12). These changes may be generally attributed to man's activities. The ambient metal supply to these reservoirs may be augmented

1. Department of Geology and Geophysics, Yale University, New Haven, Connecticut 06520.

by an increased man-induced burden created by the following processes: (a) aggravated erosion and leaching of soil, (b) injection into streams and the atmosphere of industrial waste and fuel-combustion products, (c) release to sediment of metal-enriched sewage sludge, either directly or via dredge-spoil dumping.

Particularly in the marine environment the effect of this increased burden of metal supply may have influences far beyond the point of injection. An important question for both marine geochemists and environmentalists is the extent to which metals deposited in estuaries, from whatever source, are remobilized and released to the water column.

Simple chemical arguments based on the reducing nature of estuarine sediments imply that the presence of sulfide ions in the pore waters of the sediments determines the mobility of metals. Metals that form highly insoluble sulfide compounds have little chance of leaving the sediment. On the other hand, metals that have relatively soluble sulfides can be expected to migrate up the sediment column until they encounter a new chemical environment which then sets the limits on mobility. On the basis of the solubility product constants for their sulfide compounds, Pb, Co, Ni, Hg, Ag, Cu, and Zn, for example, remain essentially fixed in reducing marine sediments whose pore waters have a sulfide ion activity of 10^{-9} moles per liter (Table 1). Under these conditions manganese and iron are readily mobilized, since MnS and FeS have fairly high solubilities (Table 1).

TABLE 1

Expected concentrations of metals in pore waters whose sediments have a sulfide ion activity of 10^{-9} moles per liter at 25°C.

Metal	Log concentration (moles/liter)
Iron	− 6.4
Manganese	− 2.6
Copper	−26.0
Zinc	−14.1
Cadmium	−16.2
Mercury	−43.7
Lead	−16.6

In order to determine the time scale of events that influence the observed metal distribution in the sediment column, a reliable chronology is required. This can sometimes be obtained by consulting historical arguments or by observing any exotic material in the column that has a unique time association. Chronological methods based on Pb^{210} (half life 22.26 yr) and Th^{228} (half life 1.91 yr) have also been shown to be useful (6, 7, and 8). Ideal sedimentary environments for dating by these radioactive methods are ones in which sedimentation is continuous and has neither major surges of sediment supply nor acute losses of sediment. In estuaries the effects of climate – or man-induced episodic deposits or erosion are often visible. If these events are frequent in relation to the time scale measurable by the particular radionuclide used, an average sedimentation rate for a core can be obtained, but the data will show greater scatter than would the ideal continuous sediment accumulation pattern.

On the other hand, a large-scale episodic deposition that results from natural or man-induced causes provides an opportunity to trace the chemical history of a deposited sediment pile almost simultaneously with its adjustment to the physical and bioligical environment of the estuary. In this case, members of the uranium and thorium decay series are useful not so much for dating as for assessing the behavior of metals during the readjustment process.

In an attempt to understand the accumulation pattern of metals in some sediments of Long Island Sound and their release patterns during the operation of typical estuarine biological, chemical, and physical processes, we have combined chemical, radiochemical, and sedimentological data. The region of Long Island Sound that we studied is typical of the silty-clay sedimentary environment of that estuary. Farther to the east of this location the bottom becomes primarily sandy. The results we obtained cannot be assumed to be typical of all estuarine sediment types in Long Island Sound or elsewhere.

METHODS

To ensure that we had complete sections with intact sediment-water interfaces, divers collected three immediately adjacent cores 12 km directly south of New Haven (Figure 1). The cores were deep-frozen soon after recovery, and the longest of the three (18 cm) was completely used up for the chemical and radiochemical work. This core was cut into seven 2.5–cm sections. Each segment of measured volume was weighed after drying at 110°C to obtain densities, and again after ignition at 500°C to obtain a rough measure of organic content.

Aliquots of 8 g from each section were leached with 300 ml of hot $1\bar{N}$ HNO_3 for three hours. The filtered leachate was split, one portion being taken for radiochemical analysis, and a smaller portion for Mn, Cu, Zn, Cd, and Pb determination by atomic absorption. U^{238} and U^{234}, Th^{232} and Th^{230} were

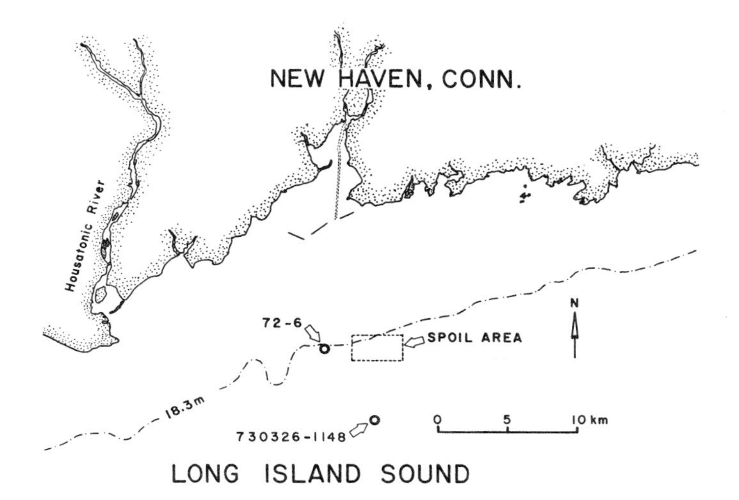

Figure 1. Long Island Sound, showing where the short cores obtained by divers (72-6) and the long-gravity core (730326-1148) were taken.

determined by the U^{232} - Th^{228} isotopic dilution α -spectrometer technique (9). A separate aliquot of each section was treated the same way, but without the U^{232}-Th^{228} spike, to determine the Th^{228}/Th^{232} isotope ratio. Ra^{226} was measured on the spiked aliquots by gas phase α scintillation counting of the Ra^{222} daughter, and Pb^{210} by low-level anticoincidence β^- counting (11). A concordant uraninite was used in cross calibration of spike specific activities and counting efficiencies for the various isotopes (D. P. Kharkar and J. Thomson, unpublished data).

The second longest diver core (13 cm) was X-rayed to reveal sedimentary structures presumed typical of the general area of the three cores, and then preserved for further sedimentologic studies.

A gravity core, 76 cm long, was raised by conventional shipboard methods some 6 km southeast of the site of the short cores (Fig. 1). Two-centimeter segments were processed and analyzed for Mn, Cu, Zn and Pb by the method cited above, and separate samples of the original core material were analyzed for Hg by the method of Applequist et al. (1). A comparison of 1N HNO_3 leached samples and totally dissolved samples indicated that 85-90% of the total concentration of the trace metals analyzed were released by the 1 $\overline{N}$ HNO_3 method, with no trend in release pattern such as might occur through diagenesis with depth.

Starting with the 2-to-4-cm segment every fourth segment of the samples of the original core material was analyzed for Pb^{210} (using 6N HC1 leaching) to obtain the data for determining the average rate of sediment accumulation of this core. A few Ra^{226} measurements at several depths were also made.

In the case of the shipboard conventional gravity core, there is no certainty that the absolute top of the core was obtained but the tops of the diver-driven cores were assured by visual observation at the time of collection.

RESULTS

The X-radiography of the long gravity core revealed shells, shell fragments, and occasional layering in the upper 30 cm or so, and occasional layering with few shells below that depth. In this it is representative of many cores taken in central Long Island Sound (D. C. Rhoads, personal communication). The data for this core are presented in Table 2. As shown in Figure 2, a mean (Pb^{210}) excess sedimentation rate of 0.45 cm/yr indicates that the shell-rich zone is a phenomenon of the last 70 years. The Cu, Zn, Hg, and Pb data along the core (Fig. 3) show that these metals also increase from low values in the bottom 45 cm to higher values towards the present time over the top 30 cm. The change does not appear to be continuous but rather to approximate a stepped increase towards the surface – a feature compatible with the concept of episodic sediment supply. The manganese concentration has a uniform low concentration from depth to about 12 cm, where it steps to a uniform higher concentration. This difference between the behavior of manganese and that of other metals will be discussed later.

The analytical data from the 18-cm short core are presented in Table 3. At face value, the (Pb^{210}) excess data indicate a sediment accumulation rate of 0.64 cm/yr (Fig. 4). In spite of this approximation to the rate observed for the long core, we suspect that this simple interpretation of the Pb^{210} data is deceptive because of the physical characteristics of the short cores.

On collection, the upper 2.54 cm of the cores were grey and indicative of oxidation, while the remainder were black and indicative of reducing conditions. An X-radiograph of the next longest diver core (Fig. 5) reveals stratification between 10 cm and the base at 13 cm and, in the upper section, horizontal layering that can be attributed to bioturbation. The 10-cm break is presumed to be present in the analyzed core as well. The analytical results can be viewed in light of this boundary between the upper bioturbated 10 cm and the lower stratified 8 cm.

The Ra^{226}/Th^{230}, Th^{230}/Th^{232} and U^{234}/U^{238} activity ratios are constant over the entire length of the core with average values of 0.35, 0.80 and 1.00, respectively. The concentrations of Th^{232} (and Th^{230}) and the trace elements Cu, Zn, Cd and Pb also show no trend related to the 1-cm break.

As we have seen, the Pb^{210} decreases from the top of the core. The two deepest samples in the stratified region of the core have the lowest Pb^{210} activities and, within two sigma errors, are identical. The anomolously high

TABLE 2

76 cm Gravity Core (41°06'N, 72°53.1'W; depth 28 m.)

depth cm	Mn ppm	Cu ppm	Zn ppm	Pb ppm	Hg ppm	Pb^{210} dpm/g	$(Pb^{210})^*_{ex}$ dpm/g	Ra^{266} dpm/g
0-2	619.	67.	176.	59.	0.24			
2-4	627.	67.	175.	57.	0.32	5.08	4.08	0.31
4-6	640.	70.	174.	56.	0.27			
6-8	552.	73.	160.	51.	0.21			
8-10	544.	75.	175.	55.	0.24			
10-12	403.	67.	158.	46.	–	2.92	1.92	
12-14	358.	45.	123.	29.	0.22			
14-16	349.	44.	120.	31.	0.24			
16-18	315.	35.	113.	25.	0.24			
18-20	326.	28.	98.	22.	–	2.14	1.14	
20-22	350.	29.	102.	23.	–			
22-24	336.	25.	91.	21.	0.21			
24-26	325.	14.	72.	13.	–			
26-28	332.	17.	78.	9.	–	1.85	0.85	0.17
28-30	317.	20.	81.	5.	–			
30-32	327.	8.	56.	4.	–			
32-34	336.	10.	63.	5.	0.06			
34-36	337.	11.	64.	7.	–	1.39	0.39	
36-38	302.	8.	55.	6.	–			
38-40	322.	8.	59.	8.	–			
40-42	294.	8.	49.	7.	–			
42-44	305.	6.	54.	10.	0.04	1.07	–	
44-46	297.	7.	55.	11.	–			
46-48	298.	8.	51.	12.	–			
48-50	306.	6.	49.	6.	–			
50-52	312.	5.	50.	7.	–	0.70	–	0.56
52-54	303.	5.	48.	7.	0.03			
54-56	307.	5.	49.	8.	–			
56-58	306.	5.	49.	7.	–			
58-60	284.	4.	46.	6.	–	1.13	–	
60-62	324.	8.	58.	10.	–			
62-64	333.	18.	77.	13.	0.10			
64-66	305.	8.	58.	8.	–			
66-68	–	–	–	–	–	1.00	–	
68-70	296.	6.	49.	8.	–			
70-72	304.	6.	50.	8.	–			
72-74	263.	6.	54.	8.	0.03			
74-76	301.	6.	55.	7.	–	1.06	–	

All measurements were made and reported on material ashed at 500°C except Hg, which was analyzed dry (90°C) and reported on 500°C basis. Organic content is 5-6% based on loss of weight on ignition.

*corrected for a background level of 1.0 dpm/g.

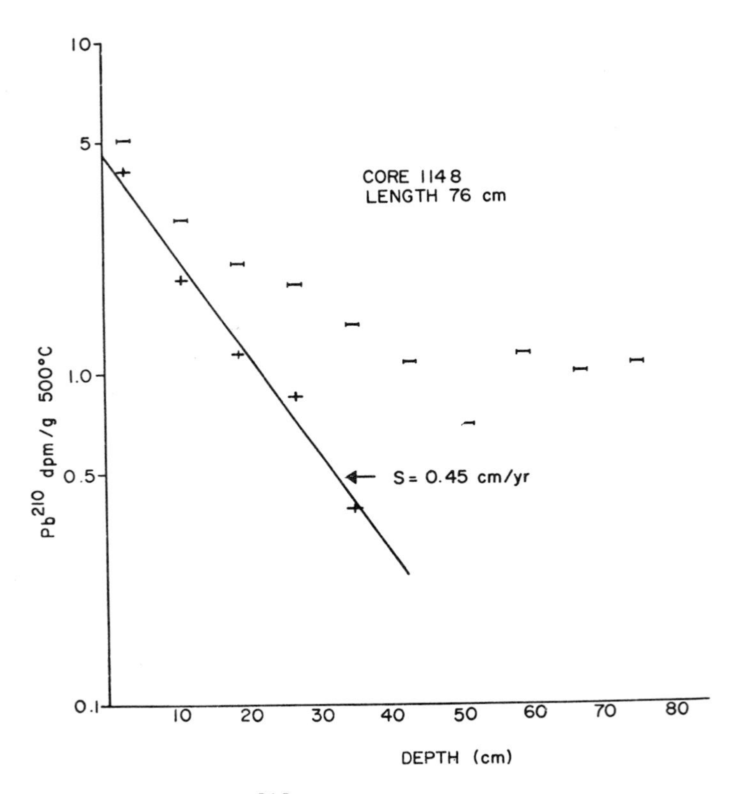

Figure 2. The distribution of Pb^{210} in the long core and the inferred sedimentation rate of sediments at this site. The correction for supported Pb^{210} is made using the virtually constant value of 1 dpm/g found in the deeper parts of the core where the excess unsupported Pb^{210} is presumed to have decayed away. The actual Ra^{226} measurements at three sites in this core are variable and lower than 1 dpm/g. Pb^{210} excess lines corrected for supported Pb^{210} based on Ra^{226} activities equal to 0.3 and 0.6 dpm/g (not shown) yield accumulation rates of 0.72 and 0.61 cm/yr respectively. Because the Pb^{210} data (determined by β-counting) may have calibration and background problems detracting from accuracy we prefer the rate calculated from the Pb^{210} data alone.

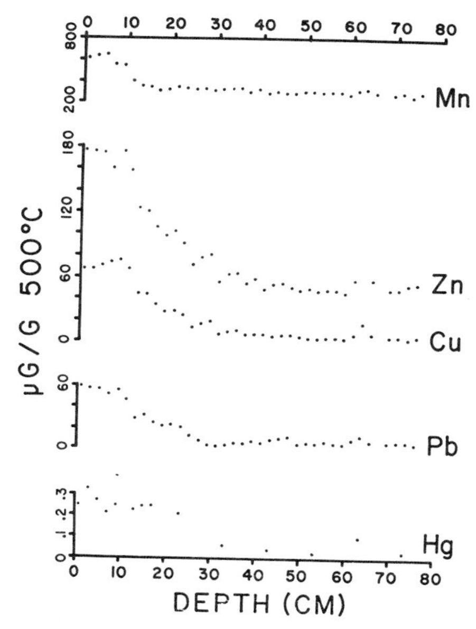

Figure 3. The distribution of Cu, Zn, Hg, Pb and Mn with depth in the long core.

Pb^{210} activity of the 10-12.5 cm interval fits neither the almost monotonic decrease with depth of the bioturbated layer nor the constant low value of the stratified layer.

The two nuclides that do show interesting variations with depth are U^{238} and Th^{228}. Uranium concentration increases monotonically with depth to the 10-cm boundary, after which a constant value typical of the stratified layer is attained. The behavior of Th^{228} is more complex and is best understood when normalized against the Th^{232} activity. Th^{232} decays to Ra^{228} (half life 6.7 yr), which then decays to Th^{228} (half life 1.9 yr). Ra^{228} was not measured. Th^{228} can be added to a sediment as the result of production from Ra^{228} in the water (whether surface run-off or sea water) at one time or another associated with the sediment (7 and 5). In time, this unsupported Th^{228} will become equilibrated with Ra^{228}, as the latter is produced from Th^{232}, if Ra^{228} does not escape relative to the thorium isotopes. The results on the Th^{228}/Th^{232} activity ratios shown in Figure 6 must represent a combination of the features mentioned above and will be discussed later.

TABLE 3

17.8 cm Diver Core ($41^{o}.09.9'N$, $72^{o}55.6'W$; depth 20 m)

Depth cm	Density g/cm^3	U^{234*}/U^{238}	Th^{230*}/Th^{232}	Th^{228*}/Th^{232}	Th^{232} dpm/g	U^{238} dpm/g	Th^{230} dpm/g
0- 2.54	0.43	1.02±0.03	0.80±0.02	1.11±0.02	2.1±0.2	0.73±0.02	1.7±0.2
2.54- 5.08	0.54	1.02±0.03	0.81±0.01	1.05±0.03	1.9±0.2	0.88±0.02	1.5±0.1
5.08- 7.62	0.57	1.05±0.02	0.83±0.02	0.77±0.03	1.9±0.2	1.09±0.03	1.5±0.1
7.62-10.16	0.74	1.05±0.02	0.78±0.01	0.73±0.02	1.8±0.1	1.22±0.03	1.4±0.1
10.16-12.70	0.66	1.03±0.02	0.76±0.01	0.74±0.02	1.8±0.1	1.32±0.03	1.3±0.1
12.70-15.24	0.72	1.00±0.02	0.79±0.01	0.88±0.02	1.7±0.1	1.14±0.03	1.3±0.1
15.24-17.78	0.71	0.00±0.04	0.79±0.01	0.91±0.02	1.7±0.1	1.04±0.04	1.3±0.1

Dept cm	Ra^{226} dpm/g	Pb^{210} dpm/g	$(Pb^{210})_{ex}$ dpm/g	Mn ppm	Cu ppm	Zn ppm	Cd ppm	Pb ppm
0- 2.54	0.68±0.03	6.8±0.3	6.1±0.3	758	81.9	197	2.01	63.4
2.54- 5.08	0.56±0.02	6.4±0.3	5.8±0.3	507	81.8	199	1.60	64.0
5.08- 7.62	0.64±0.02	5.4±0.3	4.8±0.3	458	92.0	203	1.77	55.3
7.62-10.16	0.53±0.02	4.8±0.3	4.3±0.2	479	95.9	191	1.40	54.8
10.16-12.70	0.43±0.02	4.9±0.3	4.5±0.3	544	125	190	1.43	58.3
12.70-15.24	0.32±0.01	3.0±0.2	2.7±0.2	515	89.5	185	1.16	46.9
15.24-17.78	0.40±0.01	3.6±0.2	3.2±0.2	505	97.3	192	1.41	50.8

All measurements made on core material ashed at 500°C, leached with $1\overline{N}$ HNO_3. Organic content is around 6%. Density refers to 500°C material.

*Activity ratio

DISCUSSION

The record in the long core

The average rate of accumulation of 0.45 cm/yr (Fig. 2) deduced from the Pb^{210} data can be used to estimate when man's activity began to cause a major change in the rates of metal supply to Long Island Sound. This time, determined

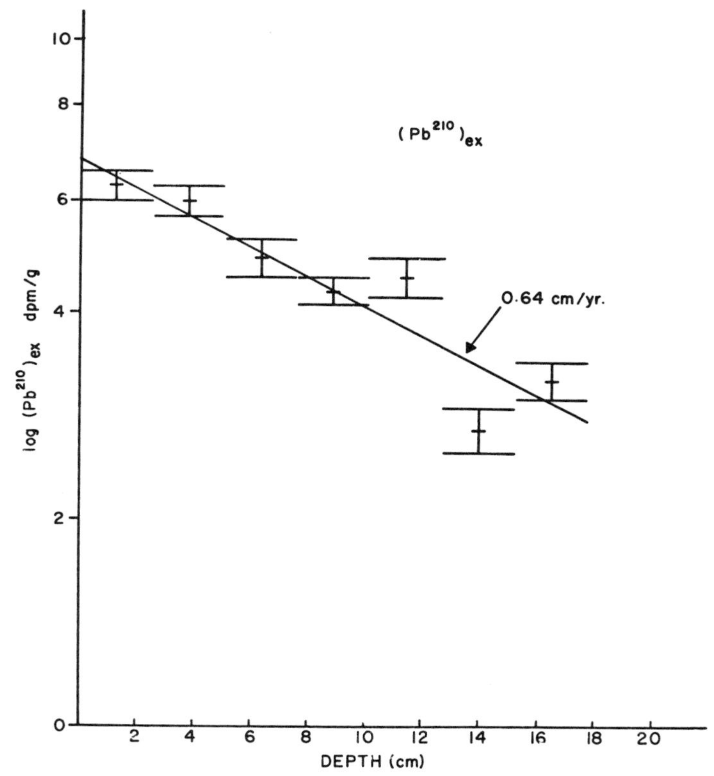

Figure 4. A plot of the excess Pb^{210} activity per gram for the short core. A simple treatment of the data yields an "accumulation rate" of 0.64 cm/yr. Other radiometric and sedimentologic data for the core indicate that this is an artifact and that the major part of the material at this site was dumped in one episode. The measured Ra^{226} activity was used to correct for supported Pb^{210}.

by assuming the top of the core to be zero age and visually spotting the changing metal concentration pattern at 50 cm (Fig. 3), appears to be 110 years ago. The sharpest increase in concentration upward starts at a depth of about 30 cm, or about 70 years ago. There is a pulse of high metal concentration at around 68 cm, corresponding to an extrapolated time of 140 years ago.

The pattern of manganese is a unique pattern. Constant and low up to 12 cm, with the high values limited to the top 12 cm, it might reflect the effects of manganese migrating up the sediment column.

The record in the short core

From the X-radiograph of one of the three short cores we inferred that the

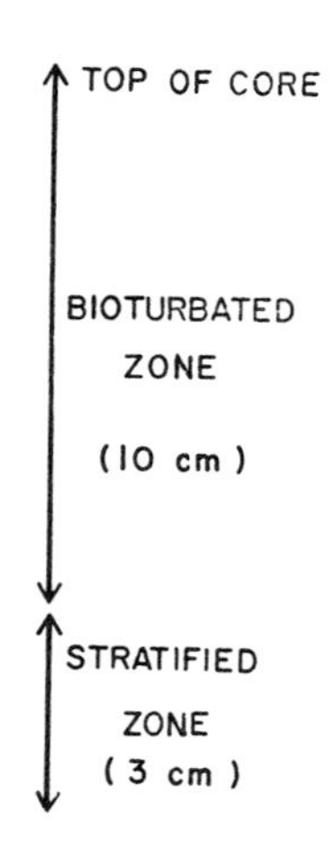

Figure 5. X-radiograph of the 13-cm diver core adjacent to the 18-cm core subjected to analysis. The bioturbation layer extends down to 10 cm, the bottom 3 cm showing stratification relict from the original deposition event. The "cross hatching" is an artifact caused by ice-crystal formation that results from storage.

surface sediments of the site from which the cores were obtained are dominated by material from an episode of gross sediment deposition, and that slower rate of sedimentation must have continued thereafter. Ra^{226}, Th^{230}, and Th^{232}, and the trace metals (except Mn) do not show any concentration differences that cannot be explained by dilution with a phase such as $CaCO_3$ or quartz. This implies that any added material must be closely similar in composition to that of the original slump.

A striking feature of the profile is that the uranium concentration decreases monotonically upward from the base of the bioturbation zone to the sediment-water interface (Fig. 6). The U^{234}/U^{238} activity ratio is close to unity over the entire length of the core, so that the source for uranium would seem to be terrestrial rather than marine, as the sea-water value of the ratio is 1.15. Thus, the implication is that under estuarine conditions typified by Long Island Sound, uranium is being released from the sediments to sea water as a result of oxidation. The mechanism for this is thought to be formation of the soluble $UO_2(CO_3)_3^{4-}$ complex in the oxidized layer of the sediment. It is to be expected that other oxidizable metalic species forming soluble anions, such as chromium and molybdenum, may also be released from estuarine sediments whose overlying water column does not become anoxic.

We were able to obtain only a crude estimate for the age of the slump. Certainly sufficient time has elapsed for well-defined profiles in the U^{238}/Th^{230} and Th^{228}/Th^{232} activity ratios and Pb^{210} to be developed. If we consider the (Pb^{210}) excess specific activity of the gravity core (4.6 dpm/g at zero depth) to

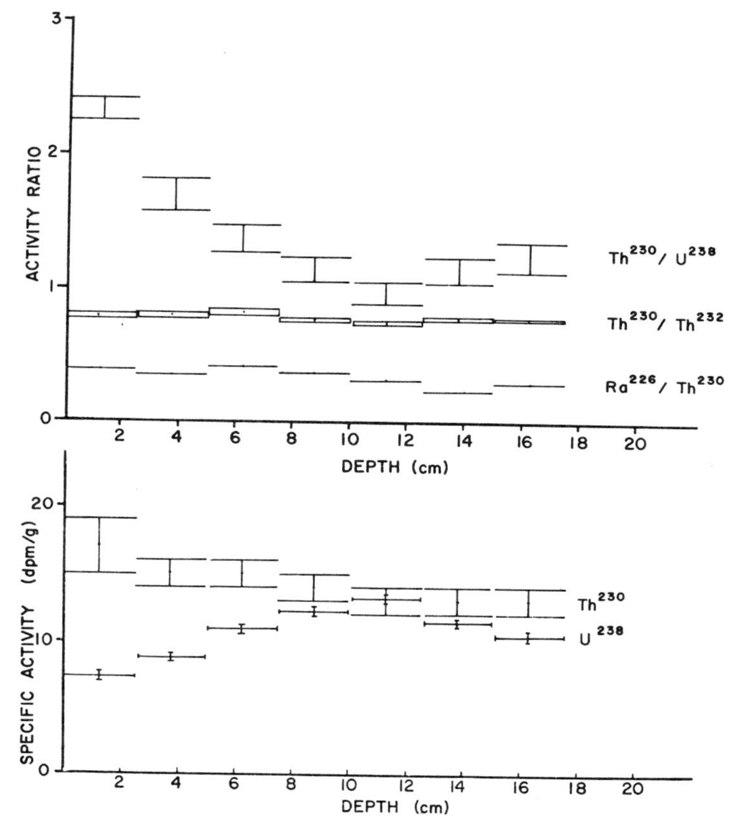

Figure 6. The decrease in the uranium concentration from the top of the stratified zone of the short core to the sediment-water interface is compared to the behavior of Th^{232} and Th^{230}. The Th^{230}/Th^{232} activity ratio is constant over the length of the core.

be representative of the contemporary specific activity of sediment in the central Sound, then the Pb^{210} activity in the stratified portion of the short core (3 dpm/g) yields an age of 14 years for the slump. The higher Pb^{210} values in the upper parts of the core are then caused by Pb^{210} in new sediment being mixed into the slumped material.

The Th^{228}/Th^{232} activity ratio of 0.90 in the stratified part of the core equals the Ra^{228}/Th^{232} activity ratio if no diffusive loss of Ra occurs. If we assume that the Ra^{226}/Th^{230} activity ratio of 0.35 is an index of the Ra^{228}/Th^{232} activity ratio of emplacement at the time (because both are subject to radium loss during transport and deposition), then the increase from a Th^{228}/Th^{232} (or Ra^{228}/Th^{232}) activity ratio of 0.35 to 0.90 means the passage of 18 years since deposition. This is an upper estimate since it is clear from the Pb^{210} and Ra^{226} data on the acid-leached fraction of the long core that acid-refractory Ra^{226} sites can release its daughters (seen finally as Pb^{210}) into the acid-labile sites thus yielding a Pb^{210}/Ra^{226} activity ratio greater than unity in the acid-leached sites at depths where secular equilibrium has clearly been established. Although Th^{228} release by decay of Ra^{228} in acid-refractory sites can be expected to be less than the release of the Ra^{226} daughters the effect probably exists and thus the Th^{228}/Th^{232} obtained by acid-leaching of the segments of the short core will be higher than that expected for simple growing in of Th^{228} from the Ra^{228} in the acid-leachable site. Our best estimate of the slump event is thus that it occurred less than 18 years before the collection of the core, and probably considerably less than that.

We interpret the Th^{228}/Th^{232} activity ratio profile (Fig. 7) in the bioturbation zone as the result of two competing phenomena: a) addition of excess authigenic Th^{228} relative to both Ra^{228} and Th^{232} to the sediment surface where the Th^{228}/Th^{232} activity ratio exceeds unity, and b) loss of Ra^{228} out of the core with increasing efficiency towards the sediment-water

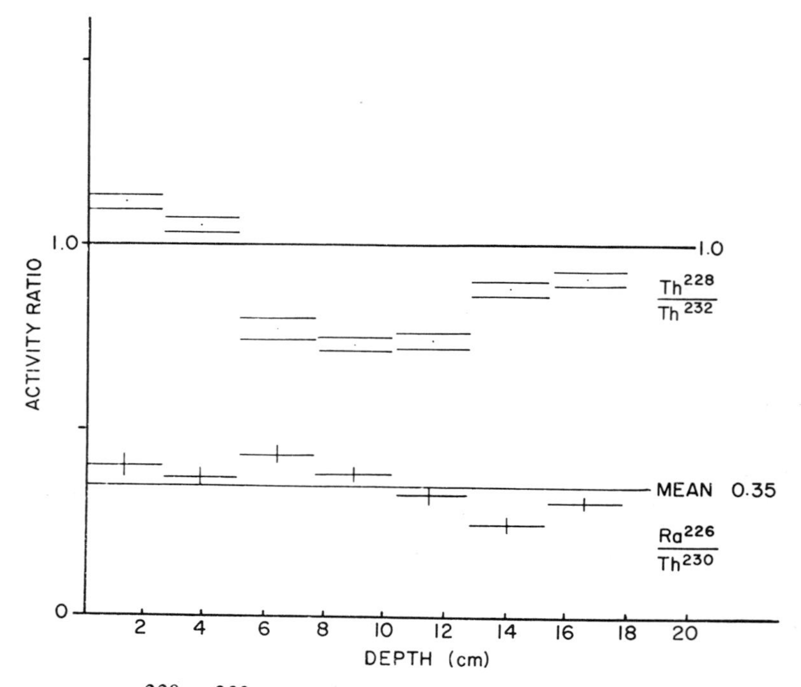

Figure 7. The Th^{228}/Th^{232} activity ratio and Ra^{226}/Th^{230} activity ratio variation with depth in the short core.

interface resulting in a decrease in the amount of Th^{228} production and, thus, in the Th^{228}/Th^{232} ratio. Integration of the idealized profile in Figure 8 indicates a Ra^{228} deficiency of 4.4 dpm/cm^2 in this core. If this deficiency was distributed over about 16 years, the flux of Ra^{228} to the overlying water would be 2.7 atoms/cm^2/min. The Ra^{228} activity of a sample of water from Long Island Sound has been determined to be 0.10$\pm$0.01 dpm/l. This implies a residence of 260 days for the water in the central Sound, assuming an average depth of 20 meters. From the idealized profile in Figure 8, we calculate a Th^{228} excess of 4.1 dpm/cm^2 over that presumed to be supported by Ra^{228}. Assuming 16 years since the original sediment was deposited, this implies that steady state for Th^{228} has been established and the downward flux is 4.1 atoms/cm^2/min. which is higher than the 0.2 atoms/cm^2/min. production rate from Ra^{228} in the water column. An additional source from the land is required to sustain this excess activity.

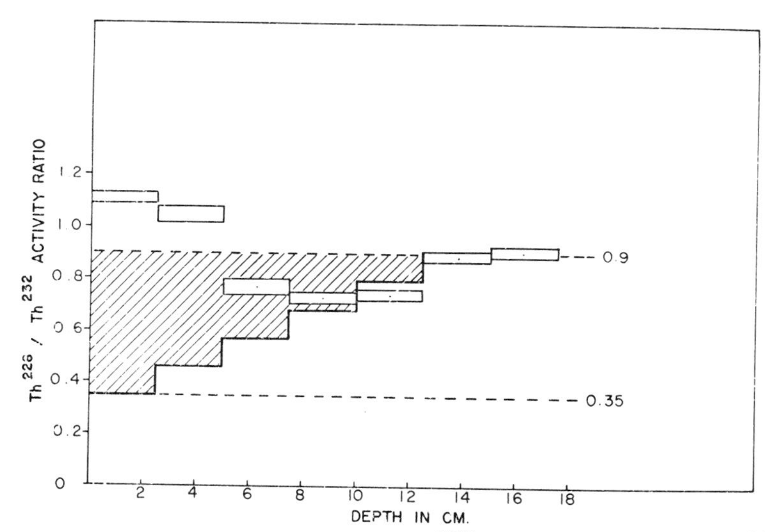

Figure 8. An interpretation of the fluxes of Ra^{228} out of the sediment and Th^{228} into the sediment at the site of the short core. It is assumed that the Th^{228}/Th^{232} activity ratio in the stratified part of the core (0.90) represents the Ra^{228}/Th^{232} ratio as the result of ingrowth from Th^{232} without diffusion of Ra^{228}. The initial Ra^{228}/Th^{232} activity ratio at the time of deposition is assumed to be the same as the Ra^{226}/Th^{230} activity ratio (0.35). The lost Ra^{228} is then approximated by the stepped triangle shown and is equal to 4.4 dpm/cm^2. If this loss occurred over 16 years then the Ra^{228} flux is 2.7 atoms/cm^2/min. The Th^{228} input flux is calculated as the difference between the measured Th^{228} less the amount assumed supported by the Ra^{228}. This is approximated by the line connecting 0.35 to 0.90. This yields a Th^{228} flux into the sediment of 4.1 atoms/cm^2/min.

Finally we note that manganese appears to be conserved in the sediment column under study (Fig. 9). The manganese deficiency in the deeper part of the reduced bioturbated sediments is more than balanced by the excess manganese found in the surface-oxidizing sediments. Calculation of the material balance actually shows 10% excess Mn in the top 2.5 cm relative to the deficiency in the lower 7.5 cm of the bioturbated zone. The excess may be the result of normal diffusion from deeper in the core or of additional sediment carrying excess manganese and excess Th^{228} and Pb^{210}. Metals released by the oxidation of metal sulfide compounds brought to the sediment-water interface by bioturbation are probably sequestered by coprecipitation with the freshly formed manganese oxide and result in the constant metal concentration along the core.

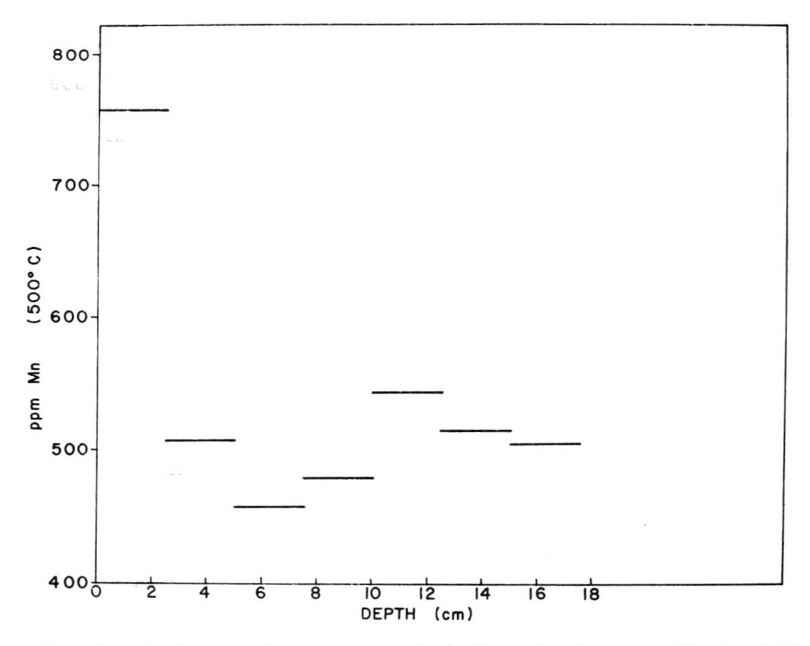

Figure 9. The distribution of manganese with depth in the short core. The top 2.54 cm oxidized layer contains manganese which is about 10% in excess of that expected from the deficiency of manganese in the reduced part of the bioturbated zone.

CONCLUSIONS

In summary we note the following points:

a) The sediments of Long Island Sound show that increases in many heavy metals have been going on for at least the past 70 years (and probably longer), a situation we ascribe to human habitation.

b) Our study of the behavior of metals in a wedge of slumped sediment shows no evidence of remobilization of any of the trace metals which form highly insoluble sulfides.

c) The release of uranium as the result of oxidation at the sediment-water interface is clearly seen. This indicates that estuaries such as Long Island Sound, whose water column never goes completely anoxic, release uranium rather than accumulate it. This situation may apply to such analogous metals as chromium and molybdenum, but no data are yet available for these elements.

d) The evidence indicates that although manganese is remobilized in the reducing portion of the sediment column and is subject to transport to the sediment-water interface, little of it escapes the sediment column because of oxidation and precipitation as an oxide. In this process, other metals may be extracted from the estuarine water.

ACKNOWLEDGMENTS

This research is part of a larger program at Yale University, supported by the United Illuminating Company, to explore the fate of dredge-spoil dumping in Long Island Sound. The physical and biological aspects of the program have been executed under the direction of Professors Robert Gordon and Donald Rhoads, and we wish to acknowledge their assistance in many aspects of this work. The U. S. Atomic Energy Commission has also been a supporter of this research, especially those aspects of it that deal with the uranium decay series nuclide distribution in sediments.

REFERENCES

1. Applequist, M. D., Katz, A., and Turekian, K. K.
1972 Distribution of mercury in the sediments of New Haven (Conn.) Harbor. **Environ. Sci. Tech.,** 6: 1123-1124.

2. Aston, S. R., Bruty, D., Chester, R., and Padgham, R. C.
1973 Mercury in lake sediments: a possible indicator of technological growth. **Nature,** 241: 450-451.

3. Bortleson, G. C. and Lee, G. F.
1972 Recent sedimentary history of Lake Mendota, Wisconsin. **Environ. Sci. Tech.,** 6: 799-808.

4. Chow, T. J., Bruland, K. W., Bertine, K. K., Soutar, A., Koide, M., and Goldberg, E. D.
1973 Lead pollution: records in southern California coastal sediments. **Science,** 181: 551-552.

5. Joshi, L. U. and Ganguly, A. K.
1973 Anomalous Th-228/Th-232 and Th-230/Th-232 activity ratios in backwater sediments along the west coast of India. In **Radioactive Contamination of the Marine Environment,** p. 747-755. I.A.E.A. Vienna.

6. Koide, M., Soutar, A., and Goldberg, E. D.
1972 Marine geochronology with Pb-210. **Earth Planet. Sci. Lett.,** 11: 407-414.

7. Koide, M., Bruland, K. W., and Goldberg, E. D.
1973 Th-228/Th-232 and Pb-210 geochronologies in marine and lake sediments. **Geochim. Cosmochim. Acta.,** 37: 1171-1187.

8. Krishnaswami, S., Lal, D., Amin, B. S., and Soutar, A.
1974 Geochronological studies in Santa Barbara Basin. **Limnol. Oceanogr.** (In press).

9. Ku, T-L.
1966 Uranium series disequilibrium in deep sea sediments. Doctoral dissertation, Columbia Univ., New York.

10. Lee, J. A. and Tallis, J. H.
1973 Regional and hisorical aspects of lead pollution in Britain. **Nature,** 245: 216-218.

11. Rama, Koide, M., and Goldberg, E. D.
1961 Lead-210 in natural waters. **Science,** 134: 98-99.

12. Young, D. R., Johnson, J. N., Soutar, A., and Isaacs, J. D.
1973 Mercury concentrations in dated varved marine sediments collected off southern California. **Nature,** 244: 273-275.

ROLE OF JUVENILE FISH IN CYCLING OF Mn, Fe, Cu, AND Zn IN A COASTAL-PLAIN ESTUARY

F. A. Cross, J. N. Willis, L. H. Hardy, N. Y. Jones
and J. M. Lewis[1]

ABSTRACT

The daily flux of Mn, Fe, Cu, and Zn was estimated for Atlantic menhaden, spot, and pinfish on both an individual-fish basis and a population basis for the summer months in the Newport River estuary. Flux estimates were determined from the equation I = A + E where I is the rate of ingestion of the metal, A is the assimilation rate of the metal, and E is the egestion rate of unassimilated metal. "I" was determined by multiplying the mean rate of food consumption times the mean trace metal concentration in stomach contents. "A" was determined by summing the rate of increase in body burden of the metal by the fish during the growth period under study and the rate of metal loss in biological turnover during the same time period. "E" was estimated from "I" - "A".

Assimilation efficiencies were highly variable and depended on the trace-metal concentration of the material ingested by the fish which, in turn, was determined by the distribution of the fish within the estuary. Because a significant fraction of the trace metal ingested by the fish is not assimilated, daily defecation of unassimilated trace metal is about 1 kg for Zn, 56 kg for Fe, 1 kg for Mn, and 0.2 kg for Cu in populations of these three species in the Newport River estuary during the summer months. This may be a major biological process in the cycling of trace metals in highly productive coastal-plain estuaries.

INTRODUCTION

Because estuaries are highly productive ecosystems, they are important nursery areas for many recreational and commercial species (e.g., fish, crabs, and shrimp), as well as permanent homes for other economically important species (e.g., clams, oysters, and scallops). The use of coastal rivers and estuaries as

1. National Marine Fisheries Service, Atlantic Estuarine Fisheries Center, Beaufort, North Carolina 28516.

receptacles for a variety of industrial and municipal wastes containing trace metals demands that the rates at which these metals move along specific pathways within the estuary be known.

The flux of trace metals into and through biota is important from both a health physics and an ecological point of view. It is essential to understand not only the deleterious effects that anthropogenic additions of trace metals can have on biota, and ultimately on man, but also the role of biota in determining the pathways along which trace metals move in estuaries.

A project to determine the cycling of trace metals and flow of energy in a small, but highly productive, coastal-plain estuary in North Carolina (Newport River estuary) has been underway for several years. Its objective is to conduct an intensive investigation of a single estuarine ecosystem and then to test the applicability of the resultant basic understanding to the many similar estuaries that exist along the southeast coast of the United States.

Wolfe (17) has reviewed the research conducted in Newport River estuary relative to energy flow and cycling of trace metals. Papers by Ferguson and Murdoch (6), Peters and Kjelson (11), and Thayer, Adams, and LaCroix (14) also discuss various aspects of energy flow in this estuary.

The first attempt to construct a conceptual model of Zn flux within the Newport River estuary was presented by Wolfe (17). This model points out the significant effect biological processes have on the cycling of Zn within the estuary. In order to refine this model and proceed to a dynamic mathematical model of trace-metal cycling in the estuary, it is necessary to identify and quantify the specific pathways along which metals move. It is the purpose of this paper to identify the role of one sub-compartment of the macrobiota--juvenile fish--in the cycling of Mn, Fe, Cu, and Zn in the Newport River estuary.

Review of Related Work

Two general approaches have been used to determine the flux of stable trace metals in marine organisms. One approach was to evaluate the role of biota in the horizontal or vertical translocation of trace metals in the marine environment either by actual migrations of the organisms themselves or by the elimination of fecal pellets, cast exoskeletons, and so forth (9, 12, 18). The other, which was to determine the flux of elements into and through individual organisms, has been defined by Small et al (13) as “physiological flux”.

Most of the research conducted to date on the flux of trace metals in the ocean has been directed towards quantifying the vertical movement of these metals from the surface waters of the ocean caused by the activities of zooplankton. Kuenzler (8) obtained elimination rates of Co in zooplankton from radiotracer experiments carried out aboard ship in the eastern Pacific Ocean, and used these data along with other published information to estimate the flux of

Co through the thermocline from diurnal migrations of zooplankton. Similarly, Lowman, et al (9), using Kuenzler's data, other published information, and assumed Fe assimilation efficiencies, calculated the daily flux of Fe, Mn, Co, Zn, Zr, and Pb from the upper mixed layer of the northeastern Pacific Ocean caused by zooplankton activities. Estimates were obtained for biological transport due to migrating zooplankton as well as for cast molts, dead organisms, fecal pellets, and particulate excretion.

In an effort to quantify more precisely the vertical flux of metals from zooplankton, Small et al (13) conducted a laboratory study on the flux of Zn through the euphausiid *Meganyctiphanes norvegica.* The authors built their model of Zn flux from the equation $K_e = \mu_e + \lambda_e$, where K_e is the rate of ingestion of particulate Zn, μ_e is the net accumulation of Zn in new tissue during growth, and λ_e is the rate of Zn elimination (feces, molts, dead carcasses, non-viable eggs, and excretion of dissolved metabolic products). Values were obtained experimentally for μ_e and λ_e, and the Zn concentrations in the food needed to satisfy the equation were compared with measurements of Zn in oceanic plankton. Parameters determined experimentally in this study were rates for feeding, growth, turnover of Zn, and rates of production of feces, molts, dead carcasses, and non-viable eggs, as well as average Zn concentrations in each component.

Using much of the information derived from the study described above, Small and Fowler (12) calculated the vertical flux of Zn in the Ligurian Sea from *M. norvegica* as the sum of dissolved Zn excreted at depth plus the concentrations of Zn remaining in feces, molts, and carcasses after they had sunk to the specified depth. Allowances were made for loss of Zn from these components as they descended.

Vanderploeg (15) developed a dynamic model to evaluate the important factors affecting ^{65}Zn specific activities of benthic fishes in Oregon coastal waters. In this model, he derived the concept of α, which is the daily rate of Zn input, as a fraction of Zn body burden in the fish. α varies as a function body weight and temperature.

Flux of Mn, Fe, Cu, and Zn in Estuarine Fish

Fish, shrimp, and crabs spawn in coastal waters and larvae migrate into the shallow coastal-plain estuaries of the southeastern United States in high numbers from January to April. These organisms feed in these productive ecosystems during the summer months and migrate back to sea in October and November. The most abundant fish species involved in these seasonal migrations to the Newport River estuary are bay anchovy (*Anchoa mitchilli*), Atlantic silversides (*Menidia menidia*), striped mullet (*Mugil cephalus*), pinfish (*Lagodon rhomboides*), spot (*Leiosotomus xanthurus*), croaker (*Micropogon undulatus*), and Atlantic menhaden (*Brevoortia tyrannus*). These seven species represent about 85% of the total fish biomass in the estuary and the estimated standing crop is about 170,000 kg (18), more than 90% of which consists of juvenile fish (Kjelson and Johnson, unpublished information). Besides the ecological importance that all these species have, adult mullet, spot, croaker, and menhaden are harvested directly by man for recreational and commerical purposes.

The trace metals that these fish accumulate in their tissues during their growth period in the estuary are lost from the estuary during the autumn emigrations. The impact of these migrations on the total flux of Mn, Fe, and Zn through the Newport River estuary has been evaluated and found to be small relative to transport by physical processes (17, 18).

Menhaden, spot, and pinfish were chosen for the purpose of quantifying their role in the flux of Mn, Fe, Cu, and Zn while residing in the estuary. These species were selected not only because they are among the most abundant, but also because there is more ecological information on them than there is on croaker, silversides, anchovy, or mullet (11).

If we set up a basic equation that I = A + E, where I is the rate of ingestion of a metal, A is the rate of assimilation of the metal by the organism, and E is the rate of egestion of unassimilated metal, then, when any two parameters are known, the flux of a metal through an organism can be described. Using Zn as an example, we will estimate the flux of Mn, Fe, Cu, and Zn through these fish in the following manner:

Rate of Zn ingestion = (rate of food consumption) x (Zn concentration in food)

Rate of Zn assimilation = (rate of increase in Zn body burden during growth period under study) + (rate of Zn loss in biological turnover during the same time period)

Zn assimilation efficiency = (Zn assimilated/Zn ingested)

Rate of Zn egestion = ingestion rate - assimilation rate.

This procedure assumes that food consumption is the primary means by which Mn, Fe, Cu, and Zn are assimilated into the internal tissues of the fish (7, 10).

Before these calculations can be made, growth rates and feeding rates during a specific time period must be known, and these are given in Table 1. Flux of trace metal has been calculated for menhaden from 1 July - 1 September, and for spot and pinfish from 1 May to 1 August.

Ingestion of trace metals

The quantity of food consumed daily was determined by multiplying the average weight of the fish during the growth period specified in Table 1 for each species times the feeding rate. Feeding rates, which vary according to

TABLE 1.

Data on growth and feeding rates used to calculate trace metal fluxes. Information on growth is from Kjelson and Johnson (unpublished) and feeding rates from Peters and Kjelson (11).

Species	Growth period	Weight change	Average weight	Feeding rate
		---- g dry wt. ----		
Menhaden	1 July - 1 Sept.	0.10 - 1.0	0.32	13
Spot	1 May - 1 Aug.	0.28 - 1.1	0.55	10
Pinfish	1 May - 1 Aug.	0.30 - 0.90	0.52	9

[1]Percent dry body weight day^{-1}

temperature, also are representative of the time periods shown.

Because it is extremely difficult to determine the specific diet of these young fish, we elected to measure the concentration of trace metal directly in their stomach contents. If these measurements were made on partially digested food, however, our estimates of trace metal consumed might be low due to assimilative losses. To minimize this problem, we dissected only fish that had been collected between daybreak and 9:00 a.m. to obtain recently consumed food. The small quantity of food taken from each fish made it impossible to measure trace metals on individual stomach contents. The results of pooled analyses are given in Table 2. The analytical methodology was the same as that described in Cross and Brooks (3) for trace metal determinations on whole fish. The fish used for stomach analyses were within the growth periods under study but they were larger than the mean weights listed for the entire growth period (Table 1).

TABLE 2.

Concentrations of trace metals in stomach contents of juvenile fish

Species	N	$\bar{X}$ wt. (g)	Zn	Fe	Mn	Cu
			----------- μg/g dry wt. ---------			
Menhaden	180	1.1	110	22,000	270	25
Spot	60	2.0	620	26,000	520	92
Pinfish	180	2.7	75	5,200	55	13

Assimilation of trace metals

For the past several years, juvenile fish have been sampled routinely for biomass studies in the Newport River estuary and aliquots of these samples were analyzed for concentrations of trace metals, as discussed by Cross and Brooks

(3). As a result of these analyses, measurements of trace-metal concentrations are available for fish over a large range of body weights. In order to determine the net accumulation of trace metals during growth of the fish, we have plotted the total µg of trace metal in individual fish against body weight as shown for Zn in menhaden, spot, and pinfish for a change in dry weight of nearly five orders of magnitude (Fig. 1-3). When the growth rate is known, the net accumulation rate of the metal into tissue can be obtained.

Regression analyses showed that the total amounts of Zn, Fe, Mn, and Cu were highly correlated with body weight in all three species, and the slopes of the regression line were very close to 1 (Table 3). R^2 values ranged from 0.83 to 0.99 and all slopes were highly significant ($P < 0.001$). Although we did not test differences among intercepts, menhaden appear to contain substantially more Zn and Fe than either spot or pinfish during this early stage in their life cycle. The fact that juvenile menhaden have a greater affinity for Zn and Fe than do four other estuarine species, including spot and pinfish, is discussed by Cross and Brooks (3). Intercepts for Mn and Cu, however, are quite similar in these same species.

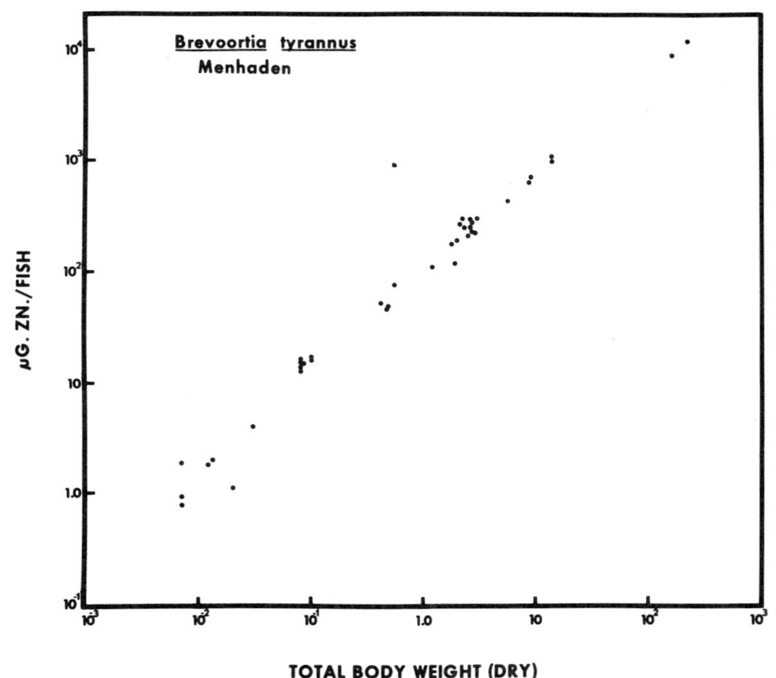

Figure 1. Relationship between total µg Zn/fish and dry body weight in menhaden. Each point represents the analysis of an individual fish.

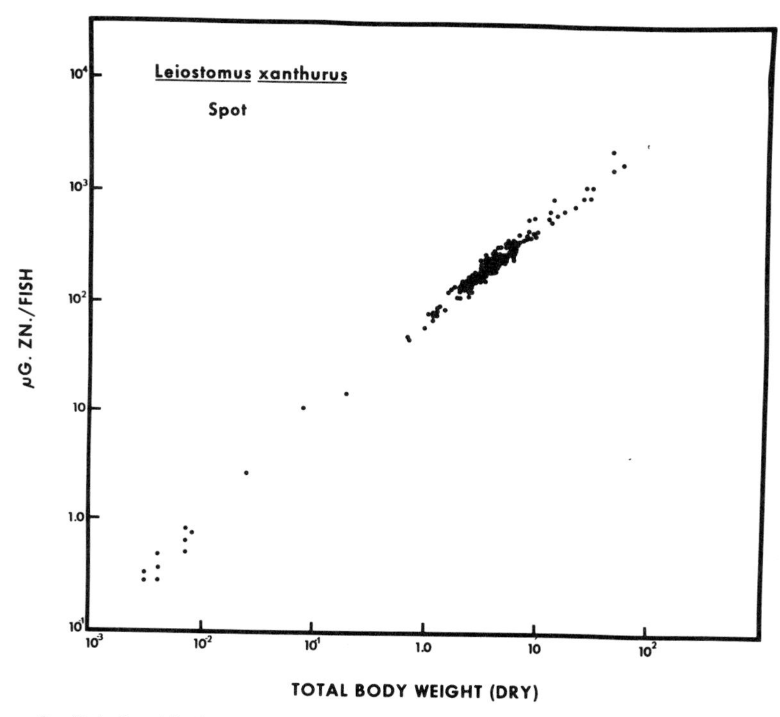

Figure 2. Relationship between total μg Zn/fish and dry body weight in spot. Each point represents the analysis of an individual fish.

As discussed previously, when determining the amount of trace metal accumulated by rapidly growing juvenile fish, the amount lost to biological turnover should be added to that which has gone into new tissue. During the past 20 years many radioecologically-oriented experiments have untilized radiotracer studies to determine turnover rates of elements in aquatic organisms. Some of these studies failed to label all compartments of the organisms completely with the radio-tracer, and others failed to maintain constant specific activities throughout the experiment. Turnover rates obtained from these experiments are applicable only to the radiotracer and are dependent upon the experimental procedures under which the organism was labeled.

Under proper experimental conditions, rate constants for uptake and loss curves of a radionuclide should be similar and, if constant specific activities are maintained, the kinetics of the radiotracer can be used to determine the rate constants of its stable counterpart. A discussion and mathematical proof of the experimental conditions necessary to obtain similar rate constants for uptake and loss curves using radiotracers and thus properly to "trace" stable elements in experimental aquatic ecosystems are made by Willis et al (16).

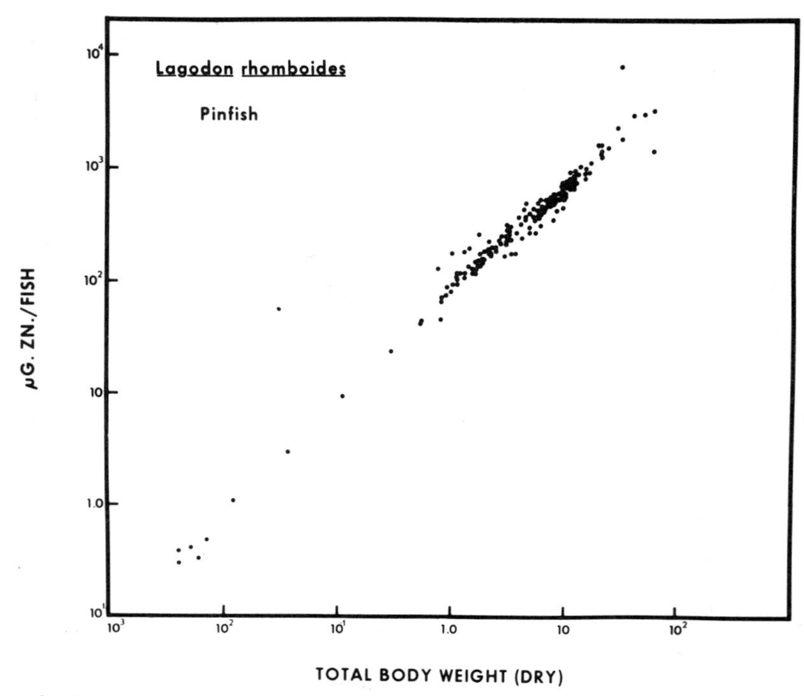

Figure 3. Relationship between μg Zn/fish and dry body weight in pinfish. Each point represents the analysis of an individual fish.

Cutshall (5), using single-compartment kinetics, demonstrated that ^{65}Zn rate constants obtained from data on uptake and loss of ^{65}Zn in oysters from Willapa Bay were similar under conditions of chronic exposure in the field. Data on uptake and loss of ^{65}Zn in these organisms were fitted to a first-order differential equation and observed rate constants were $5.49 \times 10^{-3} d^{-1}$ from loss data and $4.93 \times 10^{-3} d^{-1}$ from uptake data.

Estimates of excretion rates for stable Zn in euphausiids using ^{65}Zn were obtained by Small et al (13). Five euphausiids were labeled with ^{65}Zn through both food and water for 20 days when the animals appeared to be approaching a steady-state condition with respect to ^{65}Zn. Loss of ^{65}Zn was then followed for 71 days and the loss curve was resolved into three compartments. The percent daily loss from each compartment was then summed and the assumption was made that the loss of ^{65}Zn was equivalent to the loss of stable Zn. This loss of ^{65}Zn totaled 10.9%/day and was used to represent the total daily loss of stable Zn from the organism in the form of excretion products.

In recent months research has been underway at our center to determine exchange of stable Zn in estuarine fish using ^{65}Zn (16). Our first objective is to

TABLE 3.

Relationship between total trace metal (μ g) and dry body weight (W) in juvenile estuarine fish

Species	N [A]	Zn (S.E.E.) [B] (r^2)	Fe (S.E.E.) (r^2)	Mn (S.E.E.) (r^2)	Cu (S.E.E.) (r^2)
Menhaden	40	$110W^{0.8716}$ (±0.0147) (0.99)	$650W^{1.1662}$ (±0.0717) (0.88)	$18W^{1.1117}$ (±0.0325) (0.97)	$5.8W^{1.0921}$ (±0.0311) (0.99)
Spot	163	$72W^{0.9137}$ (±0.0086) (0.99)	$360W^{1.0567}$ (±0.0371) (0.83)	$20W^{1.0343}$ (±0.0322) (0.86)	$5.9W^{1.0335}$ (±0.0210) (0.98)
Pinfish	155	$87W^{0.9451}$ (±0.0166) (0.95)	$210W^{1.0681}$ (±0.0335) (0.87)	$25W^{1.0775}$ (±0.0249) (0.92)	$6.5W^{1.0337}$ (±0.0331) (0.86)

[A] Number of individuals analyzed

[B] Standard error of the exponent

obtain similar rate constants for uptake and loss of total Zn experimentally in these fish as Cutshall (5) was able to do for ^{65}Zn in oysters. Briefly, these experiments have involved raising the mosquitofish, *Gambusia affinis,* to adulthood in similar ecosystems in the laboratory, with the exception that one experimental system contained ^{65}Zn. Upon reaching adulthood, each population of fish was switched to the other ecosystem and simultaneous uptake and loss of ^{65}Zn was followed in each tank. Because ^{65}Zn specific activities were held constant in one tank and concentrations of stable Zn were the same in both tanks, rate constants were obtained for exchange of stable Zn in both uptake and loss experiments. A curve with three components was obtained with both the uptake and loss curves, which are shown in Figures 4 and 5. We are

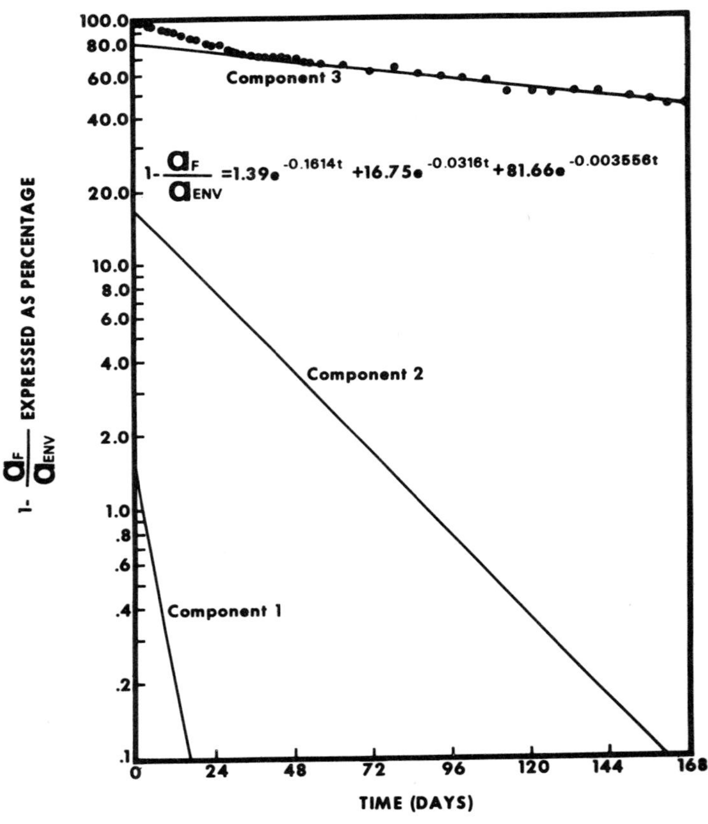

Figure 4. Mean uptake curve for five fish resolved into three components. A_F is ^{65}Zn specific activity of the fish and A_{ENV} is the ^{65}Zn specific activity of the experimental environment (from Willis et al (16)).

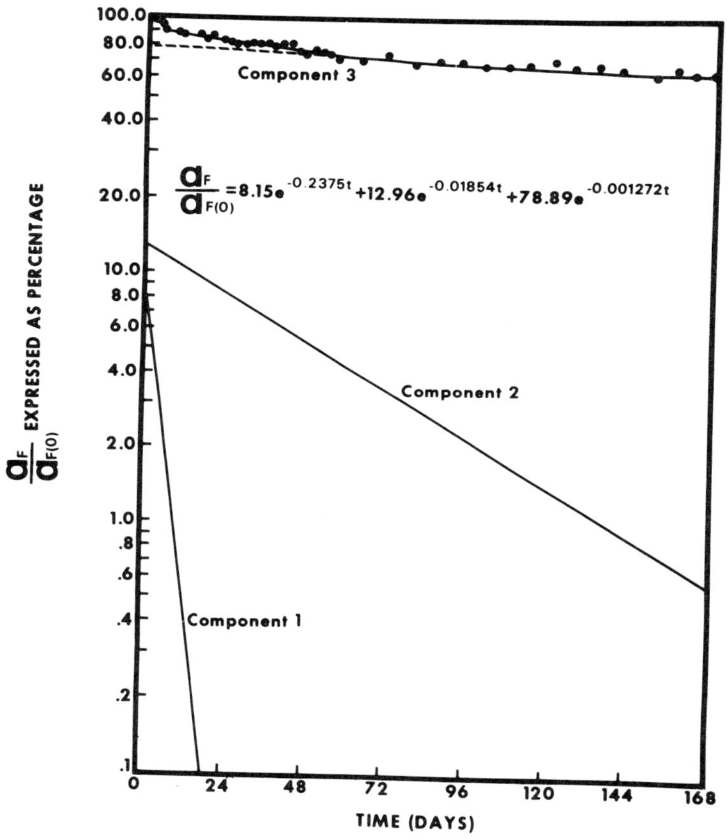

Figure 5. Mean loss curve for five fish. A_F is the ^{65}Zn specific activity of the fish at any time and $A_{F\,(0)}$ is the ^{65}Zn specific activity of the fish at zero days (from Willis et al (16)).

assuming that the long-lived component (component 3) represents the rate of exchange of internally deposited Zn.

Although Willis et al (16) did not obtain similar exchange rates from both uptake and loss experiments, we averaged the exchange rates from the two experiments for the purpose of estimating exchange of Zn in juvenile fish in this study (Table 4). We have used the mean rate constant of the long-lived component (r_3) on a per-gram basis (0.1427 μg Zn/g fish day) which represents about 80% of the total Zn in the fish (C_3). These turnover values can serve only as first approximations because the data were obtained by using a different and much smaller species under constant environmental conditions in the laboratory (20° C and $30^o/oo$ salinity). Similar experiments using other species under varying experimental conditions are planned.

TABLE 4.

Mean estimates of relative component sizes (C), rate constants (K), actual component sizes (Q), exchange rates (r), and total Zn body burden and concentration in the whole fish (Q_F) for mosquitofish, calculated from Willis et al (16).

C_1			4.77%
C_2			14.86%
C_3			80.37%
K_1			0.1994 Day^{-1}
K_2			0.0251 Day^{-1}
K_3			0.002414 Day^{-1}
Q_1	0.7134 μg Zn/Fish	=	3.463 μg Zn/g Fish
Q_2	2.494 μg Zn/Fish	=	10.82 μg Zn/g Fish
Q_3	13.30 μg Zn/Fish	=	58.51 μg Zn/g Fish
r_1	0.1596 μg Zn/Fish Day	=	0.7835 μg Zn/g Fish Day
r_2	0.06677 μg Zn/Fish Day	=	0.2809 μg Zn/g Fish Day
r_3	0.03439 μg Zn/Fish Day	=	0.1427 μg Zn/g Fish Day
Q_F	16.51 μg Zn/Fish	=	72.79 μg Zn/g Fish

Since we have no information on rates of exchange of stable Mn, Fe, or Cu in estuarine fish, we cannot estimate this parameter in our calculations of metal flux in fish.

Estimated Fluxes of Trace Metals in Juvenile Fish

Having estimated the ingestion and assimilation of trace metals in juvenile menhaden, spot, and pinfish, we can now describe the flux of Zn, Fe, Mn, and Cu in these species during their residence in the estuary (Tables 5-8).

Although the quantity of trace metal ingested varied substantially among the three species, the net rate of accumulation of trace metal in the tissues did not. The exceptions to this are that menhaden accumulate Zn and Fe at a greater rate than do the other two species (Tables 5 and 6).

For rapidly growing juvenile fish the loss of Zn caused by biological turnover is insignificant compared with the net accumulation of Zn caused by growth. As these fish approach adulthood, however, the difference between the net accumulation of Zn and the biological turnover of Zn should decrease substantially until assimilation = turnover, as pointed out by Vanderploeg (15). Although no estimates of turnover rates for Mn, Fe, and Cu are currently available, we are assuming that they are on the same order of magnitude as our

TABLE 5.

Estimates of flux of Zn (μ g/day) through juvenile estuarine fish during summer months in Newport River estuary

Species	Zn ingested	Zn accumulated in tissues	Zn loss in turnover	Zn assimilation efficiency (%)	Zn egested
Menhaden	4.6	1.6	0.04	36	2.96
Spot	34	0.60	0.08	2	33.32
Pinfish	3.5	0.58	0.07	19	2.85

TABLE 6.

Estimates of flux of Fe (μ g/day) through juvenile estuarine fish during summer months in Newport River estuary

Species	Fe ingested	Fe accumulated in tissues	Fe loss in[A] turnover	Fe assimilation efficiency (%)	Fe egested
Menhaden	920	10	–	1	910
Spot	1,400	3.6	–	0.3	1,396
Pinfish	240	1.5	–	0.6	238.5

[A] No estimates available

TABLE 7.

Estimates of daily flux of Mn (µ g/day) through juvenile estuarine fish during summer months in Newport River estuary

Species	Mn ingested	Mn accumulated in tissues	Mn loss in[A] turnover	Mn assimilation efficiency (%)	Mn egested
Menhaden	11	0.28	--	3	10.72
Spot	29	0.19	--	0.7	28.81
Pinfish	2.3	0.17	--	7	2.13

[A] No estimates available

TABLE 8.

Estimates of flux of Cu (µg/day) through juvenile estuarine fish during the summer months in Newport River estuary

Species	Cu ingested	Cu assimilated in tissues	Cu loss in[A] turnover	Cu assimilation efficiency (%)	Cu egested
Menhaden	1.0	0.09	--	9	0.11
Spot	5.1	0.05	--	1	5.05
Pinfish	0.14	0.04	--	2.9	0.10

[A] No estimates available

estimates for Zn and, therefore, inconsequential relative to the estimated rates of accumulation and egestion for these metals.

These data have demonstrated that the rate of ingestion of metal varies substantially among species, and consequently, assimilation efficiencies are not constants but vary according to the quantity of trace metal ingested. Because the ash content of the material in the gut is between 40%-50% of the dry weight of the gut contents (11) these species are ingesting a considerable amount of inorganic matter along with their food.

A major influence on the quantity of trace metal ingested by these fish is their distributional pattern within the estuary. Menhaden and spot, upon entering the estuary as larvae, migrate into low-salinity water near the head of the estuary. Pinfish, on the other hand, tend to remain in the lower portion of the estuary during their nursery period. In earlier work, we showed that concentrations of Mn, Fe, and Zn in 0.1 N HC1 extracts of sediment samples and concentrations of Mn and Fe in water of the Newport River estuary decrease in a seaward direction (4). Thus fish near the head of the estuary would ingest the greatest quantity of trace metals associated with particulate material. The juvenile menhaden and spot that were collected for measurements of trace metals in stomach contents were taken near the head of the estuary while the pinfish were collected in the lower estuary. Other measurements of juvenile spot taken from the lower estuary have shown that they have considerably lower concentrations of trace metals in the stomach contents than the values that are listed in Table 2.

The quantity of trace metals egested from these fish was estimated by subtracting assimilation from ingestion. Estimates of ingestion and assimilation were based on two significant figures and egestion values are listed in three and four significant figures only to make the equations balance (Tables 5-8). The possible ecological significance of this egested material will be discussed in the following section.

Role of Juvenile Fish in Cycling Trace Metals in Coastal-Plain Estuaries

When considered in terms of populations of juvenile fish within the estuary, the importance of the flux estimates listed in Tables 5-8 is apparent. We took the biomass estimates of these species listed in Wolfe et al (18), converted the values to dry weight (3), and assumed that the average weights of the fish were equal to the average weights listed in Table 1, in order to obtain estimates of population size. We could then convert the daily rates of ingestion, assimilation, and egestion of Zn, Fe, Mn, and Cu from a per-fish basis (Tables 5-8) to a population basis (Table 9) for these fish in the Newport River estuary. Because the population sizes were shown to be on the order of 10^7 fish, the effect of these

TABLE 9.

Flux of trace metals through populations of juvenile fish in Newport River estuary during summer months.

Species	Zn (g/day)			Fe (g/day)			Mn (g/day)			Cu (g/day)		
	Ing.[A]	Assim.[B]	Eg.[C]	Ing.	Assim.	Eg.	Ing.	Assim.	Eg.	Ing.	Assim.	Eg.
Menhaden ($1.8x10^7$)[D]	83	30	53	17,000	170	16,830	200	6	194	18	1.6	16.4
Spot ($2.8x10^7$)	950	20	930	39,000	120	38,880	810	6	804	140	1.4	139.6
Pinfish ($0.9x10^7$)	32	6	26	2,200	13	2,187	21	1.5	19.5	1.3	0.4	0.9

A Ingestion

B Assimilation

C Egestion

D Estimated population size of juvenile fish

fish on the total flux of metals could be significant, especially since these three species represent about 50% of the total fish biomass in the Newport River (18).

Having made those measurements, we had enough information to evaluate the impact of just one species, menhaden, on trace-metal budgets in the Newport River estuary. Using estimates of Wolfe on the total reservoir of metal in the water, both dissolved and particulate (17), we estimated that filter-feeding menhaden ingest daily about 0.3% of the Zn, 0.7% of the Fe, and 0.06% of the Mn in the water. These estimates are probably low because estimates of reservoir size include both water and the suspended fraction and menhaden feed only on a specific range of particle sizes. No estimate for reservoir size for Cu in water is available.

The exercise performed in the above paragraph, although quite crude, points to menhaden as an important agent in the cycling of Zn and Fe in the Newport River estuary. Wolfe (17) also discusses the importance of biological processes in the cycling of Zn in the Newport River estuary. Because the assimilation efficiencies of metal by these fish are quite low, unassimilated metal that is egested from these fish must play an important role in the cycling of trace metals in the Newport River estuary. Approximately 1 kg of Zn, 56 kg of Fe, 1 kg of Mn, and 0.2 kg of Cu are egested daily during the summer months from these three species of fish. Because of the shallowness of the estuary, this process could result in metal being rapidly moved to the surface sediments in a form where it would be subject to ingestion by other organisms. Considering the remaining fish species as well as the invertebrate populations (shrimp, crab, mollusks, polychaetes, tunicates, etc.), defecation of unassimilated trace metals may be a major process controlling the cycling of these metals in shallow estuaries. The possible effect of fecal deposition on the cycling of trace metals in the marine environment has been discussed by Lowman et al (9), Boothe and Knauer (2), Andrews and Warren (1), and Small et al (13).

In this paper, we have attempted to show how analytical measurements of different-sized juvenile fish collected from the estuary, radiotracer experiments in the laboratory, and basic ecological information can be combined to determine the pathways and fluxes of trace metals in estuaries. Results of our studies thus far point to several areas of research that need to be studied: quantifying the importance of fecal deposition, determining turnover rates of these metals in a variety of marine organisms under different environmental conditions, determining the biological availability of different chemical forms of metals, and evaluating the effect of microbes on trace-metal cycling. This information, coupled with additional knowledge of growth rates, feeding patterns, and biomass of dominant organisms, will allow us to better understand the biological processes that affect the cycling of trace metals in the Newport River estuary.

Acknowledgments

We thank Dr. M. A. Kjelson and Mr. G. N. Johnson for use of unpublished data and Dr. G. W. Thayer for supplying some of the samples of juvenile fish used in this study. This research was supported jointly by the National Marine Fisheries Service and the U. S. Atomic Energy Commission through agreement no. AT(49-7)-5.

REFERENCES

1. Andrews, H. L., and Warren, S.
1969 Ion scavenging by the eastern clam and quahog. **Health Phys.,** 17:807-810.
2. Boothe, P. N., and Knauer, G. A.
1972 The possible importance of fecal material in the bioligical amplification of trace and heavy metals. **Limnol. Oceanogr.,** 17:270-274.
3. Cross, F. A., and Brooks, J. H.
1973 Concentrations of manganese, iron and zinc in juveniles of five estuarine-dependent fishes. In **Proceedings of the Third National Symposium on Radioecology,** p. 769-775. (ed. Nelson, D. J.) USAEC CONF-710501-P2. N.T.I.S., Springfield, Virginia.
4. Cross, F. A., Duke, T. W., and Willis, J. N.
1970 Biogeochemistry of trace elements in a coastal plain estuary: Distribution of manganese, iron and zinc in sediments, water and polychaetous worms. **Chesapeake Sci.,** 11(4):221-234.
5. Cutshall, N.
Turnover of ^{65}Zn in oysters. **Health Phys.** (In Press.)
6. Ferguson, R. L., and Murdoch, M. B.
Microbial biomass in the Newport River estuary. (This symposium.)
7. Hoss, D. E.
1964 Accumulation of zinc-65 by flounder of the genus **Paralichthys. Trans. Am. Fish. Soc.,** 93:364-368.
8. Kuenzler, E. J.
1969 Elimination of iodine, cobalt, iron and zinc by marine zooplankton. In **Symposium on Radioecology,** p. 462-473. (ed. Nelson, D. J., and Evans, F. C.) USAEC CON F-670503. Oak Ridge, Tennessee.
9. Lowman, F. G., Rice, T. R., and Richards, F. A.
1971 Accumulation and redistribution of radionuclides by marine organisms. In **Radioactivity in the Marine Environment,** p. 161-199. National Academy of Sciences, Washington, D. C.
10. Pentreath, R. J.
1973 The roles of food and water in the accumulation of radionuclides by marine teleost and elasmobranch fish. In **Radioactive Contamination of the Marine Environment,** p. 421-436. I.A.E.A., Vienna.

11. . Peters, D. S., and Kjelson, M. A.
Consumption and utilization of food by various post-larval and juvenile estuarine fishes. (This symposium).

12. Small, L. F., and Fowler, S. W.
1973 Turnover and vertical transport of zinc by the euphausiid **Meganyctiphanes norvegica** in the Ligurian Sea. **Mar. Biol.,** 18:284-290.

13. Small, L. F., Fowler, S. W., and Keckes, S.
1973 Flux of zinc through a macroplanktonic crustacean. In **Radioactive Contamination of the Marine Environment,** p. 437-452. I.A.E.A., Vienna.

14. Thayer, G. W., Adams, S. M., and LaCroix, M. W.
Structural and functional aspects of a recently established **Zostera marina** community. (This symposium.)

15. Vanderploeg, H. A.
1973 Rate of zinc uptake by Dover sole in the northeast Pacific Ocean: Preliminary model and analysis. In **Proceeding of the Third National Symposium on Radioecology,** p. 840-848. (ed. Nelson, D. J.) USAEC-CONF-710501-P2. N.T.I.S., Springfield, Virginia.

16. Willis, J. N., Cross, F. A., and Lewis, J. A.
1973 Kinetics of Zn exchange in the mosquitofish, **Gambusia affinis.** In **Ann. Rept. to the Atomic Energy Commission,** p. 38-47, July 1, 1973. N.M.F.S. Atlantic Estuarine Fisheries Center, Beaufort, North Carolina.

17. Wolfe, D. A.
Modeling the distribution and cycling of metallic elements in estuarine ecosystems. (This symposium.)

18. Wolfe, D. A., Cross, F. A., and Jennings, C. D.
1973 The flux of Mn, Fe and Zn in an estuarine ecosystem. In **Radioactive Contamination of the Marine Environment,** p. 159-175. I.A.E.A., Vienna.

GEOCHEMISTRY OF MERCURY IN THE ESTUARINE ENVIRONMENT

Steven E. Lindberg[1]

Anders W. Andren[2]

Robert C. Harriss[3]

ABSTRACT

The transport and deposition of mercury in the estuarine environment is largely controlled by its interaction with natural organic matter. Field and laboratory studies indicate that 54% to 82% of the total dissolved mercury in the coastal waters of the Gulf of Mexico is associated with a fulvic acid type material primarily of less than 500 molecular weight. The mercury-organic matter association is not significantly decreased by exposure to increased salinity, meaning mercury transport in estuaries is characterized by ideal dilution. Possible exceptions are estuaries with very high dissolved organic carbon where some precipitation of high-molecular-weight organic material containing mercury may occur over the salinity gradient.

Mercury in sediments is strongly associated with particulate organic matter. Decomposition of estuarine plant material produces detrital organic material

1. Department of Oceanography, Florida State University, Tallahassee, Florida 32306.

2. Ecological Sciences Division, Oak Ridge National Laboratory, Oak Ridge, Tennessee 37830.

3. The Marine Laboratory, Florida State University, Tallahassee, Florida 32306.

that is richer in mercury than is the living tissue. Mercury in estuarine sediments from the Gulf of Mexico is not significantly redistributed by postdepositional processes active in unconsolidated recent sediment. Sediment pore water in coastal sediments is enriched in both dissolved organic matter and dissolved mercury relative to overlying water with the mercury occurring primarily in a natural fulvic acid type complex of less than 500 molecular weight. Simulated dredging of coastal sediment results in a slight increase in total dissolved mercury in the overlying water caused by mixing with mercury-enriched sediment pore water and short-term release of a small fraction of mercury associated with particulate material.

INTRODUCTION

In recent years the quantity of mercury being discharged to the natural environment from anthropogenic sources has been almost equal to the quantities estimated to be derived from natural weathering processes (54). Considerable increases in the dissolved and particulate mercury flux in marine areas adjacent to point discharges have produced ecological problems ranging from the consumption by humans of contaminated seafood in Japan (56) to the closing of extensive fisheries because high levels of mercury were found in tissues of living fish (6). As a result of several well-documented cases of acute mercury contamination in Japan and Sweden (50, 60), most large industrial sources have now been identified and the discharge of mercury to natural waters reduced to negligible flux rates relative to natural sources.

A lot of the mercury released before the pollution-control regulations were implemented has accumulated in the sediments of estuaries associated with watersheds that receive discharges. The combination of high reactivity with particulate material (14, 21, 33) and physicochemical conditions conducive to high rates of sedimentation provides a mechanism for trapping large quantities of mercury in nearshore sediments (2, 7, 15, 40, 54, 115). This paper will summarize existing information on the transport, deposition, and cycling of mercury in estuarine systems. Questions of particular ecological and geochemical significance to be considered are: (a) What factors control the concentration and chemical speciation of dissolved mercury in the estuarine environment? The uptake and toxicity in estuarine biological communities are related to the chemical form of mercury (13, 35); (b) How mobile is mercury associated with unconsolidated estuarine sediments? It is particularly important to determine whether the large quantities of anthropogenic mercury will be recycled by post-depositional processes and consequently represent a long-term ecological hazard. (c) How do such man-induced modifications as dredging, coastline modification, and reductions in marsh area, which are commonly associated with land development, affect the fluxes of mercury between various compartments of the ecosystem?

Reviews on the geochemistry of mercury in terrestrial and fresh-water environments have been published (45, 98, 111, 113).

MERCURY-WATER INTERACTIONS

Using stability constant data from the literature, several authors have investigated a range of reactions that mercury might undergo in aqueous media (5, 18, 37). From these data the authors postulated that a combination of pH and pCl governed the speciation of mercury in natural oxygenated waters. Thus in fresh-water systems that $Hg(OH)_2$, $HgOHCl$ and $HgCl_2$ should predominate, whereas in sea water at least 80% of the dissolved mercury should exist as $HgCl_4^{=}$. If other halogens are included in the calculations the HgI_2 species will also be important in the fresh-water system. Hem (37) constructed Eh-pH stability diagrams for aqueous species of mercury in 35 mg/1 chloride solutions and came to the conclusion that almost all of the mercury should exist as Hg^o.

Solubility calculations by Goldberg et al (31) and Anfält et al (5) suggest that mercury should have a concentration of from a few micrograms per liter to several hundred micrograms per liter, if equilibrium conditions are present. From the dissolved-mercury data presented in Table 1, it can be seen that concentrations found are from four to five orders of magnitude less than the calculated solubilities. It is obvious that phenomena other than solubility equilibria determine the concentration of mercury in natural waters. The most likely explanations for the observed deviations are: (a) slow dissolution kinetics of mercury from the source rocks, which would keep mercury undersaturated with respect to its least soluble salts; (b) the formation of organic complexes; and (c) the rapid adsorption of mercury by suspended matter.

Organo-mercury complexes

Recent literature contains a considerable amount of speculation on the role of organic matter and its association with metals in marine and fresh-water systems (e.g., 10, 17, 23, 39, 100, 102).Conclusions from these studies indicate that association with dissolved organic matter can both increase and decrease the solubility of trace metals. Schutz and Turekian (96) found that in areas of high productivity concentrations of silver, cobalt, and nickel increased with depth. Also they postulate that in the near-shore areas of Long Island Sound precipitation of organic matter and its associated metals can lower the concentration of these elements to values below that of open-ocean water. Sieburth and Jensen (99), Matson (70), and Szalay and Szilagyi (106) hypothesize that humic and fulvic material could precipitate at the zone of

TABLE 1

RIVER AND ESTUARINE DATA ON DISSOLVED- AND SUSPENDED-MERCURY CONCENTRATIONS

Location	Dissolved Mercury (μg/1)		Particulate Mercury (μg/kg)	
	Range	Average	Range	Average
Thames River[a]	0.003-0.515	0.109	1800-35,500	11,200
Le Havre estuary[b]	0.036-0.380	0.093	2040-34,400	11,800
Columbia River[c]	0.005-0.015	0.01	126- 176	150
Swedish lakes[d]	0.05 -0.5	0.05	----	—
Mississippi River[e]	0.02 -0.11	0.04	98- 1108	510
Mobile River estuary[e]	0.03 -0.10	0.05	62- 2310	1,333
Everglades[e]	0.05 -0.14	0.10	1201- 1314	1,270

[a]Smith et al. (103); [b]Cranston and Buckley (21); [c]Bothner and Carpenter (14); [d]Löfroth (68); [e]Andren (2).

mixing, with the consequent removal of the associated trace elements.

Mercury has a great affinity for organic matter in soils, peats, and sediments (1, 33, 92). Fitzgerald and Lyons (28) report that from 50%-60% of dissolved mercury in coastal waters may exist in association with organic matter. Part of this could presumably be methylated mercury that has escaped into the water column (78). Thus, one of the first questions to answer is how much of the total dissolved mercury that exists in estuarine surface waters is in the form of alkylmercury.

Attempts have been made to determine the methylmercury in fresh- and sea-water samples using specific inorganic-methylmercury resin (63) together with the neutron-activation analysis of Andren and Harriss (3). Although the standards produce clear peaks, no peaks could be detected for a group of six samples collected over the salinity wedge in the Mississippi River. The precision of four standard additions at the 5 ng level was only 50%. This poor precision was caused by the incomplete uptake by the resin. Nevertheless, if any methylmercury was present in the samples its concentration must have been less than one part per trillion, which is the approximate limit that can be detected by the present technique. Accordingly, using 40 ng/1 as an approximate average for the total dissolved mercury in the Mississippi River (Table 1), less than 3% of the dissolved mercury is methylated. Two samples from Mobile Bay and one from the Florida Everglades were similarly run with the same results. The Everglades water had a total dissolved-mercury concentration of 140 ng/1 , indicating that the methylmercury content was less than 0.7%. Dissolved methylmercury must be extremely low in these estuarine areas, playing an insignificant role in the transport of mercury to the oceans.

In an effort to study further the interactions between mercury and organic matter, Amicon ultrafiltration was used to fractionate dissolved organic matter from estuarine surface waters into five ranges of molecular weight. Each range was then analyzed for dissolved mercury and dissolved organic carbon. The technique and its limitations are discussed by Andren and Harriss (3). Surface waters of four different salinities collected from the salt wedge of the Mississippi River delta and of two salinities from the Everglades were studied with this method. The data for the total dissolved mercury, total dissolved organic carbon (DOC) and salinity are presented in Table 2. Data from the molecular weight fractionations are presented in Figures 1 and 2. In the Mississippi River the less than 500 molecular weight cut-off fractions represent 88%, 85%, 92%, and 95% of the total dissolved organic matter. Mercury associated with this dissolved-organic-matter fraction is 64%, 54%, 73%, and 82% of the total dissolved mercury for the respective samples. There is an increasing amount of both dissolved organic carbon and dissolved mercury in the low molecular weight cut-off fractions as higher salinities are reached. A considerable amount of mercury is associated with the larger than 500 molecular weight cut-off

TABLE 2

Total dissolved mercury, total dissolved organic carbon, methyl mercury, and salinity data from the Mississippi River (Miss) and the Everglades (Ev).

Sample	Dissolved Mercury (ng/1)	MeHg (ng/1)	DOC (mg/1)	Salinity ($^o/oo$)
Miss-1 total	40	< 1	5.0	1
Miss-11 total	70	< 1	5.2	11
Miss-22 total	70	< 1	2.3	22
Miss-33 total	30	< 1	3.0	33
Ev-4 total	110	< 1	8.0	4
Ev-22 total	50	< 1	6.7	22

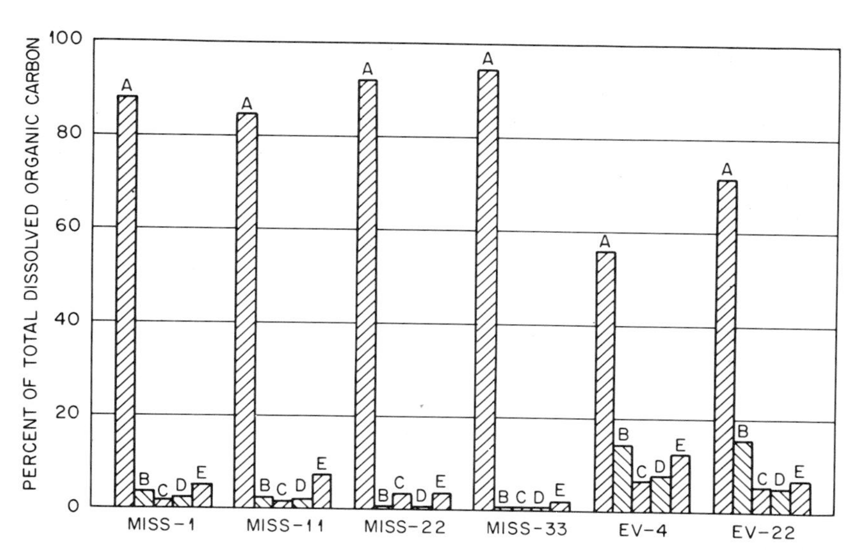

Figure 1. Molecular weight fractionations of dissolved organic carbon: A - less than 500 molecular weight; B - 500 to 10,000; C - 10,000 to 100,000; D - 100,000 to 300,000; E - greater than 300,000.

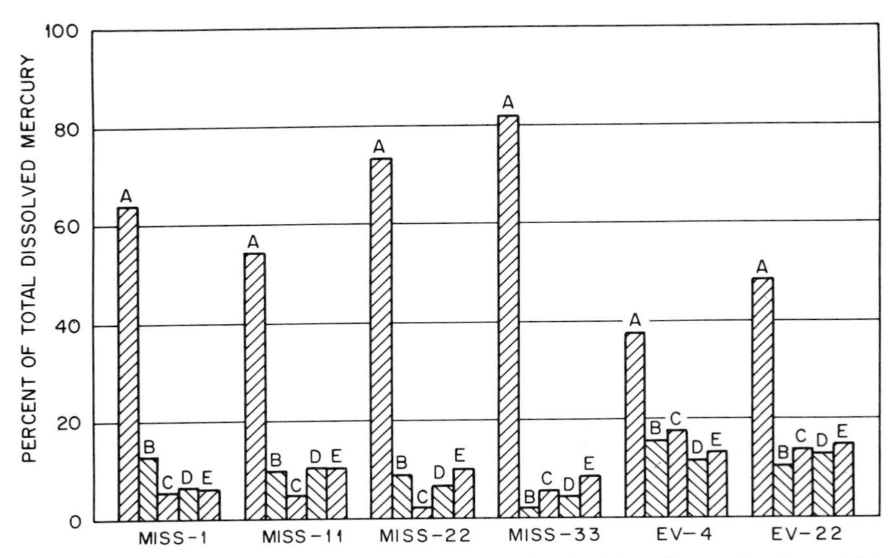

Figure 2. Fraction of total dissolved mercury associated with various molecular weight fractions of the dissolved organic carbon.

fraction of the dissolved organic matter, especially in the less saline samples. From these data it is hard to ascertain whether the relative increases in the smaller molecular weight material with higher salinity are due to precipitation of the larger molecules or are simply the effects of dilution by sea water.

The Everglades water, with its higher total dissolved organic load, exhibits a different distribution pattern. At a salinity of 4°/oo, about 42% of the dissolved organic matter exists in the greater than 500 molecular weight cut-off fraction, containing 61% of the total dissolved mercury. At a salinity of 22°/oo, the dissolved organic matter in the greater than 500 molecular weight fraction drops to 31% with 54% of the total dissolved mercury associated with it. The greater than 10,000 molecular weight dissolved organic carbon fraction decreases by 1.14 mg/l in going from low to high salinity. Since this is the approximate difference in the total dissolved organic carbon for the two stations, there may be some precipitation of higher molecular weight dissolved organic matter.

Although there may be precipitation of higher molecular weight dissolved organic matter containing mercury as more saline water is reached in the Mississippi River, the magnitude of this removal effect is probably very small compared to the one that operates in the Florida Everglades. The latter system possesses some of the highest dissolved organic carbon values recorded in literature for estuarine regions (22). It may be that this system is close to the point of saturation with respect to those materials. Thus, the salting-out effect proposed by Sieburth and Jensen (99) and Turekian (110) may be operative only in estuarine areas high in dissolved organic matter.

Mercury in the less than 500 molecular weight fraction is not necessarily organically bound, as inorganic mercury compounds will also occur in this fraction. However, when the percentage of the total dissolved mercury associated with the less than 500 molecular weight fraction is plotted versus the percentage dissolved organic carbon in the same size fraction, a linear relationship results, as illustrated in Figure 3 for the Mississippi estuary samples.

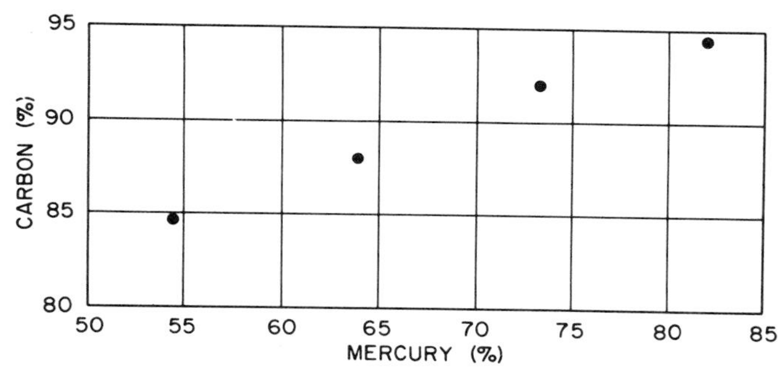

Figure 3. The percent of the total dissolved organic carbon in the less than 500 molecular weight fraction plotted versus the percent of the total mercury associated with this fraction for four Mississippi River samples.

Fulvic acids, which, according to Lamar (61), strongly resemble the dissolved organic matter in rivers and swamps, have been extracted from the sediments of Mobile Bay, the Everglades, and the Mississippi River delta by the method of Rashid and King (85). As will be discussed in a following section, the molecular weight distribution of organic carbon in the dissolved organic matter from Everglades and Mississippi River suface waters (Figs. 1 and 2) is similar to that seen for these naturally occurring sediment compounds of the fulvic acid type. This organic material was subjected to molecular-weight fractionations. In addition distribution constants, K_D, were measured between mercury and the fulvic acids by the techniques of Kruger and Schubert (59), Randhawa and Broadbent (82), and Schnitzer and Skinner (95). Table 3 shows the distribution constants for mercury (added as $HgNO_3$) and fulvic acids extracted from Mississippi River and Mobile Bay sediments. It is best to use the K_D values only to compare the stability of a mercury-fulvic-acid complex with the stability of a mercury-ion-exchange-resin complex under pH and salinity conditions simulating a estuarine environment and in this way to compare stabilities of mercury complexes with fulvic acids of different origins or molecular weights. Larger K_D values imply stronger mercury-fulvic-acid associations.

In both areas the less than 500 molecular weight cut-off fraction exhibits the highest complexing ability. In the Everglades this fraction binds 1.4 times as

TABLE 3

Distribution constants for mercury-fulvic acid complexes at 0.1 ionic strength and pH = 7.5.

Sample	Molecular Weight Cut-Off Fraction	K_D 2 mg F. A./100 ml	K_D 4 mg F. A./100 ml
Everglades	total	4110	5100
	> 300,000	2815	3270
	300,000–100,000	1370	1730
	10,000–500	4380	5470
	< 500	4768	5815
Mississippi River	total	1345	1745
	300,000	395	515
	> 300,000–100,000	412	545
	10,000–500	1410	1845
	< 500	1605	2080

much mercury per unit weight as the total fraction. In the Mississippi River this increase is by a factor of 1.3. It is also of interest that fulvic acids from the contrasting environments exhibit widely different complexing abilities. The total fulvic acid fraction from the Everglades complexes about three times as much mercury as that extracted from the Mississippi River and twice as much as that from Mobile Bay.

The strength of the mercury-fulvic-acid association is considerable. A 0.1 ionic strength solution was equilibrated with a cation exchange resin, mercury, and fulvic acid of less than 500 molecular weight and the distribution coefficient determined. KCl was then added to make the solution 0.7 ionic strength and the K_D was again measured after 4, 8, 12, and 24 hours of reequilibration. The results are shown in Table 4. Over a 24-hour period approximately 9% of the mercury is exchangeable. This fact indicates that mercury may be replaced very slowly by other cations in going from lower to higher salinities. This supports the finding of Strohal and Huljev (105) who showed that mercury was very strongly fixed to humic acids isolated from the Adriatic Sea.

Mercury suspended matter interactions

Because of the reactivity of mercury with particulate matter, the suspended load exerts a strong influence on its aqueous behaviour. The suspended load of rivers and estuaries is composed of mineral and organic material which has been shown by many investigators to adsorb trace elements (25, 32, 46, 110). Most of

TABLE 4

The effect of increased salinity on mercury-fulvic acid distribution constants.

Initial K_D	K_D 4 hrs	K_D 8 hrs	K_D 12 hrs	K_D 24 hrs
4110	4020	3940	3940	3825

organic content. Data for dissolved mercury, particulate mercury, suspended load, and salinity are presented in Table 5 for the Mississippi River. A point of interest is the relative constancy of the dissolved-mercury content. Even with the additional industrial and domestic input in New Orleans the concentration varies only from 0.03 to 0.08 μ g/1 . The fifth column in Table 5 shows the particulate-mercury variations for the stations over a salinity gradient. The mercury concentrations vary considerably but tend to decrease as salinity increases. This could be due either to desorption or mixing with low mercury-content sediments from the Gulf. However, based on the conclusions of Bothner and Carpenter (14) and the concomitant decrease in dissolved mercury as salinity gets higher, the latter hypothesis seems to be the likely one. Thus, the indication is that concentrations of both dissolved and suspended mercury are diluted upon mixing with sea water (Figs. 4 and 5)

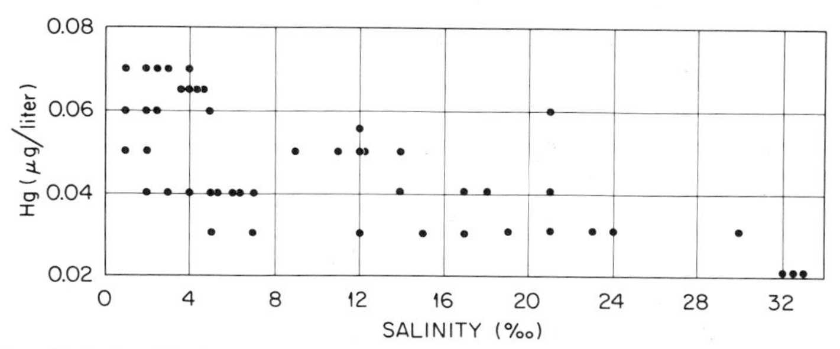

Figure 4. Relationship between dissolved mercury and salinity in the Mississippi River.

Another interesting feature apparent from the data is the relative constancy of the precent mercury on the particulate phase. In the Mississippi River from 61% to 79% of the total mercury in the water sample is associated with the particulate phase. No trends with increasing salinities are observed, but the field data agree fairly well with the experimental data of Bothner and Carpenter (14), i.e., between 50%-75% of the total mercury is associated with the particulate phase. The increase in the particulate-mercury content with distance downriver that Cranston and Buckley (21) observed in La Have estuary, Nova Scotia, is not

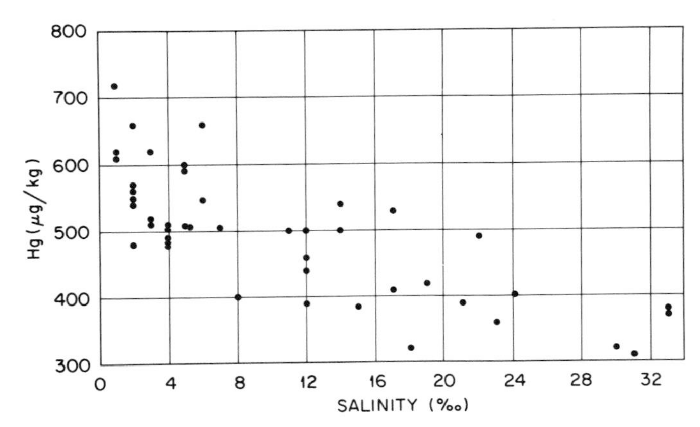

Figure 5. Relationship between mercury associated with the suspended particulate fraction and salinity in the Mississippi River.

observed in the Mississippi River. Those authors ascribed their results to the fact that as the estuary is approached, particle size decreases as a result of sedimentation of larger particles with a lower mercury content. Sedimentation in the Mississippi River estuary is negligible in the area studied by Andren (2). Also, La Have has a suspended load of 3.5 mg/1, as compared to approximately 250 mg/1 for the Mississippi estuary.

The Mobile Bay estuarine system differs in its physical characteristics from the Mississippi area because the discharge from the Mobile River passes through an enclosed bay before it enters the Gulf of Mexico. Data from the Mobile River system are presented in Table 6. The dissolved mercury concentrations exhibit once again very little variability, from 0.04 to 0.10 μg/1, and show a steady decrease as the higher-salinity waters of the Gulf of Mexico are reached. The data on the particulate-matter mercury in Mobile Bay do not show the same trend as that from the Mississippi River. The mercury content increases as salinity and distance from the point of discharge of the Mobile River increase. It is also significant that the suspended load drops drastically, indicating sedimentation of larger particles as higher salinity water is approached. Thus, it is possible that the phenomenon observed by Cranston and Buckley (21) in La Have estuary is operating in this area as well. That is, the increase in the mercury concentration is due to the increase in finer particles with higher mercury content that remain in suspension.

In general, the amount of mercury associated with particular matter in Mobile Bay is somewhat lower than that in the Mississippi River (51%-66%), but not as low as that observed in the Everglades (Table 7). One possible explanation for this difference is that dissolved organo-mercury complexes in the Everglades are not as efficiently adsorbed onto the suspended matter, and consequently leave

TABLE 5

DISSOLVED- AND PARTICULATE-MERCURY DATA FOR SALINITY GRADIENT IN MISSISSIPPI RIVER

Station	DOC mg/1	POC mg/1	Suspended Load mg/1	Particulate Matter Hg mg/kg	Particulate Matter Hg µg/1	Dissolved Hg (ppb)	Silinity °/oo	Total Hg in Water µg/1	% of Hg in Particulate Matter
1			312	443	0.14	0.05	< 1	0.19	74
2			198	414	0.08	0.05	< 1	0.13	62
3			267	1090	0.29	0.08	< 1	0.37	78
9	4.5	0.4	246	580	0.143	0.04	< 1	0.18	78
14-1 m	5.2	1.5	257	602	0.155	0.07	11	0.23	70
14-2 m	4.0	2.2	112	511	0.057	0.03	15	0.08	63
14-4 m	3.6	–	111	521	0.058	0.03	17	0.08	63
14-6 m	5.2	1.3	201	531	0.107	0.03	17	0.14	79
14-7 m	3.6	–	282	411	0.116	0.07	19	0.18	61
14-9 m	3.6	0.5	273	411	0.112	0.04	19	0.15	73
15-4 m	2.3	1.2	521	493	0.257	0.07	22	0.33	79
16-10 m	3.0	1.2	203	487	0.099	0.03	33	0.13	77

TABLE 6

DISSOLVED- AND PARTICULATE-MERCURY DATA FOR MOBILE BAY

Station	Particulate Matter mg/1	Particulate Matter Hg μg/kg	Particulate Matter Hg μg/1	Dissolved Hg, μg/1	Total Hg in Water μg/1	% Hg on Particulate Phase	Salinity $^{o}/oo$
15	20	2310	0.046	0.04	0.09	51	24
17	31	1840	0.057	0.04	0.10	57	24
19	38	1730	0.066	0.05	0.12	55	18
21	111	1170	0.130	0.08	0.21	62	12
22	130	1310	0.170	0.10	0.27	63	<6
23	84	1240	0.104	0.06	0.16	63	<1
24	41	1480	0.061	0.04	0.10	61	<1
25	46	1990	0.092	0.06	0.15	61	<1
26	48	2190	0.105	0.06	0.16	66	<1
27	52	1540	0.080	0.05	0.13	62	<1
28	48	1570	0.075	0.05	0.12	62	<1

TABLE 7

PARTICULATE- AND DISSOLVED-MERCURY CONCENTRATIONS FOR THE EVERGLADES

Station	Particulate Matter mg/1	Particulate Matter Hg μg/kg	Particulate Matter Hg μg/1	Dissolved Hg μg/1	Total Hg μg/1	% on Particulate	Salinity ⁰/oo	DOC mg/1
2	94	1295	0.122	0.14	0.26	47	<1	11.0
14	99	1314	0.131	0.11	0.24	55	4	8.0
17	62	1201	0.074	0.05	0.12	62	22	6.7

more mercury in solution.

In conclusion, it is difficult to evaluate *in situ* how much of the adsorbed mercury is exchangeable from the particulate matter in these estuaries. There is an increase in the particulate phase mercury concentration in going from the Mobile River to the Gulf of Mexico, presumably because of the higher mercury content of finer particles that remain in suspension. This phenomenon was not observed in the Mississippi River. It might be, therefore, that estuaries where considerable sedimentation takes place act as reservoirs for a large amount of mercury-laden particulate matter. Since the dissolved-mercury concentration is fairly constant within any one of the discussed estuaries, the total amount of mercury in a water sample is governed by the amount of suspended matter. That is, the higher the suspended load, the higher the mercury content of the water sample. In each estuary the percent mercury associated with the particulate phase is fairly distinctive.

MERCURY INTERACTIONS IN THE SEDIMENTARY ENVIRONMENT

Mercury-sediment interactions

Much of the mercury that enters the estuarine region eventually interacts with the sedimentary environment through adsorption on suspended sediments (21), coprecipitation with dissolved organic matter (3), concentration by degrading plant detritus (66), or a combination of the three. This section concerns the interactions and ultimate fate of mercury once it has been deposited. De Groot et al. (33) postulate that considerable post-depositional migration occurs in sediments, with mercury going into solution in the surrounding water as an organo-metallic complex. It has also been suggested that the production of hydrogen sulfide through bacterial decomposition of plant and animal matter under anaerobic conditions immobilizes dissolved mercury by precipitation as an insoluble sulfide (49, 79).

Mercury is believed to occur in the sediments in three possible forms. The predominant manner in which mercury is bound in the sediment is through association with organic material (52). Mercury shows a strong affinity for SH, NH_4, COOH, and phenol groups (20, 30, 44, 52), all of which are found in natural organic matter (42, 86, 87, 94). Statistical correlations between total sediment sulfur and mercury in lake sediments have been interpreted as evidence that mercury often occurs as the insoluble sulfide (53) or is adsorbed onto the surfaces of sulfide minerals such as FeS_2 (108). Vernet and Thomas (112) and Thomas (109) suggest that observed relationships between total Hg, Fe, and P indicate mercury to be bound to an inorganic iron-phosphate complex probably adsorbed onto a hydrated iron oxide coating on clay particles.

Within the last few years an enormous amount of data on relative mercury contents in various earth materials have appeared in the literature. A partial compilation of sediment data is presented in Table 8. From these data it is evident that mercury concentrations vary considerably both within and between these environments. For this reason such bulk sediment analyses are often useless and comparison of different systems has a limited geochemical significance unless controlling parameters are simultaneously determined. Such parameters include sediment organic content, sediment mineralogy, grain size, salinity of overlying water, suspended load, estuarine hydrology, and the sources of mercury input.

TABLE 8

COMPILED LIST OF MERCURY CONCENTRATIONS IN RECENT SEDIMENTS

LOCATION	Average (μg/gm)	Range (μg/gm)
San Francisco Bay[a]	0.29	0.02 - 2.00
Southern Lake Michigan[b]	0.14	0.03 - 0.38
Lake Wisconsin[c]	0.15	0.01 - 0.35
Rhine River[d]	6.90	1.2 - 23.3
Thames River[e]	1.80	1.0 - 3.3
James, York and Rappahannock estuaries[f]	1.14	0.4 - 2.6
La Jolla, California[g]	0.34	0.02 - 1.0
Le Havre River and estuary[h]	0.34	0.09 - 1.06
Connecticut Harbor[i]	0.83	0.04 - 2.57
Mississippi River[j]	0.33	0.07 - 1.10
Mobile River estuary[j]	0.37	0.03 - 6.14
Everglades[j]	0.60	0.04 - 1.86
Bellingham Bay[k]	4.27	0.8 - 10.7
Georgia salt marsh[l]	0.11	0.05 - 0.35

[a]McCulloch et al. (72); [b]Kennedy et al. (53); [c]Konrad (57); [d]de Groot et al. (33); [e]Smith et al. (103); [f]Huggett et al. (40); [g]Klein and Goldberg (54); [h]Cranston and Buckley (21); [i]Applequist et al. (7); [j]Andred (2), Lindberg (64); [k]Bothner and Piper (15); [l]Windom (115).

For example, Figure 6 illustrates typical mercury concentrations for surface sediments from the Mississippi River, and for sediment cores from Mobile Bay and the Everglades. Individual sample locations are described elsewhere (2, 64). A large part of the sediment mercury variability in the Mississippi River can be attributed directly to the relative proximity of the sample sites to industrial outfalls known to release mercury-laden effluents. The distribution of mercury in both cores from Mobile Bay suggests a recent increment caused by the release of mercury-containing industrial effluents to this area. The Mobile River is the site of a chemical plant that has used the mercury-cell chloralkali process since 1965. In the organic-rich environment of the Everglades, an area not known to be receiving any anthropogenic mercury inputs, total mercury in the sediments shows a significant correlation (r=.8, based on 20 surface and 16 subsurface samples) with sediment organic content. Significant mercury-organic matter correlations have previously been found in soils by Andersson (1) and Shacklette (98) in fresh-water sediments by Kennedy et al. (53) and Thomas (109), and in estuarine sediments by Bothner and Piper (15) and Applequist et al. (7).

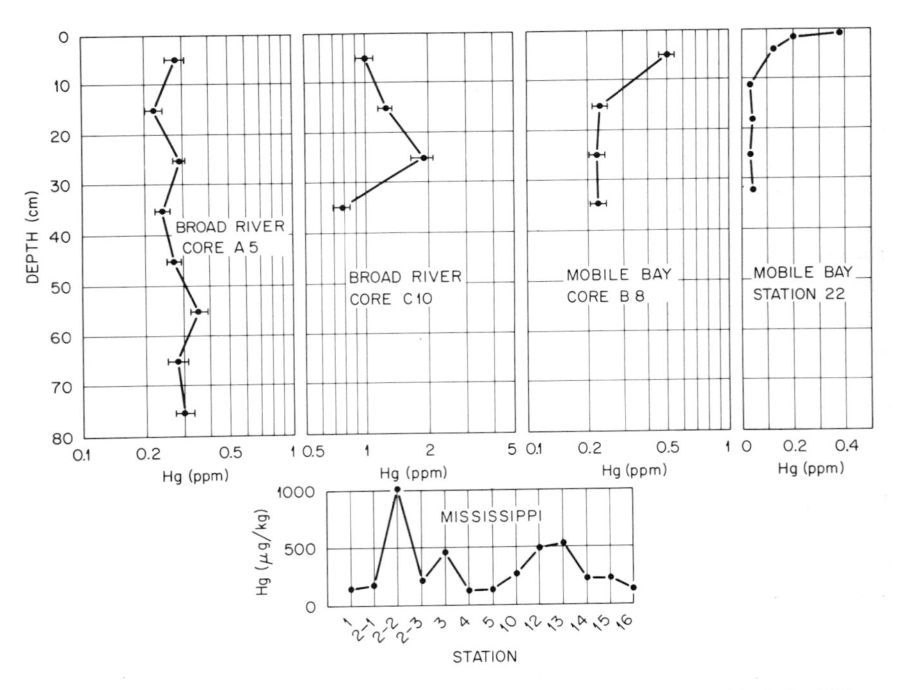

Figure 6. Total mercury concentrations for surface sediments from the Mississippi River, and for sediment cores from Mobile Bay (station 22 and core B8) and the Everglades (Broad River core A5, C10).

Distribution of mercury in interstitial water

The results of pore-water analyses on cores from the Everglades and Mobile Bay are presented in Figure 7. For the Everglades samples no consistent trends in dissolved interstitial mercury concentrations with core depth are apparent. The large variability in dissolved mercury (cores A5 and C10) suggests that distribution is influenced by a number of interacting factors in this environment. Parameters that may influence mercury distribution in these cores, such as salinity, total dissolved sulfides, and dissolved organic carbon (DOC), show similar variations (65). In contrast, the single core taken from Mobile Bay exhibits concentrations of interstitial mercury decreasing with depth to apparent background levels near 20 cm. The observed gradient corresponds to the total mercury distribution in this same core (Fig. 6).

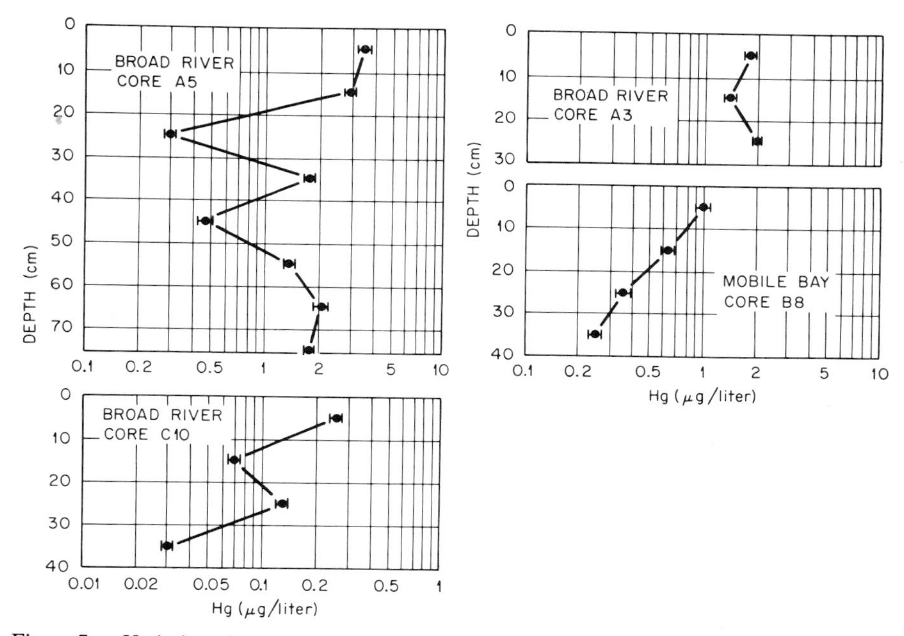

Figure 7. Variations in dissolved interstitial mercury with core depth from the Everglades (Broad River core A5, A3, C10) and Mobile Bay (core B8).

To elucidate the geochemistry of mercury in these sediments, correlation and stepwise regression analyses were applied to the data. Forthe Everglades samples, the concentration of dissolved mercury in the pore water is significantly correlated with the concentration of dissolved organic carbon (r=.55). The correlation increases significantly when calculated for the Everglades surface samples, < 10cm, (r=.81), but decreases to an insignificant level for the Everglades subsurface samples (r=.46). These results provide some insight into the processes that influence the distribution of mercury in recently deposited sediments of the Everglades. As discussed in an earlier section, there is evidence for complexing between aqueous mercury and dissolved organic matter in surface waters of the Everglades. In the process of sedimentation this complex enters the pore spaces between sediment particles. The decrease in correlation between dissolved interstitial mercury and dissolved organic carbon with increasing core depth suggests that early diagenesis reduces the integrity of the organo-mercury association. Statistical analyses do not reveal any significant relationships in the Mobile Bay core, other than the correlation between interstitial mercury and depth (r=.98).

Mercury-organic matter interactions in sediments and interstitial water

The strong evidence presented thus far for the association between mercury and organic matter in estuarine sediments and associated pore waters may be attributed to two possible mechanisms. Either the associations are the results of mercury complexing by natural organic chelators such as humic and fulvic acids (3) or they represent various alkylmercury compounds formed by microbial activity (48). Presley et al. (81) suggested that the formation of natural organic complexes with Fe, Mn, Zn, Cu, and Ni explain the observed mobility of these elements in pore waters. On the other hand, laboratory studies suggest that diagenetic processes in natural sediments can transform inorganic mercury to mono- or di-methylmercury (48).

The major evidence for the occurence of alkylmercury compounds in the environment is the discovery of the natural biological methylation of mercury (47, 48, 117). Though the exact process is not known, many workers have suggested reaction mechanisms ranging from degradation of phenylmercuric compounds to chemical methylation of inorganic mercury with methylcobalamine, a vitamin B_{12} analog (41, 62). According to Wood (116), all microorganisms capable of vitamin B_{12} synthesis are capable of methylmercury synthesis. On the basis of these findings the authors postulate that methylation of mercury in river and estuarine sediments could be the process responsible for the mobility of mercury in the aquatic environment.

Table 9 is a summary of methylmercury concentrations measured in surface sediments from the Mississippi Delta, Mobile Bay, and the Everglades, while Figure 8 illustrates the distribution of methylmercury in two sediment cores from Mobile Bay. Details of the analytical procedure are presented elsewhere (4). The most interesting aspect of the data is the fact that the methylmercury concentrations never represent more than 0.07% of the total mercury present, the average being 0.02%. The four samples from the Mississippi River have

TABLE 9

Methylmercury, Total Mercury, and Organic Matter in Estuarine Sediments.

Location	Station	Total Hg (ng g^{-1})	MeHg (ng g^{-1})	% of total Hg	% organic matter
Mobile Bay	16	210	0.06	0.03	5.5
	19	600	0.19	0.03	11.4
Mississippi River	1	140	$<$ 0.02	$<$ 0.01	0.5
	5	80	$<$ 0.02	$<$ 0.01	2.4
	10	510	0.05	0.01	6.9
	15	570	0.05	0.01	9.4
Everglades	4	120	0.06	0.05	2.0
	14	120	0.08	0.07	11.1
	17	490	0.12	0.03	69.0
	6a	370	0.05	0.01	28.8
	6b	290	0.04	0.01	33.0

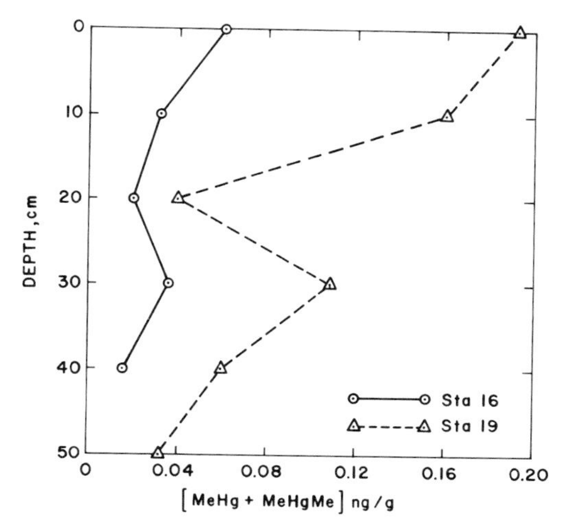

Figure 8. Variations in sediment methylmercury concentrations in two cores from Mobile Bay.

exceedingly small percentage values, where as the samples from the Everglades, where biological productivity is high, show the highest percentage values. Similar results have recently been found in Sweden where measurements of lake sediments showed that methylmercury rarely represents more than 0.10% of the total mercury present (Jernelöv, personal communication). Thus, the observed correlations between mercury and organic matter cannot be explained by alkylation.

Considerable speculation has appeared in the recent literature on the importance of the group of organic materials that possess the ability of association with metal ions in the ocean and fresh-water systems (8, 17, 104). Two recent reviews (93, 100) have discussed the role of humic and fulvic materials in this aspect. These compounds are believed to be formed by microorganism-catalyzed degradation of plant lignins, proteins, and cellulose into quinonoid and phenolic compounds (75, 80). Felbeck (27) identified approximately 50%-60% of the total humic molecule as being comprised of amino acids, hexosamines, polycylic aromatics, and oxygen-containing functional groups. Among these functional groups Rashid and King (86) found the following: carboxyl, phenolic hydroxyl, enolic hydroxyl, alcoholic hydroxyl, quinone, hydroxyquinone, other carbonyl ester, lactone, ether, and amine groups. Schnitzer (93) reported that 60% of the weight of fulvic acid molecule is in functional groups. As a consequence, fulvic acid has a strong affinity for many cations (69, 83, 95), including mercury (105). Rashid (83) also showed that as much as 90% of the cation-exchange capacity of marine sediments can be contributed by humic substances.

Because of their purported importance in determining the mercury distribution in sediments, humic and fulvic acids have been extracted from Everglades, Mobile Bay, and Mississippi River sediments for functional group analysis (total acidity, -COOH, and phenolic -OH groups), molecular weight determinations, and binding studies. The data are presented in detail elsewhere (2) but can be briefly summarized. The humic material constitutes from 42%-73% of the total organic matter. Between 50%-85% of the humic material is composed of fulvic acid. In the Mississippi River and the Everglades approximately 50% of the fulvic acid has molecular weights of less than 10,000. In the Mobile Bay most of the fulvic acid is in the 10,000-300,000 range. In all these areas most of the humic acid exists in the greater than 10,000 molecular weight fraction, a considerable amount being in the greater than 300,000 molecular weight fraction. Compared to previously measured molecular weight distributions of marine and soil humics (85), it appears that estuarine humic materials more closely resemble those of marine origin in the Mississippi delta and Mobile Bay and those of soil origin in the Everglades. The functional group distributions indicate that fulvic acid exhibits the highest acid characteristics for all areas. The less than 500 molecular weight fraction of humic and fulvic acids show the highest total acidity. The overall total acidity of the samples once again indicates that humates from the Everglades resemble soil humates, while those from Mobile Bay and the Mississippi River resemble those found in marine sediments.

Since the fulvic acid fraction constitutes most of the sedmentary organic matter, has the highest cation exchange capacity, and most closely resembles the dissolved organic matter discussed previously, this fraction has been used for laboratory studies of mercury-organic matter interactions. As discussed in a previous section, the results indicate that; (a) fulvic acid has a stronger affinity for mercury than has a strong cation-exchange resin, (b) the less than 500 molecular weight fraction of the fulvic acid has the highest complexing ability for mercury, and (c) the total fulvic fraction extracted from Everglades sediments complexes three times as much mercury as does that extracted from Mississippi River sediments and twice as much as that from Mobile Bay sediments (see Table 3). It seems that fulvic acid material is very effective in concentrating mercury in the sediment water system, especially in the organic-rich environment of the Everglades.

As was the case for mercury-sediment organic matter correlations, the mercury-dissolved organic carbon correlations in pore waters may represent mercury-humate complexes or alkylmercury compounds. In an earlier discussion a lower limit of .001 μ g/1 was set for the concentration of dissolved methylmercury in these estuarine waters. Preliminary studies on cores from Mobile Bay suggest the lower limit of CH_3HHg^+ in the interstitial water collected from 2 cm to 15 cm below the interface to be less than .1 μg/1 (Taylor, personal communication). Total dissolved mercury concentrations in these samples range from 0.7 to 1.2 μ g/1 , meaning the CH_3 Hg^+ never exceeds 15% of the total interstitial mercury and cannot entirely account for the observed mercury-dissolved organic carbon correlations.

The probable explanation for the observed relationships between dissolved mercury and dissolved organic carbon in pore waters is the formation of soluble organo-metallic complexes. Interstitial water from Mobile Bay and the Everglades has been studied with the intent of characterizing its organo-mercury complexes by molecular weight using a pressure dialysis technique termed ultra-filtration. The procedure is described and its limitations discussed in Lindberg and Harriss (65). The molecular weight distributions of pore water dissolved organic carbon and associated mercury are presented in Figure 9. The percent of the total dissolbed mercury present in any one molecular weight fraction shows a strong correlation (r=.94) with the percent of the total dissolved organic carbon in that same fraction. This suggests a significant association between dissolved mercury and organic matter in all molecular weight ranges.

The molecular weight distribution of the organo-mercury complex is

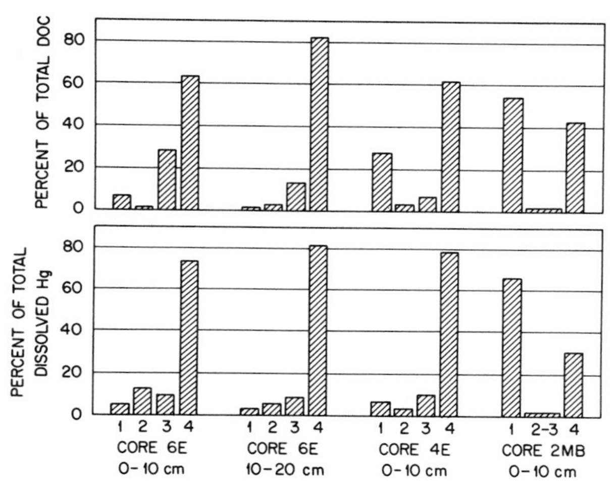

Figure 9. Pore-water molecular weight distributions of dissolved mercury and dissolved organic carbon (DOC). Molecular weight fractions are as follows: 1 - greater than 100,000; 2 - 100,000 to 50,000; 3 - 50,000 to 500; 4 - less than 500.

influenced by environmental and diagenetic factors. In pore water from Everglades surface sediments (Fig. 9, 4E and 6E, 0-10 cm) the organic fraction of less than 500 molecular weight contains a large majority of the dissolved mercury (73.7 to 78.8%) and dissolved organic carbon (61.6 to 64.1%) while the fraction greater than 50,000 molecular weight is relatively unimportant. With increasing depth in the core (from sample 6E 0-10 cm to 6E 10-20 cm) there is a general shift of mercury and dissolved organic carbon from higher to lower molecular weight ranges, suggesting that early diagenesis results in degradation of the organic material. The general molecular weight distribution of these compounds agrees quite well with the previously discussed molecular weight distribution of fulvic acids extracted from Everglades sediments. There is other evidence that dissolved organic compounds in these environments are similar to those found in soils and sediments. Fotiyev (1968, as cited in 93) reported that 85% of the dissolved organic matter found in marsh areas was fulvic acid. In addition the dissolved organic matter found in estuarine interstitial water by Nissenbaum et al. (74) was reported as very similar in composition to a fulvic acid extraced from terrestrial organic-rich waters by Black and Christman (12).

The greatest proportion of the pore-water mercury (66.4%) and organic carbon (55.0%) exists in the greater than 100,000 molecular weight fraction for the Mobile Bay sample. This is similar to the molecular weight distribution of humate compounds extracted from Mobile Bay sediments (2). Several hypotheses for the observed difference in molecular weight distributions between the Everglades and Mobile Bay include interaction between mercury and high molecular weight industrial and sewage effluents (7, 88) in Mobile Bay, as well as differences in the hydrology and characteristic microbial populations of the areas.

Studies of natural organic matter-trace metal interactions have generally concluded that the lower molecular weight fractions are the most efficient in complexing ability because of their relatively greater total acidity (carboxyl plus phenolic hydroxyl groups) (84). Table 10 was prepared by normalizing the absolute amount of mercury associated with selected molecular weight fractions to unit weight of dissolved organic carbon in that fraction. The results indicate a greater proportion of dissolved mercury complexed by the less than 500 molecular weight organic fraction than by the greater than 500 molecular weight organic fraction or parent organic fraction for the Everglades samples. This is in agreement with the relatively higher total acidity and mercury complexing ability measured for the less than 500 molecular weight fulvic acids from sediments in these estuaries.

TABLE 10

Mercury in Selected Molecular Weight Fractions Standardized to Unit Weight of Dissolved Organic Carbon.

Sample	Salinity, $^{o}/oo$	ng Hg/mg C in Molecular Weight Fraction		
		Parent Fraction	> 500	< 500
6E[a] (10-20 cm)	24.6	11.8	11.7	11.8
6E (0-10 cm)	19.7	18.1	13.3	20.8
4E (0-10 cm)	16.4	27	14.9	34
2MB[b] (0-10 cm)	2.5	178	217	127

[a]Everglades [b]Mobile Bay

The stability of the pore-water organo-mercury complex can be assessed by observing the mercury-to-dissolved-organic-carbon ratios for the parent fractions under different salinity regimes (Table 10). If the complex is resistant to competition effects resulting from major cations, the ratio should remain constant. If, however, mercury has been replaced by other cations, a decrease in the ratio would be expected as salinity increases. Though only limited data are available, it seems that increasing salinity has a negative effect on the mercury complexing capacity of the dissolved organic matter.

Referring back to the data for mercury concentrations in the sediments of the Mississippi River delta, Mobile Bay, and the Everglades (Table 8), it is interesting that in the relatively undisturbed Everglades environment, total sediment mercury occurs at a higher mean concentration than in the Mississippi River or Mobile Bay. Similarly, dissolved interstitial mercury in many Everglades samples occurs at higher concentrations than in Mobile Bay. Because of the strong evidence that mercury is primarily associated with the organic fraction, it is more significant to compare the environments after normalizing the pore-water and sediment mercury concentrations to organic contents of the samples in question. These results can be summarized as follows: Everglades mean sediment mercury concentration = 1.3 ± 0.7 nanograms mercury per milligram sediment organic matter; Mississippi River mean sediment mercury concentration = 11.8 ± 4.9 nanograms mercury per milligram sediment organic matter; Mobile Bay mean sediment mercury concentration = 2.3 ± 0.7 nanograms mercury per milligram sediment organic matter; Everglades mean interstitial mercury concentration = 0.029 ± 0.022 nanograms mercury per milligram dissolved organic carbon; Mobile Bay mean interstitial mercury concentration = 0.085 ± 0.061 nanograms mercury per milligram dissolved organic carbon. There are significantly higher relative concentrations of total sediment and dissolved interstitial mercury per unit weight of organic matter in the Mobile Bay and Mississippi River samples than in the Everglades samples. In conjunction with more extensive sampling, this may be a useful technique for evaluating the extent of mercury pollution in an area.

POSTDEPOSITIONAL MIGRATION AND REMOBILIZATION OF MERCURY

The question of whether the sediments act as a sink or a source for mercury in the estuarine environment may best be treated in terms of the stabilities of various sediment and pore-water mercury complexes. As discussed earlier, sediment fulvic acid compounds form mercury complexes with stabilities similar to those of strong cation exchange resins. Further laboratory experiments indicate that when the complex forms in solutions of varying ionic strength, increasing salinity causes a decrease in the complexing ability of fulvic acid. This agrees with the observation that increasing salinity has an adverse effect on the affinity of pore-water dissolved organic matter for mercury. However, as discussed in a previous section, the organo-mercury complex allowed to form under low salinity (4°/oo) conditions is little affected when subjected to sea water (35°/oo). The results indicate only 9% of the mercury is exchangeable after 24 hours. This suggests that, over a short time span, little remobilization of mercury occurs when sediment organo-mercury complexes are under the influence of saline overlying waters.

Methylation has been hypothesized to be an important mechanism of mercury remobilization from sediments. Methylmercury can escape from sediments in the form of dimethylmercury (47, 48). However, the low concentrations of CH_3Hg^+ and $(CH_3)_2Hg$ measured in the overlying water, sediments, and pore waters of estuaries as well as Jernelöv's finding (personal communication) that only 0.01 to 1% of the total sediment mercury is methylated per year suggest that this mechanism is relatively unimportant in these environments.

The concentration of dissolved mercury in Everglades and Mobile Bay near-surface pore waters (0-10 cm) represents enrichments of 2.3-36 times the average overlying water values measured for these areas (see Table 1). Enhanced diffusional mobility of the dissolved mercury should be an important consequence of this enrichment. To estimate the influence of ionic or molecular diffusion in the transport of mercury in pore waters, core B8 from Mobile Bay (Fig. 7) is taken as representative of the pore-water mercury distribution in that estuary. According to Berner (11), the solution to Fick's first and second laws of diffusion for a linear concentration gradient of an ion in pore water is $J_x = D(\partial c/\partial x)|_{x=o}$. The concentration gradient is calculated across the top 5 cm of the sediment useing 0.05 μg/1 for an average overlying water mercury concentration in Mobile Bay (see Table 1). The value for D, the sediment-diffusion coefficient, is 0.2×10^{-5} cm^2/sec (24) for divalent metal ions in sand. For the fine-grain clay sediments of Mobile Bay the values for mercury should be of the same order of magnitude or smaller.

The results of the diffusion calculations represent the maximum value expected to occur in Mobile Bay, because the equation assumes ideal mixing, a source with a constant concentration, and an optimum value of D for mercury, while it neglects any influence of the surfaces of solid particles on the dissolved ion mobility. The equation yields a diffusion flux of 0.012 μg Hg/cm^2 - year. Based on calculations presented later in this section concerning the average deposition rate of mercury in Mobile Bay sediments (.45 μg Hg/cm^2 - year), the diffusion of mercury out of the sediments represents less than 3.0% of the mercury entering the sediment system per year. If all of the remobilized mercury leaves the Mobile estuary, it represents approximately 0.01% of the total discharge of mercury from Mobile Bay into the Gulf of Mexico. Thus, even the maximum estimated diffusion is a negligible mechanism of mercury remobilization from these estuarine sediments.

The data discussed to this point suggest that although mercury occurs in relatively stable complexes with dissolved organic matter and is enriched in pore waters, remobilization through diffusion or alkylation is negligible. A little studied process that might lead to a more rapid release of mercury is large-scale resuspension of sediment allowing pore water and particulate matter with relatively high mercury concentration to mix with surface waters of low mercury content. Under certain conditions this could result in a significant discharge of mercury.

Nearly all engineering projects in rivers, estuaries, or near-shore marine sites involve physical changes in the quantity, suspension, or deposition of sediment by means of dredging. Extensive dredging of shipping lanes has been maintained in Mobile Bay Bay for a number of years (May 1973). A natural mechanism that has the same effect may exist in the Florida Everglades. The hydrography of the region is characterized by marked fluctuations in terrestrial run-off in response to the seasonal rainfall pattern. Heavy summer rains normally begin in June and may last until November, dropping 80% of the annual rainfall of over 60 inches during that period (36, 67). A rapid increase in the discharge of fresh water into rivers and estuaries has been known to cause extensive suspension of unconsolidated sediments.

Resuspension of bottom material by dredging, dumping, and reagitation during storms and floods is known to release nutrients and certain toxins (34, 76). Studies of effluents from pulp mills have shown suspended organic-rich sediments to be lethal to sockeye salmon smolts and walleye eggs and fry (19, 97). In each case toxicity was traced to the release of H_2S. Though the effect of released organic matter is not well known, the total amount in the water column may increase by 100,000 times over dissolved concentrations because of suspension of fine particulate matter (16). In light of the previous discussion on the association of mercury with organic matter, it is evident that the above could be an important mechanism of mercury mobilization in certain environments.

A field experiment was designed by Lindberg (64) to study the kinetics of mercury dissolution during periods of intense sediment-water mixing. The effect of resuspending freshly collected Mobile Bay near-shore marsh surface sediments or open-bay, dredge-spoil sediments in the associated overlying water was monitored with regard to the following parameters: pH, Eh, dissolved mercury, salinity, and dissolved organic carbon. In each case sediments and associated overlying water were placed in 14-liter bottles within one hour of collection and the sediment kept in constant suspension with a mechanical rotor for 5 hours. At the midpoint of the experiment suspended load was found to be 35,600 mg/1 in the marsh sediment trial, and 36,800 mg/1 in the dredge-spoil case. These compare favorably with suspended loads measured during actual dredging operations. Full details of the experimental technique are discussed elsewhere (64).

The general trend of the parameters monitored are plotted in Figure 10 (near-shore marsh sediments) and Figure 11 (dredge-spoil sediments). Table 11 is a summary of initial sediment and pore-water parameters. The major difference between the two sediments was in their organic content. The marsh sediment was approximately 20% organic matter and had a pore-water organic carbon content of 38.1 mg/1, while the dredge-spoil sediment was approximately 7% organic matter with a dissolved organic carbon content of 13.4 mg/1. In each experiment resuspension of surface sediments resulted in an initial rapid increase in the dissolved mercury concentration of the overlying water. Within 21

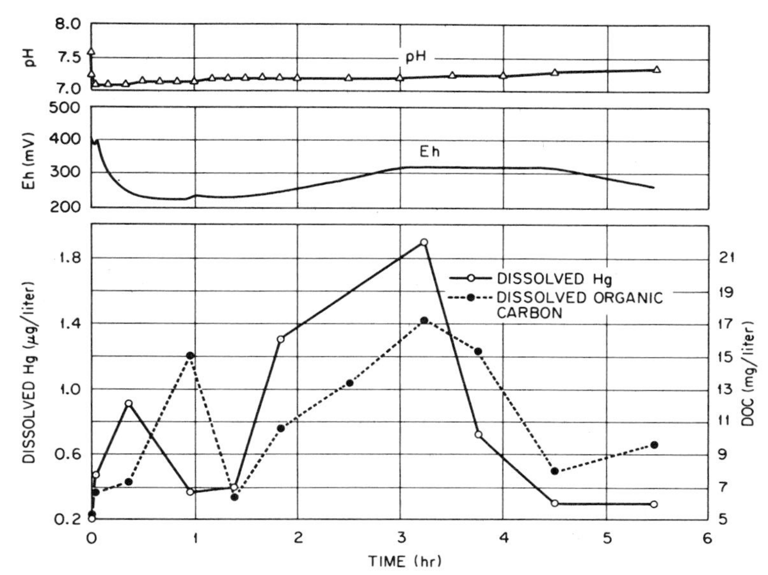

Figure 10. Release of mercury from resuspended Mobile Bay near-shore marsh sediments into the overlying water. The plot describes the time variation of the overlying water pH, redox potential (Eh), dissolved organic carbon (DOC), and dissolved mercury.

minutes for the marsh sediment case and 5 minutes for the dredge-spoil case, the dissolved mercury concentration increased from 0.2 μg/1 to appriximately 1.0 μ g/1. The concentration of dissolved mercury then decreased for approximately one hour, at which time a more substantial increase took place. The duration of this second peak was approximately 3 hours for the marsh sediment and 1.3 hours for the dredge-spoil sediment. Within 4.5 hours from the initial suspension of marsh sediment, the dissolved mercury concentration in the overlying water had reached a steady state value of .30 μg/1 which was maintained for one hour after the rotor was shut off and the sediment allowed to settle. For the dredge-spoil case a similar steady state was achieved within 2.5 hours of sediment suspension and was maintained with slight fluctuations at $.28 \pm 0.1$ μg/1 (based on 5 samples) for 2.5 hours during continued sediment suspension and for 1 hour after the sediment was allowed to settle. When the sediment was allowed to settle for a longer time (11 hours for the marsh sediment and 17 hours for the dredge-soil sediment), the mercury concentration in the overlying water remained near that of the steady state value.

For each trial a drop in the redox potential (Eh) and a fluctuation of pH of the overlying water occurred simultaneously with the initial mercury peak. Dissolved organic carbon released from the organic-rich marsh sediments

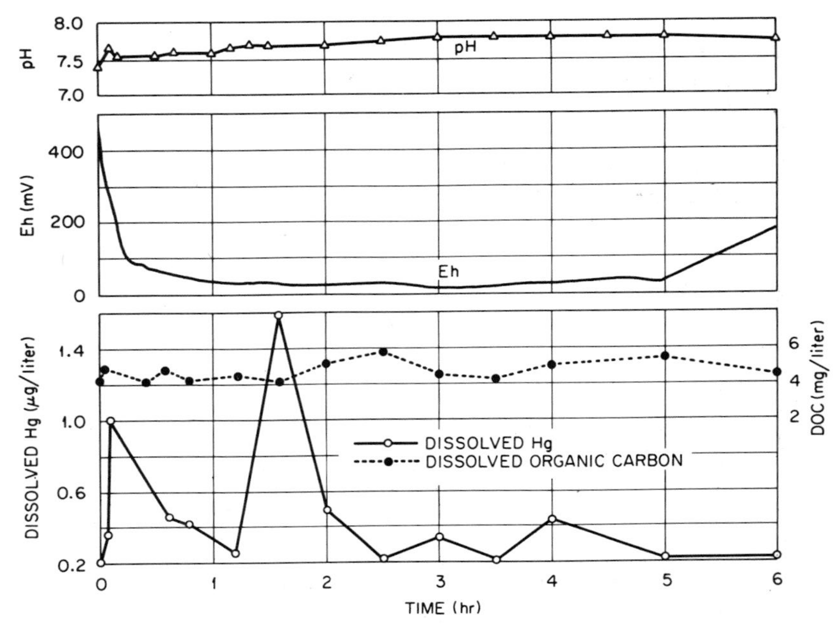

Figure 11. Release of mercury from resuspended Mobile Bay dredge-spoil sediments into the overlying water. The plot describes the time variation of the overlying water pH, redox potential (Eh), dissolved organic carbon (DOC), and dissolved mercury.

fluctuated with a pattern very similar to that of dissolved mercury. However, in the case of the less organic dredge-spoil sediment the dissolved organic carbon content fluctuated at most within 1.5 mg/1 of the initial overlying water value. In neither case did the overlying water salinity vary significantly from the value at time zero (0°/oo for the marsh sediment trial; 5°/oo for the dredge sediment trial).

In similar experiments with sediments collected from the Everglades, the observed fluctuations of dissolved mercury were reproducible in pattern but not in absolute concentrations. In two preliminary trials using Everglades estuarine sediments in overlying waters with salinities of 30.8°/oo and 2.6°/oo the dissolved mercury exhibited two distinct concentration peaks within 2 hours of sediment suspension.

It is difficult to explain the similar behavior of mercury during resuspension of sediments collected from such different environments. In the case of the organic-rich Mobile Bay marsh environment, the data on dissolved organic carbon suggest formation of a soluble organo-mercury complex. However, this

TABLE 11

Initial Sediment and Pore-Water Values for Parameters Monitored During Resuspension Experiments.

MARSH SEDIMENT TRIAL

PARAMETER	SEDIMENT	INTERSTITIAL WATER
pH	6.9	------
Eh (mv)	+339	------
Salinity (^{o}/oo)	------	0
Total Dissolved Sulfides (mg/1)	------	$<$.1
Total Hg (μ g/gm)	.26	------
Interstitial Hg (μ g/1)	------	.50
DOC (mg/1)	------	38.1
Organic Content (%)	19.9	------

DREDGE-SPOIL TRIAL

PARAMETER	SEDIMENT	INTERSTITIAL WATER
pH	7.6	------
Eh (mv)	-223	------
Salinity (^{o}/oo)	------	6.8
Total Dissolved Sulfides (mg/1)	------	$<$ .1
Total Hg (μ g/gm)	.44	------
Interstitial Hg (μg/1)	------	1.04
DOC (mg/1)	------	13.4
Organic Content (%)	6.9	------

does not explain the behavior of mercury in the dredge-spoil case. The initial peak may be controlled by redox potential, but this affects neither the subsequent drop in mercury concentration nor the second peak, as Eh remained nearly constant during this time period. A possible explanation is that mercury occurs in the sediment in two principal forms; organically complexed and adsorbed onto inorganic sediment particles. The initial mercury peak may result from loss of loosely bound exchangeable mercury. With time the mercury ions are readsorbed by some fraction of the suspended particulate matter. The second peak may be due to solubilization of sediment organo-mercury complexes. The reason for reprecipitation of these compounds is not clear. Oxidation of dissolved iron and manganese to form colloidal hydrous oxide compounds can scavenge trace metals from solution and may account for the removal of dissolved mercury from solution (43).

Knowledge of the initial mercury concentration in the pore water of the sediment used in these experiments and in the fresh and saline overlying water with which the sediment was mixed, and of the approximate volume of overlying water and pore water that were intermixed allowed another calculation. Assuming ideal dilution of the pore-water dissolved mercury by overlying waters, the resulting increase in dissolved mercury concentrations in overlying water could be calculated. The value for the marsh sediment trial was 0.21 μg/1 and for the dredge-spoil sediment 0.26 μg/1. The highest mercury concentrations measured during each trial exceeded these values by a factor of 6 to 9. However, the steady state concentrations exceeded the calculated values by less than 8% in the dredge-spoil trial and less than 43% in the marsh-sediment trial.

These results suggest that large-scale resuspension of estuarine sediments by such processes as dredging causes a sizable short-term release of mercury into the surrounding water followed by a decrease to levels close to those predicted by ideal dilution calculations. This agrees with field data indicating negligible mercury releases during actual dredging operations over a 10-day period (Windom, personal communication). In the case of organic-rich, previously undredged, marsh sediments the magnitude of the mercury peak as well as the steady state level exceeded the ambient dissolved mercury concentration of the surrounding water by a larger factor than for the dredge-spoil sediments.

With the data collected from the dredge-spoil experiment it was possible to estimate what influence dredging in the Mobile Bay ship channel had on the mercury concentrations of the overlying water. The concentration of total sediment mercury in the dredge-spoil sediment prior to resuspension was 0.44 μg/gm. At the peak mercury release, 16 μg of mercury had been released from 2.8 kg of sediment. This represents a loss of 1.3% of the total sediment mercury. Similarly, the steady state concentration represents a net mercury loss from the sediments of 0.06%. If the area of active dredging in Mobile Bay can be estimated as that of the shipping channel and adjacent spoil deposits, approximately 15% of Mobile Bay is affected annually by dredging operations. Taking 0.46 μg/gm as the average total of mercury in open-bay surface sediment, and assuming that dredging affects the sediment to a depth of 1 m below the interface, 9.0 kg mercury per year are released for a short period during the peak loss while 0.41 kg mercury per year are released and remain in solution for up to 23 hours.

This more permanent loss represents only 0.02% of the total dissolved mercury estimated to be discharged from Mobile Bay annually. The net effect of continuous dredging of open-bay sediments is negligible compared to the overall mercury budget of Mobile Bay. However, in isolated environments, especially when previously undisturbed marsh areas are dredged, the process could result in a significant discharge of dissolved mercury.

Summarizing these data leads to an overall mercury budget for Mobile Bay. Ryan and Goodell (90) have determined the sedimentation rate, the amount of suspended sediment reaching the Gulf of Mexico, and the rate of discharge from the Mobile Bay estuary. Combining these data with the average mercury contents in the overlying water and suspended load at various points in the bay yields the mercury deposition rate in Mobile Bay as well as the discharge rate of mercury into the Gulf of Mexico. Table 12 summarizes these data. The results indicate that, at most, 2%-3% of the mercury deposited in Mobile Bay is remobilized from the sediments to the overlying water annually by the combined processes of chemical diffusion, methylation, and resuspension of mechanical sediment. On the time scale of one year, the sedimentary environment of Mobile Bay is an effective sink for deposited mercury.

TABLE 12

MERCURY BUDGET FOR MOBILE BAY

Sedimentation rate[1]	3.4×10^9 kg/year
Average suspended mercury concentration	1.3 μg Hg/gm
Mercury deposition rate	4.5×10^3 kg/year (.45 μ g/cm^2 year)
Total suspended load entering Gulf of Mexico[1]	1.4×10^9 kg/year
Average suspended mercury concentration at point of entry to Gulf of Mexico	2.0 μ g Hg/gm
Suspended mercury entering Gulf of Mexico	2.8×10^3 kg/year
Mobile Bay discharge rate[1]	52.0×10^{12} 1/year
Average dissolved mercury at point of entry to Gulf of Mexico	.04 μ g Hg/1
Dissolved mercury entering Gulf of Mexico	2.0×10^3 kg/year
Total mercury entering Gulf of Mexico	4.8×10^3 kg/year
Mercury remobilization:	
methylation rate	.01 - 10% of total sediment mercury
diffusion flux	.012 μ g Hg/cm^2 year
short term loss by sediment dredging	9.0 kg Hg/year
permanent loss by sediment dredging	.41 kg Hg/year

[1]Ryan and Goodell (90).

THE ROLE OF ESTUARINE VEGETATION IN THE MERCURY CYCLE

The origin of relatively high mercury concentrations in many top carnivores in aquatic food webs is an important environmental problem that remains unsolved. During the past several years there has been considerable debate in scientific literature concerning the relative contribution of anthropogenic mercury sources versus natural mercury to the high tissue levels observed in many fish and marine mammals (9, 26, 29, 38, 51, 91).

In estuarine and coastal waters particulate plant detritus is a primary energy source for many aquatic animals (89, 107). Analyses of one common marsh plant, *Spartina alterniflora*, have demonstrated an enrichment of mercury and other metals in the plant tissues relative to the associated sediment and water (114, 115). However, the concentrations of metals in other important primary producers and the fate of metals during the decomposition of plant tissue to detritus have not been investigated. A study by Lindberg and Harriss (66) is based on the hypothesis that metals may be enriched by the decomposition products of mangroves, a major primary producer in tropical estuaries, and provide a natural pathway for the accumulation of metals in food webs. The Florida Everglades were selected for study because there are no direct anthropogenic inputs into their system and because of the predominance of mangroves.

The samples reported on in the study were collected in the estuaries that drain the western part of the Everglades National Park, Florida. Studies in this area by Heald (36) demonstrated that approximately 85 percent of the plant detritus in coastal waters of the Everglades is derived from the decomposition of the red mangrove, *Rhizophora mangle.* Heald estimates that 2.4 grams/meter^2/day (dry weight) are produced by the red mangroves, 83 percent of this material being attributed to the mangrove leaves. Odum (77) demonstrated that plant detritus originating principally from red mangrove leaves is the major energy source for the highly productive estuarine and coastal aquatic ecosystem of the southeastern Gulf of Mexico. These studies substantiate the importance of elucidating the role of detritus in the metal cycling in coastal aquatic environments.

The total mercury concentrations in the plant and detritus materials sampled are summarized in Table 13. Since previous investigations on the geochemistry of mercury have demonstrated a strong correlation between mercury and organic matter (2, 15, 53, 65), the most important column in Table 13 is the one that illustrates the mercury content of the samples per dry weight of organic material. There is a 3.2 fold enrichment in mercury concentration in mangrove litter relative to undecomposed leaves, and a 10.4 fold enrichment is suspended

TABLE 13

TOTAL MERCURY AND PERCENT ORGANICS IN MANGROVE LEAVES AND THEIR DECOMPOSITION PRODUCTS

Material	Organic Content	Total Mercury Concentration μ g/gm	
		Per Dry Weight Total Material	Per Dry Weight Organic Material
Actively photosynthesizing mangrove leaves[a]	93.2 ± 0.5	0.21 ± 0.01	0.23 ± 0.01
Mangrove leaves prior to abscission[a]	92.0 ± 0.5	0.26 ± 0.01	0.28 ± 0.01
Mangrove litter[b]	88.0 ± 0.5	0.79 ± 0.02	0.90 ± 0.03
Suspended detritus[c]	35.4 ± 0.5	1.02 ± 0.03	2.9 ± 0.1
Bottom detritus[d]	56.8 ± 0.5	1.05 ± 0.03	1.8 ± 0.1
Peat surface sediments[e]	56.1 ± 0.5	1.02 ± 0.03	1.8 ± 0.1

[a]Based on 10 samples of leaves of the mangrove **Rhizophora mangle**

[b]Based on a homogenized sample of litter containing mangrove leaves and root material collected from a 1 m^2 plot within a mangrove stand

[c]Based on a homogenized sample of estuarine particulate matter retained by a No. 25, 53 μ mesh plankton net

[d]Collected from pockets of settled detritus on a limestone bay bottom

[e]Based on the mean of 4 surface samples collected in the intertidal zone directly beneath a mangrove stand

[f]Expressed as a percent of the total dry weight

detritus. The river-bottom detritus and peat sediment, which consist of larger particles than the suspended detritus, exhibit a 6.4 fold enrichment in mercury relative to the mangrove leaves, which are the primary source material.

The enrichment of mercury in decomposing plant tissue could result from a strong chemical association between mercury and the organic constituents most resistant to degradation. In this case the total mercury in the original plant tissue is selectively concentrated into the decreasing volume of solid material during decomposition, producing an increase in the mercury per unit weight of plant detritus. An alternate hypothesis is that the mercury is concentrated by the microflora which associate with the detritus particles. The latter hypothesis has been proposed by Heald (36) to explain increases in the protein content of mangrove tissue during decomposition to detritus particles.

The results of the study by Lindberg and Harriss (66) demonstrate that the natural decomposition of mangrove leaves produces particulate organic detritus enriched in mercury. The mercury concentrations reported for the detritus in this study are approximately 3 to 30 times higher than values reported for marine phytoplankton by Knauer and Martin (55) and Windom (115). Thus, animals in the detritus-based food web are subjected to a higher natural flux of mercury than animals that feed primarily on plankton. Variations in the mercury concentrations in animal tissue, both seasonally and over the life cycle, may be related to changes in feeding habits, particularly for migratory animals that spend only part of their life cycle in the estuary.

As discussed in a previous section, the levels of mercury in Everglades sediments generally exceed values reported for many estuarine and coastal areas receiving known anthropogenic mercury discharges. The results of the above study demonstrate that natural decomposition processes supply a source of organic sediment enriched in mercury. Any attempt to determine the degree of mercury contamination in sediments must normalize to account for variable organic input.

SUMMARY

The transport and deposition of mercury in the estuarine environment is largely controlled by interaction with natural organic matter. Field and laboratory studies indicate that 54% to 82% of the total dissolved mercury in Gulf of Mexico estuaries is associated with a fulvic acid type material primarily of less than 500 molecular weight. The mercury-organic association is not significantly decreased by exposure to increased salinity, which means that mercury transport in estuaries is characterized by ideal dilution. A possible exception is estuaries with very high dissolved organic carbon where some precipitation of high molecular weight organic material containing mercury may occur over the salinity gradient.

Mercury in sediments is strongly associated with particulate organic matter. Decomposition of estuarine plant material produces detrital organic material which is enriched in mercury relative to the living tissue. Mercury in estuarine sediments from the Gulf of Mexico is not significantly redistributed by postdepositional processes active in unconsolidated recent sediment. Sediment pore water is enriched in both dissolved organic matter and dissolved mercury relative to overlying estuarine water, the mercury occurring primarily in a natural fulvic acid type complex of less than 500 molecular weight. Simulated dredging results in a slight increase in total dissolved mercury in the overlying water due to mixing with mercury-enriched sediment pore water and and short-term release of a small fraction of mercury associated with particulate material.

REFERENCES

1. Andersson, A.
   1967 Kvicksilvret i marken. **Grundförbattring,** 20: 95-105.
2. Andren, A.
   1973 The geochemistry of mercury in three estuaries from the Gulf of Mexico. Doctoral dissertation, Florida State Univ., 139 p.
3. Andren, A., and Harriss, R. C.
   1973a Observations on the association between dissolved mercury and dissolved organic matter in natural waters. (In preparation.)
4. Andren, A., and Harriss, R. C.
   1973b Methylmercury in estuarine sediments. **Nature,** 245: 256-257.
5. Anfält, D., Dryssen, D., Ivanova, E., and Jagner, D.
   1968 State of divalent mercury in natural waters. **Svensk Kem. Tinskr.,** 80: 340-342.
6. Anonymous.
   1970 Mercury in the environment. **Environ. Sci. Technol.,** 4: 890-892.
7. Applequist, M. D., Katz, A., and Turekian, K. K.
   1972 Distribution of mercury in the sediments of the New Haven (Conn.) Harbor. **Environ. Sci. Technol.,** 6: 1123-1124.
8. Baker, W. E.
   1973 The role of humic acids from Tasmanian podzolic soils in mineral degradation and metal mobilization. **Geochim. Cosmochim. Acta,** 37: 269-281.
9. Barber, R. T., Vijayakumar, A., and Cross, F.
   1972 Mercury concentrations in recent and ninety-year-old benthopelagic fish. **Science,** 178: 636-638.
10. Barsdate, R. J.
    1968 Transition metal binding by large molecules in high latitude waters. In **Organic matter in natural waters, a symposium,** p. 625 (ed. Hood, D.). College, Alaska.

11. Berner, R. A.
1971 Principles of chemical sedimentology. McGraw-Hill, New York, 240 p.

12. Black, A. P., and Christman, R. F.
1963 Chemical characterisitcs of fulvic acid. **Jour. American Water Works Ass.,** 55: 897-912.

13. Boney, A. D.
1971 Sub-lethal effects of mercury on marine algae. **Mar. Poll. Bull.,** 2: 69-71.

14. Bothner, M. H., and Carpenter, R.
1972 Sorption-desorption reactions of mercury with suspended matter in the Columbia River. Paper presented at the International Atomic Energy Agency **Symposium on the Interaction of radioactive contaminants with the constituents of the marine environment.** Seattle, Washington. 10-14 July, 1972.

15. Bothner, M. H., and Piper, D.
1971 The distribution of mercury in sediment cores from Bellingham Bay, Washington. In **Proceedings of the workshop on mercury in the western environment,** (ed. Buhler, D. R.). Oregon State University Press. (In press).

16. Carricker, M. R.
1967 Ecology of estuarine benthic invertebrates: a perspective. In **Estuaries,** p. 442-437. (ed. Lauff, G. H.). Am. Assoc. Adv. Sci. Publ. 83. Washington, D. C.

17. Christman, R. F.
1968 Chemical structures of color-producing organic substances in water. In **Organic matter in natural waters, a symposium.** p. 625. (ed. Hood, D.). College, Alaska.

18. Ciavatta, L., and Grimaldi, M.
1968 Equilibrium constants of mercury (II) chloride complexes. **Jour. Inorg. Nucl. Chem.,** 30: 197-205.

19. Colby, P. J., and Smith, L.
1967 Survival of walleye eggs and fry on paper fibre sludge deposits in Rainy River, Minnesota. **Trans. Amer. Fish. Soc.,** 96: 278-296.

20. Cotton, F. A., and Wilkinson, G.
1962 Advanced inorganic chemistry. **Interscience.** 1136 p.

21. Cranston, R. E., and Buckley, E. D.
1972 Mercury pathways in a river and estuary. **Environ. Sci. Technol.,** 6: 274-278.

22. Dreyer, C.
1973 Some aspects of dissolved and particulate organic carbon in nearshore environments of the Gulf of Mexico. M. S. thesis, Florida State Univ. 87 p.

23. Duursma, E. K.
1961 Dissolved organic carbon, nitrogen, and phosphorous in the sea., **Neth. Jour. Sea Res.,** 1: 1-148.

24. Duursma, E. K.
1966 Molecular diffusion of radioisotopes in interstitial water of sediments. In **Proceedings of the symposium on the disposal of radioactive wastes into seas, oceans, and surface waters,** p.

355-371. International Atomic Energy Agency, Vienna.

25. Duursma, E. K., and Hoede, C.
1967 Theoretical, experimental and field studies concerning diffusion of radioisotopes in sediments and suspended solid particles of the sea, Part A. **Neth. Jour. Sea Res.,** 3: 423-457.

26. Evans, R. J., Bails, J. D., and D'Itri, F.
1972 Mercury levels in muscle tissues of preserved museum fish. **Environ. Sci. Technol.,** 6: 901-905.

27. Felbeck, G. T.
1971 Structural hypothesis of soil humic acids. **Soil Sci.,** 111: 42-47.

28. Fitzgerald, W. F., and Lyons, W. B.
1973 Organic mercury compounds in coastal waters. **Nature,** 242: 452-453.

29. Ganther, H. E., Goudie, C., Sunde, M., Kopecky, M., Wagner, P., Oh, Sang-Hwan, and Hoekstra, W.
1972 Mercury concentrations in specimens of tuna and swordfish. **Science,** 175: 1121-1124.

30. Gavis, J., and Ferguson, J. F.
1972 The cycling of mercury through the environment. **Water Res.,** 6: 989-1008.

31. Goldberg, E. D., Broeker, W., Gross, M. G., and Turekian, K. K.
1971 Marine chemistry. In **Radioactivity in the marine environment,** p. 137-145. National Academy of Sciences.

32. Groot, A. J. de.
1966 Mobility of trace elements in deltas. **Trans. Comm. II and IV, Int. Soc. Soil Sci.,** p. 267-279. Aberdeen.

33. Groot, A. J. de, Goeij, J. de, and Zegers, C.
1971 Contents and behavior of mercury as compared with other heavy metals in sediments from the rivers Ems and Rhine. **Geol. Mijnbouw,** 50: 393-398.

34. Gunter, G.
1969 Reef shell or mud shell dredging in coastal bays and its effects on the environment. **North American Wildlife Conf. Trans.,** 34: 51-74.

35. Harriss, R. C., White, D., and Macfairlane, R.
1970 Mercury compounds reduce photosynthesis by plankton. **Science,** 170: 736-737.

36. Heald, E. J.
1971 Production of organic detritus in a South Florida estuary. Sea Grant Technical Bulletin No. 6. 109 p.

37. Hem, J. D.
1970 Chemical behavior of mercury in aqueuos madia. U.S.G.S. Prof. Paper 713. 19-24.

38. Heppleston, P. B., and French, M.
1973 Mercury and other metals in British seals. **Nature,** 243: 302-304.

39. Hogdahl, O. T.
1963 The trace elements in the ocean: A bibliographic compilation, Publ. Central Inst. Indust. Res., Oslo. 47 p.

40. Huggett, R. J., Bender, M., and Stone, D. H.
1972 Mercury in sediments from three Virginia estuaries. **Chesapeake**

Sci., 12: 280-282.

41. Imura, N., Sukegawa, E., Pan, S., Hagao, K., Kim, T., Kwan, T., and Ukita, T.
1971 Chemical methylation of inorganic mercury with methylcobalamine, a vitamin B_{12} analog. **Science**, 172: 1248-1249.

42. Jackson, M. P., Swift, R., Posner, A., and Knox, J. R.
1972 Phenolic degradation of humic acid. **Soil Sci.**, 114: 75-78.

43. Jenne, E. A.
1968 Controls on Mn, Fe, Co, Ni, Cu, and Zn concentrations in soils and water; the significant role of hydrous Mn and Fe oxides. In **Trace inorganics in water**, p. 337-388. Advances in Chemistry Series 73. A.C.S., Washington, D. C.

44. Jenne, E. A.
1970 Atmospheric and fluvial transport of mercury. U.S.G.S. Prof. Paper 713. 40 p.

45. Jenne, E. A.
1972 Mercury in waters of the United States, 1970-1971. U.S.G.S. Open-File Report. 34 p.

46. Jenne, E. A., and Wahlberg, J. S.
1968 Role of certain stream sediment components in radio-ion sorption.U.S.G.S. Prof. Paper 433-F. 16 p.

47. Jensen, S., and Jernelöv, A.
1967 Biosyntes av kvick-silver. **Biocidinformation** 10: 3-5.

48. Jensen, S., and Jernelöv, A.
1969 Biological methylation of mercury in aquatic organisms. **Nature**, 223: 753-754.

49. Jernelöv, A., and Lann, H.
1973 Studies in Sweden on feasibility of some methods for restoration of mercury-contaminated bodies of water. **Environ. Sci. Technol.**, 7: 712-718.

50. Johnels, A. G., and Westermark, T.
1969 Mercury contamination of the environment in Sweden. In **Chemical fallout**, p. 221-239. (eds. Miller, M. W. and Berg, G. C.). Charles Thomas Publishers, New York.

51. Johnels, A. G., Westermark, T., Berg, W., Persson, P., and Sjöstrand, B.
1967 Pike (**Esox lucius L.**) and some other aquatic organisms in Sweden as indicators of mercury contamination in the environment. **Oikos**, 18: 323-333.

52. Keckes, S., and Miettinen, J.
1970 Review of mercury as a marine pollutant. From F.A.O. technical conference on marine pollution and its effects on living resources and fishing. International Atomic Energy Agency, Laboratory of Marine Radioactivity, Rome.

53. Kennedy, E. J., Ruch, R., and Shimp, N. F.
1971 Distribution of mercury in unconsolidated sediments from southern Lake Michigan. Illinois State Geol. Surv. Environ. Geol. Notes No. 44. 18 p.

54. Klein, D. H., and Goldberg, E. D.
1970 Mercury in the marine environment. **Environ. Sci. Technol.**, 4: 765-768.

55. Knauer, G. A., and Martin, J.
1972 Mercury in a marine pelagic food chain. **Limnol. Oceanogr.,** 17: 868-876.

56. Kojima, K.
1973 Summary of recent studies in Japan on methylmercury poisoning. **Toxicology,** 1: 43-62.

57. Konrad, J. G.
1971 Mercury content of various bottom sediments, sewage treatment plant effluents, and water supplies in Wisconsin. Dept. Natural Resources, Research Report No. 74. Madison, Wisconsin. 5 p.

58. Krauskopf, K. B.
1956 Factors controlling the concentrations of thirteen·rare metals in sea water. **Geochim. Cosmochim. Acta,** 9: 1-32.

59. Kruger, P., and Schubert, J.
1953 The stability of a complex ion. **Jour. Chem. Ed.,** 30: 196-198.

60. Kurland, L. T., Faro, S. N., and Siedler, H.
1960 Minimata disease. **World Neurol.,** 1: 370-395.

61. Lamar, W. L.
1968 Evaluation of organic color and iron in natural surface waters. U.S.G.S. Prof. Paper No. 600-D. 24 p.

62. Landner, L.
1970 Restoration of mercury contaminated lakes and rivers. Swedish Air and Water Pollution Laboratory Bull. B76. 11 p.

63. Law, S. L.
1971 Methylmercury and inorganic mercury collection by a selective chelating resin. **Science,** 174: 285-287.

64. Lindberg, S. E.
1973 Mercury in interstitial solutions and associated sediments from estuarine areas on the Gulf of Mexico. M.S. thesis, Florida State Univ. 125 p.

65. Lindberg, S. E., and Harriss, R. C.
1973a Mercury - organic matter associations in estuarine sediments and interstitial water. **Environ. Sci. Technol.,** 8: 459-462.

66. Lindberg, S. E., and Harriss, R. C.
1973b Mercury enrichment in estuarine plant detritus. **Mar. Pollution Bull.** (In Press.)

67. Little, J. A., Schneider, R., and Carroll, B. J.
1970 A synoptic survey of limnological characteristics of Big Cypress Swamp, Florida. U. S. Dept. of the Interior, Federal Water Quality Administration. Southeast Water Laboratory Technical Services Program.

68. Löfroth, G.
1969 Methylmercury. Swedish Natural Science Research Council, Ecological Research Committee, Bull. No. 4. 38 p.

69. Malcolm, R. L., Jenne, E. A., and McKinley, P. W.
1968 Conditional stability constants of a North Carolina soil fulvic acid with Co^{++} and Fe^{+++}. In **Organic matter in natural waters, a symposium,** p. 479-483. (ed. Hood, D.) College, Alaska.

70. Matson, W. R.
1968 Organic matter - trace metal interactions in the aqueous

environment. Doctoral dissertation, Massachusetts Institute of Technology.

71. May, E. B.
1973 Extensive oxygen depletion in Mobile Bay, Alabama. **Limnol. Oceanogr.,** 18: 353-366.

72. McCulloch, D. S., Conomos, T., Peterson, D., and Leong, K.
1971 Distribution of mercury in surface sediments in San Francisco Bay estuary, California. San Francisco Bay Region Environment and Resources Planning Study. Basic Data Contribution No. 14. 1 p.

73. National Academy of Science.
1971 Radionuclides in the marine environment. Washington, D. C. 272 p.

74. Nissenbaum, A., Baedecker, M., and Kaplan, I. R.
1971 Studies on dissolved organic matter from interstitial water of a reducing marine fjord. In **Advances in organic geochemistry.** p. 427-440. Pergamon Press, Oxford, England.

75. Nissenbaum, A., and Kaplan, I. R.
1972 Chemical and isotopic evidence for the **in situ** origin of marine humic substances. **Limnol. Oceanogr.,** 17:570-582.

76. Odum, H. T.
1963 Productivity measurements in Texas turtle grass and the effects of dredging an intracoastal channel. **Publications of the Inst. of Mar. Sci.,** 9: 48-58.

77. Odum, W. E.
1971 Pathways of energy flow in a South Florida estuary. Univ. Miami Sea Grant Tech. Bull. No. 7. 162 p.

78. Olsson, M.
1968 Discussion: Disappearance of mercury from lakes. From Kvicksilver Symposium, Stockholm, Sweden, Oct. 10-11. 179 p.

79. Peakall, D. B., and Lovett, R. J.
1972 Mercury: Its occurrence and effects on the ecosystem. **Bioscience,** 22: 20-25.

80. Prakash, A., and Rashid, M. A.
1968 Influence of humic substances on the growth of marine phytoplankton: Dinoflagellates. **Limnol. Oceanogr.,** 13: 598-606.

81. Presley, B. J., Kolodny, Y., Nissenbaum, A., and Kaplan, I.
1972 Early diagenesis in a reducing fjord, Saanich Inlet, British Columbia. Part II. Trace element distribution in interstitial water and sediment. **Geochim. Cosmochim. Acta,** 36: 1073-1090.

82. Randhawa, N. S., and Broadbent, F. E.
1965 Soil organic matter - metal complexes: 5. Reactions of zinc with model compounds and humic acid. **Soil Sci.,** 99: 295-300.

83. Rashid, M. A.
1969 Contribution of humic substances to the cation exchange capacity of different marine sediments. **Maritime Sediments,** 5: 44-50.

84. Rashid, M. A.
1971 Role of humic acids of marine origin and their different molecular weight fractions in complexing di- and trivalent metals.

Soil Sci., 111: 298-306.

85. Rashid, M. A., and King, L. H.
1969 Molecular weight distribution measurements on humic and fulvic acid fractions from marine clays on the Scotian Shelf. **Geochim. Cosmochim. Acta,** 33: 147-151.

86. Rashid, M. A., and King, L. H.
1970 Major oxygen-containing functional groups present in humic and fulvic acid fractions isolated from contrasting marine environments. **Geochim. Cosmochim. Acta,** 34: 193-201.

87. Rashid, M. A., and King, L. H.
1971 Chemical characteristics of fractionated humic acids associated with marine sediments. **Chem. Geol.,** 7: 37-43.

88. Rebhun, N., and Manka, J.
1971 Classification of organics in secondary effluents. **Environ. Sci. Technol.,** 5: 606-609.

89. Riley, G. A.
1970 Particulate organic matter in seawater. **Advan. Mar. Biol.,** 8: 1-118.

90. Ryan, J. J., and Goodell, H. G.
1972 Marine geology and estuarine history of Mobile Bay, Alabama. Part I. Contemporary sediments. In **Environmental framework of coastal plain estuaries,** p. 517-554. Geol. Soc. Amer. Memoir 133.

91. Saperstein, A. M.
1973 Mercury and other metals in British seals. **Nature,** 243: 302-304.

92. Saukow, A. A.
1953 **Geochemie.** Veb Verlag Technik, Berlin.

93. Schnitzer, M.
1971 Metal - organic matter interactions in soils and waters. In **Organic compounds in aquatic environments,** 638 p. (eds. Faust, S. J. and Hunter, J. V.). Marcel Dekker, New York.

94. Schnitzer, M., and Skinner, S. I. M.
1965 Organo - metallic interactions in soils: 4. **Soil Sci.,** 99: 278-284.

95. Schnitzer, M., and Skinner, S. I. M.
1967 Organo - metallic interactions in soils: 7. **Soil Sci.,** 103: 247-251.

96. Schutz, D. F., and Turekian, K. K.
1965 The investigation of the geographical and vertical distribution of several trace elements in seawater using neutron activation analysis. **Geochim. Cosmochim. Acta,** 29: 259-313.

97. Servizi, J. A., Gordon, R., and Martens, D.
1969 Marine disposal of sediment from Bellingham Harbor as related to sockeye and pink salmon fisheries. International Pacific Salmon Fisheries Commission Report No. 23.

98. Shacklette, H. T., Boerngen, J., and Turner, R. L.
1971 Mercury in the environment – Surficial materials of the coterminous United States. U. S. Geol. Surv. Circular 644. 5 p.

99. Sieburth, J. M., and Jensen, A.
1968 Studies on algal substances in the sea. I. Gelbstoff in terrestrial and marine waters. **Jour. Exp. Mar. Biol. Ecol.,** 2: 174-189.

100. Siegel, A.
1971 Metal - organic interactions in the marine environment. In **Organic compounds in aquatic environments,** 638 p. (eds. Faust, S. J. and Hunter, J. V.). Marcel Dekker, New York.

101. Sillén, L. G., and Martell, A. E.
1964 Stability constants of metal - ion complexes. Special publication no. 17. The Chemical Society, Burlington House, London.

102. Slowey, J. F., and Hood, D. W.
1966 Ann. Rept. AEC - contrib. no. AT-40-1-2799. Texas A&M Univ. Report 66-2F.

103. Smith, J. D., Nicholson, R., and Moore, P. J.
1971 Mercury in water of the tidal Thames. **Nature,** 232: 393-394.

104. Stevenson, F. J., Krastanov, S. A., and Ardakani, M. S.
1973 Formation constants of Cu^{++} complexes with humic and fulvic acids. **Geoderma,** 9: 129-141.

105. Strohal, P., and Huljev, D.
1970 Investigation of mercury pollutant interaction with humic acid by means of radiotracers. In **Nuclear techniques in environmental pollution, a symposium.** International Atomic Energy Agency Proceedings, Salzburg.

106. Szalay, A., and Szilagyi, M.
1967 The association of vanadium with humic acids. **Geochim. Cosmochim. Acta,** 31: 1-6.

107. Teal, J. M.
1962 Energy flow in the salt marsh ecosystem of Georgia. **Ecology,** 43: 614-624.

108. Thomas, R.
1972 The distribution of mercury in the sediments of Lake Ontario. **Can. Jour. Earth Sci.,** 9: 636-651.

109. Thomas, R.
1973 The distribution of mercury in the surficial sediments of Lake Huron. **Can. Jour. Earth Sci.,** 10: 194-204.

110. Turekian, K. K.
1971 Rivers, tributaries, and estuaries. In **Impingement of man on the oceans,** p. 9-73. (ed. Hood, D. W.). John Wiley and Sons, New York.

111. U. S. Geological Survey
1970 Mercury in the environment. U. S. Geol. Surv. Prof. Paper 713. 67 p.

112. Vernet, J. P., and Thomas R.
1972 The occurrence and distribution of mercury in the sediments of the Petit-Lac. **Eclogae Geol. Helv.,** 65: 307-316.

113. Wallace, R. A., Fulkerson, W., Schults, W. D., and Lyon, W. S.
1971 Mercury in the environment: the human element. Oak Ridge National Laboratory Report ORNL-NSF-EP-1. 61 p.

114. Williams, R. B., and Murdock, M.
1969 The potential importance of **Spartina alterniflora** in conveying zinc, manganese, and iron into estuarine food chains. Proc. 2nd Natl. Symp. Radioecology. p. 431-439.

115. Windom, H. L.
1973 Mercury distribution in the estuarine - nearshore environment. **Jour. Waterways, Harbors and Coastal Eng. Div., Amer. Soc. Civil Eng.,** 99: 257-264.

116. Wood, J. M.
1972 A progress report on mercury. **Environment,** 14: 33-39.

117. Wood, J. M., Kennedy, F., and Rosen, C. G.
1968 Synthesis of methylmercury compounds by extracts of methanogenic bacterium. **Nature,** 220: 173-174.

PHOSPHORUS FLUX AND CYCLING IN ESTUARIES[1]

by

David L. Correll, Maria A. Faust, and David J. Severn[2]

ABSTRACT

Examples of experiments and analyses of their results are given for the measurement of phosphorus flux rates and phosphorus cycling *in situ* in tidal marsh, mud-flat periphyton, and plankton communities of the Rhode River subestuary of Chesapeake Bay. Techniques covered include phosphorus-32 orthophosphate uptake and chase kinetics, analysis of specific and total activity in various metabolically meaningful phosphorus fractions, detailed column chromatographic fractionation, continuous-flow pulse-labeling of plankton, direct microscopic examination of microbial communities, and phosphorus-33 microautoradiography.

INTRODUCTION

There have been so few field studies of phosphorus cycling and flux in estuaries that I will take the liberty of referring to field studies done in other aquatic habitats and to laboratory studies of special pertinence. In most studies of phosphorus cycling, the approach has been to add radiophosphorus to the water as dissolved orthophosphate, then follow in various ways what happens to it. Thus, in a large-scale example of this approach, Pomeroy et al. (22, 23)

1. Research was supported in part by a grant from the program for Research Applied to National Needs of the National Science Foundation and by the Smithsonian Institution's Environmental Sciences Program. Published with the approval of the Secretary of the Smithsonian Institution.

2. Radiation Biology Laboratory, Smithsonian Institution, Rockville, Maryland 20852.

attempted to trace the pathways and rates of movement of phosphorus in a salt marsh. McRoy and Barsdate (16) and McRoy et al. (17) studied the rates of phosphorus uptake and release by eelgrass from sea water in special chambers and containers. They concluded that the eelgrass plants act as phosphorus pumps between bottom sediments and the water mass. Odum et al. (18) measured relative rates of phosphate uptake from sea water by excised pieces of thallus from various species of macroscopic benthic algae. Pomeroy et al. (21) studied the rates of exchange of phosphate uptake from sea water by excised pieces of thallus from various species of macroscopic benthic algae. Pomeroy et al. (21) studied the rates of exchange of phosphate from water with suspended sediments and with sediment cores under various conditions. Pomeroy (19) has summarized a number of studies of the rates of uptake and turnover of dissolved orthophosphate from various water masses. It should be noted that most of these studies, omitted careful detailed kinetic analysis and consequently their conclusions as to rates of phosphate uptake, turnover times, and movement vary by many orders of magnitude. Field studies of fresh water, such as those undertaken by Rigler (27, 28) and Lean (15), emphasized the importance of very small particulate fractions and of a very detailed time series with many points in the first few minutes, if rates are to be estimated accurately in an isotope-uptake experiment. In some situations recycling, exhausting the pool of available phosphate in small enclosed volumes, reduced turbulence, and artifacts due to changes in many physical, chemical, and biological parameters cause apparent uptake rates to change rapidly.

A few estuarine studies have introduced radio-phosphorus to the sediments at various depths in order to observe its movement. Reimold (25) used this technique to study the movement of phosphorus into the leaves of *Spartina alterniflora,* and concluded that it originated from the sediments at depths on the order of 100 cm. McRoy and Barsdate (16) and McRoy et al. (17) studied movement into eelgrass roots, stems, and leaves from the sediments at a depth of 5 cm. These studies were in many respects qualitative, because the radiophosphorus was applied as a point source and specific activities were not determined.

Several field studies have been directed at determining the principal routes of biological phosphorus cycling in estuaries. For example, the amounts and forms of phosphorus released by large zooplankton were studied (20, 30). Laboratory work with cultures of invertebrates has demonstrated the lack of significant direct uptake of dissolved orthophosphate by these animals (11, 13). Laboratory studies of competition for phosphate between algae and bacteria indicate that, in estuarine conditions where surfaces for bacterial attachment are abundant and dissolved orthophosphate concentrations are seldom limiting, bacteria can take up a large proportion of the phosphate (9, 26). Johannes' laboratory studies (12, 13) with protozoans led him to propose that the major pathway of biological phosphorus cycling in tidal marshes is the uptake of phosphate by bacteria, the consumption of the bacteria by filter-feeding protozoans and small metazoans, and the release of a mixture of dissolved orthophosphate and organic phosphorus by these filter feeders. The organic phosphorus is then hydrolyzed enzymically by phosphatases, which are on cell surfaces and also dissolved in the water. An overview of phosphorus flux in an estuary constructed by Pomeroy et al. (24) included some results of an attempt mathematically to model phosphorus flux in estuaries.

One general point of weakness in our current knowledge of phosphorus cycling and flux in estuaries is the almost total lack of quantitative field studies of the role of the periphyton, a community composed of diverse microorganisms that uniformly coat all undisturbed underwater surfaces. This community would be expected to be very important in the functioning of an estuary, since it is at the interface between the water mass and the bottom sediments or marsh sediments. It is characterized by a very high metabolic rate and an ability to respond quickly to a change in environmental conditions by a shift in the species composition of the microbial community.

Another general comment upon the status of the field is the lack of studies that make use of radiophosphorus chase kinetics. This method offers the advantage of accuratly measuring rates of flux or cycling over long time periods without introducing serious artifacts. For the biologist, the use of phosphorus-33 and the technique of microautoradiography could also be powerful tools for the unraveling of nutrient pathways in microbial communities, as reported by Fuhs and Canelli (8). A question of fundamental importance and widespread concern is the relative role of biological mechanisms as contrasted to physical and chemical binding of phosphates to sediments and detritus in estuaries. A certain amount of information has been obtained by attempts to inhibit all biological processes, with such general inhibitors as Formalin (21), but better, more selective methods are needed.

In this paper we report attempts to introduce a series of experimental approaches to a more accurate and complete understanding of phosphorus flux and cycling in estuaries. In our view, phosphorus cycling should, when possible, be examined in terms of metabolically meaningful phosphorus fractions, their pool sizes and turnover rates. These methods evolved from earlier experiments, which included a study of planktonic phosphorus metabolism in the open ocean (4) and a discussion of approaches to making field measurements of phosphorus metabolism in microbial communities (6). All the experimental data presented in this article were obtained from studies of the Rhode River arm of Chesapeake Bay. However, the intent is not to describe the Rhode River, but to discuss phosphorus flux and cycling in estuaries. The experimental data are used only to provide examples of the application of some experimental approaches to the measurement of these phenomena.

The Rhode River is a small subestuary of Chesapeake Bay. It has an open-water surface area of 4.8×10^6 m^2. For our purposes it can be divided into three spatial categories: tidal marshes, mud flats, and a deep-water basin. In the first two categories phosphorus from the water mass was trapped primarily by surface-dwelling microbial communities called periphyton. In the basin section phosphorus was cycled by a complex interaction between the planktonic microbial community and the bottom sediment community. In the tidal marshes and mud flats, the principal importance of the higher plant communities for phosphorus cycling is the fact that they provide increased surface area for periphyton growth.

MATERIALS AND METHODS

Tidal Marshes

Ten millicuries of carrier-free orthophosphate were mixed with 10 liters of surface water and the solution was sprinkled evenly over a square plot 2.4 m on a side. The plot was in a tidal marsh dominated by *Typha angustifolia* in the Muddy Creek part of the Rhode River adjacent to station 8 (Fig. 1). The experimental area was sampled at two-week intervals for four months.

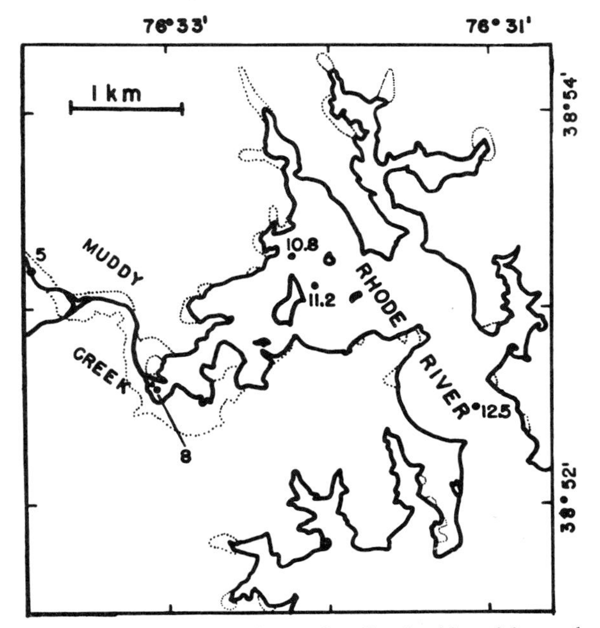

Figure 1. The Rhode River and the stations referred to in this article, marked by number. The areas within dotted lines are tidal marshes.

Composite samples were taken of (a) either young *T. angustifolia* leaves or the new growth at the base of older leaves; (b) such detritus on the surface of the sediments as decaying plant materials and what adhered to them; (c) sediment core sections taken vertically with a soil-coring tube. Composite samples were taken either scattered within the labeled plot or along transect lines parallel to the outer edges of the plot.

Periphyton

The study of periphyton communities on clear plexiglas plates (artifical substrates) was first described by Grzenda and Brehmer (10). Plexiglas plates (6 cm x 12 cm x 3 mm) were mounted on wooden bars with the 12 cm edge directed upward. The bars were positioned so that the plates were just above the surface of the bottom sediments at stations 5 and 8, both of which are shallow-water tidal mud flats (Fig. 1). The plates were exposed to the air only a few times a year, when unusual meterorological conditions caused exceptionally low water. Phosphorus -32 experiments were conducted in an apparatus designed to maintain the currents and turbulence to which the periphyton is accustomed in the estuary, thereby preventing suspended sediments from being deposited on the bottom of the apparatus during the experiment. The principle of this apparatus has already been described (6). It consists of a 50 1 capacity polyethylene tank, 72 cm long by 29 cm wide x 31 cm deep (open at the top), adapted with a high-capacity, centrifugal pump (Cole-Parmer, model 7006-20, Chicago, Illinois) powered by a gasoline engine to recirculate the water in the tank. Clear plexiglas baffles installed at both ends of the tank with a grid of various-sized holes ensure relatively equal flow rates near the walls of the tank and in the center of the water mass. All parts in contact with the water are plastic. The tank apparatus was placed in a small, flat-bottomed boat and taken to the station where the plates had been allowed to become colonized with periphyton. The tank was filled with water from the site, spiked with about 40 μc of carrier-free ^{32}p-phosphate, and allowed to mix a short time. A water sample was taken, an aliquot was filtered through a 0.45-μm pore size HA membrane filter (Millipore Corp., Bedford, Massachusetts), and the plates were then transferred into the tank. From time to time, plates and water samples were removed. Plates were placed in plastic bags and frozen on dry ice. An aliquot of each water sample was filtered as before. In the case of chase experiments, labeled plates were returned to their supports in the estuary and, at intervals, sets of two plates were removed and frozen.

Plankton-Bottle Experiments

Clear glass cylindrical bottles of 250 ml or 1000 ml capacity were equipped with rubber stoppers through which two 16-gauge syringe needles had been placed. One needle was short and one extended well into the bottle. The bottle was attached by a length of line to a float. Dark bottles were made by dipping clear bottles in black epoxy paint. The first sample of plankton was taken from the desired depth with a peristaltic pump (see Plankton Pulse-Labeling section) and at that time water samples were also taken for chemical analysis. Carrier-free orthophosphate (2 μc/1) was then added to the bottle of plankton and the contents were mixed thoroughly, the two syringe needles were capped and the bottle was incubated at the depth from which the sample was obtained. All experiments described in this paper were made with plankton from a depth of one meter. At appropriate time intervals the bottle was recovered, mixed by swirling, the needle caps were removed, and a 10 ml aliquot was removed through the long needle with a 20-ml glass syringe. About 10 ml of air was pulled into the syringe and the sample was passed through a set of three interlocked 25 mm diameter Swinex filter holders (Millipore Corp., Bedford, Massachusetts). The first holder contained Nitex screening with a pore size of 5 μm (Tobler, Ernst, and Traber, Inc., Elmsford, New York) which was coated around the edges with silicone rubber to facilitate sealing. The second and third holders contained 1.2 μm and 0.45 μm pore size membrane filters, respectively (Millipore Corp., Bedford, Massachusetts). Enough air was forced through after the sample to remove any chance of excess moisture on the filters. Each filter and a one ml aliquot of the final filtrate were also counted. Metabolically inhibited controls were run by dissolving iodoacetic acid to a final concentration of 0.05 M, five minutes prior to adding the ^{32}P-phosphate.

Plankton Pulse-Labeling Experiments

Sampling methods for this experiment have been described schematically (6). A large peristaltic pump head (Model LG-300, Little Giant Pump Co., Oklahoma City, Oklahoma) was coupled to a 12 volt, d.c. motor for automobile windows. The resulting pump operated at 60 r.p.m. and pumped at a rate of 2.1 liter/min. The pump was used to draw plankton through a clear plexiglas chamber that housed a nylon net with a 20 μm pore size. The filtrate was drawn through a mixing "tee" where it was mixed with 100 ml/min. of a solution of carrier-free ^{32}P-phosphate. A second smaller peristaltic pump was used to pump the isotope solution. The plankton was then drawn through a vertical cylindrical coil of clear Tygon tubing made by winding 30 meters of tubing, 2.54-cm inside diameter, around a plexiglas tube, 30-cm outside diameter and one meter in length. The coil and the filtering chamber were mounted under water parallel to each other and extending from 40 cm below the surface to a depth of 1.4 M. The plankton then passed through the pump and into a continuous-flow centrifuge (Cepa, Model LE, with a clarification rotor, Carl Padberg, Lahr/Baden, West Germany).

The centrifuge speed was maintained at 40,000 rpm and the rotor was first filled with phenol containing 12 percent water. Thus the plankton-bearing water flowed over the phenol and all cells were centrifuged into the phenol phase, which stops all metabolism. The residence time of the plankton from the point of radioisotope injection to the centrifuge was determined by dyes.

Extraction And Column Chromatography

Water-soluble phosphorus compounds of low molecular weight were extracted from periphyton and pulse-labeled plankton samples and prepared for column chromatography by a phenol extraction procedure (3). DEAE-cellulose chromatography was carried out as described by Correll (5). Other details of the methods used are given in the appropriate figure legends.

Radioisotope Counting

Samples were counted in a Packard, model 3320, liquid scintillation spectrometer. The scintillation cocktail was composed of 5.5 g 2,5-diphenyloxazole and 0.5 g p-bis--[2-(4-methyl-5-phenyloxazoyl)] benzene dissolved in 800 ml toluene and 200 ml Triton X-100 (Rhom and Haas, Philadelphia, Pensylvania). In the case of aqueous samples one-ml aliquots were added to 10 ml of cocktail. For isobutanol samples, up to two ml were added to 10 ml of cocktail. Isobutanol gave a consistent amount of chemical quench. When counts obtained on aqueous samples were compared with those obtained on isobutanol samples, the isobutanol values were multiplied by 2.5 to correct for this quench.

Water Chemistry

Samples were analyzed for orthophosphate colorimetrically by reaction with ammonium molybdate and reduction with stannous chloride. The same procedure was used for total phosphorus, after potassium persulfate digestion (1).

Dry Weights And Ash-Free Dry Weights

Dry weights were determined by bringing samples to constant weight at 60°. For ash-free dry weight determinations, samples previously brought to constant weight at 60° in tared Gooch cruicibles were ashed in an electric furnace at 600° for two hours.

Fractionation Of Periphyton And Marsh Samples

Samples were homogenized with a Waring Blender in 1 N HC1 and a 5-ml aliquot was transferred to a tared 10-ml beaker to determine dry weights. For orthophosphate determination, a 10-ml aliquot was added to 10-ml, 88 percent phenol and the mixture was shaken vigorously. After centrifugation the upper aqueous layer was drawn off and saved. Ten ml of water was added to the lower, phenol layer, and the sample was shaken and centrifuged again. The upper layer was combined with the first aqueous extract and diluted with water to 25 ml. For Δ_7-P determinations a 5-ml aliquot was brought to 10 ml with 1 N HC1 and was placed in a vigorously boiling water bath for seven minutes, then cooled rapidly in an ice bath. Sample preparation then proceeded as for orthophosphate. For total phosphorus determinations, a 5-ml aliquot was placed in a 30-ml microkjeldahl flask. Two ml of 72 percent perchloric acid and one drop of concentrated nitric acid were added. The sample was carefully boiled in a fume hood behind a safety shield on a microkjeldahl digestion rack until it became colorless or pale yellow. After cooling, the sample was rinsed into a 25-ml graduate and diluted with water to 25 ml. Appropriate subaliquots of the prepared orthophosphate, Δ_7-P, and total phosphorus samples were then colorimetrically analyzed for phosphate. The overall procedure for total phosphorus is a modification of King's method (14). A total phosphorus subaliquot plus 4 ml 72 percent perchloric acid were diluted with water to 25 ml and 4 ml of 5 percent ammonium molybdate in water was added and mixed. Four ml of sulfonic acid reagent were added and a timer was started. After five minutes, 20 ml of isobutanol were added and the sample was shaken vigorously. The isobutanol phase was then separated and adjusted to 25 ml with ethanol. The absorbance was then read in the far-red region of the spectrum (700-750). The absorbance should be read within 20 minutes of the end of color development. An orthophosphate or Δ_7-P sample subaliquot was diluted to 25 ml and two ml of 2.5% acidified ammonium molybdate (7) was added and mixed. Two ml of sulfonic acid reagent were added and a timer was started. After 10 minutes isobutanol was added and the analysis proceeded as for total phosphorus. Standard curves were constructed by processing various levels of phosphate standard in the same manner. In this article acid-labile phosphorus (ALP) refers to (Δ_7-P) minus orthophosphate and acid-stable phosphorus (ASP) refers to (total-P) minus (Δ_7-P).

Plankton Autoradiography

One-liter-capacity light and dark plankton bottles were filled with water as for ^{32}P-uptake experiments. Two μc of carrier-free ^{33}P-orthophosphate were added to each bottle and mixed. The bottles were incubated at a depth of one meter. At 15-minute and 30-minute intervals after the start of the experiment, 500 ml aliquots were removed and fixative was added to a final concentration of 0.4 percent gluteraldehyde and 0.01 M sodium phosphate buffer, pH 7. Fixed samples were then processed in the laboratory. First they were centrifuged for 30 min at 2000 x g. The pellet was washed with 12 ml of 0.002 N HC1 to remove any phosphate that was bound only to surfaces. After centrifuging, the pellet was washed twice more with 12 ml distilled water and finally suspended in 3 ml distilled water. Each sample was then sonicated for from 5 to 6 seconds to break up any clumps of cells (Model W 140 C, Heat Systems Co., at 60 watts output). Slides were prepared from subsamples by placing 5 drops of cell suspension on each of three microscope slides. Cells on the slides were quickly frozen on a block of dry ice and dried in a vacuum desiccator for about 6 hours. Kodak NTB-2 (nuclear track) emulsion was diluted with an equal volume of distilled water and was liquified in a water bath at 40 ^{o}C. Two drops of emulsion added to the end of each slide were spread with a stainless-steel *"raclette"* to obtain a constant thickness throughout the slide (2). After one hour's drying at room temperature, the slides were transferred to a light-proof box and stored in the refrigerator for from 3 to 5 days. After this exposure the emulsion was developed in Dektol developer at 59° for 2 minutes, rinsed in distilled water for 10 seconds, placed in an acid fixer for 5 minutes, then rinsed in running tap water for from 5 to 10 minutes. After drying, the emulsion was cleared for one hour in methylsalicylate and permanently mounted with a cover slip. The grains around individual cells whose identity could be determined were counted at 1000 x magnification in bright field. Some of the cells had no grains around them.

Biomass Determinations

The number of microorganisms was determined by the direct-count and measurement method of Rodina (29) on samples fixed in buffered glutaraldehyde, as for autoradiography. This method determines all microorganism cells present in the sample but has the disadvantage of including dead as well as living cells. Bacteria were determined on slides stained with Gram stain. Rod-shaped forms and cocci were counted separately, the number of cells in differing size classes being registered. Algal cells were counted in a similar manner but no staining was done. The quantity of cell suspension used for bacterial and algal determinations was slightly different. For the bacterial count, 0.01 ml of fixed-water sample was used, and for the algal count determinations 0.15 ml of fixed-water sample from the 3 ml final solution (*see* Plankton autoradiography section) was used. The bacterial sample was spread on a 100 mm^2 area. One hundred fields were used to count the number of microorganisms in each sample. From the mean diameter of each size group and the number of microorganisms present, the biomass was calculated as described by Rodina (29). Taking the dilution into account, the biomass per unit of water was then determined.

RESULTS AND DISCUSSION

Tidal Marsh

Examples of data acquired in studies of a *Typha angustifolia* marsh adjacent to station 8 (Fig. 1) are shown in Figures 2-7. The results of two experiments in the spring of the year are shown in Figure 2. In one experiment the changes in radioactivity under natural loading conditions were followed, while in a repeat experiment the next spring the experimental plot and the nearby marsh area (a square 9 m on a side) were sprinkled with a solution of dibasic ammonium phosphate and ammonium nitrate three times a week. The mixture contained six atoms nitrogen per atom of phosphorus and two moles ammonium ion per mole nitrate. The loading rate was 100 mg phosphorus per m^2 per application or 43 mg $P/m^2/day$. As can be seen in Figure 2, the added loading brought about a more rapid decrease in both parameters of radioactivity. It is also apparent that even under natural loading, both parameters decreased at a greater rate than would be expected as a result of the decay of ^{32}P. Figure 3 shows the equations used for the calculation of flux rates into and out of the detrital community within the experimental plot. The data from the two experiments in Figure 2 were used to calculate the results in Figure 3. The spring detrital standing crop per surface area of sediment was used to convert the data to rates per surface area, making the assumption for the time being that in the overall time course, rate into the community would be equal to rate out of the community. The flux rates thus calculated were 3.0 and 8.8 $\mu g\ P/cm^2/day$ and the difference of 5.8 is close to the calculated rate of increased loading (4.3).

The next question is where was the phosphorus going? If it had been moving very rapidly into the tidal waters, it would have disappeared rapidly from the experimental area. Figure 4 summarizes the results of vertical sampling within the labeled plot. Although cores were taken to a depth of 25 cm, the upper 5 cm contained all the significant radioactivity under natural loading conditions. The calculation was made by taking the total cpm/g dry wt. for a given section of core, correcting for isotope decay to day zero, and dividing by the initial community specific activity (intercept from Figure 2). This gives the $\mu g\ P/g$ dry wt. in the core segments, which had been derived from the detrital community. This value divided by the gram dry wt./cc of that type core segment, multiplied by the number of cc/cm^2 surface area in the core, and divided by the number of days elapsed when the sample was taken gives the final vertical flux rate. This was done for several sampling times and the results were averaged to give 4.2 $\mu g/cm^2/day$. It is obvious that a considerable amount of flux is vertical movement into the sediments and that under natural loading conditions the

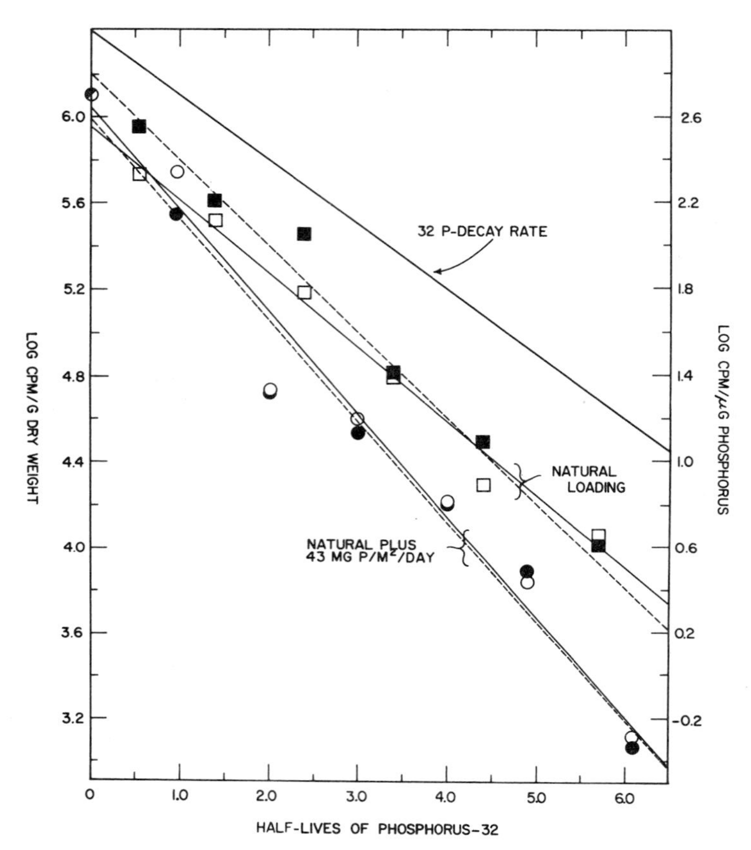

Figure 2. Kinetic analyses of the rates of change of total phosphorus specific activity and total phosphorus radioactivity per dry weight in a ^{32}P-labeled detrital community plot on the surface of the sediments in a **Typha angustifolia** tidal marsh adjacent to station 8. The half-life of ^{32}P is 14.3 days. The experimental lines are linear least-square regressions. The natural loading experiment began on 17 March 1972. The increased loading experiment began on 6 April 1973. Open symbols and solid lines are for cpm/g dry weight; solid symbols and broken lines are for cpm/μg phosphorus. Squares are data points from the natural loading experiment and circles are from the increased loading experiment.

depth of penetration is not very great (less than 5 cm in 20 days). In Figure 5 the amount of horizontal flux from the plot in the detrital community plus the surface 3 cm of sediments are summarized. In these experiments the amount of phosphorus that had been derived from inside the plot per cm^2 surface area at various transect distances was found to be a simple linear inverse function of distance from the border of the plot. Thus samples taken along a transect 60 cm

PHOSPHORUS CHASE KINETICS FOR TYPHA MARSH DETRITAL LAYER

RATE INTO COMMUNITY (μG P/G DRY WT/DAY) = $\frac{(2.303)(\text{NET SPECIFIC ACTIVITY SLOPE})(\mu\text{G P/G DRY WT})}{14.3}$

RATE OUT FROM COMMUNITY (μG P/G DRY WT/DAY) = $\frac{(2.303)(\text{NET TOTAL ACTIVITY SLOPE})(\text{INITIAL TOTAL ACTIVITY})}{14.3\ (\text{INITIAL SPECIFIC ACTIVITY})}$

	NATURAL LOADING	NATURAL + 4.3 μG P/CM^2/DAY
RATE INTO	36	62
RATE OUT	11	80
AVERAGE	24	71
AVERAGE RATE/CM^2	3.0	8.8

Figure 3. Method of calculation of flux rates in detrital communities from the data in Figure 2. Additional data used included 2720 $\pm$1329 μg P/g dry weight of detritus (n = 27) and 0.124 $\pm$0.047 g dry weight of detritus/cm^2 sediment surface area (n = 24). (Values are means $\pm$one standard deviation).

ANALYSIS OF VERTICAL MOVEMENT OF PHOSPHORUS IN TYPHA MARSH

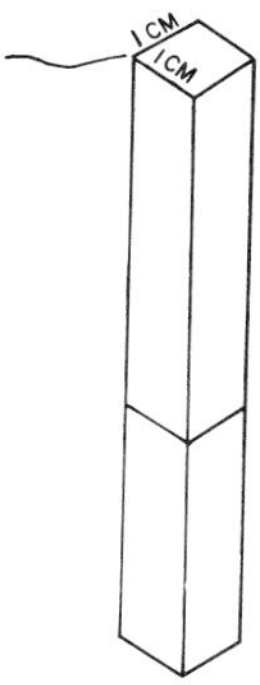

	INITIAL CPM/G DRY WT. AFTER 20 DAYS	μG P DERIVED FROM DETRITUS PER G DRY WT.	μG P/ZONE/DAY
0-3 CM SECTION	21,780	34.5	2.2
3-5 CM SECTION	3,680	5.8	0.3
			2.5

AVERAGE OF DETERMINATIONS = 4.2 μG P/CM^2/DAY AT FOUR DIFFERENT DATES

Figure 4. Analysis of total vertical phosphorus movement down from surface detrital community into sediments under natural loading conditions. Data are from the same experiment as the unloaded condition in Figure 2.

away had twice as much cpm/g dry wt. as samples taken along a 120-cm transect. The end result of the analyses shown in Figures 4 and 5 is that approximately half of the total flux calculated in Figure 3 was due to movement vertically within the plot and half due to horizontal movement out of the plot. It should be noted that the bulk of this horizontal component was found in the sediment 0-3 cm deep rather than in the detrital community. All the data in

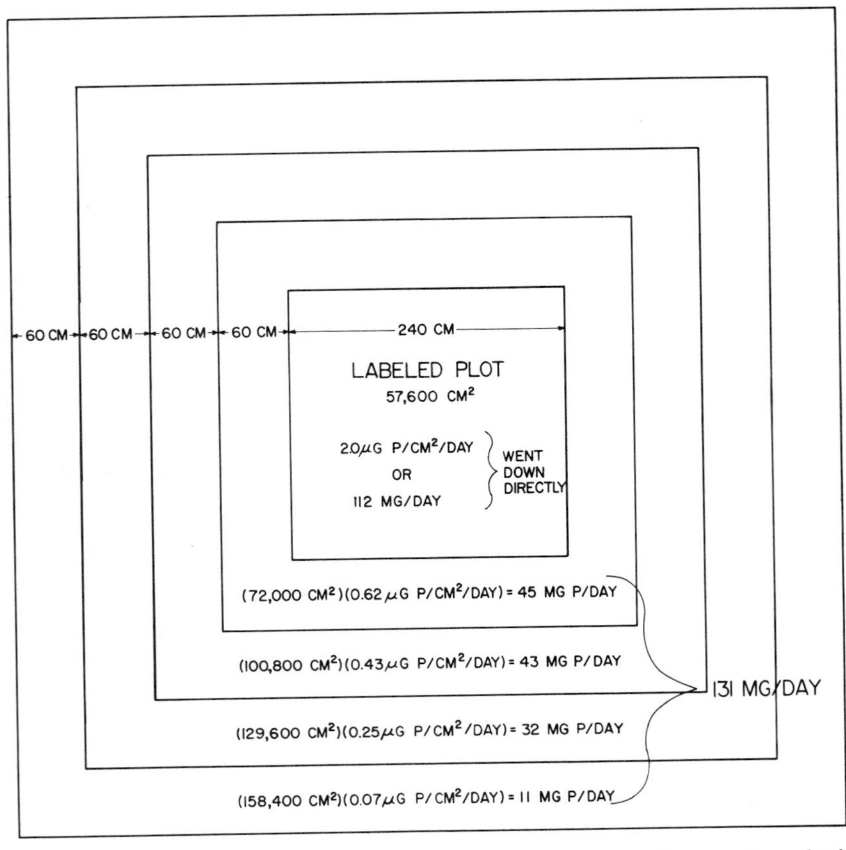

Figure 5. Analysis of total horizontal phosphorus movement from surface detrital community. Values per day were averaged for analysis 20 and 35 days from the beginning of the experiment and are compared with values for vertical movement for the same analyses. Data are from same experiment as the unloaded condition in Figure 2.

Figures 2-5 was for total phosphorus. In Figure 6, data on specific activities are summarized for day 20 in the detrital community and the surface sediments under natural loading conditions. What is apparent immediately, is the discrepancy between the high specific activities of the acid-labile fractions at all locations outside the plot and the low specific activities of the orthophosphate and total phosphorus fractions. This type of pattern has been found many times and appears to be typical for this stage of the experiment. Later in the experiment labeling closely approaches uniformity for these fractions within a given sample. The radiophosphorus is applied as orthophosphate and the purity has been checked by column chromatography. The explanation of this pattern

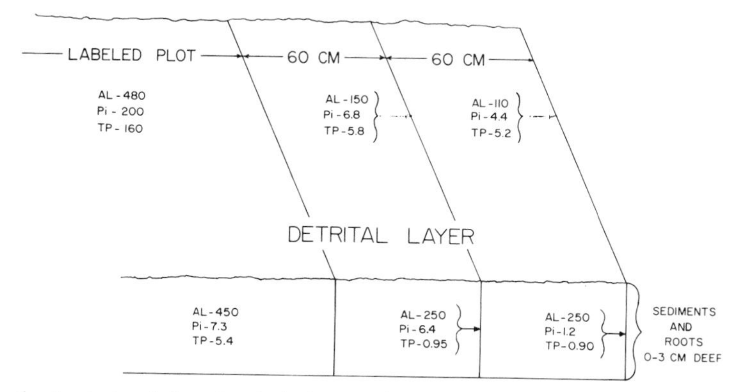

Figure 6. Analysis of the chemical characteristics of the phosphorus that is moving in a **Typha** marsh, by consideration of the specific activities of different phosphorus fractions under natural loading conditions. Specific activities are designated for acid-labile phosphorus (AL), orthophosphate (P_i), and total phosphorus (TP) at various locations along sampling transects twenty days after the experiment was initiated.

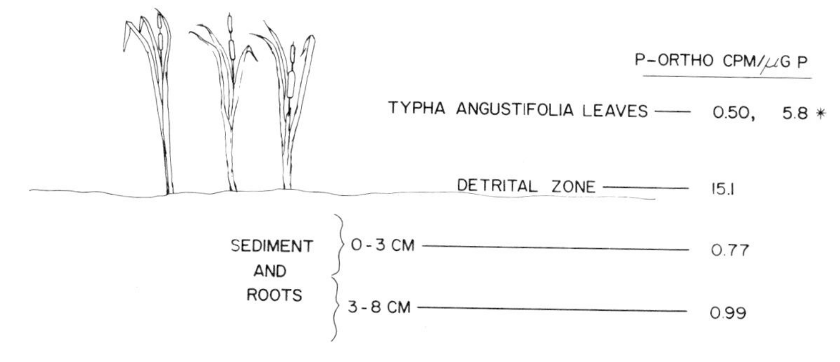

Figure 7. Analysis of the source of phosphorus for growth of **Typha angustifolia** under natural loading conditions by consideration of specific activities of phosphorus fractions 35 days from the intitiation of the experiment. All values are for orthophosphate, except the value marked with an asterisk, which is for acid-labile phosphorus.

would seem to be that the microbial cells in the detrital periphyton community take up the ^{32}P-orthophosphate inside the plot quickly and begin to approach uniform labeling. This takes considerable time and the acid-labile fraction specific activity peaks quickly (for a description of the reasons *see* Correll, 6). Apparently some of these microbial cells are constantly being transported or carried down into the sediments and along the surface film. As cells with high specific activity acid-labile phosphorus are carried along they are exposed only to low-activity orthophosphate, once they leave the experimental plot. Thus a temporary zone or front of microbial cells that has this labeling pattern can be sampled.

What can be learned from these data? It is very evident that the phosphorus that is being moved in this community is in microbial cells. Microbial cells have the currect phosphorus composition and metabolism to give these results, while phosphorus in solution or bound to particles and moved by physical processes could not behave in this manner (4, 5, 6). In Figure 7 some data are summarized relevant to the source of phosphorus for *Typha* growth. The specific activity of the acid-labile fraction in the *Typha* leaves (5.8 cpm/μg P), which is the most rapidly labeled fraction in higher plants, was higher than the specific activities of the orthophosphate fraction of the sediments at various depths (< 1) even at a depth of a few centimeters. The deeper sediments had no detectable activity. Therefore, the roots of the *Typha* were taking up phosphate from the orthophosphate in the detrital community and the very surface zone of the sediments since the specific activity of the orthophosphate there was higher, i.e. 15. When these values were compared at increasingly longer times from the initiation of the experiment, the specific activities of the leaf acid-labile phosphorus and the detrital orthophosphate began to converge. Thus, it would seem that the root hairs of the *Typha* compete on the surface with the microbial community for phosphate and, at present, are not trapping phosphate from tidal waters directly or mobilizing phosphorus from deep-sediment deposits. This result contrasts sharply with the findings for other higher plants in other locations of McRoy and Barsdate (16), McRoy et al. (17), and Reimold (25).

Periphyton

Plexiglas artificial substrates were used to study periphyton under conditions that were both easy to quantify and convenient for year-round intensive study. The kinetics of uptake of ^{32}P-phosphate by periphyton were measured at stations 5 and 8 (Fig. 1) for over a year. Typically uptake rates of phosphate varied from about 0.5 mg P/hr./m^2 in the winter to about 4 mg P/hr./m^2 in early summer. Rates of uptake of phosphate in terms of biomass ranged from about 0.2 to 1.4 mg P/hr/g ash-free dry weight. The rates of incorporation of phosphorus into various fractions of periphyton phosphorus were also measured. These values corresponded to turnover times that ranged from 2 to 600 hours, but were usually between 10 and 100 hrs. A series of papers is in preparation detailing these and related findings on the periphyton of Rhode River. Several experiments of another kind have been done to examine the meaning of these data. In one such experiment a large set of periphyton plates were labeled with ^{32}P and the phosphate compounds formed were examined chromatographically to see in more detail what had been labeled under those conditions. Figure 8 shows the results for water-soluble compounds from 16 plates labeled for 30 minutes in one such run. In the first or upper column chromatogram, the

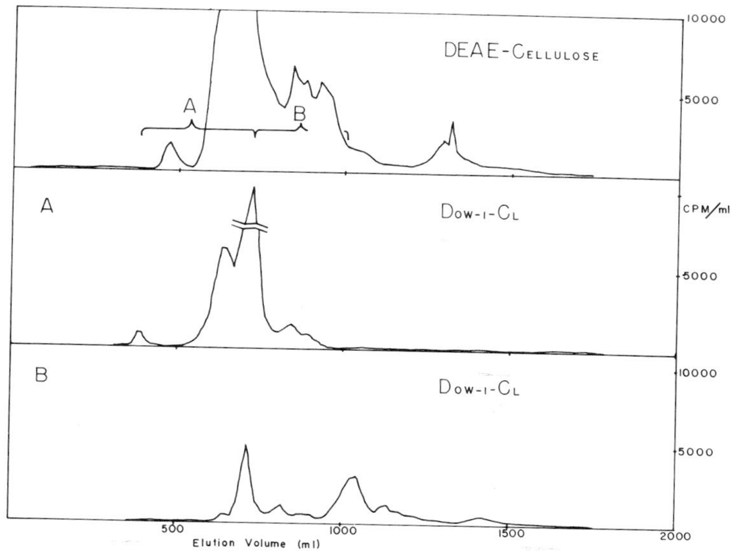

Figure 8. Column chromatograms of periphyton phosphorus fractions extracted from 16 artificial substrates at station 5. Plates had been allowed to incorporate ^{32}P-phosphate for 30 minutes near midday in clear sky conditions on 17 April 1972. The DEAE-cellulose column was 2 x 27 cm, the mixing flask was 300 ml and was initially filled with buffered 0.05 M NaCl. Elution was by 300 ml each of 0.05 M, 0.5 M, 1.0 M, 2.0 M, and 4.0 M buffered NaCl. Areas A and B from this chromatogram were separately concentrated by flash evaporation and desalted on a 2.5 x 94 cm column of Bio-Gel P-2 (100-200 mesh) equilibrated with 0.01 M sodium arsentate, pH 7.5. After desalting, fractions A and B were applied to 1.5 x 90 cm columns of Dowex-1-X8 (minus 400 mesh) in the chloride form at 40°. A closed 500 ml mixing flask was initially filled with 0.01 M cacodylate, pH 7.0. Consecutively, 500 ml portions of 0.25 M, 0.5 M, 1.0 M, and 2.0 M NaCl, buffered to pH 7.0 with cacodylate, were introduced into the mixing flask. The columns were eluted at 5 ml/min. For further details, see Methods section.

material beyond areas A and B was polymeric, predominantly polyphosphates. The materials in areas A and B were rechromatographed to further characterize the compounds present. The main point to be made here is that the phosphate taken up by the periphyton had been metabolically converted into a large number of organic phosphorus compounds. The apparent uptake was not just orthophosphate that had bound to sediment particles or to cell-wall surfaces. Instead, the rate of labeling of individual compounds was about the same as that found in uptake experiments in which the phosphorus taken up was determined as acid-labile, acid-stable, and orthophosphate fractions. The results of another way of looking at phosphorus cycling in periphyton are given in Figures 9 and 10. A 'chase' experiment was done by labeling a number of periphyton plates, then putting them back onto the supports where they had colonized in Rhode River in order to measure their rates of release of phosphorus. The chase data for total phosphorus (Fig. 9) and for orthophosphate (Fig. 10) gave about the same slopes. This means that these two fractions are being cycled at roughly the same rates, which would not be expected if a large proportion of the orthophosphate were merely bound to sediments, et cetera. Such a bound fraction would be expected to exchange very rapidly with the orthophosphate in the tidal water and would all be gone within a few minutes. Another interesting aspect of these results is the fact that the experiment is much longer in duration than uptake experiments. It includes day and night periods, low and high temperatures, et cetera, and therefore gives a better estimate of long-term uptake and release rates. The rate of phosphorus uptake in this experiment by the uptake method was 350 μg P/hr/g dry wt. and by the chase method it was 95 μg P/hr/g dry wt. Finally it becomes apparent that in this experiment the rates of phosphorus uptake were about equal to the rates of release (95 versus 99 μg P/hr/g dry wt.). This does not mean that the midday uptake rate was not really 350, rather than 95, but it says that the average rate for 66 hours was 95 μg P/hr/g dry wt.

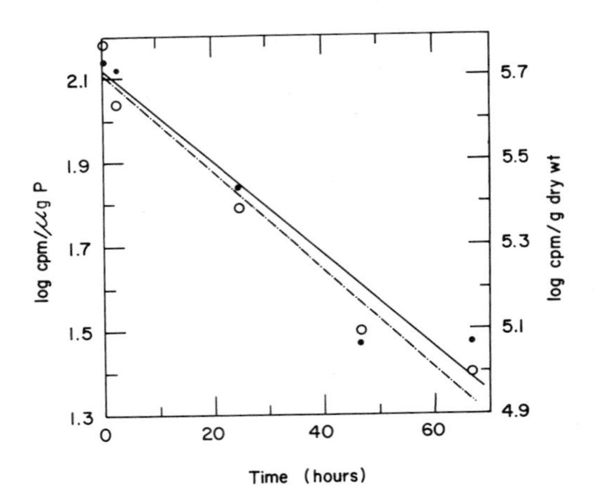

Figure 9. Kinetics of decrease in the radioactivity of periphyton total phosphorus in a chase experiment at station 8, initiated on 25 June 1973. Plates were labeled for 2 hours, then transferred back onto their support racks in the Rhode River (time zero in this Fig.). After various times of chase, pairs of plates were removed for analysis. Solid points and the solid line are cpm/μg P. Open circles and the broken line are cpm/g dry weight. The lines are linear least-squares regression fits of the data.

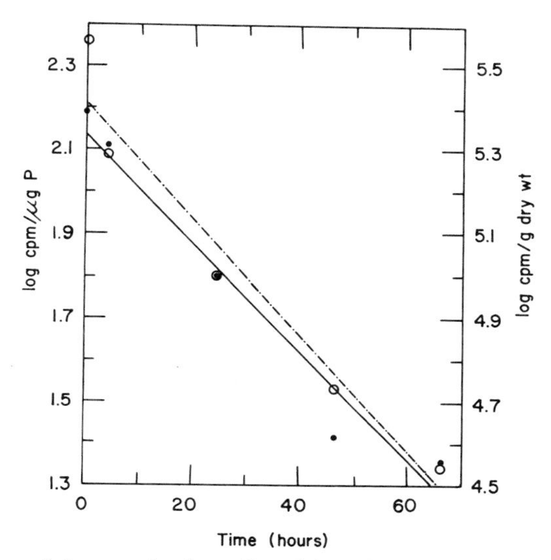

Figure 10. Kinetics of decrease in the radioactivity of periphyton orthophosphate in the same experiment shown in Figure 9, plotted the same way.

Plankton

In measuring phosphate uptake rates by plankton, two extremes of conditions may arise. In one case a relatively large pool of available orthophosphate in the water mass and a relatively low plankton standing crop or metabolic rate lead to a situation where the uptake of ^{32}P-phosphate is best estimated by measuring the appearance of radioactivity in the particulate fraction. The other type of situation is characterized by a smaller pool of available orthophosphate and a high metabolic activity rate in the plankton community. In this latter case the available orthophosphate pool turns over very rapidly and it is best to estimate uptake of ^{32}P-phosphate by measuring the disappearance of radioactivity from the water. We routinely carry out experiments in such a way that we can calculate the uptake rates by either method. An example of how helpful this can be is shown in Figures 11 and 12. In Figure 11 the data from an experiment on Rhode River at station 10.8 are plotted by the first method. On the same day at station 12.5 on Rhode River the water mass was so different in its characteristics that such a plot would have been very misleading. Figure 12 shows the data from this second location plotted by the second method. Phosphorus cycling was so rapid and the pool of available phosphate so small (8 μg P/1) that uptake apparently came to a halt within 20 minutes. This is not an extreme case. Some Rhode River samples have reached this situation within the first five minutes. In order to analyze uptake rates in this type of situation it is absolutely essential to

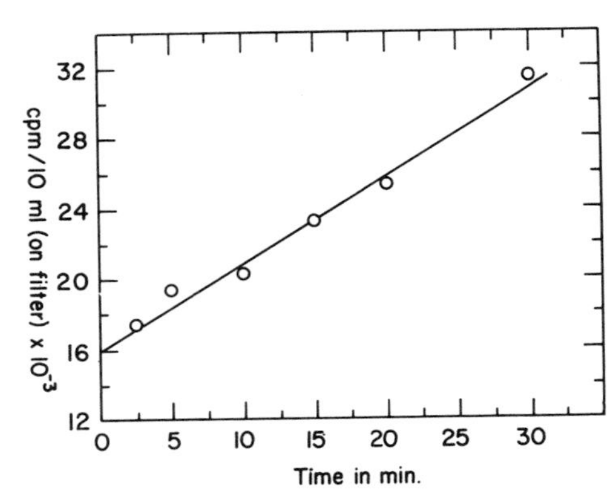

Figure 11. Analysis of the rate of plankton phosphate uptake by the measurement of the appearance of radioactivity in the particulate fraction. Experiment was run on 20 July 1972 at station 10.8 in the Rhode River at a depth of one meter. Dissolved orthophosphate was 50 μg P/1; uptake rate was calculated to be 0.042 μg P/min./1.

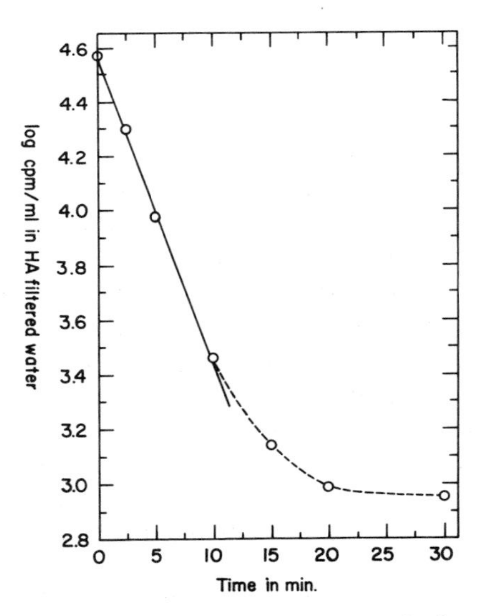

Figure 12. Analysis of the rate of plankton phosphate uptake by the measurement of the disappearance of radioactivity from the soluble fraction. Experiment was run on 20 July 1972 at station 12.5 in the Rhode River at a depth of one meter. Dissolved orthophosphate was 8 μg P/1; uptake rate was calculated to be 2.0 μg P/min./1.

assay the orthophosphate pool size and its initial specific activity as well as to take very frequent data points in the early phase of the experiment. In this method the initial rate of uptake is estimated from the slope and intercept of a plot like that in Figure 12. These experiments are now routinely run in light bottles, dark bottles, and in light bottles in the presence of the biological orthophosphate uptake inhibitor iodoacetic acid (IAA). An experiment of this type is shown in Figure 13. The rates of uptake in the light and dark bottle are about the same in Rhode River as illustrated in this experiment (*see also* Table 1). It would seem that the effects of light in stimulating algal uptake are about cancelled by the light inhibition of heterotrophic uptake. This indicates that most of the uptake is not driven by photosynthesis directly and is therefore not carried out by autotrophs. It is also apparent that IAA is an effective inhibitor of the uptake. This is strong support that the uptake is biological, since it is unlikely that IAA would interfere with physical binding to sediments and detritus. The next point to examine is the relationship of particle size to rate of phosphate uptake. Figure 14 is a typical set of data for Rhode River. Most of the phosphate uptake is carried out by particles which are in the 1.2 to 5 μm size class, whether in the light or the dark. Of course it is likely that any loosely floculated aggregates of particles would be broken up and would pass through the 5 μm pores in the first screen. The relationship between phosphate uptake rates (by ^{32}P in bottles) and bacterial biomass is examined in Table 1. A range of from 0.04 to 3.3 μg P/min./mm^3 of bacterial biomass was found in Rhode River. Table 2 gives the sizes, numbers, and biomass estimates for various sizes of bacteria and algae for the plankton of Rhode River on two occasions. The first is a normal picture for this estuary in March, while the second was during a dense dinoflagellate bloom associated with a very high population of bacteria. Even in the case of the dinoflagellate bloom of 16 May the size-class analysis of phosphate uptake looked very much like that shown in Figure 14. This was true despite the fact that about 98 percent of the biomass was dinoflagellates, which were retained by the 5μm pore size screen. It should also be noted that the uptake of phosphate by these dinoflagellates was measured by authoradiography and was only occurring in the light, yet the overall uptake was essentially the same in the light as in the dark.

The technique of microautoradiography of ^{33}P-labeled plankton has been used at Rhode River to trace the path of phosphorus cycling from dissolved orthophosphate to various locations in the particulate fraction. Table 3 gives the results from one such experiment. A rough index of rates of phosphorus assimilation per biomass is given in the last column. There is an obvious relationship between organism size and rate of uptake. The nannoplankton rate was 300 to 500 times that of the large dinoflagellates, which had the lowest rates. An inverse correlation exists between surface area per cell and grains per cell in the Table 3 data. If grain number per cell is multiplied by cell number per

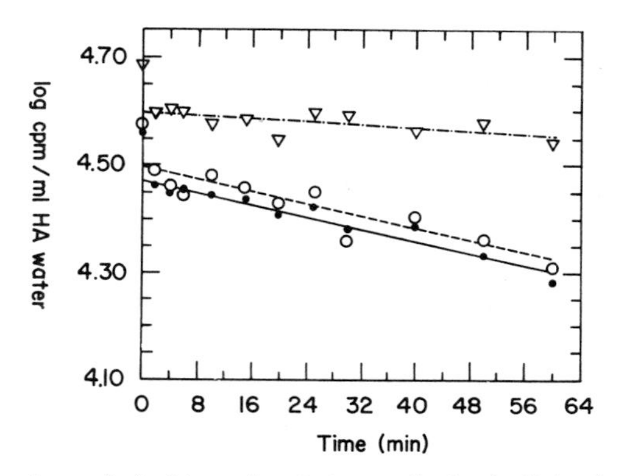

Figure 13. Comparison of plankton phosphate uptake in the light, in the dark, and in the presence of iodoacetic acid in the light. Solid points and solid line are from the light bottle; open circles and dashed line are from the dark bottle; triangles and broken line are from the light bottle containing 0.05 M IAA. Experiment was run 14 June 1973 in the Rhode River, station 11.2, at a depth of one meter.

TABLE 1

The relationship of phosphate uptake rates to bacterial biomass in plankton from Rhode River, station 11.2 at a depth of one meter.

date of experiment	^{32}P uptake rate (μg P/min./1) light	dark	bacterial cell number (cells/ml)	bacterial biomass (mm^3/1)	μg P/min. in dark per mm^3 bacteria
28 March, 1973	1.1	0.70	5.4×10^5	0.214	3.3
16 May, 1973	11.9	11.3	1.8×10^7	6.31	1.8
9 July, 1973	1.13	0.65	8.3×10^6	4.48	0.15
7 Aug., 1973	0.25	0.37	9.5×10^6	8.71	0.042
5 Sept., 1973	0.17	0.32	4.9×10^6	2.56	0.12

ml, it is evident that while *Gymnodinum nelsoni* and *G. splendens* constituted 90 percent of the phytoplankton biomass, they accounted for only 171 x 10^3 grains per ml. In contrast, the nannoplankton plus the small flagellates constituted only 1.0 percent of the phytoplankton biomass, but accounted for 625 x 10^3 grains. Thus it may be concluded that, while the bacteria carry out the bulk of the orthophosphate uptake, the next most important group is the nannoplankton. Plate 1 illustates the appearance of typical autoradiograms of Rhode River plankton. At the present stage of development of autoradiography in our laboratory, bacteria are near the limit of resolution and while they are present and sometimes aggregated into sheets, they also are found on the surfaces of algal cells and add to the "noise" level of the method. The emulsion

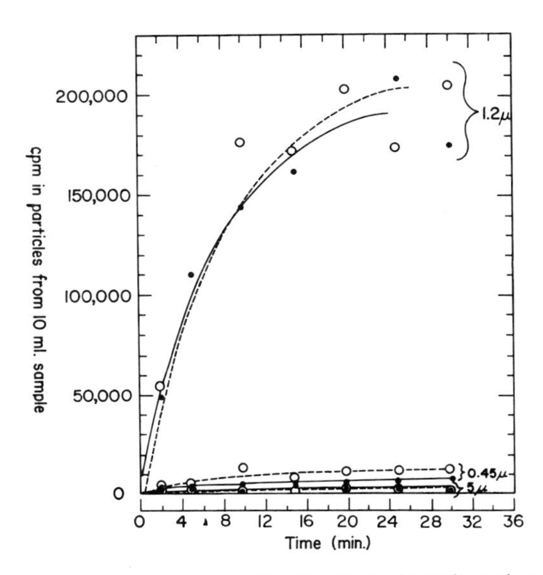

Figure 14. Time course of the appearance of radioactivity in various sizes of particles in a plankton sample. Experiment was run in the Rhode River, station 11.2, at a depth of one meter on 28 March 1973. Solid points and solid line are from the light bottle, open circles and dashed line are from the dark bottle.

lowers the resolution of the microscope by blurring the view. Species identification is therefore more difficult even for the algae. However, we believe that autoradiography is a powerful new tool deserving more effort and development.

Pulse-Label Experiments

A great deal of concern has often been expressed that ^{32}P-bottle experiments have given rise to inaccurate conclusions due to (a) binding of the phosphate to detritus, sediments and cell surfaces, which is not true biological uptake; (b) artifacts induced by containing a plankton population inside a bottle. The pulse-label experiment described in this article was designed to help evaluate the validity of these concerns as well as to enable us to begin examining the chemistry of phosphorus cycling. An example of results is given in Figure 15. In this experiment the labeling time was only six minutes, which puts it well within the time zone where exchange processes have already occurred and recyling is not yet a problem. Since relatively large volumes can be processed, it is possible to examine in detail what chemical from the phosphate is in after six minutes. For this article, the main point of Figure 15 is that a large number of

TABLE 2

Sizes and biomass estimates of bacteria and phytoplankton in the Rhode River, station 11.2, at a depth of one meter.

date (1973)	Organism	size (μm)	cells /ml	biomass ($mm^3/1$)	% of total biomass
28 March	coccoid bacteria	2.5	8.5×10^3	0.069	3.8
		1.0	8.1×10^4	0.042	2.3
		0.75	6.9×10^4	0.015	0.8
		0.25	2.8×10^4	0.0002	0.0
	bacteria (rods)	2 x 0.5	6.1×10^4	0.024	1.3
		1.5 x 0.5	2.4×10^4	0.007	0.4
		2.5 x 0.5	6.7×10^4	0.033	1.8
		2.5 x 0.25	2.0×10^5	0.024	1.3
	algae: nanno- -plankton	3.5	1.4×10^4	0.314	17.2
	small flagellates	7.5	1.4×10^3	0.309	16.9
	Chlorella sp.	4.0	2.5×10^3	0.084	4.6
	euglenoid sp.	3.0 x 12.5	1.1×10^3	0.097	5.3
	Cryptomonas sp.	4.5 x 8.5	6.0×10^3	0.811	44.3
				total - 1.83	
16 May	coccoid bacteria	1.0	1.6×10^6	0.837	0.1
		0.5	8.2×10^6	0.536	0.1
	bacteria (rods)	1 x 2	6.7×10^6	10.5	1.3
		1.5 x 2	1.1×10^6	3.89	0.5
	algae: Prorocentrum sp.	20 x 16	2.0×10^5	804	97.8
	Cryptomonas sp.	25 x 18	3.0×10^2	1.91	0.2
				total - 821.7	

TABLE 3

Relative phosphorus uptake by phytoplankton as estimated by ^{33}P-autoradiography on 9 July 1973 in Rhode River, station 11.2, at a depth of one meter. Uptake was for 30 minutes in a light bottle.

organism	size (μm)	cells /ml x10^3	volume /cell (μm^3)	% of phyto-plankton biomass	grain number* per cell	per μm^3 cell vol.
Gymnodinium nelsoni	70 x 50	3.1	1.37 x 10^5	67.9	44.8	0.00033
Gymnodinium splendens	60 x 50	1.2	1.18 x 10^5	22.7	26.4	0.00022
Peridinium sp.	70 x 60	0.17	1.98 x 10^5	5.4	—	—
euglenoid sp.	85 x 8	0.30	4270	0.2	—	—
euglenoid sp.	55 x 10	0.90	4320	0.6	—	—
Prorocentrum sp.	20 x 16	1.8	4020	1.2	14.5	.0036
Gymnodinium punctatum	12 x 5	24	236	0.9	8.3	0.035
Amphidinium globosum	10 x 5	6.2	196	0.2	—	—
nannoplankton	7 x 4	54	88	0.8	9.4	0.107
small flagellates	5 x 4	23	63	0.2	5.1	0.081

*Total grains in 100 cells were counted.

compounds are labeled in this time and a large amount of the total radioactivity is no longer orthophosphate. Some of the truly biologically assimilated phosphate would be expected to be in the form of cellular orthophosphate, but if we ignore this fraction it is possible to compare the uptake rate assayed here as nonorthophosphate after six minutes with the rate that was measured simultaneously on the same water mass by the bottle technique. The light-bottle experiment gave an estimate of 6.9 μg P/hr/1, while the pulse-label experiment gave a value of 7.5 μg P/hr/1. We feel these two values are essentially the same and verify in another way that the uptake measured is primarily biological. In most experiments of this type, the two methods have about the same rate of uptake.

Our concept of the major pathways of phosphorus cycling in estuarine plankton is summarized in Figure 16. The heavy arrows are believed to be the main pathways. Microbiological data that have been gathered as well as the size classing and inhibitor data seem to support this picture. Thus, most orthophosphate is taken up by bacteria which are mainly on the surfaces of suspended sediments and detritus, but phytoplankton also take up some

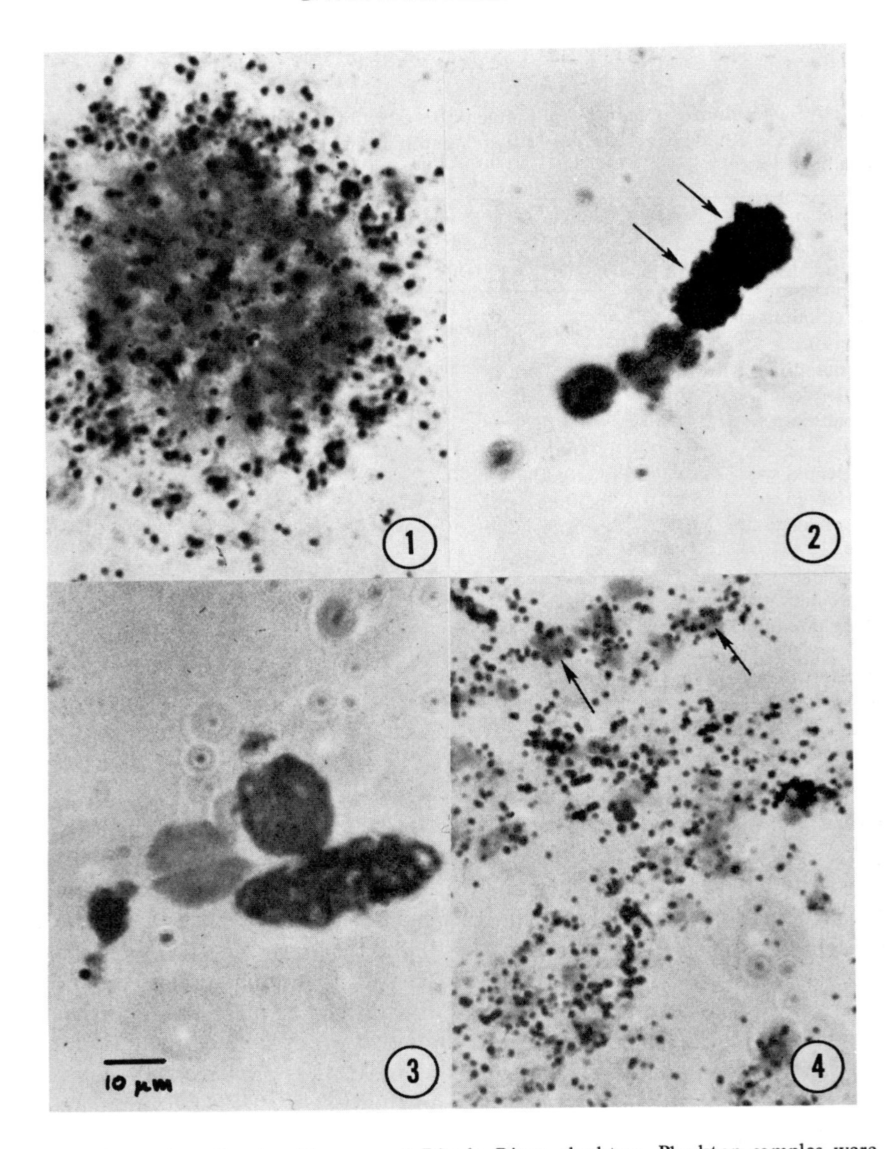

Plate 1. Examples of autoradiograms of Rhode River plankton. Plankton samples were exposed **in situ** to ^{33}P-orthophosphate in light bottles (no. 1 and no. 2) and in dark bottles (no. 3 and no. 4) for 30 minutes. In the light bottles, numerous grains are visible around and within a large **Gymnodinium splendens** cell (no. 1). In no. 2 several smaller algae cells are visible; only the two **G. punctatum** cells (arrows) assimilated ^{33}P-orthophosphate and are covered by numerious grains. In the dark bottles **Prorocentrum danicum, Glenodinium occulatum, Euglena** sp., and a

Plate 1 (Continued)

nannoplankton cell are shown without grains (no. 3). Numerous grains are found above clumps of bacteria (arrows, no. 4).

All the above representative bright field photographs were taken with a Carl Zeiss light microscope at a magnification of 1000x. All the above-mentioned species were identified on separate, non-autoradiographed preparations. The bar in no. 3 represents 10 μm.

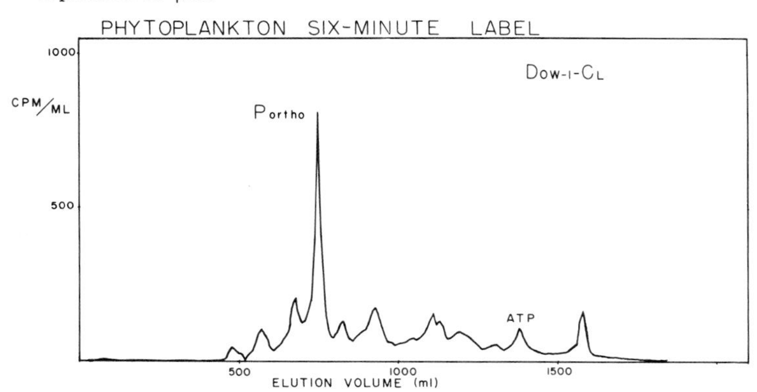

Figure 15. Column chromatographic analysis of the low-molecular weight water-soluble phosphorus fraction labeled in a continuous flow, six minute, pulse-label of plankton from a depth of one meter in Rhode River, station 11.2, on 20 September 1972. The Dowex-1-chloride column was run in the same manner as described in Figure 8. Forty liters of plankton were pulse labeled. Two hundred and sixty liters of cold plankton were also harvested by the same technique and combined with the labeled plankton to serve as "carrier". For further details see Methods section.

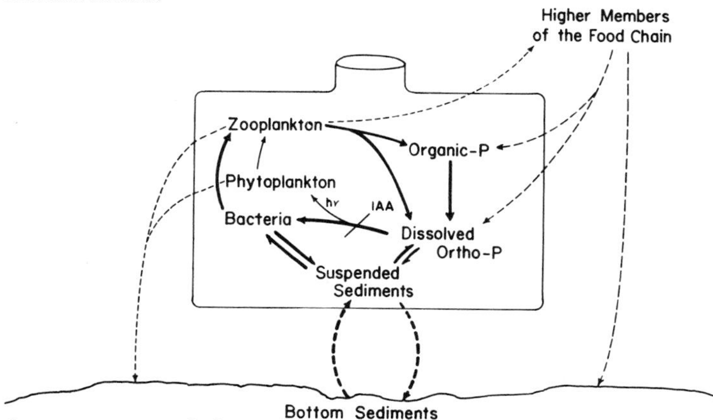

Figure 16. Current concept of the pathways of estuarine plankton phosphorus cycling. Processes stopped by enclosing a sample in a bottle are indicated as dashed arrows. Heavy lines indicate major processes. Phosphate uptake by phytoplankton requires light energy, and the presence of iodoacetic acid (IAA) inhibits direct biological uptake of orthophosphate.

orthophosphate in the light. The bacteria and phytoplankton are then eaten by filter feeders especially ciliate protozoans (31). These in turn release most of the phosphorus as dissolved orthophosphate and organic phosphorus. This is in agreement with the cycle proposed for tidal marsh communities by Johannes (12, 13).

REFERENCES

1. American Public Health Association
   1971 **Standard Methods for the Examination of Water and Waste Water,** 13th ed., APHA, New York, New York, p. 518-32.
2. Bogoroch, R.
   1972 Liquid emulsion autoradiography. In **Autoradiography for Biologists,** p. 65-94. (ed. Gahan, P. B.). Academic Press, New York.
3. Correll, D. L. and Tolbert, N. E.
   1962 Ribonucleic acid-polyphosphate from algae I. Isolation and physiology. **Plant Physiol.,** 37: 627-36.
4. Correll, D. L.
   1965a Peagic phosphorus metabolism in antarctic waters. **Limnol Oceangr.,** 10: 364-70.
5. Correll, D. L.
   1965b Ribonucleic acid-polyphosphate from algae III. Hydrolysis studies. **Plant and Cell Physiol.,** 6: 661-9.
6. Correll, D. L.
   1973 The measurement of phosphorus metabolism in natural populations of microorganisms. In **Bioassay Techniques and Environmental Chemistry,** p. 425-31. (ed. Glass, G. E.). Ann Arbor Science Publishers, Ann Arbor, Michigan.
7. Fiske, C. H. and Subbarow, Y.
   1925 The colorimetric determination of phosphorus. **Jour. Biol. Chem.,** 66: 375-400.
8. Fuhs, G. W. and Canelli, E.
   1970 Phosphorus-33 autoradiography used to measure phosphate uptake by individual algae. **Limnol. Oceangr.,** 15: 962-6.
9. Fuhs, G. W., Demmerle, S. D., Cannelli, E., and Chen, M.
   1972 Characterization of phosphorus-limited plankton algae. In **Nutrients and Eutrophication: the limiting nutrient controversy, Special Symp.** vol. I, p. 113-33. American Society of Limnology and Oceanography.
10. Grzenda, A. R. and Brehmer, M. L.
    1960 A quantitative method for the collection and measurement of stream periphyton. **Limnol. Oceangr.,** 15: 190-4.
11. Harris, E.
    1957 Radiophosphorus metabolism in zooplankton and microorganisms. **Canada Jour. Zool.,** 35: 769-82.

12. Johannes, R. E.
1964 Uptake and release of dissolved organic phosphorus by representatives of a coastal marine ecosystem. **Limnol. Oceangr.,** 9: 224-34.

13. Johannes, R. E.
1965 Influence of marine protozoa on nutrient regeneration. **Limnol. Oceangr.,** 10: 434-42.

14. King, E. J.
1932 The colorimetric determination of phosphorus. **Biochem. Jour.,** 26: 292-7.

15. Lean, D. R. S.
1973 Phosphorus dynamics in lake water. **Science,** 179: 678-80.

16. McRoy, C. P. and Barsdate, R. J.
1970 Phosphate absorption in eelgrass. **Limnol. Oceangr.,** 15: 6-13.

17. McRoy, C. P., Barsdate, R. J., and Nebert, M.
1972 Phosphorus cycling in an eelgrass (**Zostera marina** L.) ecosystem. **Limnol. Oceangr.,** 17: 58-67.

18. Odum, E. P., Kuentzler, E. J., and Blunt, M. X.
1958 Uptake of P^{32} and primary productivity in marine benthic algae. **Limnol. Oceangr.,** 3: 340-5.

19. Pomeroy, L. R.
1960 Residence time of dissolved phosphate in natural waters. **Science,** 131: 1731-2.

20. Pomeroy, L. R., Mathews, H. M., and Min, H. S.
1963 Excretion of phosphate and soluble organic phosphorus compounds by zooplankton. **Limnol. Oceangr.,** 8: 50-5.

21. Pomeroy, L. R., Smith, E. E., Grant, G. M.
1965 The exchange of phosphate between estuarine water and sediments. **Limnol. Oceangr.,** 10: 167-72.

22. Pomeroy, L. R., Odum, E. P., Johannes, R. E., and Roffman, B.
1966 Flux of ^{32}P and ^{65}Zn through a salt-marsh ecosystem. In **Disposal of radioactive wastes into seas, oceans, and surface waters,** p. 177-88, Internatl. Atomic Energy Agency, Vienna.

23. Pomeroy, L. R., Johannes, R. E., Odum, E. P., and Roffman, B.
1969 The phosphorus and zinc cycles and productivity of a salt marsh. In **Symp. Radioecology, Proc. 2nd Natl. Symp.** p. 412-9. (eds. Nelson, D. J. and Evans, F. C.). Ann Arbor, Michigan.

24. Pomeroy, L. R., Shenton, L. R., Jones, R. D. H., and Reimold, R. J.
1972 Nutrient flux in estuariess. In **Nutrients and Eutrophication: the limiting nutrient controversy, Special Symp.** vol. I, p. 274-91. American Society of Limnology and Oceanography.

25. Reimold, R. J.
1972 The movement of phosphorus through the salt marsh cord grass, **Spartina alterniflora** Loisel. **Limnol. Oceangr.,** 17: 606-11.

26. Rhee, G.-Y.
1972 Competition between an alga and an aquatic bacterium for phosphate. **Limnol. Oceangr.,** 17: 505-14.

27. Rigler, F. H.
1956 A tracer study of the phosphorus cycle in lake water. **Ecology,** 37: 550-62.

28. Rigler, F. H.
1964 The phosphorus fractions and the turnover time of inorganic phosphorus in different types of lakes. **Limnol. Oceangr.,** 9: 511-8.

29. Rodina, A. G.
1972 **Methods in Aquatic Microbiology,** p. 149-77. (trans., ed., and rev. Colwell, R. R. and Zambruski, M. S.). University Park Press, Baltimore, Maryland.

30. Satomi, M. and Pomery, L. R.
1965 Respiration and phosphorus excretion in some marine populations. **Ecology,** 46: 877-81.

31. Small, E. B.
1973 Protozoan-bacterial interrelationships in the estuarine environment. In **2nd Internatl. Estuarine Research Conference,** Myrtle Beach, South Carolina. Oct. 15-18. (This symposium.)

HEAVY METAL FLUXES THROUGH SALT-MARSH ESTUARIES

by

Herbert L. Windom[1]

ABSTRACT

The flux of heavy metals through salt-marsh estuaries is controlled by processes which occur at river-estuary and salt-marsh-sediment boundaries. These processes include adsorption-desorption reactions, flocculation, precipitation, and sedimentation. Along with estuarine circulation they effectively control the residence time of metals in estuaries. Some biological processes, such as uptake by marsh vegetation, may serve as efficient mechanisms to recycle metals otherwise lost to estuarine sediments.

Upon entry into estuaries in river run-off a large portion of dissolved iron and manganese precipitates out of solution and accumulates in sediments along with particulate fractions of these metals. Most cadmium and copper in solution is transported through salt-marsh estuaries while particulate fractions are apparently lost to the sediments. All dissolved mercury and apparently a desorbable fraction of it in particulate matter is transported through the salt-marsh system.

INTRODUCTION

Heavy metals transported by river run-off from continents must pass through estuaries before entering the ocean. Some of these metals never reach the open

1. Skidaway Institute of Oceanography, P. O. Box 13687, Savannah, Georgia 31406.

ocean but are, instead, deposited in estuarine sediments. Others may be transferred directly to the ocean and be little influenced by the estuarine environment.

The fate of heavy metals in estuaries is controlled mainly by processes that take place at two interfaces. Those occurring at the river-estuary boundary control the form and rate of input of metals. At the estuarine-sediment boundary metals may be lost from the system because of precipitation and accumulation, and previously accumulated metals may be released by chemical reaction and biological activity.

Whether a given metal is ultimately accumulated in estuarine sediments or is transported to the open ocean depends also on the residence time of the metal in the estuarine environment. This time is relatively short for heavy metals in the open ocean compared with other elements (5). In salt-marsh estuaries higher concentrations of organic matter in the water and sediments, greater biological productivity, and increased concentrations of sulfides may lead to even shorter residence times for heavy metals.

Most of the research done in estuaries has been confined to river-estuary boundaries, and its primary emphasis has been to determine the load of metals in river water and their supply to the ocean. Kharkar, et al (7), who studied the supply of dissolved metals to the ocean from many east coast rivers, determined that cobalt, silver, and selenium are desorbed from particulate matter as they enter marine waters. The supply of metals in particulate matter by several east coast rivers has also been studied (17), but, their fate in the estuaries was not considered. Turekian (14) concluded that a large part of the silver supplied by rivers must be accommodated in estuarine sediments, which would account for its relatively low abundance in marine sediments.

These studies and others (16, 20) have emphasized the role of river run-off on heavy metals in the open-ocean environment. A few have dealt with the fate of given metals in estuaries in the broad sense (8, 9). More specific studies have determined the distribution of mercury (19) and scandium (1) and their budgets in salt-marsh estuaries and the near-shore environment. Other studies have shown the importance of the marsh grass *Spartina alterniflora* in controlling the availability of such elements as Fe and Zn (10, 18) and Zn, Cu, Co, Fe, and Mn (2) to the estuarine food web. More recently, a detailed study of the role of *S. alterniflora* in the fate of Hg in salt marshes indicated that this element can be accumulated by the roots of the plant, transferred to its leaves, and subsequently to the overlying water, effecting a release of the metals from the sediment (11). No large-scale detailed studies have been carried out to budget the net fluxes of metals through estuaries.

The salt-marsh estuaries of the southeastern Atlantic coast are typical of those throughout the world, containing over 400,000 hectares of marshes covered predominantly with one macrophyte species (*S. alterniflora*). The area receives a large amount (approx. 60 billion cubic meters annually) of river run-off that contains high concentrations of metals. During the past several years information has been obtained on the input of metals to this area by rivers. The rivers studied account for 96% of the total discharge to the salt marshes along the coasts of South Carolina, Georgia, and north Florida (Figure 1). In addition, the estuaries and salt marshes of the region have been studied to determine the patterns of metal accumulation and distribution. Processes that are responsible for fate of metals in this environment can therefore be evaluated, providing a broad picture of the important aspects of metal flux through salt-marsh estuaries. Although data may not indicate, in detail, the characteristics of those processes, they do suggest the importance of various boundaries and processes that control heavy-metal transfer through the system.

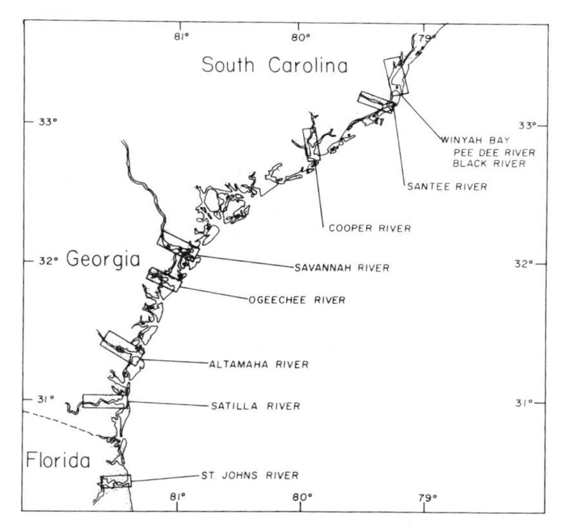

Figure 1 Southeastern Atlantic salt marsh estuarine system, indicating estuaries and rivers studied.

HEAVY METAL BUDGETS IN SALT-MARSH ESTUARIES

If it is assumed that a salt-marsh estuary is in steady state where input of metals is equal to loss, then the flux of metals can be expressed by the following equation:

$$K_i = K_d + K_p = K_s + K_f \qquad (1)$$

where K_i is the rate of total metal input by rivers, K_d and K_p are the rates of dissolved and particulate metal input by rivers, respectively, K_s is the rate of metal loss in salt-marsh sediments, and K_f is the net flux of metals through the estuarine system. If the total input of metals to an estuarine system and total losses to estuarine sediments are known the net flux of metal through the system can be calculated. If this information is known in fair detail and to a fairly accurate degree, some processes affecting the transfer of metals through the system can be suggested.

Estuaries such as those of the southeastern United States are sediment traps which apparently retain a large percentage of the particulate matter entering from rivers (3, 21). Metals supplied by rivers occur in both dissolved and particulate phases. If a metal is not redistributed between these phases at the river-estuary boundary, then all the metal contained in sediments entering the estuary will be lost to the sediments. If, also, the dissolved component of the metal undergoes no reactions which bring about its precipitation in the estuary, the net flux of the metal through the system will be approximately equal to the total input in solution ($K_f = K_d$). If the particulate matter releases a fraction of its metals, the net flux of metals through the system will be greater than the amount supplied by the river in the dissolved form ($K_f > K_d$). If on the other hand, metals are either adsorbed on particulate matter or precipitated upon entry into the estuaries, the net flux would be less than this amount ($K_f < K_d$).

The information required for equation 1 has been obtained for salt-marsh estuarine systems of the southeastern Atlantic coast. The dissolved and suspended loads of Fe, Mn, Cd, Cu, and Hg in nine major rivers emptying into the southeastern Atlantic salt-marsh environment were determined by bimonthly sampling of the rivers near their months over a one-year period. With the exception of Fe, the concentrations of the metals did not vary to a great degree in any one river over the period studied (Tables 1 and 2). In the case of Fe a clear seasonal pattern occurred with concentrations in solution increasing with increasing discharge (Fig. 2). Using the average metal concentrations and the average discharge rates for each of the rivers, it was possible to calculate the annual fluvial transport of dissolved and particulate metals (Table 3). The calculated annual rate of iron discharge, which requires more detailed information for each river, is probably significantly higher than indicated because average concentrations were used rather than concentrations integrated over variations in the discharge rates. These data supply the values for K_d and K_p for the southeastern Atlantic system.

The average concentrations of the metals in marsh sediments were determined at 10-cm intervals in 25 cores taken between Winyah Bay, South Carolina, and the St. Johns River, Florida (Table 4), an area where metal levels are relatively uniform. From these data it is possible to calculate an average accumulation rate for salt-marsh sediments in the area, if the rate of deposition is known. Obviously the rate of accumulation of salt-marsh sediments varies from place to place, depending on conditions. Rusnak's (12) estimated rate of 1 mm/yr based on eustatic changes in sea level, however, can be used as an average for the entire

TABLE 1.

Mean and Range in Dissolved Metal Concentrations in Southeastern Rivers[1]

River		ppb				
		Fe	Mn	Cd	Cu	Hg
Pee Dee	Mean	210	21	1.0	6.4	0.06
	Max	450	40	1.4	8.2	0.10
	Min	90	12	0.4	4.9	0.02
Black	Mean	240	18	1.0	7.0	0.06
	Max	410	53	3.6	13.0	0.14
	Min	60	8	0.1	1.2	0.02
Santee	Mean	170	12	0.4	2.8	0.05
	Max	360	53	0.7	4.0	0.08
	Min	30	3	0.1	1.7	0.02
Cooper	Mean	80	3	0.7	3.3	0.04
	Max	240	8	2.6	6.5	0.08
	Min	10	1	0.1	1.8	0.02
Savannah	Mean	170	21	0.3	3.0	0.07
	Max	380	34	0.6	5.5	0.13
	Min	70	6	0.2	0.8	0.02
Ogeechee	Mean	310	33	1.0	4.2	0.07
	Max	490	63	2.0	9.0	0.14
	Min.	140	15	0.4	1.0	0.02
Altamaha	Mean	150	17	0.8	3.5	0.05
	Max	280	39	2.6	5.8	0.06
	Min	30	2	0.2	1.1	0.02
Satilla	Mean	370	42	0.8	5.7	0.07
	Max	610	85	1.6	10.0	0.14
	Min	280	22	0.3	4.0	0.04
St. Johns	Mean	70	4	0.8	3.5	0.05
	Max	90	8	1.5	5.0	0.08
	Min	40	2	0.1	0.6	0.03

1 Based on results of a total of from 6 to 30 samples collected from each river bimonthly, May 1972 to July 1973.

Analyses made by atomic absorption (Fe, Mn, and Hg) and anodic stripping (Cu and Cd) µ techniques on samples filtered through 0.45 µ Millipore filters.

TABLE 2

Average Metal Concentration in Suspended Sediment of Southeastern Rivers[1]

River	Av. Suspended Sediment Load (mg/1)	Fe %	Mn	Cd	Cu	Hg
		ppm				
Pee Dee	9	6.6	1050	5	44	0.5
Black	11	5.0	1490	7	115	0.6
Santee	53	5.7	1400	3	48	0.2
Cooper	7	5.7	830	15	125	1.0
Savannah	22	8.2	1120	8	76	0.7
Ogeechee	7	5.0	1300	13	130	0.6
Altamaha	16	5.2	1600	26	100	0.7
Satilla	26	6.2	330	10	35	0.4
St. Johns	12	4.4	500	20	96	0.6

1 Based on results of a total of from 6 to 30 samples collected from each river bimonthly, May 1972 to July 1973.

Analyses made by atomic absorption after HF digestion (Fe, Mn, Cd, and Cu) or wet digestion with $H_2SO_4 - HNO_3$ (Hg).

salt-marsh system. Using this value, the rate of accumulation of metals in salt-marsh sediments was determined (Table 4) making it possible then to caluclate the annual flux of each metal through the entire 400,000-hectare salt-marsh area of the southeastern United States (Table 5). It is clear that essentially all the Fe transported in solution into the estuary precipitates out and accumulates in the salt-marsh sediments. Apparently a part of the dissolved manganese experiences a similar fate, since its net flux is less than its total dissolved input. The rates of Cu and Cd input in solution are equal to their net fluxes through the system, suggesting that these metals are relatively non-reactive in the salt-marsh estuarine system. The net flux of Hg through the estuary is greater than the net input in the dissolved form, suggesting desorption from particulate matter.

In summary, it can be said that, upon entry into the estuarine environment from river run-off, all the particulate and a fraction of the dissolved iron and manganese are precipitated out and lost to salt-marsh sediments. No significant

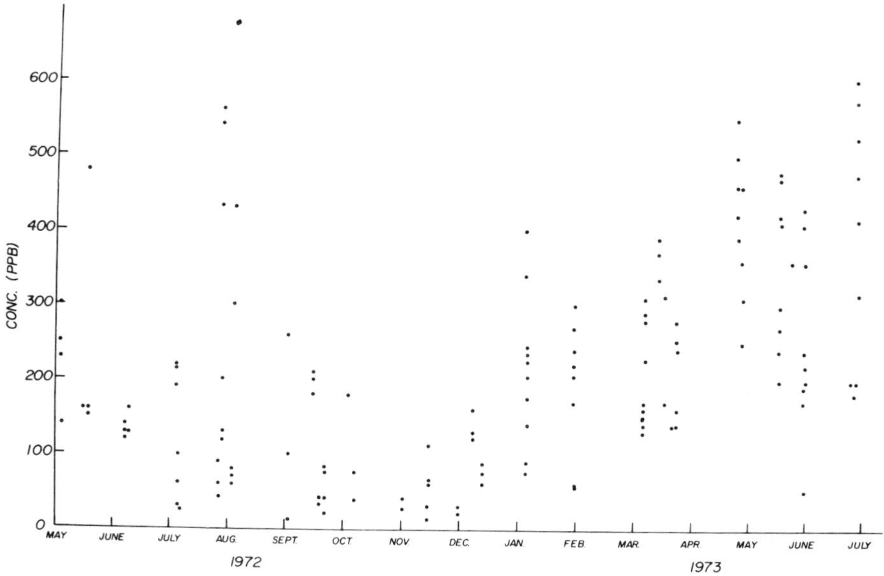

Figure 2 Seasonal variation in dissolved iron in southeastern rivers. Lowest discharge rates occur in October, November, and December. Highest discharges occur in summer.

amount of Fe escapes the estuarine zone. The accumulation of cadmium and copper in sediments is equal to that supplied by the rivers in particulate form. The dissolved fraction of these metals apparently is not affected by reactions that reduce their residence time in the salt-marsh environment. It also appears that no net transfer between dissolved and particulate phases takes place. Some release of mercury from particulate phases following entry into estuaries may occur, leading to an increase of this metal in estuarine waters. Other studies support these conclusions (6, 19).

HEAVY-METAL VARIATIONS IN ESTUARINE WATERS

If the conclusions drawn above are valid, they should be supported by the distribution of metals in estuarine waters. Concentrations of heavy metals in rivers are generally much higher than those found in marine waters (5, 15). Even coastal waters show levels of metals higher than those of the open ocean, indicating an increasing concentration gradient from offshore to inshore (13, 22). If metals transported in solution by rivers undergo no reaction that leads to their precipitation in the estuarine zone, variations in their concentrations would be a linear function of salinity, showing a decrease caused by dilution with

TABLE 3

Metal Transport to Southeastern Atlantic Salt-Marsh Estuaries by Rivers

	Average Annual Discharge (10^{12} 1)	Annual Metal Transport ($10^6 K_g$) Fe		($10^3 K_g$) Mn		Cd		Cu		Hg	
		D	P	D	P	D	P	D	P	D	P
Pee Dee	11.7	2.5	7.0	240	110	11.4	0.6	75	5	0.64	0.06
Black	0.8	0.2	0.5	14	13	0.9	0.1	5	1	0.09	0.01
Santee	1.9	0.4	5.6	20	140	0.7	0.3	5	5	0.08	0.02
Cooper	12.7	1.0	5.1	41	74	8.6	1.4	42	11	0.52	0.08
Savannah	10.4	1.9	18.7	230	250	3.2	1.8	17	32	0.65	0.15
Ogeechee	2.0	0.6	0.7	66	18	1.8	0.2	8	2	0.09	0.01
Altamaha	11.9	1.8	9.5	195	305	9.1	4.9	41	19	0.58	0.12
Satilla	2.0	0.8	3.1	83	17	1.5	0.5	11	2	0.18	0.02
St. Johns	5.0	0.3	2.5	20	30	3.8	1.2	18	5	0.26	0.04
Total	58.4	9.5	52.7	909	957	41.0	11.0	222	82	3.09	0.51
Grand Total		62.2		1866		52.0		304		3.60	
Percent Particulate		85		51		21		37		14	

TABLE 4

Average Concentrations and Accumulation Rate of Metal in Marsh Cores (Dry Weight Basis)

Station Number	% Fe	ppm Mn	Cu	Hg	Cd
1	3.9	103	30	0.12	3.4
2		250	16	0.09	5.0
3	3.9	166	26	0.16	2.9
4	5.0	142	25	0.05	3.3
5		209	24	0.08	4.4
6		158	10	0.01	1.0
7	1.2	139	4	0.04	0.4
8	1.8	114	4	0.02	0.8
9	4.1	309	16	0.04	2.1
10	1.9	174	3	0.04	0.1
11	2.4	151	8	0.06	0.5
12	4.8	260	12	0.09	0.6
13	4.6	353	20	0.16	0.8
14	3.4	319	10	0.10	0.4
15	4.0	173	10	0.11	0.8
16	2.9	366	8	0.02	0.8
17	2.7	262	6	0.05	0.2
18	3.2	273	12	0.16	0.8
19	1.1	175	4	0.03	1.6
20	3.7	168	8	0.06	0.8
21	3.1	244	8	0.07	1.2
22	4.7	132	7	0.06	0.8
23	2.9	128	10	0.08	0.7
24	0.9	205	5	0.05	0.8
25	1.5	74	2	0.05	0.4
Mean	3.1	202	11	0.07	1.4
Average Accumulation Rate* ($mg/m^2 \cdot yr$)					
	45×10^3	303	16	0.10	2.1

* Assuming an average sedimentation rate of 1 mm/yr and an average specific gravity of marsh sediments of 1.5g/cc.

Analytical techniques similar to those for suspended sediment.

TABLE 5

Budget for the Annual Flux of Metals through
Southeastern Atlantic Salt-Marsh Estuaries

	Fe (10^6 Kg)	Mn	Cd	Cu (10^3 Kg)	Hg
Total Input (k_i)	62	1866	52	304	3.6
Sedimentation Loss (K_s)	210	1200	9	66	0.4
% Sedimentation Loss	> 100	64	17	22	11
Net Flux through Estuary (K_f)	0	666	43	238	3.2

metal-poor marine waters. If, on the other hand, the metal concentrations versus salinity were more pronounced, the influence of processes other than dilution would be revealed. If metals are desorbed from particulate matter at the river-estuary boundary, their concentrations in estuarine waters should increase reflecting this addition. Subsequently, upon mixing with offshore waters, a decrease in the metal concentration would take place.

Bimonthly analyses of water samples taken from eight estuaries along the southeastern Atlantic coast (Fig. 2) indicate variations of metal concentrations across salinity gradients. These variations support what is predicted from the above budget calculations. For example, iron decreases exponentially with increasing salinty (Fig. 3) owing to its precipitation upon entry into the estuarine zone. Manganese shows a similar but less pronounced trend. These decreases in iron and manganese at the river-estuary boundary may be due to the formation of hydrated iron oxide which flocculates, precipitates, and accumulates in estuarine sediments. The increased electrolyte concentration of estuarine waters cna also influence the fate of very fine (< 0.45 microns) particulate iron and manganese associated with organic matter, also leading to their flocculation.

Variations in the other metals with salinity (Fig. 4) also appear to support the budget calculations above (Table 5). Copper and cadmium generally decrease in concentration with increasing salinity because of dilution with sea water containing lower levels of these metals (13, 22). Mercury concentrations are relatively increased in estuarine waters as compared to either river or more saline waters. This supports earlier observations from the same area (19).

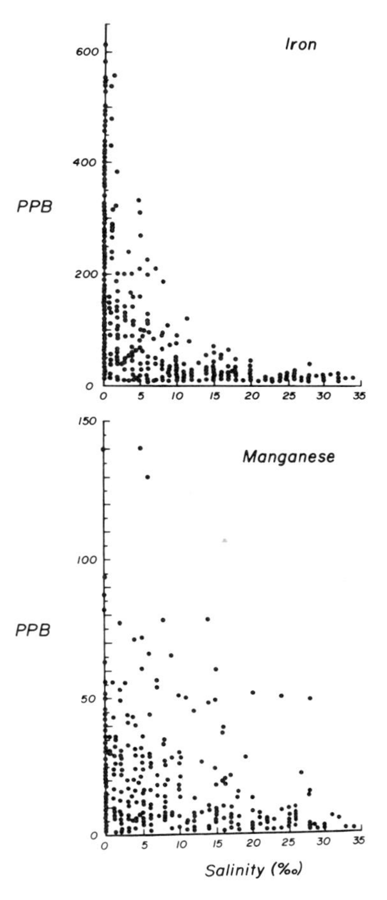

Figure 3 Dissolved iron and manganese concentration as a function of salinity in southeastern estuaries.

The two lines of argument above support the same conclusions regarding the types or, at least, the direction, of processes that affect heavy-metal transfer through salt-marsh estuaries. Specific characteristics of these processes, however, cannot be evaluated from these data alone. To determine the significance of the role played by various components of the estuarine system, each individual component must be evaluated to determine the magnitude of its effect and to assess its abiliby to influence the patterns and budgets described above.

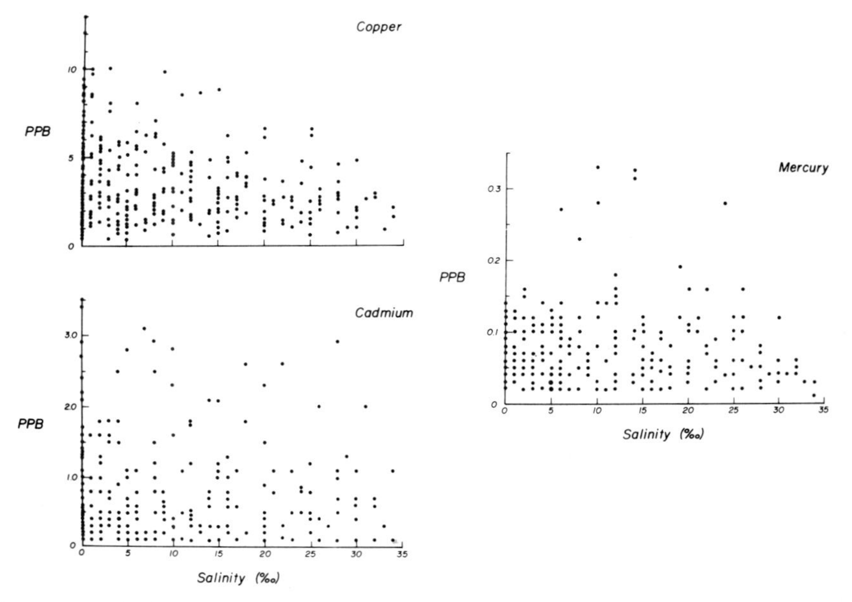

Figure 4 Dissolved copper, cadmium and mercury concentrations as a function of salinity in southeastern estuaries.

THE ROLE OF VEGETATION IN METAL TRANSFER THROUGH SALT-MARSH ESTUARIES

The dominant vegetation of salt marshes along the southeastern Atlantic coast is *S. alterniflora* which has an annual production rate of approximately 700 g per square meter. Its intimate relationship with salt-marsh sediments, which provide the largest reservoir of metals in the estuarine system, suggests its potential as a conveyor of metals from this reservoir to estuarine waters. This is further supported by the fact that *S. alterniflora* concentrates metals to a high degree. The leaves and stalks of *S. alterniflora* are harvested annually from salt marshes by the tides forming detritus, which is the base of the salt-marsh estuarine food web (4). The detritus retains metals that have been, to a large extent, extracted from the salt-marsh sediments. Using the average concentrations of the leaves and stalks of *S. alterniflora* and the average rate of production, and assuming this to be constant over the 400,000 hectare salt-marsh environment of the southeastern United States, the average annual uptake rate of metals through the *S. alterniflora* can be calculated (Table 6).

TABLE 6

Average Concentration in and Annual Uptake of Metals
by **Spartina alterniflora**
(Leaves and Stalks)

	ppm Dry Weight				
	Fe	Mn	Cd	Cu	Hg
Mean	750	50	0.5	3.7	0.20
Max.	2100	300	1.9	7.8	0.37
Min.	100	17	0.1	0.5	0.07
Average Annual Uptake (10^3kg)	2.1×10^3	140	1.4	10	0.6
Percent of Total Input	3	8	3	3	17

Analyses accomplished by atomic absorption on wet digested samples.

When the percentage of the total input to the system taken up annually by the *S. alterniflora* (Table 6) is compared to the annual loss of metals by sedimentation (Table 5), it is clear that, with the exception of mercury, the annual supply of metals to salt marshes is more than enough to maintain their uptake by *S. alterniflora.* If all the metals taken up by *S. alterniflora* were released to the water column there would probably be no significant increase in the metal levels, again with the exception of mercury. In the case of mercury, the marsh grass uptake is sufficient to more than offset the sedimentation loss. If this amount were cycled back to the water column, the total input of mercury to the salt-marsh estuaries could be transported through the system to the open ocean. Since such is obviously not the case, some of the metal contained in the *S. alterniflora* is apparently recycled to the salt-marsh sediments in the form of plant detritus. This would suggest that a large part of the mercury in the marsh sediment should be associated with organic matter (Fig. 5). Such association has been found in other estuarine environments (6).

Although the above results and conclusions do not describe the details of heavy-metal cycles and fluxes through salt-marsh estuaries, they at least indicate the general processes involved. This information provides a logical basis for further investigations into the characteristics of metal-transfer cycles in salt-marsh estuaries.

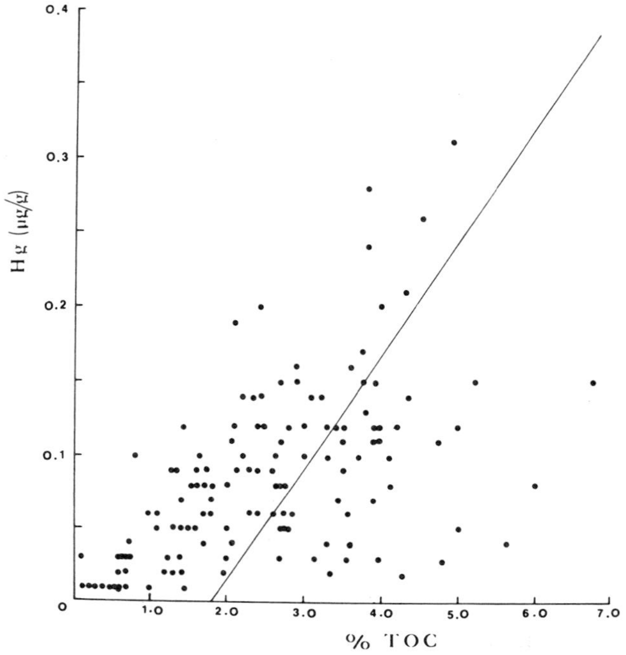

Figure 5 The relation between mercury and total organic carbon. Linear regression curve shown for significant correlation.

ACKNOWLEDGEMENTS

This research was partially supported by the U. S. Environmental Protection Agency (project no. R-800372) and the National Science Foundation, Office of the International Decade of Ocean Exploration (grant no.. GX-33615).

The author wishes to thank R. Smith, F. Taylor, and E. Waiters for their assistance in this research.

REFERENCES

1. Barnes, S. S., Craft, T. F., and Windom, H. L.
1973 Iron-scandium budget in sediments of two Georgia salt marshes. **Bull. Gn. Acad. Sci.,** 31 (1): 23-30.

2. Bhate, U.
1972 **Trace Metal Distributions in Natural Salt Marsh Sediments,** M. S. thesis, Georgia Institute of Technology.

3. Bigham, G. N.
1973 Zone of influence – inner continental shelf of Georgia. **Jour. Sed. Pet.,** 43 (1): 207-214.

4. de la Cruz, A. A. and Odum, E. P.
1967 Particulate organic detritus in a Georgia salt marsh-estuarine ecosystem. In **Estuaries,** p. 383-388, (ed. Lauff, G. H.) Am. Assoc. Adv. Sci. Publ. 83. Washington, D. C.

5. Goldberg, E. D.
1965 Minor elements in sea water. In **Chemical Oceanography** Vol. I, p. 163-196, (eds. Riley, J. P. and Skirrow, G.) Academic Press, London.

6. Harriss, R. C., Anders, A., and Dion, E.
1971 The distribution of mercury in rivers and estuaries of the Northern Gulf of Mexico. Progress Rpt. to EPA, FSU, Tallahassee, Florida.

7. Kharkar, D. P., Turekian, K. K., and Bertine, K. K.
1968 Stream supply of dissolved silver, molybdenum, antimony selenium, chromium, cobalt, rubidium and cesium to the oceans. **Geochim. Cosmodrim. Acta.,** 32: 285-298.

8. Parker, P. L.
1962 Zinc in a Texas Bay. **Texas Univ. Mar. Sci. Inst.,** 8: 75-79.

9. Parker, P. L., Gibbs, A., and Lawler, R.
1963 Cobalt, iron and manganese in a Texas Bay. **Texas Univ. Mar. Sci. Inst.,** 9: 28-32.

10. Pomeroy, L. R., Johannes, R. E., Odum, E. P., and Roffman, B.
1969 The phosphorus and zinc cycles and productivity of a salt marsh. **In Proc. 2nd Nat. Symp. on Radioecology,** p. 412-419 (eds. Nelson, D. J. and Evans, F. C.) N.B.S. Springfield, Virginia.

11. Rahn, W. R.
1973 The role of **Spartina alterniflora** in the transfer of mercury in a salt marsh environment. M. S. thesis, Georgia Institute of Technology, 61p.

12. Rusnak, G. A.
1967 Rates of accumulation in modern estuaries. In **Estuaries,** p. 180-184 (ed. Lauff, G. H.) Am. Assoc. Adv. Sci. Publ. 83. Washington, D. C.

13. Spencer, D. W. and Brewer, P. G.
1969 The distribution of copper zinc and nickel in seawater of the Gulf of Maine and the Sargasso Sea. **Geochim. Cosmochim. Acta.,** 33: 325-339.

14. Turekian, K. K.
1968 Deep-sea deposition of barium, cobalt and silver. **Geochim. Cosmochim. Acta.,** 32: 603-612.

15. Turekian, K. K.
1971 Rivers, tributaries, and estuaries. In **Impingement of Man on the Oceans,** p. 9-74 (ed. Hood, D. W.) Wiley Interscience, New York.

16. Turekian, K. K., Harriss, R. C., and Johnson, D. G.
1967 The variations of Si, Cl, Na, Cu, Sr, Ba, Co and Ag in the Neuse River, North Carolina. **Limnol. Oceanogr.,** 12: 702-706.

17. Turekian, K., and Scott, M. R.

1967 Concentrations of Cr, Ag, Mo, Ni, Co and Mu in suspended material in streams. **Environ. Sci. Tech.,** 1: 940-942.

18. Williams, R. B. and Murdoch, M. B.
1969 The potential importance of **Spartina alterniflora** in conveying zinc, manganese and iron into estuaric food chains. In **Proc. 2nd Nat. Symp. on Radioecology,** p. 431-439 (eds. Nelson, D. J. and Evans, F. C.), Springfield, Virginia.

19. Windom, H. L.
1973 Mercury distribution in estuarine-nearshore environment. **Jour. of the Waterways, Harbors and Coastal Eng. Div.,** ASCE, 99 (WW2) 257-264.

20. Windom, H. L., Beck, K. C., and Smith, R.
1971 Transport of trace metals to the Atlantic Ocean by three southeastern rivers. **Southeast. Geol.,** 12 (3): 169-181.

21. Windom, H. L., Neal, W. J., and Beck, K. C.
1971 Mineralogy of sediments in three Georgia estuaries. **Jour. Sed. Pet.,** 41 (2): 497-504.

22. Windom, H. L. and Smith, R. G.
1972 Distribution of cadmium, cobalt, nickel and zinc in southeastern United States continental shelf. **Deep-Sea Res.,** 19: 727-730.

PROCESSES CONTROLLING THE DISSOLVED SILICA DISTRIBUTION IN SAN FRANCISCO BAY

by

D. H. Peterson
T. J. Conomos[1]
W. W. Broenkow[2]
E. P. Scrivani[3]

ABSTRACT

Dissolved silica (DS) is supplied to San Francisco Bay primarily by river inflow and, to a lesser extent, by ocean water. Its distribution in the bay differs from that of other micronutrients (e.g., phosphate, nitrate, and ammonia) because man's input to the estuary and recycling by dissolution of siliceous phytoplankton within the estuary appear to be relatively minor sources of DS.

Major variations in the DS distribution are seasonal and are related to the variations in rates of river supply and silica utilization by phytoplankton. When the rate at which river inflow supplies silica to the bay is large compared with the rate of which silica is used within the estuary, the decrease in DS as salinity increases is controlled primarily by mixing of river and ocean waters. When the DS utilization rate increases significantly relative to the supply rate, the DS concentration in the estuary is considerably less than that predicted by simple

1. U. S. Geological Survey, 345 Middlefield Road, Menlo Park, California 94025.

2. U. S. Geological Survey, 345 Middlefield Road, Menlo Park, California 94025, and Moss Landing Marine Laboratory, Moss Landing, California 95039.

3. U. S. Geological Survey, 345 Middlefield Road, Menlo Park, California 94025, and Department of Botany, University of California, Berkeley, California 94720.

mixing, and is often lower in concentration than DS in the near-surface ocean water.

The influence of phytoplankton on DS concentrations in the bay appears to depend on how long the water remains in the estuary. Because river discharge is the fundamental parameter of estuarine circulation as well as the major source of high DS concentrations, river discharge modulates both seasonal variations in the estuarine water replacement time and the DS supply. Thus, when river discharge is high, even rapid DS utilization do not have a significant influence on the DS distribution.

INTRODUCTION

Nitrogen and phosphorus distributions in the open ocean have been the object of considerable interest and study since the inception of rapid colorimetric chemical analyses (e.g., 55). These micronutrient-element studies provide a framework for describing and understanding ocean circulation and plankton ecology. In the study of estuarine processes, however, their usefulness has not yet been completely exploited because the complex sources and sinks of nitrogen and phosphorus resulting in temporal and spatial distributions are generally difficult to describe.

By contrast, in the San Francisco Bay estuary, dissolved silica (DS) has a relatively simple spatial and temporal distribution. This simple distribution suggests that DS, as a non-conservative property, may provide an excellent reference for comparison with distributions of other less simple, non-conservative properties.

This paper describes major features of the San Francisco Bay estuarine system and demonstrates the relation between the seasonal DS variations and the effects of insolation and river discharge (used as rate terms herein). Particular emphasis is placed on the longitudinal (river to ocean) DS variations which are associated with estuarine circulation.

METHODS

Field and Laboratory Determinations

The San Francisco Bay estuary is defined as the northern reach of the San Francisco Bay system located between Rio Vista and Golden Gate. The southern reach, from Golden Gate to San Jose, has only a small fresh-water source at its southern end.

Water chemistry observations were made within one- to two-day periods at as many as 20 stations in the mid-channel of the estuary between the fresh-water boundary (as far landward as Rio Vista) and the estuary mouth at Golden Gate

(Fig. 1; *see* McCulloch et al. (46) for station locations). The hydrographic surveys have been repeated on a near-monthly basis from April 1969 to the present.

Most of the analyses were made with a continuous-flow system in which the water was pumped from 2-m depth to the ship through a towable salinity-temperature-depth pumping system (7). On shipboard the water was filtered through a silver membrane filter (0.45 μ pore diameter) and analyzed continuously for dissolved silica, phosphate, nitrate plus nitrite, and ammonia using a Technicon AutoAnalyzer®[1] calibrated at 4-hour intervals. Simultaneous

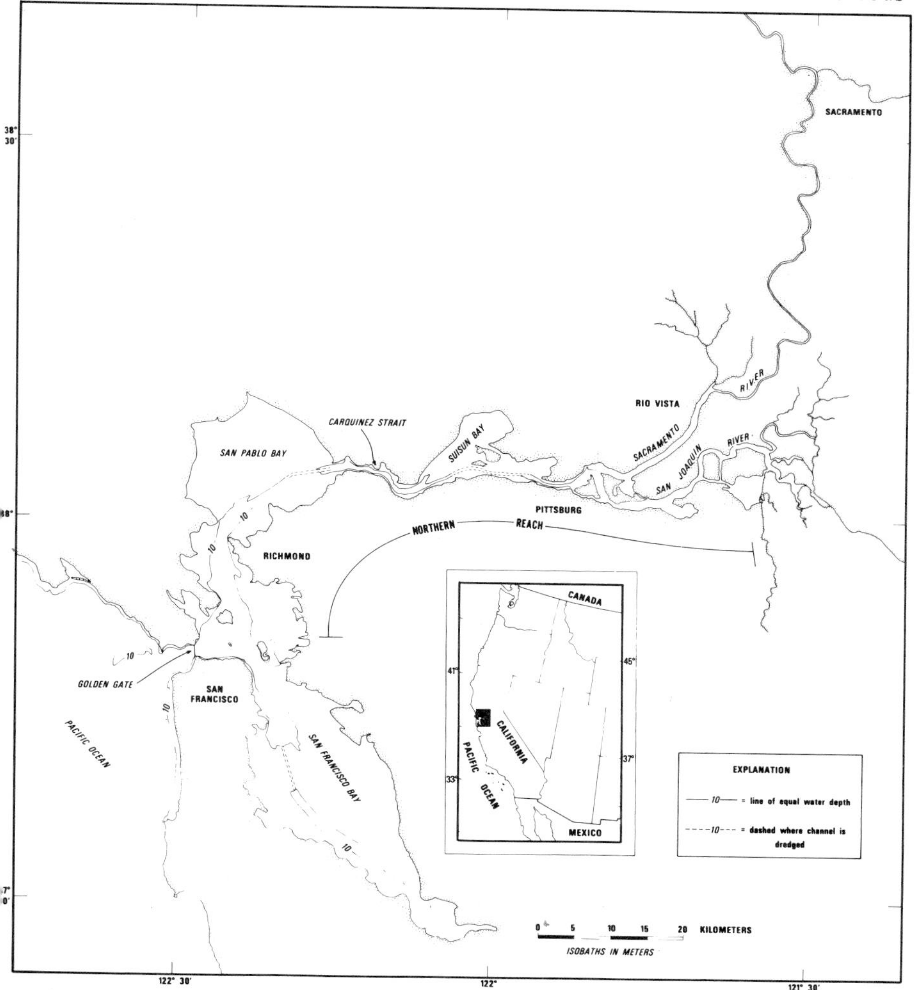

Figure 1. The San Francisco Bay system and environs. The northern reach is the part of the system between Rio Vista and Golden Gate, the entrance to the ocean.

continous determinations of pH, dissolved oxygen, turbidity, and *in vivo* chlorophyll *a* were made. Because the DS analyses were made immediately following collection, special freezing and thawing procedures (13, 37) were not necessary. The accuracy of our silicomolybdate method (2) is estimated to be within ± 2 μ g-atoms SiO_2-Si/liter. Discrete water samples were drawn for studies of phytoplankton cell counts and productivity; the procedures used were similar to those of Strickland (67), Vollenweider (70), and Schindler and Holmgren (57).

Additional Data Sources

To extend our data through a longer time base, we used DS, phytoplankton, and suspended particulate matter (SPM) data from the investigations of McCarty et al. (45); Storrs et al. (65, 66); Bain and McCarty (4). Data sources for various parameters are indicated in figure captions. River discharge data from U. S. Geological Survey are published annually in Surface Water Records, California; data used from U. S. Bureau of Reclamation sources are unpublished; insolation data were furnished by the Bay Area Air Pollution Control Board, San Francisco, California.

Mathematical Modelling

Because the major DS variations can be represented in one dimension, we have modeled the effects of DS advection, diffusion, and biological utilization with an equation for one-dimensional distribution of variables (68 and Doherty, written communication, 21, 22). In this model, diffusion coefficients were estimated from the salinity distribution during a known river discharge (advection) and the DS utilization rates were assumed.

SAN FRANCISCO BAY ESTUARY

Sacramento-San Joaquin River System

The Sacramento-San Joaquin river system is the dominant feature controlling the major seasonal hydrologic processes in the San Francisco Bay estuary. This river system, which supplies more than 90 percent of the fresh-water discharge to the San Francisco Bay system, is the major source of riverborne dissolved and particulate matter (including DS) and the basic feature controlling nontidal water circulation in the estuary. River discharge influences both the supply rates of substances to the estuary and the removal rates of substances from the estuary.

The combined discharge of the Sacramento and San Joaquin rivers varies annually between approximately 100 m^3/sec and 2,000 m^3/sec. The seasonal cycle (Fig. 2-A) is characterized by a high winter discharge (December-April) and a low summer discharge (June-October). Seasonal riverborne suspended particle concentrations vary with the discharge levels (Fig. 2-C).

Salinity Regime

The salinity values in the estuary respond closely to variations in river discharge (Fig. 2-B); the salinities are depressed in the winter and approach ocean values in the summer. During extremely high river discharge, surface-to-bottom salinity differences in the estuary can exceed 10 ^{o}/oo, but during normal discharge conditions the estuary is partially mixed or well-mixed by strong tidal currents (69).

During winter, salt water intrudes landward to San Pablo Bay and during summer to Rio Vista on the Sacramento River (Fig. 1). In both seasons the estuarine part of the bay system is long relative to its width. Within this system, sea-bed drifter movements have shown that the near-bottom, landward-flowing, density-driven (salinity) current, a typical feature of estuaries, persists throughout the year (16).

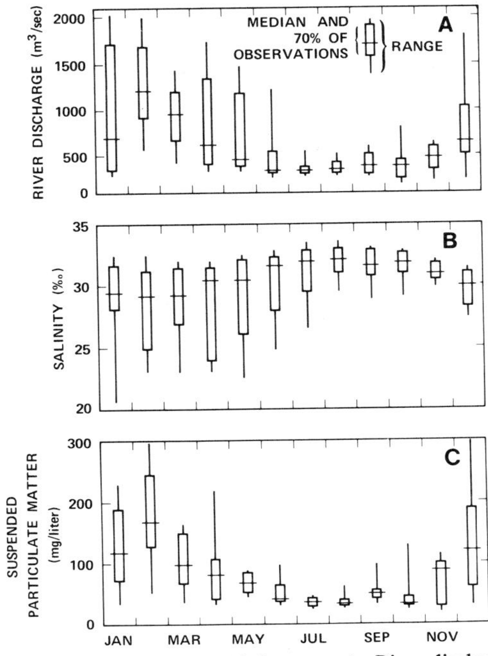

Figure 2. Monthly means 1960-1970. (A) Sacramento River discharge at Sacramento. (B) Surface salinity at Golden Gate. (C) Suspended particulate matter at Sacramento.

Dissolved Silica

Seasonal variations of salinity and DS in the estuary can be illustrated by comparing typical winter and summer longitudinal distributions in the near-surface (2 m) water (Fig. 3). Several features are evident: (a) an inverse relation between salinity and DS concentrations during winter (Fig. 3, A and C); (b) higher DS concentrations during winter than during summer (Fig. 3, C and D); (c) a rapid decrease in DS concentration with distance in the upper portion of the estuary during summer (Fig. 3-D); and (d) lower DS concentrations during summer within the estuary than at the seaward boundary (Fig. 3-D).

To eliminate the large tidally-caused variations in salinity and DS and to make the above features more apparent, the subsequent discussion relates DS to salinity instead of geographic position (Fig. 4). Plots of DS as a function of the corresponding salinity define three river-ocean mixing situations within estuaries. A linear distribution (Fig. 4-A) indicates that horizontal mixing rates between river and ocean water dominate non-conservative processes. When DS utilization rates become significant, DS concentrations are less than those produced by a simple mixing process (Fig. 4-C). In both these instances the concentrations in the estuary are higher than at the ocean boundary, but in the latter instance the estuary is both a sink for river-supplied DS and a DS source to the ocean. When utilization rates are pronounced, the estuary is a sink for both river- and

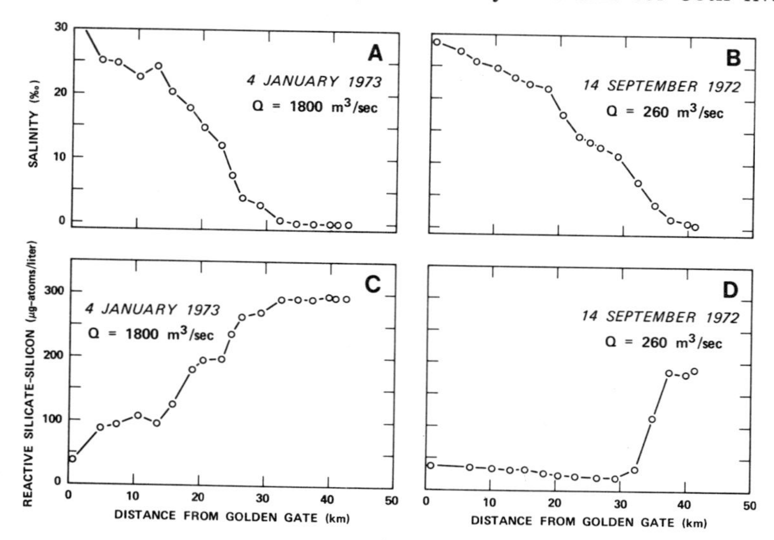

Figure 3. Longitudinal distribution of salinity and silicate-silicon at 2-m depth during typical winter and summer conditions in the northern reach of San Francisco Bay. Q indicates combined mean monthly discharge from the Sacramento and San Joaquin rivers.

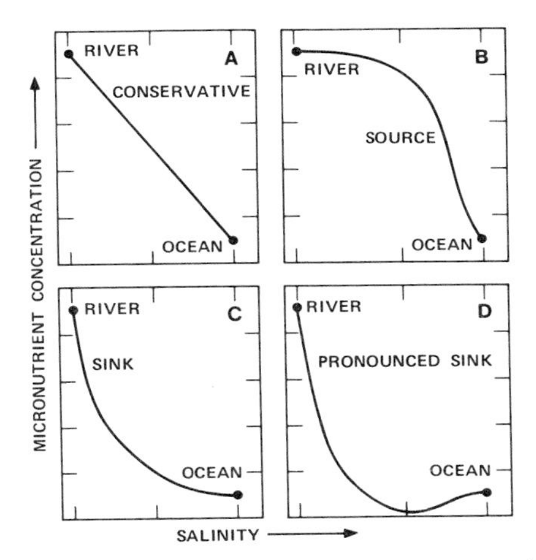

Figure 4. Idealized longitudinal micronutrient-salinity relations showing and mixing of nutrient-rich river water with nutrient-poor sea water in the estuary.

ocean-supplied DS (Fig. 4-D). These three examples typify seasonal DS behavior in the bay (Fig. 3). A clear example of a DS distribution corresponding to that shown in Figure 4-B has not been observed in the bay or, to our knowledge, in any other estuary.

Two major temporal features and one major spatial feature can be identified in these DS distributions. The first temporal feature (during 1973) is illustrated by an annually recurring winter and summer variation in the distribution of DS relative to salinity (Fig. 5): the near-linear DS-salinity distribution during winter becomes progressively non-linear towards late summer. The decrease in DS concentrations in the estuary during summer is similar to the decrease observed in coastal waters (3, 18). The second temporal feature is illustrated by comparing a summer of high discharge, 1971, with a summer of lower discharge, 1972 (Fig. 6): the summer level of DS in the estuary in 1971 is demonstrably higher than that in 1972 (Fig. 7).

The two seasonal features above indicate basic processes that are not unique to estuaries. Perhaps more appropriate to the investigation of estuaries, but also more difficult to explain, is the pronounced departure in linearity of the DS-salinity correlations in the inner (landward) part of the estuary (Fig. 7). We believe that in partially-mixed estuaries this non-linear distribution is controlled largely by the hydrodynamics of estuarine circulation, which controls water column residence time, and in turn the turbidity and phytoplankton maxima. For this reason, we will discuss longitudinal variations before seasonal variations.

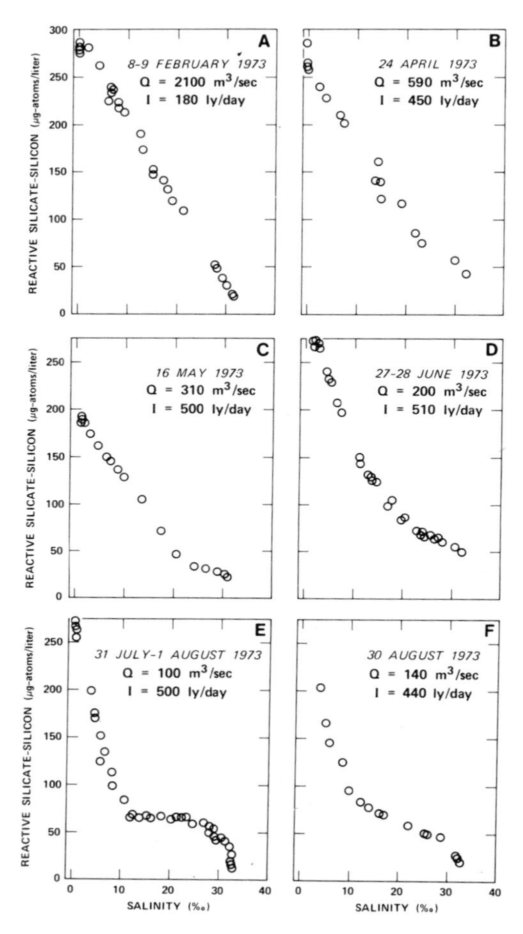

Figure 5. Silicate-salinity relations at 2-m depth in the northern reach of San Francisco Bay, 1973. Q indicates combined mean monthly discharges from the Sacramento and San Joaquin rivers, and I indicates mean monthly insolation at Richmond.

PRIMARY FACTORS AFFECTING DISSOLVED SILICA DISTRIBUTIONS

Longitudinal Variations

In the inner part of the estuary the depression of DS concentrations persists throughout the summer and cannot be explained by the patchy short-term events that are typically associated with phytoplankton blooms. The depression

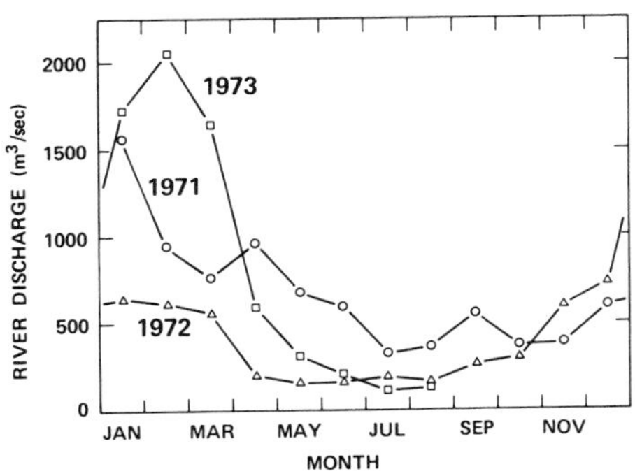

Figure 6. Combined mean monthly discharges from the Sacramento and San Joaquin rivers, 1971-1973.

appears to be best explained by the constant and ordered characteristics of estuarine circulation.

Estuarine circulation and null zone

When measurements of currents in estuaries are averaged over one or more tidal cycles, a nontidal mean flow is commonly observed. This flow may consist of river-, density-, and wind-induced components. The density component, a basic feature of estuaries, is caused by the density difference between river and sea water. It produces a constant net landward bottom flow of dense sea water in opposition to the net seaward flow of less dense river water (Fig. 8, A and B) (30). The presence of this landward-flowing current distinguishes an estuary from a tidal river system, in which tidal variations are imposed only on the river-flow regime. The landward-flow regime in an estuary contains a vertical advective component (not indicated in Figure 8-A), which is also absent from tidal river systems. The geographic area where the landward-flowing density current and the seaward-flowing river current have equal and opposite effect on the nontidal flow is termed the *null zone* (29).

Turbidity maximum

Although the inflow and outflow of water are nearly equal in an estuarine system, the inflow and outflow of suspended particulate matter (SPM) are typically not equal. Riverborne SPM does not necessarily escape to the sea but may settle by gravity from the seaward-flowing surface layer to the landward-flowing bottom layer. This inward advective transport supplies SPM to the null zone and often maintains a higher concentration of SPM there (turbidity maximum) than in either the river or the lower estuary (33, 47, 52, 59, 62).

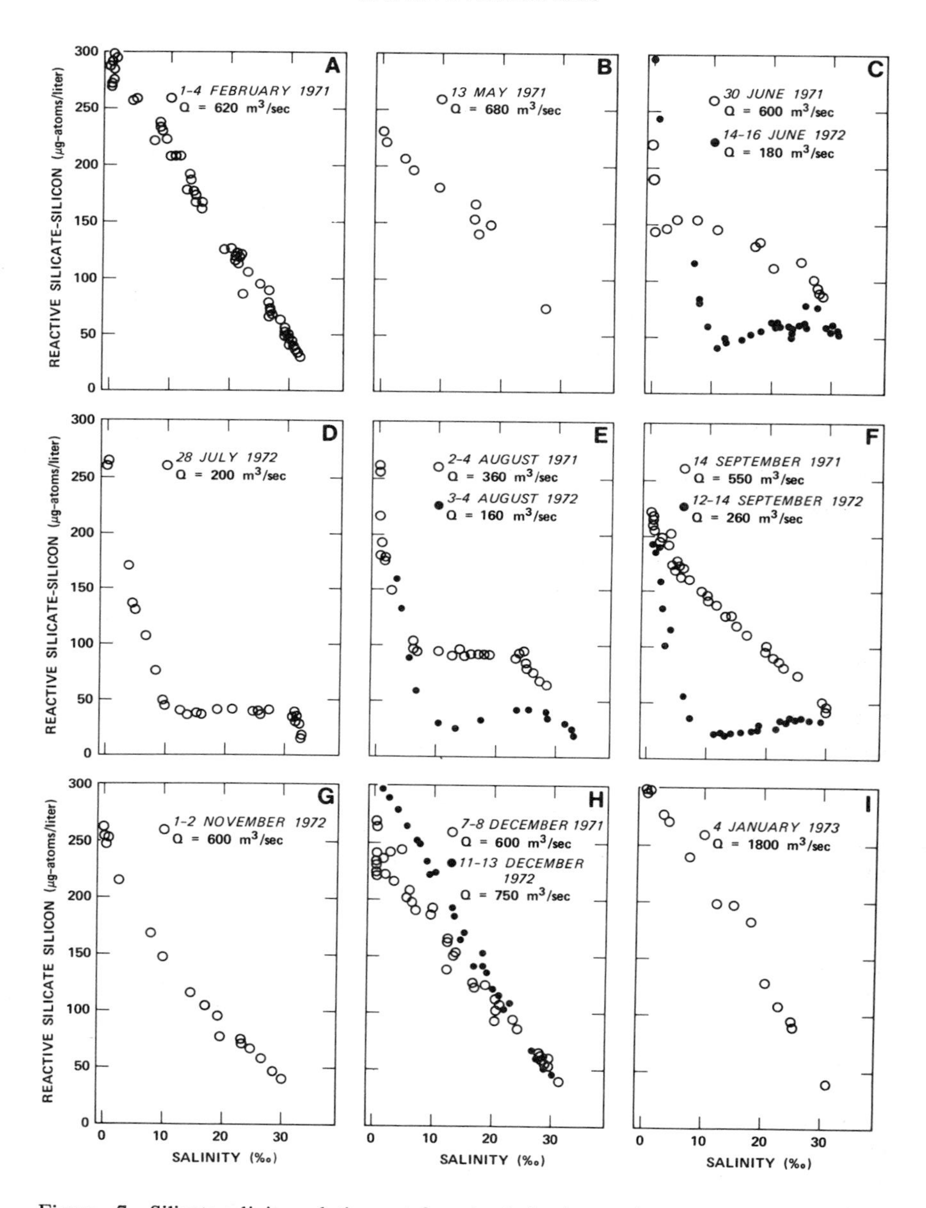

Figure 7. Silicate-salinity relations at 2-m depth in the northern reach of San Francisco Bay as functions of river discharge. Q indicates combined mean monthly discharges from the Sacramento and San Joaquin rivers.

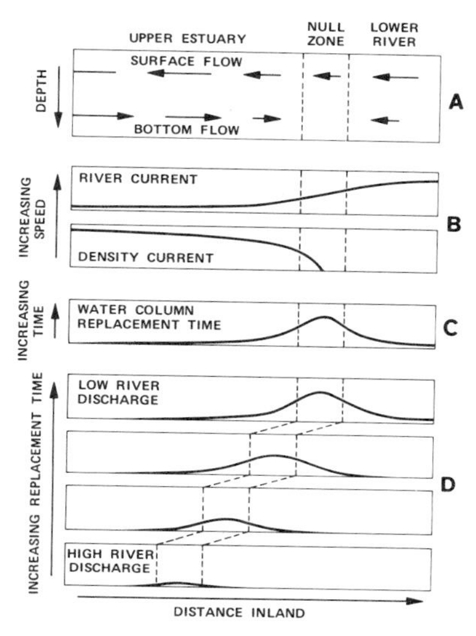

Figure 8. Schematic representation of (A) net drift in vertical section through a river-estuarine system, in which the length of the arrows indicates relative current strength; (B) longitudinal variation in average river and density currents within the river, null zone, and estuary; (C) generalized longitudinal variation in average advective water-column replacement time; and (D) seaward movement of null zone and diminishing water column replacement time with increasing river discharge.

Near-monthly SPM samples taken at 1-meter depth in the main channel of the San Francisco Bay estuary show that concentrations are generally highest in winter, apparently in response to the higher concentrations found in the river during that period (Fig. 9, A and B). Also, SPM concentrations are higher in the null zone (Fig. 9-C) than in either the upper or the lower part of the estuary (Fig. 9-A). This existence of a null-zone-associated turbidity maximum in San Francisco Bay demonstrates the typical response of the longitudinal distribution of SPM to estuarine circulation (47). Furthermore, a living photosynthetic fraction of this SPM, the phytoplankton, is also influenced by this circulation.

Phytoplankton maximum

The summer composition of the turbidity maximum suggests that estuarine circulation controls SPM distribution including phytoplankton: the phytoplankton cell concentrations illustrate the same relation with salinity as do the SPM concentrations (51). This pattern of phytoplankton concentration has been observed in several surveys; 1961-1963 (45, 65, 66), 1964 (4), and our

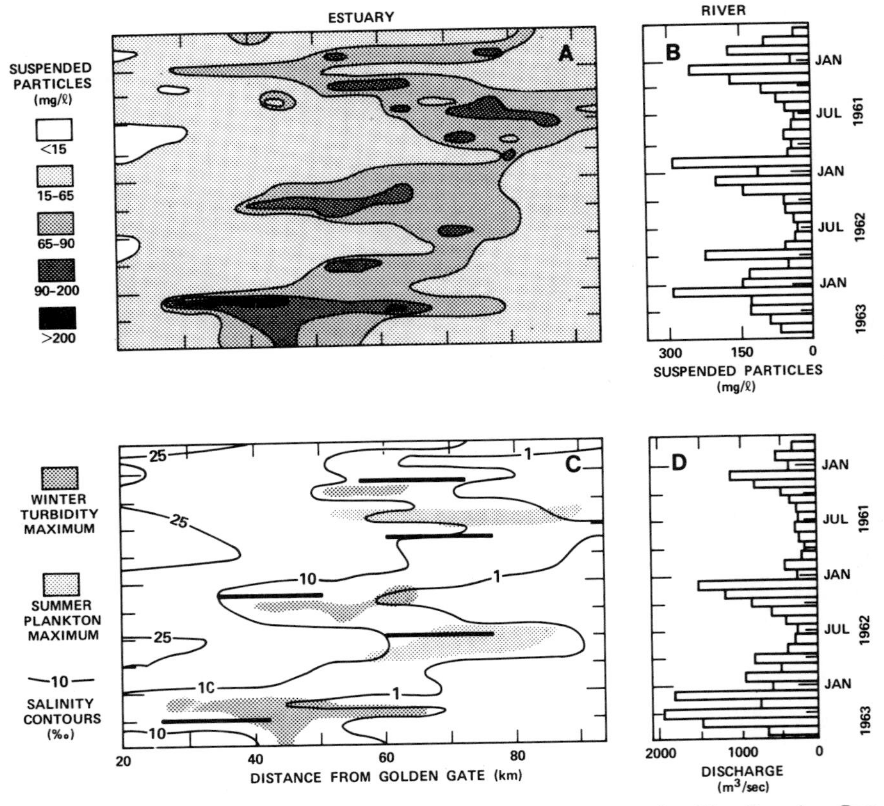

Figure 9. Seasonal distribution of properties in the northern reach of San Francisco Bay, 1961-1963: (A) Suspended particle concentration in the estuary; (B) suspended particle concentration in Sacramento River; (C) salinity at 1-m depth in the estuary, compared with location of suspended particle (turbidity) maxima and approximate location of nontidal current null zone (solid black lines); (D) combined discharge from the Sacramento and San Joaquin rivers. Suspended particle and salinity data in estuary from Storrs et al. (65, 66).

studies from 1971 to 1973. In addition, the location of both phytoplankton and SPM maxima correspond with the winter and summer estuarine null zone locations (Fig. 9-C), and, during summer, phytoplankton often comprises a significant weight fraction of the turbidity maximum (17).

Thus, as river discharge and riverborne SPM concentrations decrease during the summer (Fig. 9, B and D), a null-zone-associated phytoplankton maximum appears. This maximum coincides with the null-zone-associated turbidity maximum that typifies partially mixed and well-mixed estuaries.

Residence time and phytoplankton growth

In addition to the advective processes described above, the SPM concentrations within estuaries may be increased by *in situ* phytoplankton production (19, 26, 71). Significant *in situ* production necessitates a minimum period during which the water containing phytoplankton must remain in the estuary (6, 34, 56). For this reason a longitudinal variation in abundance of continuously reproducing phytoplankton may reflect the longitudinal variation of water-column residence time as well as of estuarine circulation. To explore the dependence of phytoplankton production on water movements, we have qualitatively evaluated, by inference, the longitudinal variation in water-residence time.

Water replacement, residence, or flushing time of estuaries is generally defined as the average time required for an entering parcel of water to pass through the estuary. The length of this period is controlled by river discharge, density currents, tidal mixing, and wind-induced currents (9). Residence time or water-column replacement time as used in this paper differs from the foregoing. It describes the variation in average longitudinal advective water flow regardless of the flow direction. As such, it can be expressed as the average time required for the water to flow through a unit length of the estuarine system, or

$$\left[\frac{1}{A} \int_z^0 w|v|\, dz \right]^{-1},$$

where A is the cross-channel area, w is the depth-variable width of the channel, $|v|$ is the absolute landward or seaward current speed, and z is water depth.

In most estuaries, the river discharge per unit cross-channel area (river current) decreases in a seaward direction owing to the seaward increasing cross-channel area, and the density current must decrease to zero in the landward direction. From purely advective considerations, this circulation pattern and geometry cause the maximum water-column-replacement time to be found within the null zone when the density currents are stronger than the river currents. A typical contribution of river and density currents to average advective replacement time of the water column is illustrated for the river, null zone, and estuary (Fig. 8-C). In this representation wind effects are assumed to be contributing primarily to turbulent processes, cross-channel variations in current speeds have been averaged, and cross-channel areas are considered to be decreasing in the landward direction. As the river current increases in strength relative to the density current, the replacement time of the water column decreases and the maximum is reduced and shifted in a seaward direction (Fig. 8-D; 51) At very high river discharges the river component may dominate estuarine circulation throughout the entire estuary. A similar pattern has been described for the Escaut estuary (73).

Although this model provides only a simplified description of estuarine circulation in San Francisco Bay, it reflects the flow rates measured by current meters (51) and surface and sea-bed drifters (16). These measurements indicate that estuarine circulation is relatively strong compared with the summer river-discharge component. Strong estuarine circulation is typical of low-flow conditions in many partially- and well-mixed estuaries (9, 10, 54). The advective residence time reaches a miximum in the null zone during low-flow period because both the volume transport and speed of estuarine currents are large relative to river discharge per unit cross-channel area.

A plankton community will be afforded a longer period for growth in the null zone of the estuary than in waters landward or seaward of the null zone. Thus, both landward advective transport to the null zone and longer advective residence time within the null zone are suggested as favorable hydrodynamic conditions for development of the large seasonal phytoplankton community in the upper reaches of the San Francisco Bay estuary. Because the major fraction, both by cell number and biomass, of the phytoplankton community generally consists of diatoms, it follows that highest rates of DS utilization per unit volume occur and are maintained in the upper reaches. These high utilization rates are reflected in the marked departure of DS concentrations from the near-linear river and sea-water mixing relation (compare Fig. 7, A and E). These DS departures closely parallel the average increases in phytoplankton productivity (Fig. 10).

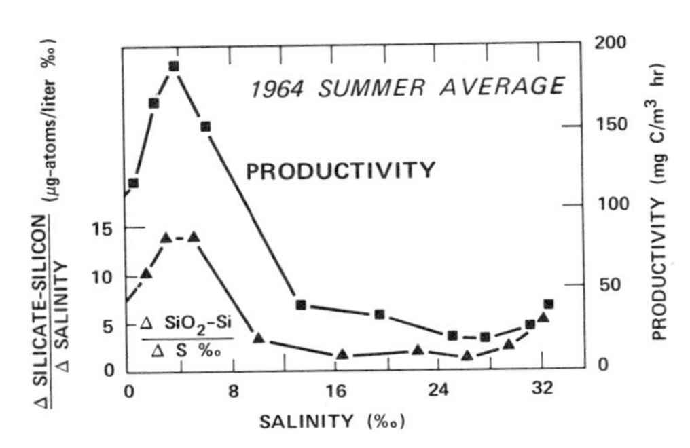

Figure 10. Averaged longitudinal Δ(silicate-silicon) / Δ salinity rate of change and phytoplankton productivity at 1-m depth in summer 1964. Data from Bain and McCarty (4).

Seasonal Variations

Large changes in the DS concentrations in the estuary result from seasonal variations in supply and removal rates. In this section we describe the relation between seasonal DS concentrations and the effects of seasonal variations in insolation and river discharge.

Insolation and primary production

Insolation appears to be the dominant factor limiting phytoplankton productivity in the bay because the nutrient supply seems sufficient throughout most of the year (4). Coincident with increased insolation rates during summer, phytoplankton abundance increases over that during winter (Fig. 11; 4, 5, 45, 65, 66). Our studies indicate that more than 80 percent of these phytoplankton cells are living and that the near-surface productivity of phytoplankton per unit volume of water nearly doubles with a doubling of the cell concentration (Fig. 12). Diatoms comprise a major fraction of the plant cell numbers and biomass; the observed summer depression in DS largely reflects the increase in the size of the phytoplankton population.

Despite the fact that insolation appears to be the dominant factor limiting phytoplankton productivity on a seasonal basis, its control is gross. There is no

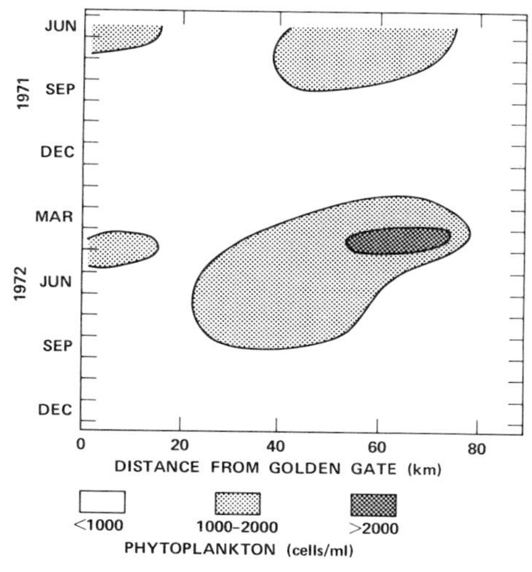

Figure 11. Phytoplankton abundance at 2-m depth in the northern reach of San Francisco Bay, 1971-1972.

close correspondence between levels of DS utilization and insolation when considered on shorter time increments during the summer months (April-September). For example, during a period of high insolation (Fig. 13-A), the DS distribution for June 1972 (Fig. 14-A) was similar to DS distributions during periods of similar discharge but lower insolation (Figs. 14, B and C and 15-A). This suggests that the difference in insolation rate between June and August 1972 had only a minor influence on the DS distribution. Although the

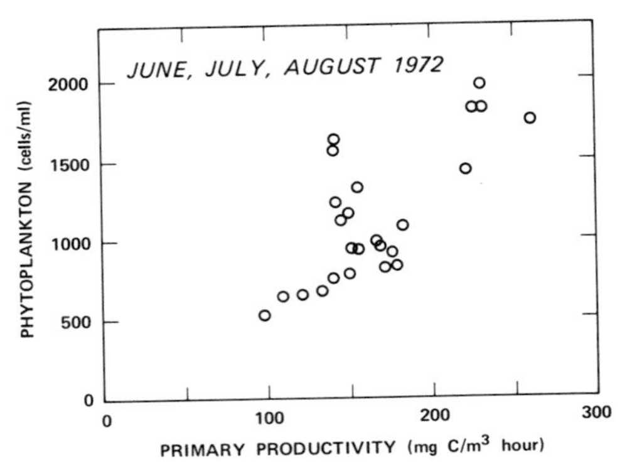

Figure 12. Phytoplankton cell count at 2-m depth as a function of primary productivity in the northern reach of San Francisco Bay, summer 1972.

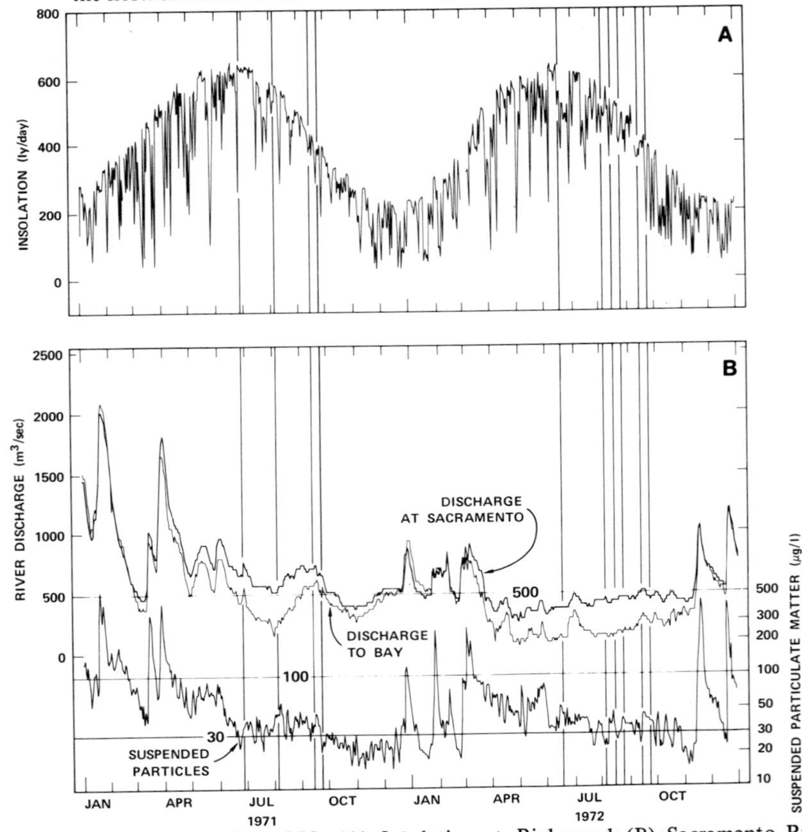

Figure 13. Daily means 1971-1972. (A) Insolation at Richmond; (B) Sacramento River discharge (at Sacramento) compared with combined Sacramento River and San Joaquin River discharges to the bay; suspended particulate matter concentration in Sacramento River at Sacramento. Cruise periods are indicated by vertical lines.

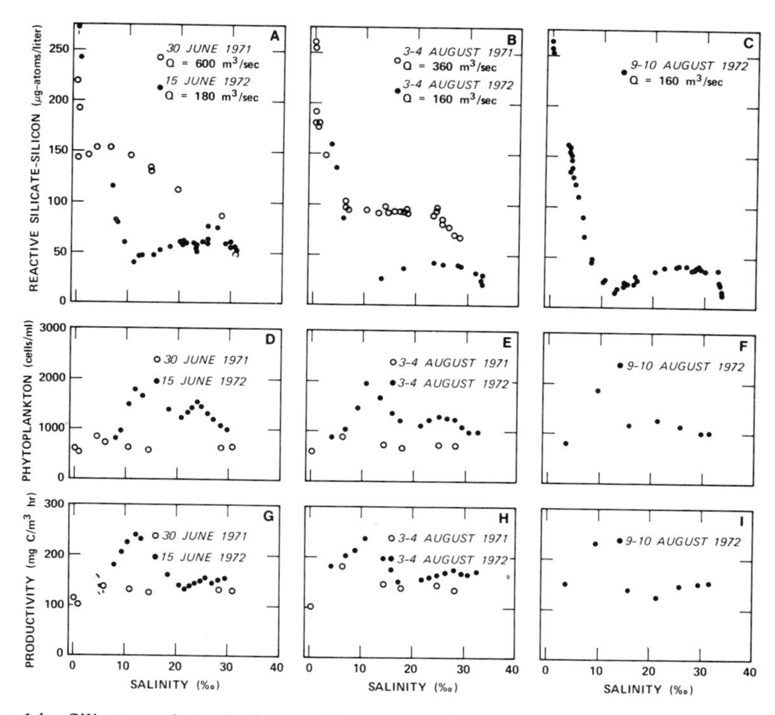

Figure 14. Silicate-, phytoplankton cell count- and productivity-salinity relations at 2-m depth in the northern reach of San Francisco Bay, mid-summer 1971-1972.

discharge was approximately the same in September as in August 1972 (Fig. 13-B), the DS distribution had changed very little (Fig. 15, B and C) despite the lower September insolation rate (Fig. 13-A).

We believe, therefore, that effects of seasonal variations in insolation on DS distributions are less complicated to evaluate than effects of river discharge. The mean monthly insolation rate for a given month of the year may be considered more or less constant from year to year, while the mean monthly river discharge rate is highly variable (Fig. 16).

River discharge and dissolved silica supply

We noted above that insolation is an important factor limiting phytoplankton productivity relative to nutrient availability in the San Francisco Bay system. Because increasing SPM concentrations tend to reduce water-column-light intensity by decreasing water transparency, the potential effects of river discharge variations on turbidity and, hence, on the availability of light in water should be considered (38).

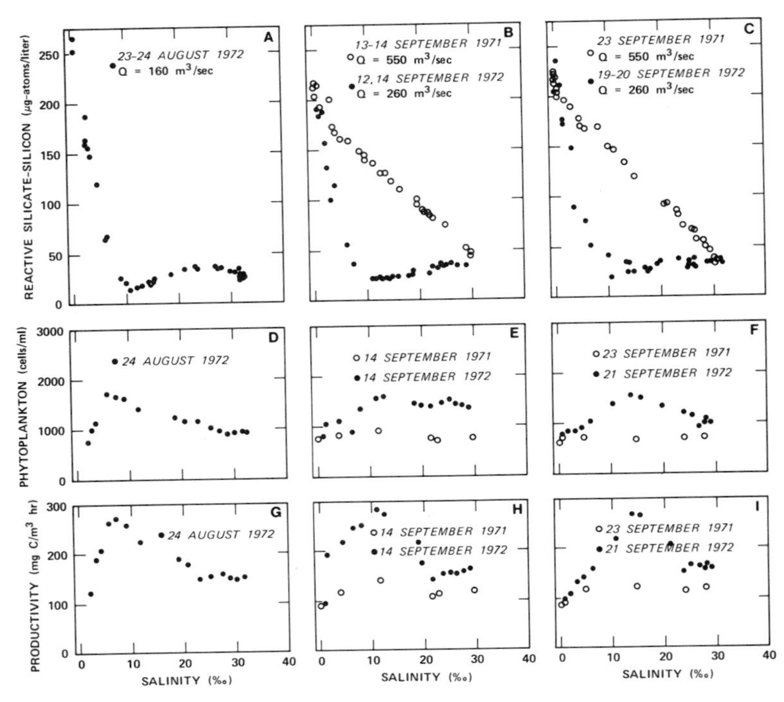

Figure 15. Silicate-, phytoplankton cell count- and productivity-salinity relations at 2-m depth in the northern reach of San Francisco Bay, late summer 1971 and 1972.

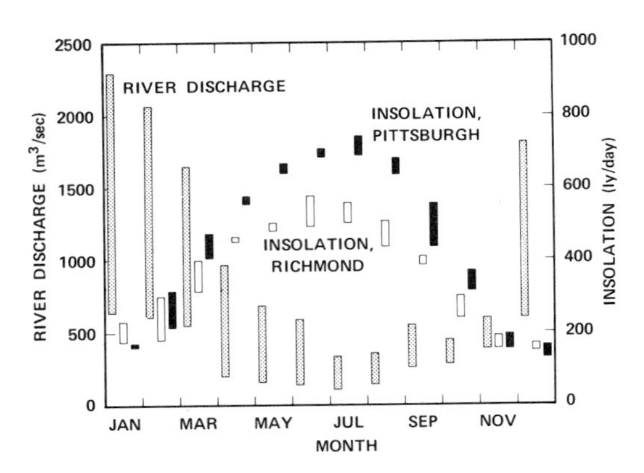

Figure 16. Range of combined mean monthly discharges from the Sacramento and San Joaquin rivers and mean monthly insolation at Richmond and Pittsburg, January 1970-August 1973.

As river discharge increases, both river-current speed and riverborne SPM concentrations generally increase. High SPM concentrations are evident immediately following the first storm-flow conditions if the storms are separated by at least a month (Fig. 13-B). The mean monthly SPM concentrations during three relatively high-discharge summer periods, 1967, 1969, and 1971, were higher than during three low-discharge summer periods, 1964, 1966, and 1968, in the Sacramento River (Fig. 17). However, SPM concentrations during the low flow of late summer were unexpectedly higher than those of high-flow September concentrations (Fig. 17). We suspect that the differences in the effects that riverborne SPM have on the availability of light in the estuary are significant seasonally only when large changes in discharge occur (e.g., in the early summer April 1971 vs. April 1972; Fig. 13). In the general case, turbidity in the estuary must be controlled by a number of factors in addition to river discharge; the effects of a difference in summer discharge on water transparency appear as only a secondary factor in controlling DS distributions.

The influence of the river discharge rate on the DS supply rate and on estuarine water residence time is much more obvious than is its impact on the DS utilization rate. It is relatively simple to estimate the increase in DS supply with increasing discharge, and the decrease in advective water residence time with increasing river discharge is also conceptually simple. Using the estimated river current (river discharge per unit cross-channel area) as a useful approximation of the strength of the river component in the estuary, we have compared the highest (1971) and lowest (1961) rates of summer flow during the period from 1960 to 1973 (Fig. 18). It may be assumed that this seasonally varying river current represents the average advective water transport in the estuary landward of the null zone and is the primary source of DS. If river discharge rate is doubled (and the DS concentration held constant) the DS

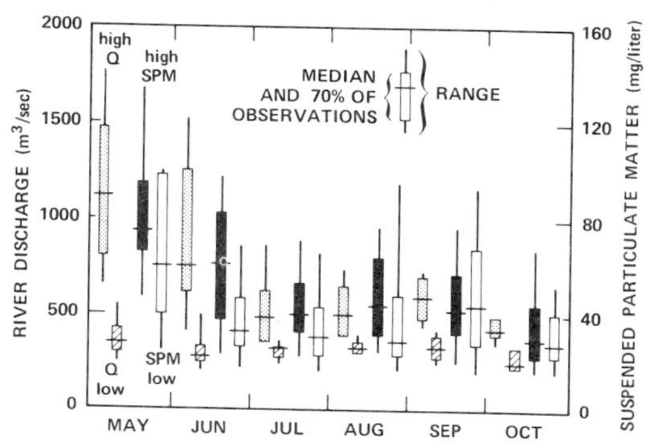

Figure 17. Comparison of daily Sacramento River discharge, Q, and suspended particulate matter concentrations, SPM, during summer months with high discharge (1967, 1969, 1971) and low discharge (1964, 1966, 1968).

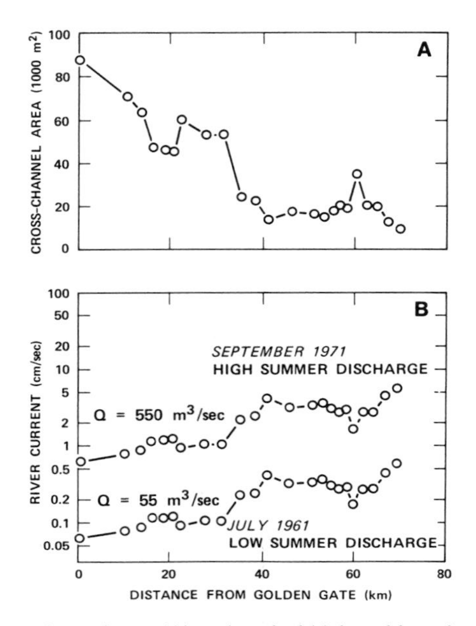

Figure 18. Mean cross-channel area (A) and typical high and low river current speeds during summer (B) in the northern reach of San Francisco Bay. Q indicates combined mean monthly discharges from the Sacramento and San Joaquin rivers. Cross-channel areas from Glenne and Selleck (27).

supply rate is doubled, but the advective residence time is reduced to one-half the initial value. It follows that doubling the discharge rate reduces to one-fourth the time available per unit silica for *in situ* utilization of a DS supply equal to the initial supply. Therefore, river discharge, as a fundamental parameter of estuarine circulation, controls not only the estuarine water-residence time, but also the DS supply rate to the estuary.

Linear and non-linear distributions

The predictable behavior of DS distributions relative to seasonal insolation and river discharge permits us to determine approximately the relative DS supply and removal rates that produce linear or near-linear DS distributions, or that result in apparent phytoplankton growth-limiting DS concentrations. For these estimates we assume that the (a) mean monthly river discharge is a measure of the DS supply rate, (b) month of the year is a measure of the DS utilization rate, and (c) mean monthly insolation for a given month is relatively constant from year to year. The DS distribution that is near-linear for the lowest discharge rate would represent the lowest supply rate for which DS still exhibits conservative behavior (Fig. 19). All DS measurements made during months having a higher

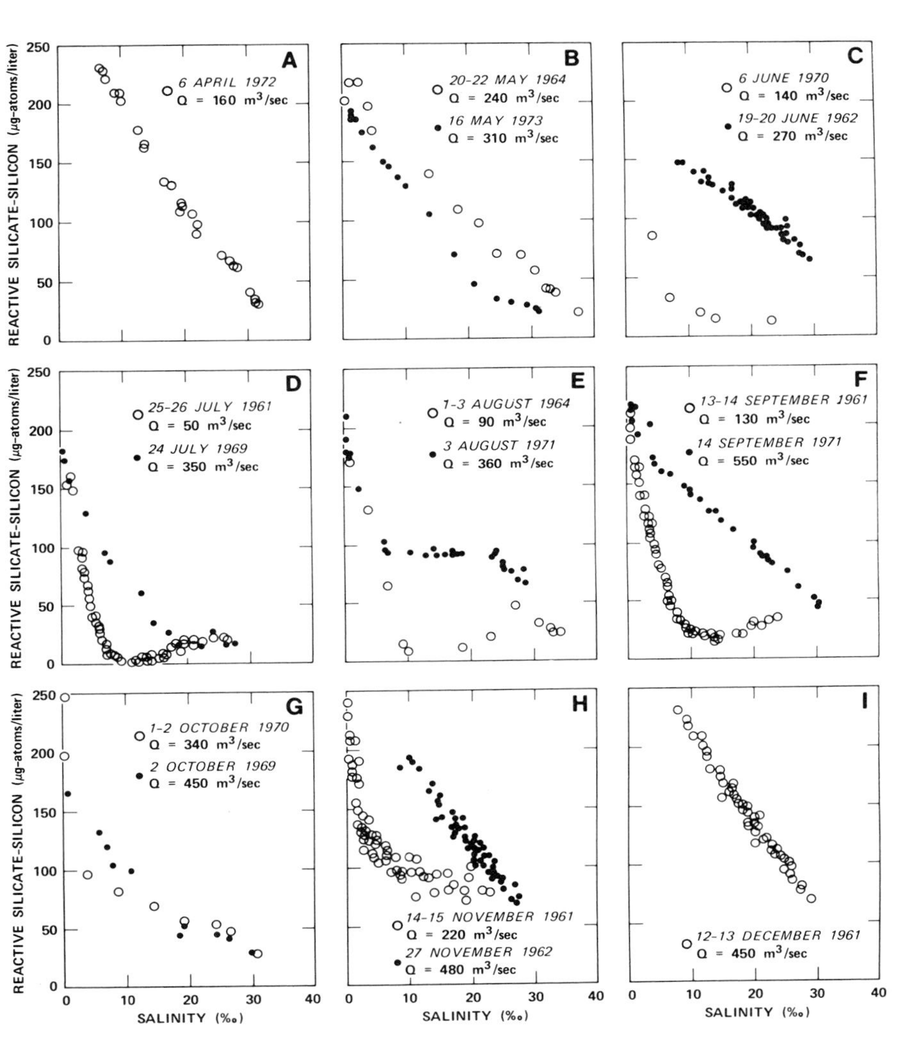

Figure 19. Linear and non-linear silicate-salinity relations at 2-m depth representative of high and low river discharge in the northern reach of San Francisco Bay. Q indicates combined mean monthly discharges from the Sacramento and San Joaquin rivers.

discharge than those given in Figure 19 have described near-linear relations with salinity. Alternatively, with one exception, June 1964, all DS measurements made during the months having the lowest discharge showed DS-salinity distributions that were most non-linear.

DS distributions at particular monthly rates of discharge and insolation are not generally in a steady state. During late summer, the effects of non-conservative processes that occurred during the preceding months may be observed over several months, particularly after a summer of low river discharge. This lag effect during the summer of 1961, which was one of particularly low discharge, may explain the non-linear DS-salinity relation observed in November 1961 (Fig. 19-H). Conversely, the dominance of conservative processes during winter may explain the near-linear relation during April for a low-discharge condition (Fig. 19-A). Persistence of near-linear DS-salinity relations into early summer following near-linear relations between December and April or May is probably a consequence of the difference in timing of the seasonal discharge and insolation cycles. No winter discharge was sufficiently low to allow comparison between the response of DS during low winter insolation and low summer discharge. For lowest discharge levels we have no data to determine how DS-salinity relations vary with insolation. The lowest summer DS concentration occurred during July 1961, the period of maximum insolation as well as minimum discharge (Fig. 19-D).

We have discussed two primary factors that control the seasonal distribution of DS in an estuary. These are (a) the river discharge rate that seasonally modulates both the DS and SPM supplies and the water residence time; and (b) the seasonal variation in insolation that controls phytoplankton DS utilization rates.

Primary factors can be distinguished from such secondary factors as variable DS concentrations in either the river or at the ocean boundary. The importance of one primary factor relative to another is, however, not easily determined. For example, bioligical DS utilization may lower DS concentrations below the levels estimated from conservative mixing, but the particular DS concentration at which this biologically-reduced concentration is maintained may be influenced by varying either the river supply or the phytoplankton DS utilization rate.

SECONDARY FACTORS AFFECTING DISSOLVED SILICA DISTRIBUTIONS

DS variations discussed thus far are large-scale temporal and spatial features. There are, however, additional factors that are assumed to be of secondary importance but may relate to some of the minor variations observed in the distribution of DS.

Small-Scale Variations

Diurnal variations, which may be large in some estuarine systems, are not notable in San Francisco Bay DS distributions. Data have shown a slight diurnal time dependence, but these effects could explain only the most minor variations in the basic DS distribution pattern.

Depth variations in DS utilization rates due to vertical light attenuation have been sought, but have not been observed. The absence of this feature suggests that water-column vertical mixing rates are large relative to the potential effects of the vertical variation in DS utilization rates. How effective mixing is in reducing vertical DS variations can be illustrated. If the average daily DS utilization in the upper meter of a 10-m-deep water column is 10 μM/liter per day and the water column is mixed vertically in less than 24 hours, the average DS utilization in the upper meter would be only 1 μM/liter per day. Such an apparent utilization rate is generally too small to be detected by measuring the DS-salinity relation on two successive days, or by comparing the DS-salinity distribution at the surface with that at the bottom, because random variations in DS concentrations are greater than ±1 μM/liter.

Similar inferences can be made for cross-channel variations. Cross-channel mixing reduces potential differences in the cross-channel DS-salinity distribution. Summer 1964 data (4) indicate that cross-channel variations in phytoplankton cell numbers relative to salinity are nearly similar to the longitudinal variations. These data suggest that carbon uptake rates per unit phytoplankton cell tend to be slightly lower in the shoal areas, but this causes no major differences in the DS-salinity relations.

In summary, only the large-scale longitudinal variations in DS utilization are demonstrable. Vertical and cross-channel variations in DS utilization appear so small relative to vertical and cross-channel mixing that they can be ignored.

Additional Dissolved Silica Sources

The paucity of data available from the adjacent ocean (with the exception of CalCOFI stations 60.52 and 60.80; 60, 61) has prevented elaboration of the potential effects of seasonal variations in DS concentration and exchange rate at the ocean-estuary boundary. Although these boundary effects on the DS distributions are probably small, the total mass transport of ocean-derived DS to San Francisco Bay by the inward-flowing bottom density current may be quite large in comparison to the river supply during low-flow conditions (Table 1).

Wind-induced upwelling (from < 200 m) may produce minor DS variations at the mouth of the estuary during early summer, but at Golden Gate these effects are small in comparison with variations produced by river discharge. During high or low winter discharge conditions or during high summer inflow, DS levels are

TABLE 1

Estimated dissolved silica sources to San Francisco Bay

Source	Winter	Summer
	(metric tons/day)	
River Discharge[1]	2,000	100
Bottom Density Current[2]	400	400
Waste Discharge[3]	30	30

[1]Assumed winter discharge is 1,400 m^3/sec; summer discharge is 80 m^3/sec; and the river DS concentration is 300 μg-atoms/liter.

[2]The average cross-sectional area of the landward flowing current is assumed to be one-half of the total cross-channel area at Golden Gate (area estimated from Glenne and Selleck, 27); the net landward current speed is assumed to be 5 cm/day (Conomos et al., 16) and the average DS concentration is 30 μg-atoms/liter.

[3]Using a waste discharge of 2 x 10^6 m^3 per day (63) and typical DS concentration of 300 μg-atoms/liter (12).

higher in the seaward-flowing upper layer than in the landward-flowing lower layer. Observations indicate that only after a period of very low summer flow (e.g., September 1972; Q = 250 m^3/sec) is San Francisco Bay a net sink for both river- and ocean-supplied DS; during such periods DS concentrations are higher in the inflowing near-bottom (ocean) water than in the outflowing surface (estuary) water.

Estimates of the importance of municipal and industrial sources of DS indicate that sewage effluent is a relatively minor DS source in San Francisco Bay. These effluents supply less than 5 percent of the annual budget (Table 1).

The DS resupply by dissolution of biogenic silica is exceedingly difficult to evaluate because there is no obvious method to obtain reliable estimates of dissolution rates. Studies of DS in the open ocean indicate that silica dissolution of biogenic silica must be considered an important source of DS. Vertical DS gradients in the ocean, which indicate effects of a dissolution process, are developed on a time scale considerably longer than exists in San Francisco Bay (1, 32). An evaluation of silica dissolution in the bay has been attempted with a one-dimensional mathematical model, in which measured values of salinity, river discharge and river and ocean boundary DS concentrations, and assumed values of DS utilization were used to simulate DS concentrations.

The relative effects of eddy diffusivity, K, and river discharge, Q, on the salinity distribution in the model are first illustrated (Fig. 20-A). These distributions can be compared with the similar distributions in Figure 3 by reversing the position of the horizontal (distance) axis. The relative effects of diffusivity, river discharge, and DS utilization, R, are then illustrated (Fig. 20-B) in which both the diffusivity and utilization rates are considered to be constant throughout the estuary. The model is further refined by introducing a variable eddy diffusivity (Fig. 20-C) which represents typical values calculated from the observed salinity. It seems significant that by using a spatially variable utilization rate in the model (Fig. 20-D) a more realistic DS distribution is obtained in the model (Fig. 20-E, distribution VII) than a constant rate (Fig. 20-E, distributions V and VI). This is the case because the observed carbon-production rate is also spatially variable (Fig. 10). With only minor adjustments in the selected DS utilization rates, the resulting DS distribution (Fig. 20E, distribution VII) could duplicate the observed distribution during a river discharge of 50 m^3/sec (Fig. 19-D). In this case, however, such refinement would serve no useful purpose because the values of eddy diffusivity represent a more gross approximation. The fact that distribution VII (Fig. 20-E) is similar to the observed distribution (Fig. 19-D) illustrates that the river DS supply during low summer-flow conditions is an adequate source for support of typical DS utilization rates (20, 31, 48, 49). These results suggest that the DS production rate by silica dissolution is small relative to the DS river supply. It would not be surprising, however, if the biogeneous silica dissolution rates in the estuary proved to be significant, at least during the periods of extremely low DS concentrations.

Biological Factors

Evaluation of temperature and salinity effects on diatom growth rates and related DS utilization rates is beyond the scope of this paper. For two reasons, however, it is clear that seasonal and longitudinal variations in phytoplankton abundance and productivity are not controlled solely by temperature and (or) salinity. First, longitudinal temperature variation in the estuary is influenced by water depth as well as by mixing of river and ocean water; the bay is considerably shallower than the potential depth of vertical mixing. As a result, this longitudinal temperature variation during summer is small in comparison with the observed longitudinal variation in phytoplankton productivity. Second, temperature differences at comparable times of year between 1971 and 1972 were too insignificant to explain the obvious differences in phytoplankton abundance and productivity in 1971 compared with 1972 (e.g., Figs. 14 and 15). These annual differences would also be difficult to explain on the basis of variations in the salinity field.

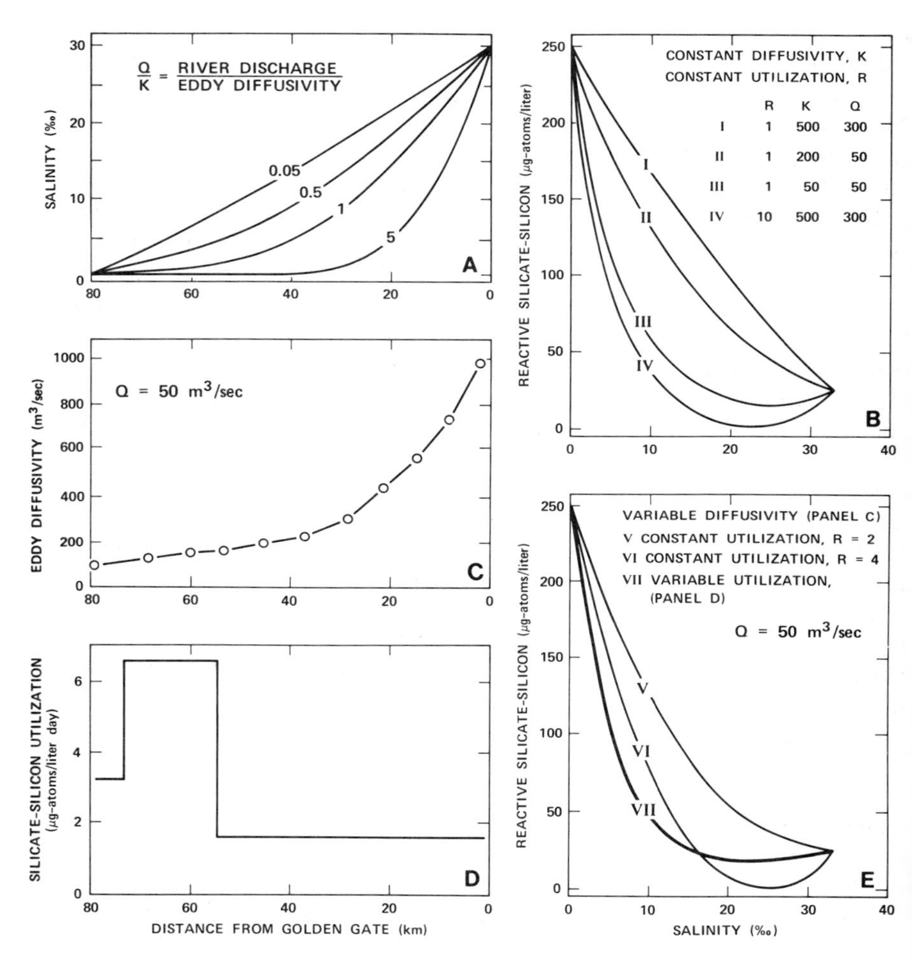

Figure 20. One-dimensional silicate-salinity model in northern reach of San Francisco Bay: (A) Salinity versus distance for various ratios of river discharge, Q, to constant longitudinal eddy diffusivity, K; (B) silicate-salinity relation from the model with constant eddy diffusivity and constant utilization; (C) variation of longitudinal eddy diffusivity from the observed salinity distribution at Q = 50 m^3/sec; (D) parametric representation of variable silicate utilization rate; and (E) silicate-salinity relation from the model with variable diffusivity and silicate utilization rate. R is given in g-atoms Si-SiO_2/liter/day, K in m^2/sec and Q in m^3/sec.

The circulation factors we have discussed emphasize that high DS utilization rates per unit volume exist in the phytoplankton maximum primarily because of an increased phytoplankton abundance per unit volume, rather than because of an increased temperature- or salinity-induced DS utilization rate per unit phytoplankton. The increased turbidity of this maximum further supports our interpretation.

IMPLICATIONS

Ketchum (34) emphasized the importance of water-residence time to biological systems in estuaries. Probably the most important feature implied in DS behavior relative to salinity is that physical (conservative) processes in San Francisco Bay have a strong influence on the biological (non-conservative) distributions. River discharge provides a seasonal modulation of residence time in the estuary. Estuarine circulation imposes a spatial variation. Thus the effects of time-dependent biological processes are partly controlled by the physically-controlled variations in water-residence time.

Because the importance of residence time cannot be clearly separated from insolation, we cannot predict what phytoplankton concentrations would develop in the estuary during low winter insolation if the river discharge in winter were as low as in summer. High winter river discharge in addition to the low winter insolation may be an important factor controlling both the near-linear winter DS-salinity relation and the lower winter abundance of plankton in the estuary. Alternatively, if high river discharge prevailed throughout summer, lower phytoplankton concentrations and linear DS distributions might be maintained (i.e., Fig. 9-F). In an extreme case, such as in the Orange River estuary (11), an exceptionally small residence time may result in a paucity of biota.

Longitudinal variations in strength of estuarine circulation imply longitudinal variations in advective residence time, particularly during low river-discharge conditions. The importance of such variations to the distribution of substances in the estuary, however, must depend on wind- and tidally-induced diffusive processes as well. Advective processes are known to have an important influence on zooplanktonic distributions in estuaries (6, 8, 28, 44, 53). The fact that the seasonal DS variations seem to be strongly associated with river discharge implies that advective processes influence the seasonal distributions of substances in the estuary. Similarly, observed longitudinal variations in the non-conservative (non-linear) distributions of SPM, phytoplankton, and DS suggest that the effects of advective processes influence the longitudinal distributions of these substances in the estuary. In contrast, diffusive processes may tend to develop uniform distributions.

Cycles of nitrogen and phosphorus and other biologically-reactive substances in estuaries are more complex than that of DS (15, 35, 36). San Francisco Bay

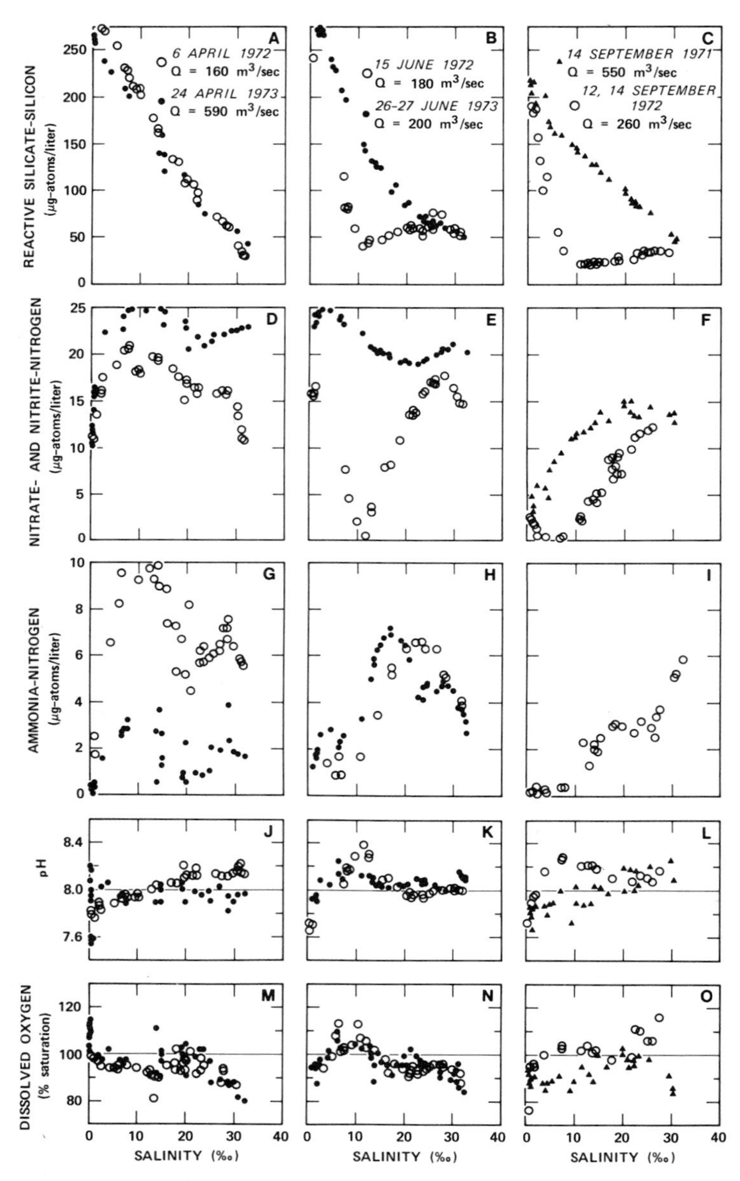

Figure 21. Longitudinal salinity relations with (A-C) silicate-silicon, (D-F) nitrate- and nitrite-nitrogen, (G-I) ammonia-nitrogen, (J-L) pH and (M-O) dissolved oxygen at 2-m depth during summers representing various river discharges. Q indicates combined mean monthly discharges from the Sacramento and San Joaquin rivers.

exhibits similar complexity. Some of the processes controlling the DS distributions apparently control nitrate and ammonia distributions which appear less simple than those of DS (Fig. 21). During early winter (December-March), river discharge is typically a major source of nitrate to the estuary. The relatively high nitrate concentration observed in April 1972 followed a winter of high discharge (Gig. 21-D). In contrast, ammonia concentrations may result largely from sources within the estuary. Thus, higher concentrations are achieved during periods of longer water-residence time, as suggested by the higher levels in April 1972 compared with those in April 1973 (Fig. 21-G). Such interpretations obviously must be made with caution. Relative to DS, rapid ammonia and nitrate regeneration times would require much smaller residence time scales to clearly influence their seasonal distributions, and factors other than residence time may influence significantly their distributions.

It is apparent then that biological factors are responsible for the non-linear DS-salinity relation during summer in San Francisco Bay; bioligical and physical factors are responsible for its seasonal variation; and physical factors appear primarily responsible for its longitudinal variation. Too few observations have been made to conclude whether the major factors identified in San Francisco Bay apply to most estuaries. The few available observations in other estuaries, however, indicate that DS behavior is not unusually different (14, 24, 25, 37, 39, 40, 41, 42, 43, 50, 58, 64, 72, 73). Thus, the changes and the rates of changes in the seasonal DS-salinity relation probably can be used as a framework for the initial investigation of other physical and biological processes in many estuaries.

ACKNOWLEDGMENTS

We thank S. W. Hager, L. E. Schemel, R. E. Smith, and S. W. Wienke for conducting the shipboard observation and performing the chemical analyses, and D. S. McCulloch, F. H. Nichols, K. V. Slack, and J. A. West for reviewing the manuscript.

REFERENCES

1. Armstrong, F. A. J.
   1965 Silicon. In **Chemical Oceanography,** Vol. I, p. 409-432. (eds. Riley, J. P. and Skirrow, G.). Academic Press, New York.

2. Armstrong, F. A. J., Stearns, C. R., and Strickland, J. D. H.
   1967 The measurement of upwelling and subsequent biological processes by menas of the Technicon AutoAnalyzer® and associated equipment. **Deep-Sea Res.,** 14: 381-389.

3. Atkins, W. R. G.
   1926 Seasonal changes in the silica content of natural waters in relation to the phytoplankton. **Jour. Mar. Biol. Ass. U. K.,** 14: 89-99.

4. Bain, R. C., Jr., and McCarty, J. C.
   1965 Nutrient-productivity studies in San Francisco Bay. U. S. Pub. Health Serv., Cent. Pacific Basins Water Poll. Cont. Admin. 116 p.

5. Bain, R. C., Jr., and Pintler, H. E.
   1968 Effects of the San Joaquin Master Drain on water quality of the San Francisco Bay and Delta, Appendix C, Nutrients and Biological Response. Fed. Water Poll. Cont. Admin. 116 p.

6. Barlow, J. P.
   1955 Physical and biological processes determining the distribution of zooplankton in a tidal estuary. **Biol. Bull.,** 109: 211-225.

7. Beers, J. R., Stewart, G. L., and Strickland, J. D. H.
   1967 A pumping system for sampling small plankton. **Jour. Fish. Res. Bd. Canada,** 24(8): 1811-1818.

8. Bousfield, E. L.
   1955 Ecological control of the occurrence of barnacles **(Crustacea: Cirripedia)** in the Miramichi estuary. **Bull. Nat. Mus. Canada,** 137: 1-70.

9. Bowden, K. R.
   1967 Circulation and diffusion. In **Estuaries,** p. 15-36. (ed. Lauff, G. H.). Am. Assoc. Adv. Sci. Publ. 83. Washington, D. C.

10. Bowden, K. R. and Gilligan, R. M.
    1971 Characteristic features of estuarine circulation as represented in the Mersey estuary. **Limnol. Oceanogr.,** 16: 490-502.

11. Brown, A. C.
    1959 The ecology of South African Estuaries Part IX: Notes on the estuary of the Orange River. **Trans. of the Royal Soc. of South Africa,** 35(5): 463-473.

12. Brown, R. L. and Varney, G.
    1971 Algal assays of selected waste discharges; Vol. VII of A study of toxicity and biostimulation in San Francisco Bay-Delta Waters. California Dept. of Water Resources. 41 p.

13. Burton, J. D., Leatherland, T. M., and Liss, P. S.
    1970 The reactivity of dissolved silicon in some natural waters. **Limnol. Oceanogr.,** 15: 473-476.

14. Burton, J. D.

1970 The behavior of dissolved silicon during estuarine mixing. II. Preliminary investigations in the Vellar Estuary, southern India. **Jour. Cons. perm. int. Explor. Mer.,** 33: 141-148.

15. Carpenter, J. H., Pritchard, D. W., and Whaley, R. C.
1969 Observations of eutrophication and nutrient cycles in some coastal plain estuaries. In **Eutrophication: Causes, Consequences, Correctives.** p. 210-221. Natl. Academy of Sciences, Washington, D. C.

16. Conomos, T. J., McCulloch, D. S., Peterson, D. H., and Carlson, P. R.
1971 Drift of surface and near-bottom waters of the San Francisco Bay system: March 1970 through April 1971: U. S. Geol. Survey open-file map.

17. Conomos, T. J. and Peterson, D. H.
1974 Biological and chemical aspects of the San Francisco Bay turbidity maximum; **Symposium Internatl. Relations sedimentaires entre estuaries et plateaux continentaux,** Institut de Geologic du Bassin d'Aquitaine, 15 p.

18. Cooper, L. H. N.
1933 Chemical constituents of biological importance in the English Channel November 1930 to January 1932. Part I. Phosphate, silicate, nitrate nitrite, ammonia. **Jour. Mar. Biol. Ass. U. K.,** 18: 677-728.

19. Cronin, L. E. and Mansueti, A. J.
1971 The biology of the estuary. In **A symposium on the Biological Significance of Estuaries,** p. 14-39. (eds. Douglas, P. and Stroud, R.). Sport Fishing Institute, Washington, D. C.

20. Davis, C. O., Harrison, P. J., and Dugdale, R. C.
1973 Continuous Culture of Marine Diatoms Under Silicate Limitation. I. Synchronized Life Cycle of **Skeletonema costatum. Jour. Phycology,** 9: 175-180.

21. Doherty, P. C.
1970 One-dimensional model of solute concentration in an estuary. U. S. Geol. Survey Computer Contribution C423; Menlo Park, California. 36 p.

22. Doherty, P. C.
1971 Diffusion and source functions by residual minimization. U. S. Geol. Survey Computer Contribution C937; Menlo Park, California. 34 p.

23. Dugdale, R. C.
1967 Nutrient limitation in the sea: dynamic identification and significance. **Limnol. and Oceanogr.,** 12(4): 685-695.

24. Fa-Si, L., Yu-Duan, W., Long-Fa, W., and Ze-Hsia, C.
1964 Physiochemical processes of silicates in the estuarial region. I. A preliminary investigation on the distribution and variation of reactive silicate content and the factors affecting them. **Oceanol. Limnol. Sinica,** 6: 311-322.

25. Fanning, K. A. and Pilson, M. E. Q.
1974 The lack of inorganic removal of dissolved silica during river-ocean mixing. **Geochimica et Cosmochimica Acta.** (In press.)

26. Flemer, D. A.
1970 Primary Production in the Chesapeake Bay. **Chesapeake Sci.,** 11(2): 117-129.

27. Glenne, G. and Selleck, R. E.
1969 Longitudinal estuarine diffusion in San Francisco Bay, California. **Water Res.,** 3: 1-20.

28. Graham, J. J.
1972 Retention of larval herring within the sheepscot estuary of Maine. **Fishery Bull.,** 70: 299-305.

29. Hansen, D. V.
1965 Currents and mixing in the Columbia River Estuary. **Ocean Sci. and Ocean Eng.,** Trans. Joint Conf. Marine Tech. Soc. Am. Soc. Limnol. and Oceanogr. p. 943-955.

30. Hansen, D. V. and Rattray, J., Jr.
1966 New dimensions in estuary classification. **Limnol. Oceanogr.,** 11: 319-326.

31. Harrison, P. J.
1974 Continuous culture of the marine diatom **Skeletonema costatum** (Grev.) Cleve under silicate limitation. Doctoral dissertation. Univ. Washington, Seattle, 140 p.

32. Hurd, D. C.
1972 Factors affecting solution rate of biogenic opal in seawater. **Earth and Planetary Science Letters,** 15: 411-417.

33. Inglis, C. C. and Allen, F. H.
1957 The regimen of the Thames Estuary as affected by currents, salinities, and river flow. **Inst. Civil Engrs. Proc.,** (London), 7: 827-868.

34. Ketchum, B. H.
1954 Relation between circulation and planktonic populations in estuaries. **Ecology,** 35: 191-200.

35. Ketchum, B. H.
1967 Phytoplankton nutrients in estuaries. In **Estuaries,** p. 329-335. (ed. Lauff, G. H.). Am. Assoc. for the Adv. of Science, Publ. 83. Washington, D. C.

36. Ketchum, B. H.
1969 Eutrophication of estuaries. In **Eutrophication: Causes, Consequences, correctives,** p. 197-209. Natl. Academy of Sciences, Washington, D. C.

37. Kobayashi, J.
1967 Silica in fresh water and estuaries. In **Chemical environment in the aquatic habitat,** p. 41-55. (eds. Golterman, H. L. and Clymo, R. S.) N. V. Noord-Hollandesche Vitgeveps Maatschapping, Amsterdam.

38. Krone, R. B.
1966 Predicted suspended sediment inflows to the San Francisco Bay system. Prepared for Cent. Pac. Basins Comprehensive Water Poll. Cont. Proj., Fed. Water Poll. Cont. Admin., 33 p.

39. Kühl, H. and Mann, H.
1953 Beiträge zur Hydrochemie der Unterelbe. **Veröffentlichungen Inst. fur Meeresforschung Bremerhaven,** 2: 236-268.

40. Kühl, H. and Mann, H.
1954 Über die Hydrochemie der untersen Ems. **Veröffentlichungen Inst. für Meeresforschung Bremerhaven,** 3: 126-158.

41. Kühl, H. and Mann H.
1957 Beiträge zur Hydrochemie der unteren Weser. **Veröffentlichungen Inst. fur Meeresforschung Bremerhaven,** 5: 34-62.

42. Liss, P. S. and Pointon, M. J.
1973 Removal of dissolved boron and silicon during estuarine mixing of sea and river waters. **Geochim. Cosmochim. Acta 37.** (In press.)

43. Maéda, H. and Takesue, K.
1961 The relation between chlorinity and silicate concentration of waters observed in some estuaries. **Records of Oceanogr. Works in Japan,** 6(1): 112-119.

44. Massmann, W. H.
1971 The significance of an estuary on the biology of aquatic organisms of the middle Atlantic region. In **A symposium on the biological significance of estuaries,** p. 96-109. (eds. Douglas, P. and Stroud, R.). Sport Fishing Institute, Washington, D. C.

45. McCarty, J. C., Wagner, R. A., Macomber, M., Harris, H. S., Stephenson, M., and Pearson, E. A.
1962 An investigation of water and sediment quality and pollutional characteristics of three areas in San Francisco Bay 1960-61. Sanit. Eng. Research Lab., Univ. California, Berkeley. 571 p.

46. McColloch, D. S., Peterson, D. H., Carlson, P. R., and Conomos, T. J.
1970 Some effects of fresh water inflow on the flusing of South San Francisco Bay. A preliminary report. U. S. Geol. Surv. Circ. 637A: A1-A27.

47. Meade, R. H.
1972 Transport and deposition of sediments in estuaries, p. 91-120. **Geol. Soc. Amer. Memoir,** 133.

48. Paasche, E.
1973a Silicon and the ecology of marine plankton diatoms. I. **Thalassiosira pseudonana** (**Cyclotella nana**) grown in a chemostat with silicate as limiting nutrient. **Mar. Biol.,** 19: 117-126.

49. Paasche, E.
1973b Silicon and the ecology of marine plankton diatoms. II. Silicate-uptake kinetics in five diatom species. **Mar. Biol.,** 19: 262-269.

50. Park, P. K., Osterberg, C. L., and Forester, W. O.
1972 Chemical budget of the Columbia River. In **Columbia River Estuary and Adjacent Ocean Waters: Bioenvironmental Studies,** p. 123-134. (eds. Pruter, A. T. and Alverson, D. L.) Univ. Washington Press, Seattle, Washington.

51. Peterson, D. H., Conomos, T. J., Broenkow, W. W., and Doherty, P. C.
Location of the nontidal current null zone in northern San Francisco Bay. **Estuarine and Coastal Mar. Sci.** (In press.)

52. Postma, H.
1967 Sediment transport and sedimentation in the estuarine environment. In **Estuaries,** p. 158-179. (ed. Lauff, G. H.). Am.

Assoc. Adv. Sci. Publ. 83. Washington, D. C.

53. Pritchard, D. W.
1953 Distribution of oyster larvae in relation to hydrographic conditions. **Proc. Gulf Caribbean Fisheries Inst., 5th Session,** 1952: 123-132.

54. Pritchard, D. W.
1967 Observations of circulation in coastal plain estuaries. In **Estuaries,** p. 37-44. (ed. Lauff, G. H.). Am. Assoc. Adv. Sci. Publ. 83. Washington, D. C.

55. Redfield, A. C., Ketchum, B. H., and Richards, F. A.
1963 The influence of organisms on the composition of sea water. In **The sea,** p. 26-77. (ed. Hill, M. N.). Interscience, New York.

56. Rogers, H. M.
1940 Occurrence and retention of plankton within the estuary. **Jour. Fish. Res. Bd. Canada,** 5: 164-171.

57. Schindler, D. W. and Holmgren, S. K.
1971 Primary production and phytoplankton in the Experimental Lakes area, northwestern Ontario, and other low-carbonate waters, and a liquid scintillation method for determining C^{14} activity in photosynthesis. **Jour. Fish. Res. Bd. Canada,** 28: 189-201.

58. Schink, D. R.
1963 Budget for dissolved silica in the Mediterranean Sea. **Geochim. Cosmochim. Acta,** 31: 987-999.

59. Schubel, J.R.
1968 Turbidity maximum of northern Chesapeake Bay. **Science,** 161: 1013-1015.

60. Scripps Institution of Oceanography
1965 Data report–Physical and Chemical data: CCOFI Cruise 6401. Univ. California (**also** SIO Ref. 65-7).

61. Scripps Institution of Oceanography
1966 Data report–Physical and Chemical data: CCOFI Cruise 6404 and 6407. Univ. California (**also** SIO Ref. 65-20).

62. Simmons, H. B.
1955 Some effects of upland discharge on estuarine hydraulics. Proc. Am. Soc. Civil Eng. 81 (Separate 792).

63. State of California
1971 Interim water quality control plan for the San Francisco Bay Basin. Calif. Regional Water Quality Control Board, San Francisco Bay Region. 72 p.

64. Stefansson, U. and Richards, F. A.
1963 Processes contributing to the nutrient distributions of the Columbia River and the Strait of Juan de Fuca. **Limnol. Oceanogr.,** 8: 394-410.

65. Storrs, P. M., Selleck, R. E., and Pearson, E. A.
1963 A comprehensive study of San Francisco Bay, 1961-62: second annual report. Univ. California San. Eng. Res. Lab., Rept. No. 63-4, 323 p.

66. Storrs, P. M., Selleck, R. E., and Pearson, E. A.
1964 A comprehensive study of San Francisco Bay, 1962-63: third annual report. Univ. California San. Eng. Res. Lab., Rept. No. 64-3, 227 p.

67. Strickland, J. D. H.
1966 Measuring the production of marine phytoplankton. Fish. Res. Bd. Canada Bull. No. 122, 172 p.

68. Sverdrup, H. U., Johnson, W., and Fleming, R. H.
1942 **The oceans, their physics, chemistry, and general biology.** Prentice-Hall, Inc., Englewood Cliffs, New Jersey. 1087 p.

69. U. S. Army Corps of Engineers.
1963 Comprehensive survey of San Francisco Bay and tributaries, California. Appendix H. **Hydraulic model studies,** Vol. I, U. S. Army Engr. District, San Francisco, California. 339 p.

70. Vollenweider, R. A.
1969 A manual on methods for measuring primary production in aquatic environments. **Intl. Biological Program Handbook No. 12,** Blackwell, Oxford.

71. Williams, R. B.
1966 Annual phytoplankton production in a system of shallow temperate estuaries. In **Some Contemporary Studies in Marine Science,** p. 689-716. (ed. Barnes, H.). George Allen and Unwin Ltd., London.

72. Wollast, , R. and De Broeu, F.
1971 Study of the behavior of dissolved silica in the estuary of the Scheldt. **Geochim. Cosmochim. Acta,** 35: 613-620.

73. Wollast, R.
1972 Circulation, accumulation et bilan de mass dans l'estuaire de l'Escaut: Programme Nat'l. sur l'Environment physique et biologique. Pollution des Eaux. Rapport de synthése, tome II, Chapitre V: 231-264.

PROCESSES AFFECTING THE COMPOSITION OF ESTUARINE WATERS (HCO_3, Fe, Mn, Zn, Cu, Ni, Cr, Co, and Cd)

by

J. H. Carpenter[1], W. L. Bradford[2] and V. Grant[3]

ABSTRACT

Consideration of data for alkalinity and the concentrations of several metals in the Susquehanna River and northern Chesapeake Bay shows that variations in the composition of the river water that mixes with sea water to produce such estuarine waters are a major cause of temporal changes in the estuarine-water composition. Biogenic or physical-chemical processes in the estuary are extremely difficult to distinguish from processes in the watershed as causes of concentration variations.

For several transition metals, the maximum concentrations of filer-passing (soluble) forms in the river water were found during late fall and winter, and input from decaying vegetation on the watershed during these seasons appears to be important. However, the influences of biogenic and physical-chemical processes in the estuary are identifiable as concentration variations larger than the source fluctuations for some metals. Substantial release of manganese from the sediments was found during the summer when concentrations of pH and dissolved oxygen in the overlying water were low. The release of zinc from

1. Rosenstiel School of Marine and Atmospheric Science, University of Miami, 10 Rickenbacker Causeway, Miami, Florida 33149.

2. URS Research Corporation, 155 Bovet Road, San Mateo, California 94402.

3. Chesapeake Bay Institute, The Johns Hopkins University, Baltimore, Maryland 21218.

recently deposited sediments was observed as the chlorinity of the overlying water increased in late spring. Apparently zinc is actively cycled between the sediments and the water, as indicated by observed vertical gradients in the distribution of zinc that reflect probable turnover times of one to two weeks. Most of the soluble zinc appears in a complex form during late summer and these transitory organic complexes may have an influence on the zinc cycles in estuaries.

Sometimes it is of interest to wonder how the "water" (the liquid with all its dissolved and suspended components, including living organisms, retrieved by samplers) at some position in an estuary came to have its particular chemical composition, i.e., to evaluate the relative importance of all the conceivable processes. This paper is an attempt to explore the extent to which some of such questions are tractable. During the Conference on Estuaries at Jekyll Island, Georgia, in 1964, definitions of the word estuary were discussed, various authors made proposals according to their own sciences, physical, biological, or geological. With respect to chemical composition, the definitions "a body of water in which the river water mixes with and measurably dilutes sea water" (10) and " . . . within which sea water is measurably diluted with fresh water derived from land drainage" (19) are obviously pertinent, and the caution that our desire "to determine common features – a dangerous approach if we forget that reality in nature is individuality" (4) is especially appropriate. However, emphasis may be placed on the fact that "the body of water" is being continuously formed and transported out of the estuary and the estuary, perhaps, is most importantly a geomorphic feature where the process occurs.

The same "water" is never present again at the position where it was observed on some particular occasion and nearly all the "water" is lost from the estuary during periods of time ranging from days to at most a year, bodies of water the size of the Baltic Sea being excluded from consideration as estuaries. The obvious approach to evaluating processes by observing at two successive points of time is complicated by this inherent feature of estuaries.

Complete enumeration of possible estuarine processes and representation of the various reactions produces such complex diagrams that the viewer is unduly perplexed. Figure 1 shows the simplest, realistic approximation of the processes that affect the composition of estuarine waters, processes that, for crude analysis, may be grouped as follows:

1. Physical flow and mixing of river waters with coastal and oceanic waters.
2. Biogenic Reactions.
   a. Uptake and release processes involving living and detrital materials.
   b. Synthesis of new compounds and degradation.
3. Secondary Biogenic Reactions.
   a. Change of proton equilibria.
   b. Change of oxidation – reduction equilibria.

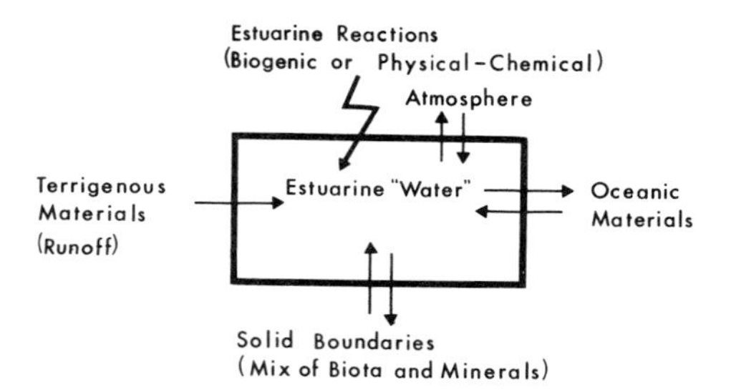

Figure 1. Schematic representation of the processes that affect the composition of estuaries.

4. Physical-Chemical Reactions.
   a. Changes of equilibria due to temperature changes.
   b. Changes of equilibria due to non-specific ion interactions.
   c. Ion exchange with solid minerals.

FLOW AND MIXING

In the absence of any biogeochemical processes, the composition of an estuary would be determined by the composition of the ocean and of the particular river that discharged into it. This "simple mixture" forms a reference for considering the magnitude of other processes. The concentration of a component of a solution that results from the mixture of two other solutions may be found in the following manner:

Let X_1 = concentration in solution 1
X_2 = concentration in solution 2
A_1 = amount of solution 1 in mixture
A_2 = amount of solution 2 in mixture
X = concentration in the resulting mixture of solutions 1 and 2

then

$$X = \frac{A_1X_1 + A_2X_2}{A_1 + A_2}$$

or

$$\frac{A_1}{A_1 + A_2} X_1 + \frac{A_2}{A_1 + A_2} X_2 \qquad (1)$$

$$\text{Let } F = \frac{A_1}{A_1 + A_2}\text{; so that } (1 - F) = \frac{A_2}{A_1 + A_2};$$

and substituting in (1)

$$X = FX_1 + (1 - F)X_2$$

or

$$F = \frac{X - X_2}{X_1 - X_2} \qquad (2)$$

Equation 2 would apply to any conservative component and the concentration of one element may be calculated from the concentration of another by eliminating F from the individual equations. Since chloride is the most abundant element in sea water, it is frequently used as a reference element for the fraction of sea water present in an estuarine water sample. Neglecting the chloride in river water, the resulting equation has the following form:

$$X_{sample} = X_{river} + \frac{(X_{sw} - X_r)}{Cl_{sw}} Cl_{sample}$$

For a conservative element, there would be a linear variation with chlorinity and a zero chloride intercept corresponding to the river = water concentration. The river-water concentration is implicit in the slope term, and, since it can be directly measured, the following equation is also useful for analysis of the distributions of the conservative elements:

$$X_{river} = \frac{X_{sample}Cl_{sw} - X_{sw}Cl_s}{Cl_{sw} - Cl_s}$$

This equation provides the basis for a comparison between observed concentrations in an estuary in terms of a computed "river-water concentration" and the observed concentrations in the river=water source, for elements that have a constant ratio to chlorinity in sea water.

The compositions of rivers that mix with sea water are variable and appear to reflect (a) the character of the drainage basin and (b) the rate of discharge from the basin. The differences between drainage basins produce the greatest differences in composition between their respective estuaries. The basis for this statement is the fact that the variation in river compositions amounts to many milligrams per liter for the major ions and there are no known biogeochemical processes that produce changes in composition of more than a few milligrams per liter.

Variations in river-water composition frustrates efforts to demonstrate the other processes in an estuary. Even though the most pronounced difference between estuaries caused by the character their drainage basins is a constant for a given estuary, the river-water composition varies with stream flow (and perhaps with season) to an extent approaching the differences between many drainage basins. The flow-rate dependence is the most important of the processes that produce changes in estuarine relative composition, with the exception of a few trace metals and the major biological elements. Our studies of the Chesapeake Bay and review of the analyses of the San Francisco estuary and several Canadian estuaries have shown that the standard "water quality" components (Na, K, Ca, Mg, Cl, SO_4, HCO_3) can be calculated from the simple-mixture equation (3) with agreement approaching the accuracy and precision of the analytical values.

The distribution of alkalinity in the upper Chesapeake Bay may be cited, as an example. Table 1 shows the best-fit equations for the results of monthly cruises made in 1959 and 1960. Alkalinity was determined by adding an excess of acid, driving off the carbon dioxide, and back-titrating the solution to the bromothymol blue end point (8). The standard deviation computed from analysis of replicate samples was 0.01 meq/liter. The year 1959 was dry and steam flow was below average throughout the year. During the first half of 1960, runoff was above average. Seasonal and yearly variations in run-off result in a large variation in the alkalinity of the Bay's fresh-water source, as shown by the intercept term in the equations (Table 1). The extremes of this effect were observed in April 1960, when water of 5°/oo salinity had an alkalinity of 0.62 meq/liter, and in December 1960, when water of the same salinity had an alkalinity of 0.86 meq/liter.

Alkalinity is a convenient parameter for illustrating the effects of variations in the river-water composition. Other investigators who have measured alkalinity in estuaries have presented the results in terms of the variation of specific alkalinity with chlorinity or salinity, for example Mitchell and Rakestraw (15), Brust and Newcombe (3), and Turekian (23). This form of presenting the data is equivalent to the following transform of the simple-mixture equation:

$$\frac{X_{sample}}{Cl_{sample}} = \frac{X_{river}}{Cl_{sample}} + \frac{X_{sw} - X_r}{Cl_{sw}}$$

This hyperbolic function is more difficult to analyze using simple statistical procedures than is the linear equation 3, and does not seem to be the best way of looking at the data.

The alkalinity equations in Table 1 have not been examined for "surplus alkalinity" in the manner of Gripenberg (9), because the variations in alkalinity appear to be entirely within the range of river-water variations and it does not appear to be necessary to invoke anions of organic acids that are metabolized to account for them. A reaction that might involve substantial quantities of protons is the formation of iron sulfides from solids containing ferric iron and dissolved sulfate, but the river-water variations frustrate any attempt to discern this process as a change in alkalinity in upper Chesapeake Bay. Similarly, the reverse weathering reactions of sodium bicarbonate with clay minerals were not

TABLE 1

ALKALINITY – UPPER CHESAPEAKE BAY

Year	Month	
1959	January	Alk(meg/L) = .46 + .55S^{o}/oo
	February	.38 + .57S
	March	.49 + .54S
	April	.50 + .54S
	May	.43 + .56S
	June	.45 + .55S
	July	.42 + .56S
	August	.37 + .57S
	September	.40 + .56S
	October	.42 + .56S
	November	.49 + .54S
	December	.39 + .57S
1960	January	.39 + .57S
	February	.36 + .58S
	March	.35 + .58S
	April	.32 + .59S
	May	.43 + .56S
	June	.49 + .54S
	July	.51 + .53S
	August	.52 + .53S
	September	.53 + .53S
	October	.59 + .51S
	December	.61 + .50S

detected by Turekian (23) in Long Island Sound because of the scatter in the alkalinity that most probably originates in the watersheds rather than in Long Island Sound.

Another feature of the physical processes that affect estuarine composition appears when inputs from human activities are considered. Consider an input to the river or near the head of an estuary and the resulting concentration distribution as shown in Figure 2, which might be a plot of a single survey. Measurement of chlorinity makes it possible to compute the effects of dilution with sea water and such plots have been used to estimate the effects of processes other than dilution. However, simple interpretation is not possible; for example, the difference labeled with a question mark in Figure 2 might be due to (a) transformation of the constituent into a form not determined in the analytical procedure, (b) loss to the sediments, or (c) the flow rate of the river having varied with time. The last possibility always exists and the curve in Figure 2 may simply indicate that the water in the more seaward part of the estuary passed the point of contaminant input at a high rate of flow and never had the concentration that was being produced at the source point at

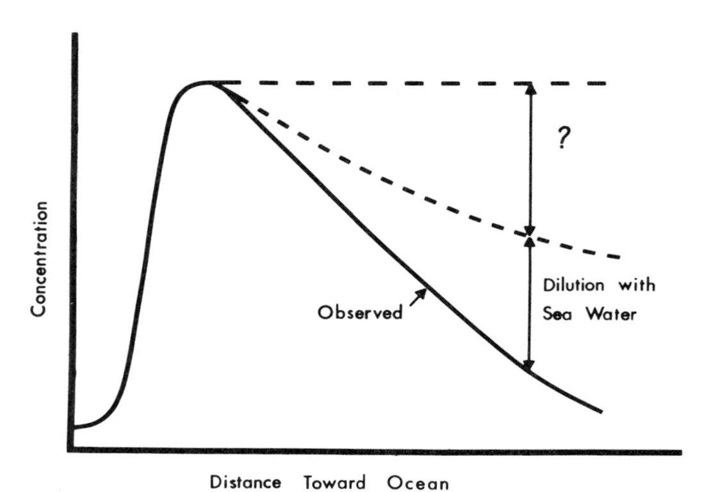

Figure 2. Example of the longitudinal variation of concentrations from a point source input (contamination), solid line. Small dash curve represents estimate of concentration that would have been found if no dilution had been present. Difference between small-dash curve and large-dash curve requires identification of additional process from several possibilities.

the time of the survey. Evaluation of this effect requires adequate time-varying "models" (equations) for estuarine flow and mixing and a comprehensive set of data for variations as a function of both position and time. Because of these features, our understanding of the fate of contaminants has not progressed very far, and progress awaits simultaneous developments in theory and observation.

BIOGENIC AND SECONDARY BIOGENIC REACTIONS

Iron and Manganese in northern Chesapeake Bay.

It was hoped that estuarine processes clearly distinct from physical mixing would be observable for elements whose concentrations in estuarine waters are relatively low. In particular, the transition metals should be involved in such processes. Iron and manganese are the most abundant of these metals and potentially may influence the occurrence of many others such as zinc, copper, nickel, cadmium, cobalt, and chromium through coprecipitation reactions and competition in coordination (chelate) reactions. The results of an exploratory survey of iron and manganese at 6 stations located at 12=mile intervals along the center line of Chesapeake Bay running south from the mouth of the Susquehanna River are shown in Figure 3. The observations made during March, 1961, show a simple pattern of decreasing concentrations of both iron and

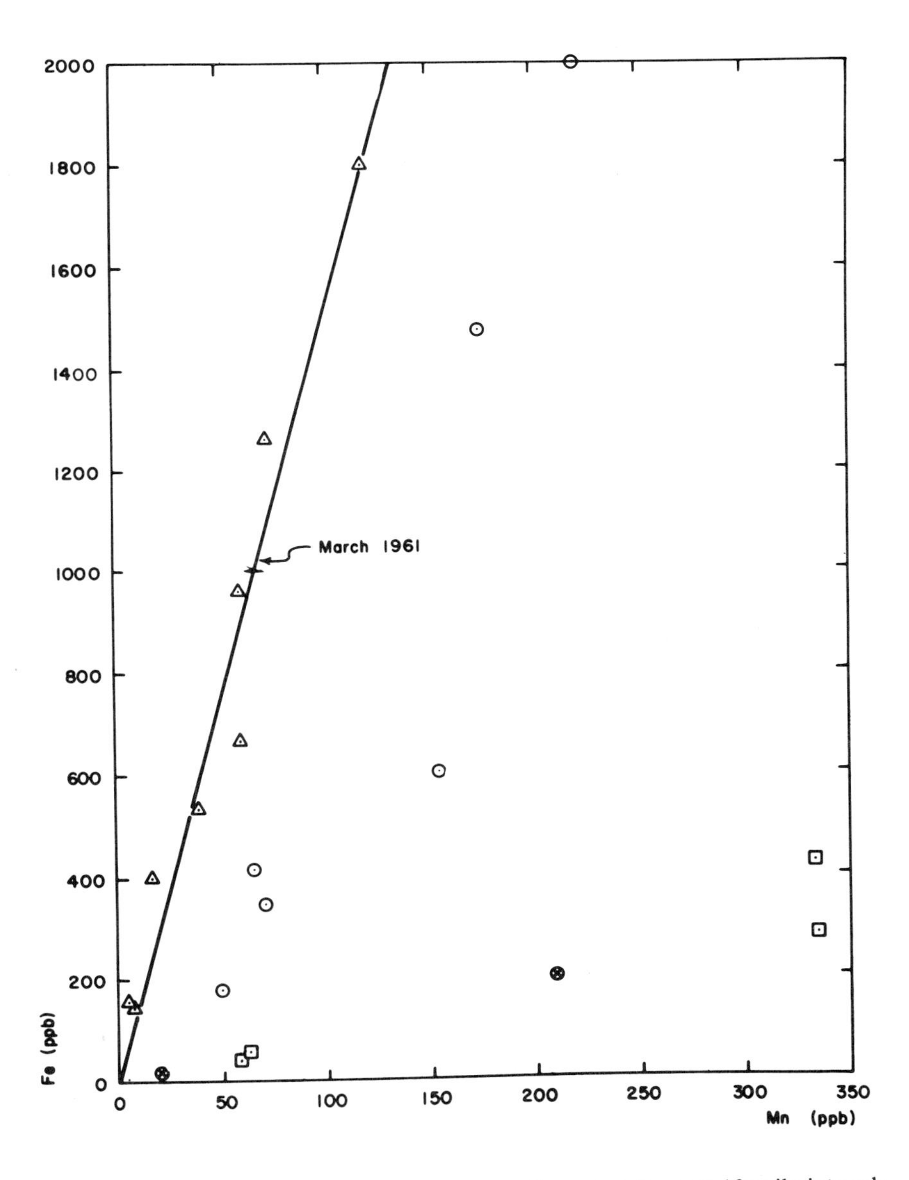

Figure 3. Observed iron and manganese concentrations at stations at 12=mile intervals along Chesapeake Bay from the mouth of the Susquehanna River. The triangles represent data points for the March, 1961, survey. Other points are for the July, 1961, survey. (See text for details)

manganese with distance from the mouth of the Susquehanna River. The decrease is much greater than simple dilution would cause and a possible interpretation is that the Susquehanna River was delivering material that had an Fe/Mn ratio of approximately 17/1 and this material was being deposited in the upper Chesapeake Bay.

The other data points in Figure 3 are observations made during July, 1961, and are for the surface and bottom data at each station. The circles are for the stations nearest the Susquehanna River, the circles with crosses are for the next most distant stations, and the squares are for the most distant stations in the region that is persistently estuarine (salinities 5-15°/oo). Much of the manganese in the deeper samples was filter-passing, while that at the surface was not. The striking feature of these data is that, at the three most seaward stations, a relatively uniform Fe/Mn ratio of 1/1 was found at all depths. It might be inferred that the local processes of selective solubilization of iron and manganese and subsequent precipitation had produced within the estuary a new material and the vigor of this process had completely obliterated the influence of the source material that was delivered during March, 1961. Even though this area of Chesapeake Bay is strongly stratified with respect to density and the near-bottom waters develop low oxygen and pH conditions during the summer, there had been sufficient vertical exchange so that the flux from the sediments was determinate of the Fe/Mn ratios in the surface waters.

This interpretation does not appear to be complete, since further study (see below) shows that the materials delivered by the Susquehanna River have Fe/Mn ratios of 8/1 during high flow and 4/1 during low flow and the material found in March, 1961, did not all appear to have come from the Susquehanna River. Analysis of bottom sediments collected along a transect across the Bay, near Annapolis, Maryland, shows Fe/Mn ratios ranging from 25/1 to 40/1. Some of these sediments are derived from material supplied by the Susquehanna River and the relative enrichment with iron appears to be the result of selective loss of manganese. The existence of this process is shown in the data for July, 1961, where the bottom waters were greatly enriched in manganese. In addition, analyses of the interstitial waters of the sediments along the Annapolis transect showed iron contents of 1-3 ppm and manganese contents of 5-20 ppm, with a resulting average Fe/Mn ratio of 0.2 that confirms the water-column measurements of selective solubilization. Apparently the material observed in Chesapeake Bay during March, 1961, was a mixture of material recently delivered by the Susquehanna River and material that had been resuspended following sedimentation and manganese loss. The observed Fe/Mn ratio of 17/1 could have been produced by a mixture of approximately equal parts of recent river-derived and sediment-derived materials. If this is so, both sources are important and must be considered in describing the dynamics of iron and manganese in northern Chesapeake Bay. The biogenic processes in the water column are completely masked by the biogenic sediment reactions and physical resuspension, as well as by the time-varying characteristics of the river input.

Observations of the Concentrations of Iron, Manganese, Zinc, Nickel, Copper, Cobalt, Chromium, and Cadmium in the waters of the Susquehanna River near Conowingo, Maryland.

The Susquehanna River, which drains the central part of Pennsylvania and New York, contributes 90% of the fresh-water input to the Chesapeake Bay north of the entrance of the Potomac River, and 42% of the total fresh-water input to the bay. It is thus a major source, as noted above. There have been few investigations of the trace-element composition of the Susquehanna River or of any other large river. As pointed out by D. A. Livingstone (11), "the principal gaps in geochemical data for lakes and rivers are long-term downstream averages for the general composition of large tropical rivers and trace-element analyses for large rivers everywhere." Analyses of Susquehanna River samples have been reported by Durum, Heidel and Tison (57) and the four samples studies showed great variability. Turekian and Scott (22) reported an analysis of one sample taken at Conowingo, Maryland, in June, 1966, for which Cr, Ag, Ms, Ni, Co, and Mn concentrations in the suspended material were measured. Both of these papers present estimates of the transport of various metals into Chesapeake Bay. Results for one sample at Conowingo that was examined for As, Cd, Cr, Co, Pb, Hg, and Zn content were reported by Durum, Hem and Heidel (6) but only As and Zn were above their detection limit.

Observations of the compositions of water samples collected at intervals of approximately seven days from April 1965 through August 1966, are summarized below. The Susquehanna River was sampled at Lapidium, Maryland, approximately 1 mile downstream from the Conowingo dam. Duplicate samples were collected from January 1966 through August 1966.

Sample treatment

The samples, of 100 liters volume, were collected with a nylon pump and polyethylene hose into covered polyethylene tanks. They were processed into three fractions; (a) the solid material that could be collected by gravity settling, (b) the solid material that could be collected by filtration, and (c) the filter-passing elements that could be collected on a chelating ion exchange column.

(a) Settled solids – Turbidity measurements on the samples showed that settling was essentially complete after approximately 11 days, and the samples were allowed to settle for 10 to 14 days. Most of the supernatant water was transferred to another tank and the settled solids were resuspended in the remaining liquid. The resulting slurry was centrifuged and the solids collected and weighed. The water content of the centrifuged material was determined on an aliquot and the remaining material was transferred to glass bottles for chemical treatment and analysis. The coefficient of variation for 22 pairs of duplicate determinations of the concentration of settled solids was 12 percent.

Chemical treatment of the collected solids consisted of making the slurry to a standard volume (100 ml) with hydrochloric and acetic acids so that the final suspension was 1 molar HCl and 1.5 molar acetic acid. The acidified suspension was gently agitated at 60C for 48 hours and the samples stored at room temperature.

(b) Filtered solids – A 50-liter aliquot of the sample that had been separated from the settleable solids was filtered through acid-washed cellulose acetate membrane filters having a 0.2 micron pore size. Another aliquot of the sample was passed through a continuous centrifuge and the collected material weighed to determine the weight concentration of the filterable solids.

The collected material on the filters was treated with hydrochloric and acetic acids in the manner used for the settled solids.

(c) Filter-passing material – The "soluble" fractions of the samples were collected by passing 50 liters of filtrate through a cation exchange column of Chelex 100 resin. Column performance was tested with radioactive manganese as a tracer and complete retention with the maximum total cation content (Ca and Mg) was observed. The collected material was eluted from the column with 1 molar hydrochloric acid. The eluate was evaporated to dryness, oxidized with nitric acid, and made to volume with 0.1 molar hydrochloric acid.

Analysis of the collected materials

The three fractions prepared as described were analyzed for iron, manganese, zinc, copper, nickel, cobalt, chromium, and cadmium. The analytical techniques are summarized in Table 2. The atomic-absorbance measurements were made with a Jarrel-Ash 82-360 spectrophotometer, using either a total-consumption burner or a laminar-flow, slot burner. Instrumental conditions that produced the best combination of sensitivity and range of linearity were determined for each element. Solution matrix effects and inter-element interference were minimized by sample dilution, extraction of the interfering element (most notably iron), or selective extraction of the element being measured. The standard addition technique was used for all the analyses, which required that the linear range of the absorbance-versus-concentration relationship be known, and the test solutions and spiked-test solutions were diluted so that their concentrations were within the linear range. The average deviation between replicate determinations ranged from a low of 1 percent for zinc to a high of 2.2 percent for cobalt and cadmium. Sample matrix absorbance was significant in the zinc procedure and

TABLE 2

Summary of analytical techniques for Conowingo samples

Metal	Additional Treatment	TERMINAL ANALYSES conc-ppm	Method
Fe	Dilution with distilled water and reagents.	.1-5	Calorimetric measurement of o-phen. complex in aqueous solution.
Mn & Zn	Dilution with D. W. and/or std Mn & Zn solns.	Mn .05-1.0 Zn .03-.35	At. Abn. of aqueous soln with total consumption burner using air & H_2.
Cu, Ni & Co	Simultaneous extraction of PAD Carbamate complexes into MIBK - usually followed by further diln with MIBK for Cu & Ni.	Cu .05-.6 Ni .0-1.5 Co .1-1.4	At. Abn. of organic solns with T. C. burner using air and H_2.
Cr	Extraction of Cr complex in acetyl acetone and dilution with MIBK.	.05-5	At. Abn. of mixed organic soln. using laminar flow burner with C_2H_2 and N_2O.
Cd	Extraction of APDC complex from aqueous phase after Cr extraction using acetylaceton as extractant.	.03-.3	At. Abn. of organic solution using T. C. burner with air & H_2.

the correction for background absorbance was based on measurements at the nonabsorbed wavelength of 2288 A from a cobalt lamp, which is close to the analytical wavelength used for zinc.

Results and discussion

The discharge rate of the Susquehanna River at Conowingo and the observed concentrations of suspended solids are plotted in Figure 4. Some correlation between the discharge rate and the suspended-solids concentration is visually obvious. Because high discharge rate is associated with high suspended solids, most of the suspended material is delivered to the Chesapeake Bay during high flow. For example, our sample on 17 February 1966 indicates that 5.1 percent of the total solids that were discharged to the Bay during the 12-month period, 1 September 1965 through 31 August 1966, were discharged on that day. Since the flow was very high for the previous 3 days and the following 3 days, roughly 30% of the suspended solids may have been discharged during this week.

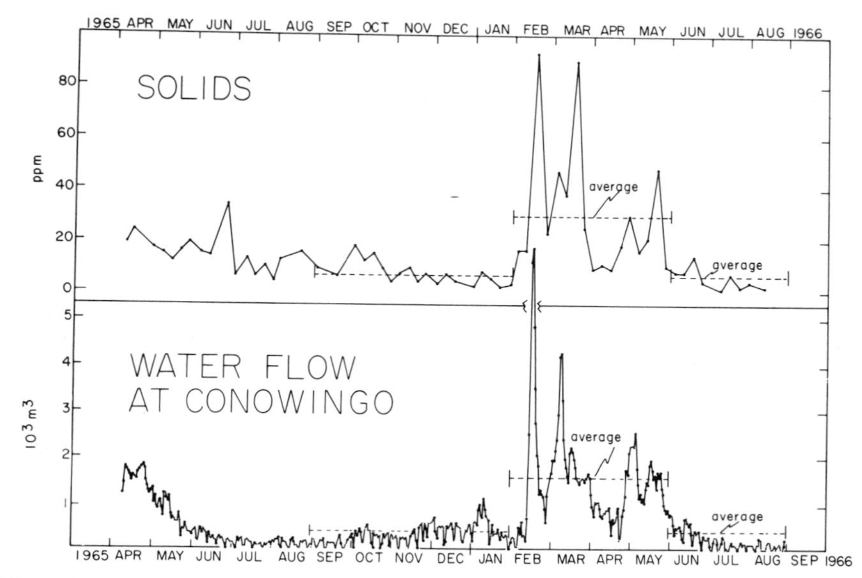

Figure 4. Flow of the Susquehanna River and the concentrations of suspended solids in the samples.

Samples during low and moderate flow are not representative of the bulk of material discharged to the Bay.

The total concentrations of iron, manganese, zinc, nickel, copper, cobalt, chromium, and cadmium are plotted in Figures 5, 6 and 7. There was marked covariance of the iron concentrations with flow and the concentration of suspended solids. From May through November, nearly all the iron was present in settleable solids. The occurrence of substantial (several hundred ppb) amounts of iron in the non-setteable and filter-passing fractions during winter suggests that much of the iron was introduced into the Susquehanna River in either soluble, reduced form or as a colloidal dispersion stabilized by increased concentrations of organic materials from decaying vegetation.

The manganese concentrations varied with time in a way that showed some qualitative similarities to the iron variations. There are a number of dams and reservoirs along the Susquehanna River and it was expected that some soluble, reduced manganese might occur during the summer from reducing environments in the reservoirs. This was not found and, during the summer and early fall nearly all the manganese was in the settleable fraction. The striking increases in non-settleable and filter-passing forms of manganese in December and January were not associated with any large increase, in river-flow rate, and flushing of stagnant areas of the watershed does not appear to be the principal cause. Input from decaying vegetation seems more plausible and is supported by the

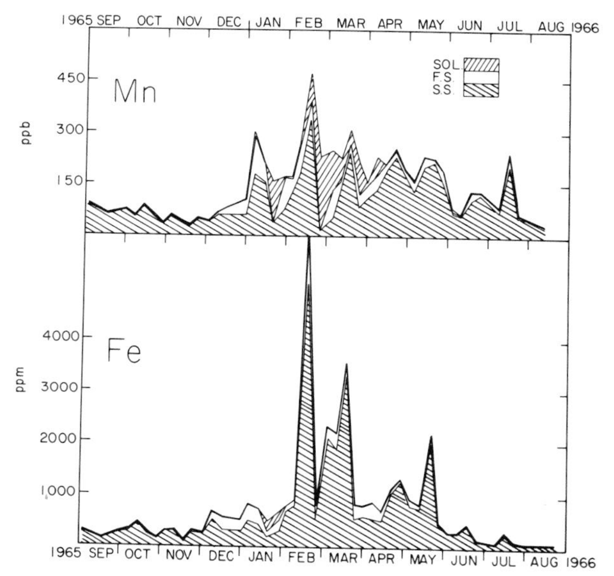

Figure 5. Iron and manganese concentrations in the soluble, filtered-solids, and settled-solids fractions of the samples.

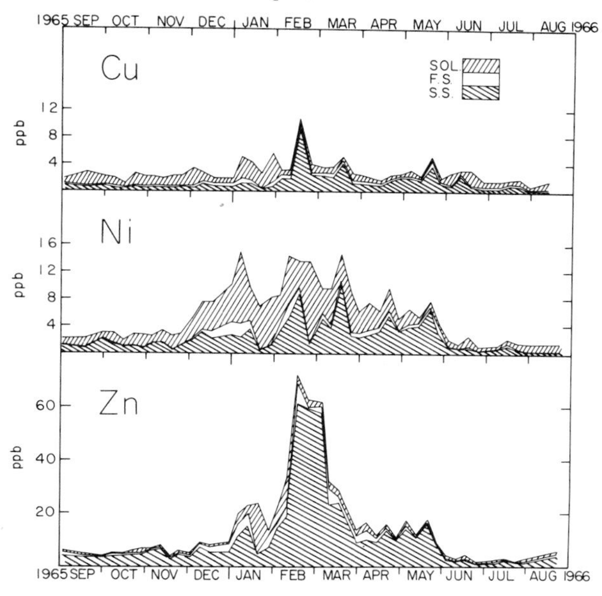

Figure 6. Zinc, nickel, and copper concentrations in the soluble, filtered-solids, and settled=solids fractions of the samples.

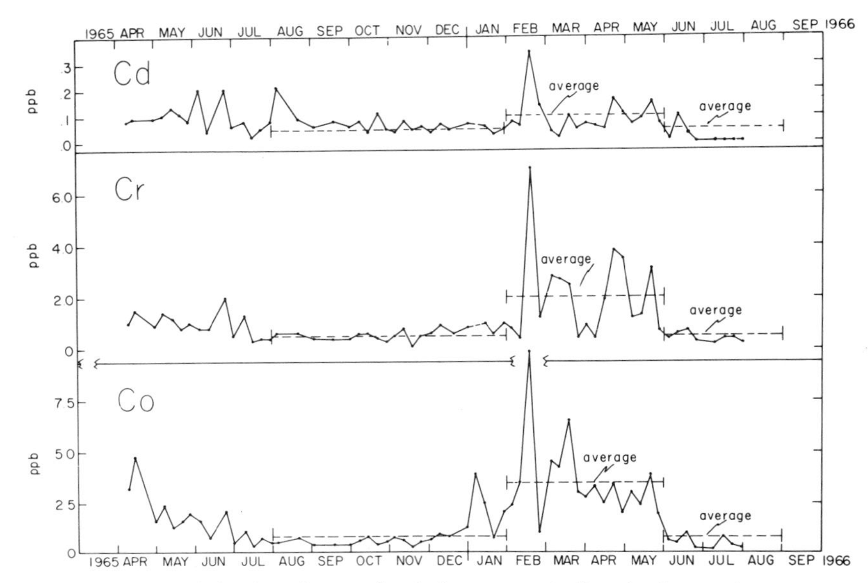

Figure 7. Total cobalt, chromium, and cadmium concentrations in the samples.

observations of Slack and Feltz (21) on a tributary of the Potomac River that iron and, particularly, manganese stream concentrations increased in response to increased leaf-fall rate, with soluble manganese reaching as high as 5000 ppb in late October. If this feature is general, previous estimates of the flux of manganese to the ocean through estuaries have probably been extreme underestimates.

The data for copper, nickel, and zinc are plotted in Figure 6. Zinc was predominately associated with settleable solids, but does not correlate very well with the variations in iron concentrations. Week-to-week variations of twofold or more were found, with a pulse of soluble zinc in January. Copper and nickel were roughly equally distributed between the solid phase and filter-passing or soluble phase. The winter pulse of nickel might be a vegetative input, but the possibility of inputs from the burning of fossil fuels needs to be studied using rain-water samples from the watershed. However, the lack of large metropolitan areas on the Susquehanna watershed may rule this out as a quantitatively important source. This lack of relative increase in the copper and nickel concentrations during February, March, and May in the manner shown by iron, manganese, and zinc suggests that the solids carrying the copper and nickel have a different source and character.

The concentrations of cadmium, chromium, and cobalt in the filter-passing and non-settleable solids were below our useful analytical detection limits. Results for the settled solids that contained nearly all of these elements in the

Susquehanna water are plotted in Figure 7. All three of the metals appear to vary with the total suspended-solids concentration, but not in a simple manner. Although the increased cobalt concentrations in January are mysterious, they correlate with the high soluble-zinc occurrence.

The effects of variation in the quantity of suspended solids may be removed from the data by computing the weight concentrations of the various elements in the solids. The concentrations of the eight elements in the separated solids are plotted in Figures 8, 9 and 10. As shown in Figure 8, the solids have iron concentrations of from 5 percent to 10 percent and the filterable solids may range up to 25 percent iron. The sample collected on 30 December 1965 appears to be unusual in that iron, copper, nickel, zinc, chromium, and cadmium are all relatively high in it – suggesting that the sample solids were missing in the normal concentration of materials low in these elements. The lack of covariance of iron and manganese is apparent. The most striking feature of these plots is the seasonal variation in the composition of the solids, metal-rich materials being more abundant during December and January. The non-settleable materials, i.e., either small-size or low-density, probably are high in organic and aluminosilicate content and frequently had much higher concentrations of metals, but they are extremely variable.

Because of the large size of the Susquehanna watershed and the opportunity for the mixing of water and material during passage through its naturally tortuous channels and artificial reservoirs, it was hoped that relatively uniform solutions and suspended solids were being introduced into Chesapeake Bay and

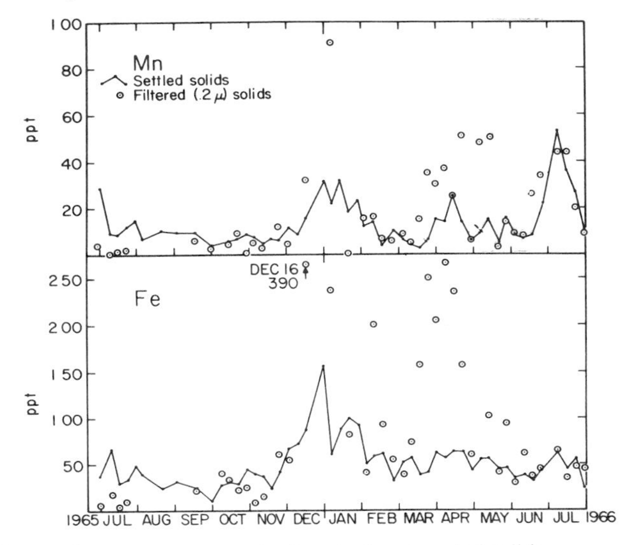

Figure 8. Iron and manganese concentrations in the separated solids.

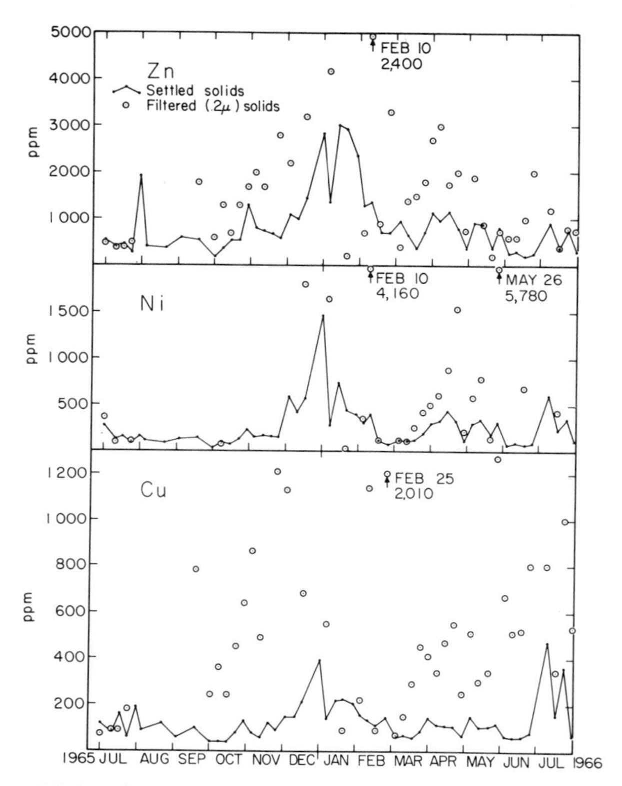

Figure 9. **Zinc, nickel, and copper concentrations in the separated solids.**

that such solutions and materials could be identified by characteristic ratios of one metal to another. That is to say it was hoped that observations of metal concentrations in the Susquehanna estuary or northern Chesapeake Bay could be "normalized," perhaps by reference iron for the river-derived input, and estuarine processes could thus be distinguished from the time variations in the gross concentrations of the input from the Susquehanna River. The data show that this is not true to a useful extent and only pronounced processes (e.g. solubilization of manganese) can be observed above the river-input variability.

The data have been used to compute conventional statistics, as listed in Table 3. There does not seem to be much use in comparing these data with other estimates in the literature that were based on a very few samples, as discrepancies of severalfold would come from any limited selection of our data. Since the year 1965 was extremely dry and the average flow throughout the period sampled was roughly 10 percent below the long-term average, it is difficult to estimate how representative the data are for more extended periods

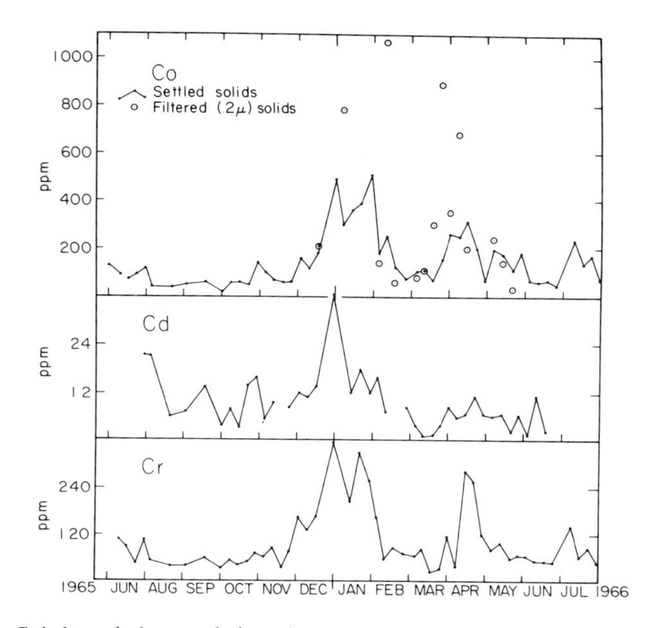

Figure 10. Cobalt, cadmium, and chromium concentrations in the separated solids.

of time. Other workers are making surveys to collect "baseline data" and our data suggest that intensive sampling over several years will be required to obtain numerical values of one significant figure.

Zinc in the northern part of Chesapeake Bay.

In spite of the difficulties presented by the variations in the river-water compositions, we attempted to resolve approximately some estuarine processes for zinc. The concentrations of zinc at three positions in Chesapeake Bay were observed in samples collected monthly from March through December, 1971: station 914S, (39°14'06", 76°14'20") with a depth of 7 meters is located near the head of the Bay, in an area that is occupied by Susquehanna River water during high flow and brackish water (3-5°/oo Cl) during low flow; station 858C (38°57'20", 76°23'04") with a depth of 30 meters is located off Annapolis and is never freshened completely; and station 818P (38°18'48", 76°17'42") with a depth of 33 meters is located off the mouth of the Patuxent River and is approximately at the center of the length of the Bay. Attention was directed to the "soluble" zinc and the chemical form of this zinc.

TABLE 3

Susquehanna River – September 1965 through August 1966

	Time-average concentration Metric tons/day			
	High Flow 4 months	Low Flow 8 months	Annual Input- in Metric tons (weighted by flow)	Volume - average concentration (=ppm)
Water	141×10^6 (57×10^3 C.F.S.)	34×10^6 (14×10^3 C.F.S.)	25×10^9 (82×10^9 C.F.)	
Total Solids	5,900	240	770×10^3	31
Total Fe	325	15	40×10^3	2
Total Mn	37	3.6	5×10^3	.2
Total Zn	4.6	.3	600	.03
Total Ni	1.4	.2	200	.008
Total Cu	.7	.1	100	.004
Total Co	.7	.04	90	.004
Total Cr	.4	.02	50	.002
Total Cd	.014	.002	2	.00008

Sampling and treatment

Samples were taken from 1, 2, and 4 meters above the bottom using a 3-liter Van Dorn bottle that was well leached from having been used for several years. Possible sources of zinc from components of the sampling system and the ship were examined and appear to be negligible. The samples were transferred to 1-liter Teflon® storage bottles and immediately refrigerated at 3C. They were stored for 5-7 days in the dark at 3C, then the upper 800 ml were siphoned off. The bottles were flushed with distilled water to remove the settled material and the supernate samples were returned to each for storage. Material with an equivalent Stokes diameter greater than 0.1 microns should settle out in 5 days. Filtration of similarly treated samples showed only a small amount of material remaining, which appeared under microscopic phase contrast examination to be organic detritus at concentrations of less than 1 mg/liter. We estimate that the upper limit for zinc concentrations associated with this suspended material was 1 μ g/liter. This arbitarily defined method of separating out the solids is, of course, different from filtration, and close comparison is tenuous; however, it avoids the possibility of an intolerable uncertainty in the procedural blank which can be easily introduced by filtration, as shown by Marvin, et al. (13)

The effects of storage on the aqueous-zinc concentration were examined, using Zn-65 as a tracer. Under bacteriostatic conditions of either low temperature or added formaldehyde, loss of zinc to the container was less than 5 percent during 5 weeks. Samples left at room temperature for 5 weeks showed losses averaging 45 percent, suggesting that microbial uptake can be extreme.

Aliquots of the samples were irradiated in quartz tubes for 3 hours or more with light from a 1200-watt mercury vapor lamp (Englehard-Hanovia 189A10). Subsampling during irradiation showed that the effects occurred during the first hour of irradiation. The purpose of the irradiation was to hydrolyze and oxidize organic compounds that might be forming complexes with the zinc. There was no detectable blank from the irradiation procedure.

Analysis

The treated samples were analyzed using an anodic stripping tammet (ASV) technique. Details of the equipment and analytical procedures have been described by Bradford (2). The ASV technique has the virtue of making easily measured signals obtainable from natural waters (high sensitivity), and the manipulations do not introduce substantial blanks. However, inter-element effects reduce the specificity, and we observed larger inter-element effects than had been reported previously by Matson (14) and Fitzgerald (7). In particular, copper strongly influences the zinc stripping curve and causes peak height reductions and tailing. Study of ASV curves for various solutions in which the concentrations of copper and zinc were varied suggests that copper and zinc form an intermetallic complex in the thin layer mercury electrode with a formation constant of 10^5. The interference of copper in the samples required the estimation of the copper content from its stripping curve and treatment of the data using appropriate algebraic expressions involving the formation constant. The interference of nickel was greatly reduced by adding sodium tartrate to the samples after the solutions had been deoxygenated and the plating potential applied to the electrode. The electrode responses varied with time and the technique was continually calibrated, using additions of zinc standards to the samples. The precision for Chesapeake Bay samples based on several series of 5 replications had a coefficient of variation of 3.7%. It is estimated that uncorrected inter-element effects range up to 1 μ g/liter uncertainty in the analytical results for Chesapeake Bay samples.

Results and Discussion

The observed concentrations of zinc in the settled and irradiated samples are plotted as a function of time in Figures 11, 12, and 13. The data for station 914S show a striking increase in the zinc concentrations for the May samples, an increase larger than might be expected from the observed variations in the Susquehanna River. The zinc concentrations had a strong vertical gradient, suggesting that the bottom was the source of the additional zinc. The increase occurred in parallel with an appearance of sea salts (3°/oo Cl) at this station in May, while during March and April the chlorinity was less than 0.2°/oo. The vertical gradient and coincidence with increasing salinity suggest that ion exchange release of zinc from the deposited solids was occurring. This supposition agrees with the experiments of Bachman (1) and O'Conner (16), but differs from the observations of Osterberg et al., (17) that Columbia River solids do not release zinc upon mixing with sea water. Additional evidence that zinc is released is that the sediments in the Bay have zinc contents averaging 300 ppm and material supplied by the Susquehanna River during high flow contains 1000 ppm zinc (Figure 9).

The more seaward stations did not show increases in zinc concentrations during June and July that can be identified as resulting from the seaward drift of the zinc pulse observed earlier at station 914S. During June and July, the zinc concentrations decreased at all three stations and, since phytoplankton abundance increases during these months, biogenic uptake may be the process that caused these decreases.

The zinc concentrations increased during August and September and much of the zinc was in a form that required UV irradiation to make it available to the ASV analytical technique. While no analyses for organic compounds were performed, the samples were observed to be faintly yellow-colored and, after

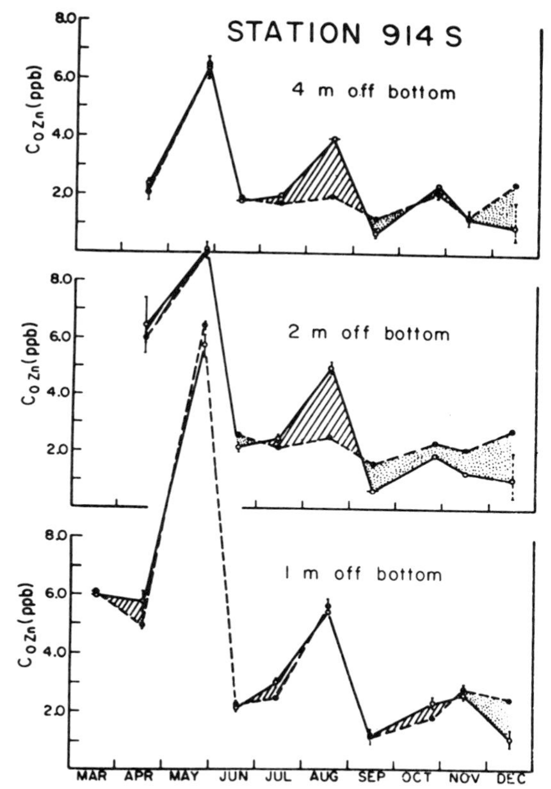

Figure 11. Zinc concentration in the samples at station 914S with solid circles for the unmodified aliquots and open circles for the irradiated (oxidized) aliquots. Lined shading designates samples for which irradiation in the analytical zinc results, and stippling designates samples for which the reverse was found. Vertical lines on some points indicate the range of replicate analyses.

bubbling during the ASV, organic material collected on the walls of the vessel used for analysis. It appears that a significant fraction of the zinc was in a complex form during these two months.

During the fall, the variations in the zinc concentrations were not interpretable. However, a new characteristic appeared; UV irradiation reduced the ASV peak height and changed the shape of the stripping curves. Our tentative interpretation is that irradiation released an unidentified interferent that was not present during other seasons of the year.

At stations 914S and 858C, the vertical distributions of zinc frequently showed increases as the bottom was approached, but were variable. At station 818P (the mid-bay position), increases were found in 7 of the 10 monthly

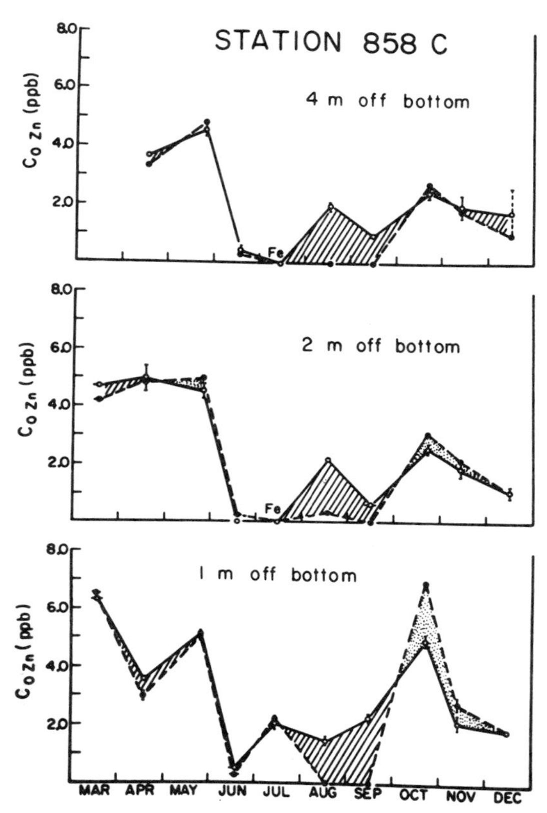

Figure 12. As for 11 but station 858C.

samplings. The average gradient was 1.4 ppb/meter and maintenance of this gradient requires input from the sediment surface. Pritchard (19) reported eddy diffusivity values of ca. 1 cm^2/sec and this value would require a zinc flux of 1.2 μ g/cm^2/day to maintain the gradient. Since the inventory of zinc at this position averaged 10 g/cm^2, the indicated flux would double the inventory in 8 days. The lack of increase suggests that zinc is continuously returned to the sediments, and we speculate that biogenic uptake by phytoplankton and return to sediments in fecal particles may be the primary process.

The possibility of biogenic zinc flux to the sediments appears to be of the required magnitude, using "typical" values. For Chesapeake Bay, the phytoplankton abundance during the summer is ca. 1 mg/liter carbon with a volume of ca. 0.02 cc/liter. Zinc concentration factors are of the order of 10^4, as given by Lowman et al. (12), so the zinc inventory in the cells is ca. 0.6 μ g/liter for waters averaging 3 μ g/liter zinc concentration. Phytoplankton productivity

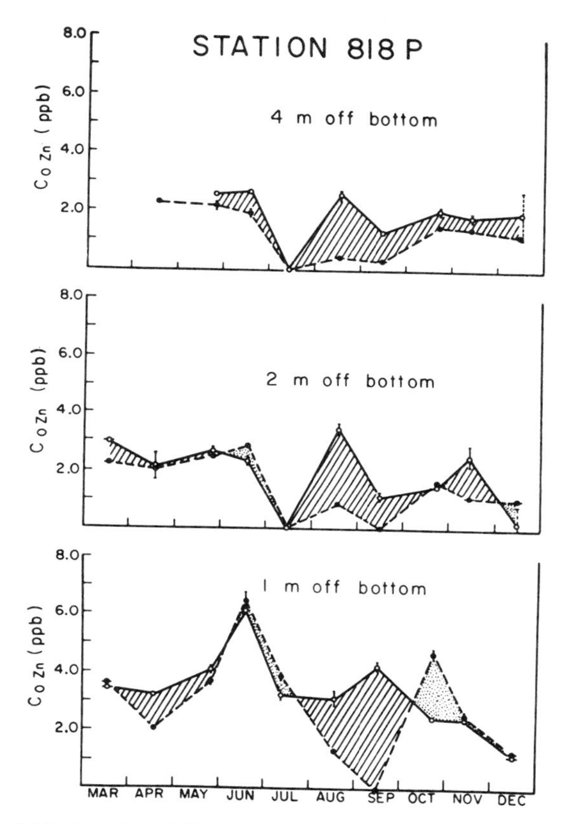

Figure 13. As for 11 but station 818P.

amounts to a doubling in a day or less, so that 5 days' production would permit incorporation of all the zinc inventory into phytoplankton cells, i.e., the sediment input amounting to 2x the inventory in 8 days is easily balanced by phytoplankton removal at a rate equivalent to loss of the total zinc inventory in 5 days.

In addition to the data above, a series of 18 samples was collected at station 914S on August 16 and 17, 1971. The time series data show a smooth variation from 1100, August 16, to 0100, August 17, with values ranging from 3 ppb to 8 ppb, indicating horizontal variations of severalfold might have been present over distances of a 6-10 mile tidal excursion. This variability in total zinc may be the result of the variations in Susquehanna River source (above) or in processes in the Bay, but did not correlate with the variations in suspended solids that increase and decrease through resuspension and deposition during the tidal cycle

(20). The lack of correlation suggests that the water and surface sediments are in short-term chemical equilibrium.

These results indicate that a quantitative description of the temporal and spatial variations in the zinc concentrations in estuaries similar to Chesapeake Bay will require a *large* quantity of data and that the following features may be anticipated:

1. Zinc in the river sources may be primarily in the particulate phases and much of this terrigenous zinc may be released near the head of the estuary during part of the year.

2. Biogenic up take in the water column and transport to the sediment is counteracted by return to the water by decompositional processes at the sediment surface, with turnover times of a week or so. Either process may by the more intense for short periods of time and the results will be short-term fluctuations in the zinc concentrations in the water that pose serious sampling and data rate problems in obtaining acceptably precise estimates of the zinc inventory in the water.

3. Zinc may be expected to appear in organic complexes and such transformations will vary seasonally. The chemical nature of the complexes is unknown, but the occurrence correlates with increasing phytoplankton standing crops and productivity. While complex formation would be expected to favor the release of zinc from minerals, the extent of such processes in increasing the estuarine total zinc concentrations can only be demonstrated with data that reliably average out the rather variable (factor of 2 or more) zinc-concentration fluctuations.

REFERENCES

1. Bachman, R. W.
1963 Zn-65 in studies of the water zinc cycle. In "Radioecology." **Proceedings of the First National Symposium on Radioecology, Colorado State University, 1961.** p. 485-496. (eds. Schultz, V. and Klement, A. W.). Reinhold Publ. Corp. and The Amer. Inst. Biol. Sciences, Washington, D. C.

2. Bradford, W. L.
1972 A study on the chemical behavior of zinc in Chesapeake Bay water using anodic stripping voltammetry. Tech. Report 76, Chesapeake Bay Institute, The Johns Hopkins University, Baltimore, Maryland.

3. Brust, H. F. and Newcombe, C. L.
1940 Observations on the alkalinity of estuarine waters of the Chesapeake Bay near Solomous Island, Maryland. **Jour. Mar. Res.,** 3: 105-111.

4. Caspers, H.
1967 Estuaries: Analysis of definitions and biological considerations. In **Estuaries** (ed. Lauff, G. H.), Am. Assoc. Adv. Sci. Publ. 83. Washington, D. C.

5. Durum, W. H., Heidel, S. G., and Tison, L. J.
1960 World-wide runoff of dissolved solids. Int. Assoc. Sci. Hydrology, I.V.G.G., Pub. No. 51, p. 618-628.

6. Durum, W. H., Hem, J. D., and Heidel, S. G.
1971 Reconnaissance of selected minor elements in surface waters of the United States, October 1970. Geo. Survey Circular 643, Washington, D. C.

7. Fitzgerald, W. F.
1970 A study of certain trace metals in sea water using anodic stripping voltammetry. Doctoral dissertation, Mass. Inst. Technology and Woods Hole Oceanogr. Inst., January 1970.

8. Gripenberg, S.
1937 The determination of excess base in sea water. Internat. Assn. Phys. Oceano., Union Géod. and Géophys. Internat., Proc-Verb., 2: 150-152.

9. Gripenberg, S.
1960 On the alkalinity of Baltic waters. **Jour. Cons. Int. Explor. Mer.,** 26: 5-20.

10. Ketchum, B. H.
1951 The flushing of tidal estuaries. **Sewage Ind. Wastes,** 23: 198-209.

11. Livingstone, D. A.
1963 Data of Geochemistry Chapter G, Chemical Composition of Rivers and Lakes, Geological Survey Professional Paper 440-G, Washington, D. C.

12. Lowman, F. G., Rice, T. R., and Richards, F. A.
1971 Accumulation and redistribution of radionuclids by marine organisms. In **Radioactivity in the Marine Environment,** National Academy of Sciences, Washington, D. C.

13. Marvin, K. T., Proctor, Jr., R. P., and Neal, R. A.
1970 Some effects of filtration on the determination of copper in freshwater and salt. **Limnol. Oceanogr.,** 15: 320-325.

14. Matson, W. R.
1968 Trace metals, equilibrium and kinetics of trace metal complexes in natural media. Doctoral dissertation, Mass. Inst. Tech., Dept. of Chemistry, January 1968.

15. Mitchell, P. H. and Rakestraw, N. W.
1933 The buffer capacity of sea water. Biol. Bull., 65, 437-442.

16. O'Conner, J. T.
1968 Fate of zinc in natural surface waters. Civil Engineering Studies, Sanitary Eng. Ser. No. 49. Dept. of Civil Eng., University of Illinois, Urbana.

17. Osterberg, C., Cutshall, N., Johnson, V., Cronin, J., Jennings, D., and Frederick, L.
1966 Some non-biological aspects of Columbia River radioactivity. In **Symposium on the Disposal of Radioactive Wastes into Seas, Oceans and Surface Waters,** pp. 321-333. International Atomic Energy Agency, Vienna.

18. Pritchard, D. W.
1967 What is an estuary: Physical viewpoint. In **Estuaries** (ed. Lauff, G. H.). Am. Assoc. Adv. Sci. Publ. 83, Washington, D. C.

19. Pritchard, D. W.
1967 Observation of Circulation in Coastal Plain Estuaries. In **Estuaries.** pp. 37-44. (ed. Lauff, G. H.), Am. Assoc. Adv. Sci. Publ. 83, Washington, D. C.

20. Schubel, J. R.
1968 Suspended sediment of the northern Chesapeake Bay. Chesapeake Bay Inst., The Johns Hopkins Univ., Tech. Rept. No. 35. Reference 68-2.

21. Slack, K. V. and Feltz, H. R.
1968 Tree leaf control on low flow water quality in a small Virginia stream. **Environ. Sci. Tech.,** 2: 126-131.

22. Turekian, K. K. and Scott, M. R.
1967. Concentrations of Cr, Ag, Co and Mn in suspended material in streams. **Environ. Sci. Tech.,** 1: 940-942.

23. Turekian, K. K.
1971 Rivers, Tributaries and Estuaries. In **Impingement of Man on the Oceans** (ed. Hood, D. W.), Wiley-Interscience, New York.

PART II.

BIOLOGY: DYNAMICS OF FOOD WEBS IN ESTUARIES

Convened By:
John Costlow, Director
Duke University Marine Laboratory
Beaufort, North Carolina 28516

DETRITUS PRODUCTION IN COASTAL GEORGIA SALT MARSHES[1]

Robert J. Reimold, John L. Gallagher,

Rick A. Linthurst and William J. Pfeiffer[2]

ABSTRACT

The change in quantity of dead material and the detritus production flux (instantaneous rate of disappearance) were measured at four-week intervals in three plant stands in Georgia salt marshes. Detritus production (areal rate of disappearance) was calculated from these data. The average standing crop of dead material was highest in *Juncus roemerianus* and lowest in short-form *Spartina alterniflora*. The detritus production fluxes were an average of 7 mg g^{-1} day^{-1} for tall-form *S. alterniflora*, 18 mg g^{-1} day^{-1} for short-form *S. alterniflora*, and 7 mg g^{-1} day^{-1} for *J. roemerianus*. The average monthly detritus production of tall-form *S. alterniflora* (197.9 g m^{-2}) was significantly greater than short-form *S. alterniflora* detritus production (113.6 g m^{-2}). The average monthly *J. roemerianus* detritus production (188.4 gm^{-2}) was significantly greater than that in short-form but not that in tall-form *S. slterniflora.* Mean annual detritus production from the aerial portion of the plants (weighted for the percentage of the watershed occupied by each stand) was 1845.8 g m^{-2} $yr.^{-1}$.

INTRODUCTION

Primary production in coastal Georgia marshes centers around *Spartina alterniflora* Loisel., salt marsh cordgrass, and *Juncus roemerianus* Scheele, black rush. These two plant species account for between 70% and 90% of the primary

1. Contribution No. 275, The University of Georgia Marine Institute.

2. The University of Georgia Marine Institute, Sapelo Island, Georgia 31327.

productivity in Georgia salt marshes (10, 11, 13). This production results in biogenic material which is a potential energy source for consumers entering the estuarine detritus food web via tidal action rather than part of the grazer food chain (1, 2, 5, 12, 18).

A recent summary of detritus in aquatic systems (8) stressed the importance of particle-size distribution. Lenz (6) considered detritus to be particles between 1 μm and 300 μm in size. In coastal systems where there is an inverse relationship between the number of particles per unit volume and particle size, detritus accounts for about 90% of the particulate matter. The detritus serves as an energy source which is consumed at a slow rate. Saunders (15) has characterized the decomposition of detritus as a bimolecular, second-order reaction where the rate of decomposition is a function of detritus concentration and fungal and bacterial enzyme concentration, while dependent on temperature and nutrient concentrations.

Approximately 90% of the aerial primary production of salt-marsh plants enters the detritus food web when the plants die (16). The production of dead material by these plants represents the potential detritus input to the estuarine system. The instantaneous disappearance rate from the marsh surface is a measure of the rate at which each gram of dead material is entering the detritus food web (detritus production) either through microbial growth, tidal harvest, or incorporation into the soil system.

The objectives of the research reported here were to: 1) measure the change in quantity of dead material in three stands of marsh plants; 2) determine detritus production flux (the instantaneous rate of disappearance); and 3) compute the detritus production (aerial rate of disappearance).

METHODS

Three areas for study were chosen in coastal Georgia. Using aerial photographic techniques (14), areas of short-form *S. alterniflora,* tall-form *S. alterniflora,* and *J. roemerianus* were selected in the Duplin Estuary watershed, Sapelo Island, Georgia (Fig. 1). In a pure stand of short-form *S. alterniflora,* a plot 50 by 10 meters was delineated; in *J. roemerianus,* the plot was 60 by 8 meters; and the tall-form *S. alterniflora* plot was a 180-meter-long band extending along the banks of a major stream. Samples were collected at four-week intervals; sample locations within each plot were chosen using random coordinates (7). All vegetation was harvested at ground level using pruning shears or dissecting scissors. Plants were dried to constant weight at 100 °C in a forced-draft oven; wet and dry weights were recorded (9). Dry plant tissue was ground in a Wiley mill (40 μm mesh) for mineral analyses by the methods of Jones and Warner (4).

The paired-plot method of measuring mortality and instantaneous

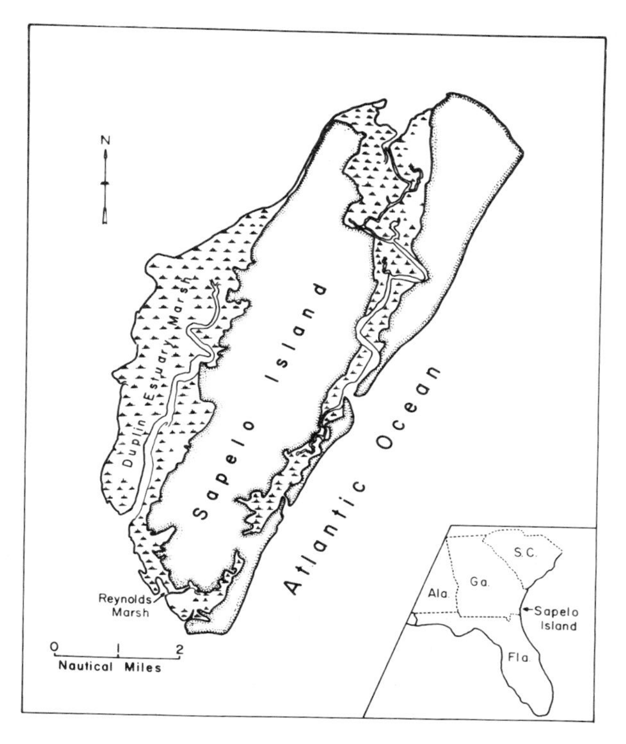

Figure 1. Duplin Estuary, Sapelo Island, Georgia.

disappearance rates (detritus production flux) was selected (20). In each plant stand at each harvest interval, six replicates were harvested. Each replicate consisted of two square quadrats side by side with a common border between them. Both living and dead material were harvested from one quadrat; only the living was removed from the adjacent quadrat. Optimum quadrat size determined previously (19) was 0.25 m^2 for short-form *S. alterniflora* and *J. roemerianus* and 1.0 m^2 for tall-form *S. alterniflora.*

The instantaneous rate of disappearance (detritus production flux) was computed using the equation:

$$r = \frac{\ln (W_0/W_1)}{t_1 - t_0} \tag{1}$$

where r = detritus production flux in milligrams per gram per day; W_0 = weight of dead material at time t_0; and W_1 = weight of dead material at time t_1 (time in days). Equation 1 is based on the assumptions that the biomass of the two paired quadrats was identical and that the rate of disappearance from the two quadrats was equal.

Detritus production per month was computed using:

$$x_i = [(a_i + a_{i-1}) / 2]\ r_i t_i \tag{2}$$

where a_{i-1} = the standing crop of dead material (in grams per m^2) at the beginning; a_i = the standing crop of dead material one month later; r_i = the instantaneous detritus production flux during the interval; and t_i = the time interval in days. Annual detritus production was computed by summing the monthly values.

All statistical computations were conducted according to the methods outlined in Steel and Torrie (17).

RESULTS AND DISCUSSION

The standing crops of both living and dead portions of the plants are depicted for short-form *S. alterniflora,* tall-form *S. alterniflora,* and *J. roemerianus* in Figures 2, 3, and 4, respectively. Regression analyses were used to compute the relationship between wet and dry weights of plant tissue. The results of this relationship are summarized in Table 1. In all instances a significant relationship was found between wet and dry weights of both living and dead material.

In short-form *S. alterniflora*, the amount of standing dead material frequently exceeded the living material (70% of the monthly observations), while the

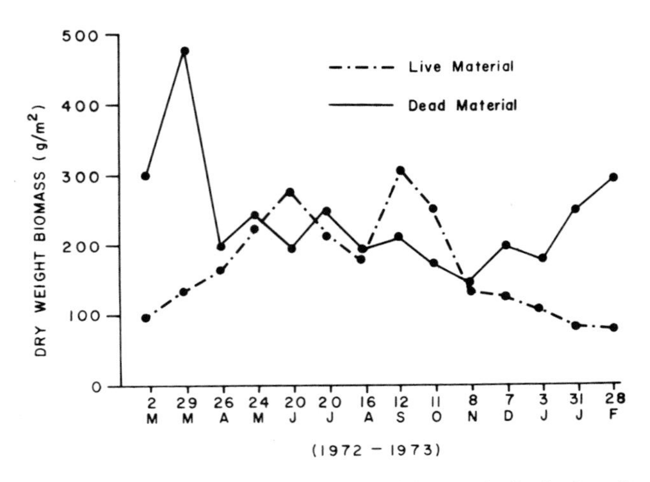

Figure 2. Mean dry weight biomass for both living and dead short-form **Spartina alterniflora.**

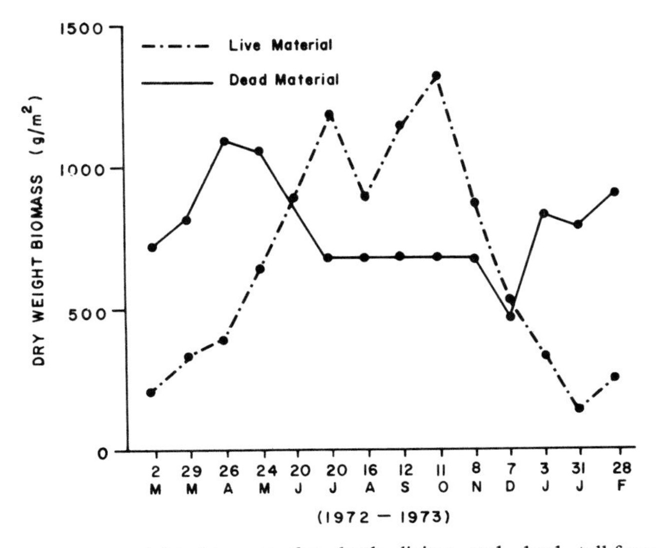

Figure 3. Mean dry weight biomass for both living and dead tall-form **Spartina alterniflora.**

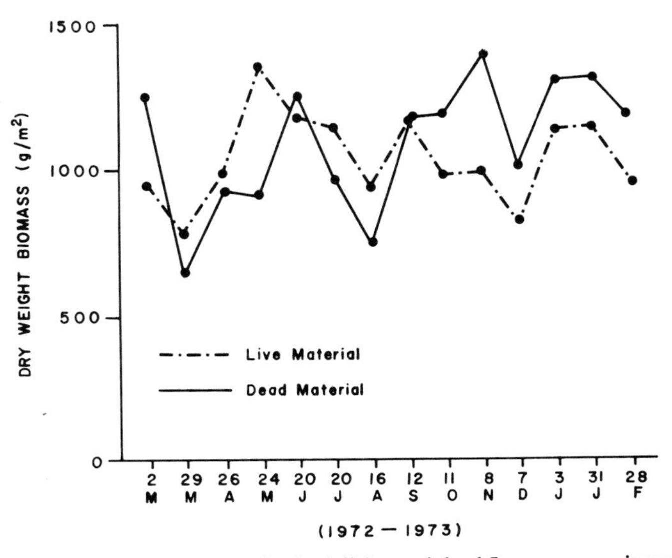

Figure 4. Mean dry weight biomass for both living and dead **Juncus roemerianus.**

TABLE 1.

Regression equations comparing dry weight (Y) and wet weight (X) for two forms of **Spartina alterniflora** and **Juncus roemerianus**. Correlation coefficients (r) are significant at the 99.9% confidence interval. For each of the plant types there are 72 observations.

Plant type and equation	Correlation coefficient
Tall-form **Spartina,** living material $Y = 167.6 + 0.20\ X$	0.83
Tall-form **Spartina,** dead material $Y = 215.8 + 0.25\ X$	0.71
Short-form **Spartina,** living material $Y = 1.00 + 0.34\ X$	0.95
Short-form **Spartina,** dead material $Y = 19.45 + 0.23\ X$	0.51
Juncus, living material $Y = 68.4 + 0.30\ X$	0.99
Juncus, dead material $Y = 137.6 + 0.20\ X$	0.70

amount of standing dead material in tall-form *S. alterniflora* and *J. roemerianus* was about equal to the living material (50% of the monthly observations).

The average standing crop of dead material was highest in *J. roemerianus* and lowest in short-form *S. alterniflora.* On a percentage basis the greatest amplitude in dead standing crop occurred where the biomass was lowest and the least where it was highest.

Figure 5 shows the monthly variation in the potential detritus production flux for all three plant types. Several of the observations (negative values) represent times when tidal action imported additional detritus into the quadrats. These detritus production fluxes were an average of 7 mg g^{-1} day^{-1} for tall-form *S. alterniflora,* 18 mg g^{-1} day^{-1} for short-form *S. alterniflora,* and 7 mg g^{-1} day^{-1} for *J. roemerianus.*

Although flux rates in *J. roemerianus* and tall-form *S. alterniflora* were low, detritus production was high (Fig. 6), and the high rates can be attributed to large biomass (Figs. 3, 4). Short-form *S. alterniflora,* on the other hand, had a relatively low biomass but a flux rate which averaged more than twice that of the other two stands.

A summary of the detritus produced and contributed to the estuary for each plant type is shown Figure 7. The average monthly detritus production of tall-form *S. alterniflora* (197.9 g m^{-2}) was significantly greater (95% confidence

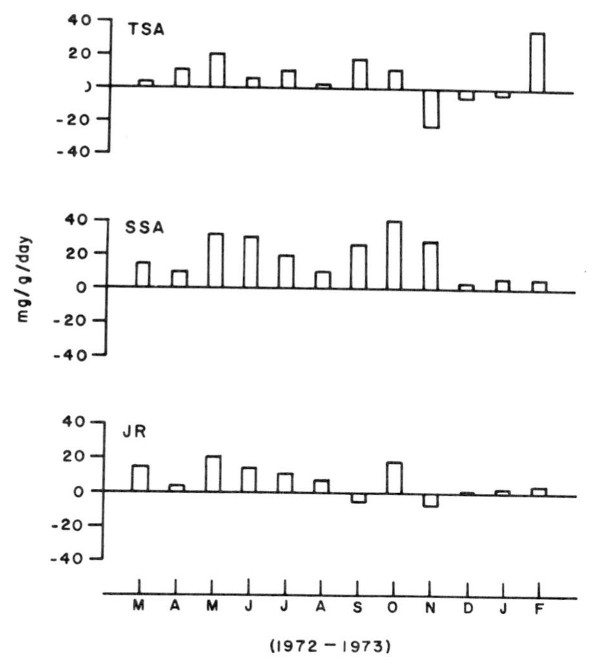

Figure 5. Average instantaneous rate of disappearance of dead material for tall-form **Spartina alterniflora** (TSA), short-form **Spartina alterniflora** (SSA), and **Juncus roemerianus** (JR) based on six observations.

interval) than the short-form *S. alterniflora* detritus production (113.6 g m^{-2}). The average monthly *J. roemerianus* detritus production (188.4 g m^{-2}) was significantly greater (95% confidence interval) than that of short-form *S. alterniflora* but not significantly different from that of tall-form *S. alterniflora.* The monthly variation in detritus production was greatest in tall-form *S. alterniflora.* In both tall and short forms of *S. alterniflora* and in *J. roemerianus,* the late fall and winter months exhibited a detritus production considerably below the monthly average.

A comparison of monthly detritus production of different plants (Fig. 6) contrasts the monthly change in rank of productivity as well as the actual production values. In early spring (March and April) tall-form *S. alterniflora* produced the most detritus per month, while in May, June, and July, *J. roemerianus* was responsible for greatest production. In late fall and early winter, detritus production was at a minimum. In several instances there was a negative value indicating that dead plant material was moved into the collection plot by extremely high spring tides and wind (P. C. Adams, personal communication).

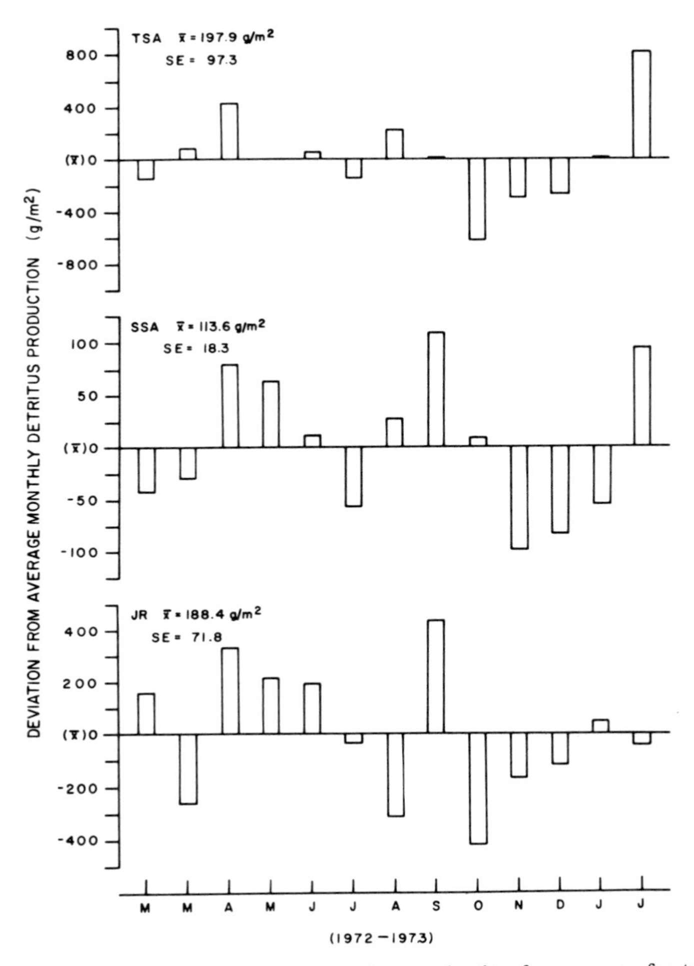

Figure 6. Deviation of individual monthly detritus production for one year for tall-form **Spartina alterniflora** (TSA), short-form **Spartina alterniflora** (SSA), and **Juncus roemerianus** (JR).

Data from Reimold et al. (14) showed that the Duplin watershed was composed of 23% tall-form *S. alterniflora,* 65% short-form *S. alterniflora,* and 12% *J. roemerianus.* The area weighted mean annual detritus production for this marsh watershed was 1845.8 g m^{-2} $year^{-1}$. On a monthly basis the integrated mean detritus production was 141.9 g m^{-2} $month^{-1}$.

In these figures for detritus production from the aerial portion of the plants, only part of the total production is considered since the soil detritus component

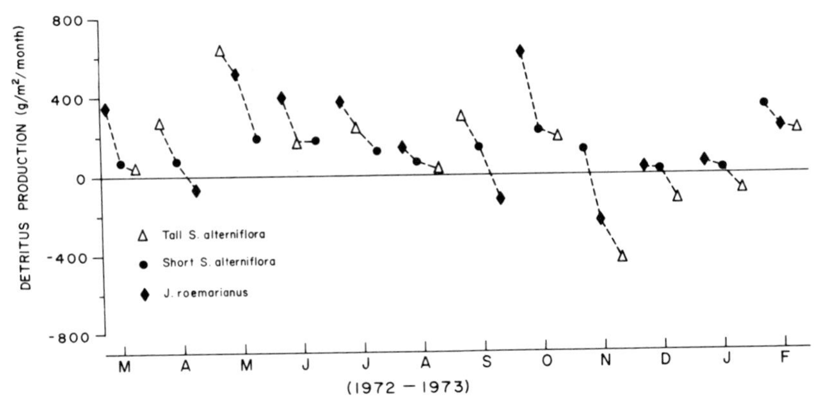

Figure 7. Detrital production on a monthly basis for tall-form **Spartina alterniflora**, short-form **Spartina alterniflora**, and **Juncus roemerianus.**

is neglected. Underground biomass (living and dead) in tall-form *S. alterniflora* is less than in both short-form *S. alterniflora* and *J. roemerianus*, which are similar (3). The pathways of soil detritus are even less well known that those of the aerial portions of the plants.

The comparison of the mineral composition of dead plants in March and August is shown in Table 2. A number of differences between the species and between the seasons are evident. Notable among the species differences are the

TABLE 2.

Comparison of mineral composition of the dead material in three stands of marsh plants in the cool and in the warm season (based on dry weight of six replicates).

Mineral	Tall-form **Spartina alterniflora**		Short-form **Spartina alterniflora**		**Juncus roemerianus**	
%	Mar.	Aug.	Mar.	Aug.	Mar.	Aug.
N	1.0	0.8	0.5	0.8	0.6	0.6
P	0.11	0.10	0.12	0.11	0.08	0.08
K	0.43	0.32	0.32	0.42	0.22	0.25
Ca	0.08	0.13	0.13	0.32	0.14	0.22
Mg	0.23	0.41	0.31	1.09	0.27	0.53
ppm						
Mn	107	109	166	102	242	288
B	25	37	33	50	35	45
Cu	8	7	9	8	6	6
Zn	17	18	14	13	10	9
Mo	12	14	14	16	4	5
Sr	14	27	20	58	22	35

lower phosphorus and molybdenum contents and the higher manganese content in *J. roemerianus.* Calcium, magnesium, boron, and strontium concentrations were higher in August than in March for all species.

Detritus produced in stands of marsh plants varies in many parameters, some of which have been reported here. Standing crops of dead plants were on the average higher in *J. roemerianus* and tall-form *S. alterniflora* than in short-form *S. alterniflora.* Detritus production flux rates were the opposite, the highest rates being found in short-form *S. alterniflora.* Detritus production was highest in tall-form *S. alterniflora* and *J. roemerianus* while short-form *S. alterniflora* produced least. Mineral composition analyses indicated that the material available for the detritus food web differed not only in quantity with species and season but also in quality.

ACKNOWLEDGEMENTS

The extensive field work required to complete this study involved numerous man-hours of work. The authors are especially indebted to: Patrick C. Adams, James R. Duerbig, Ann O. Fornes, Gwen Johnson, Helen D. Walker, Jackie A. Ulmer, Owen M. Ulmer, Jr., and Victoria C. Wray for their unlimited and uninhibited diligence in the soft marsh substrate. We dedicate this paper to them and others who actually "get their feet into the mud" of one of the planet's most productive ecosystems. We also wish to thank the Sapelo Island Research Foundation, Inc., for their support of this research.

REFERENCES

1. Darnell, R. M.
1967 The organic detritus problem. In **Estuaries,** 374-375. (ed. Lauff, G. H.). Am. Assoc. Adv. Sci., Publ. 83. Washington, D. C.

2. Darnell, R. M.
1967 Organic detritus in relation to the estuarine ecosystem. In **Estuaries,** 376-382. (ed. Lauff, G. H.) Am. Assoc. Adv. Sci. Publ. 83. Washington, D. C.

3. Gallagher, J. L.
In press Sampling macro-organic matter profiles in salt marsh plant root zones. **Proc. Soil Sci. Soc. Amer.**

4. Jones, J. B. and Warner, M. H.
1969 Analysis of plant ash solutions by spark-emission spectroscopy. In **Developments in applied spectroscopy,** Vol. 7A: 152-160. Plenum Press, New York.

5. Krey, J.
1961 Der Detritus im Meere. **Jour. Cons. Perm. Int. Explor. Mer.,** 26 (3): 262-280.

6. Lentz, J.
1972 The size distribution of particles in marine detritus. In **Detritus**

and its role in aquatic ecosystems, p. 17-35. (eds. Melchiorri-Santolini, U. and Hopton, J. W.) Mem. Ist. Italiano Idrobiol. (Proc. IBP-UNESCO Symp.) 29, Suppl. 13-16. Pallanza.

7. Lewis, T. and Taylor, L. R.
   1967 **Introduction to experimental ecology.** Academic Press, Inc., New York. 401 pp.
8. Melchiorri-Santolini, U. and Hopton, J. W., (eds.)
   1972 **Detritus and its role in aquatic ecosystems.** Mem. Ist. Italiano. (Proc. IBP-UNESCO) 29, Suppl. 13-16. Pallanza.
9. Milner, C. and Hughes, R. F.
   1968 **Methods for the measurement of the primary production of grassland.** Blackwell Scientific Publications, Oxford, England. 70 p.
10. Odum, E. P.
   1963 Primary and secondary energy flow in relation to ecosystem structure. **Proc. XVI Intern. Congr. Zool.,** 4: 336-338.
11. Odum, E. P. and de la Cruz, A. A.
   1963 Detritus as a major component of ecosystems. **AIBS Bull.,** 13: 39-40.
12. Odum, E. P. and de la Cruz, A. A.
   1967 Particulate organic detritus in a Georgia salt marsh-estuarine ecosystem. In **Estuaries,** p. 383-388. (ed. Lauff, G. H.) **Am. Assoc. Adv. Sci.** Publ. 83. Washington, D. C.
13. Pomeroy, L. R., Johannes, R. E., Odum, E. P., and Roffman, B.
   1969 The phosphorus and zinc cycles and productivity of a salt marsh. In **Proceedings of the 2nd symposium on radioecology,** p. 412-419. (eds. Nelson, D. J. and Evans, F. C.) (U. S. Atomic Energy Commission Conf. 670503.) Cleaning house for Federal Scientific and Technical Information, U. S. Dept. Commerce, Springfield, Virginia.
14. Reimold, R. J., Gallagher, J. L., and Thompson, D. E.
   1973 Remote sensing of tidal marsh. **Photogram. Eng.,** 39 (5): 477-488.
15. Saunders, G. W.
   1972 Summary of the general conclusions of the symposium. In **Detritus and its role in aquatic ecosystems,** p. 533-540. (eds. Melchiorri-Santolini, U. and Hopton, J. W.) Mem. Ist. Italiano Idrobiol. (Proc. IBP-UNESCO Symp.) 29, Suppl. 13-16. Pallanza.
16. Smalley, A. E.
   1959 The role of two invertebrate populations, **Littorina irrorata** and **Orchelimum fidicinium,** in the energy flow of a salt marsh ecosystem. Doctoral dissertation, Univ. Georgia, Athens. 126 p.
17. Steel, R. G. D. and Torrie, J. H.
   1960 **Principles and procedures of statistics.** McGraw-Hill Book Co., Inc., New York. 481 p.
18. Teal, J. M.
   1962 Energy flow in the salt marsh ecosystem of Georgia. **Ecology,** 43: 614-624.
19. Wiegert, R. G.
   1962 The selection of an optimum quadrat size for sampling the

standing crop of grasses and forbes. **Ecology,** 43: 125-129.

20. Wiegert, R. G. and Evans, F. C.
1964 Primary production and the disappearance of dead vegetation on an old field in southeastern Michigan. **Ecology,** 45 (1): 49-63.

MICROBIAL ATP AND ORGANIC CARBON IN SEDIMENTS OF THE NEWPORT RIVER ESTUARY, NORTH CAROLINA

Randolph L. Ferguson and

Marianne B. Murdoch[1]

ABSTRACT

The standing crops of carbon in heterotrophic and autotrophic microorganisms in the sediments and in the water of the Newport River estuary were estimated from adenosine triphosphate (ATP) and chlorophyll *a* measurements. ATP was extracted with boiling tris (hydroxymethyl) aminomethane buffer. The extraction efficiency of ATP from the sediment was 47 $\pm$ 15% ($\pm$ SD). ATP is strongly absorbed to acid- washed clay but the extractibility of ATP from natural sediment was independent of clay content. Instrumentation for measuring as little as 5 x 10^{-10}g ATP ml^{-1} of sample extracts and a hand-coring device for collecting shallow water sediment samples are described.

The increase in standing crop of heterotrophic microbes from winter to summer over the entire estuary, 8.4-20.9 g C m^{-2} in the upper 15 cm of sediment, was associated with a drop in detritus and an increase in the macroscopic infauna. Implications of the seasonal distribution of carbon in the detritus, microbe, and infauna compartments are discussed relative to apparent carbon flows through the detritivore food web.

1. National Marine Fisheries Service, Atlantic Estuarine Fisheries Center, Beaufort, North Carolina 28516.

2. Mention of manufacturer or trade name of a product does not imply endorsement by the Federal Government.

INTRODUCTION

Heterotrophic microorganisms are a major component of estuarine ecosystems and are found predominantly in the sediment. Included in this heterogeneous group are: bacteria; fungi; protozoans; and small metazoans. This group is often subdivided based upon arbitrary size catagories (16, 26) as well as along major taxonomic lines. Both of these approaches have been used to compartmentalize microbial species but these approaches do not necessarily yield categories that aid in our understanding of the complex functional interrelationships of these species (6). The term *heterotrophic microorganism* will be used here to refer not only to those microorganisms generally considered to be heterotrophic, such as the saprophytic bacteria and fungi, the protozoans and the microbial metazoans, but also to the chemosynthetic and phototrophic bacteria which, strictly speaking, are secondary producers that use energy originally derived from green plants (13, 34, 35). Heterotrophic microorganisms also include those benthic microalgae which are living heterotrophically at aphotic depths in the sediment (23, 25). Microalgae in the water and at photic depths in the sediment are autotrophic microorganisms. Microbes are present in the water column of aquatic ecosystems, but their biomass per unit volume of sediment can be greater by several orders of magnitude (53). In estuarine ecosystems the role of the microorganisms of the sediment is particularly important because their biomass greatly exceeds the biomass of microorganisms in the shallow overlying water.

The energy requirements of the heterotrophic microbes in the sediment are supplied by autotrophic microbes in the sediment and by allochthonous detritus, the latter being by far the most important (12). The depth into the sediment to which primary production occurs depends upon the penetration of light. This depth is 3-5 mm in most sediments (14, 39). Methodology for measuring the primary production of benthic microalgae has not been standardized and available estimates from shallow marine sediments vary greatly. They range from 4-420 g C m^{-2} yr^{-1} for submerged sediments (2, 36). Intertidal sediments of a Georgia salt marsh had a mean of 200 g C m^{-2} yr^{-1} (32), which is probably higher than the upper limit for microalgae production in continuously submerged sediments because the overlying water reduces light intensity. High estimates of production obtained from grossly disturbed subtidal sediments are more appropriately termed estimates of a potential for production (3, 31), because they might be realized occasionally but only for short periods of time while resuspension of the sediments exposes subsurface microalgae to light. Reliable measurements of the production of benthic microalgae in the sediments of the Newport River estuary have not been made, but this production might be significant (46).

In most estuarine ecosystems a major portion of total primary production is contributed by macroscopic plants, either submerged or in salt marshes, which are not used directly by estuarine macrofauna (10, 30, 47, 49). Although most high marsh production is used within the marsh (50), production from benthic and low marsh plants enters the estuarine food web as plant detritus (27, 40). Detritus includes all biogenic material in various stages of microbial decomposition representing potential energy sources for consumer species (8). Additional detritus from phytoplankton production in the estuary and detritus derived from freshwater-marsh and terrestrial production also reaches estuarine sediments.

The microorganisms interact with species in the detritivore food web of estuarine systems. This food web is important in estuaries since many macroscopic invertebrates and vertebrates consume particulate detritus (9, 29). The nutritive value of this detritus to the macrofauna varies with time in part as a function of the biomass of associated microorganisms and their activity (4, 28), and both detritus and its associated microorganisms, particularly bacteria and fungi, are probably used by detritivore species. However, fine detritus consisting of organic matter most resistant to microbial decomposition contains no microorganisms (45) and is probably not digestible by most species considered to be detritivores (30). Thus, while the impact of heterotrophic microbes on the production of commercially important estuarine fish and shellfish is profound, it is also contradictory. On one hand, heterotrophic microbes compete with macroscopic detritivores as primary consumers of the readily available protein and soluble extractives of detritus, while, on the other hand, they make plant detritus more useful to detritivores by converting its cellulose and lignin to a more readily digestible form, that is, to microorganisms.

Members of the sediment microbial community differ according to trophic function. Bacteria and fungi convert detritus into their own substance and into simpler organic and inorganic compounds (53). The other heterotrophic microbes consume, not plant detritus, but bacteria, fungi, autotrophic microbes, and each other (8). Although the interrelationships of microbes in the sediment are undoubtedly complex, the only way to approach an understanding of the functional role of microorganisms in the estuarine system may be by simplifying them. Sokolova (33), for example, has demonstrated the positive correlation between production and activity of the bacteria and the biomass of 0.5-5 mm invertebrates. But what about the protozoans and the metazoans smaller than 0.5 mm? Present standard methods for quantifying biomass in marine sediments do not measure at least 90% of the protozoans and fragile metazoans (8). Somewhat more recent discussions of sampling methodology show no major advances (44).

The standing crop of detritus in the Newport River estuary varies seasonally and is located predominantly in the sediment. The death of macroscopic plants in fall and winter supplies a large quantity of plant matter to the detritus standing crop of the estuary at these times of year (42, 49; Thayer, Adams and LaCroix, this symposium), while the detritus is used most rapidly in the spring and summer (4). Detritus that remains suspended in the water can be lost from the estuary by the flushing of estuarine water as fresh water enters the estuary and flows through it into the ocean, and by the flushing that tidal oscillations cause in water volume in the estuary.

The standing crop of detritus in the sediment has an effect upon the sediment biocenosis. Heterothrophic bacteria reduce their environment (18) and, when supplied with an excess of detritus, have the potential to use oxygen at a rate that can exceed its supply from the overlying water (53). In the absence of oxygen, fermentation reactions characterize bacterial metabolism (53), and reduced organic compounds accumulate in the sediment. As a result, anaerobic bacteria will dominate the sediment then characterized by the presence of black iron sulfide and H_2S (1). On the other hand, the autotrophic microbes, when supplied with sufficient light energy, tend to oxidize their environment by releasing oxygen to the sediment. Thus, the rate of input of detritus relative to the rate of input of oxygen and the penetration of light into the sediment determines the type of biocenosis present in the sediment.

Fenchel (12) has identified two types of sediment microbial communities found in estuarine sediments: the estuarine sand microbiocenosis and the sulfuretum. The estuarine sand microbiocenosis is found in shallow estuaries in oxidized surface sediments, 1-10 cm thick, which are predominantly sand but can be rich in silt and detritus. The sulfuretum is a sediment biotope characterized by high silt and detritus content and by anaerobic and reducing conditions with only a very thin oxidized surface layer less than 1 cm in depth. Schematic representations of the energy flow through these communities and into the fauna are presented in Figure 1. Utilization of detritus in a given sediment will be predominantly aerobic or predominantly anaerobic. An understanding of the structural and functional relationships within these two microbenthic biotopes and among the communities of the estuarine ecosystem as a whole will require comprehensive determinations of biomass, metabolic rate, and trophic relationships in these two sediment communities. These data and the methods required for such comprehensive measurements are not presently available.

The purpose of the work reported here was to develop the methodology required to obtain quantitative estimates of standing crops of carbon within the microbial compartments of the simplified flow diagram of Figure 2. These data are being collected in an on-going seasonal survey of the biomass of microorganisms in the Newport River estuary. This survey is being conducted in conjunction with surveys of macroscopic infauna and nekton biomass. During the winters of 1972 and 1973 and the summer of 1973, we collected data that have provided estimates of the biomass of these compartments and first

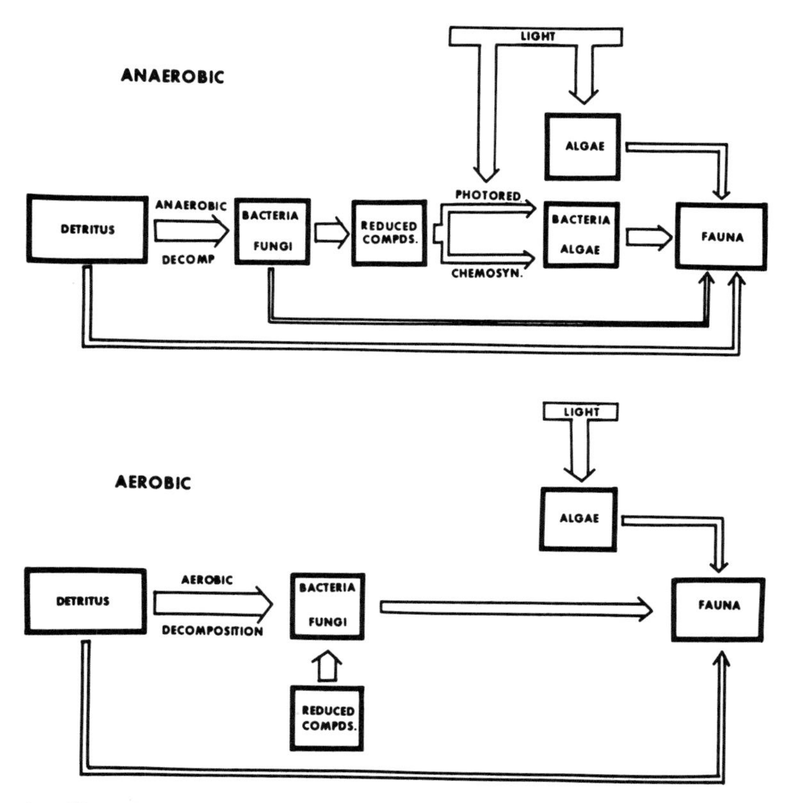

Figure 1. Flow diagram of carbon in anaerobic and aerobic sediment communities (after [12], Fig. 14, p. 47, modified).

approximations to minimum net flows through some of them for the first half of the year 1973.

METHODS OF

ESTIMATION OF MICROBIAL BIOMASS FROM MEASUREMENTS

OF ADENOSINE TRIPHOSPHATE AND CHLOROPHYLL *a*

General Discussion

Total microbial carbon was estimated by determining the adenosine triphosphate (ATP) content of estuarine water and sediment. The rationale for

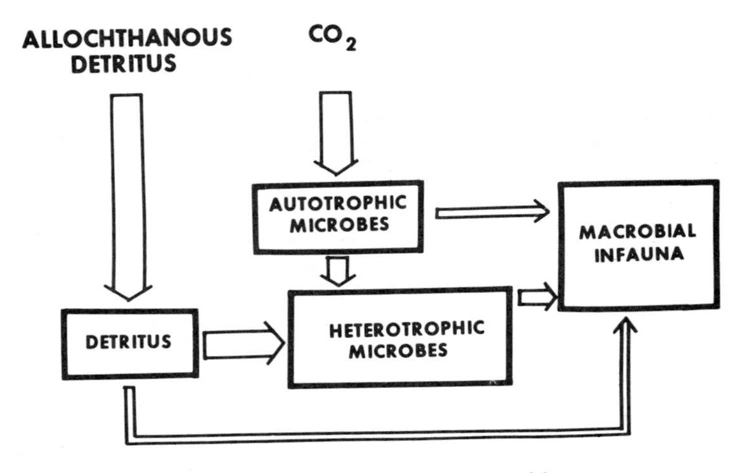

Figure 2. Simple flow diagram of carbon in sediment communities.

using ATP measurement included the facts that ATP is correlated with cell carbon and is not associated with nonliving materials (21). The relationship between ATP and cell carbon has been determined using bacteria, unicellular algae, and zooplankton in which the weight of ATP is about 0.4% of the carbon weight (17, 19) and appears to be a satisfactory way of estimating total microbial biomass (15). Since ATP is present in all living cells, the ATP we measured was contributed by all organisms included in our samples, and the total microbial carbon estimate at the surface of the sediment included heterotrophic and autotrophic carbon.

Autotrophic carbon was estimated independently by fluorometric analysis of chlorophyll *a* (Chl *a*) and phaeophytin. Chl *a* is related to cell carbon by a factor which varies with species, light intensity, and nutrient availability. In estuarine sediment we would expect a factor near 30 would be appropriate to convert chl *a* to cell carbon, since algal populations here are exposed to relatively high nutrient and relatively low light levels (37). Heterotrophic carbon in photiczone sediments was estimated by taking the difference between total and autotrophic biomass.

Sample Collection

Sediments were collected with a small-diameter Plexiglas®[2] coring tube. We used a hand corer (Fig. 3) that can sample in water up to 4.5 m deep. The device, made of stainless steel, aluminum, and polyvinyl chloride, holds a 2.8 cm ID

Figure 3. Coring device for sampling estuarine sediment: (a) technician holding pushrod, (b) union for tubing extension, (c) core sample, (d) stopper poised at end of pushrod, (e) coring tube, (f) coring tube clamp with wing nut.

coring tube by means of a swing-out, stainless-steel clamp tightened with a wing nut. A rubber stopper is held above the tube in a steel clamp, which in turn is attached to a steel push rod. The coring tube was lowered to the sediment on the end of 1 to 3 lengths of aluminum pipe. The tube was gently screwed into the sediment and then stoppered by rotating the push rod and pressing it downward. The sampler collected a core 18-20 cm long plus 50-75 ml of overlying water. This water had an ATP content similar to that in surface-water samples. The core was pushed out of the tube onto aluminum foil, upper surface first, and immediately subsampled at the surface and at 5 and 15 cm from the surface of the core. Each subsample consisted of sediment measured into two 0.1 cm^3 aluminum boats. Two replicate subsamples for both ATP and chl *a* determinations were placed into plastic scintillation vials. All sampling of the sediment from each core was completed within 5-10 minutes from the time the coring tube was inserted into the sediment. From two or four replicate cores were collected at each station. Surface water ATP and chl *a* samples were collected with a bucket at one to three stations per transect, concentrated on glass fiber filters, and extracted.

ATP Extraction

ATP was extracted for 4 minutes using 5 ml of boiling tris (hydroxymethyl) aminomethane buffer, pH 7.75, hereafter referred to as tris. Although extraction of water samples on filters is considered quantitative, variable quantities of ATP may be retained by the sediment following release of ATP from cells killed by the hot extracting solution. In lake sediments, less than 5% of the ATP present was extracted by boiling tris (24). Our sediment samples yielded amounts of ATP which indicated that boiling tris must extract much more than 5% of the total ATP present in estuarine sediments. Furthermore, additions of reagent ATP to estuarine sediment samples during extraction were recovered.

We examined the extraction efficiency by adding reagent or bacterial ATP to mixtures of acid-washed sand and clay, American Standard Kaolin. Bacterial ATP was prepared by lyophilizing the centrifugate of broth (5) cultures of local strains of estuarine bacteria. Kaolin has a strong affinity for ATP (Fig. 4). Ninety-three percent of added reagent ATP was recovered from pure sand, while less than 15% was recovered from a 1% clay-99% sand (by weight) mixture. With 10% clay, less than 1% of added ATP was recovered. Extractions of bacterial ATP did not give the same results. Although recovery dropped as clay content

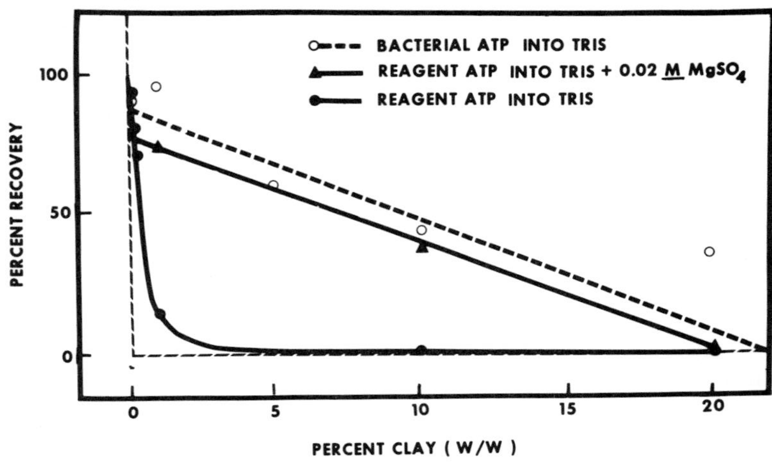

Figure 4. Extraction efficiency of ATP from mixtures of clay (American Standard Kaolin) and acid-rinsed sand.

increased, 47% was recovered with 10% clay. The recovery of bacterial ATP could have been enhanced by competition for adsorption sites on the clay particles by the salts in the preparation, 40% by weight. The bacteria had not been rinsed prior to freezing and drying. We checked the recovery of reagent ATP from clay-sand mixtures using tris plus 0.02 M $MgSO_4$. This compound is a major component of sea water and would supply ions missing from the acid-washed sediment. Average recovery with 10% clay was 41%.

We also examined recovery of ATP additions to estuarine sediments which had been dried and then autoclaved. Ninety-seven percent of our sampling locations had a clay content within the range of 3-10% (43). The recovery of ATP additions averaged 47.3% for the 16 sediments tested (Fig. 5). Recovery could not be related to clay content and was not affected by the origin of ATP added, reagent or bacterial, or by the presence of 0.02 M $MgSO_4$ in the extracting solution. The standard deviation of percent recovery of ATP was 15.

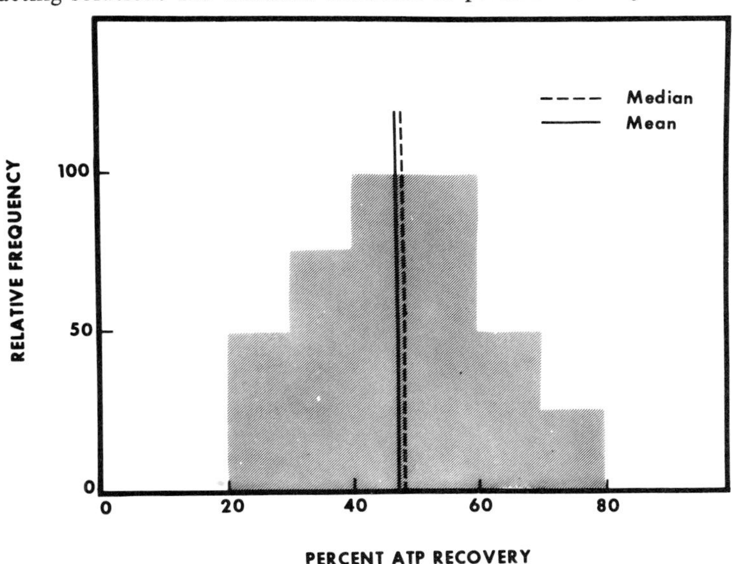

Figure 5. Relative frequency distribution of percent recovery of ATP from estuarine sediments using boiling tris (hydroxymethyl) aminomethane buffer.

Therefore, if we assume a factor of 2.1 to convert extractable ATP to actual ATP present, 97% of the time we will estimate total ATP within a factor of 2 of the true value.

ATP Analysis

ATP was quantified by monitoring light emission resulting from addition of our sample extracts to centrifuged aqueous extracts of lyophilized firefly lanterns (20, 38). We used a tuberculin syringe, fitted with a 14-gauge cannula and shrouded in black cloth to prevent light leakage, to inject 0.8 ml of an ATP sample in tris into 0.2 ml of Sigma® firefly lantern extract (Fig. 6). The force of the sample injection at the bottom of the cuvette thoroughly mixed the sample and reagent. A record of photomultiplier tube output was traced on a 10-inch recorder and the area under the trace was integrated continuously for 30 seconds before and after sample injection (Fig. 7). The difference between the area under the sample trace and the area under the enzyme-extract trace prior to sample injection was the gross sample emission. A large negative gross emission was observed at the highest instrument sensitivity when tris buffer lacking ATP was injected into the firefly lantern extract due to dilution of the extract by the tris. Net sample emission was computed from gross sample emission by subtracting the gross blank emission, which corrected for the reduction in light emission resulting, from dilution of the enzyme extract by the sample.

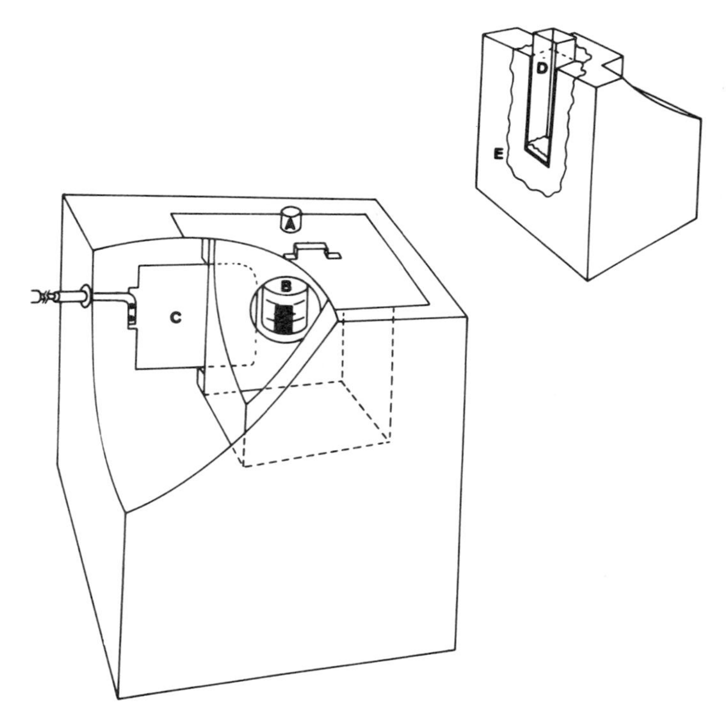

Figure 6. Sample holder for ATP photometer: (a) injection port, (b) photomultiplier tube, (c) shutter, (d) cuvette containing firefly lantern extract, (e) cuvette holder with aluminum foil lining.

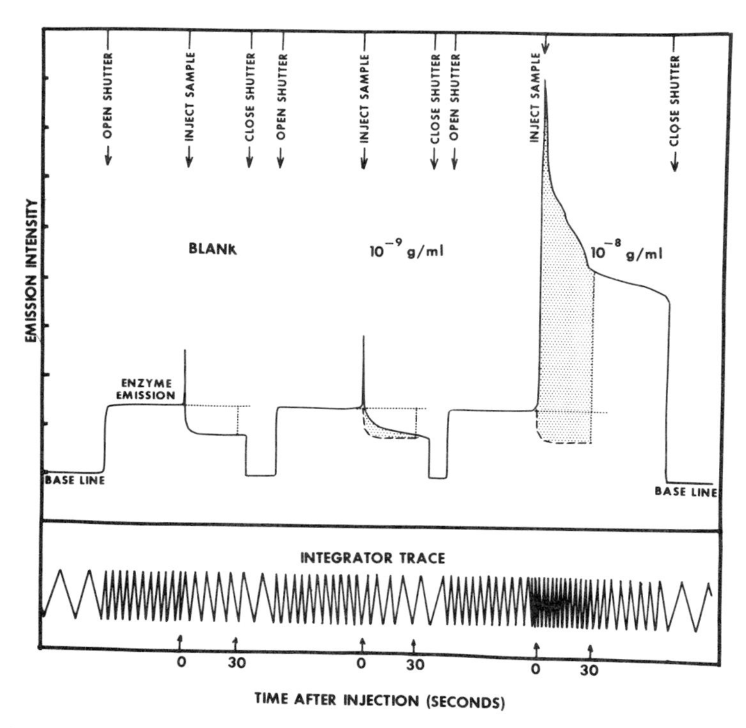

Figure 7. Sequential events and observed light emissions and integrator tracings of ATP additions to firefly lantern extracts.

The logarithm of net sample emission was linearly related to the logarithm of ATP concentration over the range 10^{-9} - 10^{-6} g ATP ml^{-1}. The slope and intercept of this line was dependent upon the particular batch of firefly lanterns and upon the pretreatment of the enzyme extract. The extract was aged 16 hours at 0 °C or for 3-4 hours at room temperature. Uncentrifuged, fresh extracts gave high enzyme traces and were not stable through time. Aged extracts which had been centrifuged gave lower enzyme traces and were relatively stable for 3-6 hours at 0 °C. The calibration curve for each enzyme extract was computed by regressing the logarithm of net emission on the logarithm of ATP concentration at 10^{-9}, 10^{-8}, 10^{-7} and 10^{-6} g ml^{-1}. The mean intercept of 66 log-log (base 10) calibration curves was 13.3 with a standard deviation of 1.04. For these curves the mean slope was 1.24 with a standard deviation of 0.15. Differences in the slope and intercept of these calibration curves were mainly a function of differences in enzyme preparations. Our theoretical limit of detection was equal to 2×10^{-11} g ATP ml^{-1}. Our practical limit of detection with two replicate injections was about 5×10^{-10} g ml^{-1} of extract.

Chlorophyll *a* Extraction

Samples of chl *a* for analysis were placed in 5 ml of 90% acetone which contained suspended $MgCO_3$. Filters were extracted with grinding in a glass-teflon tissue homogenizer. Winter sediment samples were ground in the tissue homogenizer or disrupted in an 80-watt sonic bath for 1 minute at room temperature. We found no difference in the amounts of pigment extracted by the two methods. Both of these extraction procedures are subject to criticism concerning completeness of extraction (7). The summer sediment samples were sonicated 3 minutes using a 1.9 cm ultrasonic probe using 40% full power from a 300-watt amplifier (Artek no. 300 sonic dismembrator). The sample was maintained below 10° C using a salt-ice-water bath.

Chlorophyll *a* Analysis

Chlorophyll *a* was quantified by measuring fluorescence of sample extracts in 90% acetone (22, 38). Fluorescence of the winter samples was measured with a Turner (model 111) fluorometer equipped with a high-sensitivity door, a F4T4-BL lamp, a CS 5-60 excitation filter, and a CS-2-64 emission filter (52). Summer samples were measured with similar instrumentation but a Wratten 47-B excitation filter was used. The ratio Fo/Fa, fluorescence before acidification divided by fluorescence following acidification, was used to compute the relative proportion of chl *a* and its breakdown product phaeophytin. For actively growing phytoplankton cultures this ratio was near 2.1 or 2.0 with the different excitation filters, and near 1.0 if only phaeophytin was present.

Sampling the Estuary

Sediment and water samples were collected in the Newport River estuary along 6 approximately evenly spaced shore-to-shore transects which were sampled at 5 or 6 stations each at 100-300 m intervals across the estuary. Three of these transects were predominantly sulfuretum. The rest were predominantly estuarine sand biocenosis. We sampled these transects from 28 November 1972 to 16 March, 1973, the winter survey, and from 26 June to 18 July 1973, the summer survey.

Biomass Estimation

The ATP and chl *a* data were converted to estimates of total microbial carbon and of autotrophic carbon respectively. The extractible ATP per cm^3 of sediment was converted to total microbial carbon per cm^3 by multiplying by the ATP to carbon factor of 250 and the ATP extraction factor of 2.1. These data were plotted as microbial carbon versus depth for each core and integrated by planimetry to yield g C m^{-2} . Data points were connected by assuming an exponential decrease of carbon between sampled depths for the winter sediment samples. The mean of factors relating mg C m^3 at the sediment surface to g C m^{-2} was 45 $\pm$ 7($\pm$ SD). Data for surface sediments in the summer were converted to microbial carbon per m^2 by this factor. Our limit of detection for total microbial carbon was approximately 0.01 mg C cm^{-3} and 0.1 g C m^{-2}. The estimate of autotrophic carbon per cm^3 at the sediment's surface was converted to carbon per m^2 by assuming that autotrophic carbon was present uniformly to a depth of 5 mm. Heterotrophic microbial carbon per m^2 equaled total microbial carbon per m^2 minus autotrophic carbon per m^2.

Statistical Treatment

The winter and summer ATP data and the chl *a* and Fo/Fa summer data were analyzed using output from the computer program BMDO8V (11) which computed the *F* statistic for determining the significance of factor effects and the estimates of the variance components of the factors. ATP data were converted to logarithms before statistical computations were performed.

RESULTS AND DISCUSSION

All of the factors, transect, depth in sediment and station, produced significant effects ($F, \alpha = 0.01\%$) in both surveys. With the ATP data 73% of the total variation in the data could be attributed to these factors at 15%, 45% and 13% respectively. Total random error resulting from variation between replicate cores, samples, and analyses accounted for only about 4% of the total variation. Total random error in chl *a* and Fo/Fa data were 6% and 2% of the respective total variations. Transect, depth, and station accounted for 59% of the total variation in the chl *a* data or 28%, 21% and 10%, respectively, while these factors accounted for 79% of the total variation in the Fo/Fa data or 57%, 1% and 21%, respectively.

A rough demarcation of the extent of the estuarine sand biocenosis and the sulfuretum of the Newport River estuary is shown in Figure 8. The outlines of these sediment communities are only approximate because they are based upon observations taken at 34 stations or about 1 station per 0.9 km^2. The extent of the sulfuretum in the lower portion of the estuary was based upon one transect in Calico Creek and upon data presented by Fenchel and Riedl (9). Data were not collected within Core Creek. The sulfuretum and sand biocenosis each occupy approximately 50% of the estuary, exclusive of Core Creek.

ATP, chl *a* and Fo/Fa data from the 20 stations within the sand biocenosis were pooled and compared to pooled data from the 14 stations within the

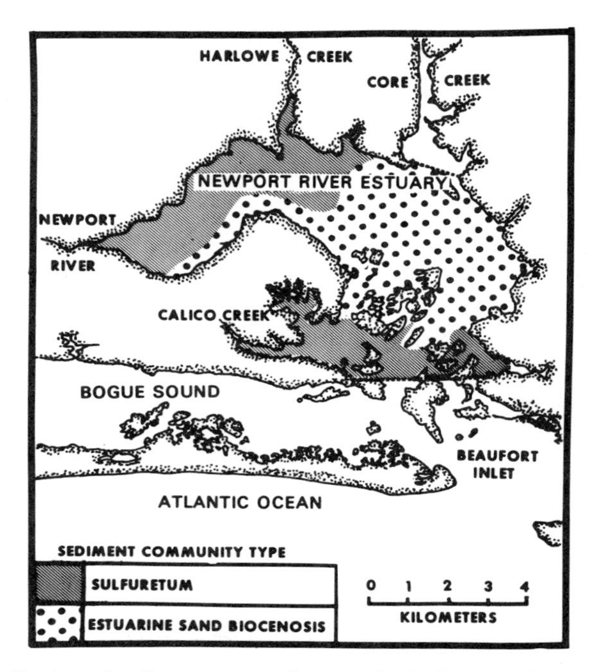

Figure 8. Distribution of sediment community types in the Newport River estuary, North Carolina.

sulfuretum (Table 1). All of the parameters-ATP, chl *a* and Fo/Fa - decrease as depth in sediment increases and are higher in the sand biocenosis than in the sulfuretum. With the change of season ATP increased markedly, particularly in the sulfuretum where ATP values in the winter were lower than the ATP values at 5 cm depth in the sand biocenosis. The surface ATP in summer in the sulfuretum was at about the concentration of the ATP in the surface sediment of the sand biocenosis in winter. The concentration of chl *a* in the surface sediments of the estuary was approximately 1,000 times that observed in the water. Chl *a* in the water was in the range of 2.0-9.3 μ g l^{-1} observed previously (41, 48). However, the Fo/Fa ratios in the sand biocenosis at 15 cm in the sediment and at all depths in the sulfuretum are all well below 1.40, which may indicate some low potential for primary productivity (51) at these depths in these sediments. The ratio observed in surface sediments of the sand biocenosis in winter is just below the range of the ratio commonly observed in the upper photic zone in stratified ocean water (51). The ratios observed in the sand biocenosis at 0 and 5 cm are somewhat below those ratios observed by Pamatmat (31) in the upper 10 cm of sediment in intertidal sand flats. If one estimates the concentration of autotrophic microbial carbon at the surface of

TABLE 1.

Extractible ATP, chlorophyll **a** and Fo/Fa by depth in subtidal sediments of the estuarine sand biocenosis and the sulfuretum of the Newport River estuary during the periods November, 1972 through March, 1973 and during June, July 1973[1]

Depth in sediment cm	Winter			Summer		
	ATP $\mu g/cm^3$	Chl a $\mu g/cm^{-3}$	Fo/Fa	ATP $\mu g/cm^{-3}$	Chl a $\mu g/cm^{-3}$	Fo/Fa
			Estuarine Sand Biocenosis			
0	0.64	7.45	1.75	1.18	3.77	1.66
5	0.16	2.00	1.51	0.32	2.01	1.63
15	0.04	0.60	1.27			
			Sulfuretum			
0	0.13	3.02	1.31	0.62	1.49	1.27
5	0.01	1.21	1.22	0.10	0.67	1.21
15	0.00	0.38	1.18			

[1] ATP data are geometric means; chl **a** and Fo/Fa data are arithmetic means. All comparisons of means wthin two of the factors, season, sediment community and depth in sediment and across the levels of the third factor differ by more than the sum of the 2 respective standard errors, not included in the table, except chl **a** in the sand biocenosis at 5 cm which did not change from winter to summer.

the sediment from the chl *a* concentration and also estimates total microbial carbon from the ATP concentration, the estimate of carbon in autotrophic microbes exceeds the estimate of cell carbon of the total living microorgansims in only one instance, and that is in the surface sediment of the sulfuretum in winter. This sediment had a very low Fo/Fa ratio, 1.31, and given this ratio, the factor of 30 used to convert estimates of chl *a* to estimates of living autotrophic carbon in the sulfuretum may be too great.

The ATP and chl *a* data, converted to estimates of heterotrophic and autotrophic microorganism carbon on a m^2 basis, are shown in Table 2 along with estimates of detrital and macroscopic infaunal carbon. The detritus and macroscopic infauna data were obtained from Thayer, Price, LaCroix, and Montgomery (43, and unpublished data). Infauna data were converted from carbonate-free, ash-free dry weight to carbon weight by multiplying by 0.5. Almost all of the organic carbon in estuarine sediment is present in detritus, and there is about four times as much detritus in the sulfuretum as in the sand

TABLE 2.

Carbon in detritus and in living organisms in the surface 15 cm of subtidal sediments of the Newport River estuary during the periods November 1972 through March 1973 and June-July 1973.[1]

	Estuarine Sand biocenosis		Sulfuretum	
	Winter g C/m^{-2}	Summer g C/m^{-2}	Winter g C/m^{-2}	Summer g C/m^{-2}
Detritus	1152.	888.5	4965.	3478.
Heterotrophic microorganisms	14.2	27.4	2.62	14.4
Autotrophic microorganisms	1.10	0.55	0.45	0.20
Macroscopic infauna				
a. sand biocenosis	2.30	6.55		
b. Upper estuary sulfuretum			0.44	0.34
c. Calico Creek sulfuretum			5.54	20.7

1 Detritus and macroscopic infauna data computed from Thayer, Price, LaCroix and Montgomery (43, and unpublished data).

biocenosis. The largest amount of carbon in living organisms is found in the heterotrophic microorganisms which contain, on the average, about 25 times the carbon present in the autotrophic microorganisms and about 2.6 times the carbon present in the macroscopic infauna.

Measurements of *in situ* primary production by microalga in the surface sediments were not made. Even in the estuarine sand biocenosis the contribution of this production is probably small relative to detritus utilization. Pamatmat (31, Table 4, p. 231) reported chl *a* concentrations in the upper 1 cm of intertidal sand flats in False Bay, Washington, which average 44.2 μg cm^{-3}, a value equal to 8 times the average concentration observed at the surface of subtidal sand in the Newport River estuary (Table 1). The gross primary production in the intertidal sand flats averaged 144 g C m^{-2} yr^{-1} (31, Table 15,

p. 259) assuming a PQ of 1. An estimate of net productivity of autotrophic microbes of the sand biocenosis based on relative chl *a* concentration, assuming net production equals 50% gross production, is 9 g C m^{-2} yr^{-1}, which on the basis of an equal length of time is less than 2% of the net detritus loss of 248 g C m^{-2} that occurred in 183 days in these subtidal sediments (Table 2). Similarly, an estimate of net production by subtidal benthic microalgae in the Newport River estuary is 6.3 g C m^{-2} yr^{-1}, an addition of only 3% to the net annual production in the Newport River estuary (Wolfe, this symposium, Table 1).

The heterotrophic microorganisms increased and the detritus content of the sediment decreased from winter to summer throughout the estuary. In the sand biocenosis the microbes increased 93% and detritus decreased 23%, while in the sulfuretum the microbes increased by 450% and detritus decreased by about 30%. The estimates of minimum trophic efficiency of detritus use are 5.0% and 0.8%, respectively.

In the sand biocenosis and in the embayment of the lower estuary sulfuretum zone into which Calico Creek empties, macroscopic infauna also increased from winter to summer. Overall minimum trophic efficiency of detritus use including the infauna are 6.6% and 1.8%, respectively. If estimated net production by phytoplankton and benthic microalgae are included in these computations, the trophic efficiency estimates decrease to 5.3% and 1.8%. For computation of primary production between surveys we used average chl *a* values for surface sediments and an average of 152 and 183 station-days between surveys in the sulfuretum and sand biocenosis, respectively.

The decrease in the macroscopic infauna in the upper estuary sulfuretum from winter to summer might be due to cropping by bottom-feeding nekton. The biomass of nekton in water overlying the sulfuretum, particularly in the upper estuary, exceeds that of the sand biocenosis (M. Kjelson, personal communication). The apparent trophic efficiency of the sediment community of the upper estuary is only 0.8%. The feeding habits of the juvenile fish and invertebrates which dominate the upper estuary (Peters and Kjelson, this symposium), make it seem probable that these organisms are using the detritus, microorganisms, and members of the macroscopic infauna particularly in this zone.

The difference in trophic efficiency of the two sediment community types might be the result of the relative efficiencies of the detritus-to-microbe link of the sediment community food chain caused by aerobic versus anoerobic breakdown of organic matter. On the other hand these differences might be apparent and might be the result of nekton using sulfuretum benthos to a greater extent.

Assuming that detritus use proceeds at the observed rates for 6 months of the year and that there is no net annual change, the net increase of detritus in estuarine sediments for the remaining 6 months must be about 898.5 g C m^{-2}.

Net production by phytoplankton, *Spartina* and *Zostera* in the estuary is only 204.92 g C m^{-2} yr^{-1} (Wolfe, this symposium), or about 23% of the above value. These data indicate that perhaps as much as 77% of the detrital input to sediments of the Newport River estuary may be derived from primary production in the adjacent sounds (Wolfe, this symposium), from terrestrial production, and from the outfalls of the Morehead City and town of Newport sewage treatment plants which enter the estuary via Calico Creek and the Newport River, respectively.

ACKNOWLEDGEMENTS

The authors thank J. Crosswell, J. Devane, D. Evans, J. Mosser, C. Shelton, C. Turner, and R. Turner, who participated in collecting samples or making analyses during the development or execution of this survey. We also thank G. Thayer, T. Price, M. LaCroix, and G. Montgomery for making available to us unpublished data on detrital carbon and macroscopic infauna biomass in sediments. This research was supported jointly by NMFS and USAEC Agreement No. AT(49-7)-5.

REFERENCES

1. Baas Becking, L. G. M.
1925 Studies on the sulfur bacteria. **Ann. Bot.,** 39: 613-650.
2. Bunt, J. S., Lee, C. C., and Lee, E.
1972 Primary productity and related data from tropical and subtropical marine sediments. **Mar. Biol.,** 16: 28-36.
3. Bunt, J. S., Lee, C. C., Taylor, B., Rost, P., and Lee, E.
1972 Quantitative studies on certain features of Card Sound as a biological system. **Univ. Miami Tech. Rept.,** UM-RSMAS-72011. 13 p. (Unpublished manuscript.)
4. Burkholder, P. R. and Bornside, G. H.
1957 Decomposition of marshgrass by aerobic marine bacteria. **Bull. Torrey Bot. Club,** 84(5): 366-383.
5. Colwell, R. R. and Wiebe, W. J.
1970 "Core" characteristics for the use in classifying aerobic, heterotrophic bacteria by numerical taxonomy. **Bull. Georgia Acad. Sci.,** 28: 165-185.
6. Coull, B.
1973 Estuarine meiofauna: A review, trophic relationships and microbial interactions. In **Estuarine microbial ecology,** p. 499-511. (eds. Stevenson, L. H. and Colwell, R. R.) (Belle W. Baruch Coastal Res. Inst.) Univ. South Carolina Press, Columbia.
7. Daley, R. J., Gray, B. J., and Brown, S. R.
1973 A quantitative, semiroutine method for determining algal and sedimentary chlorophyll derivatives. **Jour. Fish. Res. Board**

Canada, 30(3): 345-356.

8. Darnell, R. M.
   1967 The organic detritus problem. In **Estuaries,** p. 374-375. (ed. Lauff, G. H.) Am. Assoc. Adv. Sci. Publ. 83. Washington, D. C.
9. Darnell, R. M.
   1967 Organic detritus in relation to the estuarine ecosystem. In **Estuaries,** p. 376-382. (ed. Lauff, G. H.) Am. Assoc. Adv. Sci. Publ. 83. Washington, D. C.
10. Dillon, C. R.
   1968 Distribution and production of the macrobenthic flora of a North Carolina estuary. M. A. thesis, Univ. North Carolina, Chapel Hill.
11. Dixon, W. J. (ed.)
   1968 **Biomedical computer programs.** Univ. California Publ. Automatic Computation 2. Univ. California Press, Berkeley and Los Angeles. 600 p.
12. Fenchel, T. M.
   1969 The ecology of marine microbenthos. IV. Structure and function of the benthic ecosystem, its chemical and physical factors and the microfauna communities with special reference to the ciliated protozoa. **Ophelia,** 6: 1-182.
13. Fenchel, T. M. and Riedl, R. J.
   1970 The sulfide system: a new biotic community underneath the oxidized layer of marine sand bottoms. **Mar. Biol.,** 7(3): 255-268.
14. Fenchel, T. and Straarup, B. J.
   1971 Vertical distribution of photosynthetic pigments and the penetration of light in marine sediments. **Oikos,** 22: 172-182.
15. Ferguson, R. L. and Murdoch, M. B.
   1973 On the estimation of microbial ATP in estuarine sediments and water. **Bull. Ecol. Soc. Am.,** 54(1): 27. (Abstract.)
16. Gomoiu, M. T.
   1971 Ecology of subtidal meiobenthos. **Smithson. Contrib. Zool.,** 76: 155-160.
17. Hamilton, R. D. and Holm-Hansen, O.
   1967 Adenosine triphosphate content of marine bacteria. **Limnol. Oceanogr.,** 12: 319-324.
18. Hewitt, L. F.
   1950 **Oxidation-reduction potentials in bacteriology and biochemistry.** E. and S. Livingston Ltd., Edinburgh. 215 p.
19. Holm-Hansen, O.
   1973 Determination of total microbial biomass by measurement of adenosine triphosphate. In **Estuarine microbial ecology,** p. 73-89. (eds. Stevenson, L. H. and Colwell, R. R.) (Belle W. Baruch Coastal Res. Inst.) Univ. South Carolina Press, Columbia.
20. Holm-Hansen, O.
   1972 Total microbial biomass estimation by ATP method. In **Techniques for the assessment of microbial production and decomposition in fresh waters,** p. 71-76 (eds. Sorokin, Y. I. and Kodota, H.) IBP Handbook 23. Blackwell Scientific Publications, Oxford.

21. Holm-Hansen, O. and Booth, C. R.
1966 The measurement of adenosine trophosphate in the ocean and its ecological significance. **Limnol. Oceanogr.,** 11: 510-519.

22. Holm-Hansen, O., Lorenzen, C. D., Holmes, R. W., and Strickland, J. D. H.
1965 Fluorometric determination of chlorophyll. **Jour. Cons. Int. Explor. Mer.,** 30(1): 3-15.

23. Hutner, S. H. and Provasoli, L.
1951 The phytoflagellates. In **Biochemistry and physiology of the protozoa,** Vol. 1, p. 27-121. (eds. Hutner, S. H. and Lwoff, A.) Academic Press, New York.

24. Lee, C. C., Harris, R. F., Williams, J. D. H., Armstrong, D. E., and Syers, J. K.
1971 Adenosine triphosphate in lake sediments: I. Determination. **Soil Sci. Soc. Am. Proc.,** 35: 82-86.

25. Lewis, J. C.
1963 Heterotrophy in marine diatoms. In **Symposium on marine microbiology,** p. 229-235. (ed. Oppenheimer, C. H.) Thomas, Springfield, Ill.

26. Mare, M. F.
1942 A study of a marine benthic community with special reference to the micro-organisms. **Jour. Mar. Biol. Assoc., U.K.,** 25(3): 517-554.

27. Odum, E. P. and De la Cruz, A. A.
1963 Detritus as a major component of ecosystems. **AIBS Bull.,** 13: 39-40.

28. Odum, E. P. and De la Cruz, A. A.
1967 Particulate organic detritus in a Georgia salt marsh-estuarine ecosystem. In **Estuaries,** p. 383-388. (ed. Lauff, G. H.) Am. Assoc. Adv. Sci., Publ. 83. Washington, D. C.

29. Odum, W. E.
1969 Pathways of energy flow in a south Florida estuary. Doctoral dissertation, Univ. Miami, Miami, Florida. 162 p.

30. Odum, W. E., Zieman, J. C., and Heald, E. J.
1973 The importance of vascular plant detritus to estuaries. In **Proceedings of coastal marsh and estuary management symposium,** p. 91-135. (eds. Chabreck, R. H.) Louisiana State Univ., Div. Continuing Education, Baton Rouge.

31. Pamatmat, M. M.
1968 Ecology and metabolism of a benthic community on an intertidal sandflat. **Int. Rev. gesamt. Hydrobiol.,** 53(2): 211-298.

32. Pomeroy, L. R.
1959 Algal productivity in salt marshes. In **Proceedings salt marsh conference 1958,** p. 88-95. (eds. Ragotzkie, R. A., Pomeroy, L. R., Teal, J. M., and Scott, D. C.) (Univ. Georgia Mar. Inst.) Univ. Georgia Printing Dept., Athens.

33. Sokolova, M. N.
1970 Weight characteristics of meiobenthos from different parts of the deep-sea trophic regions of the Pacific Ocean. **Oceanology,** 16(2): 266-272. (Trans. **Okeanologiyi, Akad. Nauk. USSR,** Scripta Technica.)

34. Sorokin, Y. I.

1965 On the trophic role of chemosynthesis and bacterial biosynthesis in water bodies. In **Primary production in aquatic environments,** p. 188-205. (ed. Goldman, C. R.) Univ. California Press, Berkeley.

35. Sorokin, Y. I.
1969 Sampling problems, filtration apparatus and other devices. In **A manual on methods for measuring primary production in aquatic environments, including a chapter on bacteria,** p. 128-151. IBP Handbook 12. Blackwell Scientific Publications, Oxford.

36. Steele, J. H. and Baird, I. E.
1968 Production ecology of a sandy beach. **Limnol. Oceanogr.,** 13: 14-25.

37. Strickland, J. D. H.
1960 Measuring the production of marine phytoplankton. **Bull. Fish. Res. Board Canada,** 122: 1-172.

38. Strickland, J. D. H. and Parsons, T. R.
1968 A practical handbook of sea water analysis. **Bull. Fish. Res. Board Canada,** 167: 1-311.

39. Taylor, W. R.
1964 Light and photosynthesis in intertidal benthic diatoms. **Helgolander wiss. Meeresunters.,** 10: 29-37.

40. Teal, J. M.
1962 Energy flow in the salt marsh ecosystem of Georgia. **Ecology,** 43: 614-624.

41. Thayer, G. W.
1971 Phytoplankton production and the distribution of nutrients in a shallow unstratified estuarine system near Beaufort, N. C. **Chesapeake Sci.,** 12(4): 240-253.

42. Thayer, G. W.
1974 Identity and regulation of nutrients limiting phytoplankton production in the shallow estuaries near Beaufort, N. C. **Oecologia,** 14: 75-92.

43. Thayer, G. W., Price, T. J., LaCroix, M. W., and Montgomery, G. B.
1973 Structural and functional aspects of invertebrate communities in the Newport River estuary. In **Ann. Rept. to the Atomic Energy Commission,** p. 98-139. NMFS Atlantic Estuarine Fisheries Center, Beaufort, North Carolina. (Unpublished manuscript.)

44. Wells, J. B. J.
1971 A brief review of methods of sampling the meiobenthos. **Smithson. Contrib. Zool.,** 74: 183-186.

45. Wiebe, W.
1972 Microorganisms and their association with aggregates and detritus in the sea: a microscopic study. In **Detritus and its role in aquatic ecosystems,** p. 325-352. (eds. Melchiorri-Santolini, U. and Hopton, L. W.) Mem. Ist. Italiano Idrobiol. (Proc. IBP-UNESCO Symp.) 29, Suppl. 13-16. Pallanzo.

46. Williams, R. B.
1970 A general evaluation of fishery production and trophic structure in estuaries near Beaufort, North Carolina. In **Ann. Rept. to the Atomic Energy Commission,** p. 178-216. NMFS Atlantic

Estuarine Fisheries Center, Beaufort, North Carolina. (Unpublished manuscript.)

47. Williams, R. B.
1973 Nutrient levels and phytoplankton productivity in the estuary. **Second coastal marsh and estuary symposium,** p. 59-89. (ed. Chabreck, R. H.) Louisiana State Univ., Div. Continuing Education, Baton Rouge.

48. Williams, R. B. and Murdoch, M. B.
1966 Phytoplankton production and chlorophyll concentration in the Beaufort Channel, N. C. **Limnol. Oceanogr.,** 11(1): 73-82.

49. Williams, R. B. and Murdoch, M. B.
1969 The potential importance of **Spartina alterniflora** in conveying zinc, manganese, and iron into estuarine food chains. In **Symposium on radioecology,** p. 431-439. (eds. Nelson, D. J. and Evans, R. C.) (U. S. Atomic Energy Commission 670503.) Oak Ridge, Tenn.

50. Williams, R. B. and Murdoch, M. B.
1972 Compartmental analysis of the production of **Juncus roemerianus** in a North Carolina salt marsh. **Chesapeake Sci.,** 13(2): 69-79.

51. Yentsch, C. S.
1965 The relationship between chlorophyll and photosyntheitc carbon production with reference to the measurement of decomposition products of chloroplastic pigments. In **Primary productivity in aquatic environments,** p. 323-346. (ed. Goldman, C. R.) Univ. California Press, Berkeley.

52. Yentsch, C. S. and Menzel, D. W.
1963 A method for the determination of phytoplankton chlorophyll and phaeophytin by fluorescence. **Deep-Sea Res.,** 10: 221-231.

53. ZoBell, C. E.
1946 **Marine microbiology.** Chronica Botanica Co., Waltham, Mass. 240 p.

PRELIMINARY STUDIES WITH A LARGE PLASTIC ENCLOSURE

BY

J. M. Davies, J. C. Gamble and J. H. Steele[1]

ABSTRACT

A cylindrical plastic enclosure 3 m diameter x 17 m deep was used to follow temporal changes within a particular body of water for a period of 46 days. The enclosed water column behaved in much the same way as the outside water, although after about 25 days there was evidence of wall-effects.

The large zooplankton bloom that occurred in the bag implies good growth and survival of the herbivores but raises questions about the lack of predators.

The contrast between detritus data from inside the bag and that from outside it suggests that the zooplankton may be reworking the detrital material in the bag.

INTRODUCTION

The use of plastic enclosures in aquatic ecology provides time-series data of changes within a particular body of water rather than changes at a single position or in the neighborhood of a drogue. Two main artificialities are introduced. The first is that enclosures prevent the lateral mixing of populations, which could be an important process in maintaining the general equilibrium of these populations in the open sea. Since the question of the relevance of spatial heterogeneity and dispersal to population stability needs elucidation, the basic experiment with "bags" is to determine the importance of diffusion by studying the consequences of its removal. The second artificiality comes from the wall-effect caused by fouling on the interior of the bag. This cannot be eliminated and we need to determine what effects it has on the interior of the system. Both these problems are related to the size of the bag, through changes in the ratio of

1. Marine Laboratory, PO Box 101, Victoria Road, Aberdeen AB9 8DB, Scotland.

volume-to-surface area. Bags in the sea, such as those used by Strickland and his colleagues (2, 5), can be used to follow the relatively rapid changes in a phytoplankton bloom. At the other extreme, large columns, 45 m diameter and 11 m deep, stretching from surface to bottom, have been used in fresh water by the FBA in Blelhan Tarn (4). The flora and fauna inside remained reasonably similar to those outside for periods of 18 months.

We wished to study changes at several trophic levels in an inshore but relatively marine, rather than estuarine, environment. Both tidal excursion and wave action preclude the type of structure used in Blelhan Tarn. Further, we were particularly interested in exchange and recycling processes between the water and the bottom; thus we wished to isolate the bag from the bottom but to collect the "fall-out" for concurrent experiments at the mud surface. We also wanted to follow the flow of energy inside the bag from primary production to its final removal from the system and to compare it with the outside water column, which had been studied in previous years (6). These factors, together with cost, led to the design of an enclosure which was used for preliminary studies in Loch Ewe during April and May, 1973. The aims were to find how the bag operated as a structure and for how long its contents, particularly the zooplankton, would function as an ecosystem similar to the populations outside it. The results, although not replicated, may be of value to others involved in this type of research in inshore or estuarine environments.

EXPERIMENTAL DESIGN

Ideally, the plastic enclosure should be constructed from a nontoxic, strong yet flexible material with better than 95% light transmission. To minimize possible wall-effects the volume-to-surface area ratio should be as large as possible; however, a compromise must be reached. Since for this experiment we wanted to enclose a sample water column, we used a cylindrical enclosure of 3 m diameter and 17 m depth. It was moored in 30 m of water in an area, Loch Thurnaig, which has been studied previously (6).

The final design was a completely closed system. At the top the cylinder was about one meter below the surface, and a funnel (46 cm diameter) protruded from the surface and was used for sampling. The bottom of the bag was conical, reducing to 46 cm diameter; a replaceable bottom bucket was used to collect the detrital material (Fig. 1). The bag was held in place inside a semirigid frame by elastic straps located radially at one-meter intervals down the length of the bag. The semirigid frame consisted of metal rings, one at the top holding a flotation ring and three at the bottom, connected by seven ropes, the whole being held rigid by 112 kg weights at the end of each rope (Fig. 1).

Both bags were constructed of plastic materials reinforced with nylon mesh: one was made of PVC; the other polyethylene. A first trial bag made of PVC was

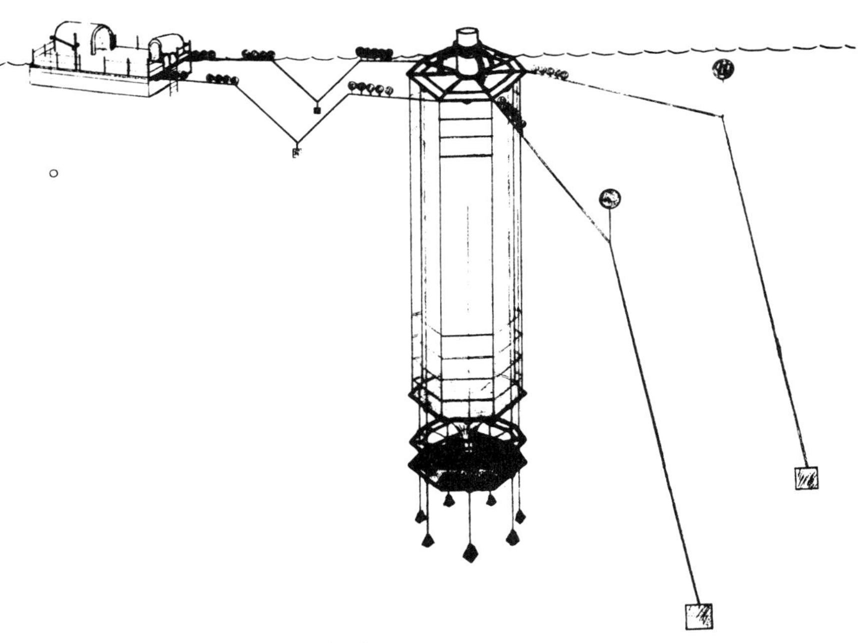

Figure 1. Diagram of the bag and raft

much stronger and easier to work with than the polyethylene, although its light transmittance (≃85%) was not as good as the polyethylene (> 95%). Moreover, PVC is not suitable material biologically, since the phthalates, used as plasticisers, leach out from the material and contribute significantly to the dissolved organic carbon pool. This caused a microbial bloom, measured by heterotrophic production, 7-10 times that outside the bag. Thus polyethylene was used for the experiment described here.

The sampling procedures inside the bag and out were as follows. Profiles of nitrate, nitrite, ammonia, phosphate, silicate, salinity, temperature, chlorophyll, and phaeopigment were measured every day. Zooplankton hauls were taken every second day and particulate carbon and nitrogen measurements, C^{14} production, light meter readings and heterotrophic production measurements were carried out every five days. Detritus that settled out was collected once per week. Radiation, wind speed, and direction were recorded continuously.

The most convenient way to take water samples from the bag would have been to use a pump. However, snatch-bottles were used because the C^{14} production of pumped samples was about 30% lower than in those taken with the bottles, and the heterotrophic production was up to thirty times higher, depending on the phytoplankton biomass.

During the course of the 46-day experiment there were a number of severe storms with winds exceeding 150 km hr^{-1} and waves of 2 m. During one of these storms on 1-2 May, the top of the bag ripped. The tears were repaired as far as possible two days later, but the bag could no longer be considered a completely closed system, since some leakage could occur in the top 3 m. However, the chlorophyll *a* and zooplankton data show that the water exchange was probably fairly small.

The bag was filled initially by pumping into the top; then, with the bottom open, a volume equal to that of the bag was pumped out of the top in an attempt to fill the bag with subsurface plankton populations.

In general, leakage or exchange of water could occur at the top through a joint in the sampling column which was not completely watertight, as well as through small holes along the repaired tears. Leakage also could occur at the bottom seal which, because of the need to remove the sediment bucket, was also not completely watertight. The salinity results (Fig. 2) show that some exchange occurred at the surface but was confined to a shallow layer (the 1.5 sampling position in the bag was situated at the lower end of the sampling column). At the bottom there was also evidence of intrusion of water during the first 10 days, as shown by an increase in salinity. This probably was caused by the lower density of water in the bag indicating that the filling process had not equalized the inside and outside densities.

Changes of temperature in the bag were similar to those outside, implying relatively efficient heat exchange with the outside and resulting in comparable density changes (Fig. 3). There was complete vertical homogeneity at the end of

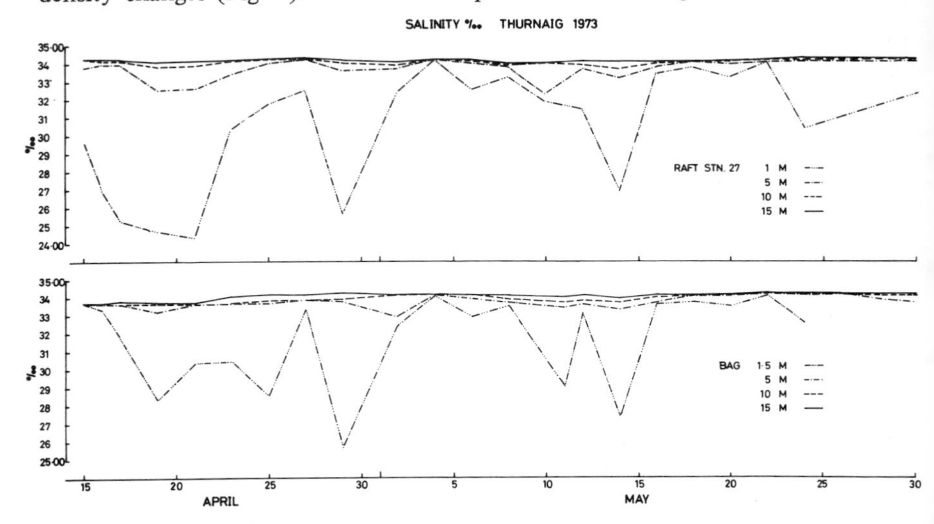

Figure 2. Salinity values in the bag and outside water

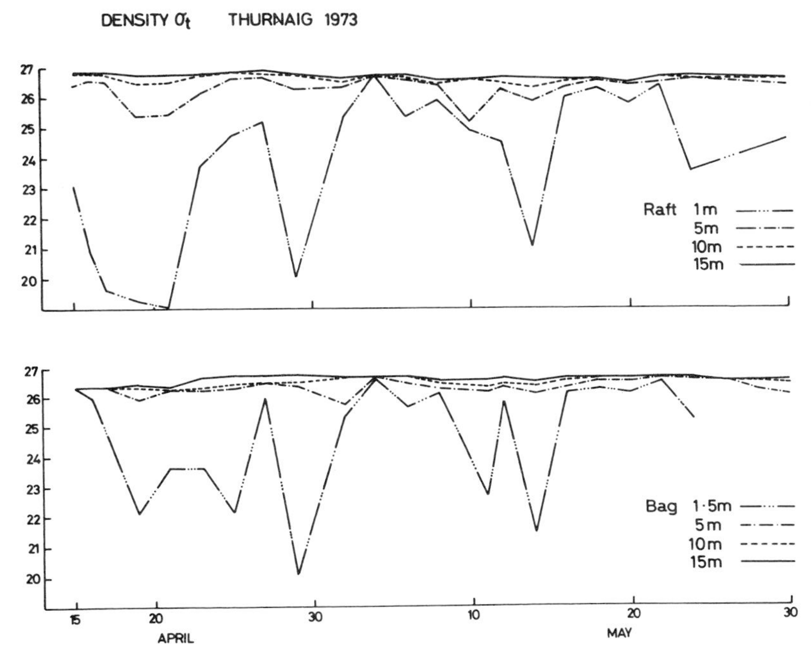

Figure 3. Density changes in the bag and outside water

a storm (4 May) in both environments. As will be seen, this had a considerable effect on the biota.

RESULTS

The biological data will be presented first and then compared with physical and chemical measurements. The vertical chlorophyll profiles showed two main periods occurring in both the bag and outside water (Fig. 4). From 15 April - 3 May outside and 15 April - 4 May inside, a bloom started at the surface and was followed by a midwater maximum at 5 m. The storm conditions on 1 and 2 May, which broke down stratification in both systems, started a second cycle from 4 May onwards, with a general trend from a small surface maximum to a large midwater peak at 10 m during the period 15-22 May outside (corresponding to a peak in radiation), followed by a similar peak 17-26 May inside.

There were marked fluctuations within these general trends. The changes outside the bag could be attributed to lateral movements, but the changes inside are inexplicable even by leakage. Since any leakage occurred near the surface or

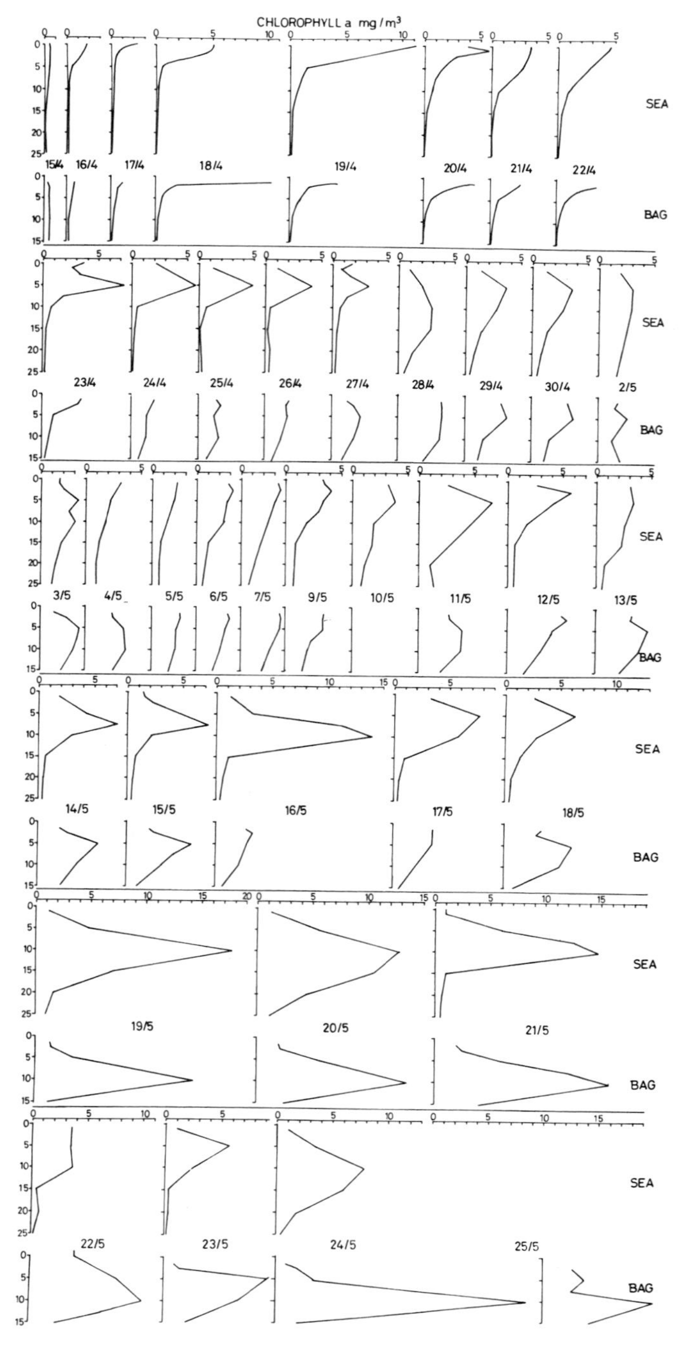

Figure 4. Vertical profiles of chlorophyll **a** in bag and outside water

the bottom of the bag, midwater peaks could not be included by exchange with the outside. These changes in chlorophyll provide the best proof that the bag was responding in a manner not very different from the outside environment. The main changes at the beginning of May could have been associated with the strong vertical mixing which occurred at that time. There was an increase in nitrate in the upper layers of water outside the bag following the heavy swell conditions of 30 April, and the nutrient profile suggests that nutrients were released from the bottom sediment. Similar increase occurred inside but may have been due, in part, to water exchange when the bag was ripped.

The primary production (per m^2 of water surface) in the bag was significantly less than that outside, in part owing to exclusion of the top 1.5 m by the neck of the bag. There was also some production below 15 m outside the bag since the 1% light level in this area was usually found at 20-25 m. The maximum decrease in production, on 9 May, was associated with a heavy growth on the outside of the bag, which significantly reduced light inside the bag. At the beginning of the experiment and after the outside was cleaned, light levels inside and outside did not differ significantly. In general, decreased production within the bag does not seem to have led to significantly lower chlorophyll concentrations or to decreases in herbivore populations.

The numbers of the larger zooplankton in the bag were significantly different from those outside (Fig. 5). For the larger animals there was only one occasion when total numbers inside were less than those outside. This occasion (2 May) occurred just before the bag ripped. It was followed by a rapid increase in numbers in the bag which did not occur outside. The species composition was qualitatively similar throughout the whole period with two species, *Acartia clausi* and *Pseudocalanus elongatus,* forming the major part of the copepod population and two other species making up the remainder. However, in the bag there was a switch on 2 May, *Acartia* being replaced by *Psuedocalanus,* which was the species responsible for the later larger populations in the bag. The large population increase inside the bag was mirrored in the silicate values (Fig. 6) which peaked on 6 May and fell steadily during the rapid zooplankton increase to a minimum on 14-15 May, suggesting very heavy grazing of the phytoplankton. This trend was not evident outside the bag. The qualitative similarity in species composition indicates that, during the period of the experiment, there was no drift to a completely different community as happened in tanks on-shore in Loch Ewe (7). The quantitative differences support the conclusions from the chlorophyll data that there was no significant exchange of water between the bag and the outside except possibly around 2 May.

Differences were also found in the small-mesh collections which showed larger numbers of nauplii in the bag after 28 April. Since there were many gravid females in the bag during the first two weeks of the experiment, it is possible

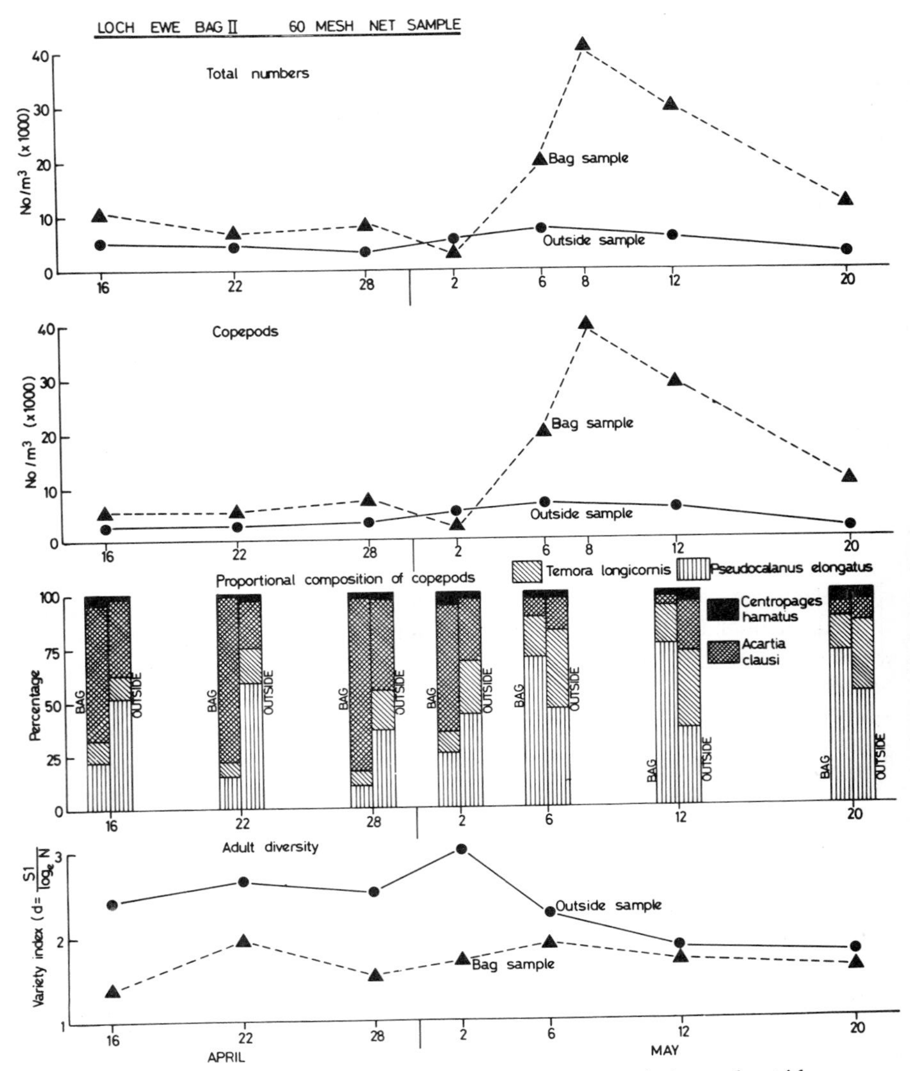

Figure 5. Zooplankton (250 μ mesh net) numbers and species in bag and outside

that the naupliar increase was from eggs produced in the bag during the first few days. However, if, as according to Corkett (3), the developmental time from nauplius I to adult of *Pseudocalanus minutus* is 35 days at 12 °C, the dramatic population increase in copepods caught in 250/μ mesh cannot be totally explained in this way.

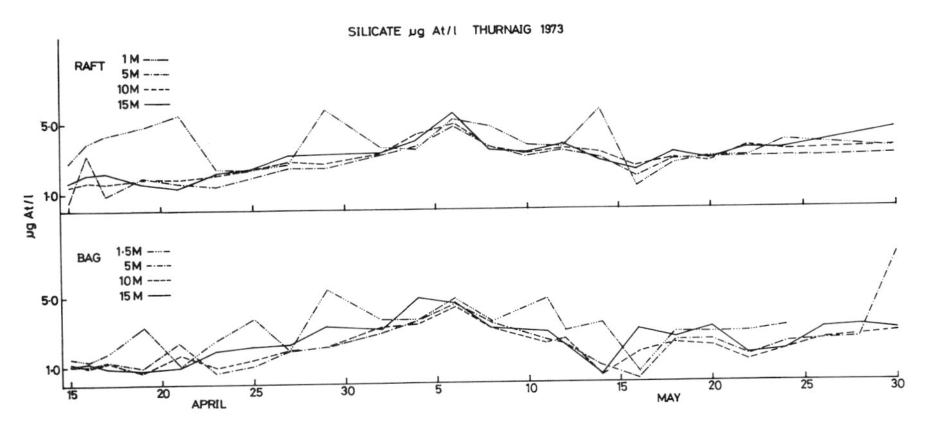

Figure 6. Silicate values in bag and outside water

Although total numbers were small, there were significantly more cuphausiids in the bag than outside, but, since these were at the furcilia stage, they can be considered as herbivores. The other major difference was in numbers of animals that could be considered as predators (ctenophores, medusae and *Sagitta* spp.). Except for the first day of sampling, there were significantly fewer in the bag than outside. The difference indicates that predation on the copepods in the bag probably was at a much lower level than that on the natural populations outside. Low predation could account for the survival of a large number of nauplii leading to a large adult population. The reason for the small numbers of predators is not clear. It may have resulted from avoidance during filling of the bag, or from problems of sampling these animals adequately with a slow-moving net, but the one sample on 16 April suggests that "inside" and "outside" were the same. Another possibility is that these organisms did not survive well in the bag, although the numbers tended to increase with time. This aspect of predator survival is a main subject for future studies.

A further problem is the decrease in numbers of copepods at the end of the experiment. One feature of the detrital material (to be discussed in the next section) was the marked increase in crustacean fragments in the second half of the experiment. This material was not usually found in detrital collections outside the bag (6) and suggests that, in the absence of predators, there may have been natural mortality.

The feeding experiment (Fig. 8) showed that particulate matter inside the bag depressed filtering rates, particularly for larger particles which normally are grazed on most efficiently, as shown by outside animals in outside water. Further, the animals taken from the bag were in poor condition, judging by their lower filtering rates even in outside water. Microscopic examination of the material in the bag water showed that some of the diatom populations (mainly

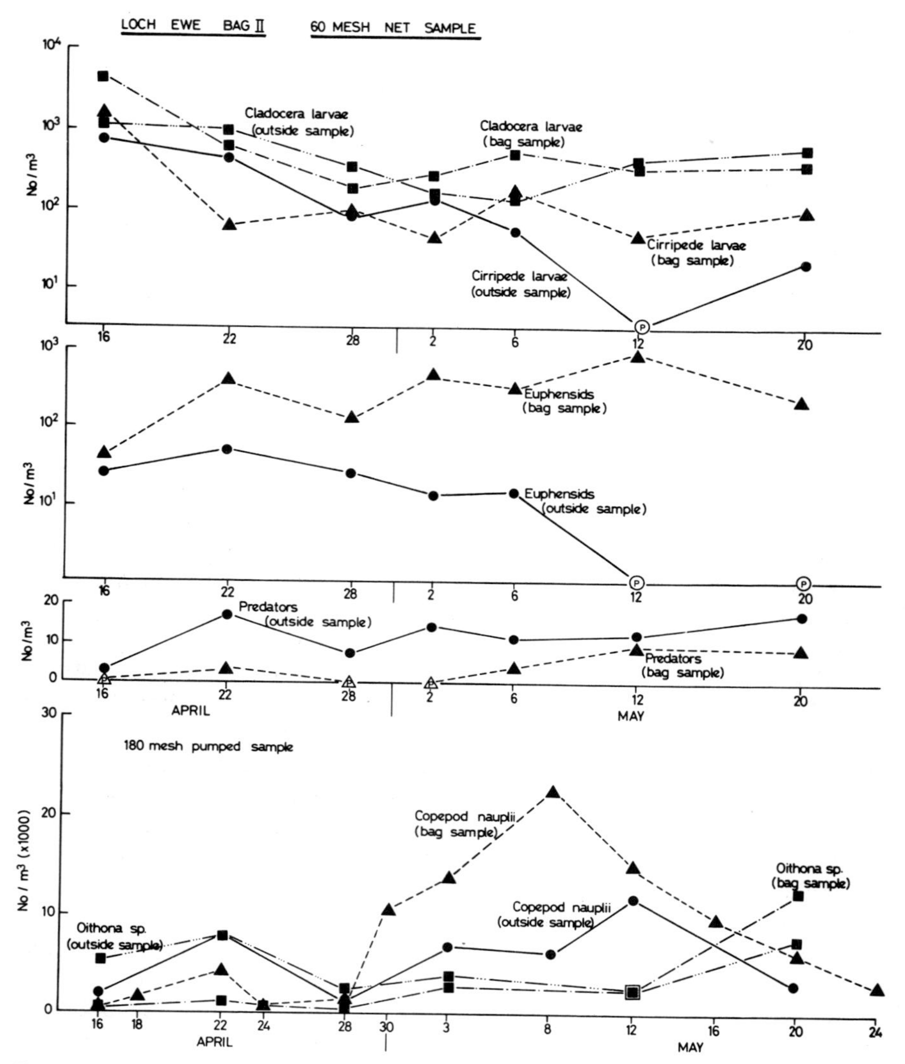

Figure 7. Continuation of 250 μ mesh net data and results of 68 μ mesh net data

Skeletonema and *Chaetoceros*) were mixed in a matrix of gelatinous material which may have originated from the walls of the bag. Although the ripping of the bag immediately afterwards may have exchanged some of the zooplankton, the later development of large populations of both nauplii and copepods suggests that the effect found in this one feeding experiment could not have persisted to the detriment of the plankton.

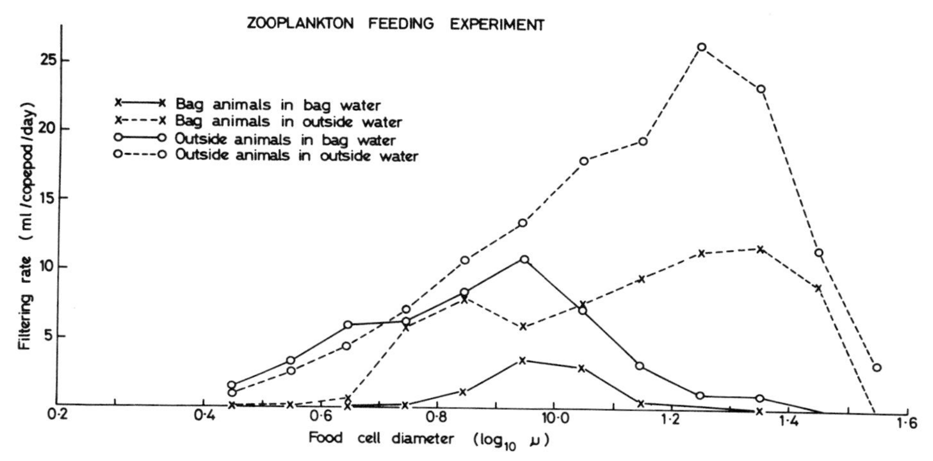

Figure 8. Zooplankton feeding experiment

The detrital material settling out of the water column in the bag was collected once per week, and a string of three sediment jars was used to collect detrital material outside the bag over the same period.

The most interesting aspect was the change in composition of detrital material during the course of the experiment. For the first 25 days the detritus contained about 85% faecal pellets and small amounts of crustacean fragments and benthic diatoms and was very similar in composition to the material collected outside the bag (6). However, after about 25 days the percentage faecal pellets dropped to about 50%, the crustacean fragments increased from 5% to 25% and the percentage benthic or sessile diatoms also increased 10% to 20%. About 5% pelagic diatoms were recovered ungrazed (Fig. 9).

This change in detrital composition within the bag followed the zooplankton peak (8 May) and its subsequent decline, which suggests that about this time events in the bag were being influenced by either wall-effects or the lack of predators, or both.

The other striking difference between the inside and outside detrital material was the percentage of primary production carbon recovered as detrital faecal material. Outside the bag about one-third of the production was recovered as faecal pellets, whereas inside the recovery was as low as 2% and ranged up to 22% only toward the end of the experiment (17-25 May), when the zooplankton numbers had declined. It is unlikely that this was due to inefficient collection, since there was no evidence that material stuck to the sides of the cone. One possible explanation is that the zooplankton reworked the faecal pellets, and there is some evidence to support this idea; in bad weather the material in the bucket was partially resuspended, and on one occasion when the bucket was changed there were large numbers (almost 20 times the natural density) of zooplankton in the bucket.

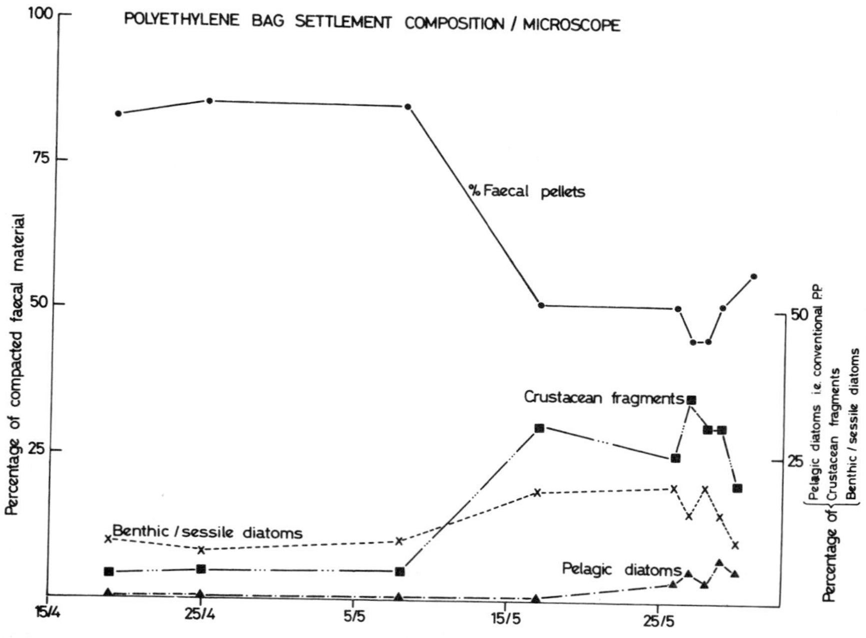

Figure 9. Composition of the detrital material in the bag

There is no reason why zooplankton could not rework fragmented faecal pellets and in fact *Centropages hamatus* and *Acartia tonsa* are both known to be omnivorous (1). Moreover, when we consider the large zooplankton bloom 30 April - 8 May, when copepod numbers increased by 36 x 10^3 m^{-3}, and take an average dry weight of *Pseudocalanus elongatus* as 10 µg and 50% of the dry weight as carbon, then this represents growth of 2,800 mg C m^{-2} of the water column. Since the production for this period was only 3,600 mg C m^{-2} of the water column, this would require a conversion efficiency of 77%.

Normal assimilation efficiency is between 25% and 50% suggesting that the zooplankton might have been reworking faecal pellets, which would help explain the low percentage recovery of detrital material. The fate of the detritus, representing about 30 g C m^{-2} y^{-1} (an average of 3 years' data, [6; Davies, unpublished]), has been studied by measuring the oxygen uptake of the sea bed using translucent plastic chambers. The sea bed in this area (30 m) is compacted, jelly-like mud with a flocculent surface layer. Respiration measurements were made once per month throughout a year, and an integrated oxygen consumption for this period was converted to an equivalent carbon figure using an RQ value of 0.7. The carbon requirement per m^2 per year worked out in this way is 34 g C m^{-2} y^{-1} (Davies, unpublished data), which is not significantly different from the input.

During the course of the experiment, the percentage of benthic and sessile diatoms appearing in the bucket increased, and so at the end of the experiment squares of wall material were taken from three areas (0-6 m, 6-12 m, and 12-16 m) and inspected for fouling. The upper region on the outside was relatively clear of algae but characterized by large numbers of gastropods and bivalves. The gastropods, being grazers, may have helped to keep the sides of the bag clean. The central region outside showed dense amorphous algal growth, probably naviculoid diatoms, with little multicellular growth. The lower region showed less abundant algal growth but numerous colonies of the bryozoan *Electra pilosa*, which are often the starting point for macrophyte growth.

The inside walls of the bag had much less growth($<20\%$) probably because of the smoother inside surface of the polyethylene and the lack of water-flow over the surface.

CONCLUSIONS

Considering the extremely adverse weather conditions, the bag performed reasonably well. Improvements in design are required and a more sheltered site would be an advantage, but the general conclusion is that enclosures of this size can be operated without too great difficulty in marine environments. The main problem was deciding what exchange had occurred when the bag ripped. For this and for possible exchanges at other times, the biological, rather than the physical, data prove better indicators.

Both chlorophyll measurements and zooplankton counts indicate that, except for a short period around the beginning of May, the bag contained separate populations which were adequate representations of marine conditions even though the dominant zooplankton species differed. Because of the marked changes immediately after 2 May, it may be better to regard these trials as two successive experiments, particularly as the zooplankton declined in numbers at 2 May, which could have resulted from inadequate feeding as shown by the feeding experiment. The later outburst in the zooplankton population would suggest that this was not a dominant effect for the second part of the trial, but more experimental work is required.

The large number of zooplankton in the bag, while implying generally good survival of the herbivores, also raises questions about the low number of predators. At present, the most likely cause appears to be escape of these organisms when the bag was filled, but this is a second major problem for future work.

REFERENCES

1. Anraku, M.
   1963 A review of feeding habits of copepods **Inform. Bull. Planktol.,** (Japan), 9: 10-35.
2. Antia, N. J., McAllister, C. D., Parsons, T. R., Stephens, K. and Strickland, J. D. H.
   1963 Further measurements of primary production using a large volume plastic sphere. Limno. Oceanogr., 8 (2): 166-184.
3. Corkett, C. J.
   1970 Techniques for breeding and rearing marine calanoid copepods. **Helgolander wiss. Meeresunters.,** 20: 318-324.
4. Lund, J. W. G.
   1972 Preliminary observations on the use of large experimental tubes in lakes. **Verh. internat. Verein. Limnol.,** 18: 71-77.
5. McAllister, C. D., Parsons, T. R., Stephens, K. and Strickland, J. D. H.
   1961 Measurements of primary production in coastal sea water using a large volume plastic sphere. **Limnol. Oceanogr.,** 6 (3): 237-258.
6. Steele, J. H., and Baird, I. E.
   1972 Sedimentation of organic matter in a Scottish sea loch. In **Detritus and its role in aquatic ecosystem,** p. 73-88. (ed. Melchiorri-Santolini, U. and Hopton, J.) Mem. Ist. Italiano Idrobiol. (Proc. IBP-UNESCO Symp.) 29, Suppl. 13-16. Pallanza.
7. Trevallion, A., Johnstone, R., Finlayson, D. M., and Nicoll, N. T.
   1973 Studies on the bivalve **Tellina tenuis** da Costa. IV. Further experiments in enriched sea water. **Jour. exp. mar. Biol. Ecol.,** 11 (2): 189-206.

THE DETRITUS-BASED FOOD WEB OF AN ESTUARINE MANGROVE COMMUNITY

by William E. Odum[1] and Eric J. Heald[2]

ABSTRACT

This paper is an attempt to construct a conceptual model of the food web of an estuarine mangrove community; the objective is to ascertain the importance of vascular plant detritus to the heterotrophic community.

From stomach-content data most of the important organisms of the estuary were arranged into the following groups: herbivores, omnivores, and three levels of carnivores. Most species function at more than one trophic level.

Of about 120 species examined, roughly one-third can be classified as detritus consumers. These are defined as organisms whose digestive-tract contents averaged at least 20 percent vascular plant detritus by volume on an annual basis. These detritus consumers included herbivorous and omnivorous species of crustaceans, mollusks, insect larvae, nematodes, polychaetes, and a few fishes. They appearred capable of digesting algae, portions of vascular plant detritus particles, microorganisms and, perhaps, dissolved substances sorbed upon inorganic particles.

A schematic diagram of the North River food web suggests that the principal flow of energy is along the route: mangrove leaf detritus → bacteria and fungi → detritus consumers → lower carnivores → higher carnivores.

1. Department of Environmental Sciences, University of Virginia, Charlottesville, Virginia 22903.

2. Rosensteil School of Marine and Atmospheric Science, Univeristy of Miami, Rickenbacker Causeway, Miami, Florida 33149.

INTRODUCTION

The importance of vascular plant detritus to estuarine consumers has been emphasized by a number of authors (summarized by W. E. Odum et al. [40]). Others have questioned the validity of the concept (54). Critical arguments have contended that direct usage of vascular plant detritus by estuarine aquatic consumers has not been demonstrated satisfactorily. In this paper we present evidence from an estuarine mangrove community which refutes these arguments.

In previous studies (13, 38) we showed that detritus production from a south Florida mangrove forest was high, in excess of 8 tonnes dry weight per hectare per year, and that colonization of decaying mangrove leaves by fungi and bacteria created a plentiful food source with a high caloric value and relatively great protein content. In addition, we examined the feeding habits of most of the fish and invetebrate species that inhabit the estuary (39).

The present paper is based upon the findings of these earlier publications and is an attempt to construct a conceptual model which represents the food web of the mangrove estuary; the objective is to ascertain the importance of vascular plant detritus to the heterotrophic community. If the principles relating vascular plant production to aquatic secondary production can be demonstrated in a simple estuarine system, then the same principles should operate in more complex estuaries.

PROCEDURES

A complete account of the methods which we employed in our field research and a description of the Everglades estuarine region have been presented elsewhere (13, 38). Briefly, it should be mentioned that this estuary consists of a belt of seasonally flooded mangrove forests in which the surface fresh-water flow of the Everglades mixes with the Gulf of Mexico in a system of tidal rivers, small streams, ponds, and coastal embayments. Our study was focused upon the North River Basin, a 21.7 km^2 drainage basin which is two-thirds covered by red mangroves *Rhizophora mangle* and one-third by open water.

Two characteristics of this mangrove-dominated estuary make it particularly suitable for a study of this type. First, relatively few species are able to adapt to the annual salinity range of from near zero to $30^{o}/oo$ which results from the seasonal rainfall pattern. This produces a food web of relatively simple dimensions. Second, and most important, the region is characterized by high levels of vascular plant production and low rates of algal production. The reasons for the latter are related to the high tannin content of the water, shading from the mangroves, and the low nutrient content of water draining the Everglades. The result is that vascular plant detritus forms a major component of the diet of primary consumers and its importance to the total food web is readily apparent.

NORTH RIVER TROPHIC GROUPS

In order to understand the North River food web, it was necessary first to understand the trophic relationships of the individual organisms. This was accomplished by analyses of diets and arrangement of the organisms into a trophic sequence.

Larkin (23) and Regier (44) considered that the concept of trophic levels has limited direct application to aquatic consumers because so many species function at two or more trophic levels. However, as emphasized by E. P. Odum (personal communication) the trophic-level concept pertains to the step-wise flow of energy in the community and not so much to individual species. Therefore, the concept must apply equally to aquatic and terrestrial ecosystems even though a higher proportion of terrestrial species function at only one level.

In order to obtain a relative measure of the trophic position of individual species in the North River food web we calculated a numerical index, similar to that adopted by Williams and Murdock (unpublished data). This method assigns values of 1.0, 2.0, 3.0, and 4.0 to successive heterotrophic levels: herbivores, carnivores feeding upon herbivores, and carnivores feeding on carnivores (two levels). Organisms with mixed diets are assigned intermediate index values based on the proportion of the energy intake derived from different sources. For instance, consumers whose diet was evenly split between plant material and herbivorous animals were assigned a value of 1.5; in other words this consumer functioned half at the 1.0 level and half at the 2.0 level. An organism whose diet indicated that it functioned 20% at the 1.0 level, 20% at 2.0, 40% at 2.6 and 20% at 3.0 would be assigned an index value of 2.2.

By converting our quantitative stomach-content data (39) to numerical indices, it was possible to arrange the North River animal community into a trophic sequence of primary to top consumers (Table 1). In this manner organisms which had trophically similar diets could be grouped together as functional units in the ecosystem.

Herbivores (mean values from 1.1-1.3)

Included among the herbivores were organisms which are primarily plant eaters but on occasion derive some nourishment from animal tissues. The stomach contents were normally composed of plant detritus and smaller amounts of algal filaments and cells. Additional nourishment must come from microorganisms adsorbed on ingested particles and from dissolved organic substances sorbed upon fine inorganic particles.

The herbivores were subdivided into three groups based upon the extent to which macro-animal material such as copepods and insect larvae was occasionally ingested. The species in group A (mean value of 1.1) are nearly strict herbivores.

TABLE 1

Trophic position of North River herbivores, omnivores, and carnivores based on function trophic indices[1] (For more detailed information regarding stomach contents see Odum and Heald, 39.)

HERBIVORES	MEAN VALUE	RANGE OF VALUES	PRINCIPAL COMPONENTS OF DIET
(GROUP A)			
Heterotrophic bacteria, yeasts & fungi	1.1	1.0-4.0	plant material
Ciliates, **Frontonia marina** and **Strombidium** sp.	1.1	1.0-1.1	microalgae
Copepods (at least 5 species)	1.1	1.0-2.0	microalgae, detritus
False mussel, **Congeria leucophaeata**	1.1	1.0-1.1	microalgae, detritus
Scorched mussel, **Brachidontes exustus**	1.1	1.0-1.1	microalgae, detritus
Eastern oyster, **Crassostrea virginica**	1.1	1.0-1.1	microalgae, detritus
Chironomid midge larvae (3 species)	1.1	1.0-1.1	detritus, microalgae
Nematodes (undetermined number of species)	1.1	1.0-1.1	detritus, microalgae
Ostracods (undetermined number of species)	1.1	1.0-1.1	microalgae, detritus
Cumaceans, **Cyclaspis varians** & **Oxyurostylis** sp.	1.1	1.0-1.1	microalgae, detritus
Isopods (at least 3 species)	1.1	1.0-1.1	microalgae, detritus
(GROUP B)			
Striped mullet, **Mugil cephalus**	1.2	1.0-2.0	microalgae, detritus
Sailfin molly, **Poeciliia latipinna**	1.2	1.0-2.0	detritus, microalgae
(GROUP C)			
Sheepshead minnow, **Cyprinodon variegatus**	1.3	1.0-2.0	detritus, microalgae
Diamond killifish, **Adinia xenica**	1.3	1.0-2.0	detritus, microalgae
Amphipods, **Melita nitida, Grandidierella bonnieri, Corophium lacustre**	1.3	1.0-2.0	detritus

TROPHIC POSITION OF NORTH RIVER OMNIVORES

TABLE 1 (Cont'd)

OMNIVORES			
(GROUP A)			
Polychaetes, **Neanthes succinea, Nereis pelagica**	1.4	1.0-2.0	detritus, microalgae
Crab, **Rhithropanopeus harrisii**	1.4	1.0-3.0	detritus, small animals
Snapping shrimp, **Alpheus heterochaelis**	1.4	1.0-2.0	detritus, small animals
Goldspotted killifish, **Floridichthyes carpio**	1.5	1.0-2.0	small animals, detritus
Caridean shrimp, **Palaemonetes intermedius, P. paludosus**	1.5	1.0-3.0	small animals, detritus
Nematodes (undetermined number of species)	1.5	1.0-4.0	plant & animal material
(GROUP B)			
Pink shrimp, **Penaeus duorarum**	1.8	1.0-3.0	small animals, detritus
Mosquito fish,. **Gambusia affinis**	1.8	1.0-2.0	small animals, algae
Least killifish, **Heterandria formosa**	1.8	1.0-2.0	small animals, algae
Crested goby, **Lophogobius cyprinoides**	1.8	1.0-3.0	small animals, detritus

TABLE 1 (Cont'd)

LOWER CARNIVORES			
Most larval & post-larval fishes	2.0	1.0-3.0	zooplankton, eggs
Copepods (undetermined number of species)	2.0	1.0-3.0	small zooplankton
Pinfish, **Lagodon rhomboides**	2.3	1.0-3.0	crustaceans
Sheepshead, **Archosargus probatocephalus**	2.4	1.0-3.0	crustaceans, mollusks
Juveniles of most middle and top carnivores	2.5	2.0-3.0	crustaceans, mollusks
Blue crab, **Callinectes sapidus**	2.5	1.0-4.0	mollusks, crustaceans
Scaled sardine, **Harengula pensacolae**	2.5	2.0-3.0	crustaceans
Bay anchovy, **Anchoa mitchilli**	2.5	1.0-3.0	crustaceans
Marsh killifish, **Fundulus confluentus**	2.5	1.0-3.0	crustaceans, midge larvae
Gulf killifish, **F. grandis**	2.5	2.0-3.0	crustaceans
Rainwater killifish, **Lucania parva**	2.5	1.0-3.0	crustaceans, midge larvae
Bluefin killifish, **L. goodei**	2.5	2.0-3.0	crustaceans, ostracods
Spotted sunfish, **Lepomis punctatus**	2.5	2.0-3.0	assorted small animals
Silver jenny, **Eucinostomus gula**	2.5	2.0-3.0	crustaceans
Spotfin mojarra, **E. argentius**	2.5	2.0-3.0	crustaceans
Striped mojarra, **Diapterus plumieri**	2.5	2.0-3.0	crustaceans
Code Goby, **Gobiosoma robustum**	2.5	1.0-3.0	crustaceans, midge larvae
Clown goby, **Microgobius gulosus**	2.5	1.0-3.0	crustaceans
Frillfin goby, **Bathygobius soporator**	2.5	2.0-3.0	crustaceans
Tidewater silversides, **Menidia beryllina**	2.5	2.0-3.0	crustaceans, insects
Hogchoker, **Trinectes maculatus**	2.5	2.0-3.0	assorted small animals
Lined sole, **Achirus lineatus**	2.5	2.0-3.0	assorted small animals
Common eel, **Anguilla rostrata**	2.6	2.0-3.0	assorted small animals
Skilletfish, **Gobiesox strumosus**	2.6	2.0-3.0	crustaceans
Silver perch, **Bairdiella chrysura**	2.8	2.0-3.5	crustaceans, small fishes
Gulf toadfish, **Opsanus beta**	2.8	2.0-3.5	crustaceans, mollusks
Sea catfish, **Arius felis**	2.8	1.0-4.0	assorted animals

TABLE 1 (Cont'd)

MIDDLE CARNIVORES			
Wood stork, **Mycteria americana**	3.1	2.0-3.5	fishes, crustaceans
White ibis, **Eudocimus albus**	3.1	2.0-3.5	fishes, crustaceans
Great blue heron, **Ardea herodias**	3.1	2.0-3.5	fishes, crustaceans
Little blue heron, **Florida caerulea**	3.1	2.0-3.5	fishes, crustaceans
Louisiana heron, **Hydranassa tricolor**	3.1	2.0-3.5	fishes, crustaceans
Green heron, **Butorides virescens**	3.1	2.0-3.5	fishes, crustaceans
Great white heron, **Ardea occidentalis**	3.1	2.0-3.5	fishes, crustaceans
Common egret, **Casmerodius albus**	3.1	2.0-3.5	fishes, crustaceans
Inshore lizardfish, **Synodus foetus**	3.1	2.0-3.5	fishes
Needlefishes, **Strongylura** spp.	3.1	2.0-3.5	fishes, crustaceans
Leatherjacket, **Oligoplites saurus**	3.2	2.0-4.0	crustaceans, fishes
Ladyfish, **Elops saurus**	3.2	2.0-4.0	crustaceans, fishes
Gafftopsail catfish, **Bagre marinus**	3.2	2.0-4.0	fishes, crustaceans
Snook, **Centropomus undecimalis** and **C. pectinatus**	3.2	2.0-4.0	fishes, crustaceans
Spotted seatrout, **Cynoscion nebulosus**	3.2	2.0-4.0	crustaceans, fishes
Red drum, **Sciaenops ocellata**	3.2	2.0-4.0	crustaceans, mollusks
Juveniles of certain top carnivores	3.2	2.0-4.0	fishes, crustaceans
Jewfish, **Epinephelus itajara**	3.2	2.0-4.0	fishes, crustaceans
Crevalle jack, **Caranx hippos**	3.2	2.0-4.0	crustaceans, fishes
Florida gar, **Lepisosteus platyrhincus**	3.3	2.0-4.0	fishes, crustaceans
TOP CARNIVORES			
Tarpon, **Megalops atlantica**	3.5	2.0-4.0	fishes, crustaceans
Barracuda, **Sphyraena barracuda**	3.5	2.0-4.0	fishes, crustaceans
American alligator, **Alligator mississipiensis**	3.6	2.0-4.0	fishes, reptiles
Bull shark, **Carcharhinus leucas**	3.6	2.0-4.0	fishes, crustaceans
Bald eagle, **Haliaeetus leucocephalus**	3.7	2.0-4.0	fishes
Osprey, **Pandion haliaetus**	3.7	2.0-4.0	fishes

1. 1.0 = strict herbivore
   2.0 = carnivore feeding on herbivores
   3.0 = carnivore feeding on 2.0 carnivores
   4.0 = carnivore feeding on 3.0 carnivores

The only animal material which was normally ingested was in the form of microorganisms. The two fishes which comprised group B (1.2) fed rarely upon small crustaceans, insect larvae and nematodes. Animals in group C (1.3) included animal material in their diet more frequently. Over the entire year, however, macro-animal material contributed less than 30 percent of the volume of the digestive tract contents.

Omnivores (mean values from 1.4 to 1.8)

These species exhibited a catholic choice in the particles which they ingested. Their diet seemed to be dictated by the availability of different types of food. The animals in omnivore group A (1.4-1.5) usually ingested more plant material than animal matter, although they can be grown well in captivity on a diet of animal tissue. The four species which made up group B (1.8) were able to exist on a diet of either plant or animal material, but showed a perference for animal tissue when it was available. In most habitats group B omnivores feed upon a mixture of the two components with small animal forms such as insects and crustaceans predominating.

Lower Carnivores (mean values from 2.0-2.8)

These species, predominantly small fishes, derived their nourishment primarily from the preceding herbivorous and omnivorous groups. Algal strands and mangrove detritus were occasionally ingested in small quantities (less than five percent of the total stomach-content volume), but their presence in the digestive tract was probably the result of accidental ingestion during capture of benthic animals. Exceptions were pinfish and sheepshead, which intentionally ingest quantities of plant material.

Middle Carnivores (mean values from 3.1 to 3.3)

This group included most of the wading birds and gamefishes. Their food was derived from all of the animals of the lower trophic groups but was dominated by small fishes from the lower carnivore group. Juvenile forms of most of the middle carnivore fishes functioned as lower (2.5) carnivores and are preyed upon by the higher carnivore groups.

Top Carnivores (mean values from 3.5 to 3.7)

These species formed the top group of the North River food web. Their food was derived from all of the animals of the lower trophic groups, but the most commonly ingested organisms were those from the middle (3.2) carnivore group.

ONTOGENETIC DEVELOPMENT

As pointed out by Regier (44) and others, individual fish species often function at different trophic levels during successive stages of development. For instance, *Mugil cephalus* of less than 25 or 30 mm are 2.0 carnivores feeding upon zooplankton; at larger sizes they become 1.1 herbivores feeding upon microalgae, detritus and sediments. Therefore, a single taxonomic species may exist as several functional species during different stages of its life history.

IMPORTANCE OF DETRITUS AS FOOD

More than 20 percent of the material contained in the digestive tracts of all of the organisms classified here as herbivores and omnivores was detritus of a vascular plant origin, usually mangrove leaf. It was present in much greater quantities than were algal tissues. Furthermore, field observations revealed sparse growths of benthic algae, low phytoplankton standing crops (10^5 cells per liter or less), and low rates of phytoplankton net production, indicating that vascular plant detritus is the most important element of the energy base for the North River food web. To comprehend fully the mechanism by which the energy from this detritus is transferred to higher trophic levels, it is necessary to examine the nature of detritus and how the material is produced.

Definition of Detritus

The word "detritus" originated from the Latin verb *deterere* which means "to rub away" or "to wear off". Although the term was used originally by geologists to denote material resulting from the disintegration of rock, it has been in use in biology at least since the time of Petersen's classic paper (42). As defined by E. P. Odum and De La Cruz (35), organic detritus refers to particulate material that was formerly part of a living organism. Particles ranging from the freshly dead bodies of plants and animals through finely disintegrated particles of these organisms to fecal pellets and even aggregates of colloidal-size particles are included in the definition. Detritus also includes materials which are sorbed upon the basic particle: bacteria, fungi, and protozoans, along with adsorbed dissolved organic and inorganic compounds. The entire particle and its sorbed load should be considered as a single unit – a small ecosystem within a larger system.

Organic detritus forms a significant fraction of the available food particles in many ecosystems. Its importance has been recognized in shallow estuarine embayments (6, 7, 8, 28, 36, 37, 42, 43), in salt marshes (25, 48, 51, 52), in soil communities (24, 41, 55), in old field communities (35), in forests (3, 45), in temperate lakes (26), in tropical lakes (11, 14) and in rivers and streams where it often is of a terrestrial origin (4, 9, 15, 17, 21, 22, 27, 30, 31). Organic detritus is present in the open sea and deep embayments, but recent investigations (2, 33, 46, 50) indicate that most of the detritus particles present are not derived directly from the breakdown of living cells, but are organic aggregates which have been synthesized from dissolved organic matter by physico-chemical adsorption. Such aggregates are present in shallow estuaries, but are insignificant in comparison to the great bulk of decaying plant particles. It is the annual production of tons per acre of this plant detritus that provides the energy to drive the biological machinery of the shallow esturine system.

Production of Vascular Plant Detritus in the North River System

To understand the trophic relationships of the North River animal and plant communities and before attempting to draw any conclusions, it is necessary to summarize our results concerning detritus production within the basin.

We estimated (unpublished data) that over 85 percent of the total production of vascular plant detritus in the North River basin was derived from approximately 1050 hectares of the red mangrove, *Rhizophora mangle;* the remaining percentage of detritus came from 160 hectares of *Juncus* and sawgrass marshes. The contribution of aquatic algae to the detrital pool is unknown, but based on the observed small standing crop of algae within the system, it was probably an unimportant source of detritus. Utilizing leaf catchment devices, surveys of mangrove density and size, and aerial photographs, we estimated that the North River mangroves produced 12,400 tonnes dry weight per year of leaves, twigs, leaf scales and flowers or 8.8 dry tonnes per hectare per year (2.41 dry grams per m^2 per day). Of this total 83 percent was composed of dead leaves.

The next step was to monitor the fate of this mangrove debris once it entered the water. Using litter bags with two-millimeter mesh openings, we found that the degradation rate of leaf material was highly variable, depending upon where the leaf fell. Those falling on dry ground decomposed slowly; those landing in fresh water broke down somewhat more rapidly; and the greatest degradation rate was in full strength sea water, presumably because of the larger populations of grazing organisms present there. After four months from the time of leaf fall, only 9% of the original leaf remained in the litter bags placed in seawater, while 39% was still present in those in brackish water and 54% in fresh water.

The disintegrating leaf material was monitored closely to determine its value as a potential food source. During the first six months after the leaf entered the water, the relative percent of protein present in the remains of the leaf rose from 5 to 21%. This does not mean that there was an actual increase in protein, but that there was relatively more protein present. As the leaf particles disintegrated, fats, carbohydrates, and plant protein were lost through a combination of autolysis and microbial activity, and might have been replaced to some extent by microbial protein. The end result was a smaller particle which, even though it had been reduced in all of its components, contained relatively more protein than the original leaf. Accompanying this increase in the relative amount of protein was an increase in the caloric value of the leaf material from 4.7 to 5.3 kcal per ash-free gram (19.7-22.2 J).

We analyzed the water flowing out of the North River and computed the amounts of suspended detritus which were lost from the North River. Of the detrital material produced in the North River basin roughly one half was exported from the system into the surrounding bays and inshore waters in particulate form; less than two percent of the leaf material remained permanently in the system to form peat. The exported detritus was primarily in the form of particles between 50 and 350 μm in size. These particles were composed of 21% protein on an ash-free basis and contained numbers of bacteria, fungi and yeasts.

Export of dissolved organic material from the North River was not estimated, but must be considerable. Only a few measurements were made, but these indicated dissolved concentrations as high as 10-15 mg C per liter.

Utilization of Detritus as Food

The tonnes of mangrove leaf material produced in each hectare of the North River mangrove community present a potential food source of great magnitude. There are at least four ways by which this freshly fallen leaf organic material may be utilized by the heterotrophic community: (1) dissolved organic substances → microorganisms → higher consumers, (2) dissolved organic substances → sorption on sediment particles → higher consumers either directly or by way of microorganisms, (3) leaf material → higher consumers, (4) leaf material → bacteria and fungi → higher consumers.

The first two routes of energy exchange are based upon the rapid loss of water-soluble organic substances (e.g., simple sugars, organic acids, starches) which occurs during the first few weeks after the leaf enters the water. Nykvist (34) found that ash leaves lost 22 percent of their original dry weight during the first few days in water. Soluble organic substances of mangrove leaf origin either may be used by bacteria and other microorganisms directly from the water or may become sorbed upon fine organic and inorganic particles in suspension or in the surface sediments. These particles, in turn, may be ingested by fishes and invertebrates and the sorbed substances removed in the digestive tract and assimilated by the animal. Inorganic particles with their loads of dissolved organic compounds are potentially a valuable food source.

The third route, that of leaf material serving directly as a food source for higher consumers, may be important early in the degradation process when the leaf still retains significant amounts of digestible plant proteins, fats, and carbohydrates. Only a few of the macroorganisms in the North River system appear capable of utilizing leaf material directly (e.g., the crab *Rhithropanopeus harrisii* and the amphipod *Melita nitida*). As the process of degradation proceeds, little digestible material remains and this third pathway probably is of less importance.

The fourth means of energy transfer in a detritus system such as the North River depends upon the ability of bacteria and fungi to break down and assimilate resistant plant substances such as celluloses and lignins while at the same time removing dissolved nutrients from the water. As plant detritus decomposes into decreasingly smaller particles, little remains except for empty cell walls composed of these resistant materials. When such a particle is ingested by a detritus feeder there is little labile material available for digestion except for adsorbed bacteria and fungi, along with Protozoa which may be feeding upon the bacteria.

There is little doubt that bacteria can furnish an adequate nutritional source. Shapira and Mandel (49) in experiments with rat growth found that bacteria fed at low levels provided an adequate sole source of protein. Baier (1) was one of the first to hypothesize that the nourishment from detritus particles comes from the bacteria involved in decomposition. Ivlev (16) was able to grow chironomid larvae on filter paper with the help of bacteria which were decomposing the paper. Zobell and Felthan (57) demonstrated that certain marine invertebrates could live almost indefinitely on an exclusive diet of bacteria. Zobell (56) suggested that bacteria were useful as food in two ways: (1) they were important directly as nourishment, and (2) they assisted the organism's digestion.

The value of bacteria as food has been emphasized by Teal (51) for fiddler crabs, *Uca* sp.; Fish (11) for *Tilapia;* Thompson (53), Darnell (6) and W. E. Odum (37) for the striped mullet *Mugil cephalus;* Fredeen (12) for blackfly larvae; and Marzolf (29) for the sediment-feeding amphipod *Pontoporeia affinis.* Newell (32) has shown that two marine deposit feeders, the prosobranch *Hydrobia ulvae* and the bivalve *Macoma baltica,* were unable to alter or utilize ingested detritus particles but were capable of digesting decomposer microorganisms adsorbed on the particles.

The importance of fungi as food has not been as widely recognized as it has for bacteria, primarily because, as Cooke (5) has pointed out, it is difficult to analyze quantitatively the presence of fungi in the aquatic environment. Although Rodina (47) could find no fungi associated with detrital particles in Russian lakes, fungi played an important role in the breakdown of many plant materials, particularly the more resistant vascular plants. Kaushik and Hynes (20) found most of the protein increase in decaying elm leaves to be due to fungal growth rather than bacterial activity. Our mangrove detritus samples from the North River system were heavily colonized by fungi, among which were species of *Fusarium, Nigrospora, Dendryphaiella, Cladosporium, Alternaria, Phomopsis,* and *Mucor.* Stained thin sections indicated that the relative amounts of fungi present in these particles was high. Subsequent work on mangrove fungi has been done by Fell and Masters (10).

One variation of this detritus → microorganism pathway involves the possible importance of digestive-tract microbes in breaking down detritus material which has been ingested by the host animal. This is a situation which parallels the microorganism's function in the ruminant stomach. Johannes and Satomi (19) showed that bacteria in the hindgut of a crustacean were able to reduce and assimilate the previously undigestible portion of *Nitzschia closterium* in only 30 minutes. W. E. Odum (37) noted that there were large numbers of bacteria and Protozoa present in the digestive tracts of the striped mullet, *Mugil cephalus,* a fish which eats large quantities of detritus. These intestinal Protozoa may be of importance judging from the finding of Johannes (18) that bacteria reduced detritus more rapidly in the presence of Protozoa since they were kept in a prolonged state of "physiological youth" by grazing Protozoa.

The Process of Detritus Formation

The microbial conversion of detritus and the subsequent utilization of this material as food by consumers was summarized by W. E. Odum (37) in the following manner. Initially the detritus particle, which may range in size from a few microns to an entire leaf, is attacked by fungi and bacteria, which begin oxidation, hydrolysis and assimilation of the basic carbon structure of the detritus particles while at the same time removing dissolved nutrients from the water. During the process of microbial breakdown the bacteria are continuously grazed by Protozoa creating a rich Protozoa-bacteria-fungi-detritus complex with great potential food value. At intervals the entire complex may be ingested by a larger organism such as a crab or mysid and most of the bacteria, fungi, and Protozoa digested off of the particle. In addition, there may be intestinal microbes which further reduce the particle and provide the host organism with nutrition in the form of excreted organic substances or of the microorganism itself. Once the particle, or fragments of the original particle is released as fecal material into the water, the entire process begins again. A single particle may be ingested and reingested in this manner by a number of different detritus feeders with the size of the particle decreasing with the completion of each cycle. Eventually it reaches a very fine size, becomes joined together with a number of other fine particles to form a conglomerant, and the process begins again.

DETRITUS CONSUMERS: THE KEY TO THE NORTH RIVER FOOD WEB

The decaying mangrove particles with their increased caloric value, high protein content and microbial loads present a rich food source for detritus consumers. The organisms from the North River whose digestive tract contents averaged at least 20 percent vascular plant detritus particles by volume were as follows: sheepshead killifish, *Cyprinodon variegatus;* gold spotted killifish, *Floridichtyes carpio;* diamond killifish, *Adinia xenica;* sailfin molly, *Poecilia latipinna;* crested goby, *Lophogobius cyprinoides;* striped mullet, *Mugil cephalus;* polychaete, *Nereis pelagica;* polychaete, *Neanthes succinea;* cummaceans, mysids, harpacticoid and planktonic copepods, amphipods, ostracods, caridean shrimp, penaeid shrimp, snapping shrimp, *Alpheus heterochaelis;* crab, *Rhithropanopeus harrisii;* chironomid midge larvae. This group of detritus consumers was dominated by invertebrates, chiefly crustaceans; only six species of fish were considered to be detritivores.

Feeding Mechanisms

Detritivores feed in a variety of ways with the result that different sized detrital material is used by different organisms. Figure 1 shows the types of detritus particles ingested by some of the North River detritivores. This information suggested that there are at least three major types of detritivores: (a) grinders, (b) deposit feeders and (c) filter feeders. For example, *Rhithropanopeus* and *Melita* are grinders which fed to a great extent upon large pieces of leaf material which were masticated into smaller particles. *Cyprinodon* and *Lophogobius* selected intermediate sized detritus particles from the benthos, while *Adinia* and *Palaemonetes* selected much finer benthic deposits. The finest particles were ingested by filter feeders such as *Brachidontes* and *Congeria.* All of the organisms, with the possible exception of the filter feeders, routinely ingested fecal pellets, thus accounting for the occurrence of conglomerates, even in the digestive tracts of large-particle feeders.

Comparison with Lake Pontchartrain

Our findings from the North River are in marked contrast to the situation reported by Darnell (6, 7, 8) for Lake Pontchartrain, Louisiana. He considered a detritus consumer to be any organism which contained more than five percent "detritus" in its digestive tract. Since he lumped most materials of an organic origin in the category of detritus, well digested organisms or their intestinal contents may have been included in this classification. His list of detritus consumers was dominated by 21 species of fishes and relatively few invertebrates. In our opinion only two of these fishes, *Mugil cephalus* and *Brevoortia patronus,* derive much nourishment from the detritus particles which they ingest. The remaining fishes are not equipped with the extremely long digestive tract characteristic of detritus consuming species. The following species

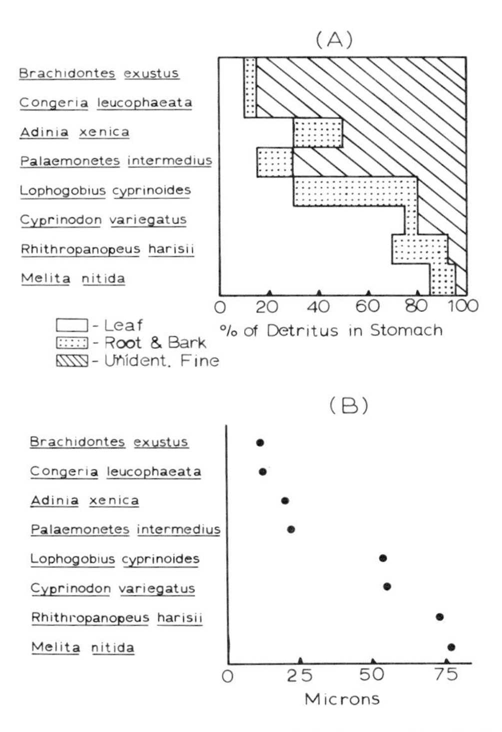

Figure 1. A. Types of mangrove detritus particles ingested by detritivores expressed as mean percent volume of stomach or buccal cavity contents on an annual basis. B. Mean particle size in microns of detritus particles from stomach and buccal cavity contents of devrivores.

which Darnell classified as obtaining nourishment from detritus rarely contained appreciable amounts in the North River samples: the spotted seatrout, *Cynoscion nebulosus;* the silver perch, *Bairdiella chrysura;* the ladyfish, *Elops saurus;* the red drum, *Sciaenops ocellata;* the needlefish, *Strongylura notata;* the bay anchovy, *Anchoa mitchilli;* and the hogchoker, *Trinectes maculatus.*

The Principle of Detritus Utilization Applied to All Shallow Estuaries

We agree with Darnell (6) when he suggested that detritus is important to consumers in the estuary, but we do not believe that it is important as food in significant quantities for the more carnivorous species. The important principle, both for the North River and most shallow estuaries, is that there is a group made up of a few species but many individuals of herbivorous and omnivorous crustaceans, mollusks, insect larvae, nematodes, polychaetes and a few fishes, all of which derive their nourishment primarily from a diet of vascular plant detritus and small quantities of fresh algae. Moreover, fecal material extruded by one organism in this group may be reingested a short time later by another species and the entire process of microbial enrichment and subsequent digestion by the detritus consumer be repeated. This concept of detritus cycling is depicted in a simple form in Figure 2.

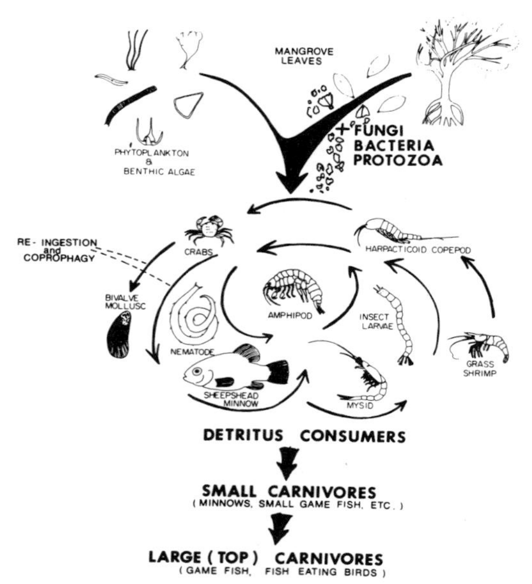

Figure 2. A schematic diagram of the detritus-consuming omnivorous organisms of the North River estuary. The cyclical nature of the diagram depicts the utilization and reutilization of detritus particles in the form of fecal material.

CONSTRUCTION OF A FOOD WEB

Utilizing the information presented in this study, we structured a schematic representation of the North River food web (Fig. 3). Although many food chains were incorporated into this food web, the principal flow of energy is along the route: mangrove leaf detritus → bacteria and fungi → detritus consumers → lower carnivores → higher carnivores. The detritus cycle of Figure 2 is represented in this diagram by the central box containing herbivorous and omnivorous organisms, all of which appear capable of digesting algae, detritus, microorganisms, and dissolved substances sorbed upon inorganic particles.

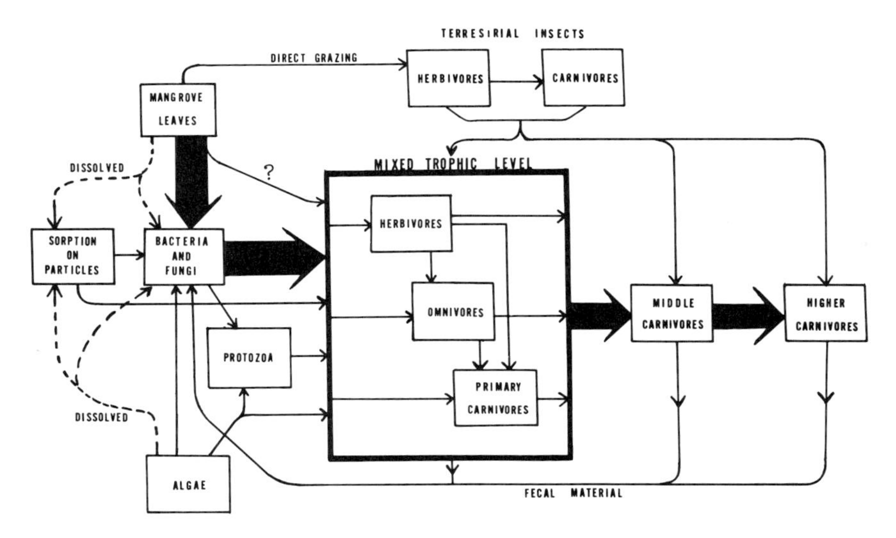

Figure 3. A conceptual model of the North River food web showing the most important flow of energy as a broad arrow, less important food chains as narrow arrows and the pathway of dissolved leaf material as a dotted line.

A GENERALIZED THEORY OF ESTUARINE FOOD WEBS

From consideration of our results and other published studies, there appear to be two basic types of estuarine food webs. The first is based primarily upon vascular plant detritus, but always with a component of fresh benthic micro-algae. The key organisms in this food web are a group of detritus consumers – herbivorous and omnivorous crustaceans, mollusks, insect larvae, nematodes, polychaetes and a few fishes. This type of food web is characteristic of shallow, muddy estuaries with extensive plant communities of marsh grasses, seagrasses, mangroves or macro-algae. The second type of food web is found in deeper estuaries with clearer water. This is a grazing food web based upon phytoplankton and dependent upon a key group of zooplankton and zooplankton grazers (anchovies, herring, etc.).

Usually an estuary is dominated by one of these food webs, although the situation may be reversed at specific sites within the estuary. Examples of detritus based estuaries include most of the shallow sounds behind barrier islands on the south Atlantic coast. Much of the Everglades estuary is characterized by this type of food web, although direct-grazing food webs become more important in the larger coastal embayments. The grazing food web is most important in large, deep estuaries such as Long Island Sound and Chesapeake Bay. Even in the Chesapeake system, however, the detritus based food web is of local importance where the wetlands to water ratio is high such as along the York River.

In summary, although each estuary probably has a tendency to be dominated by one type of food web, both types will always be present.

ACKNOWLEDGEMENTS

This paper is based, in part, upon research performed by the authors in partial fulfillment of the requirements for their Ph.D. degrees at the University of Miami, Florida. Thanks have been given elsewhere (13, 39) to a large number of people who made this work possible. Particular appreciation is extended to E. J. F. Wood, C. P. Idyll and D. C. Tabb. The project was supported by the National Park Service (Grant No. PO-160-713), by a Biomedical Sciences Support grant from the National Institutes of Health to the University of Miami, and by a National Science Foundation grant to the senior author while at the University of Virginia (Grant No. GA-34100).

REFERENCES

1. Baier, C. R.
1935 Studien zur Hydrobakteriologie stehender Bennengewasser. **Arch. Hydrobiol.**, 29: 183-264.

2. Baylor, E. R., Sutcliffe, W. H., Jr., and Hirschfield, D. S.
1962 Adsorption of phosphate onto bubbles. **Deep-Sea Res.**, 9: 120-124.

3. Bray, J. R. and Gorham, E.
1964 Litter production in forests of the world. **Adv. Ecol. Res.**, 2: 101-157.

4. Chapman, D. W.
1966 The relative contributions of aquatic and terrestrial primary producers to the trophic relations of stream organisms. **Spec. Publ. Pymatuning Lab.**, 4: 116-130.

5. Cooke, W. B.
1961 Pollution effects on the fungus population of a stream. **Ecology**, 42: 1-18.

6. Darnell, R. M.
1958 Food habits of fishes and larger invertebrates of Lake Pontchartrain, Louisiana, an estuarine community. **Pub. Inst. Mar. Sci.** Texas, 5: 353-416.

7. Darnell, R. M.
1961 Trophic spectrum of an estuarine community based on studies of Lake Pontchartrain, Louisiana. **Ecology**, 42(3): 553-568.

8. Darnell, R. M.
1967 The organic detritus problem. In **Estuaries**, p. 374-375. (ed. Lauff, G. H.) Am. Ass. Adv. Sci. Publ. 83, Washington, D. C.

9. Egglishaw, H. J.

1964 The distributional relationship between the bottom fauna and plant detritus in streams. **Jour. Animal Ecol.**, 33: 463-476.

10. Fell, J. and Masters, I. M.
    1973 Fungi associated with the degradation of mangrove (**Rhizophora mangle** L.) leaves in south Florida. In **Estuarine microbial ecology**, p. 455-465. (eds. Stevenson, L. H. and Colwell, R. R.) (Belle N. Baruch Coastal Res. Inst.) Univ. South Carolina Press, Columbia.
11. Fish, G. R.
    1955 The food of **Tilapia** in East Africa. **Uganda Jour.**, 19(1): 78-91.
12. Fredeen, F. J. H.
    1964 Bacteria as food for black fly larvae (Diptera: Simulidae) in laboratory cultures and in natural streams. **Can. Jour. Zool.**, 42: 527-548.
13. Heald, E. J.
    1969 The production of organic detritus in a south Florida estuary. Doctoral dissertation, Univ. Miami, Miami, Florida. 110 p.
14. Hickling, C. F.
    1961 **Tropical inland fisheries.** John Wiley and Sons, Inc. New York. 287 p.
15. Hynes, H. B. N.
    1963 Imported organic matter and secondary productivity in streams. **Proc. XIV Int. Cong. Zool.**, 4: 324-329.
16. Ivlev, V. S.
    1945 [The biological productivity of waters.] **Adv. Mod. Biol. (Usp. Sovrem. Biol.)**, 19: 98-120. (In Russian.)
17. Ivlev, V. S. and Ivassik, W. M.
    1961 [Data on biology of mountain rivers of Soviet Carpathian Ruthonia.] **Trudy Vses. Gidrobiol. Obshch.**, 11: 171-188. (In Russian.)
18. Johannes, R. E.
    1965 Influence of marine Protozoa on nutrient regeneration. **Limnol. Oceanogr.**, 10: 434-442.
19. Johannes, R. E. and Satomi, M.
    1966 Composition and nutritive value of fecal pellets of a marine crustacean. **Limnol. Oceanogr.**, 11: 191-197.
20. Kaushik, N. K. and Hynes, H. B. N.
    1968 Experimental study on the role of autumn-shed leaves in aquatic environments. **Jour. Ecol.**, 56: 229-243.
21. Kawanabe, H., Mori, W. and Mizuno, N.
    1957 Modes of utilizing the riverpools by a salmon-like fish, **Plecoglossus altivelir. Japanese Jour. Ecol.**, 7: 22-26.
22. Keast, A.
    1966 Trophic interrelationships in the fish fauna of a small stream. **Univ. Michigan, Great Lakes Res. Div. Publ.**, 15: 1-79.
23. Larkin, P. A.
    1956 Interspecific competition and population control in fresh-water fish. **Jour. Fish Res. Bd. Canada**, 13: 327-342.
24. MacFadyen, A.
    1961 Metabolism of soil invertebrates in relation to soil fertility. **Ann. Appl. Biol.**, 45: 215-218.

25. Marples, T. G.
1966 A radionuclide tracer study of arthropod food chains in a **Spartina** salt marsh ecosystem. **Ecology,** 47(2): 270-277.

26. McConnel, W. J.
1968 Limnological effects of organic extracts of litter in a southwestern impoundment. **Limnol. Oceanogr.,** 13: 343-349.

27. Mann, K. H.
1967 The cropping of the food supply. In **The biological basis of freshwater fish production,** p. 243-257. (ed. Gerking, S. D.) John Wiley and Sons, Inc., New York.

28. Mann, K. H.
1972 Macrophyte production and detritus food chains in coastal waters. In **Detritus and its role in aquatic ecosystems,** p. 325-352. (eds. Melchiorri-Santalini, U. and Hopton, J.) Mem. Ist. Italiano Idrobiol. (Proc. IBP-UNESCO Symp.) 29, Suppl. 13-16. Pallanzo.

29. Marzolf, G. R.
1966 The trophic position of bacteria and their relation to the distribution of invertebrates. **Spec. Publ. Pymatuning Lab.,** 4: 116-130.

30. Minshall, G. W.
1967 Role of allochthonous detritus in the trophic structure of a woodland spring-brook community. **Ecology,** 48: 139-149.

31. Nelson, D. J. and Scott, D. C.
1962 Role of detritus in the productivity of a rock-outcrop community in a Piedmont stream. **Limnol. Oceanogr.,** 7: 396-413.

32. Newell, R.
1965 The role of detritus in the nutrition of two marine deposit feeders, the prosobranch **Hydrobia ulvae** and the bivalve **Macoma baltica. Proc. Zool. Soc. London,** 144(1): 25-45.

33. Nishigawa, S.
1966 Suspended material in the sea; from detritus to symbiotic microcosmos. **Inform. Bull. Planktol. Japan,** 13: 1-4.

34. Nykvist, N.
1959 Leaching and decomposition of litter. **Oikos,** 10: 190-224.

35. Odum, E. P. and De La Cruz, A. A.
1963 Detritus as a major component of ecosystems. **AIBS Bulletin,** 13(3): 39-40.

36. Odum, E. P. and De La Cruz, A. A.
1967 Particulate organic detritus in a Georgia salt marsh-estuarine ecosystem. In **Estuaries,** p. 383-388. (ed. Lauff, G. H.) Am. Assoc. Adv. Sci. Publ. 83. Washington, D. C.

37. Odum, W. E.
1970 Utilization of the direct grazing and plant detritus food chains by the striped mullet **Mugil cephalus.** In **Marine food chains,** p. 222-240. (ed. Steele, J. H.) Univ. California Press, Berkeley and Los Angeles.

38. Odum, W. E.
1970 Pathways of energy flow in a south Florida estuary. Doctoral dissertation, Univ. Miami, Miami, Florida. 162 p.

39. Odum, W. E. and Heald, E. J.
1972 Trophic analyses of an estuarine mangrove community. **Bull. Mar. Sci.,** 22(3): 671-738.

40. Odum, W. E., Zieman, J. C., and Heald, E. J.
1973 The importance of vascular plant detritus to estuaries. In **Proceedings of the coastal marsh and estuary management symposium,** p. 91-135. (ed. Chabreck, R. H.) Louisiana State Univ. Div. Continuing Education, Baton Rouge.

41. Overgaard-Nielsen, C.
1962 Carbohydrates in soil and litter invertebrates. **Oikos,** 13: 200-215.

42. Petersen, C. J. G.
1918 The sea bottom and its production of fish food. The survey of work done in connection with the valuation of the Danish waters from 1883-1917. **Rept. Danish Biol. Sta.,** 25: 1-82.

43. Petersen, C. J. G. and Boysen-Jensen, P.
1911 Havets Bonitering. I. Havbundens Dyreliv, dets Naering og Maengde. **Rept. Danish Biol. Sta.,** 20: 1-81.

44. Regier, H. A.
1972 Community transformations – some lessons from large lakes. In **Proc. 50th Anniversary Symposium** (Univ. Washington, College of Fisheries.) Univ. Washington Press, Seattle.

45. Reiners, W. A.
1968 Energy and nutrient dynamics of detritus pools in three Minnesota forests. (AIBS Interdisciplinary Meeting, Madison, Wisconsin, June 1968.) (Unpublished abstract.)

46. Riley, G. A.
1967 Organic aggregates in sea water and the dynamics of their formation and utilization. **Limnol. Oceanogr.,** 8: 372-381.

47. Rodena, A. G.
1963 Microbiology of detritus in lakes. **Limnol. Oceanogr.,** 8: 388-393.

48. Schelskie, C. L. and Odum, E. P.
1961 Mechanisms maintaining high productivity in Georgia estuaries. **Proc. Gulf Carib. Fish. Inst.,** 14: 75-80.

49. Shapira, J. and Mandel, A. D.
1968 Nutritional evaluation of bacterial diets in growing rats. **Nature,** 217: 1061-1062.

50. Sutcliffe, W. H., Baylor E. R., and Menzel, D. W.
1963 The measurement of dissolved organic and particulate carbon in sea water. **Limnol. Oceanogr.,** 9: 139-142.

51. Teal, J. M.
1958 Distribution of fiddler crabs in Georgia salt marshes. **Ecology,** 39(2): 185-193.

52. Teal, J. M.
1962 Energy flow in the salt marsh ecosystem of Georgia. **Ecology,** 43: 614-624.

53. Thompson, J. M.
1954 The organs of feeding and the food of some Australian mullet. **Australian Jour. Mar. Fresh. Res.,** 5: 469-485

54. Walker, R. A.

1973 Wetlands preservation and management on Chesapeake Bay: The role of science in natural resource policy. **Jour. Coastal Zone Manag.,** 1(1): 75-101.

55. Wiegert, R. G., Coleman, D. C., and Odum, E. P.
1970 Energetics of the litter-soil subsystem. In **Methods of study in soil ecology,** p. 93-98. UNESCO, Paris.

56. Zobell, C. F.
1942 The bacterial flora of a marine flat as an ecological factor. **Ecology,** 23: 69-78.

57. Zobell, C. F. and Felthan, C. B.
1938 Bacteria as food for certain marine invertebrates. **Jour. Mar. Res.,** 4: 312-327.

SOURCES AND FATES OF NUTRIENTS OF THE PAMLICO RIVER ESTUARY, NORTH CAROLINA

J. E. Hobbie, B. J. Copeland, and W. G. Harrison[1]

ABSTRACT

Nutrient studies on the Pamlico River of North Carolina, an estuary extending some 65 km from fresh water to Pamlico Sound, were initiated in 1966. Phosphorus was always abundant and in concentrations far above algal needs. Nitrogen was relatively abundant all year as ammonia. In contrast, nitrate nitrogen was abundant only during winter months. Two peaks of algae (dinoflagellates) occurred each year, one in winter and one in late summer. These blooms correlated well with the appearance of nitrate in the middle reaches of the estuary. Additional evidence that the algae were using nitrate was the direct correlation of the numbers of *Peridinium triquetrum* with the quantity of nitrate reductase.

The estuary is a sink for nutrients. Over one year, 60% of the total phosphorus and 50% of the nitrate nitrogen remained in the estuary, presumably attached to sediments. Ammonia output was approximately equal to input, but complete turnover of ammonia is likely at least twice a day. Although total dissolved organic nitrogen plus particulate nitrogen was as great as total inorganic nitrogen, this organic material did not appear to be biologically active. Uptake and cycling experiments with nitrogen-15 are necessary for direct measurements of the fate of nitrogen in this estuary.

1. Department of Zoology, North Carolina State University, Raleigh, North Carolina 27607.

INTRODUCTION

The Pamlico River of North Carolina is a large estuary extending from fresh water to Pamlico Sound. Salinity ranges up to 21 ppt and tidal influence is slight. In recent years there has been an increase in use of fertilizers in the drainage basin, in the population in the drainage basin, in the number of persons on sewage treatment plants, and in manufacturing. One industry in particular has the potential to affect the estuary; Texas Gulf Industries operates a major phosphate mine and fertilizer plant in Aurora, North Carolina.

With all this potential for change in the river, studies were initiated in 1966 to follow eutrophication in the river and to study the ecology of a North Carolina estuary. A number of reports, papers, and theses have been produced, and we are now at the stage where we have some preliminary information on the total budget of nutrients in the estuary.

DESCRIPTION OF THE RIVER

The Pamlico River is a shallow estuary (average depth 3.5 m) extending for 65 km from Washington, North Carolina, on the west to Pamlico Sound on the east (Fig. 1). The chief tributary of the estuary is the Tar River which has a watershed of 8008 km^2. The basin is sparsely populated with a total population of 300,000, but only 40% of this is urban. Some 71% of the area is farm land,

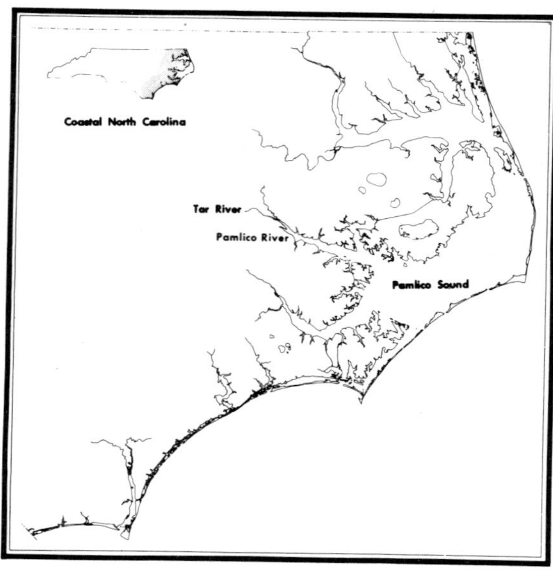

Figure 1. Map of eastern North Carolina, showing the location of the Pamlico River.

only 40% of which is in crops. Tobacco is the chief agricultural product. The climate is mild, with an average temperature of 16.5 °C and moderate rainfall of 122 cm per year. An additional land area of approximately 3120 km^2 drains into the estuary from the north and south which is mostly swamps and woodland, although there are a few large farms. Most of the tributaries draining this land have low flows and contribute little water in comparison to the flow of the entire river. However, actual flows are difficult to measure. One large river, the Pungo, drains a total area of 2210 km^2, but the actual flow has not been measured because of tidal problems. These tributary streams are not polluted, but high waters can flush much organic matter out of swamps.

As a result of low population density and lack of industrialization, the water entering Pamlico River is not polluted with large quantities of organic matter and has adequate oxygen.

The lunar tide in this estuary, only 15 cm, is greatly damped by Pamlico Sound and is much less important than frequent wind tides up to 1 m. Because of the shallowness, strong stratification is infrequent; there is usually a very weak stratification that is easily broken down by moderate winds. As a result, the estuary is usually well mixed and the turbulence keeps particulate matter in suspension and the waters quite turbid. This turbidity, along with a high humic content of water, causes the 1% light level to lie at 1 m in the upper and middle parts of the estuary and at 4 m in the lower estuary. Salinity, temperature, and oxygen data for 1965 to 1971 have been already reported (3, 6, 11). Salinity ranged from 0-to 20 ppt and temperature from 3 to 34 °C. When stratification lasted for more than a week, the bottom waters usually became anaerobic.

The yearly phytoplankton cycle is dominated by a massive dinoflagellate bloom during January, February, and March. The dominant form was found to be *Peridinium triquetrum* but other dinoflagellates were also important (4). A late summer dinoflagellate peak of *Gymnodinium, Polykrikos,* and *Gyrodinium* was not as large as the winter-spring peak but was metabolically more active as the temperature was so much higher. Both these peaks occurred in the middle reaches of the estuary. The dominant zooplankter was a copepod, *Acartia tonsa* (9), and the benthic system was dominated by *Rangia cuneata* and *Macoma balthica* (11). Blue crabs, shrimp, clams, and oysters are commonly harvested from the estuary. Marshes along the estuary are all of the high marsh type (*Juncus* sp.) and probably do not contribute much organic material. There are extensive beds of widgeon grass (*Ruppia maritima*) and *Potamogeton* in the shallows.

METHODS

Sampling was carried out at 34 stations every 2 weeks for one year, August 1971-August 1972, weather permitting. Samples for nitrogen and phosphorous determinations were frozen in the field with dry ice. Chemical methods followed closely those in a standard methods book (10). Nitrogen was reduced on a copper-cadmium column. Ammonia was oxidized with hypochlorite, and total nitrogen was measured after UV oxidation. Total phosphorous was measured after potassium persulfate treatment. Chlorophyll *a* was calculated after concentration of the plankton on glass fiber filters, extraction with acetone and spectrophotometric measurement at three wave lengths. Results were corrected for pheophytin.

Contour maps for individual sampling dates as well as graphs of concentration over the entire year were produced with a SYMAP program.

RESULTS AND DISCUSSION

A typical surface salinity distribution is shown in Figure 2. Freshwater enters at the extreme left and Pamlico Sound lies at the extreme right. Salinity in this estuary is mainly determined by the quantity of inflowing water. On the date illustrated here, the estuary was well mixed and salinity differences from top to bottom were nowhere greater than 1 ppt. On the immediately preceding and

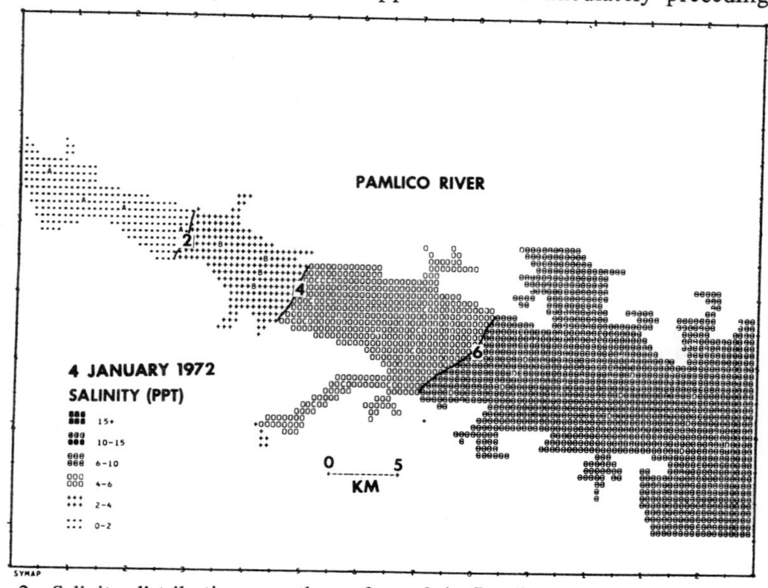

Figure 2. Salinity distribution over the surface of the Pamlico River on 4 January 1972. The river mouth is at the right edge and north is towards the top.

succeeding sampling dates large amounts of fresh water were entering the estuary and moving down almost to the middle section by flowing on top of the denser, more saline water. As a result, on those dates, there were salinity differences of up to 7 ppt between the top and bottom.

Another way of presenting the sampling data is given in Figure 3. Here, the transects across the estuary have been averaged and the resulting thirteen values plotted on a single line with the fresh-water head of the estuary at the top. This

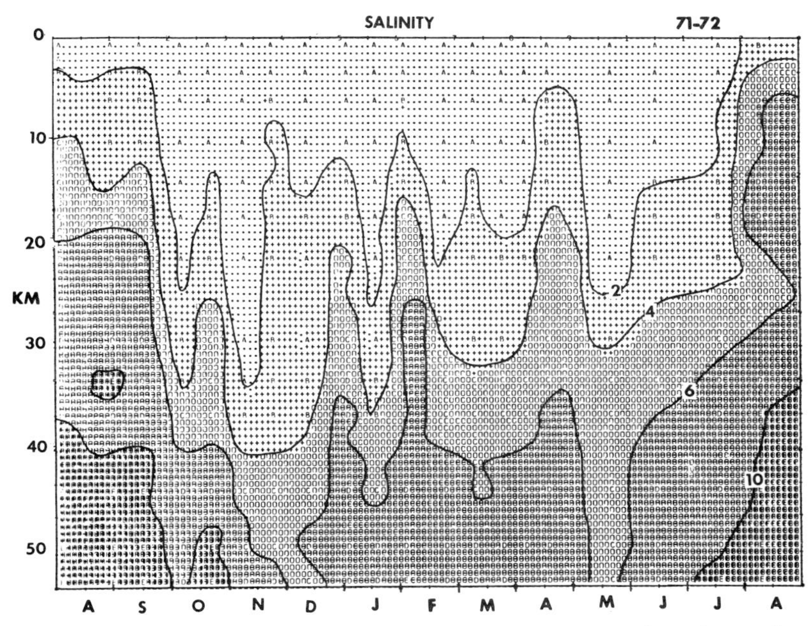

Figure 3. Salinity distribution in the Pamlico River (concentration in ppt) over the year August 1971 through August 1972. Upstream is at the top of the vertical axis, the mouth at the lower edge.

was done for a year (August 1971 through August 1972) and the isohalines plotted. In that particular year, August and September were dry and the salinity of the river was greatest at this time. The tremendous amount of fresh water in late September was the result of a hurricane. However, the entire fall had high amounts of rainfall and salinities were abnormally low.

Total phosphorous concentrations were always relatively high in the river (Fig. 4). Most of the phosphorous came from upstream where concentrations were always between 2 and 6 μg-atoms l^{-1}. There was, however, some input from the phosphate mining operations in the middle reaches of the estuary. Typically, This small patch of phosphorous-rich water stayed close to the south shore of the estuary and did not mix across to the north side.

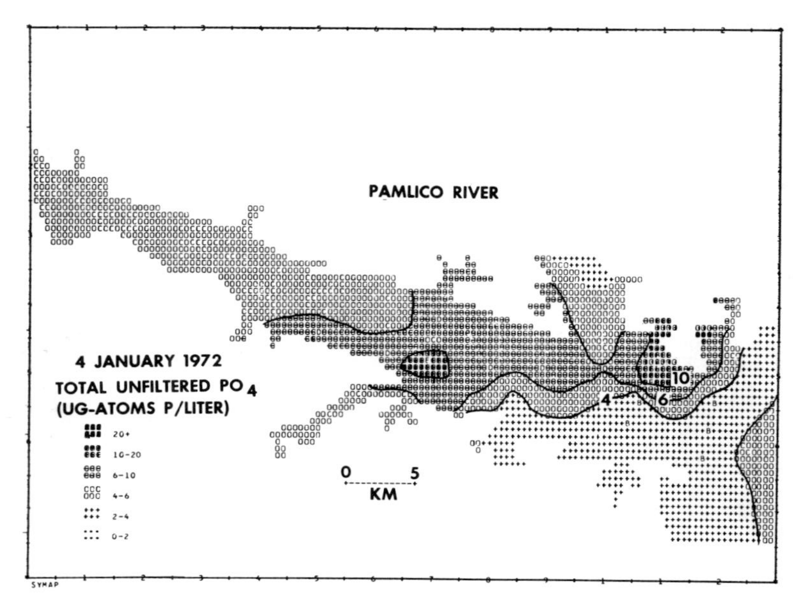

Figure 4. Total unfiltered phosphorus (µg-atoms P per liter) distribution over the surface of the Pamlico River on 4 January 1972.

The graph of concentration over the year (Fig. 5) also illustrates that there were slightly higher concentrations of total phosphorous in the middle parts of the estuary. The times of the highest average concentrations were August, September, and October while the lowest average concentrations were in April, May, June, and July.

Based on such concentration data as these, on photosynthesis measurements, and on nutrient addition experiments, it was concluded (5) that phosphorus was always abundant in the Pamlico River and did not limit primary productivity. Additional evidence for these conclusions comes from a statement (7) that 2.8 µg-atoms P l^{-1} is an upper level for unpolluted coastal waters. Levels were always higher than this in the Pamlico River.

Ammonia was similar to phosphorus in that high concentrations were always present (Fig. 6) both in the influents and in the estuary. It was only rarely, such as in October and November of 1971, that high quantities entered in the fresh water. As noted, these sampling dates followed a hurricane. Experiments that have been carried out (2) show a frequent photosynthetic response to added ammonia and also that ammonia is being continually produced within the planktonic part of the system. Presumably there are even greater amounts of ammonia being regenerated from the sediments. In view of the fact that many algae use ammonia, it is difficult to understand why there was not a greater and

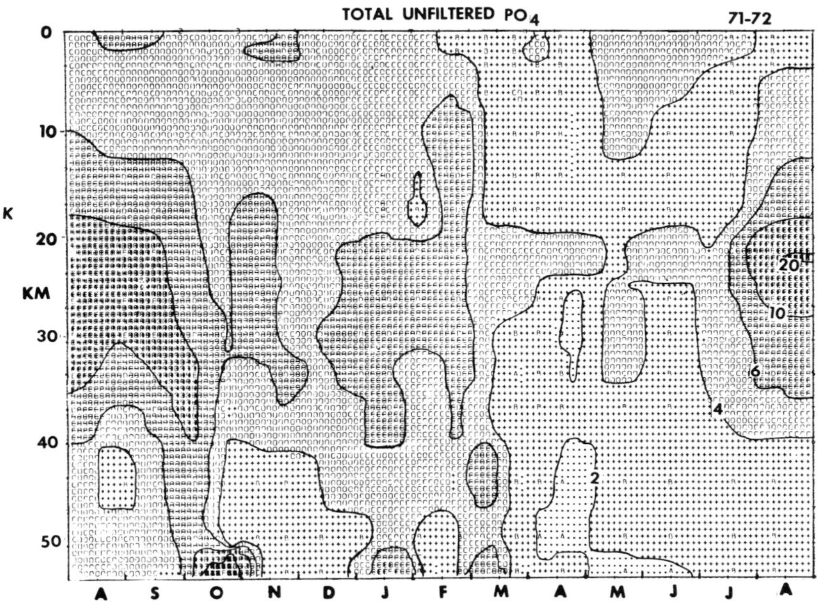

Figure 5. Total unfiltered phosphorus distribution over the year August 1971 through August 1972.

more extensive algal bloom in Pamlico River throughout the year. There is a possibility that because the chemical method used is sensitive to amino groups as well as to ammonium, it overestimated the concentration. However, the Solórzano technique has been tested and gave similar results (2).

A key to an understanding of the phytoplankton cycles is the concentration of nitrate. This began to increase in the rivers and streams in late fall and continued at a high level until spring (Fig. 7). This increase of nitrate in the rivers correlated with increased amounts leaving farm land after the removal of crops and the recharging of the water table (8). When high concentrations of nitrate reached the middle part of the estuary in early winter, a bloom occured and the chlorophyll content increased (Fig. 8). It is difficult, however, to use a correlation approach for a causative analysis as there were lags in the growth of algae. Thus, by the time the algae populations had built up, the nitrate might have been reduced to such a low concentration that there would be little correlation between nutrient and bloom. Figures 9 and 10 show that there were moderate-to-high nitrate concentrations but low chlorophyll concentrations in the middle part of the estuary on 13 December. By 4 January, however, the algae had built up to bloom conditions in the middle estuary but the nitrate was reduced almost to zero (Figs. 11 and 12).

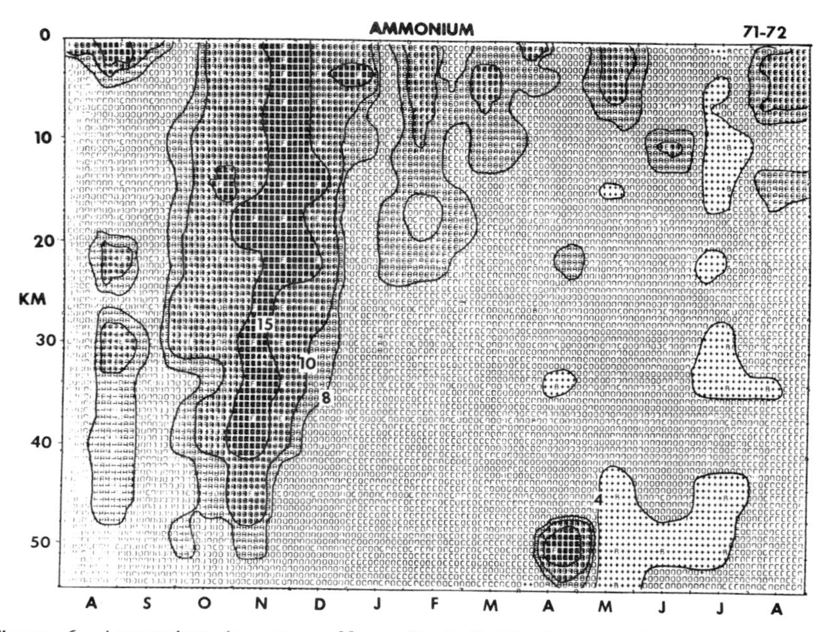

Figure 6. Ammonium (μ g-atoms N per liter) distribution over the year August 1971 through August 1972.

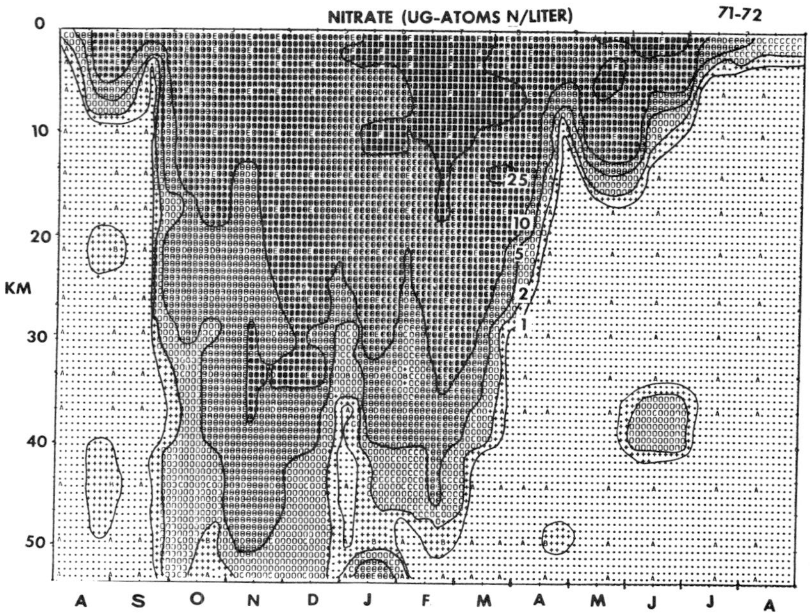

Figure 7. Nitrate distribution over the year August 1971 through August 1972.

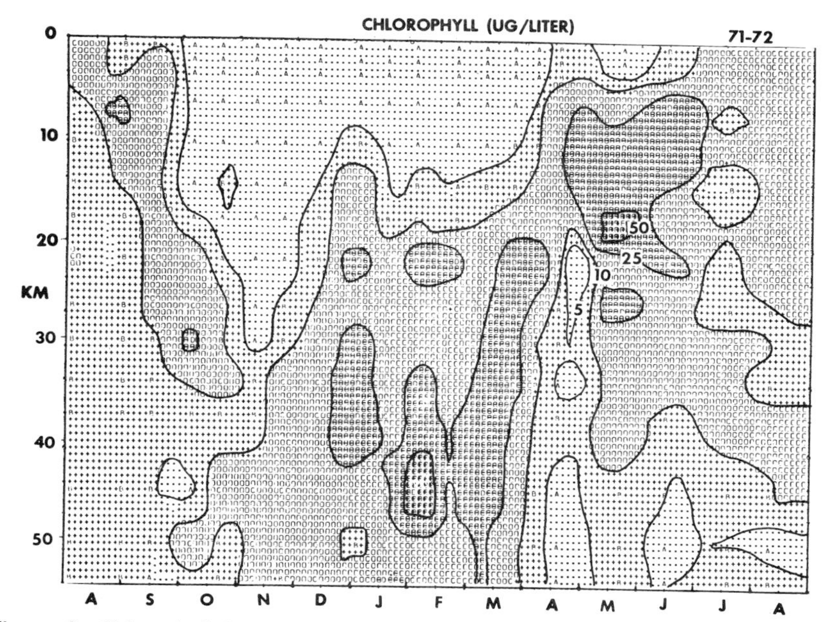

Figure 8. Chlorophyll distribution over the year August 1971 through August 1972.

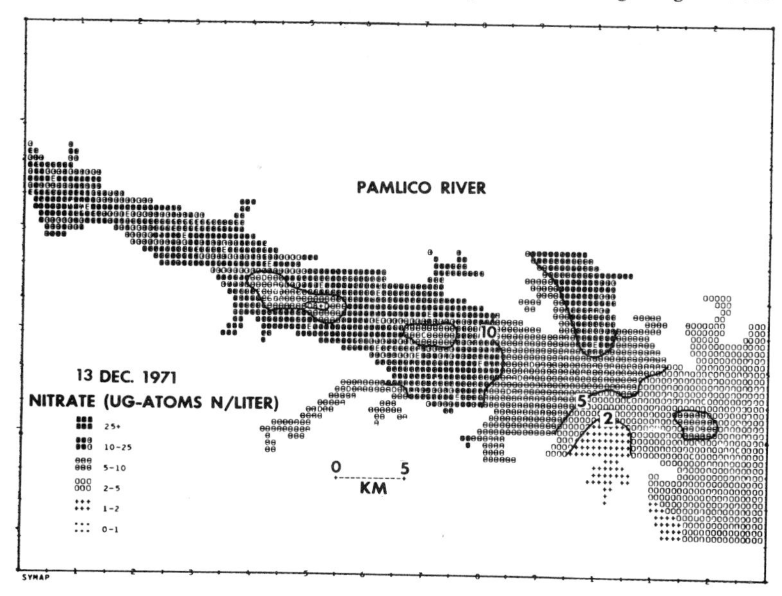

Figure 9. Nitrate distribution over the surface of the Pamlico River on 13 December 1971.

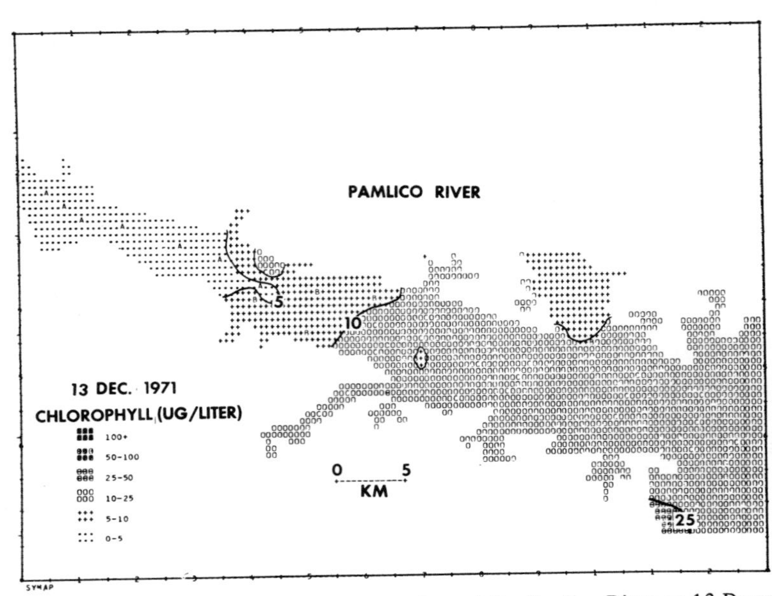

Figure 10. Chlorophyll distribution over the surface of the Pamlico River on 13 December 1971.

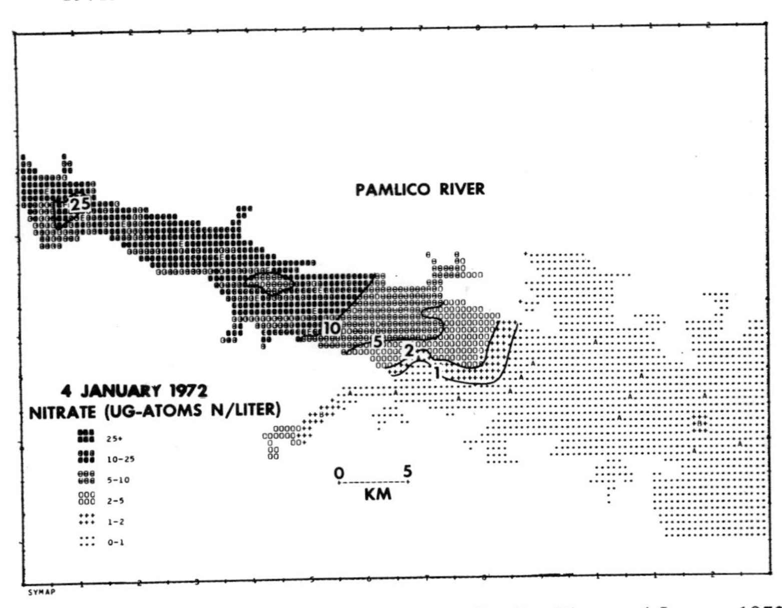

Figure 11. Nitrate distribution over the surface of the Pamlico River on 4 January 1972.

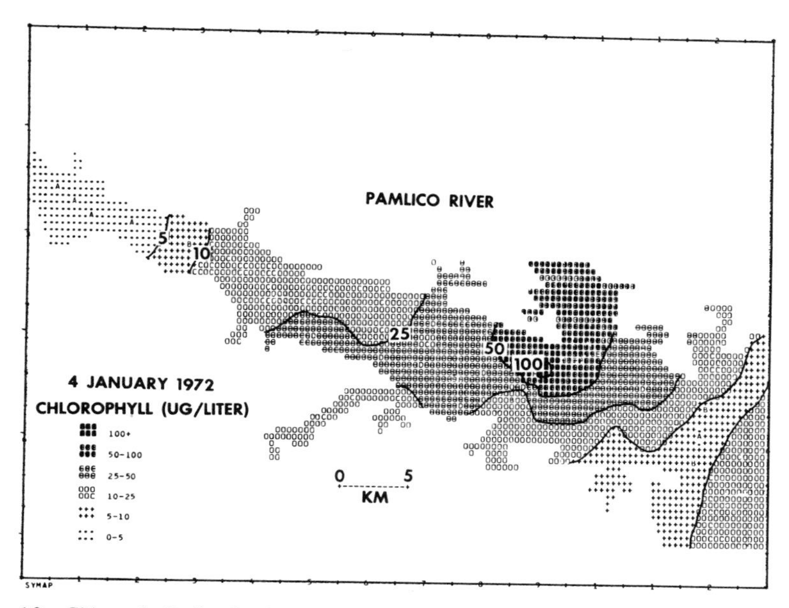

Figure 12. Chlorophyll distribution over the surface of the Pamlico River on 4 January 1972.

The winter dinoflagellate bloom has been investigated in more detail (1). In this study, the concentration of nitrate reductase was found to correlate well with the presence of *Peridinium triquetrum* and with the presence of high nitrate concentrations. Thus, it appears to be well documented that the algae are using the nitrate. The question why they should use nitrate when ammonia is so abundant and can be used with less expenditure of energy remains unanswered.

The next step is our study was to quantify the inputs and outputs of nutrients to the Pamlico River. Details of the calculations for nitrogen are given in Harrison (2) and the calculations for phosphorus are similar. Monitored hydrologic data came from the U. S. Geological Survey which maintains two gauges on the lower Tar River system, one gauge on a stream on the north, and one on a stream on the south side of the Pamlico River. Since most of the short tributaries that enter the Pamlico River are affected by the tide, direct flow was impossible to measure and, instead, flows were calculated from area and from the flow at the nearest gauge. Overall, the Tar River system flow was probably estimated quite well (75% of the total flow) while the remaining flow estimates contained considerable errors ($\pm$ 35%) . Chemical data consisted of measurements of nitrogen in the Tar River and at a series of stations where bridges crossed tributaries, a total of 8 samples, and of measurements of phosphorous in the Tar River only. Total input was calculated from total flow on a given day and from concentrations in the tributaries. Total output was calculated from input flow and from concentrations at the last downstream estuarine station. Evaporation and precipitation were assumed to be in balance. Total amounts in the estuary were calculated from water volumes in subsections of the river adjacent to each station and from the concentration of nutrients in these subsections.

The total quantity of nutrients present in the estuary was relatively constant for ammonia, total unfiltered phosphorous, and dissolved organic nitrogen, but the quantity of nitrate varied seasonally (Table 1). Thus, nitrate nitrogen was very low (5-17 tonnes) during July and August but was high (103-142 tonnes) from November through March. Ammonia and total unfiltered phosphorus were always quite high. Both dissolved organic nitrogen and particulate nitrogen were also quite high and approximated the sum of ammonia plus nitrate. Unfortunately, the precision of these organic measurements was not as good as for the inorganic forms, so output data cannot be calculated (note the extremely high dissolved organic nitrogen for May, for example).

The uncertainties increase in calculating total nutrient input and output and the results must be considered as the first approximation only. For all inorganic nutrients, input to the estuary exceeded output (Table 2). However, while 36% of the total phosphorus and 50% of the nitrate remained in the estuary, only 7% of the ammonia remained. In view of our knowledge of nitrogen cycling in

TABLE 1

Average amounts of nutrients present in the Pamlico River as N or P from August 1971 through August 1972 in tonnes.

	NO_3	NH_4	Total unfiltered	Dissolved organic	Particulate
August	16.9	116.2	261.5	215	48
September	8.8	105.4	240.2	264	50
October	85.7	144.3	294.6	222	46
November	137.0	190.6	178.0	161	31
December	139.2	132.9	156.1	194	46
January	117.5	100.2	214.6	176	51
February	141.5	99.4	180.6	174	43
March	102.6	93.3	146.2	163	57
April	78.3	137.9	86.8	159	35
May	43.7	119.9	114.7	422	83
June	32.6	111.1	107.0	174	65
July	7.03	71.5	123.7	214	37
August	4.9	97.8	195.7	199	45

TABLE 2

Total inputs and outputs of nutrients as N or P to the Pamlico River August 1971 through July 1972 in tonnes.

	PO_4	NO_3	NH_4
Input	715	2,804	795
Output	459	1,425	744
Net gain to estuary	+256	+1,379	+51

aquatic systems, it is obvious that much recycling is occurring, for example, in photosynthetic uptake of nitrogen and in decomposition alone. Incubation experiments in the river (2) showed that ammonia always decreased (up to 4 μg-atoms N l^{-1} day^{-1}) and nitrate usually increased (up to 8 μg-atoms N l^{-1} day^{-1}) during incubation. While these were net changes, they do indicate a potentially rapid cycling of nitrogen.

It should be pointed out that the entire input was assumed to have come from the streams. One additional source, the phosphate mine, exists for the phosphate. Dr. C. O'Melia (personal communication) has provided us with the figure of about one tonne of phosphorus added per day by this operation (90 cubic feet per second, 11 mg P_2O_5 per liter). This would amount to 366 tonnes additional annual input of phosphorus (in Table 2) and would decrease to 42% the percentage leaving the estuary.

Another way to examine losses and gains within the estuary is by measuring some of the processes that are at work. For example, release of nitrogen from the sediments has been measured (2) in long-term experiments where estuary water flowed over sediments in shallow troughs. Photosynthesis rates in the estuary were also measured and nitrogen uptake was calculated as carbon:nitrogen = 100:16. The results for February and August are given in Table 3 along with gains from rainfall and the measured gains from water movement. It is obvious that sediment release is unimportant but that recycling at least twice each day is necessary. For example, the August data suggest a loss from the estuary of 231 tonnes of nitrogen, but there were only 103 tonnes present.

There are several sources of error in these input and output calculations. The largest error is due to inadequate flow measurements in the streams entering the Pamlico River. This involved 25% or so of the total flow so the error is much less than this. Another related error is due to inadquate chemical data for the Tar River and inflowing streams. Here, it is very important to have good flow data as the flow varies by orders of magnitude while the concentrations vary by only a

TABLE 3

Net gains of nitrogen from the Pamlico River estuary on a single day in February and in August, 1972, in tonnes per day.

	February	August
Sediment release	+ 0.5	+ 3.4
Rainfall	+ 0.1	+ 0.7
N Assimilation from photosynthesis	– 6.7	- 231
Calculable gain to estuary	- 6.1	-227
Measured net gain to estuary	+ 6.9	+ 0.1

factor of two or three. Still another error is due to removal of nutrients by fish migration from the estuary. However, this amount was probably less than 0.5% of the total output for both phosphorus and nitrate plus ammonium (Table2). Finally, we have not considered the loss of nutrients resulting from turbulent mixing at the mouth of the estuary. This error would likely be very small because the salinity gradients are so small. In addition, much of the error is accounted for by the method of calculation.

CONCLUSIONS

Nutrients were abundant in the inflowing water in the estuary itself. Phosphorus was plentiful throughout the year while the presence of large amounts of nitrate in the early winter apparently triggers the phytoplankton bloom. There may be interactions with turbidity, salinity, and temperature as these large blooms are always in the middle parts of the estuary.

The estuary acts as a trap for phosphorus and nitrate, as half of the input of these compounds remained in the estuary. The exact mechanisms are not known but presumably the phosphate is trapped in the sediments while the nitrate is denitrified, assimilated, and also trapped in the sediments. In contrast, there was no net loss of ammonia from the system. Since it is unlikely that ammonia was not being used, and since some experiments proved that it was being removed by plankton, it is obvious that ammonia was being regenerated. Preliminary indications are that turnover times for ammonia may be as short as one-half day during the summer. The organic forms of nitrogen were almost as abundant as the inorganic forms. However, there was no evidence that these were greatly changed in their passage through the estuary. Thus, although there was some additional dissolved organic and particulate nitrogen formed, most of this organic nitrogen was not biologically active. Unfortunately, the evidence for this conclusion is somewhat shaky.

The next step in this study is to use nitrogen-15 as a tracer to investigate the rates of uptake of nitrogen and ammonia into the plankton, regeneration rates of ammonia, and nitrification. Only then will we completely understand the fate of nutrients in this estuary.

ACKNOWLEDGMENTS

Funds for this research were provided by the Office of Water Resources Research Matching Grants Program via the University of North Carolina Water Resources Research Institute. Matching funds and other support were provided by Texas Gulf Incorporated and by the North Carolina State University Agriculture Experiment Station and Department of Zoology.

REFERENCES

1. Harrison, W. G.
   1973 Nitrate reductase activity during a dinoflagellate bloom. **Limnol. Oceanogr.,** 18: 457-465.
2. Harrison, W. G.
   1974 Certain aspects of the nitrogen cycle of the Pamlico River estuary, North Carolina. Doctoral dissertation, Dept. Zool., North Carolina State Univ., Raleigh.
3. Hobbie, J. E.
   1970 Phosphorus concentrations in the Pamlico River estuary of North Carolina. **Rept. Water Resour. Res. Inst., Univ. North Carolina,** 33: 1-47. (Unpublished manuscript.)
4. Hobbie, J. E.
   1971 Phytoplankton species and populations in the Pamlico River estuary of North Carolina. **Rept. Water Resour. Res. Inst., Univ. North Carolina,** 56: 1-147. (Unpublished manuscript.)
5. Hobbie, J. E.
   In press Effect of phosphorus on estuarine systems. In **Coastal ecological systems of the U. S.** (eds. Odum, H. T. and Copeland, B. J.) Conservation Foundation, Washington, D. C.
6. Hobbie, J. E., Copeland, B. J., and Harrison, W. G.
   1972 Nutrients in the Pamlico River estuary, N. C., 1969-1971. **Rept. Water Resour. Res. Inst., Univ. North Carolina,** 76: 1-242. (Unpublished manuscript.)
7. Ketchum, B. H.
   1969 Eutrophication of estuaries. In **Eutrophication: causes, consequences, correctives,** p. 197-209. Natl. Acad. Sci., Washington, D. C.
8. Kilmer, V. J., Gilliam, J. W., Joyce, R. T., and Lutz, J. F.
   1972 Loss of fertilizer nutrients from soils to drainage waters. Part I.

Studies on grassed watersheds in western N. C. Part II. Nitrogen concentration in shallow groundwater of the N. C. coastal plains. **Rept. Water Resour. Res. Inst., Univ. North Carolina,** 56: 1-25. (Unpublished manuscript.)

9. Peters, D. S.
1968 A study of relationships between zooplankton abundance and selected environmental variables in the Pamlico River estuary of Eastern North Carolina. M. S. thesis, Dept. Zool., North Carolina State Univ., Raleigh.

10. Strickland, J. D. H. and Parsons, T. R.
1968 A manual of seawater analysis. **Bull. Fish. Res. Bd. Canada,** 167: 1-311.

11. Tenore, K. R.
1972 Macrobenthos of the Pamlico River estuary, North Carolina. **Ecol. Monogr.,** 42: 51-69.

NUTRIENT INPUTS TO THE COASTAL ZONE: THE GEORGIA AND SOUTH CAROLINA SHELF

Evelyn Brown Haines[1]

ABSTRACT

The concept of "outwelling" of nutrients from salt-marsh estuaries to coastal waters as proposed by E. P. Odum is evaluated for the Georgia and South Carolina shelf. The distribution of inorganic nutrients, particulate organic carbon, and chlorophyll *a* was determined in shelf waters between Charleston, South Carolina, and Fernandina Beach, Florida. Nearshore, lower salinity waters were enriched in phosphate, organic carbon, and chlorophyll *a* but contained little nitrate or ammonia. A budget of the nitrogen influx from terrestrial discharge, intrusion of nutrient-rich deep water at the edge of the shelf, and precipitation was calculated for the shelf area studied. The annual nitrogen input was less than five percent of the calculated yearly uptake of nitrogen by primary producers on the shelf suggesting that regeneration is the most important factor in maintaining high rates of production in the coastal waters of Georgia and South Carolina.

INTRODUCTION

The inshore waters of the Georgia continental shelf are highly productive. Thomas (30) observed an annual primary production in waters off the Altamaha River equal to the annual primary productivity of the salt-marsh estuaries that border the Georgia coast. The suggestion that estuaries contribute to the productivity of coastal waters has been formalized in the *outwelling* concept

1. Duke Marine Laboratory, Beaufort, North Carolina 28516.

originally presented by E. P. Odum (18). Inorganic nutrients unused in the estuaries because of light limitation were hypothesized to stimulate coastal production when flushed to clearer waters offshore. The study reported here was designed to determine to what extent the process of estuarine outwelling affects production in Georgia shelf waters. This mechanism of nutrient influx to shelf waters was examined in the more general context of the total nutrient input to the coastal zone. The potential sources of nutrients considered were fresh-water drainage, intrusion of nutrient-rich deep water onto the edge of the shelf, and precipitation.

In 1971-72, the distributions of salinity, inorganic phytoplankton nutrients, organic carbon, and chlorophyll *a* in shelf waters from Charleston, South Carolina, to Fernandina Beach, Florida, were investigated to determine the extent to which materials were exported from estuaries to the coastal zone. The coasts of Georgia and South Carolina are bordered by highly productive *Spartina* salt-marsh estuaries which are well flushed by river flow and by tides with a mean range of 2 m. Estuarine water in this region contains significant concentrations of phosphate (about 1.0 μg at l^{-1}) (30), organic carbon (2-20 mg l^{-1}) (18), and chlorophyll *a* (2-6 mg l^{-3}) (30). Southeastern rivers flowing through the estuaries are rich in nitrate (32, 33) and silicate (Haines, unpublished data). The hydrography of the Georgia shelf has received much less attention than that of the shelf waters farther north. Salinity and temperature data along a few transects off the Georgia coast are available from work aboard the M/V *GILL* (1). Subsequent studies on the Georgia shelf have been limited to nearshore waters (11, 30).

To evaluate the importance of river and estuarine discharge as a source of phytoplankton nutrients for the coastal zone, a nitrogen flux budget for Georgia and South Carolina shelf waters was calculated. Nitrogen was chosen because it has been considered the inorganic nutrient most frequently limiting to marine primary production (24). The computed influx was compared both to the annual nitrogen uptake by producers on the shelf and to the amount of nitrogen in particulate matter in shelf waters.

AREA STUDIED

The continental shelf off the Georgia and South Carolina coast is a broad, gently sloping platform, varying from 90 to 120 km in width and from 10 to 50 m in water depth over most of the shelf. The shelf begins to slope steeply at water depths of 50 to 100 m. The approximate inner edge of the Gulf Stream lies along the edge of the shelf. The general seasonal direction of circulation in Georgia shelf waters summarized by Kuroda and Marland (11) is usually southerly, except in the spring, when the general tendency of circulation is northerly. The wind blows from the north in the fall and winter and from the south in the spring and summer. Circulation is strongest in the winter (19-28 cm s^{-1}) and weakest in the summer (1.9-3.7 cm s^{-1}).

The region of shelf considered in this study is shown in Figure 1; the boundaries of the area are the transects over the shelf normal to the coast at Charleston and Fernandina Beach and the 200 m depth contour at the edge of the shelf. This region has a surface area of 29,270 km^2 and a water volume of 105.9 x 10^{10} m^3 (Table 1).

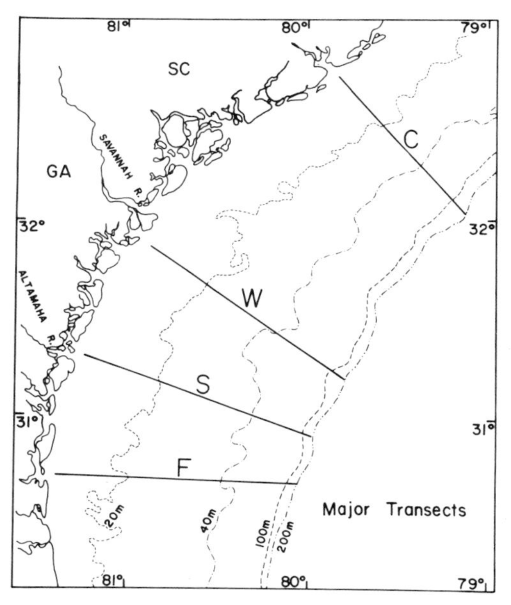

Figure 1. Transects between Charleston, South Carolina, and Fernandina Beach, Florida. C: Charleston, W: Wassaw Sound, S: Sapelo Sound, F: Fernandina Beach.

TABLE 1

Volume of water contained in various depth contours for the continental shelf between Charleston, South Carolina and Fernandina Beach, Florida.

Depth range m	Surface area km^2	Water volume 10^{10} m^3
0-20	11200	11.2
20-40	10300	30.9
40-100	6600	46.2
100-200	1170	17.6
Total	29270	105.9

METHODS

Sampling was carried out on four cruises of R/V *EASTWARD* between September, 1971, and October, 1972. In September, 1971, a survey was made of continental shelf waters from Miami, Florida, to Cape Hatteras, North Carolina. In November, 1971, and June and October, 1972, cruises were jointly conducted with investigators from the University of Georgia. Four sections over the shelf between Charleston, South Carolina, and Fernandina Beach, Florida, were established for these three cruises (Fig. 1). Water was collected with 1.5-liter Nansen bottles in September and with 8-liter Niskin bottles in the following months at depths of 1 m and at intervals of 5, 10, or 20 m to the bottom.

Analyses of reactive phosphate (29), ammonia (10), and chlorophyll *a* (12) were carried out at sea. Salinity was determined conductometrically with a Hytech salinometer. Sea-water samples for the determination of nitrate and nitrite (29) were collected in 250 ml polyethylene bottles, quick frozen at -90 ^{o}C, and stored at 0 ^{o}C until thawed and analyzed ashore. Volumes of 0.1 to 1.0 liters of sea water for particulate organic carbon analysis and 1.0 to 3.0 liters of sea water for particulate nitrogen analysis were filtered through two Gelman-type A glass fiber filters which had been precombusted to 500 ^{o}C. The bottom filter of each pair was analyzed to obtain a blank value which was subtracted from the top filter value (2). The filters were dried and stored in a 60 ^{o}C oven. Particulate organic carbon was determined by high temperature combustion according to a modified Menzel and Vaccaro (17) method. Particulate nitrogen was determined by the Micro-Dumas method with a Coleman Model 29 Nitrogen Analyzer.

OBSERVATIONS

The September, 1971, survey of the southeastern continental shelf showed that lowest salinities and highest concentrations of phosphate occurred off the Georgia coast (Fig. 2). Highest concentrations of chlorophyll *a* (2.4 mg m^{-3}) and particulate organic carbon (0.8 mg l^{-1}) were also observed off the Georgia coast. Concentrations of nitrate, nitrite, and ammonia were low in shelf waters, except for near the bottom at the edge of the shelf where there were concentrations of nitrate up to 1.5 μg at l^{-1}. This was presumed to indicate intrusion of nutrient-rich deep water.

During subsequent cruises off the Georgia and South Carolina coasts, lowest salinities and highest concentrations of reactive phosphate, chlorophyll *a*, and particulate organic carbon were observed in nearshore waters between Savannah and Fernandina Beach, with a peak usually off the Altamaha River (Figs. 3-6). The passage of hurricane Agnes over Georgia during the June 1972 cruise

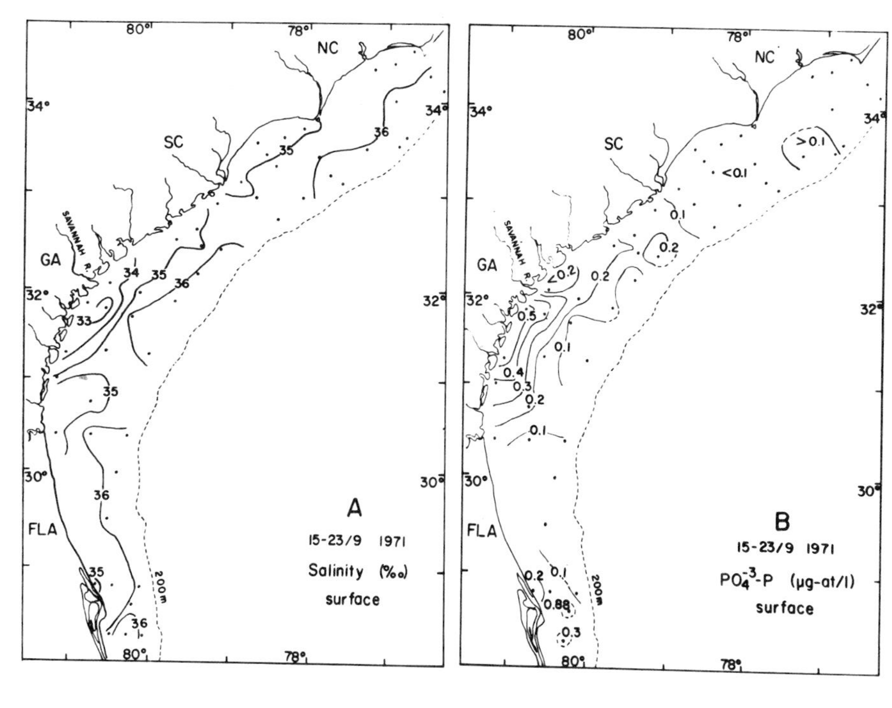

Figure 2. Distribution of salinity and phosphate in southeastern U. S. continental shelf waters, September, 1971.

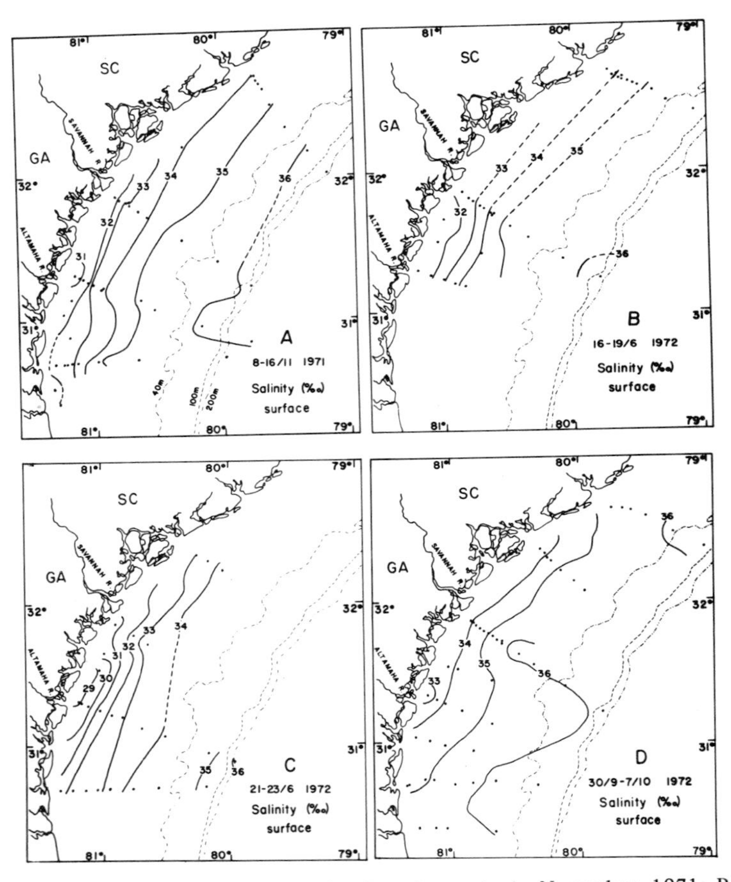

Figure 3. Distribution of salinity off the Georgia coast. A: November, 1971; B: June, 1972, before hurricane Agnes; C: June, 1972, after hurricane Agnes; D: October, 1972.

required a thirty-six hour syspension of sampling. Distribution patterns for salinity, phosphate, and chlorophyll *a* were sufficiently different before and after the storm to warrant separate distribution charts for the two parts of the cruise.

The concentration of inorganic nitrogen nutrients was very low during all months. Small quantities of nitrate were occasionally found in nearshore waters (Fig. 7). No intrusions of nitrate-rich water were observed at the edge of the shelf in November, June, or October. The concentration of ammonia showed no discernible distribution pattern. Nearshore mean ammonia concentrations were similar to offshore mean concentrations (Table 2).

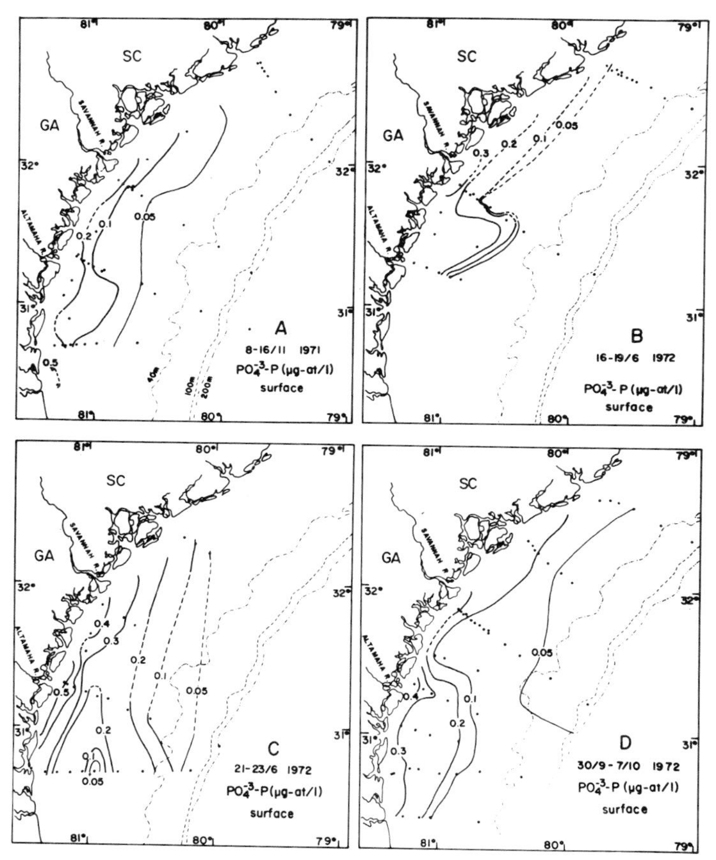

Figure 4. Distribution of phosphate off the Georgia coast. A: November, 1971; B: June, 1972, before hurricane Agnes; C: June, 1972, after hurricane Agnes; D: October, 1972.

The vertical distributions of salinity, phosphate, and nitrate (Figs. 8-10) show the vertical homogeneity of Georgia coastal waters.

The concentrations of phosphate, chlorophyll *a*, and particulate organic carbon found in Georgia coastal waters in this study are similar to levels reported by other workers. Thomas (30) found phosphate concentrations ranging from 0.85 μg at l^{-1} near the mouth of the Altamaha River to 0.03 μg at l^{-1} 28 km offshore. The chlorophyll *a* concentrations reported by Thomas were somewhat higher than those found in this study. Concentrations were generally 3-9 mg chl *a* m^{-3} in nearshore waters and below 3 mg m^{-3} at distances greater than 15 km offshore (30). Manheim et al. (13) reported concentrations of combustible

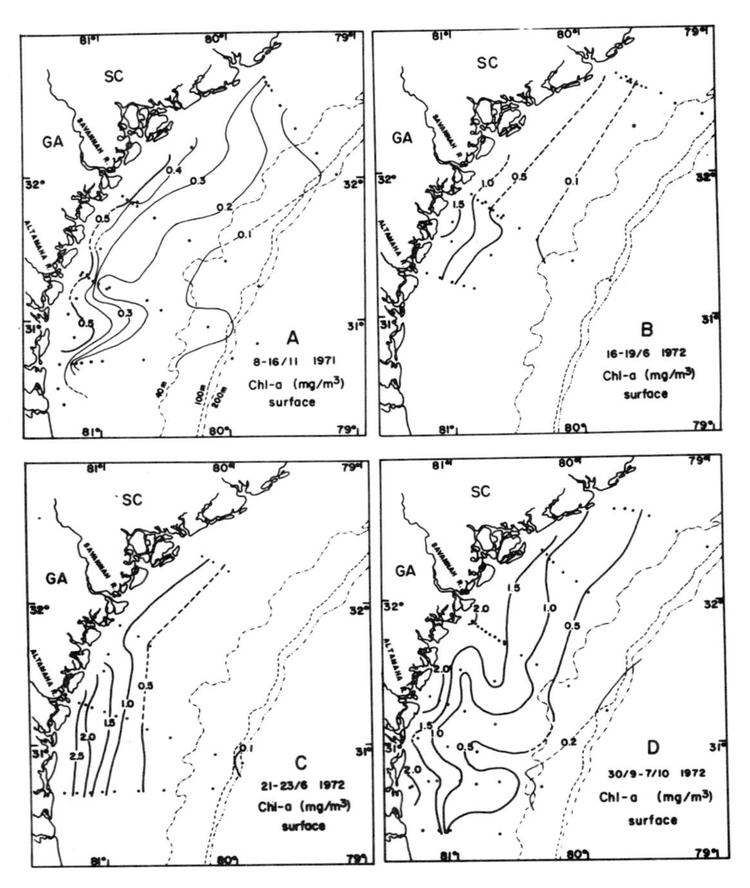

Figure 5. Distribution of chlorophyll a off the Georgia coast. A: November, 1971; B: June, 1972, before hurricane Agnes; C: June, 1972, after hurricane Agnes; D: October, 1972.

organic suspended matter of 0.4-1.6 mg C 1^{-1} in nearshore southeastern shelf waters. Water containing more than 1 mg 1^{-1} suspended matter was generally restricted to river mouths and a narrow nearshore zone less than 10 km wide. Surface waters over the rest of the shelf generally contained less than 0.125 mg 1^{-1} seston, of which 60-90% was combustible organic material.

The distribution patterns of salinity, inorganic nutrients, and organic materials found in this study indicate that there is a zone adjacent to the coast which is greatly influenced by river and estuarine discharge. The potential importance of the export of nitrogen from rivers and estuaries to the shelf waters is examined in the next section.

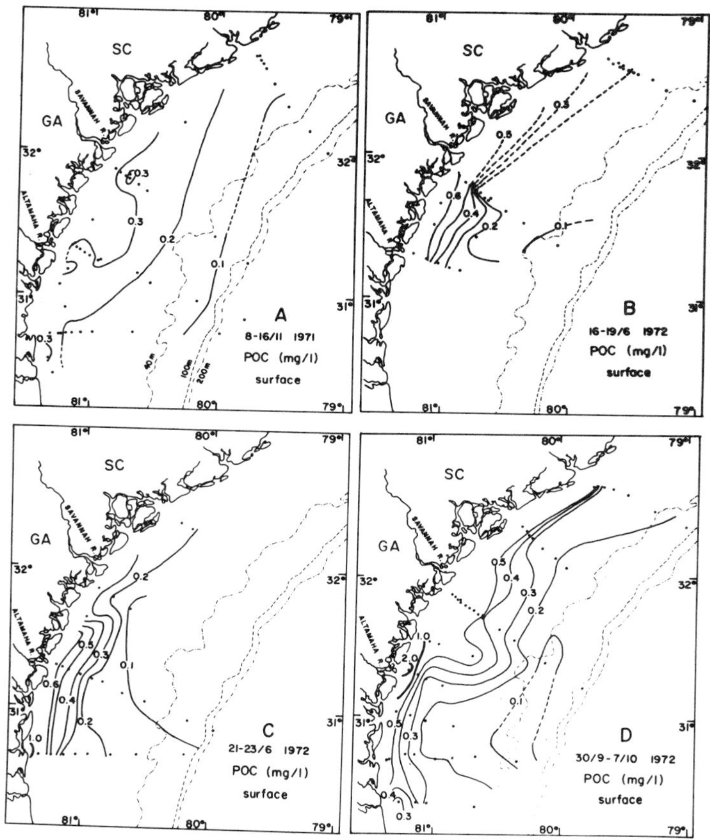

Figure 6. Distribution of particulate organic carbon off the Georgia coast. A: November, 1971; B: June, 1972, before hurricane Agnes; C: June, 1972, after hurricane Agnes; D: October, 1972.

NITROGEN BUDGET

A budget of the nitrogen flux from various sources into shelf waters between Charleston, South Carolina, and Fernandina Beach, Florida (Fig. 1, Table 1) is considered here. Calculated nitrogen influx from fresh-water drainage, deep-water intrusions, and precipitation is compared to nitrogen uptake by primary producers and to standing stock of particulate nitrogen. Nitrogen fixation is assumed to be an unimportant source of new nitrogen relative to the other sources. *Trichodesmium* blooms have been observed in this area (14), but no quantitative estimates of *Trichodesmium* abundance are available. Measured rates of nitrogen fixation by *Trichodesmium* have generally been low (4, 5).

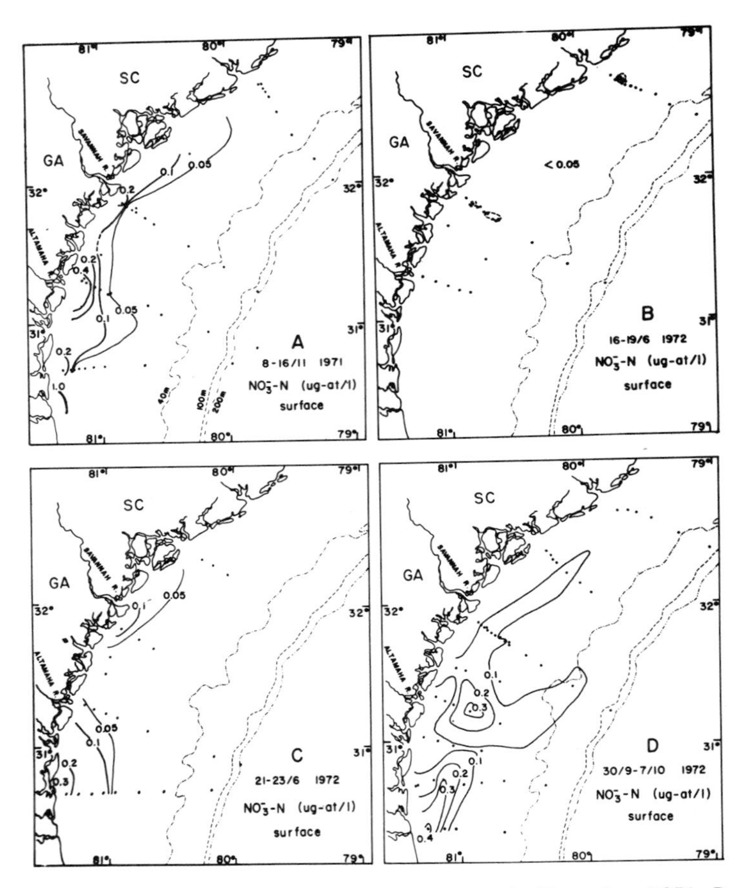

Figure 7. Distribution of nitrate off the Georgia coast. A: November, 1971; B: June, 1972, before hurricane Agnes; C: June, 1972, after hurricane Agnes; D: October, 1972

TABLE 2

Mean ammonia concentrations (μg at 1^{-1})

Depth	Sept. 71	Nov. 71	June 72a	June 72b	Oct. 72
0-20 m	0.31 ±0.17	0.12 ±0.09	0.15 ±0.10	0.33 ±0.34	0.27 ±0.21
20-200 m	0.22 ±0.14	0.12 ±0.08	0.09 ±0.09	0.35 ±0.23	0.25 ±0.13

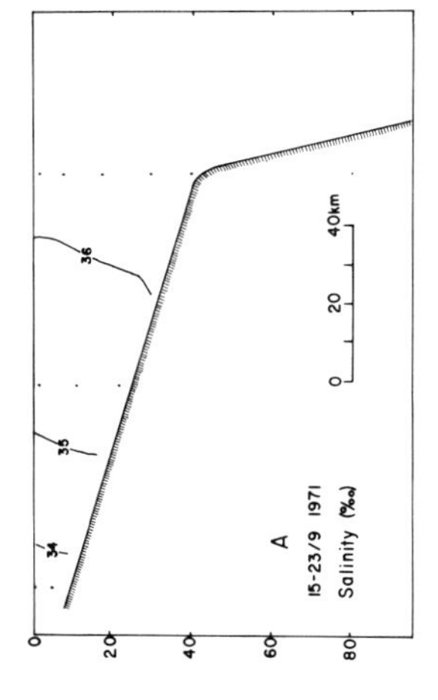

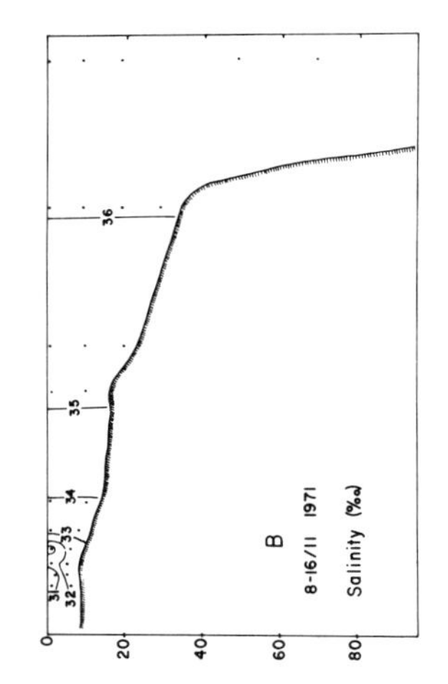

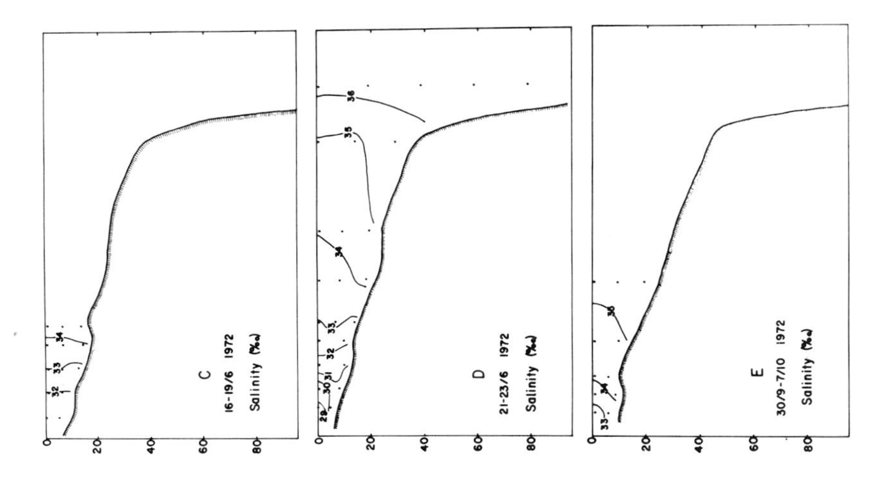

Figure 8. Vertical distribution of salinity, Sapelo Transect.

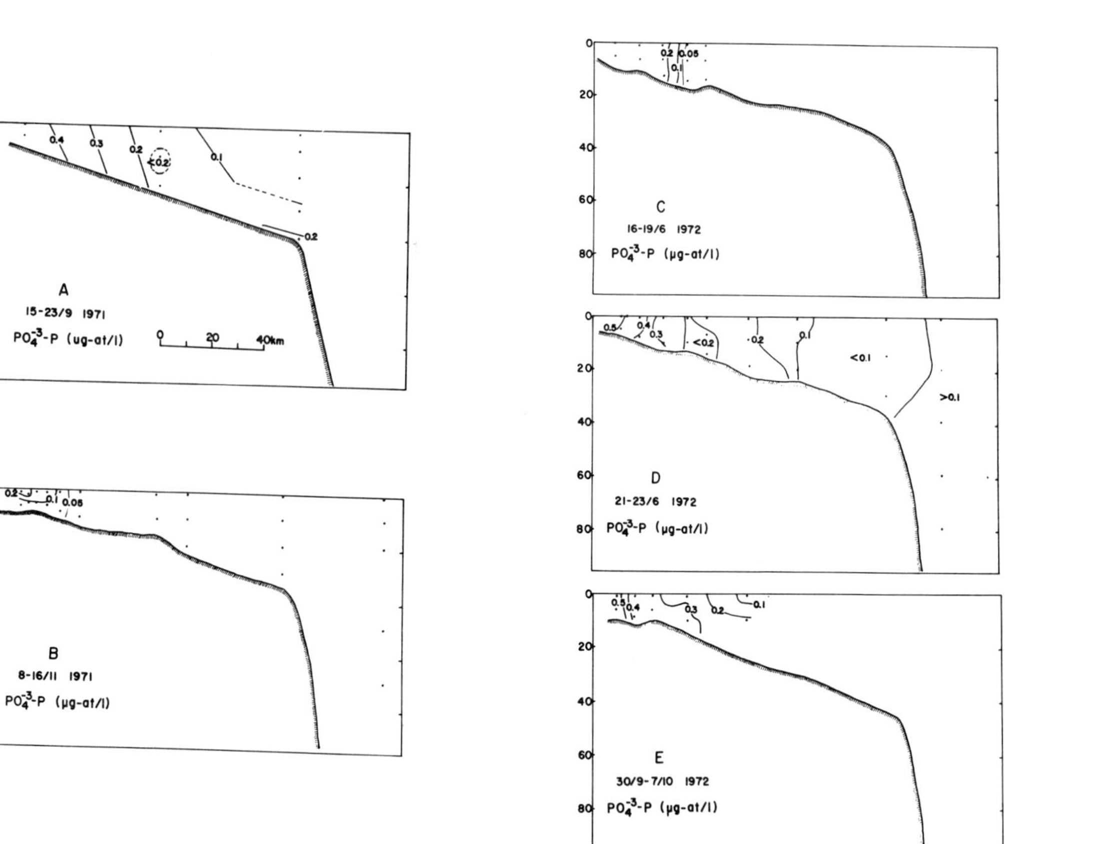

Figure 9. Vertical distribution of phosphate, Sapelo Transect.

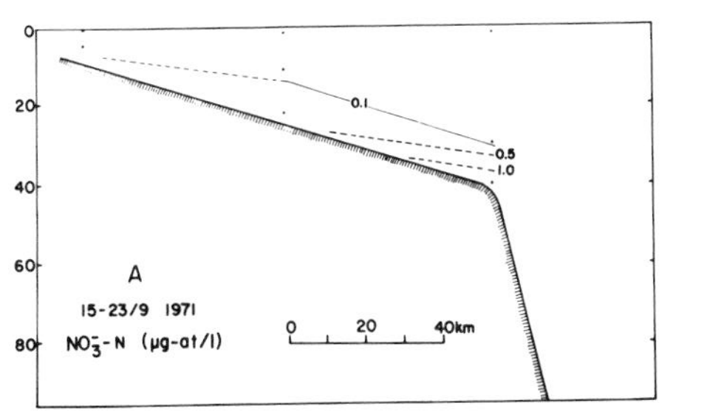

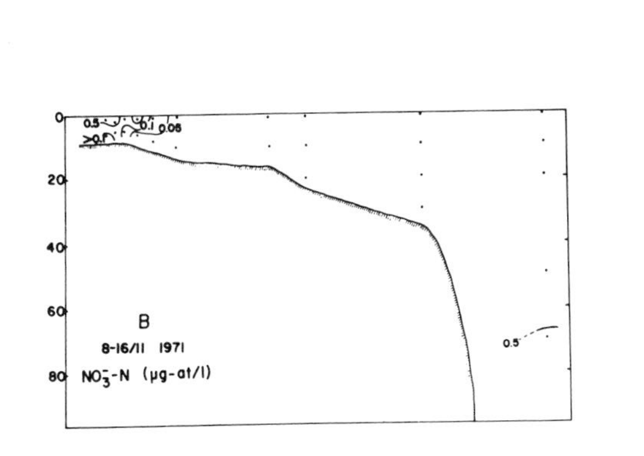

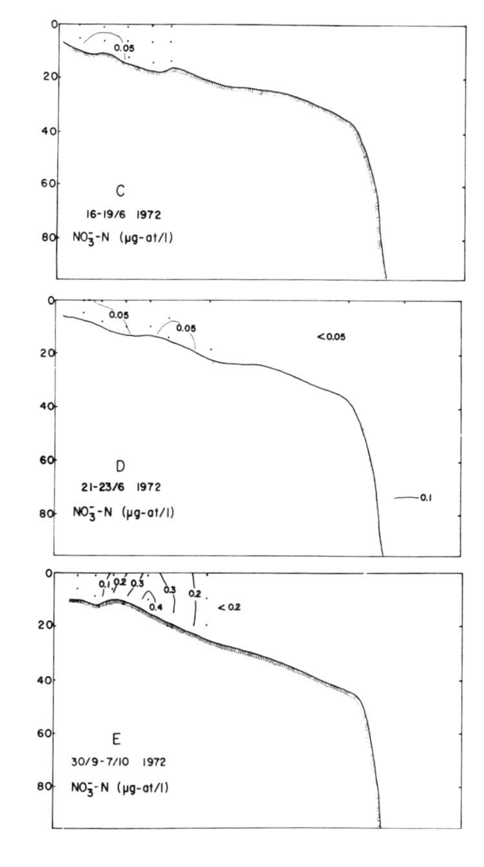

Figure 10. Vertical distribution of nitrate, Sapelo Transect.

Nitrogen inputs

Fresh-water drainage

The amount of nitrogen contributed to offshore water by river discharge is assumed to depend only on the inorganic nitrogen content of river water. Georgia salt-marsh estuaries probably do not add nitrogen to river water and may in fact remove most of the nitrate in river water (Haines, unpublished data).

Major gauged rivers discharging into the shelf area considered are listed in Table 3. The total drainage area is estimated to be 118,500 km^2. Fresh-water drainage from this area was computed by multiplying the drainage area by the ratio of total gauged river flow to total gauged drainage area as given in Table 3. The calculated mean discharge is 46,800 cubic feet per second, and the volume of fresh-water contributed annually by river flow is 4.17×10^{10} m^3.

The average inorganic nitrogen content (nitrate plus ammonia) of river water was estimated from U. S. Geological Survey water resources data (32, 33) to be 0.1 mg N l^{-1}, or about 7 μg at l^{-1}. The calculated amount of nitrogen discharged annually to coastal waters is 0.42×10^{10} g.

TABLE 3

Major gauged rivers discharging into shelf waters between Charleston and Fernandina Beach; from U. S. Geological Survey data for 1971 (32, 33).

River	Years of Record	Drainage area mi^2	(km^2)	Mean Discharge cfs	(m^3s^{-1})	cfs mi^{-2}	(m^3s^{-1} km^{-2})
Edisto	32	2730	(7071)	2617	(74)	0.97	0.01
Coosawhatchie	20	203	(526)	177	(5)	0.87	0.01
Savannah	38	9850	(25511)	11730	(332)	1.19	0.01
Ogeechee	34	2650	(6863)	2333	(66)	0.88	0.01
Canoochee	34	555	(1437)	435	(12)	0.82	0.01
Altamaha	40	13600	(35224)	13460	(381)	0.99	0.01
Satilla	41	2790	(7226)	2136	(60)	0.77	0.01
St. Marys	45	700	(1813)	677	(19)	0.97	0.01
Total		33078	(85672)	33583	(950)	1.02	0.01

Deep-water intrusions

The influx of nitrogen through eddy diffusion of nutrient-rich deep water at the edge of the shelf as described by Riley (23) is difficult to quantify from available information. Calculations of the inputs from this source are based on the following three assumptions: (a) A slope-water intrusion which extends 10 km onto the shelf occurs 100 days each year. Stefansson et al. (28) observed slope-water intrusions onto the North Carolina shelf in July, August, and September, but not in winter or spring months. Conditions which appeared to favor this phenomenon were southerly winds and stratification of surface layers. In this study an apparent intrusion onto the Georgia shelf was found only once, in September, 1971. (b) Nitrogen content of the slope water is 15 μg at l^{-1}. Nitrate concentrations in Atlantic deep water are generally equal to or greater than this concentration. Riley assumed in his model a nitrate concentration of 15 μg at l^{-1} in deep water at the edge of the shelf. The nitrate concentrations in acutal deep-water intrusions may not be so high, however. Stefansson and Atkinson (27) found nitrate concentrations of about 5 μg at l^{-1} in apparent intrusions, and the maximum nitrate concentration observed at the edge of the shelf in this study was 1.5 μg at l^{-1}. (c) There is a 2% per day diffusion or exchange of the top 10 meters of intruding deep water with shelf water. Riley (23) postulated a 1% per day exchange of water, based on rates of eddy diffusion, between stratified layers on the outer shelf. The computed amount of nitrogen contributed to shelf waters from deep-water intrusions is 0.76×10^{10}g N yr.$^{-1}$.

Precipitation

Junge (7) has summarized data on nitrate and ammonia content of rainfall at coastal stations around the world. The range of nitrate concentrations was 0.15-0.60 mg l^{-1}, with an average concentration of 0.29 mg l^{-1} (4.7 μg at N l^{-1}). Ammonia concentrations ranged from 0.04-0.29 mg l^{-1} with an average of 0.15 mg l^{-1} (8.3 μg at l^{-1}). Junge's data (8) on southeastern U. S. rainfall showed ammonia concentrations consistently lower than 0.02 mg l^{-1} (1.5 μg at l^{-1}) and nitrate concentrations generally lower than 0.20 mg l^{-1} (3.2 μg at N l^{-1}). These observations appear to be low compared to more recent analyses. Menzel and Spaeth (16) found an average ammonia concentration in oceanic rain near Bermuda of 4.97 μg at l^{-1}. Dugdale and Goering (3) suggested that the concentration of nitrate in precipitation near Bermuda may equal or exceed the ammonia concentration. Pomeroy (personal communication) measured nitrate concentrations of 15 and 17 μg at l^{-1} and ammonia concentrations of 10 and 56 μg at l^{-1} in rainfall at Sapelo Island, Georgia. I have measured ammonia concentrations in outer shelf rainfall of 0.7, 2.5, and 8.0 μg at l^{-1} (unpublished data).

The average concentration of nitrogen in rain water chosen for the estimate of nitrogen influx in rainfall is 0.18 mg N l^{-1}, or 13 μg at l^{-1}. This value corresponds to the sum of Junge's (7) figures for average nitrate and ammonia concentrations in coastal rainfall. This concentration is hopefully close to, or an overestimation of, average concentrations in shelf rainfall. Nitrogen concentrations may be much higher in rainfall near the coast. The average rainfall on the Georgia coast is about 1.3 m yr^{-3} (31). The calculated nitrogen input to the shelf area considered in rainfall is 0.75 x 10^{10} g yr^{-1}. For the shelf region inshore of the 20 m depth contour, the annual nitrogen input is about 0.29 x 10^{10}g.

Nitrogen uptake

To estimate annual nitrogen requirements of primary producers in the shelf region considered, it is necessary to know the annual carbon production on the shelf and the average C:N ratio of phytoplankton. Thomas (30) reported an annual phytoplankton primary production of 547 g C m^{-2} yr^{-1} in Georgia coastal waters near the Altamaha River. Productivity along the coast above and below this region might not be so high. Annual production in nearshore waters from the coast to the 20 m depth contour is estimated to be 250 g C m^{-2}. There are no good data for production over the rest of the Georgia shelf. Productivity in New England coastal waters of 25 to 50 m depth is about 160 g C m^{-2} yr^{-1} (25). Annual production of Georgia shelf waters of depths from 20 to 200 m is assumed to be similar and is set at 150 g C m^{-2}. Shelf production in the nearshore area is then 280 x 10^{10} g C m^{-2} yr^{-1}, and over the rest of the shelf, 270 x 10^{10} g C m^{-2} yr^{-1}. Total calculated shelf production is 550 x 10^{10} g C m^{-2} yr^{-1}.

Menzel and Ryther (15) measured C:N ratios in the North Atlantic ranging from means of 12.5:1 to 5.3:1. Redfield's (21) value of 7:1 for average C:N ratio in the sea falls within this range and will be used here. The total nitrogen requirement for production on the shelf is 93.7 x 10^{10} g N yr^{-1}. Calculated nitrogen uptake in the nearshore zone to the 20 m depth contour is 46.7 x 10^{10} g yr^{-1}.

Particulate nitrogen standing crop

The average concentration of particulate nitrogen observed in this study in coastal waters inshore of the 20 m contour is 0.043 mg l^{-1}, and over the rest of the shelf, 0.015 mg l^{-1}. Amount of nitrogen in particulate matter is about 0.48 x 10^{10} g in the near shore zone and 1.40 x 10^{10} g for the rest of the shelf. The total amount is about 1.88 x 10^{10} g N.

Ratio of inputs to uptake

Calculated nitrogen inputs, uptake, and standing stock are summarized in Table 4 and in Figure 11. The total amount of influx of new nitrogen is much less than the computed annual nitrogen uptake by primary producers. For the nearshore zone, the ratio of total influx to uptake is 4.1%, for the shelf, the ratio of input to uptake is 2.3%. The annual input of new nitrogen into shelf waters is, however, about equal to the amount of nitrogen in particulate matter in shelf waters at any one time.

TABLE 4

Calculated nitrogen inputs, uptake, and standing stock in shelf waters, between Charleston, South Carolina and Fernandina Beach, Florida (10^{10} g N).

	Nearshore Zone 0-20 m	Total for Shelf 0-200 m
Annual inputs		
Fresh-water drainage	–	0.42
Deep-water intrusions	–	0.76
Precipitation	0.29	0.75
Total		1.93
Annual uptake	46.7	93.7
Standing stock	0.48	1.88

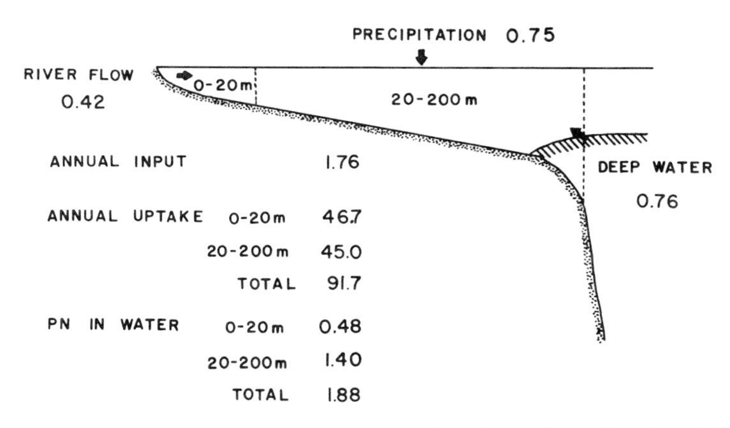

Figure 11. Summary of the nitrogen fluxes on the Georgia and South Carolina shelf.

DISCUSSION AND SUMMARY

There is no conclusive evidence based on previous work that any potential source of new nutrients provides significant amounts of nutrients to coastal waters. Ketchum and Keen (9) calculated that annual river discharge into New England coastal waters was only about 1% of the total volume of water on the shelf, and therefore represented a negligible source of phytoplankton nutrients. Plumes from major rivers have various effects on the productivity of adjacent sea water, and may decrease as well as increase production in the zone of mixing (6). Although apparent intrusions of nutrient-rich deep water onto the shelf have been observed off North Carolina (28), there has been only speculation about the importance of this process to shelf production. Riley (23) developed a mathematical model showing that shoreward transport of nutrients from deep water at the edge of the shelf may account for the observed pattern of increasing shelf production in a shoreward direction. Precipitation has been suggested as a source of enrichment for both open ocean (16) and estuarine (21) environments, but there are no comparable studies for the coastal zone. Nitrogen fixation has been shown to be an important source of new nitrogen in the ocean only in *Trichodesmium* blooms (3, 4, 5). If new information quantifying nitrogen inputs to shelf waters by nitrogen fixation becomes available, the model presented here can be revised to include it.

The budget of nitrogen inputs to Georgia and South Carolina shelf waters which has been presented indicates that the total input of new nitrogen to shelf waters is small compared to the amount of nitrogen taken up in primary production. Regeneration must account for 95% or more of nitrogen uptake by primary producers in shelf waters. Dugdale and Goering (3) have made independent estimates of the ratio of uptake of new nitrogen (nitrate) to total nitrogen uptake (nitrate plus regenerated nitrogen, or ammonia) by measuring uptake of ^{15}N-labeled nitrate and ammonia in the euphotic zone of various parts of the world ocean. Mean ratios varied from 8.3%, characteristic of subtropical regions, to 39.5%, typical of temperate and coastal areas. The calculated ratio of new to total nitrogen uptake of less than 5% for Georgia shelf waters is lower than Dugdale's and Goering's (3) estimates for coastal areas based on ^{15}N uptake. The high regeneration rates required do not seem unreasonable for the shallow, well-mixed waters of the Georgia coastal zone, where nutrients are not lost from the euphotic zone by sinking or grazing as in open-ocean waters. R. E. Turner (personal communication) has observed respiration rates in Georgia coastal waters equal to a turnover of 80% of the daily production.

Although the amount of new nitrogen contributed annually appears small compared to the annual uptake, the nitrogen influx is of the same order of magnitude as the amount of nitrogen in particulate matter on the shelf. The calculated input of new nitrogen can support an annual turnover of the nitrogen in the particulate pool.

Although the influx of nitrogen from rivers and estuaries does not appear to be an important factor in maintaining high rates of nearshore production in Georgia coastal waters, river and estuarine discharge may affect offshore productivity in a number of other ways. The annual fresh-water discharge into the shelf area considered is equal to 39% of the total water volume in the nearshore zone from the coast to the 20 m depth contour. A water influx of this magnitude is certain to influence the chemistry and production of offshore waters. The higher concentrations of phosphate in nearshore water is certainly maintained by discharge of phosphate-rich estuarine water (30). River water may also contribute significant amounts of trace metals (34) as well as silicate to the coastal zone. Organic nutrients such as vitamin B_{12} which are plentiful in estuaries (26) may stimulate productivity when flushed offshore. Prakash (20) has suggested that organic compounds of terrestrial origin, such as humic acids, which act as chelators, may enhance coastal production.

Export of organic materials from the salt-marsh estuaries may also add directly to offshore secondary production. Odum and de la Cruz (19) reported concentrations of organic detritus of 2-20 mg C l^{-1} in water discharged from a salt-marsh tidal creek. If the average concentration in discharged estuarine water is 20 mg C l^{-1}, then the annual carbon influx to the shelf waters would be 83 x 10^{10} g yr^{-1}, but should be considered in the carbon budget.

The observations and analysis reported here indicate that for the Georgia and South Carolina shelf, nutrient influx to the coastal zone via outwelling is of minor importance, mixing of deep water across the edge of the shelf is of minor importance, and *in situ* regeneration is the most important process in maintaining high rates of nutrient flux and hence high rates of biological productivity in the shelf waters.

ACKNOWLEDGMENTS

The author acknowledges the support of National Science Foundation grant no. GA-28742 and the support of the Oceanographic Program of Duke University Marine Laboratory for use of R/V *EASTWARD* on cruises E-22c-71, E-26-71, E-9-72, and E-18b-72. The Oceanographic Program is supported by National Science Foundation grant nos. GA-27725, GD-28333, and GD-32560. Many thanks to all the people who helped in the sample collection of my cruises. I am grateful to Dr. R. T. Barber for criticizing this manuscript, and to my long-suffering husband, Bruce.

REFERENCES

1. Anderson, W. W., Gehringer, J. W., and Cohen, E.
   1956-1959 Physical oceanographic, biological, and chemical data. South Atlantic coast of the United States, M. V. THEODORE GILL Cruises 1-9. **Spec. Scient. Rept. U. S. Fish Wildl. Serv. Fish.** 178, 198, 210, 234, 265, 278, 303, 313.
2. Banoub, M. W. and Williams, P. J. le B.
   1972 Measurements of microbial activity and organic material in the western Mediterranean Sea. **Deep-Sea Res.,** 19: 433-444.
3. Dugdale, R. C. and Goering, J. J.
   1967 Uptake of new and regenerated forms of nitrogen in primary productivity. **Limnol. Oceanogr.,** 12: 196-206.
4. Dugdale, R. C., Goering, J. J., and Ryther, J. H.
   1964 High nitrogen fixation rates in the Sargasso Sea and the Arabian Sea. **Limnol. Oceanogr.,** 9: 507-510.
5. Dugdale, R. C., Menzel, D. W., and Ryther, J. H.
   1961 Nitrogen fixation in the Sargasso Sea. **Deep-Sea Res.,** 7:298-300.
6. Goldberg, E. D.
   1971 River-ocean interactions. In **Fertility of the Sea,** Vol. 1, p. 143-156. (ed. Costlow, J. D.) Gordon and Breach, New York.
7. Junge, C. E.
   1958 Atmospheric chemistry. In **Advances in Geophysics,** Vol. IV, (ed. Landsberg, H. E., and Van Miegham, J.) Academic Press, New York.
8. Junge, C. E.
   1958 The distribution of ammonia and nitrate in rainwater over the United States. **Trans. Amer. Geophs. Union,** 39: 241-248.
9. Ketchum, B. H. and Keen, D. J.
   1955 The accumulation of river water over the continental shelf between Cape Cod and Chesapeake Bay. **Deep-Sea Res.,** 3: 346-357.
10. Koroleff, F.
    1970 ICES. Information on techniques and methods for sea water analysis. **Interlab. Rept.,** 3: 19-22.
11. Kuroda, R. and Marland, F. C.
    1973 Physical and chemical properties of the coastal waters of Georgia. Georgia Inst. Tech. Environmental Resources Center, ERC 0373. Atlanta. (Unpublished manuscript.)
12. Lorenzen, C. J.
    1966 A method for the continuous measurement of **in vivo** chlorophyll concentration. **Deep-Sea Res.,** 13: 223-227.
13. Manheim, F. T., Meade, R. H., and Bond, G. C.
    1970 Suspended matter in surface waters of the Atlantic continental margin from Cape Cod to the Florida Keys. **Science,** 167: 371-376.
14. Marshall, H. G.
    1971 Composition of phytoplankton off the southeastern coast of the United States. **Bull. Mar. Sci.,** 21: 806-825.
15. Menzel, D. W., and Ryther, J. H.

1964 The composition of particulate organic matter in the western North Atlantic. **Limnol. Oceanogr.,** 2: 179-186.

16. Menzel, D. W. and Spaeth, J. P.
1962 Occurrence of ammonia in Sargasso sea waters and in rain water at Bermuda. **Limnol. Oceanogr.,** 7: 159-162.

17. Menzel, D. W. and Vaccaro, R. F.
1964 The measurement of dissolved organic and particulate carbon in seawater. **Limnol. Oceanogr.,** 9: 138-142.

18. Odum, E. P.
1968 A research challenge: Evaluating the productivity of coastal and estuarine water. **Proc. 2nd Sea Grant Conf.,** Grad. School Oceanography, University of Rhode Island, Newport. pp. 63-64.

19. Odum, E. P. and de la Cruz, A.
1967 Particulate organic detritus in a Georgia salt-marsh estuarine ecosystem. In **Estuaries** (ed. Lauff, G. H.) Am. Assoc. Adv. Sci. Publ. 83. Washington, D. C., pp. 383-388.

20. Prakash, A.
1971 Terrigenous organic matter and coastal phytoplankton fertility. In **Fertility of the Sea,** Vol. II, p. 351-368. (ed. Costlow, J. D.) Gordon and Breach, London.

21. Redfield, A. C., Ketchum, B. H., and Richards, F. A.
1963 The influence of organisms on the composition of sea-water. In **The Sea,** (ed. Hill, M. N.), Vol. II, pp. 26-77. Interscience Publ., New York.

22. Reimold, R. J.
1968 Euthrophication of estuarine areas by rainwater. **Chesapeake Sci.,** 8: 132-133.

23. Riley, G. A.
1967 Mathematical model of nutrient conditions in coastal waters. **Bull. Bingham Oceanogr. Coll.,** 19: 72-80.

24. Ryther, J. H. and Dunstan, W. M.
1971 Nitrogen, phosphorus, and eutrophication in the coastal marine environment. **Science,** 171: 1008-1012.

25. Ryther, J. H. and Yentsch, C. S.
1958 Primary production of continental shelf waters off New York. **Limnol. Oceanogr.** 2: 281-286.

26. Starr, T. J.
1956 Relative amounts of vitamin B_{12} in detritus from oceanic and estuarine environments near Sapelo Island, Georgia. **Ecology,** 37: 658-664.

27. Stefansson, U. and Atkinson, L. P.
1967 Physical and chemical properties of the shelf and slope waters off North Carolina. **Tech. Rept.,** Duke Univ. Mar. Lab., 230 pp.

28. Stefansson, U., Atkinson, L. P., and Bumpus, D. F.
1971 Hydrographic properties and circulation of the North Carolina shelf and slope waters. **Deep-Sea Res.,** 18: 383-420.

29. Strickland, J. D. H. and Parsons, T. R.
1967 A practical handbook of seawater analysis. **Bull. Fish. Res. Bd. Canada,** 167: 1-311.

30. Thomas, J. P.
1966 The influence of the Altamaha River on primary production beyond the mouth off the river. M. S. thesis, Univ. Georgia, Athens. 88 p.

31. U. S. Department of Commerce
1968 U. S. Climatic Atlas. Washington, D. C.

32. U. S. Geological Survey
1972 1971 Water resources data for Georgia. Atlanta. 295 p.

33. U. S. Geological Survey
1973 1971 Water resources data for South Carolina. Part 1. Surface water records. Part 2. Water quality records. Part 3. Ground water records. Columbia. 208 p.

34. Windom, H. L., Beck, K. D., and Smith, R.
1971 Transport of trace metals to the Atlantic Ocean by three southeastern rivers. **Southeast. Geol.,** 12: 169-181.

POPULATION DYNAMICS OF ZOOPLANKTON

IN THE MIDDLE ST. LAWRENCE ESTUARY

by E. L. Bousfield[1], G. Filteau[2], M. O'Neill[3], and P. Gentes[4]

ABSTRACT

Preliminary studies on the distribution of medium zooplankton in the Middle St. Lawrence Estuary between Ile d'Orléans and the Saguenay fiord during 1971 revealed three principal populations: (1) an essentially coastal marine group of species, further subdivisible into reproductive and nonreproductive populations, that penetrate landwards mainly in cold, high-salinity bottom waters; (2) an estuarine-endemic holoplanktonic group, dominant in medium-brackish, cool, middle waters, mainly near the surface; and (3) a fresh-water holoplanktonic group dominant in the warm tidal fresh waters of the estuary. The vertical swimming behavior of the animals and the two-layered estuarine circulation apparently combine to concentrate the marine forms in discrete, deep, channel locations near their upper estuarine limit of occurrence, in biomass sufficient to form a significant food source on which the capelin and ultimately white whales depend.

INTRODUCTION

During the period May to September 1971, a preliminary survey of the

1. National Museum of Natural Sciences, Ottawa, Canada.

2. Faculté des Sciences, Université Laval, Québec 10.

3. Deceased, June 15, 1973.

4. Département de Biologie, Université Laval, Québec 10.

5. Groupe Interuniversitaire de Recherches Océanographiques du Québec.

6. Deceased, June 15, 1973.

hydrobiology of the Middle St. Lawrence Estuary in the region from Ile d'Orléans to St. Siméon and Hare Island was undertaken (Fig. 1). The purpose of the investigation was to obtain new information on the distribution and abundance of medium zooplankton on a geographical, bathymetrical, and seasonal basis, and to delimit hydrobiological mechanisms that enable eurytopic members of the intertidal and littoral fauna, some with pelagic larvae, to penetrate broadly into this physically rigorous and changeable environment. It was hoped that some explanation would be found for the apparently low secondary productivity of the Middle Estuary, a region that paradoxically supports a food pyramid capped by white whales, but which excludes most species of groundfish, commerical shellfish and benthic crustaceans. The study also endeavored to establish a general framework and methodology on which long-range future surveys might successfully be based.

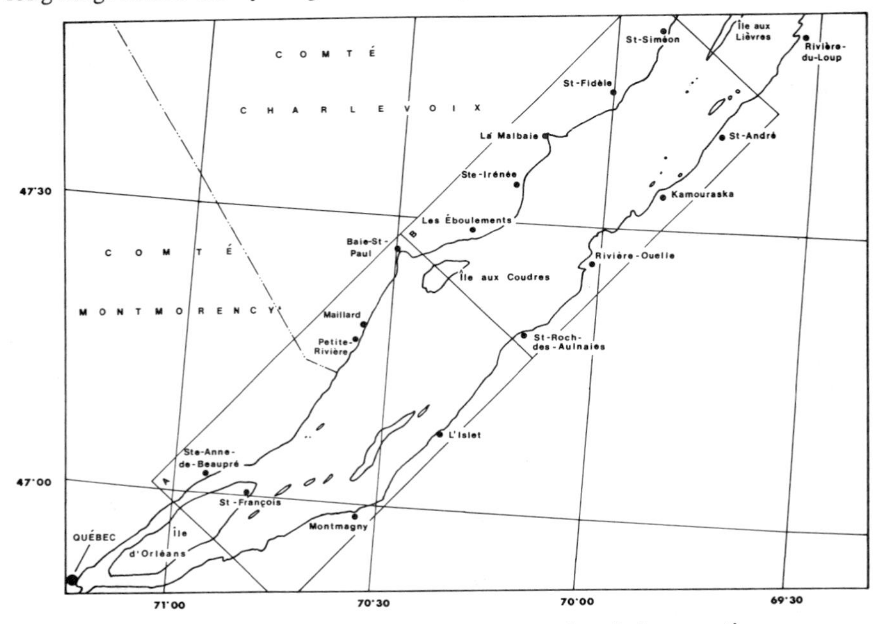

Figure 1. The Middle St. Lawrence Estuary. A: upper section; B: lower section.

PRINCIPAL PHYSICAL FEATURES OF THE MIDDLE ESTUARY

The St. Lawrence Estuary comprises the tidal waters from Lake St. Peter on the west to the Gulf of St. Lawrence on the east. It is a low-latitude, cold-water hydrobiological system unique to eastern Canada, having no counterpart elsewhere in the world. Within this 350-mile-long (560 km) estuary the entire fresh-water surface flow of the St. Lawrence drainage basin meets and mixes with salt water from the Gulf of St. Lawrence on its way to the Atlantic Ocean. For convenience of study, some authors (e.g., 4, 5) have subdivided the Estuary geographically and hydrologically into an Upper Estuary (tidal fresh-water portion above Ile d'Orléans), a brackish Middle Estuary (Ile d'Orléans to Saguenay fiord and Trois Pistoles), and a cold, salt Lower Estuary (Saguenay fiord to Pte. des Monts and Cap Chat).

Within the Middle Estuary, the greatest degree of hydrological and biological change takes place, not only horizontally, but also vertically and seasonally. Herein major ecological factors presumably reach limiting levels for marine, brackish, and fresh-water organisms. The distribution of surface temperature and salinity in summer has been illustrated by Gaudry (7) and Nadeau (11). In the Middle Estuary may be observed the greatest effect of the St. Lawrence freshet on the marine flora and fauna, the greatest extremes of horizontal and vertical mixing and sedimentation, and the maximum lateral effect of the coriolis force on the circulation of the water and dispersal of endemic plankton organisms.

The upper section of the Middle Estuary (Fig. 2A) is relatively broad and shallow, studded by many rocky islands and shallow sand bars treacherous to navigation. Mean tidal amplitudes are large, ranging to about 16 feet (5 m) at the east end of Ile d'Orléans. The main channel is located close to the steep north coast, with normal depths of 60-100 feet (18-30 m), dredged to 30 feet (9 m) in shoalest channel areas. The lower section of the Middle Estuary (Fig. 2B) is slightly narrower but considerably deeper, with channel depths everywhere greater than 150 feet (45 m), and more than 300 feet (90 m) in seaward areas. A central axial ridge, shoaling to the surface at (Morin Shoal and Hare Island Bank, separates north and south Channels. The north coast is very steep, almost precipitous in places, with a narrow mid-tidal shelf that drops abruptly into deep water close inshore. The lower terrace of the south coast slopes gently offshore, forming broad tidal flats and shoals before dropping sharply at the south channel. The high tidal amplitudes and boulder-strewn mid-tidal shelf, characteristic of the Middle Estuary, prevent access to the Estuary from small wharves except during the high-water part of the tidal cycle.

The first significant studies on the zooplankton of the Middle Estuary are those of Herdman et al. (8). Between Quebec City and the Saguenay fiord these authors recorded the following copepod species: *Calanus finmarchicus* (Gunnerus), *Pseudocalanus elongatus* (Boeck), *Acartia longiremis* (Lilljeborg), *Eurytemora herdmani* (Thompson & Scott), and *E. affinis* (Poppe). The last-named species, a well-known brackish-water form, was noted in considerable abundance, prompting Herdman et al. to conclude that the Middle Estuary was a region of mixing of fresh and salt water. They also recorded as rare *Oithona spinifrons* (Boeck) and *Ectinosoma sarsi* (Boeck). Willey (18) compiled a list of copepod species occurring in the estuary off Trois Pistoles, Prefontaine (13) listed some invertebrate species of the lower Middle Estuary, and Tremblay (17)

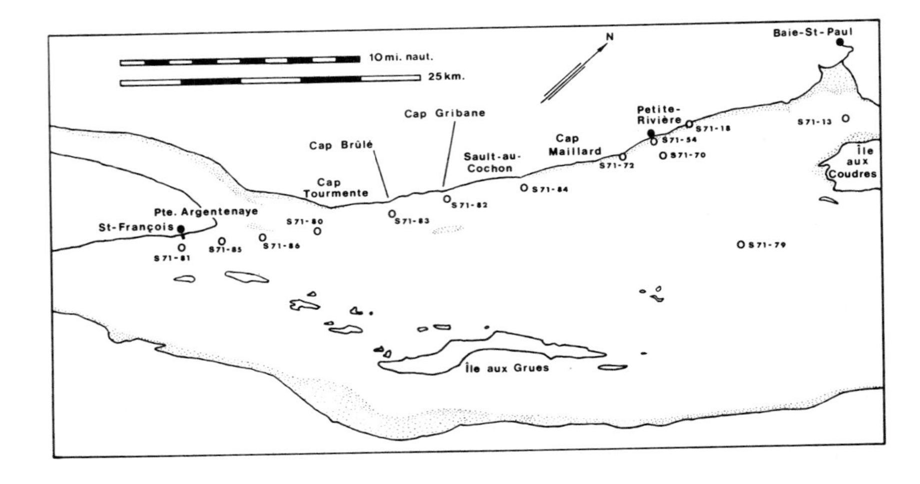

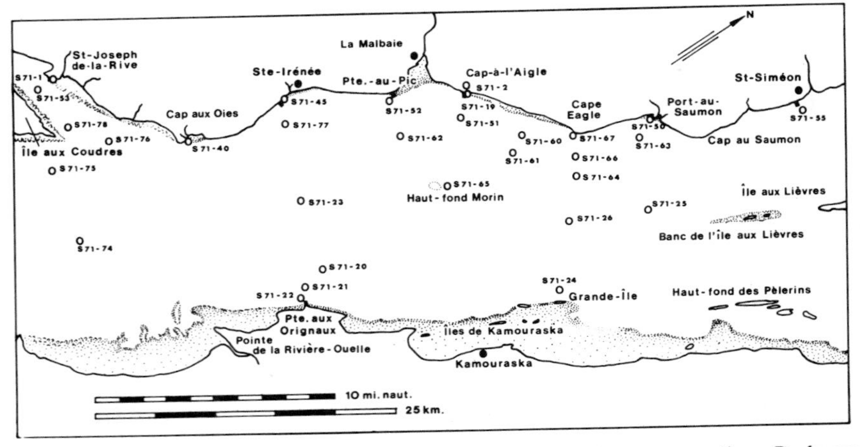

Figure 2. Stations visited, Middle St. Lawrence Estuary. A: upper section; B: lower section.

provided a list of regional benthonic harpacticoids (identified by A. G. Nicholls) that included *Ectinosoma intermedia* Nicholls, *E. littoralis* Nicholls, and an unidentified species of *Ectinosoma*. Tremblay's principal contribution was an annotated summary of planktonic copepods taken throughout the St. Lawrence Estuary; it included *Euchaeta norvegica* Boeck, *Metridia longa* Lubbock and *Calanus hyperboreus* Krøyer in waters of high salinity and negative temperatures. Characteristic of less salt and warmer estuarine waters were *Calanus finmarchicus, Pseudocalanus elongatus, Oithona similis* Claus and *Oncea borealis* Sars. However, in the brackish Middle Estuary, Tremblay found three common species of *Eurytemora*, viz. *E.herdmani*, *E. hirundoides* Norquist (= *E. affinis*) and *E. americana* Williams. These were "replaced" in the tidal fresh-water upper estuary by species of *Epischura* Forbes.

These previous qualitative accounts of the dominant crustacean zooplankton provided little information on regions of dominance, on distributional limits, and on depth distribution and seasonal occurrence. Clues to the seasonal occurrence of immature or larval stages of barnacles (*Balanus crenatus* Bruguière), ostracods, harpacticoids and amphipods were provided by Bourget (1) in a study of attachment and settlement of fouling organisms on special collectors set off Pte. aux Orignaux. Relative to populations in the extreme Lower Estuary and Gulf of St. Lawrence (see 3, 9, 10), densities of the Middle Estuary were found to be very low, and spawning seasons much delayed. Bourget's work, in part, stimulated the present studies on the regional zooplankton and subsequent studies.

MATERIALS AND METHODS

A grid of collecting stations was established in the Middle Estuary, concentrated mainly along the north channel from the east end of Ile d'Orléans to St. Siméon. Coverage was more complete in the middle and south sides of the lower part of the Middle Estuary then elsewhere (Fig. 2). All stations were reached by means of a small outboard-powered figerglas boat capable of speeds up to about 20 knots under favorable operating conditions. The vessel could be launched at only three locations, all on the north side of the estuary; at St. François (Ile d'Orléans), Baie St. Paul, and Cap à l'Aigle. All station circuits had to be completed during a maximum launch-and-recovery time of about 5-6 hours, depending on location, lest the expedition be "locked out" of the docking areas during the low-tide period. Unpredictable and sudden changes in meteorological conditions, especially wind, and heavy steamship traffic in the main ship channels posed appreciable risk to these small-boat operations.

Plankton samples were obtained by means of a portable pump powered by a 2 HP Briggs-and-Stratton gasoline engine. The rubber intake hose, of one-inch inside diameter, reached depths of 225 feet (70 m). The intake nozzle was fitted with a check valve. Plankton was strained by means of a standard Clarke-Bumpus net of No. 10 mesh size held in the pump outflow. Because of flow friction in the hose, the rate of flow varied from about 25 gallons (115 1) per minute (surface sample) to about 5 gallons (25 1) per minute (225-foot [70 m] sample). The duration of sampling was adjusted to a total delivery of about 100 gallons (450 1). With the boat drifting freely, the nozzle was weighted to maintain the hose in a nearly vertical position. Pump delivery rate was calculated by timing the filling of a 3-gallon pail immediately after each plankton sample was taken. Water temperature was measured with a small alcohol thermometer accurate to the nearest 0.5 °C. Salinities were determined in the laboratory by electric conductivity instruments, accurate to 0.01 °/oo, and later with a Beckman RS5 salinometer. The scientific validity of pump samples was checked by means of 3-minute surface tows using a standard Clarke-Bumpus plankton sampler and No. 10 net mesh. Despite an 8-fold increase in total catch, relative proportions of medium zooplankton species remained approximately the same using both methods.

Plankton samples were preserved in the field in approximately 5% sea-water formalin. In the laboratory, total counts were made on about 75% of the nearly 200 samples obtained, since most of these were made in impoverished surface waters of the lower Middle Estuary where less than 500 organisms were usually taken per 100-gallon (450 1) sample. Aliquot portions (to 1/16 in a few cases) of richer collections, especially from the upper Middle Estuary, were obtained by means of a standard Folsom plankton splitter. All animal species and stages were recorded using standard denominator counters and a binocular microscope. Copepod identifications were based mainly on Sars (14) and barnacle larvae on Crisp (6). Counts were transferred to special mimeographed sheets according to station, date, and depth. Biomass conversion factors were worked out for each species and stage group by (1) micromeasurement and (2) decantion methods, with remarkably close agreement. Original and summary data sheets are available from the Department of Biology, Université Laval (*see also* preliminary report by O'Neill and Gentes [12]).

OBSERVATIONS AND RESULTS

From field data accompanying plankton samples taken June 15 to September 1, it is possible to plot a composite picture of the average summer distribution of salinity (Fig. 3) and temperature (Fig. 4) during the high-tide period. Essentially cold (< 5 °C), salt (> 25°/oo) water penetrates up-estuary along the bottom to the Coudres Channel off Baie St. Paul. Warm (> 19 °C), fresh water meets and mixes thoroughly with the salt water at all depths (absence of thermo- or hali-clines) below Ile d'Orléans to the Coudres Channel. From there the cool, high-brackish mixture spreads seaward in a relatively thin surface layer, mainly along the south side owing to the Coriolis effect. From *a priori* considerations, the residual (nontidal) circulation appears to be seawards at the surface and landwards below about 40 feet (12 m). Major regions of upwelling apparently occur along the north channel off Cap à l'Aigle and in the Coudres Channel; at these localities, salinities are persistently higher and temperatures lower than in surface waters at immediately adjacent localities. Judging from the few salinities taken from wharves at low water and from the average speed of tidal currents, the isopleths of salinity and temperature at times of low water would be shifted about ten miles (16 km) seaward, but the structural relationships within the

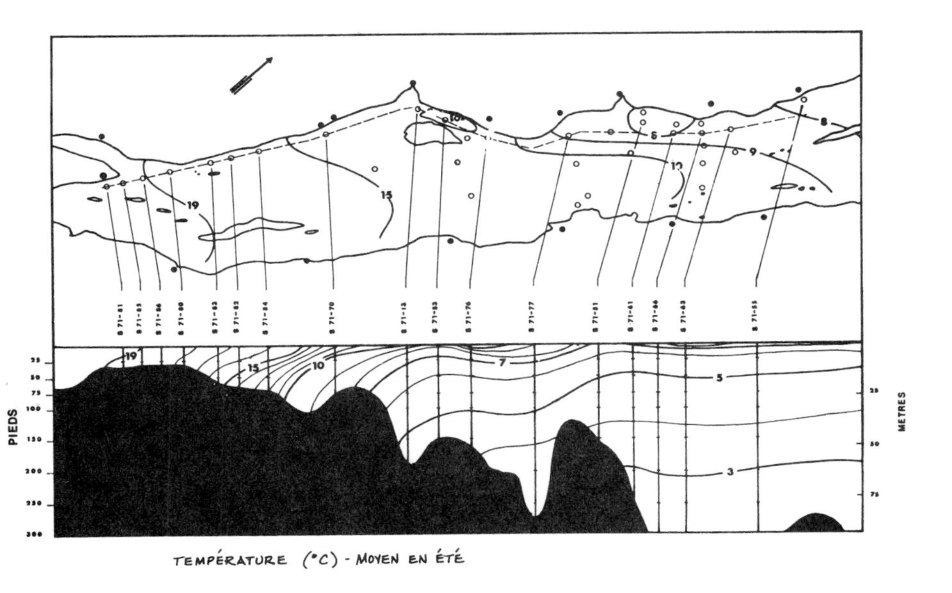

Figure 3. Average distribution of salinity (°/oo) in summer along the Middle Estuary.

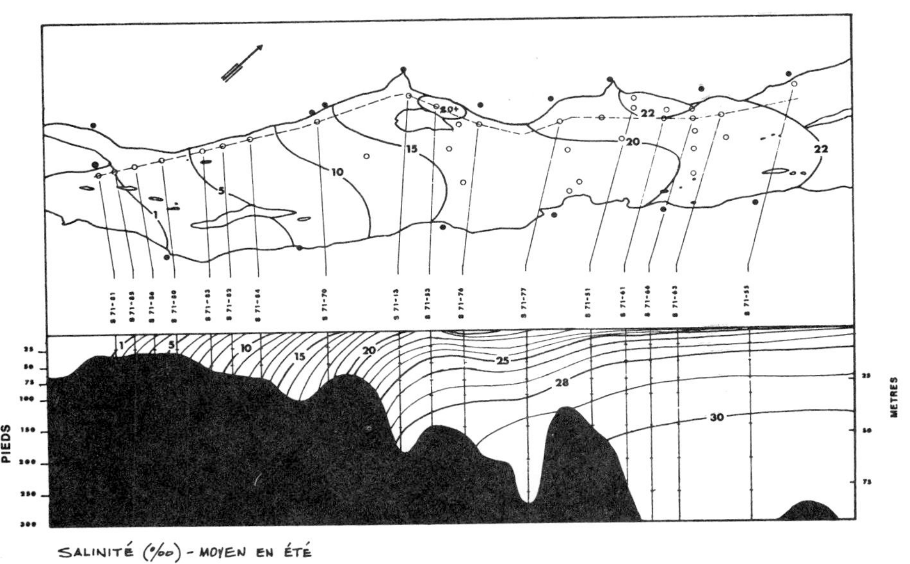

Figure 4. Average distribution of temperature (°C) in summer along the Middle Estuary.

water column would remain nearly constant.

Although no Secchi disc readings were obtained, visual observations from the boat suggest that water turbidity is highest in the upper section of the Middle Estuary and along the south side of the lower section. The turbidity appears to be caused mainly by fine silt and organic detritus maintained in suspension by the strong tidal currents that scour the extensive muddy tidal flats and submerged banks. The fresh-water discharge of the St. Lawrence Upper Estuary and the landward-moving, cold salt water of the Lower Estuary appear to have higher transparencies than the mixed waters of the intervening zone.

The medium zooplankton of the Middle Estuary is distinguished by its low species diversity and low total abundance. Although more than 25 species of all phyla and classes were identified, only sixteen species of medium crustacean zooplankton were analyzed. The Copepoda was the dominant group. Of ten calanoid species recorded (Fig. 5), only three (*Eurytemora affinis, E. herdmani* and *Acartia longiremis*) were present in relatively large numbers (frequently more than 100 per sample) and, during summer, were reproducing significantly in the Middle Estuary.

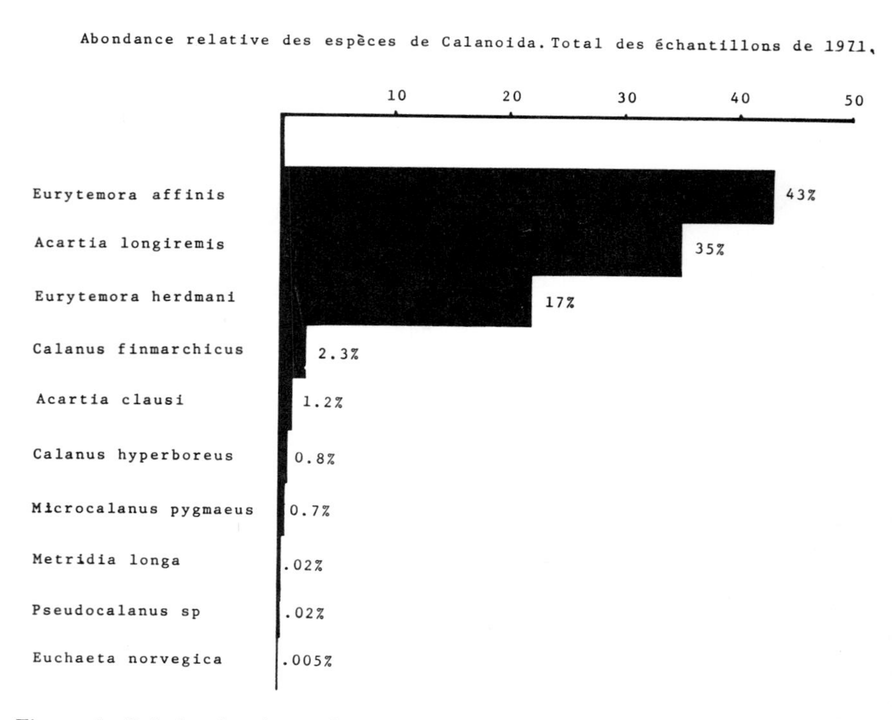

Figure 5. Relative abundance of calanoid copepod species in total plankton catches, 1971.

The dominant estuarine medium zooplankton species may be grouped in three principal subdivisions, viz.:

(1) Marine coastal species that require relatively cold, high salinity waters. These may be further grouped into: (a) Nonreproducing species, of which only very late larval stages and adults were taken in the Middle Estuary during summer months. Species included *Calanus finmarchicus, C. hyperboreus, Acartia clausi* Giesbrecht and miscellaneous rare Calanoid species (Fig. 5), as well as *Oithona* sp. and occasional specimens of *Thysanoessa* spp. (Euphausiacea). (b) Reproducing species, of which all copepodite and nauplius stages and adults were taken in the Middle Estuary during summer months. Species included the holoplanktonic copepods *Acartia longiremis* and *Eurytemora herdmani,* and the meroplanktonic larvae of the benthonic barnacle *Balanus crenatus.* Occasional larvae and subadults of the euphausiid *Thysanoessa* sp. were taken, especially in Clarke-Bumpus plankton tows. However, *not* taken were the zoeal stages of *Crangon,* an epibenthic decapod shrimp not uncommon along the south shore of the Middle Estuary.

(2) Estuarine-endemic species that occur essentially in cool-to-warm brackish waters. Species included the holoplanktonic copepod *Eurytemora affinis,* the epibenthic and planktonic harpacticoid *Ectinosoma curticorne* Sars, and the hyperbenthic and planktonic opossum shrimp *Neomysis americana* (Smith). Other benthic harpacticoids, specimens of *Mysis gaspensis* Tattersall and *M. stenolepis* Smith, and all stages of the amphipod *Gammarus tigrinus* Sexton were taken sporadically. A second harpacticoid species occurred commonly in samples from the zone of recovery and lower section in 1972.

(3) Fresh-water species that are characteristic of the Upper Estuary. They penetrate seawards only into very low salinities and are rapidly eliminated in the Middle Estuary. Species included *Bosmina longirostris* (O. F. Müller), occasionally *Daphnia* sp., *Epishchura,* and fresh-water diaptomid copepods.

Marine Coastal Species

Calanus finmarchicus (Gunnerus), a boreal marine species, is the dominant large copepod in the Gulf and Estuary of the St. Lawrence. It extends into the Middle Estuary as late copepodites and adult stages, mainly in waters deeper than 50 feet (15 m) (Figs. 6, 11). Populations are relatively low, reaching maximum densities of 30-40 specimens per 100 gallons (450 l) in middle depths below Cap à l'Aigle and in the deep trough off Ste. Irénée (Fig. 6). Population densities of higher than 10 per 100 gallons (450 l) were obtained where summer temperatures are continuously below about 6 °C and salinities above 25‰. The species penetrates only slightly above Ille aux Coudres on the north side, to maximum temperatures of about 8 °C and minimum salinities of 21‰. Populations are generally higher on the north than on the south side of the

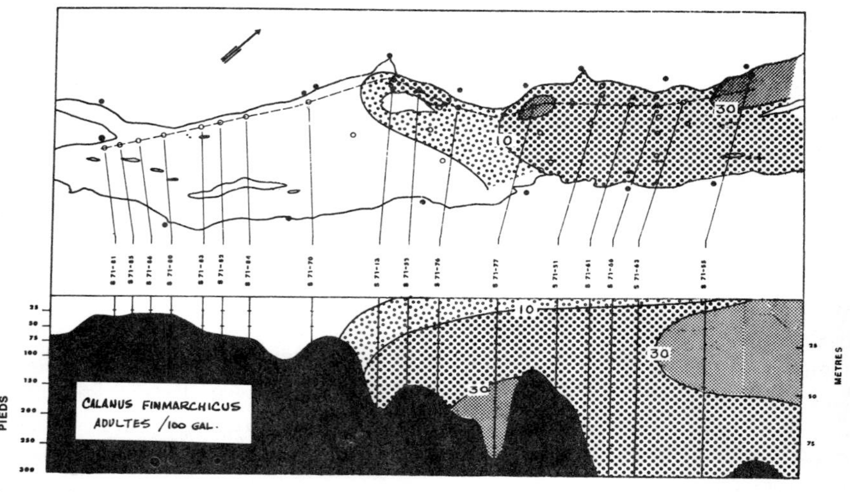

Figure 6. Average distribution of **Calanus finmarchicus** (adults) along the Middle Estuary.

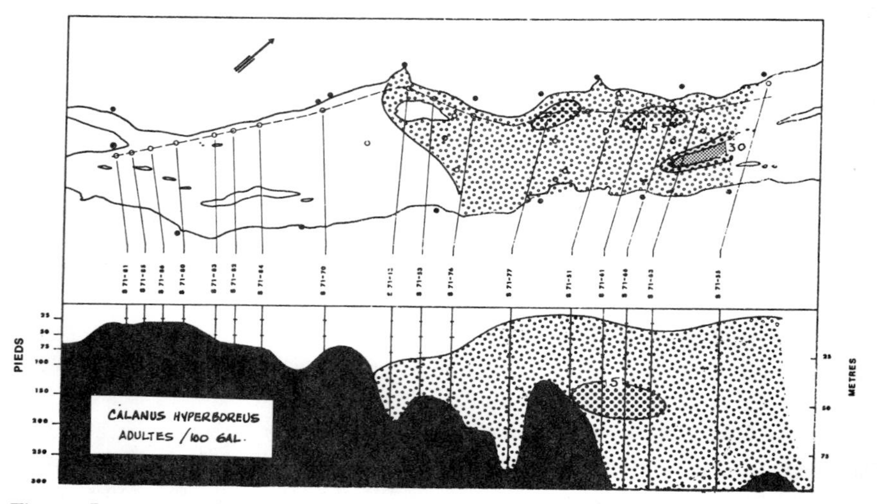

Figure 7. Average distribution of **Calanus hyperboreus** (adults) along the Middle Estuary.

Estuary (Fig. 11 A, B. C).

Calanus hyperboreus Krøyer is a very large arctic and subarctic coastal marine species that occurs widely (but not in great abundance) in the Gulf and Estuary of the St. Lawrence, especially in deeper and colder regions. Although immature stages have been taken Previously off Trois Pistoles (15), the species penetrates into the Middle Estuary only as terminal copepodite and adult stages (Fig. 7). It occurs in quite low numbers, even in channels along the north side, almost exclusively at depths below 50 feet (15 m). Maximum concentrations of more than 30 per 100 gallons (450 l) were attained in the deep channel off Grande Ile, in salinities higher than 25°/oo and temperatures lower than 5°C. The species penetrated in very small numbers to the Coudres Channel at about the same limits of temperature and salinity as for *C. finmarchicus.*

Acartia clausi Giesbrecht is a small- to medium-sized boreal marine species widely distributed and regionally dominant in coastal waters of eastern Canada and the Gulf of St. Lawrence. It penetrates the Middle Estuary, mainly in deeper waters, in small numbers (up to 25 per 100 gallons [450 l]) to channels off Grande Ile and Ste. Irénée, in salinities of more than 28°/oo and temperatures lower than 5 °C (Fig. 8). Its extreme upper limit is reached at about Petite Rivière in salinities of 16°/oo and temperatures of 10 °C. No early copepodite stages of *Acartia* were attributed to this species, indicating reproduction is proceeding at very low levels, if at all, in this physically unfavorable environment.

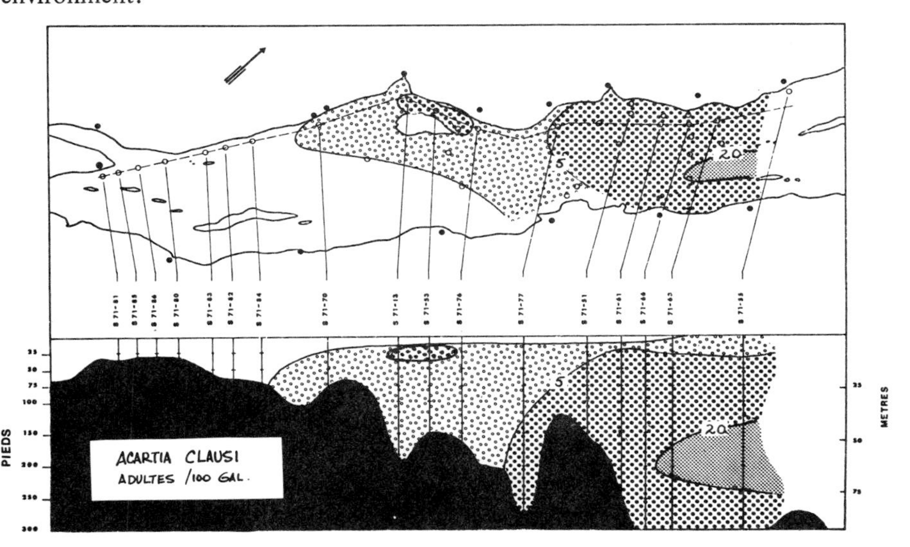

Figure 8. Average distribution of **Acartia clausi** (adults) along the Middle Estuary.

Acartia longiremis (Lilljeborg), a subarctic-boreal species, was the dominant planktonic crustacean in surface waters of the lower Middle Estuary. All life stages tended to be concentrated along the north channel, with adult stages especially numerous in the deep off Ste. Irénée, mainly in salinities of 20-28°/oo and temperatures of 4-8 °C (Figs. 9, 10). A few animals occurred up-estuary to Petite Riviere in salinities as low as 15°/oo and temperatures as high as 12 °C. Slightly higher concentrations occurred over south-side channels than in mid-estuary, although none were taken near the bottom at most landward

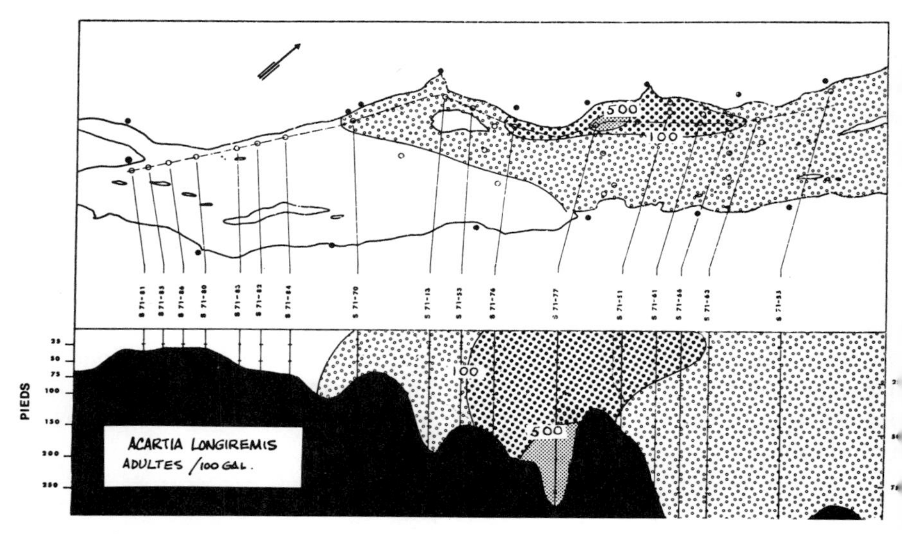

Figure 9. Average distribution of **Acartia longiremis** (adults) along the Middle Estuary.

south-side stations (Fig. 10). Under certain meteorological conditions and light intensities, adults tend to congregate near the bottom, as at Station S71-77. All nauplii and early copepodites identifiable as *Acartia* were attributed to *A. longiremis.*

Estuarine-endemic species

Eurytemora herdmani Thompson & Scott, a boreal coastal marine species, is dominant in the Gulf of St. Lawrence but less abundant than *Acartia longiremis* in the Estuary. Within the Middle Estuary, however, adult specimens were actually quite scarce (maximum of one per gallon [4.5 l]), were relatively less abundant than *Acartia longiremis,* and occurred at somewhat greater depths, on the average (Fig. 12). Adults tended to be concentrated in deep pockets near the upper estuarine limit of occurrence that coincided with salinity levels of about 15°/oo and summer temperatures of about 10 °C. Some copepodites of *Eurytemora* were identified as this species, all well down-estuary.

Balanus crenatus Bruguière is the dominant subtidal cirriped species in the Gulf and Estuary of the St. Lawrence. It occupies depths mainly in the upper 300 feet (90 m) extending to 600 feet (180 m) (2). Judging from previous studies on the ecology of this species (3) and from distribution of newly spawned larvae, reproducing adults extend up-estuary along the bottom at least to the Coudres Channel. The spawning period extends from May (possibly earlier) at least through September. The shallow-water adult populations presumably spawn before those of deeper water.

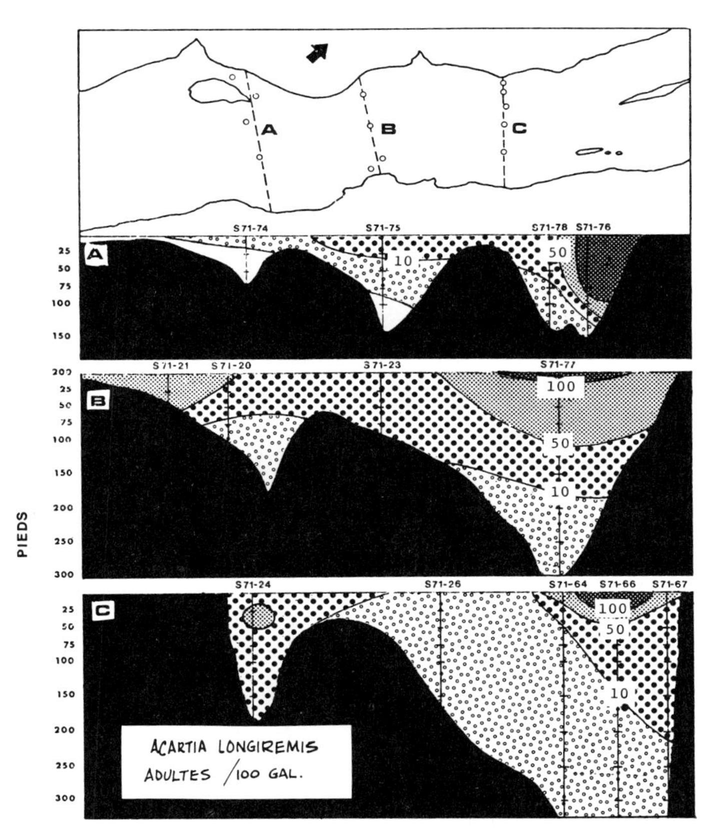

Figure 10. Average distribution of **Acartia longiremis** (adults) across the Middle Estuary.

Nauplius stage II, having a planktonic period of about one week, is abundant in surface waters below Cap à l'Aigle and Pte. aux Orignaux and in the channel between Petite Rivière and Ile aux coudres (Fig. 13A). Upper (landward) limits of occurrence were noted at salinities of approximately 10°/oo and temperatures of 15 °C. Nauplius stages III and IV, each lasting about three days, occurred at greater depths, mainly down-estuary, with a small pocket persisting below Ile aux Coudres, in lowest salinities of about 20°/oo and temperatures as high as 10 °C (Fig. 13B). Nauplius stages V and VI, with similar duration, occurred at about the same localities but at somewhat greater depths (Fig. 14A). The small up-estuary population was lacking. The final, deep-swimming cyprid with a long-lasting pelagic stage (two weeks or more, failing attachment) was not only abundant in down-estuary localities but was taken in apparently greater concentrations in middle and bottom waters in the north channel above Ile aux Coudres to Sault au Cochon (Fig. 14B). Accompanying lowest salinities were

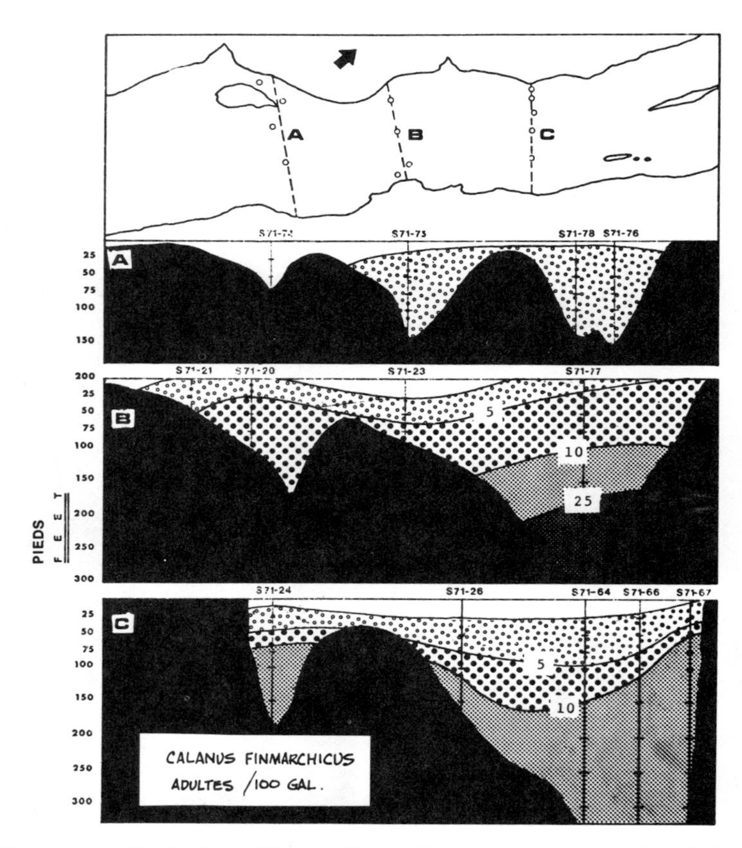

Figure 11. Average distribution of **Calanus finmarchicus** (adults) across the Middle Estuary.

about 6⁰/oo and temperatures up to 18 ⁰C.

The explanation of this anomalous distribution pattern may be essentially similar to that for larval dispersal in *Balanus improvisus* Darwin in the Miramichi estuary (3). All larval stages apparently survive, develop, and settle normally in the cold, high-salinity water below Goose Cape and Pte. aux Orignaux. Newly spawned nauplius II from uppermost estuarine adult populations in the Coudres Channel are probably killed by low salinities in surface waters or drift seawards in nontidal flow before metamorphosing to stage III. Because of natural mortality (10% per day?) the four stages III-VI are present in progressively smaller numbers but maintain a horizontal position by sinking into landward-flowing, deeper waters. The very low actual numbers of these late nauplius stages taken is attributable to their short duration in the plankton. The long-lasting, deep-swimming cyprid, virtually absent from surface waters except in areas of upwelling, is carried up-estuary and concentrated in deep channels

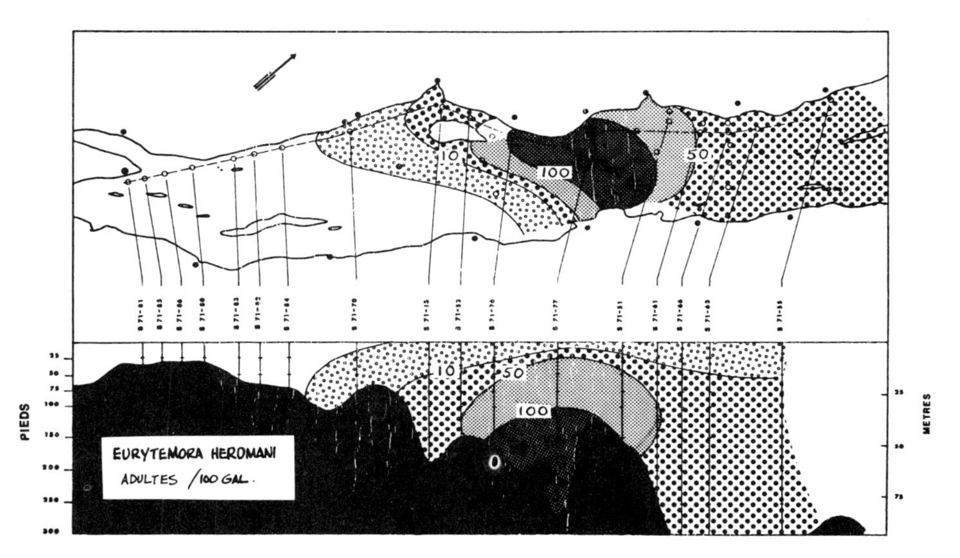

Figure 12. Average distribution of **Eurytemora herdmani** (adults) along the Middle Estuary.

above Ile aux Coudres to its limit of low salinity tolerance. Moribund or partly decomposing cyprids were noted among specimens from uppermost stations. Interestingly, small barnacle spat-of-the-year were noted in small numbers on buoy cylinders and chains lifted from the estuary off Petite Rivière and Ile aux Coudres in December, 1971. Although settlement, attachment, and metamorphosis obviously do take place at these lower near-surface salinities, feeding and growth are apparently inhibited.

Barnacle larvae were more numerous in south-side than in north-side channels. This phenomenon may reflect the greater initial abundance of spawning adults in the shallow, more extensive substrata of south-shore localities.

Eurytemora affinis (Poppe) was the dominant copepod species, and indeed the dominant medium zooplankton species, in the Middle Estuary. The species occurred over a salinity range of 0-30°/oo (mainly less than 25°/oo) and a summer temperature range of 5-20 °C. Adult animals apparently formed two main populations, the principal concentration being located in the low-brackish (1-10°/oo), warm (15-20 °C), turbid region above Petite Rivière, and a second concentration (especially in June) in the Coudres Channel in salinities of 20-25°/oo and temperatures 6-10 °C (Fig. 15C). Copepodites were concentrated in the same area as the upper adult population, suggesting that the latter is the source of most, if not all, of the regional reproduction in the species (15A, B). The bulk of adults and copepodites occurs at mid-depths and near the bottom, apparently enabling the animals to maintain an up-estuary position and avoid being swept seaward by the net outflow of brackish waters at the surface.

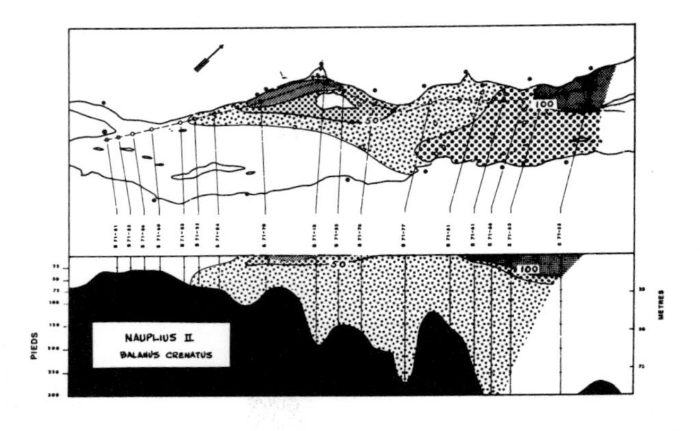

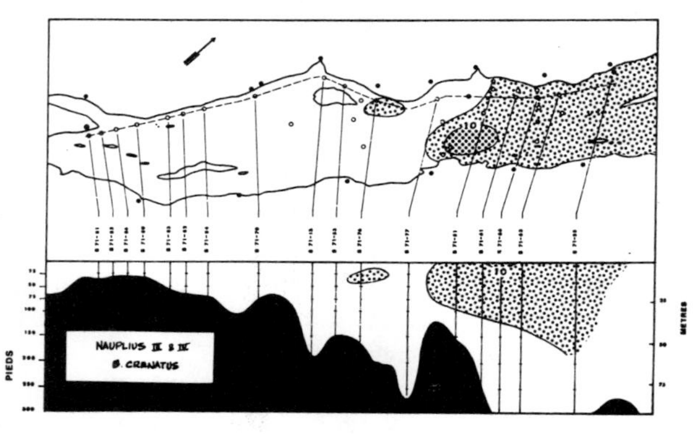

Figure 13. Average distribution of larval stages of **Balanus crenatus** along the Middle Estuary. A: nauplius II; B: nauplius III, IV.

However, other species of *Eurytemora* (e.g., *E. lacustris* Forbes, *E. americana* Williams) have previously been recorded from the Saguenay and Middle St. Lawrence regions (15). The possibility that these species comprise at least a small part of present material, but were not detected, would bear further study.

Although *Neomysis americana* (Smith) has been recorded intertidally throughout the estuary (16), the present records are apparently the first pelagic and epibenthic records from the Middle Estuary. All stages, from newly hatched young to adult, occur most abundantly in the low brackish (0-20°/oo), summer warm (10-20 ^{o}C) turbid waters above Petite Rivière (Fig. 16). Concentrations are highest at mid-depth and near the bottom; very few were taken near the surface. Although concentrations of specimens are only about 5% of *Eurytemora affinis,* their much larger size results in a greater biomass of *Neomysis* in that

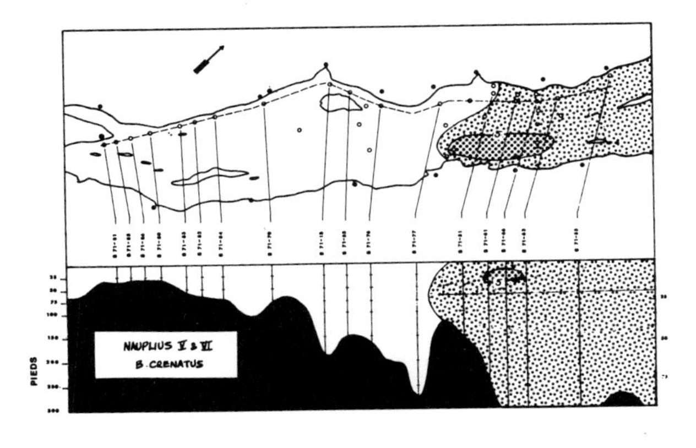

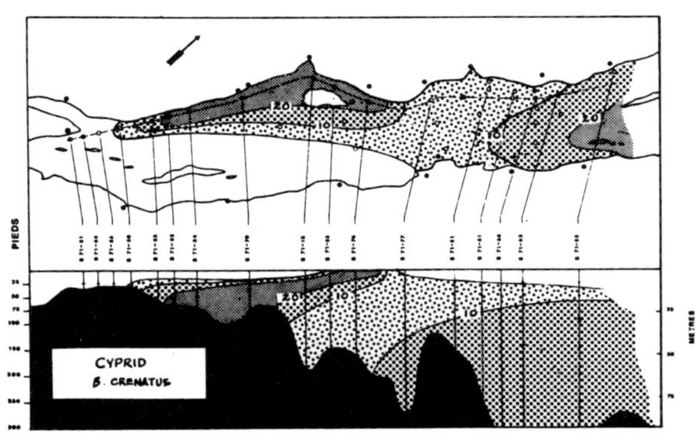

Figure 14. Average distribution of larval stages of **Balanus crenatus** along the Middle Estuary. A: nauplius V, VI; B: cyprid.

region.

In early spring and late fall, a few planktonic specimens of *Mysis stenolepis* Smith and *M. gaspensis* Tattersall were captured. Very large specimens of the former species have been observed (December, 1971) on buoy chains, either clinging to, or caught in, the chain links as they are pulled out of the water by the ship's winch.

Ectinosoma curticorne Sars may have previously been reported from this region as *E. sarsi* Boeck by Herdman et al. (8) and as *Ectinosoma* sp. by Willey (18) and others. Planktonic specimens of this benthic and epibenthic harpacticoid species are abundant in low brackish (0.2-10°/oo) and summer warm (15-20 ^{o}C) turbid waters above Petite Rivière (Fig. 17). Some specimens were taken seawards, mainly at mid-depths, to about the same limits as *Neomysis,* but less far than all stages of *Eurytemora affinis.* The species

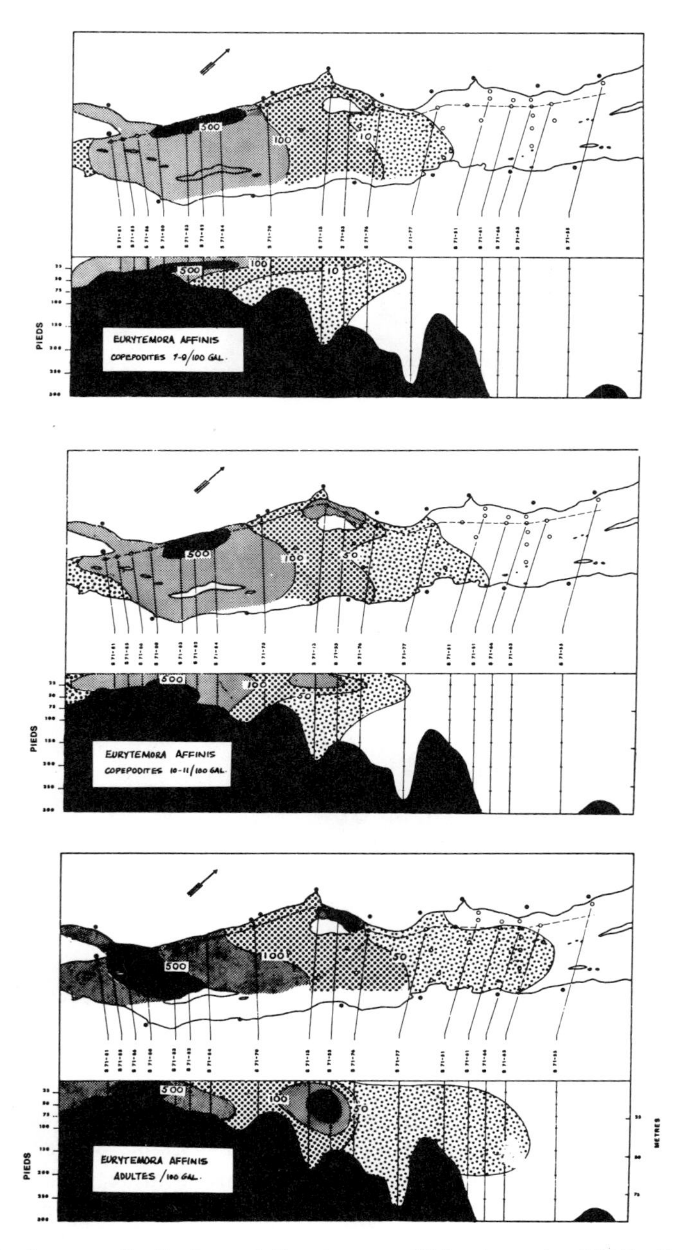

Figure 15. Average distribution of **Eurytemora affinis** along the Middle Estuary. A: copepodites VII-IX; B: copepodites X, XI; C: adults.

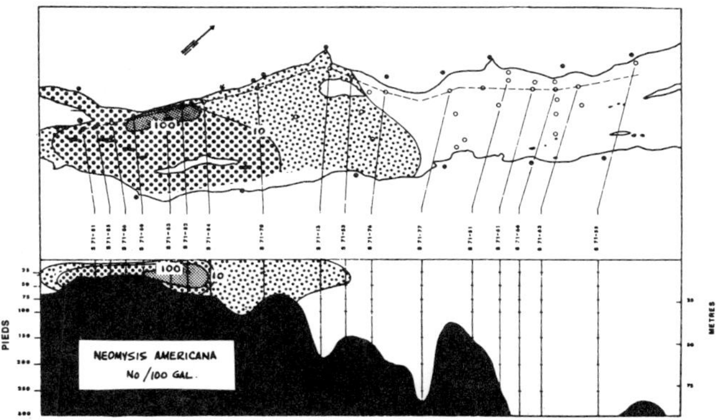

Figure 16. Average distribution of **Neomysis americana** along the Middle Estuary.

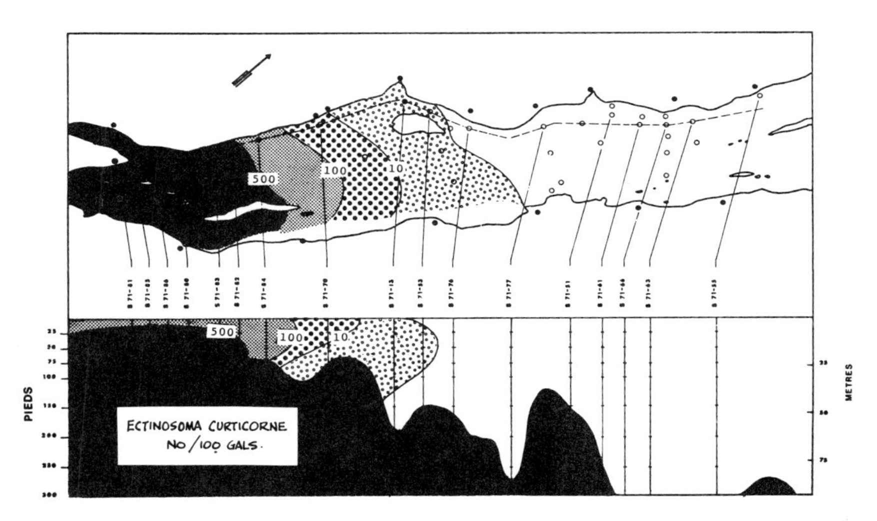

Figure 17. Average distribution of **Ectinosoma curticorne** (all stages) along the Middle Estuary.

apparently occurs only in low concentrations in purely fresh, less turbid waters, above the extreme limit of salt-water penetration. Further information on the taxonomy, life history, and ecology of this interesting species is needed.

Fresh-water species

The small cladoceran *Bosmina longirostris* (O. F. Müller) was superabundant in the strictly fresh-water portion of the Estuary, above the limit of salt penetration and mainly near the surface (Fig. 18). Although abundantly present

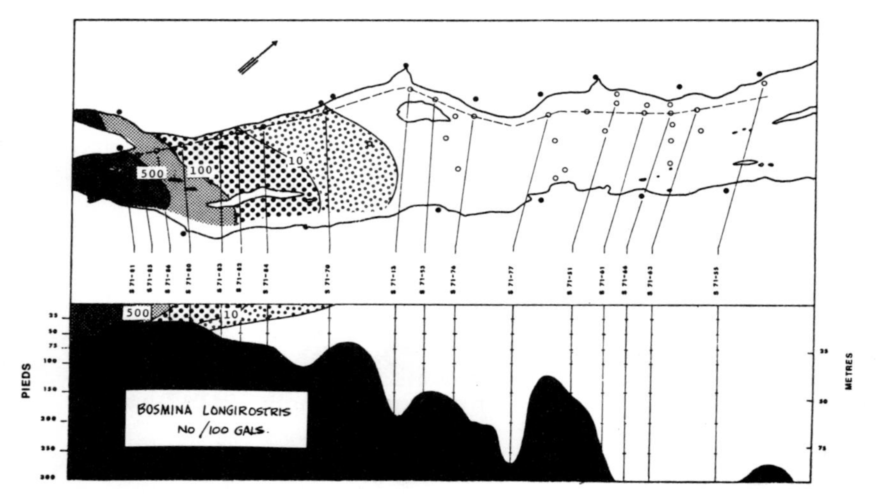

Figure 18. Average distribution of **Bosmina longirostris** (all stages) along the Middle Estuary.

also in extreme low-brackish (0-2°/oo) and summer warm (above 19 °C) turbid waters, the species rapidly diminished down-estuary in salinities greater than 5°/oo. A very few specimens, in surface samples only, were taken at salinities of 15°/oo.

A few specimens of *Daphnia* sp., relatively very large, occurred in some fresh-water but not in brackish-water samples.

DISCUSSION

Distribution

These preliminary surveys of the medium zooplankton in the Middle St. Lawrence Estuary have established the overall composition, abundance, and three-dimensional distribution of the principal crustacean species during the summer period. However, we yet have little information on vertical swimming behavior in response to diurnal and tidal cycles, or distribution and composition during fall, winter, and early spring. The plankton analyses confirm previous observations on the generally very low abundance of zooplankton, especially in near-surface samples, and especially from stations along the north side of the Middle Estuary. As noted previously, more than half of all samples, corrected to 100-gallon (450 l) volume, had fewer than 500 specimens of all species and stages, and counts could easily be made on total samples without subsampling. These numbers are calculated to be less than 1/50 the concentration of zooplankton in equivalent parts of the Miramichi Estuary (3).

Richest zooplankton concentrations were located in two or three main areas. The down-estuary concentrations corresponded with observed areas of concentrations of capelin (*Mallotus*) and white whales (*Delphinapterus*). The mechanism of retention of copepods in these regions is not yet clear. However, it would seem that these marine species are carried progressively landwards (up-estuary) in deep, salt, cold water, not reproducing, to the limit of survival. By controlling vertical swimming behavior, terminal (adult) stages may seek refuge in deep salt "holes", selectively swimming down and avoiding death or seaward loss in salt water that is advected to the surface and mixed with outflowing brackish surface water. Their source of food is also not clear. The animals may feed by night during vertical migration into detritus-rich middle layers, or exist on a "rain" of organic detritus sinking into deeper layers from outflowing surface layers.

Comparative abundance and secondary productivity

Assessing the total abundance of medium zooplankton throughout the region and at various depths, utilizing the present limited three-dimensional and seasonal sampling results, is a difficult task. Nonetheless, some indication of overall differences can be expressed in terms of total numbers of plankters or in terms of approximate biomass of all plankters combined. Gross inspection reveals two prime areas of plankton abundance, namely (1) the upper fresh-water and low-brackish region and (2) the deep channels on either side of the lower Middle Estuary. On a more quantitative basis, i.e., total number of all species and larger stages of medium crustacean zooplankton per 100 gallon (450 l) sample, the numbers generally conform to the two areas of concentration (Table 1). Numbers of zooplankton from the upper (fresh-water) section are three to ten times greater than from the lower section. Within the latter, numbers of animals are two to four times greater at north-side stations (especially at S71-77) than elsewhere, although numbers on the south side channel are greater than in mid-estuary (e.g., Figs. 10, 11). In all cases, the lowest concentration of zooplankton species is in the zone of greatest salinity change in the region to the west and south of Ile aux Coudres, where fewer than 200 specimens (average maximum per 100 gallons [450 l]) were netted.

Total maximum zooplankton biomass can be calculated with the aid of a conversion factor that accounts for the different size of species and life stages (Table 2). Included in the smallest biomass group are *Bosmina, Ectinosoma,* nauplii of *Balanus crenatus,* and copepodites of *Eurytemora* and *Acartia longiremis.* Adult stages of *Eurytemora* and *Acartia* have five to ten times the body mass of their larvae and adults of *Calanus* may have from two hundred to five hundred times the mass of the smaller species and stages. *A priori* considerations of body length and diameter (thickness) and gross inspection of

TABLE 1

Average maximum numbers of principal zooplankton species and stages per 100-gallon (450 l) sample at principal collecting stations. (Estimated from base data and Figs. 6-18.)

Species/Stage	Station Number, Upper Middle Estuary									
	81	80	83	84	70	13	73	76	75	74
Bosmina longirostris (all stages)	700	90	30	8	1	–	–	–	–	–
Ectinosoma curticorne (all stages)	2000	1000	700	250	25	3	1	–	2	5
Neomysis americana (all stages)	25	100	150	9	6	2	1	–	2	4
Eurytemora affinis (copepodites)	250	1100	1500	1400	180	150	125	90	65	65
Eurytemora affinis (adults)	450	600	400	200	80	100	500	80	70	80
Balanus crenatus (nauplii)	–	–	–	20	125	125	90	60	50	2
Balanus crenatus (cyprid)	–	1	12	20	25	25	25	20	12	1
Eurytemora herdmani (adults)	–	–	–	–	2	30	45	100	50	2
Acartia clausi (adults)	–	–	–	–	1	5	7	5	2	1
Acartia longiremis (copepodites)	–	–	–	–	3	17	34	67	10	3
Acartia longiremis (adults)	–	–	–	–	5	25	50	100	15	4
Calanus finmarchicus (adults)	–	–	–	–	–	10	20	25	5	–
Calanus hyperboreus (adults)	–	–	–	–	–	1	1	2	1	1
Totals	3425	2900	2800	1910	450	495	950	550	285	165

TABLE 1 (cont'd)

Species/Stage	Station Number, Lower Middle Estuary								
	77	23	20	51	65	67	64	26	24
Bosmina longirostris (all stages)	–	–	–	–	–	–	–	–	–
Ectinosoma curticorne (all stages)	–	–	–	–	–	–	–	–	–
Neomysis americana (all stages)	–	–	–	–	–	–	–	–	–
Eurytemora affinis (copepodites)	2	4	15	–	–	–	–	–	–
Eurytemora affinis (adults)	10	25	40	–	5	0	1	4	2
Balanus crenatus (nauplii)	25	40	75	40	50	90	90	70	75
Balanus crenatus (cyprid)	7	7	6	7	8	9	10	15	18
Eurytemora herdmani (adults)	150	120	100	55	90	30	35	30	25
Acartia clausi (adults)	5	5	5	8	7	10	14	18	20
Acartia longiremis (copepodites)	340	20	34	134	47	50	50	6	34
Acartia longiremis (adults)	500	30	50	200	70	75	75	9	50
Calanus finmarchicus (adults)	35	25	15	20	20	15	13	10	15
Calanus hyperboreus (adults)	5	4	3	4	4	4	5	3	30
Totals	1075	280	345	435	300	285	295	165	270

samples in which the larger species were taken, amply confirm this point. Total numbers, converted to biomass, more strikingly reveal the concentrations of zooplankton. Marine species reach maximum levels in the deep channels off Ste. Irénée and St. Joseph de la Rive on the north side and off Grand Ile on the south. Biomass consists largely of terminal stages of the large copepods *Calanus finmarchicus* and *C. hyperboreus* that concentrate in deep cold waters in a form suitable for predation by such pelagic plankton-feeding organisms as capelin. These fishes are apparently more abundant in these very parts of the Estuary and are in turn preyed upon by white whales (beluga) that are conspicuously common here also in summer.

Maximum zooplankton biomass is attained in the extreme low salinity range (1-10°/oo) largely in the form of immature specimens of *Neomysis americana,* but to significant extent by all stages of *Eurytemora affinis* and *Ectinosoma curticorne.* To what extent it is available to, or utilized by, plankton-feeding fishes is a matter of conjecture. The mysids are relatively large and suitable prey for larger fishes, but their presence must be masked by the strong tidal currents, high turbidity, and great amounts of detritus in the water column. In addition , relatively few species of fish find this physically harsh and variable environment suitable for prolonged existence. Also, we mught expect the zone of mixing of the Upper Estuary to be, in effect, a "graveyard" of bodies of marine and

TABLE 2

Average maximum biomass of principal zooplankton species and stages, relative units per 100 gallon (450 l) sample at principal collecting stations. (From Table 1.)

Species/ Stage	Estimated Biomass Conversion Factor	Station Number, Upper Middle Estuary									
		81	80	83	84	70	13	73	76	75	74
Bosmina longirostris (all stages)	1.0	700.0	90.0	30.0	8.0	1.0	–	–	–	–	–
Ectinosoma curticorne (all stages)	0.5	1000.0	500.0	350.0	125.0	12.5	1.0	0.5	–	1.0	2.5
Neomysis americana (all stages)	25.0	625.0	2500.0	3750.0	225.0	150.0	50.0	25.0	–	50.0	100.0
Eurytemora affinis (copepodites)	0.6	150.0	660.0	900.0	840.0	108.0	90.0	105.0	54.0	54.0	39.0
Eurytemora affinis (adults)	3.5	1620.0	2160.0	1440.0	720.0	288.0	360.0	2000.0	288.0	252.0	288.0
Balanus crenatus (nauplii)	0.2	–	–	–	6.0	25.0	25.0	18.0	12.0	10.0	0.4
Balanus crenatus (cyprid)	0.8	–	0.8	9.4	16.8	21.0	21.0	21.0	16.8	9.4	0.8
Eurytemora herdmani (adults)	2.0	–	–	–	–	4.0	60.0	90.0	200.0	100.0	4.0
Acartia clausi (adults)	1.3	–	–	–	–	1.3	6.5	9.1	6.5	2.6	1.3
Acartia longiremis (copepodites)	0.38	–	–	–	–	1.10	6.4	13.0	25.5	3.8	1.1
Acartia longiremis (adults)	1.3	–	–	–	–	6.3	31.5	63.0	126.0	19.0	5.0
Calanus finmarchicus (adults)	83.0	–	–	–	–	–	830.0	1660.0	2075.0	415.0	–
Calanus hyperboreus (adults)	350.0	–	–	–	–	–	350.0	350.0	700.0	350.0	350.0
TOTALS		4095.0	5910.0	6479.0	1941.0	618.0	1831.4	4419.6	3503.8	1266.8	792.1

TABLE 2 (cont'd)

Species/ Stage	Estimated Biomass Conversion Factor	Station Number, Lower Middle Estuary								
		77	23	20	51	65	67	64	26	24
Bosmina longirostris (all stages)	1.0	–	–	–	–	–	–	–	–	–
Ectinosoma curticorne (all stages)	0.5	–	–	–	–	–	–	–	–	–
Neomysis americana (all stages)	25.0	–	–	–	–	–	–	–	–	–
Eurytemora affinis (copepodites)	0.6	1.2	2.4	9.0	–	–	–	–	–	–
Eurytemora affinis (adults)	3.5	35.0	87.5	140.0	–	17.5	3.5	3.5	14.0	7.0
Balanus crenatus (nauplii)	0.2	5.0	8.0	15.0	8.0	10.0	18.0	18.0	14.0	15.0
Balanus crenatus (cyprid)	0.8	5.6	5.6	4.8	5.6	6.4	7.2	8.0	12.0	14.4
Eurytemora herdmani (adults)	2.0	300.0	240.0	200.0	110.0	180.0	60.0	70.0	60.0	50.0
Acartia clausi (adults)	1.3	6.5	6.5	6.5	10.4	9.1	13.0	18.2	23.4	26.0
Acartia longiremis (copepodites)	0.38	129.0	7.6	12.9	51.0	18.0	19.0	19.0	2.3	12.9
Acartia longiremis (adults)	1.3	650.0	39.0	65.0	260.0	91.0	97.5	97.5	11.7	65.0
Calanus finmarchicus (adults)	83.0	2905.0	2075.0	1245.0	1660.0	1660.0	1245.0	1079.0	830.0	1245.0
Calanus hyperboreus (adults)	350.0	1750.0	1400.0	1050.0	1400.0	1400.0	1400.0	1750.0	1050.0	10500.0
TOTALS		5786.3	3871.6	2748.2	3505.0	3392.0	2863.2	3063.2	2017.4	22925.3

fresh-water zooplankton, inexorably carried there and trapped by their swimming behavior and the circulation, and killed by the extreme physical conditions. Understanding the mechanism of utilization of such a secondary source of organic production in the regional food pyramid is a further challenging problem.

ACKNOWLEDGEMENTS

This work forms part of the GIROQ[5] program of hydrobiological studies on the St. Lawrence Estuary. The authors are grateful to Drs. Guy Lacroix, André Cardinal, and Pierre Brunel and their research associates who provided much helpful advice, encouragement, and cooperation in all phases of the study. The previous experience and effective leadership provided by Mr. Edwin Bourget was vital to the successful commencement of the field operations. Dr. Gaston Moisan and Mr. Henri Savard graciously expedited field work in the Baie St. Paul region, particularly at critical sampling periods. Michel Demers[6] provided valuable assistance in determination of plankton biomass conversion factors. The senior author is especially grateful to Dr. J. -L. Tremblay for many personal kindnesses and the benefit of his long experience with the hydrobiology of the Middle St. Lawrence Estuary, and to Dr. Lucien Huôt and his secretarial and technical staffs who so pleasantly facilitated his visiting professorship in the Department of Biology at the Université Laval during the study period.

REFERENCES

1. Bourget, E.
1971 Aspects saisonniers de la fixation de l'épifauna benthique de l'étage infralittoral de l'estuarie du St. -Laurent. M. S. thesis, Univ. Laval, Quebec. 115 p.

2. Bousfield, E. L.
1954 The distribution and spawning seasons of barnacles on the Atlantic coast of Canada. **Bull. Nat. Mus. Canada,** 132: 112-154.

3. Bousfield, E. L.
1955 Ecological control of the occurrence of barnacles in the Miramichi Estuary. **Bull. Nat. Mus. Canada,** 137: 1-69.

4. Bousfield, E. L.
1955 Studies on the shore fauna of the St. Lawrence Estuary and Gaspé coast. **Bull. Nat. Mus. Canada,** 136: 95-101.

5. Brunel, P.
1970 Les grandes divisions du Saint-Laurent: 3e commentaire. **Rev. Géogr.** Montreal, 24(3): 291-294.

6. Crisp, D. J.
1962 The planktonic stage of the Cirripedia **Balanus balanoides** L. and **Balanus balanus** L. from north temperate waters. **Crustaceana** 3: 207-221.

7. Gaudry, R.

1938 Les températures de l'estuaire du St. Laurent. **Contr. Stat. Biol. St. Laurent,** 13: 1-4.

8. Herdman, W. A., Thompson, I. C., and Scott, A.
1898 On the plankton collected continuously during two traverses of the north Atlantic in the summer of 1898. **Proc. Trans. Liverpool Biol. Soc.,** 12: 33-90.

9. Lacroix, G. and Filteau, G.
1971 Les fluctuations quantitatives de zooplankton de la Baie-des-Chaleurs (Golfe St. Laurent). II. Composition des copépodes du genre **Calanus. Trav. Pêch. Québec,** 24: 711-748.

10. Lacroix, G. and Filteau, G.
1971 Les fluctuations quantitatives du zooplankton de la Baie-des-Chaleurs (Golfe St. Laurent). III. Fluctuations des copépodes autres que **Calanus. Natr. Canad.,** 98: 775-813.

11. Nadeau, A.
1938 Salinité des eaux de l'estuaire du St. Laurent. **Contr. Stat. Biol. St. Laurent,** 11: 1-19.

12. O'Neill, M. and Gentes P.
1972 Étude écologique de l'holoplankton et du méroplankton de l'estuaire moyen du St. -Laurent. Partie I & II. Initiation à la recherche, Univ. Laval, Quebec. 51 p.

13. Prefontaine, G.
1936 Additions à la liste d'espèce animales de l'estuaire du St. Laurent dans la région de Trois Pistoles. **Trans. Roy. Soc. Canada,** 3(26): 205-209.

14. Sars, G. O.
1901 **An account of the Crustacea of Norway.** Vol. 4, parts 1 & 2. Bergen Museum, 162 p.

15. Shih, C. -T., Figueira, A. J. G., and Grainger, E. H.
1971 A synopsis of Canadian marine zooplankton. **Bull. Fish. Res. Bd. Canada,** 176: 1-264.

16. Tattersall, O. S.
1954 Shallow-water Mysidacea from the St. Lawrence estuary, Eastern Canada. **Canadian Field-Nat.,** 68(4): 143-154.

17. Tremblay, J. -L.
1942 Plancton. **Rapp. Gén. Sta. Biol. Saint -Laurent, 1936-1942.** p. 11

18. Willey, A.
1931 Preliminary report on copepod plankton collection by the Station de Biologique du St. -Laurent à Trois Pistoles in July, 1931. **Rapp. Univ. Laval Sta. Biol. St. -Laurent,** 1: 82-84.

19. Wilson, C. B.
1932 The copepods of the Woods Hole Region, Massachusetts. **Bull. U. S. Nat. Museum,** 158: 1-635.

THE ECOLOGICAL SIGNIFICANCE OF THE ZOOPLANKTON IN THE SHALLOW SUBTROPICAL WATERS OF SOUTH FLORIDA[1]

Michael R. Reeve[2]

ABSTRACT

Much emphasis has been placed on the imported detrital and benthic sea grasses in the biological economy of shallow subtropical inshore waters. This report seeks to present the existing data on plankton which indicate that the shallow water column may support planktonic production at least as large as some much deeper, colder-water, inshore regions. It is suggested that plankton organisms are important in the utilization of this material in the marine ecosystem both by their rapid uptake of dissolved organics flushed out from the land and released from sediments and by the possible ability of the ultra-microzooplankton to graze on detrital-bacterial aggregations in the water column. In two adjacent areas substantial differences in plankton biomass could be correlated with fresh-water run-off. In the low biomass region summer temperatures were associated with severe depressions of macroplankton populations.

INTRODUCTION

Biological studies on the nature of inshore and estuarine systems in warm

1. Contribution No. 0000 from the University of Miami, Rosenstiel School of Marine and Atmospheric Science.

2. Rosenstiel School of Marine and Atmospheric Science, University of Miami, Rickenbacker Causeway, Miami, Florida 33149.

waters are far less advanced than those on colder northern waters, which have a history of several decades, and even those in tropical oceanic waters which saw great expansion in the sixties (e.g., Indian Ocean and Equilant expeditions). It was not until the late sixties in Miami, for instance, that systematic studies were launched in an effort to comprehend the nature of the interaction of biological and physical phenomena which comprised the ecosystem of these warm shallow seas adjacent to the laboratory. The funding generated by a rapid upsurge in public awareness of the frailty of these regions to the ravages of ongoing or threatened installation of deep-water ports, oil refineries, electrical generating plants, modification of natural land drainage patterns by urbanization, etc. provided the impetus for these studies. Unfortunately many studies suffered from their dependence on funding by one or more of a variety of mission-oriented agencies whose scientific goals and funding levels fluctuated, so that the establishment of integegrated long-term goals even by workers within the same organization was effectively prevented. Nevertheless, useful work has been forthcoming, some of which is presented in this volume, e.g., Carpenter on nutrient cycling, Odum on the utilization of detrital material of terrestrial origin by marine invertebrates and vertebrates, and Zieman on sea-grass production. In this report I hope to summarize my own studies and those of colleagues on the inshore plankton, relate them to other available biological information and indicate areas for future study.

Inspection of a map of South Florida and the adjacent Caribbean (Fig. 1) shows that there are many thousands of square miles of marine waters of depths

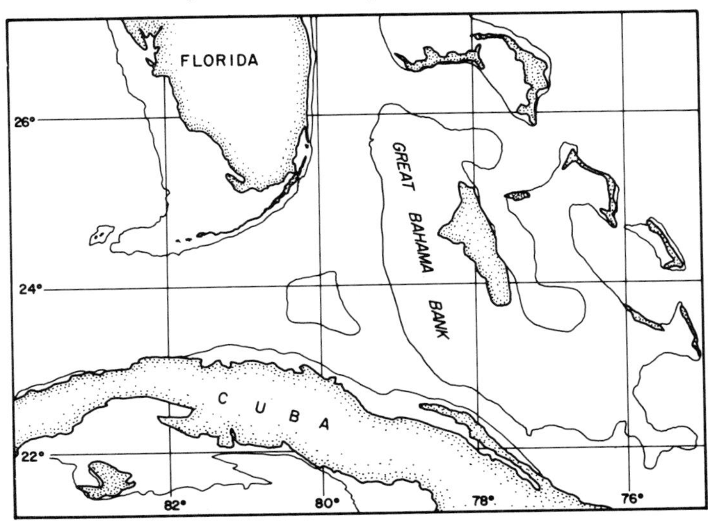

Figure 1. Sketch map of south Florida and the adjacent Caribbean showing the extent of marine waters shallower than 20 m (contour line).

no greater than 20 meters. Much of this area consists of the Bahama Banks and the gently sloping shallows of the west coast of Florida bordering the Gulf of Mexico. Even so, most of the coastline of Florida, including the southeast, despite its proximity to the deep, fast-flowing Gulf Stream, has a borderline of shallow water which, to a great extent, isolates it from offshore water masses of oceanic origin. In many places around southern Florida these waters are further isolated from offshore influences by barrier islands or bars which form such areas as Biscayne Bay and Card Sound adjacent to Miami (Fig. 2). Few of these areas can be classified as estuaries according to classical dictionary definitions (reviewed by Pritchard, 21) based on river mouths, and whether they meet the broadened definitions of Pritchard with regard to hyposalinity or of Caspers (6) with regard to instability of environmental factors depends largely on location and degree of isolation, land drainage, and time of year. The most important

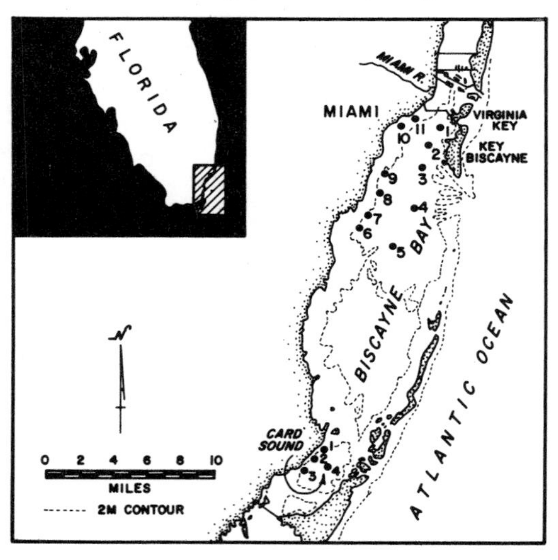

Figure 2. Sketch map of Biscayne Bay and Card Sound showing station locations.

feature which these waters have in common is their shallowness; most of them fit Emery's and Stevenson's (10) description of lagoons, in that they are "elongate parallel to shore, and lie athwart the mouths of one or more streams". They differ in respect to the nature and extent of land drainage and degree of isolation from the open sea. At one end of the scale are the regions into which flow such detritus-laden water courses emerging from wetland vegetation as the mangrove forests of southwest Florida. Intermediate in amount of suspended material and run-off are regions such as Biscayne Bay, where relatively little natural run-off is augmented by man-made drainage canals, which carry down particulate and dissolved material from agricultural and urban activities. Further down the scale are regions such as extend down from Card Sound into Florida Bay, where natural drainage may be restricted by natural geological and sometimes man-made barriers and where poor exchange with the open sea can create extended periods of hypersalinity. At the other extreme are the shallows surrounding smaller Bahamian islands, which receive virtually no land run-off and whose waters are closest to the open ocean in their salinity and water clarity.

All these environments are of commercial and often recreational importance. Most of them are experiencing a rapid upswing in both activities concomitant with the southward surge of population to the tropics and are being additionally stressed by all the other artificial influences that such population explosions bring in their wake.

Many studies of a biological nature concerning particular components of these ecosystems have accumulated over the years, but attempts to examine the overall ecology of a particular location have only recently been undertaken, particularly in response to concern over siting and operation of electrical generating plants. Much of this information is still available only in such unpublished reports as the south Biscayne Bay and Card Sound study (1, 2, 3) and the Crystal River study (13).

Qualitative seasonal studies of zooplankton were made by Smith et al. (29) and Woodmansee (32) in Biscayne Bay. Reeve (22, 23), Reeve and Cosper (25) and Baker (4) reported on quantitative seasonal surveys from the same region. Further information on the distribution and identification of copepods from Floridian neritic waters was provided by Davis (7), Fleminger (12), Grice (14), and Owre and Foyo (20). Quantitative studies in waters whose plankton populations closely resemble those of south Florida were done in inshore Bermudan waters by Herman and Beers (16) and the St. Andrew Bay system on the Gulf Coast of northern Florida by Hopkins (17). There is no quantitative seasonal information on the plankton either from waters receiving extreme loads of fresh-water and organic detrital material or from waters at the other extreme referred to above.

The seasonal pattern in the region of the inshore south Florida and adjacent waters consists of dry winters and wet summers, so that, in May, immediately prior to the beginning of the wet season, salinity gradients across a transect from land seawards show minimum variation often tending towards the hypersaline, while during the summer great salinity variations might exist across the same transect (Fig. 3). In Card Sound, on the other hand, although the same patterns exist, the extent of variation is much smaller. Winter and summer are also characterized by plateaus of temperature in the region of 19-22 ^{o}C and 28-32 ^{o}C, respectively, in Biscayne Bay, with rapid temperature changes in spring and fall. Extemely shallow areas with poor circulation can fluctuate considerably below the winter plateau, following the passage of cold fronts, and above the

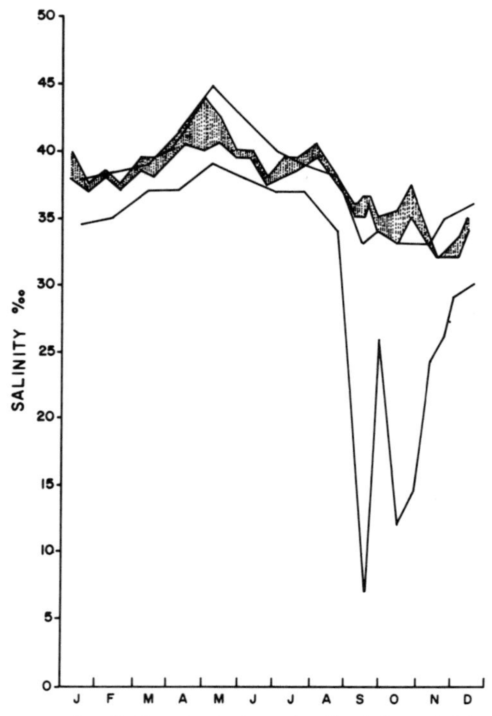

Figure 3. Seasonal range of salinity at stations in Card Sound (shaded) and central Biscayne Bay (unshaded).

summer plateau.

Unlike the open ocean, where the only significant primary input to the system is via the surface layer of endemic phytoplankton, coastal waters, in addition to their phytoplankton, support benthic primary producers in the form of macro-algae and sea grasses as well as sediment and epiphytic populations of micro-algae. Organic dissolved and particulate material is also imported into the system from both natural and man-made terrestrial sources, as described earlier.

It is notoriously easy to present an unconsciously biased view of the nature of an ecosystem, because no worker can make a detailed examination of anything but a very small part of it. Smith et al. (29), for instance, compared plankton standing crops in Biscayne Bay with northern waters and concluded that it was an "impoverished" ecosystem. Thorhaug and Stearns (31), on the other hand, on the basis of their sea-grass study, believed that the productivity of the region was based on "a benthic macro-plant driven food web", which was "at least six times more productive than the mean for upwelling areas" based on estimates of Ryther (28). Odum (this volume) makes strident claims for a direct utilization of detritus of terrestrial origin by invertebrates and vertebrates as a major food chain.

It is appropriate to review what we know regarding the ecology of the plankton in relation to the rest of the system, since heavy emphasis has recently been laid on the significance of detrital and sea-grass components.

PHYTOPLANKTON BIOMASS AND PRODUCTION

I am aware of no published data on phytoplankton production and of only one unpublished report containing a few carbon-assimilation experiments (5), which indicate an extremely high carbon assimilation to chlorophyll *a* ratio of 15. Standing-crop measurements in terms of chlorophyll *a* were reported over one year from Card Sound and averaged 0.56 with a peak bloom of 4.4 mg m^{-3} (25). Measurements over 3½ months in central Biscayne Bay indicated a mean of 2.1 and peak bloom of 4.0 mg m^{-3} (Bunt, unpublished).

Direct comparisons between various components of primary production at this stage are of dubious validity. Zieman estimated that production of the leaves of the sea grass *Thalassia* was 5 g dry weight (2 g C) m^{-2} day^{-1} (this volume), and that the beds covered at least 30% of the bay bottom (personal communication). Converting mean pigment value for central Biscayne Bay to production using Bunt's pigment-to-carbon assimilation ratio yields a value of 95 mg C m^{-2} day^{-1}, which may be compared to an average *Thalassia* value over the bay of 600 mg m^{-2} day^{-1} computed from Zieman's data reported at this conference. The other measured source of primary production, the red macro-algae complex, is over two orders of magnitude less productive than *Thalassia* (30). Nothing is known of the amount contributed by sediment and epiphytic microalgae. A crude estimate of phytoplankton production in central Biscayne Bay, therefore, might place it an order of magnitude less than the sea grasses. Since pigment values per m^3 are not uncharacteristic of other estuarine ecosystems on an annual basis, its insignificance compared to benthic production per m^2 is more a function of water depth than anything else. Card Sound, on the other hand, with an average phytoplankton standing crop four times less than central Biscayne Bay, seems by comparison to warrant the label "impoverished."

The input of organic material to these shallow seas is not, of course, confined to the endemic autotrophs since, as noted in the introduction, land drainage brings with it varying amounts of decaying terrestrial vegetation and dissolved organics as well as inorganic nutrients from a variety of sources. Apart from Heald's (15) dramatic demonstrations of the vast quantities of detritus estimated to be exported from the luxuriant mangrove forests and grass swamps of the southwest Everglades region of Florida at one extreme, we have neither real information on the export of dissolved materials from that system nor data on either particulate or dissolved materials from any other region. The imported terrestrial material is not to be regarded as a separate and independent source of primary production, as is implied by statements concerning its relative importance compared to benthic grass and algal production (e.g., 31). Nutrients derived from the breakdown of the former must be used by the latter. It is probably not accidental that the *Thalassia* beds are concentrated near the shoreline, and phytoplankton blooms often follow periods of heavy rainfall (Reeve, 22). Apart from some work of Fell and his co-workers on the sequence of mangrove seedling breakdown and utilization through fungal, bacterial, and meiofaunal food chains, and of Odum (19) on detritus-feeding invertebrates and fish and work reported in this volume, we know virtually nothing, either in terms of pathways or rates of use, of the subsequent history of these materials once they enter the marine environment. For that matter, the same statement may be made of the *Thalassia* beds since it is generally agreed that the vast majority of them are not grazed down but die and gradually disintegrate (e.g., 11). The rate of use may well be an order of magnitude slower than that of the phytoplankton.

ZOOPLANKTON

Biomass and Production

Turning to the zooplankton, to which most of my personal observations are confined, I want to consider its quantitative significance as well as its ecological relationships to the other components of the ecosystem in these waters. The most extensive study of standing stock of zooplankton in terms of dry weight was by Reeve and Cosper (25) in Card Sound. Dry weight of 20-64, 64-200, and over 200 μ m size-fractions were calculated separately and corrected for detritus and phytoplankton; mean annual surface standing stocks of 3.45, 5.28, and 2.02 mg m^{-3}, respectively, were estimated. As with phytoplankton standing-crop data, dry weight of zooplankton in Card Sound appeared to be very low compared to central Biscayne Bay, i.e., an annual mean of 2 mg and 20 mg m^{-3}, respectively (in the 200 μ m fraction), or an order of magnitude less. This value for central Biscayne Bay is quite close to the only other regional data available, those from the work of Hopkins (17) for the St. Andrew Bay system in northwest Florida (33.1 mg m^{-3}) and of Maturo (18) for the Crystal River region of west Florida (44.8 mg m^{-3} over half a year).

As a further indication of the impoverishment of Card Sound, it may be calculated from the data of DeSylva (8), who reported on settled volume from a 500 μ m mesh net over two years at 39 stations through Biscayne Bay and Card Sound, that mean volume of central Biscayne Bay stations was some five times higher than in Card Sound. Further information from the same study suggests that over nine months larval fish were about four times more abundant in central Biscayne Bay.

Even less information is available on zooplankton production from these waters, but it might be expected that compared to colder waters, life cycles would be shorter and breeding would be extended throughout the year, so that the ratio of production to biomass would be higher. Reeve and Baker (unpublished data) measured the production of the carnivores *Sagitta hispida* Conant (Chaetognatha) and *Mnemiopsis mccradyi* Mayer (Ctenophora) and derived mean daily production-to-biomass ratios of 0.31 and 0.12, respectively. Since most of the zooplankton population is much smaller than *S. hispida* in biomass, its production-to-biomass ratio is likely to at least equal that of *S. hispida.* Applying this ratio to the surface standing-stock figures for the 64-200 and over 200 μ m mesh fractions for central Biscayne Bay and Card Sound, and assuming a depth of 3 m and conversion factor of 1.52 from surface to mean biomass per m^3 over the water column (see 25), a mean daily production of zooplankton of 64 μm and over in central Biscayne Bay and Card Sound may be calculated at 46.1 and 4.6 mg C m^{-2}, respectively (carbon as 50% of organic dry weight). These figures may be compared with values provided by Riley (27) for zooplankton production in Long Island Sound, which he described as a "somewhat estuarine environment of moderately high productivity" of 27 mg C m^{-2} day^{-1} in a 20 m depth water column. In the case of Long Island Sound the production-to-biomass ratio over the year was 0.027 (i.e., a tenth of that assumed for Biscayne Bay). Both Riley's figures and mine are admitted to be gross approximations, but they serve to indicate that in central Biscayne Bay and other estuarine areas of the region there is substantial zooplankton production which may well equal that of such areas as Long Island Sound, even on a m^2 basis.

Whatever the relative importance of the zooplankton in terms of production, included at one time or another are larval stages of virtually all other animals that inhabit these shallow seas, including those of commercial importance, whose life cycles depend on completion of their planktonic stages.

The rest of the zooplankton is an essential part of the diet of the carnivorous meroplankton such as decapod and fish larvae, as well as the filter-feeding benthic invertebrates such as sponges, mollusks, and juvenile and some adult fish.

The smaller, more delicate larvae, as well as most of the copepods, presumably rely on the phytoplankton as their source of nutrition, but the presence of large quantities of detritus of a widely varying size in the water column cannot be discounted as a nutritional source.

Possibly the ultra-microzooplankton plays an important part in mobilizing this material into the animal food chain, which represents both the remains of *Thalassia* and vegetation of terrestrial origin. Bacterial accumulation by multiplication on these detrital cores, as described by Fenchel (11), might be grazed down by such organisms as tintinnids and nonloricate ciliates, which in turn would provide food for the crustacean microzooplankton and establish a pathway by which suspended detritus is incorporated directly into the zooplankton food chain.

Composition and Seasonal Cycles

Detailed information on the zooplankton of south Biscayne Bay, where populations were separated into inshore and mid-bay groups, was provided by Reeve (23). Reeve and Cosper (25) reported on the Card Sound population, and Baker (4) collected samples from 11 stations in central Biscayne Bay over the same period, which she analyzed in terms of dry weight. She kindly made available her samples, aliquots of which were subsequently pooled and counted and are reported on here. A variety of net meshes were used in these studies, but most of the comparisons drawn below are on the basis of samples from a 200 μ m mesh net (Card Sound, central Biscayne Bay) and 300 μ m mesh net (south Biscayne Bay).

The composition of the zooplankton in these subtropical inshore waters is characteristic of estuarine waters in general and not of the tropical ocean, in terms of small numbers of species, their small size and taxonomic affinities. The holoplankton is dominated by copepods, the only others of quantitative significance being larvacean tunicates, tintinnids, the chaetognath *Sagitta hispida* and the ctenophore *Mnemiopsis mccradyi.*

Acartia tonsa Dana and *Paracalanus parvus* Claus predominated in all samples taken with net meshes in the range of 200-300 μm. Davis (7) considered the former to be the most common of all copepods in Florida marine waters. The adults of the two species are approximately equal in size and weight (being about 1 mm in length and 5 μg dry weight). In south Biscayne Bay, *A. tonsa* predominated at the inshore stations and *P. parvus* was more important in the mid-bay. The latter was rarely to be found at temperatures above 30 ^{o}C or in salinities below 25°/oo. In Card Sound, salinities remained high in 1971 and temperatures did not exceed 30 ^{o}C. On this basis it would be predicted that *P. parvus* would predominate over *A. tonsa* in Card Sound, and this proved to be true, the annual mean numbers being 524 and 300 m^{-3}, respectively.

In the composite samples from central Biscayne Bay, which included stations with large and small annual salinity ranges, *A. tonsa* was more abundant than *P. parvus* throughout the year although, on the basis of the above information, it is likely that *P. parvus* predominated at the easternmost stations. Mean annual numbers in Card Sound and south and central Biscayne Bay were 300, 187, and 2,833 for *A. tonsa* and 524, 87, and 516 for *P. parvus,* respectively. Two species of *Labidocera* (*L. scotti* Giesbrecht and *L. mirabilis* Fleminger) occurred in these inshore waters and were not separated in counts. Their mean numbers in Card Sound and south and central Biscayne Bay of 47, 22, and 46 m^{-3} respectively, seriously underestimated their significance in terms of biomass because they are several times larger than *A. tonsa* and *P. parvus* (see 23).

Oithona nana Giesbrecht was overwhelmingly the major constituent of the cyclopoid copepods, with a few members of the genus *Corycaeus* in evidence at times. It is the smallest of the Biscayne Bay and Card Sound copepods, not being retained in significant numbers on a 200 μ m mesh. Since a fine mesh was not employed in central Biscayne Bay, it was not possible to make comparisons between the locations, but Reeve and Cosper (25) estimated that *O. nana* accounted for over 50% of the copepod standing stock (dry weight) in Card Sound with mean numbers of 14,000 m^{-3} in the 64 μ m mesh net.

Temora turbinata (Dana) occurred in small numbers and could be associated with regions of lower salinity (23); its mean numbers in inshore south Biscayne Bay and central Biscayne Bay were 37 and 33 m^{-3}, respectively. *Calanopia americana* Dahl and *Metis jousseaumei* (Richard) also occurred in low numbers, the latter being taken commonly in shallow water where it was presumed to have risen from the bottom.

Holoplanktonic carnivores were represented by *Sagitta hispida* Conant and *Mnemiopsis mccradyi* Mayer. *S. hispida* occurred in the 200 μ m mesh net in mean numbers of 50 and 70 m^{-3} in Card Sound and central Biscayne Bay, respectively. These numbers are not very different considering that their food supply was 10 times more abundant in the latter location. *M. mccradyi,* on the other hand, was rarely seen in south Biscayne Bay and Card Sound, although it occurred in blooms in the winter in central Biscayne Bay (Baker, 4), where there was a direct relationship between ctenophore and zooplankton biomass. Less than 3% of the recorded occurrences of ctenophores were associated with macroplankton dry weights of less than 5 mg m^{-3}. The suggestion from this data is that the biomass of the Card Sound macroplankton, which was always less than this, was not sufficient to sustain ctenophore populations, and that *S. hispida* may be more efficient at collecting food at low densities than *M. mccradyi.*

Tintinnids were not collected effectively except in Card Sound where a 20 μ m mesh was used and resulted in annual mean numbers of 121,000 m^{-3}. Larvacean tunicates, although usually absent from south Biscayne Bay and not abundant in Card Sound, were often prominent in central Biscayne Bay (mean 117 m^{-3}). The meroplankton was predominated by molluscan veligers and decapod larvae with occasional bursts of polychaete and echinoderm larvae.

The general pattern of seasonal variations of the zooplankton of Card Sound and Biscayne Bay was one of rapid fluctuation throughout the year (Fig. 4). The single most obvious fluctuation was the extreme low point in biomass and numbers of the macroplankton in the middle of the summer, in which all the major constituents of the Card Sound macroplankton share *(Acartia tonsa, Paracalanus parvus, Labidocera spp., Sagitta hispida,* decapod larvae and

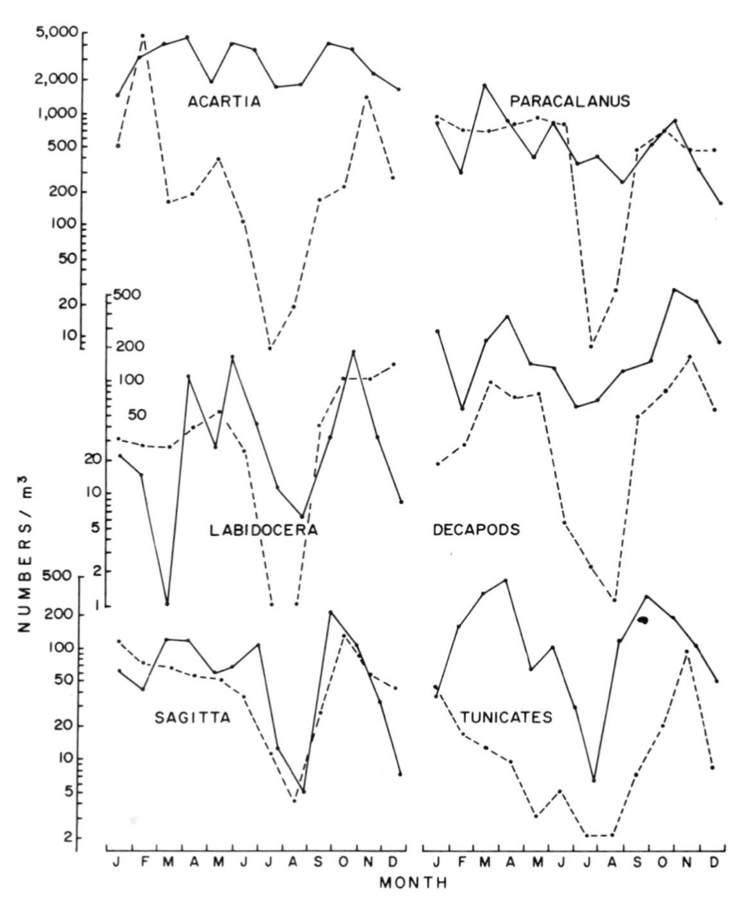

Figure 4. Seasonal variation of plankton numbers from the 200 μ m net in central Biscayne Bay (solid line) and Card Sound (broken line).

larvacean tunicates): Reeve (23) noted that a summer depression had been pointed out in all previous studies (22, 25, 29, 32). Its occurrence in Card Sound (Fig. 5) in 1971 was the most extreme example. The counts from Baker's central Biscayne Bay samples showed only a depression of similar magnitude to others in the same year. Perhaps the most striking example of this phenomenon was provided by the data on plankton volume of DeSylva (personal communication), who sampled more than 30 stations between northern Biscayne Bay and Card Sound monthly over two years (1970-71) using a 500 μ m mesh net during a larval fish survey. Averaging these volumes over all stations during both years resulted in the distribution shown at the top of Figure 5. The effect of combining all these data was to retain only major fluctuations characteristic of

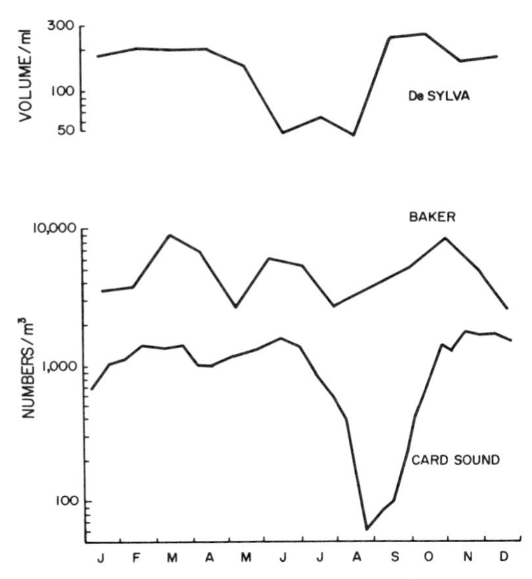

Figure 5. Seasonal changes in total numbers of macroplankton from a 200 μ m net in Card Sound (Reeve and Cosper, 25) and central Biscayne Bay (Baker, 4); and settled volume from a 500 μm net (DeSylva, 8).

most stations over both years and to minimize the contributions of more variable influences. The resulting curve was almost level during half of the year (mid-November to mid-May), indicating that actual fluctuations in individual years were not predictably seasonal. The summer minimum was, however, very obvious and extensive and was reduced to only 25% of the winter levels. This figure also demonstrated that an autumn bloom or peak follows the summer minimum, which was indicated as well by data from Card Sound and south Biscayne Bay.

Of the five species or groups named in the previous paragraph, all reached their lowest point in biomass and numbers for the year in Card Sound during July and August. In central Biscayne Bay only *Labidocera spp.* reached an annual low point, whereas the others exhibited moderate low points characteristic of the rest of the year. *Labidocera spp.,* by virtue of their size, must have been the predominant copepod in DeSylva's 500 μm samples, and this, together with the fact that his samples came from both Baker's and the south Biscayne Bay and Card Sound sampling regions, may explain why his summer minima were also low points for the year.

A correlation between the summer macroplankton minimum and temperature maximum is plausible on the basis of evidence that the critical thermal maxima of the major constituents occurred above 30 °C (24). On the other hand, in Baker's samples from central Biscayne Bay, most of the species occurred in greater numbers during their summer low point than they did in Card Sound during their peak periods. In addition, their numbers during the summer low point were no lower than those at other times and other temperatures, and the summer microzooplankton and phytoplankton of Card Sound showed no severe depression.

A second explanation for the summer macroplankton minimum may be that there is a shortage of food over the summer. If it is assumed, however, that most of the plankton are herbivores and cannot make significant use of detritus, then the phytoplankton pigment curve (Fig. 6) for Card Sound, which shows a rather stable increasing trend with little fluctuation over the summer, lends no support to this possibility.

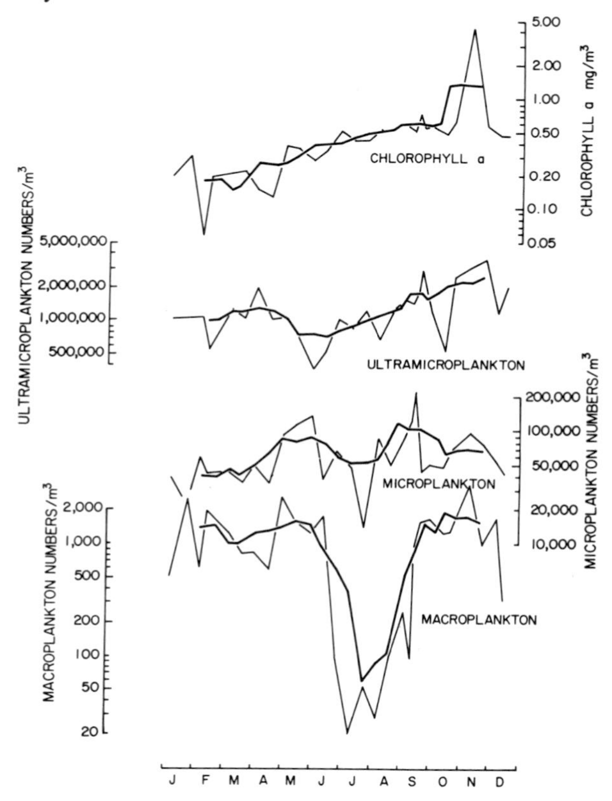

Figure 6. Seasonal changes in chlorophyll **a**, ultramicrozooplankton (20-64 μ m), microzooplankton (64-200 μ m) and macroplankton (> 200 μ m) in Card Sound, with 5-point moving averages (from Reeve and Cosper, 25).

A third explanation concerns the basic weather pattern, which is dominated by a dry season extending approximately from November to May and a wet season extending through the summer and autumn. From Figure 6 it can be seen that a significant pigment peak occurred at the beginning of May, which was sustained and continued to increase steadily throughout the summer. The average curves for the ultramicro-, micro- and macroplankton in Figure 6 show a minimum followed by an upward trend which occurred progressively later in each, culminating in the extreme depression and subsequent rapid increase exhibited by the summer minimum and autumn bloom of the macroplankton. This concept of a progression of effects upwards through size categories and, by inference, the food chain, implies that the summer macroplankton minimum is the end result of the earlier dry months. The subsequent autumn bloom is attributable to the beginning of the rains in May which flushed out the nutrients accumulating in the fringing mangroves and farmlands beyond, and the nutrients stimulated the phytoplankton, which in turn encouraged reproduction of huge quantities of copepod young which grew up during the autumn.

The main drawback to this explanation is that too long a time elapses between initial nutrient input and subsequent bloom of adult copepods, most of which probably require little more than a week to complete their life cycles at summer temperatures.

An environmental population response extending from May, beginning of the rainy season, to October, peak macroplankton, seems almost an order of magnitude too slow. Reeve and Cosper (25) concluded, therefore, that the summer macroplankton minimum, which appears to be most devastating in the low biomass regions, could be the result of the added stress of low food levels combining synergistically with such factors as high temperature when a moderate minimum would otherwise have occurred.

Interrelationships of Plankton and Benthos

It is obvious that in these shallow seas, which are often no more than 2-4 meters in depth, there must be a close physical relationship between the pelagic and benthic environments. The pronounced vertical layering and migration of the older animals must being them into very close contact with the sediments, sea grass, and other organisms associated with the bottom.

Emery (9), making a preliminary investigation of the plankton associated with coral reefs of south Florida at the edge of these shallow waters, drew attention to aggregations or swarms of planktonic copepods and mysids over and around coral growths. Such swarms may be seen visually by direct observation with Scuba and, as he observed, are not confined to coral reefs but occur also in close association with *Thalassia* beds. From my own observations, copepod swarms, which are often composed of *Acartia tonsa* and *Oithona nana*, vary in diameter from a few centimeners around grass blades and individual coral branches to several meters in deeper areas over and between large coral formations. Animal density can reach 1,000 per liter, which is equivalent to 2 mg C l^{-1}. As Emery noted, species form a behavioral continuum from species normally considered planktonic to those assumed to be epibenthic, such as the copepod *Metis jousseaumei* and the mysids. Whether this phenomenon is a specialized adaptation to permit maintenance of a specific position in relation to bottom topography in these shallow waters and reduce chances of gradually drifting out into deeper water, or whether it is an obvious expression of a more generalized phenomenon of plankton behavior, is not clear at present. Patchiness, after all, is a very-well-established phenomenon of zooplankton distribution on a larger scale, and short of direct observation and *in situ* collection, there is no means of identifying it on so small a scale. Bottom topography, such as coral and grass, helps the diver to focus closely and see clusters of tiny organisms which may exist within the general water column in deeper waters.

Adaptation to shallow bottoms by inshore zooplankton can be demonstrated in other ways. The copepod *Acartia tonsa* has been shown by Zillioux and Gonzalez (33) to have eggs that are capable of lying dormant in the sediments and hatching when conditions are favorable for the survival of the young. The chaetognath *Sagitta hispida,* which is also a characteristic and often sole representative of the genus in these inshore waters, has developed behavior intermediate between all the other planktonic members of its phylum and the benthic genus *Spadella.* It can attach itself to surfaces upon which it deposits its eggs in clusters embedded in a sticky matrix. It is probable that the reef is simply a special case of a general phenomenon characteristic of these shallow inshore waters, and further investigation of the life-history of individual species will probably show that the vast majority enter into some intimate association with the bottom and that there may be no such thing as a truly planktonic species.

Laboratory Maintenance

Laboratory maintenance and culture of zooplankton is an essential tool in understanding its ecology. This is even more important in warm waters than in cold waters, because animals pass through their life-cycles more rapidly and have a longer breeding season, often extending year-round. The synchronous breeding periods of higher latitudes initiated by such annual events as the spring bloom, which permit the identification and following through of a cohort from size-frequency analysis of monthly samples and provide data on growth and mortality rates, are not characteristic of warm waters.

A typical size-frequency distribution for *Sagitta hispida* or *Mnemiopsis mccradyi,* for instance, shows decreasing number with increasing age on virtually any sampling date throughout the year. Growth rate, therefore, can only be estimated from laboratory populations. Reeve and Walter (26) reported on the growth rate of *Sagitta hispida* which could go through its life cycle in a minimum of 18 days. Baker (4) showed that *Mnemiopsis mccradyi* could lay eggs within 22 days after hatching. Zillioux (personal communication) has data indicating that the copepod *Acartia tonsa* has a life-cycle time of 10 days, even at winter temperatures. Such rapid generation times suggest how populations can build up very quickly to take advantage of increases in food supply, and how large variations in sampled zooplankton biomass can occur from month to month. They also indicate production by inference from comparisons of biomass between these and colder water environments, where growth rates may be much slower, can seriously underestimate warm-water production.

Some care must be exercised in interpreting growth rates obtained in the laboratory, since the conditions are different from those in the natural environment. On the one hand, we have evidence that container size, even up to 400 liters, inhibits the growth rate of *Sagitta hispida,* and mortality rates in culture are often no better than in the environment, despite the absence of predators. On the other hand, food may be provided in excess of its natural availability, which would promote faster growth rates. Environmental growth rates, therefore, may be different from those in the laboratory, but not necessarily slower. Care must also be taken in interpreting adverse effects in laboratory populations as reflecting the natural environment. Long-term maintenance becomes particularly difficult near the upper limit of temperature in the natural environment (i.e., 30-32 °C in Biscayne Bay). At such temperatures, microbial blooms and other events can occur with extreme rapidity in laboratory systems.

Plankton as Indicators of Pollution

Inshore plankton in laboratory culture may prove to be a very useful tool in pollution studies. Their virtues in this connection reside in their small size, rapid growth and reproductive rates, extreme sensitivity, and amenability to laboratory culture. We can hope to establish not only acute lethal pollutant doses in short-term experiments, but the effects of longer-term, sublethal doses on growth and fecundity, and even the effects of accumulation of toxic materials through the food chain and the disruption of community structure by large-scale, long-term experiments (see Davies et al., this volume). Such an approach would enable us to develop models which could be used to predict the outcome of environmental pollution where environmental surveys fail. On the other hand, the inherent patchiness of plankton and the extreme mobility of populations in water currents render the analysis of short- or long-term sampling at fixed environmental stations a very insensitive measure of anything but the most catastrophic of polluting influences. Even when laboratory experiments as well as direct observation showed that all the plankton passing through an electrical generating plant were being killed, it was impossible to detect significant differences between populations in the bay immediately adjacent to the intake and outflow canals (Reeve and Cosper, 25).

CONCLUSION

Shallow inshore subtropical waters of Florida are much closer in their plankton composition, biomass, and extent of seasonal fluctuation to inshore regions of higher latitudes than they are to the offshore tropical waters where low biomass, high species diversity, and relatively little seasonal fluctuation is the rule. Populations fluctuate very rapidly throughout the year, probably because the extremely short life cycles that occur at these temperatures permit rapid response to increased food supplies. There are predictable summer seasonal effects such as a minimum in macroplankton biomass followed by an autumn bloom, particularly noticeable in areas of low biomass. Winter production, on the other hand, is sustained by equable water temperatures and high light intensity in contrast to higher latitudes, so that annual zooplankton production may be considerably higher than that in some inshore areas in higher latitudes, on a m^3 basis, and total production in the water column of these shallow waters may equal inshore areas of considerably greater depth.

Although there are very few comparative data available, it appears that zooplankton production is proportional to the input of nutrients from terrestrial sources, and in areas such as Card Sound, where seasonal salinity patterns indicate little land run-off, biomass may be several times lower than in such areas as central Biscayne Bay. Unfortunately, there are no data for such areas as the turbid, particle-laden waters adjacent to the western Everglades, which are likely to be much richer in plankton, or the clear waters over the Bahama Banks, where one might expect that an extremely oligotrophic situation exists.

Phytoplankton production, although probably at least an order of magnitude lower than that of benthic sources, may assume more importance in the food chain by being rapidly used rather than by entering into a slow process of detrital decomposition. A decade or two ago, our traditional ideas of primary production in marine environments prevented us from recognizing the importance of nonplanktonic sources in estuarine systems, but I believe now that current thought has swung too far in the other direction. The need over the next few years is to try to unravel the complex pathways and rates by which detrital material is utilized into higher animal trophic levels and to learn to what extent the plankton play a role. I believe the role to be significant, and even question whether there is any point in still referring to plankton and benthos in these shallow, warm waters as if they were separate major subdivisions instead of trophically and physically closely connected parts of the same system.

ACKNOWLEDGEMENTS

This review draws heavily and quotes directly from an unpublished report by Reeve and Cosper (25) and unpublished material of Linda Baker and Donald DeSylva. Technical assistance was provided by Duane Phillips, Chere Steen, and Mary Ann Walter. Funds were provided by AEC grant no. AT-(40-13)-3801 SUB 4 and NSF grant no. GA-28522X.

REFERENCES

1. Bader, R. G. (ed.)
1969 An ecological study of south Biscayne Bay in the vicinity of Turkey Point. **Prog. Rept. to USAECC.** Univ. Miami, School Mar. Atmosph. Sci. 63 p. (Unpublished manuscript.)

2. Bader, R. G. and Roessler, M. A. (eds.)
1971, 1972 An ecological study of south Biscayne Bay and Card Sound, Florida. **Prog. Rept. to USAEC and FPL.** Univ. Miami, School Mar. Atmosph. Sci. 303 p.; 296 p. (Unpublished manuscripts.)

3. Bader, R. G. and Tabb, D. C. (eds.)
1970 An ecological study of south Biscayne Bay in the vicinity of Turkey Point. **Prog. Rept. to USAEC.** Univ. Miami, School Mar. Atmosph. Sci. 81 p. (Unpublished manuscript.)

4. Baker, L. D.
1973 The ecology of the ctenophore **Mnemiopsis mccradyi** Mayer, in Biscayne Bay, Florida. **Univ. Miami Tech. Rept.,** UM-RSMAS-73016. 131 p. (Unpublished manuscript.)

5. Bunt, J. S., Lee, C. C., Taylor, B., Rost, P., and Lee, E.
1972 Quantitative studies on certain features of Card Sound as a biological system. **Univ. Miami Tech. Rept.,** UM-RSMAS-72011. 13 p. (Unpublished manuscript.)

6. Caspers, H.
1967 Estuaries: Analysis of definitions and biological considerations. In **Estuaries,** p. 6-8. (ed. Lauff, G. H.) Am. Assoc. Adv. Sci. Publ. No. 83.

7. Davis, C. C.
1950 Observations of plankton taken in marine waters of Florida in 1947 and 1948. **Quart. Jour. Florida Acad. Sci.,** 121: 67-103.

8. DeSylva, D.
1970 Ecology and distribution of postlarval fishes in southern Biscayne Bay, Florida. Prog. Rept. to EPA, Div. Water Res., Water Quality Office. 198 p. (Unpublished manuscript.)

9. Emery, A. R.
1968 Preliminary observations on coral reef plankton. **Limnol. Oceanogr.,** 13(2): 293-303.

10. Emery, K. O. and Stevenson, R. E.
1957 Estuaries and lagoons. I. Physical and chemical characteristics. In **Treatise on marine ecology and paleoecology,** Vol. I, p. 693-729. (ed. Hedgpeth, J. W.) Geol. Soc. Am. Mem. 67. Waverly Press,

Baltimore, Maryland.

11. Fenchel, T.
1973 Aspects of the decomposition of sea grasses. Paper Presented at International Sea Grass Workshop, Leiden, Netherlands. 18 p. (Unpublished manuscript.)

12. Fleminger, A.
1957 New calanoid copepods of **Pontella** Dana and **Labidocera** Lubbock with notes on the distribution of genera in the Gulf of Mexico. **Tulane Stud. Zool.,** 5: 19-34.

13. Florida Power Corporation
1973 **Environmental status report, October 1972-March 1973.** St. Petersburg, Florida. 120 p. (Unpublished manuscript.)

14. Grice, G. D.
1960 Copepods of the genus **Oithona** from the Gulf of Mexico. **Bull. Mar. Sci.,** 10: 485-490.

15. Heald, E. J.
1971 The production of organic detritus in a south Florida estuary. **Univ. Miami Sea Grant Tech. Bull.,** 6: 1-110. (Unpublished manuscript.)

16. Herman, S. S. and Beers, J. R.
1969 The ecology of inshore plankton populations in Bermuda. Part II. Seasonal abundance and composition of the zooplankton. **Bull. Mar. Sci.,** 19: 483-503.

17. Hopkins, T. L.
1966 The plankton of the St. Andrew Bay System, Florida. **Publ. Inst. Mar. Sci. Texas,** 11: 12-64.

18. Maturo, F. J. S., Jr.
1973 A supplementary zooplankton survey at the Crystal River Plant site. In **Environmental Status Report, October 1972-March 1973,** p. 36-44. Florida Power Corporation, St. Petersburg, Florida. (Unpublished manuscript.)

19. Odum, W. E.
1971 Pathways of energy flow in a south Florida estuary. **Univ. Miami Sea Grant Tech. Bull.,** 7: 1-162. (Unpublished manuscript.)

20. Owre, H. B. and Foyo, M.
1967 Copepods of the Florida current. **Fauna Caribaea** 1, **Crustacea** 1, **Copepoda.** Univ. Miami Press, Miami, Florida. 137 p.

21. Pritchard, D. W.
1967 What is an estuary: physical viewpoint. In **Estuaries,** p. 3-5. (ed. Lauff, G. H.) Am. Assoc. Adv. Sci. Publ. 83.

22. Reeve, M. R.
1964 Studies on the seasonal variation of the zooplankton in a marine subtropical inshore environment. **Bull. Mar. Sci.,** 14: 103-122.

23. Reeve, Mr.
1970 Seasonal changes in the zooplankton of south Biscayne Bay and some problems of assessing the effects on the zooplankton of natural and artificial thermal and other fluctuations. **Bull. Mar. Sci.,** 20: 894-921.

24. Reeve, Mr. and Cosper, E.
1972 Acute effects of heated effluents on the copepod **Acartia tonsa**

from a subtropical bay and some problems of assessment. In **Marine pollution and sea life,** p. 250-252. (ed. Ruivo, M.) Fishing News (Books) Ltd., London.

25. Reeve, M. R. and Cosper, E.
1973 The plankton and other seston in Card Sound, South Florida, in 1971. **Univ. Miami Tech. Rept.,** UM-RSMAS-73007. 24 p. (Unpublished manuscript.)

26. Reeve, M. R. and Walter, M. A.
1972 Conditions of culture, food size selection and the effects of temperature and salinity on growth rate and generation time in **Sagitta hispida** Conant. **Jour. exp. Mar. Biol. Ecol.,** 9: 191-200.

27. Riley, G. A.
1972 Patterns of production in marine ecosystems. In **Ecosystem structure and function,** p. 91-112. (ed. Wiens, J. A.) Proc. **3rd** Ann. Biol. Colloq. Oregon State Univ. Press, Corvallis.

28. Ryther, J. H.
1969 Photosynthesis and fish population in the sea. **Science,** 166: 72-76.

29. Smith, F. G. W., Williams, R. H., and Davis, C. C.
1950 An ecological survey of the subtropical inshore waters adjacent to Miami. **Ecology,** 31: 119-146.

30. Thorhaug, A. and Garcia-Gomez, J.
In press. Ecological investigations of the macroalgae in Biscayne Bay and Card Sound, Florida. I. Preliminary results of the red algae complex. **Jour. Phycol.**

31. Thorhaug, A. and Stearns, R. D.
In press. An ecological study of **Thalassia testudinum** in unstressed and thermally stressed estuaries. **Am. Jour. Bot.**

32. Woodmansee, R. A.
1958 The seasonal distribution of the zooplankton off Chicken Key in Biscayne Bay, Florida. **Ecology,** 39: 247-262.

33. Zillioux, E. J. and Gonzalez, J. G.
1972 Egg dormancy in a neritic calanoid copepod and its implications to over-wintering in boreal waters. In **Fifth European marine biology symposium,** p. 217-230. (ed. Battaglia, B.) Piccin Editore, Padua.

RELATIONSHIP OF LARVAL DISPERSAL, GENE-FLOW AND NATURAL SELECTION TO GEOGRAPHIC VARIATION OF BENTHIC INVERTEBRATES IN ESTUARIES AND ALONG COASTAL REGION

Rudolf S. Scheltema[2]

ABSTRACT

Both gene-flow and natural selection can play an important role in determining the amount of geographic variation between populations of marine benthic organisms. Genetic similarity between populations depends upon gene-flow. Clinal and disjunct differences between spatially separated populations may result from either nongenetic adaptations or genetically determined variations.

Among most coastal and estuarine benthic species gene-flow is largely limited to the dispersal that occurs during pelagic larval development. To understand the dispersal capability of sedentary bottom-dwelling invertebrate species it is necessary, first, to know their life history, i.e., whether development is pelagic or nonpelagic and the length of planktonic larval existence; second, to have information on the estuarine and coastal circulation which can disperse the larvae; third, to study behavior of larvae, particularly their vertical movement and small-scale distribution or "patchiness" and its relationship to local hydrography; and fourth, to know from direct evidence the geographical distribution of the larvae. However, gene-flow, even with extensive dispersal, is only successfully completed if immigrants survive to reproduce.

Genetic differences between populations result largely from natural selection of favored genotypes, although random loss of alleles from small populations and the "founder effect" near the species geographic limit can account for some genetic heterogeneity. One direct way to measure selection in natural

1. Contribution no. 3239 from the Woods Hole Oceanographic Institution.

2. Woods Hole Oceanographic Institution, Woods Hole, Massachusetts 02543.

populations is to relate variation to age structure within a population, another is to demonstrate genetic variation within a population in a heterogeneous or mosaic environment.

This review considers some of the evidence for gene-flow between and natural selection within estuaries and coastal regions of the continental shelf.

INTRODUCTION

Even though the importance of geographic variation to the understanding of animal evolution has long been recognized, remarkably little study has been devoted to those factors which contribute to the similarity and differences between populations of benthic marine animals. Geographical patterns of variation may be markedly disjunct, that is, with distinct differences between spatially separated populations, or they may be clinal, showing gradients of change from one population to the next.

Disjunct variations may result either from barriers to gene-flow or from intense differential selection. It is generally expected that if gene-flow between populations is restricted or completely absent, then genetic differences between populations will result from natural selection. In small populations or near the geographical limits of a species, genetic drift and the "founder effect" may also account for some genetic differences.

Clinal variation, unlike disjunct variation, is a gradual change in the phenotype of successive populations. Gradients of variation are related to gradual changes in the environment. The basis for clinal variation is not always immediately evident. Two possibilities present themselves. The first is non-genetic adaptations of genetically similar populations. Here environmental differences affect only the phenotype and the same genotype may produce different phenotypes under various environmental conditions (e. g., physiological acclimatization or morphological adaptation; *vide,* Mayr [27], p. 140, Table 7-1 for summary of noninherited variation). The second is adaptation in which the clinal differences in phenotype are genetically determined. Such differences can occur in the absence of significant gene- flow.

Two processes contribute importantly to the geographical variation of genetically determined phenotypes. One, natural selection, reduces variability within populations; the other, gene-flow, introduces new genetic variability. The amount of interpopulation variation will depend on the relationship of gene-flow and the intensity of natural selection. However, populations with well balanced genetic systems are known to resist changes from both selection and gene-flow. Such homeostasis is due to highly integrated, delicately balanced genetic interrelationships.

Too frequently, if populations appear to be phenotypically similar, lack of geographic variation is attributed to gene-flow. On the other hand, if variability exists, it is said to be the result of natural selection. Seldom among marine species has direct evidence for either natural selection or gene-flow been advanced. To obtain such evidence is often difficult, but not impossible.

It is in this context that I wish to review what is known about gene-flow and its relationship to natural selection among marine benthic species in estuaries and along coastal regions.

Dispersal Along the Continental Shelf

Gene-flow is determined by the dispersal of a species. Transport of many attached invertebrate forms has long been known to occur by rafting on objects floating upon the sea surface. Herbaceous plants and trees (13), marine algae (48), pumice from volcanic island, and, in recent years, solidified concentrations and lumps of petroleum (17) are invariably colonized by a wide variety of invertebrate organisms. Egg masses of species too large to be rafted have been found attached to drifting debris. However, Thorson (46) estimated that about 70 percent of all benthic species have pelagic larval development and among the relatively sedentary species of the continental shelf and in estuaries, dispersal occurs mostly by the planktonic larvae.

Dispersal of larvae along the continental shelf and nearshore waters has been mostly observed in boreal and temperate seas. Studies on the spatial distribution of zoea were made by Johnson (18), who traced the larvae of the sand crab *Emerita analoga* to a distance of 125 miles off the coast of southern California. During July and August, 36 percent of all plankton samples contained zoea of *Emerita.* These data were interpreted to show that large numbers of zoea were carried offshore to perish. However, Efford (11) observed that new recruits appeared on the beaches each year and concluded that, even though larval development was of four months' duration, long-shore currents and their corresponding countercurrents returned most of the larvae to the coast rather than dispersing them out to sea.

Knudsen (21) followed the dispersal of xanthic crab larvae also along the coast of California. He reasoned that a hydrographic mechanism for retaining zoea must exist, otherwise the crab population center would move progressively southward each year.

Makarov (26) examined the distribution of larval decapods in the plankton off the western Kamchatka shelf. The zoea were distributed in "larval belts" parallel to the continental shelf, and, Makarov remarked, "an abundance of pelagic larvae of sea bottom invertebrates in the shelf waters and their paucity in waters remote from the shore is a characteristic feature of practically all seas except those at the highest latitudes .... As a rule, only an insignificant portion of the larvae are carried out into the open sea" (p. 252, translation). "The larval belt owes its existence to . . . the system of nearshore currents usually moving along the coast, and the effect of tidal movements of the shelf waters.... Differences in the distribution of larvae of various species within the belt depend largely on the activity of the larvae themselves" (p. 257, translation).

Along the Atlantic coast of America there has been little study of meroplankton in shelf waters. An interesting exception is the analysis by Nichols and Keney (32) of brachyuran decapod larvae between Cape Hatteras, North Carolina, and Cape Kennedy, Florida (Fig. 1). Zoea were found at a distance of up to 60 miles (96.6 km) from shore, and as far out as the axis of the Gulf Stream. A total of 78 stations were taken along transects extending out to sea. Ninety percent of all stations had *Callinectus zoea;* only those stations beyond the axis of the Gulf Stream lacked these zoea. In general the number of crustacean larvae decreased as distance from shore increased. First and second zoea were abundant near the coastline, but 5th- through 8th-stage zoea were commonly found more than 40 miles offshore (Fig. 2). The length of pelagic development of *Callinectus sapidus* is known from laboratory culture to extend from 31 to 49 days (10), and the larvae, even if transported at only a very

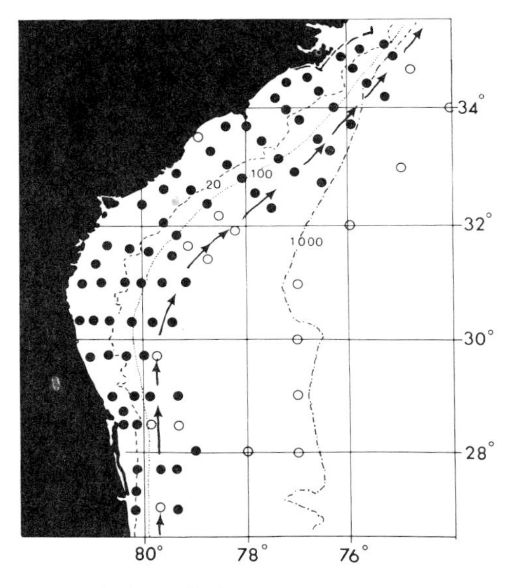

Figure 1. Stations occupied during a plankton survey over a two-year period from 1953 to 1954 in the South Atlantic Bight. Filled circles indicate locations where zoea and megalopa larvae of **Callinectes** were found; open circles are stations where crab larvae were absent. Depths are in fathoms. With a single exception larvae were found at all locations on the continental shelf (within the 100-fanthom line) at some time during the two-year period. (Modified after Nichols and Keney [32].)

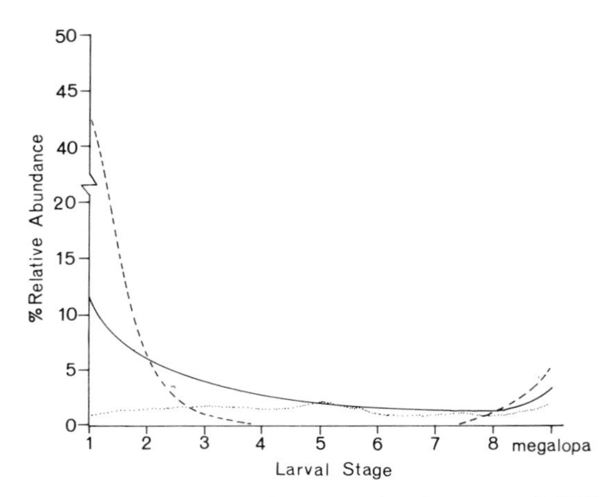

Figure 2. Percent relative abundance of various larval stages of **Callinectes** over the continental shelf at various distances off the coast of North Carolina. Solid line represents larvae along the shore; dashed line, 20 miles offshore; and dotted line, 40 miles or more offshore. (Modified after Nichols and Keney [32].)

modest rate, could theoretically be dispersed almost anywhere between the two capes. Larvae of other shoal-water decapod genera encountered by Nichols and Keney (32) included *Emerita, Hepatus, Portunus, Panopeus, Eurypanopeus, Neopanope, Menippe, Rhithropanopeus, Pinnotheres, Dissodactylus, Pinnixa,* and *Sesarma.*

Hydrographic Mechanisms that Act to Retain Larvae in Coastal Areas

Even from the relatively sparse evidence presented we see that larvae, at least those of crustaceans, are regularly dispersed over the continental shelf and frequently as far as 125 miles from shore. Williamson (50, p. 818) recently remarked on this phenomenon and has asked, "how do sufficient numbers of larvae remain in the spawning area or return to it for the stock to be replenished?". The answer to this question must be found in the hydrography of coastal regions.

Miller (30) has plotted isopleths showing the percentage drift bottles released and subsequently recovered along the mid-Atlantic Coast between Cape Cod and Cape Hatteras. Such information allows one to estimate the probability of shoreward drift of surface water from points along the coast of the Middle-Atlantic Bight. From much additional data collected over a period of ten years (7), a useful approximation can be made for odds that a larva will be retained if drifting near the sea surface at various distances from shore and at different times of the year.

Whereas drift bottles give information on sea-surface drift, data from sea-bed drifters give the direction and velocity of currents along the bottom and also the probability of shoreward dispersal (5, 23). In most regions along the Atlantic Coast between the Bay of Fundy and southern Florida there is an onshore drift along the sea floor. Particularly marked landward components of water movement occur at the mouths of such major estuaries as Delaware Bay and Chesapeake Bay (*vide,* 14). Indeed there seems to be a general convergence along the bottom toward these regions (Fig. 3) and Meade (28) remarked on the transport of offshore sediments into the mouths of these estuaries. The velocity of this bottom component is ordinarily about 0.4 nautical miles (0.74 km) per day, but has been recorded as high as 1.85 nautical miles (3.4 km) per day.

The probability of shoreward dispersal along the bottom and onto the coast or into the major estuaries can be summarized for the entire Middle Atlantic Bight (Table 1). The percentage recovery of bottom drifters is substantially higher than that of surface drift bottles, so that larvae near the bottom have a

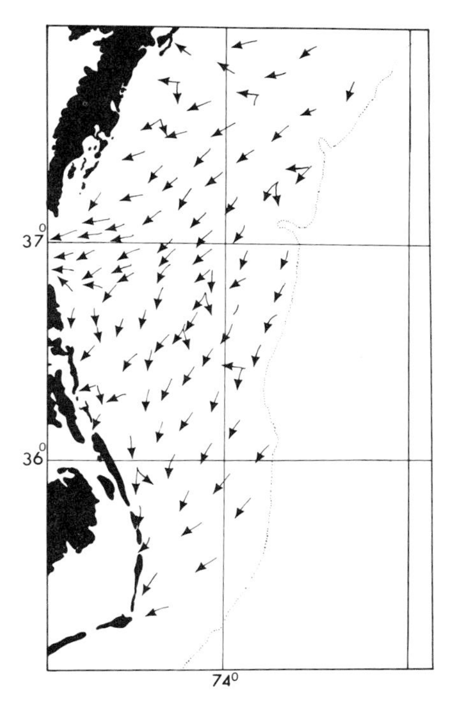

Figure 3. Bottom circulation on the continental shelf of the Chesapeake Bight. The arrows show the average direction of currents throughout the year. Returns from bottom-drifters far exceeded those from surface bottles. (Modified after Harrison et al. [14].)

TABLE 1.

Percentage drift bottles and sea-bed drifters released and subsequently recovered in the Middle Atlantic Bight of the North American coast between 36° and 40° north latitude between 1960 and 1970. (Data from D. F. Bumpus [7].)

	Number released	% recovered on coastline	% recovered in estuaries	% recovered on bottom	% recovered overseas
Surface drift bottles	76,326	12.5	none	none	0.04
Bottom sea-bed drifters	31,166	18.1	2.0	2.6	none

better chance for transport toward shore than those near the ocean surface. The likelihood of successful drift to the coast is also related to the distance from shore (Fig. 4).

Considering the hundreds of miles of shoreline along the mid-Atlantic coast, a much larger number of drifters arrived at the small entrances of estuaries than might be expected. This fact shows that bottom water from the shelf converges towards the mouths of estuaries and can entrain larvae with it.

At the mouths of estuaries there is a marked decrease in landward transport of drift bottles and none released over the shelf have been recovered subsequently within major estuaries. It is therefore inferred that larvae will not be carried into these estuaries in the surface waters.

The residual bottom drift in the South Atlantic Bight yields only slightly higher bottom drifter than surface drift-bottle returns (Table 2). This near similarity in surface and bottom shoreward transport is in contrast with the Middle Atlantic Bight and probably results from there being less fresh-water drainage per mile of coastline in the South Atlantic Bight. It is probable that a significant proportion of the zoea and megalopa larvae described by Nichols and Keney (32) would be retained in the continental shelf and that, on the average for the whole region, between 10 percent and 11.5 percent would actually be returned to the shoreline by passive drift alone. A megalopa within 10 to 20 miles from the coast has a good chance that it will not be carried out to sea.

The surface and bottom currents along the entire continental shelf of the Atlantic Coast of the United States and part of Canada are now well known from drift bottle and bottom-drifter data for each month of the year (5, 8). The current system is quite different at the surface for the spring of the year than that typically found during the late summer. In spring there is a southward surface current from Cape Cod to Cape Hatteras and a northward surface current from Florida to Cape Hatteras (Fig. 5). During late summer there is a reversal of current flow south of Cape Hatteras resulting in a strong southward current

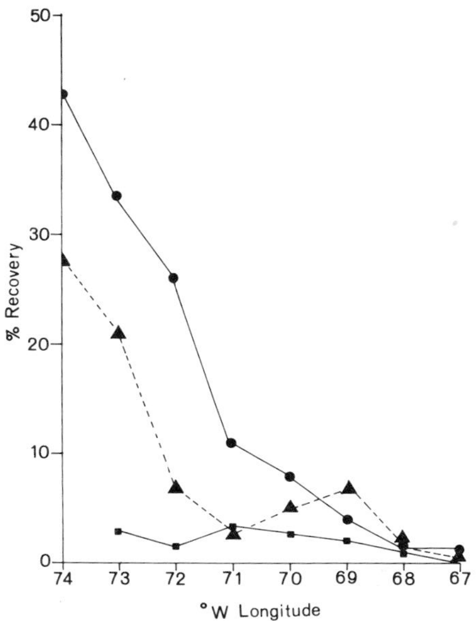

Figure 4. Drift bottle and bottom drifter returns from off the coast of New Jersey (40° north latitude) in the Middle Atlantic Bight. The abscissa indicates release points in degrees west latitude with shoreward points on the left and offshore on the right of graph. Ordinate is the average year-round percentage recovery. Circles indicate percentage of bottom drifters recovered on the coast; triangles show recovery of surface drift-bottles on shore. The squares show the number of bottom drifters retrieved in estuaries. The bottom recoveries exceeded those of the surface and the number of returns is directly related with the distance of releases from shore. No surface drift-bottles were recovered in estuaries. (Data from D. F. Bumpus.)

TABLE 2.

Percentage drift bottles and sea-bed drifters released and subsequently recovered in the South Atlantic Bight of the North American Coast between 25° and 35° north latitude between 1960 and 1970. (Data from D. F. Bumpus [7].)

	Number released	% recovered on coastline	% recovered in estuaries	% recovered on bottom	% recovered overseas
Surface drift bottles	46,460	10.0	none	none	0.1
Bottom sea-bed drifters	30,474	11.5	0.3	0.7	none

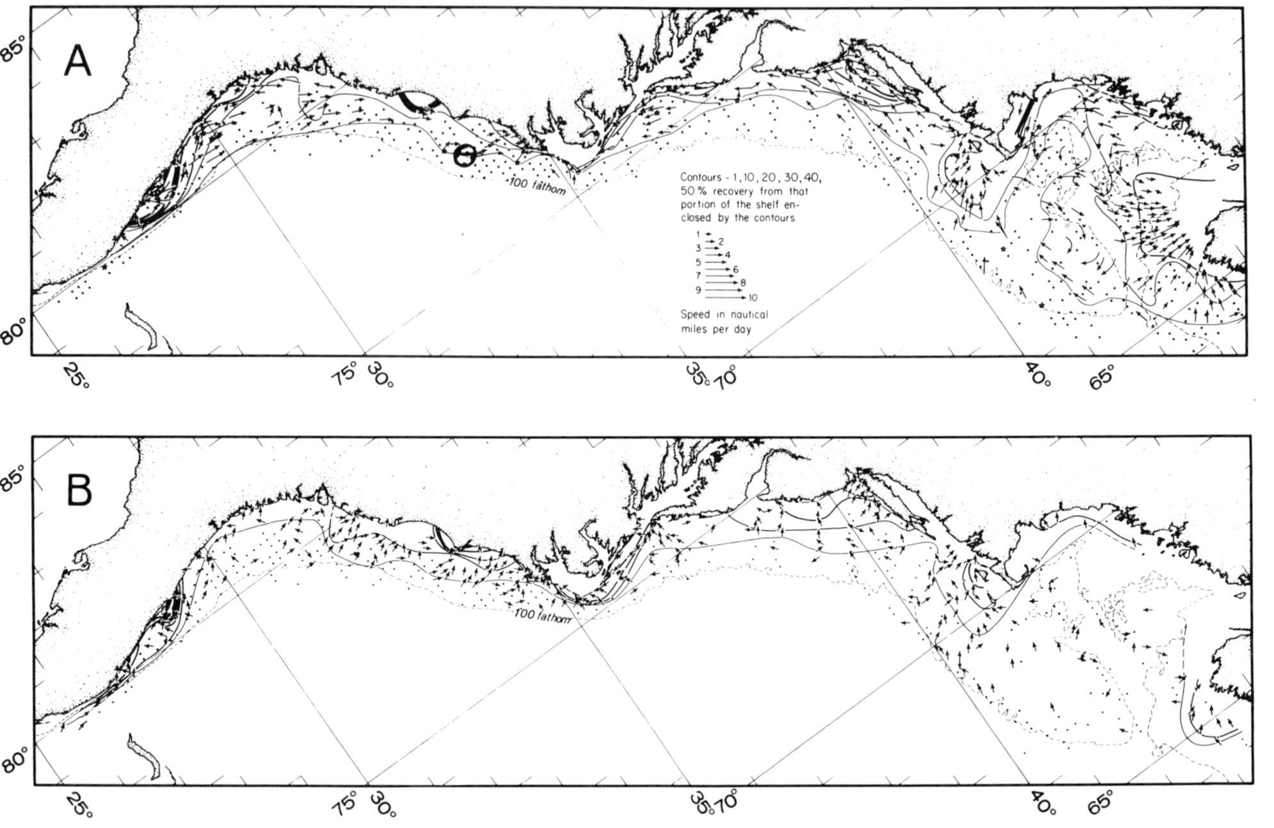

Figure 5. Inferred surface (A) and bottom (B) drift during the month of April from ten years' data. (After Bumpus [71].)

extending along the entire coastline from Cape Cod to Florida (Fig. 6). However, August surface currents north of Cape Hatteras may in certain years move northward to Cape Cod. This unusual reversal is associated with low runoff from Delaware Bay and the Hudson River (6).

Obviously the time of spawning of a benthic species can make considerable difference in the ultimate direction of its dispersal and that the behavior of larvae, particularly that affecting their vertical distribution, may make a difference in the probability of their vertical distribution, may make a difference in the probability of their retention over the continental shelf (*vide*, 47; 2, pp. 105-109).

Dispersal of Larvae in Estuaries

A study of larval dispersal in estuaries should consider three points: (a) The number of larvae retained within the estuary; these larvae will be important in maintaining an endemic adult population. (b) The number of larvae flushed out of the estuary; these larvae may be important not only in establishing gene-flow with other populations but also in maintaining and extending the species range. (c) The number of larvae entering the estuary and originating from some other region; these larvae introduce new genetic variability and maintain genetic continuity with distant adult populations.

The retention of pelagic larvae within estuaries has been considered in a number of studies. In a coastal-plain type of estuary there is a net downstream movement of less saline water flowing out along the surface and a nontidal upstream drift of more dense and more saline water moving up toward the head of the estuary along the bottom (36). Salt balance and volume continuity are maintained by vertical mixing from the bottom to the surface stratum (34). The retention of barnacle larvae in a coastal-plain estuary was shown by Bousfield (4) in the Miramichi. Plankton tows revealed that each successive nauplius stage was found to live deeper in the water column than its preceding instar. Consequently the early nauplii were carried toward the mouth, and the later stages, being below the level of no net motion, were carried upstream toward the head of the estuary. By the time that settlement was imminent most of the cyprids had been returned to the approximate point of their origin as nauplii.

The application of studies on physical hydrography in estuaries to biological problems was considered by Pritchard (33, 35), who demonstrated that oyster larvae were carried upstream along the bottom in the James River and that this dispersal accounted for the regular recruitment of spat on the oyster seed beds. The same phenomenon was further studied in the James River by Wood and Hargis (51), who compared differences in distribution between coal particles and oyster larvae. They concluded "that the bivalve larvae sampled . . . were not being transported passively, but by a process of selective swimming which

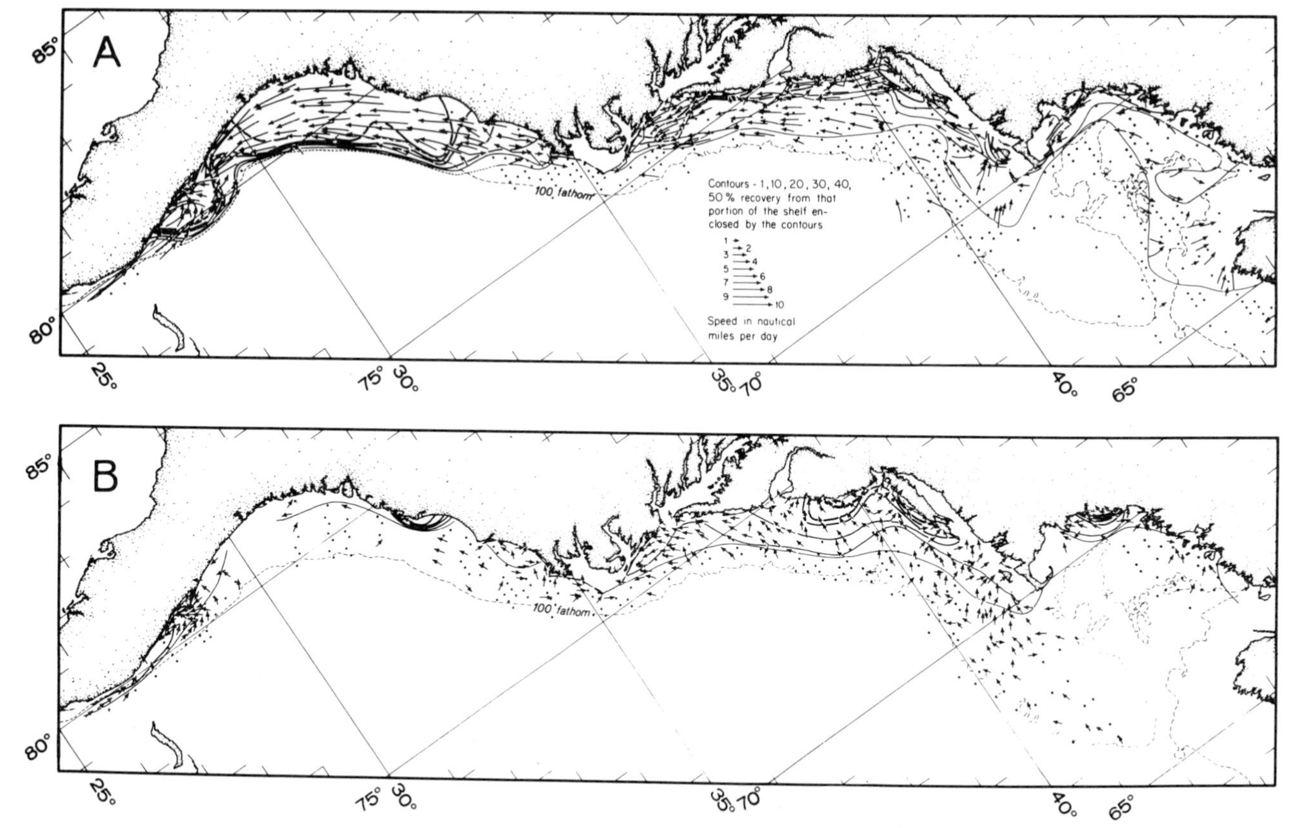

Figure 6. Inferred surface (A) and bottom (B) drift during the month of August from ten years' data. (After Bumpus [7].)

contributed actively to their upriver movement.... The data suggest that this behavior is correlated not with an increasing current speed but with increases in salinity that accompany the flood tide." (p. 41).

The dynamics of bar-built estuaries, a second type common along the mid-Atlantic coast of the United States, are usually very complex and theoretically difficult (15, 37, 38). This complexity arises from their shallow depth and small fresh-water inflow and from the fact that their mixing and exchange are largely governed by the wind regime. Although bar-built estuaries are the habitat and spawning areas of large numbers of benthic invertebrates, the dispersal of larvae them has been little studied.

The loss of larvae by mortality and by transport out of an estuary has been considered by Ketchum (19) and Ayers (1). The equations derived are limited to instances with complete vertical mixing and consequently have only restricted application. Ayers concluded that "Under estuarine . . . conditions dilution and mortality proceed simultaneously and produce a greater decrease in population than could be achieved by either process alone." (p. 28). He suggested that in estuaries having a high flushing rate there is no possibility for endemic larvae to complete development within them, and that settlement is solely from mature immigrant larvae carried in with the flooding tide.

Chamberlain (9) attempted to estimate the loss of the xanthid crab larvae of *Rhithropanopeus harrisii* from a small tributary estuary of upper Chesapeake Bay. By using rhodamine B dye as a tracer and from frequent plankton sampling, he concluded that the loss of larvae from flushing was 2.90% per tidal cycle.

Since there is no way to distinguish endemic larvae from immigrants, it is difficult to estimate the number introduced from a foreign origin. Only in those estuaries with high flushing rates, such as the special cases that were studied by Ayers (1), can the number of immigrants be roughly estimated. Strathmann (44) has pointed out that it is also necessary to know whether larvae are dispersed as individuals or in large patches, for there is probably a genetic advantage to the immigration of "sibling larvae".

Larvae near the bottom along the continental shelf of the Atlantic seaboard will tend to be swept toward the shore and also into the mouths of estuaries. From a knowledge of the abundance of larvae outside the estuary and from sea-bottom drifter data on the direction, rate, and probability of shoreward drift, it is possible to approximate very roughly the number of pelagic stages that will return shoreward or be entrained into estuaries. However, we still will not know the origin of the larvae. Although the manner in which larvae can be carried from one region of coastline to another, or from one estuary to another, can now be understood in a qualitative way, it is still not yet possible to make any quantitative estimates of the exchange of larvae between adjacent estuaries.

Natural Selection in Species With Pelagic Development

Even when gene-flow between populations is readily possible, i.e., dispersal stages such as pelagic larvae occur in the life history of a species, genetic differences may be maintained between populations as a result of local differences in genotypes favored by selection. Indeed, both disjunct and clinal genetically determined variations are known among benthic species having pelagic larval development. For example, the effect of natural selection on allele frequency has been shown in the blue mussel, *Mytilus edulis.* Milkman et al. (29) compared the leucine aminopeptidase (LAP) allele frequencies between upstream and downstream populations in a small estuary on Cape Cod, and showed that they diverged markedly as the mussels developed from settlement to maturity. They interpreted this divergence to be the result of differences in selection pressures acting upon the two populations.

Two sympatric intertidal species of mussels, *Mytilus edulis* and *Modiolus demissus,* were studied by Koehn and Mitten (20) with respect to the polymorphism of the leucine aminopeptidase locus. The data showed that there was considerable variation in gene frequencies at a single LAP locus in both species of mussels in localities only one-quarter mile apart along the marsh studied. In each local environment, different allele frequencies were found. At any single locality the frequencies of the three common and two rare alleles were similar for both species, i.e., "the rank order of relative abundance of alleles of particular electrophoretic mobility is nearly identical for both species at all locations". The evidence indicated that both species were responding to short-term selection at the LAP locus in the same way.

In a further study of *Modiolus demissus,* Koehn et al. (22) showed that populations separated vertically on the intertidal by 0.66 meters differed significantly in allele frequency and in the proportion of heterozygotes at the indolphenol oxidase locus. There also appeared to be a significant increase in the proportion of heterozygotes with increasing age (i.e., increasing shell length). The genetic differences were attributed either to site selection by larvae or more probably to differential mortality or natural selection.

The foregoing evidence based on the use of starch gel electrophoresis in identifying loci controlling enzyme polymorphism should be accepted with some reservations. Selander et al. (40) in their study of genetic diversity of the horseshoe crab *Limulus polyphemus* cautioned that "statements regarding the genetic basis of polymorphism in electrophoretic patterns . . . are to be understood as hypotheses, since for none of the polymorphic proteins do we have direct evidence from progeny studies . . . ". However, they advanced indirect evidence to support their genetic interpretation.

A study of another species with a pelagic mode of development is that of the prosobranch gastropod *Littorina picta* by Struhsaker (45), who further demonstrated how natural selection can act differently on two populations in very close spatial proximity. The intraspecific variation in shell sculpture of *Littorina picta* appeared to be genetically associated with certain physiological variations. Both the morphological and physiological viriations were in turn correlated with particular supratidal environments. Populations with smooth shells predominated on low-angle beaches with heavy horizontal wave force, while populations that had sculptured shells predominated on high-angle beaches receiving spray and slight horizontal wave force. Populations with intergrades in shell sculpture occurred in intermediate types of environments. Larvae from populations of different sculpture types were reared under identical conditions in laboratory culture. The larvae varied in shell morphology, shell size, growth rate, and mortality rate. Progeny of parents with sculptured shells had a higher growth rate and lower mortality rate than larvae from smooth parents. The progeny of sculptured parents had heavily sculptured, larger shells; the progeny of smooth parents, almost no sculpture and small shells; progeny of intermediate parents, intermediate shell sculpture and size. The rearing experiments indicated that the shell variations had a genetic basis independent of any environmental effect. The major environmental factors selecting for the shell variations are probably wave action, prolonged submersion, desiccation, high temperature, and extreme salinity. These factors appear to act most strongly against settling larvae, post-veligers, and juvenile stages within the supratidal environment. The shell sculpture variation appears to be an example of balanced, adaptive polymorphism maintained by differential selection and nonrandom mating within two major types of supratidal habitats.

An interesting example of clinal variation is the case of the minimum spawning temperature in the oyster *Crassostrea virginica,* whose geographical range extends from the St. Lawrence to the northern half of the Gulf of Mexico. Nelson (31) proposed that this species first spawns at a sea-water temperature of 20 °C. Subsequently the minimum spawning temperature was observed to vary with different populations along the North American coast, the southern populations commencing to spawn at 25 °C. Stauber (42, 43) posed the question whether this difference was genetically determined or was merely the result of physiological acclimatization. The problem was seemingly solved when Loosanoff and Nomejko (25) showed that the spawning temperature of southern oysters was not modified even when held in Long Island Sound over periods that were judged to be sufficient for physiological acclimatization.

Biochemical and serological data have given further evidence for genetic differences between populations of *Crassostrea virginica* (16, 24). Either the transport of oyster larvae between estuaries is restricted, or natural selection overrides the effect of gene-flow resulting from dispersal. It is possible that the larvae of some estuarine species will not tolerate the conditions found in open waters of the continental shelf.

There are also instances in which there is little genotypic variation among populations throughout wide areas of the species' geographical range. An explanation for this distribution may be the existence of unrestricted gene-flow between populations. Indeed, species with low genotypic variation between populations usually have pelagic larval development. Milkman et al. (29) found a remarkably uniform frequency of certain electrophoretically determined alleles in *Mytilus edulis* throughout the southern part of its range. That larvae of this species are regularly dispersed over considerable distances is attested by the recent evidence of Wells and Gray (49), who showed that *Mytilus edulis* is found periodically at Cape Hatteras even though this bivalve is not able to survive the hot summer months there.

The gastropod *Nassarius obsoletus* was collected by Gooch et al. (12) from eleven sites along the Atlantic coast from Cape Cod, Massachusetts, to Beaufort, North Carolina. The genetic variability between these populations was compared by polyacrylamide gel electronphoresis with respect to two polymorphic alleles. A remarkable geographic homogeneity of allele frequency was found, and Gooch was inclined to attribute this to "extensive gene-flow" although he did not exclude the possibility of genetic homeostasis or "balance polymorphism insensitive to local variations in the environment". Clearly, obtaining direct evidence for the dispersal of *Nassarius obsoletus* larvae would be desirable.

It has proved easier to demonstrate natural selection than to give evidence for gene-flow. Some insight into the importance of a pelagic stage to gene-flow may be obtained, however, by comparing species with planktonic larvae to those with direct development (i.e., lacking a pelagic dispersal stage). Berger (3) compared the three species of *Littorina* that are found sympatrically on intertidal fucoids from Labrador to New Jersey. Only one species, *Littorina littorea,* has pelagic larvae; the other two, *Littorina obtusata* and *Littorina saxatilis,* have developments without a planktonic phase. Fifteen populations between Prince Edward Island and the south shore of Cape Cod were examined electrophoretically for gene-enzyme variation. Gene-frequency, determined by assuming a correspondence between bands on gels and alleles, showed interesting differences between the three species. Allele frequency data for polymorphic loci seemed to reveal interpopulation heterogeneity in the species *L. saxatilis* and *L. obtusata,* but homogeneity between populations of *L. littorea.* Berger concluded that the "differences appear to be correlated with the mechanisms and capabilities for larval dispersal in these species".

Snyder and Gooch (41) showed further that the degree of differentiation between populations of *Littorina saxatilis,* as demonstrated by electrophoretically determined enzyme polymorphism, was independent of the distance between populations. Significant population differentiation occurred over distances as little as 2 km, while widely separated populations were sometimes nearly identical. By means of probability indices, the genotypic identity between populations of *Littorina saxatilis* with its ovoviviparous development was compared to that between populations of *Nassarius obsoletus* with its planktonic larval development. Snyder and Gooch concluded that there was "a strong correlation between population differentiation and dispersal ability of the organisms".

If the planktonic larval stages of benthic organisms are truly effective in establishing gene-flow, then a relationship would be expected between the intensity or frequency of such dispersal and the degree of genetic similarity between spatially disjunct populations. Such a relationship seems to exist among some tropical gastropod species (39).

SUMMARY AND CONCLUSIONS

Among marine benthic organisms with pelagic larvae, both gene-flow and natural selection play an important role in determining the amount of geographic variation between populations.

Gene-flow is limited by dispersal and among most benthic species occurs during the pelagic larval stage. It has been shown that the larvae of coastal benthic species are commonly carried out over the continental shelf. The hydrography of a coastal region will determine the direction and distance that a larva can be dispersed. Larval behavior, particularly that which determines vertical distribution, is also important in determining the direction of dispersal.

There is evidence that some benthic species will pelagic larvae will show genotypic similarity over hundreds of kilometers of shoreline. On the other hand, other benthic forms with pelagic development show genotypic differences between populations only a few hundred meters apart. Even though this review cannot pretend to be comprehensive, it should make obvious that much more data are necessary if the relationships between larval dispersal, natural selection, and genotypic variation between benthic populations are to be understood.

ACKNOWLEDGEMENTS

I wish to thank Mr. Dean F. Bumpus for free access to data collected and compiled under his direction and also for a number of stimulating discussions on coastal circulation. This work was made possible through grant no. GA 40144 from the National Science Foundation.

REFERENCES

1. Ayers, J. C.
1956 Population dynamics of the marine clam, **Mya arenaria. Limnol. Oceanogr.,** 1: 26-34.

2. Banse, K.
1964 On the vertical distribution of zooplankton in the sea. **Progr. Oceanogr.,** 2: 54-125.

3. Berger, E. M.
1973 Gene-enzyme variation in three sympatric species of **Littorina. Biol. Bull.,** 145: 83-90.

4. Bousfield, E. L.
1955 Ecological control of the occurrence of barnacles in the Miramichi estuary. **Bull. Nat. Mus. Canada,** 137: 1-69.

5. Bumpus, D. F.
1965 Residual drift along the bottom on the continental shelf in the middle Atlantic Bight area. **Limnol. Oceanogr.,** 10 (supplement): R50-53.

6. Bumpus, D. F.
1969 Reversal in the surface drift in the middle Atlantic Bight area. **Deep-Sea Res.,** 16: 17-23.

7. Bumpus, D. F.
1973 A description of the circulation on the continental shelf of the east coast of the United States. **Progr. Oceanogr.,** 6: 111-157.

8. Bumpus, D. and Lauzier, L. M.
1965 Surface circulation on the continental shelf off eastern North America between Newfoundland and Florida. **Serial Atlas of the Marine Environment** (ed. Webster, W.) Amer. Geogr. Soc. Folio 7. 4 p., 8 pl.

9. Chamberlain, N.
1962 Ecological studies of the larval development of **Rhithropanopeus harrisii** (Xanthidae, Brachyura). **Johns Hopkins Univ., Tech. Rept.** 28, Ref. 62-18. 47 p. (Unpublished manuscript.)

10. Costlow, J. D. and Bookhout, C. G.
1959 The larval development of **Callinectus sapidus** reared in the laboratory. **Biol. Bull.,** 115: 373-396.

11. Efford, I. E.
1970 Recruitment to sedentary marine populations as exemplified by the sand crab, **Emerita analoga** (Decapoda, Hippidae). **Crustaceana,** 18, 293-308.

12. Gooch, J. L., Smith, B. S., and Knapp, D.
1972 Regional survey of gene frequencies in the mud snail **Nassarius obsoletus. Biol. Bull.,** 142: 36-48.

13. Guppy, H. B.
1917 **Plants, seeds, and currents in the East Indies and Azores.** William Norgate, London. xi + 513 p.

14. Harrison, W., Norcross, J. J., Pore, N. A., and Stanley, E. M.
1967 Circulation of shelf waters off Chesapeake Bight – surface and bottom drift of continental shelf waters between Cape Henlopen, Delaware, and Cape Hatteras, North Carolina. June 1963-December 1964. **U. S. ESSA Prof. Pap.,** 3: 1-82.

15. Hela, I., Carpenter, C. A., and McNulty, J. K.
1957 Hydrography of a positive, shallow tidal, bar-built estuary (report on the hydrography of the polluted area of Biscayne Bay). **Bull. Mar. Sci. Gulf Carib.,** 7: 47-99.

16. Hillman, R. E.
1964 Chromatographic evidence of intraspecific genetic differences in the eastern oyster, **Crassostrea virginica. Syst. Zool.,** 13: 12-18.

17. Horn, M., Teal, J. M., and Backus, R. H.
1970 Petroleum lumps on the surface of the sea. **Science,** 168: 245-246.

18. Johnson, M. W.
1939 The correlation of water movement and dispersal of pelagic larval stages of certain littoral animals, especially the sand crab **Emerita. Jour. Mar. Res.,** 2: 236-245.

19. Ketchum, B. H.
1954 Relation between circulation and planktonic populations in estuaries. **Ecology,** 35: 191-200.

20. Koehn, R. K. and Mitten, J. B.
1972 Population genetics of marine Pelecypods. I. Ecological heterogeneity and evolutionary strategy at an enzyme locus. **Amer. Nat.,** 106: 47-56.

21. Knudsen, J. W.
1960 Reproduction, life history, and larval ecology of the California Xanthidae, the pebble crabs. **Pacific Sci.,** 14: 3-17.

22. Koehn, R. K., Turano, F. J., and Mitton, J. B.
1973 Population genetics of marine Pelecypods. II. Genetic differences in microhabitats of **Modiolus demissus. Evolution,** 27: 100-105.

23. Lauzier, L. M.
1967 Bottom residual drift on the continental shelf area of the Canadian Atlantic coast. **Jour. Fish. Res. Bd. Canada,** 24: 1845-1859.

24. Li, M. F., Flemming, C. W., and Stewart, J. E.
1967 Serological differences between two populations of oysters **(Crassostrea virginica)** from the Atlantic coast of Canada. **Jour. Fish. Res. Bd. Canada,** 24: 443-446.

25. Loosanoff, V. L. and Nomejko, C. A.
1951 Existence of physiologically-different races of oysters, **Crassostrea virginica. Biol. Bull.,** 101: 151-156.

26. Makarov, R. R.
1969 Transport and distribution of Decapoda larvae in the plankton of the western Kamchatka Shelf. **Okeanologyi Akad. Nauk. USSR,** 9: 306-317. (**Oceanology,** 9 (2): 251-261, Scripta Technica.)

27. Mayr, E.
1963 **Animal species and evolution.** Belknap Press, Harvard University Press, Cambridge, Mass. xvi + 797 p.

28. Meade, R. H.
1969 Landward transport of bottom sediments in estuaries of the Atlantic coastal plain. **Jour. Sed. Petrol.** 39: 222-234.

29. Milkman, R., Zeitler, R., and Boyer, J. F.
1972 Spatial and temporal genetic variation in **Mytilus edulis:** natural

selection and larval dispersal. **Biol. Bull.** 143: 470.

30. Miller, A. R.
1952 A pattern of surface coastal circulation inferred from survace salinity-temperature data and drift bottle recoveries. **Woods Hole Oceanogr. Inst. Tech. Rept.,** Ref. No. 52-28. 47 p. (Unpublished manuscript.)

31. Nelson, T. C.
1928 Relation of spawning of the oyster to temperature. **Ecology,** 9: 145-154.

32. Nichols, P. and Keney, P. M.
1963 Crab larvae **(Callinectes)** in plankton collections from cruises of M/V **Theodore N. Gill** South Atlantic coast of the United States 1953-54. **U. S. Fish Wildl. Serv. Spec. Sci. Rept. Fisheries,** 448: 1-14.

33. Pritchard, D. W.
1951 The physical hydrography of some estuaries and some applications to biological problems. **Trans. 16th North American Wildlife Conf.,** p. 368-376.

34. Pritchard, D. W.
1952 Salinity distribution and circulation in Chesapeake Bay estuarine system. **Jour. Mar. Ress.,** 11: 106-123.

35. Pritchard, D. W.
1952 Distribution of oyster larvae in relation to hydrographic conditions. **5th Ann. Sess. Gulf Carib. Fish. Inst.,** p. 123-132.

36. Pritchard, D. W.
1955 Estuarine circulation patterns. **Proc. Amer. Soc. Civil Eng.,** 81: 717-1 to 717-11.

37. Pritchard, D. W.
1960 Salt balance and exchange rate for Chincoteague Bay. **Chesapeake Sci.,** 1 (1): 48-57.

38. Roelofs, E. W. and Bumpus, D. F.
1953 The hydrography of Pamlico Sound. **Bull. Mar. Sci. Gulf Carib.,** 3: 181-205.

39. Scheltema, R. S.
1971 Larval dispersal as a means of genetic exchange between geographically separated populations of shoal-water benthic marine gastropods. **Biol. Bull.,** 140: 284-322.

40. Selander, R. K., Yang, S. Y., Lewontin, R. C., and Johnson, W. E.
1970 Genetic variation in the horseshoe crab **(Limulus polyphemus),** a phylogenetic "relic". **Evolution,** 24: 402-414.

41. Snyder, T. P. and J. C. Gooch.
1973 Genetic differentiation in **Littorina saxatilis** (Gastropoda). **Mar. Biol.,** 22: 177-182.

42. Stauber, L. A.
1947 On the possible physiological species in the oyster **Ostrea virginica. Anat. Rec.,** 99: 614.

43. Stauber, L. A.
1950 The problem of physiological species with special references to oyster and oyster drills. **Ecology,** 31: 109-118.

44. Strathmann, R.
1972 Spread of sibling larvae of sedentary marine invertebrates. **Amer. Zool.,** 12: 721-722.

45. Struhsaker, J. W.
1968 Selection mechanisms associated with intraspecific shell variations in **Littorina picta** (Prosobranchia; Mesogastropoda). **Evolution,** 22: 459-480.

46. Thorson, G.
1946 Reproduction and larval development of Danish marine bottom invertebrates. **Medd. Komm. Danmarks Fisk. Havunders., ser. Plankton,** 4 (1): 1-523.

47. Thorson, G.
1964 Light as an ecological factor in the dispersal and settlement of larvae of marine bottom invertebrates. **Ophelia,** 1: 167-208.

48. Vallentin, R.
1895 Some remarks on the dispersal of marine animals by means of seaweeds. **Ann. Mag. Nat. Hist.,** ser. 6, 16: 418-423.

49. Wells, H. W. and Gray, J. E.
1960 Seasonal occurrence of **Mytilus edulis** on the Carolina coast as a result of transport around Cape Hatteras. **Biol. Bull.,** 119: 550-559.

50. Williamson, D. I.
1967 Some recent advances and outstanding problems in the study of larval crustacea. **Proc. Symp. Crustacea, Symp. Ser. 2, Mar. Biol. Assoc. India,** Pt. II, p. 815-823.

51. Wood, L. and Hargis, W. J.
1971 Transport of bivalve larvae in a tidal estuary. In **Fourth European Marine Biology Symposium,** Cambridge Univ. Press, Cambridge. p. 29-44.

GEOGRAPHICAL DISTRIBUTION AND MORPHOLOGICAL DIVERGENCE IN AMERICAN COASTAL-ZONE PLANKTONIC COPEPODS OF THE GENUS *LABIDOCERA*

Abraham Fleminger[1]

ABSTRACT

The combined eastern and western coastal zone of the American continents support 15 morphologically and geographically distinct populations assignable to 12 known species of temperate and tropical planktonic copepods belonging to the genus *Labidocera.* Morphologically, they comprise two natural and distinctive phylogenetic groups. Each group is represented by several species along the eastern and western coastal zones.

Comparative phylogenetic-biogeographic study of these populations has been underway to determine (a) the nature and extent of divergent morphological characters in cephalic and thoracic appendages and body segments and (b) the geographical location and extent of the coastal zone occupied by each population. Comparison between species groups revealed (a) co-occurrences of pairs of species are common, (b) differences in feeding and swimming appendages are restricted to small details such as one more or less small seta per appendage, and (c) extensive qualitative divergence in sexually modified structures indicating essential differences in function. Within a species group (a) co-occurrence is more frequently limited to overlap at the edge of adjacent ranges or in coastal sectors experiencing pronounced seasonal changes in climate, (b) feeding and swimming appendages do not differ materially, and (c) divergence in sexually modified structures varies in proportion to the geographical extent of actual or potential overlap with proximal species.

The results indicate that among coastal-zone species with similar genetic and geographical heritages there is a predominance of selection for the development of prezygotic mating barriers but little or no selection for partitioning food resources by means of morphological differentiation or body size.

1. Scripps Institution of Oceanography, P. O. Box 1529, La Jolla, California 92307.

Planktonic copepods of estuaries and coastal zones share in occupying an area of transition between fresh water, the land, and the open ocean. It is an area in which conservative oceanic properties fluctuate broadly and strong advective forces meet abrupt hydrographic gradients (10). In short, this area presents an atmosphere hostile to zooplankton, and no doubt the low diversity of estuarine and coastal-zone planktonic copepods is an outcome of its rigorous demands. In addition to abiotic stresses, estuarine and coastal-zone copepods have undergone intense biotic stresses. Ample indication of past biotic selection pressures is seen in the habitat zonation and interspecific compatibility needed to sustain the stabilized zooplankton fauna found in coastal zones and large estuaries.

Of the two, coastal-zone species are a better subject for inquiry about natural selection produced by biotic interactions. Not only is there a greater diversity of species, but the reduced impact of daily fluctuations from tides, weather, and contamination in the zone provide a broader spectrum of relatively stable co-occurrences. Individual idiosyncrasies of specific estuaries have only local influence. The coastal zone also offers a more moderate response to temporal and zonal changes in climate. This makes it easier to separate persistent geographical co-occurrences of potentially interacting species from more ephemeral contacts. Related species and, to an extent, the same species of copepods are shared by the coastal zone and its adjacent estuaries. Thus, implications about evolution in coastal-zone species may also be expected to apply to estuarine species.

Divergence in external morphology and in adaptation to habitat are the raw data of evolution to the biogeographer-systematist. The extent of concordant patterns among these features provides a yardstick for judging what dominant selection pressures operate on populations. It is from this perspective that I have been examining morphology and distribution among a number of groups of closely related species of planktonic calanoid copepods.

In this paper I consider morphology relative to distribution among two species groups of the genus *Labidocera* occupying the eastern and western coastal zones of the Americas from mid-latitudes to the tropics. The term species group is used in a conventional sense; i.e., a monophyletic lineage of recently derived morphologically distinct populations inhabiting geographical regions that are often contiguous or overlapping. Their monophyly is a judgment based on morphological evidence; the criteria that they are valid species are based on the absence of intergradation in areas of geographical overlap and the predominance of morphological differences located in sexually modified structures; and their recent derivation is a judgment based on geographical distributions relative to the geological history of the region.

As in terrestrial lineages (e.g., cf. 2) recent common ancestry in plantonic copepods confers upon the daughter populations similarity in behavior, morphology, and environmental requirements. In planktonic copepods, similarity in the selection of habitat is readily apparent among the thirteen species groups I have been examining over the past few years (Table 1). Similar patterns would emerge from a consideration of their food organisms as well as from sets of diagnostic morphological characters.

The two coastal-zone species groups of *Labidocera* found in the Americas show parallel distributions correlated to continental zones of climate moderated by adjacent hydrographic circulations. Geographical overlap and opportunity for interaction among those species have probably varied in different times ranging from short-term cyclical sequences of cool and warm years to glacial and interglacial geological time periods of tens of thousands of years. Contemporary patterns of distribution and morphology reflect the outcome of primary selection pressures that have modified these populations with time.

Undoubtedly during large-scale fluctuations in climate, the range of coastal-zone species expanded and contracted, isolating daughter populations and providing ample opportunities to produce the 15 morphologically distinct populations of *Labidocera* found in the American coastal zone. It is against this historical perspective, as well as in the light of contemporary climates in the coastal zone, that we should examine geographical distribution relative to morphological divergence in *Labidocera.*

The western coastal zone of the American continents is conditioned by eastern boundary currents (17) flowing from mid-latitudes (Fig. 1). These currents, together with coastal upwelling, extend temperate conditions closer to the tropics than do to the eastern coasts of the Americas. The eastern coasts are modified by western boundary currents transporting tropical water poleward and warming the coastal zone up to mid-latitudes. Along the west coast the tropical-to-temperate coastal zone is confined to latitudes between 5°S to about 30°N and seasonal change is barely noticeable. On the east coast the tropical-to-warm temperate zones extend from latitudes 40°S to 40°N in appropriate seasons and the strong influence of seasonality is indicated by pronounced changes in the coastal location of the 15, 20 and 25 °C isotherms.

The influence of these fundamental differences in marine climates along the eastern and western American coasts may also be seen in the distribution of both planktonic and non-planktonic neritic organisms. For example, neritic species of the calanoid genera *Centropages* and *Temora* (Fig. 2) and species of the shrimp genus *Penaeus* (Fig. 3) show similar patterns of geographical distribution from mid- to low-latitudes, patterns that are repeated in the coastal-zone species of *Labidocera,* discussed below. Eastern coastal ranges are much more extensive than those of the west and the more gradual change of climate is reflected in the greater number of eastern regions showing faunal changes in coastal zooplankton (e.g. Cape Cod, Cape Hatteras, southern Florida, the northern Gulf of Mexico, the Yucatan Peninsula, Trinidad, Cape Frio, and the La Plata estuary). On the west coast there is an abrupt change from cool temperate to tropical conditions

TABLE 1.

Habitat and climate zones occupied by the species comprising each of the species groups listed by the name of a known representative of the group. Species groupings and judgments as to distribution are based on published and unpublished data. See text for further discussion.

Species-group	No. supp.	Depth range in epipelagic zone	Coastal[1]				Open neritic[1]				Eutrophic Oceanic[1]				Oligotrophic Oceanic[1]			
			P	B	Te	Tr	P	B	Te	Tr	P	B	Te	Tr	P	B	Te	Tr
Pontellina plumata	(4)	broad - 1												4				
		shallow - 3																
Ecalanus elongatus	(5)	deep										1	1	3				
Eucalanus attenuatus	(4)	deep - 2											2	2				
		shallow - 2																
Labidocera detruncata	(3)	shallow												3				
Eucalanus subtenuis	(7)	shallow							1	1			1	4				
Calanus finmarchicus	(7)	broad					1	2	4									
Centropages furcatus	(2)	shallow								2								
Temora stylifera	(2)	shallow								2								
Labidocera acuta	(2)	shallow								2								
Undinula vulgaris	(4)	shallow								4								
Eucalanus pileatus	(3)	shallow				3												
Labidocera darwinii	(13)	shallow			7	4				2								
Labidocera jollae	(6)	shallow			2	4												
13 groups	62 species																	

1. P = polar; B = boreal; Te = temperate; Tr = tropical.

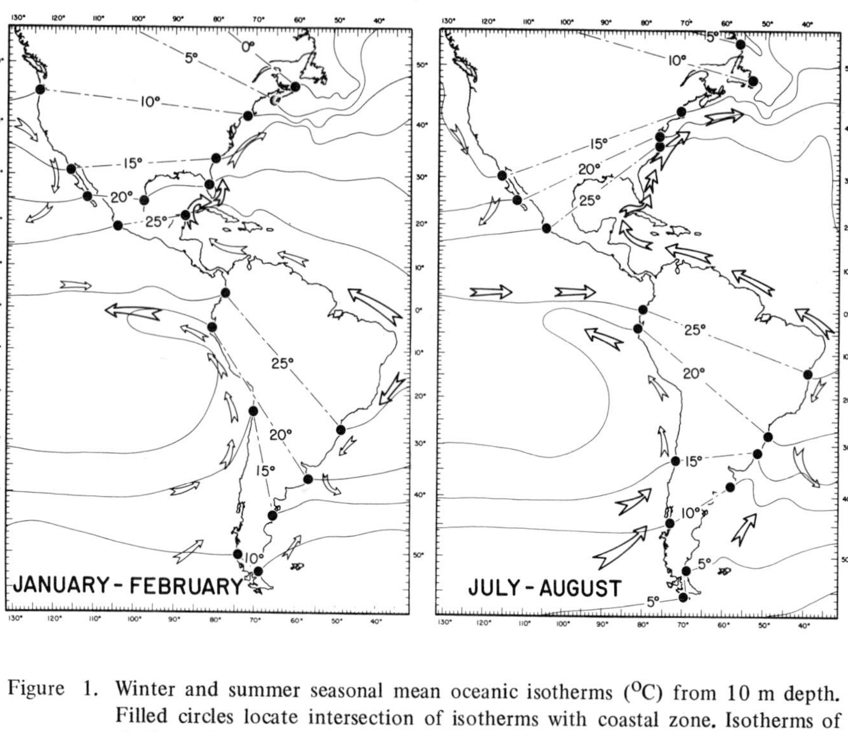

Figure 1. Winter and summer seasonal mean oceanic isotherms (^{O}C) from 10 m depth. Filled circles locate intersection of isotherms with coastal zone. Isotherms of similar value are connected across continents by dashed lines. Arrows show direction of prevailing surface currents; more strongly weighted arrows indicate seasonally stronger currents. Temperature data based on Muromtsev (13, 14). Surface currents compiled from various sources.

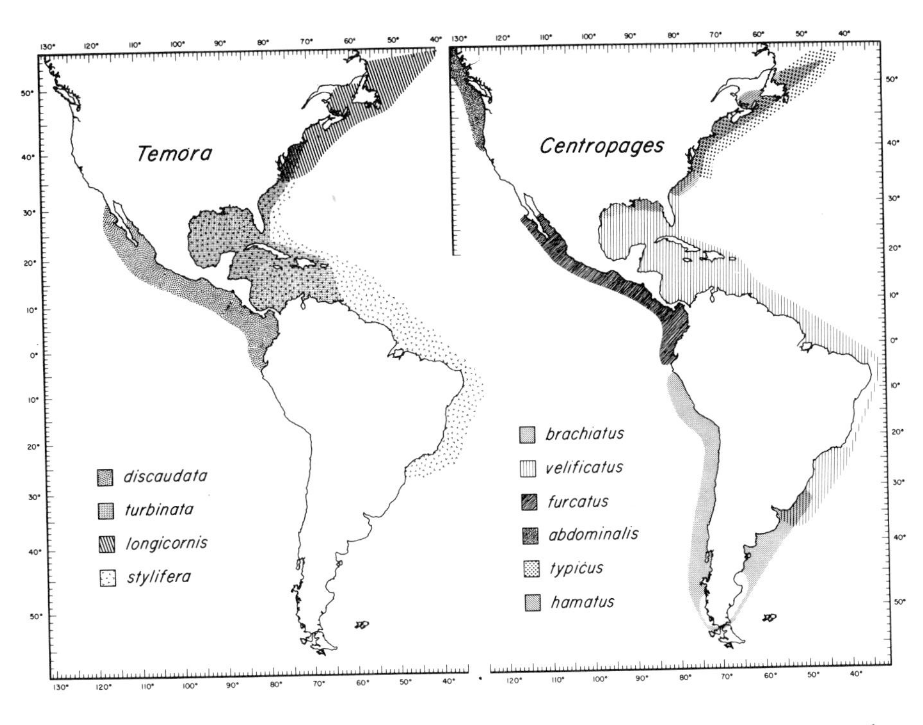

Figure 2. Schematic distribution of species of the calanoid copepod genera **Temora** and **Centropages** in American neritic waters. Distributions based on Fleminger (6), a review of published records, and unpublished data.

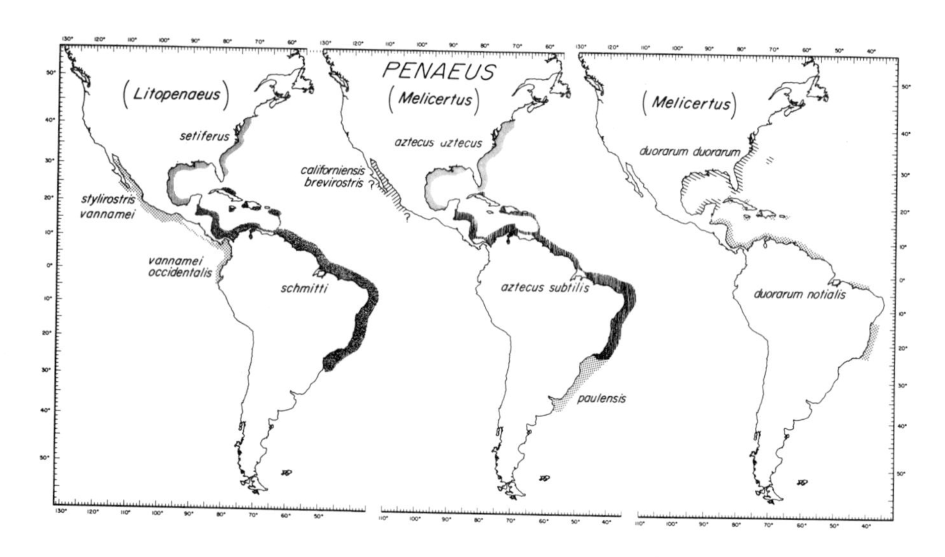

Figure 3. Distribution of penaeid shrimps in American neritic waters. Based on data from Pérez Farfante (15).

etween northern Peru and the Gulf of Guayaquil. Tropical froms extend north) the Gulf of California and the southern half of Baja California and are ›ruptly replaced by temperate species in the northern third of the Gulf of alifornia as well as on the Pacific coastline along upper Baja California and alifornia.

Unpublished records from extensive sampling within and outside of the merican coastal zone (Fig. 4) and available published records have established ıe distribution of American coastal-zone species of *Labidocera* (Fig. 5) and erified their absence away from the American continents. With each species roup represented in the eastern and western coastal zones by a number of pecies, appreciable differences in range dimensions (Table 2) as well as the elative extent of geographical overlap (Fig. 5) provide a natural experiment in volution. This natural experiment is controlled in a sense by the availability of omparative patterns for two phylogenetically independent though congeneric pecies groups that are remarkably similar in size and in their general equirements for habitat.

Diagnostic differences among the species of *Labidocera* are concentrated in he sexually modified structures of reproductively mature individuals. In the nale (Figs. 6a, b) they include the right first antenna, the fifth pair of wimming legs, and the right lateral corner of the last thoracic segment. In the female (Figs. 6c-e) the genital segment, the fifth pair of legs, and often the caudal furca are useful for separating species. The spermatophore with its attachment device, "the coupler," is also species specific, at least in American species. Presumably with the aid of the male's left fifth leg, the coupler is cemented to the female genital segment in a position also characteristic to the species, while the male holds the female with the large claw (chela) of its right fifth leg. Indications are that a species-specific key-and-lock mechanism serving as a mechanical barrier against hybrid mating is formed by a combination of the female structure held by the male's claw, the orientation of the male's body relative to that of the female, body and appendage proportions, and a combination of swellings, integumental sensory organs, and glands located in the vicinity of the female's genital pore.

The *darwinii* and *jollae* species groups, co-occurring extensively in the western coastal zone, exhibit morphological differences in sexually modified structures that suggest functionally significant differences exceeding that of a simple key-and-lock mechanism. These differences are illustrated by a comparison of *johnsoni* representing the *darwinii* group and *kolpos* representing the *jollae* group. The two species have overlapping distributions in the Gulf of California. Appreciable differences in mating behavior by males of each species group are indicated by mechanical differences in the hinging of the right first antenna (Figs. 6b, 7a, c), the chela, and the proportional length and structure of the left fifth leg (Figs. 7b, d). In females, asymmetries in the genital segment, the caudal

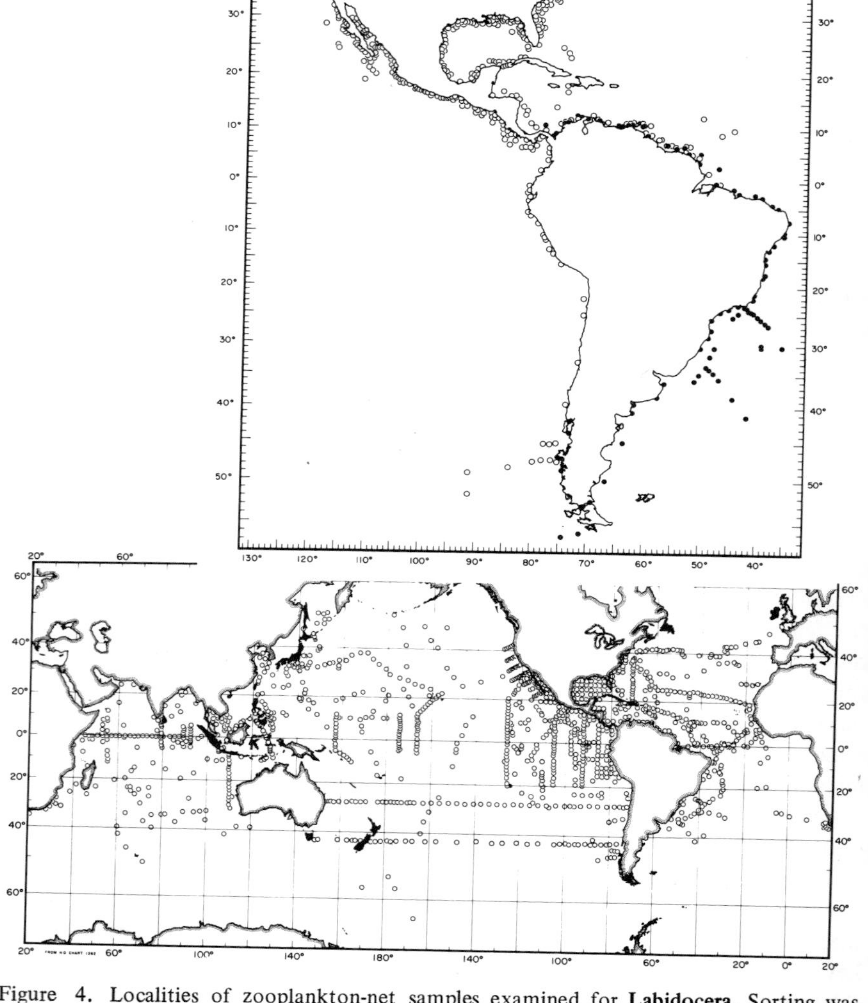

Figure 4. Localities of zooplankton-net samples examined for **Labidocera.** Sorting was done with the aid of a microscope from a randomly drawn subsample representing at least 10% of the total catch. Upper panel: inshore oblique and horizontal hauls from the uppermost 20 m or less collected routinely within one km of the coastline using a standard one-half-meter or a one-meter CalCOFI net. Many localities are represented by two to four samples taken equidistantly along a transect between the edge of the surf zone and roughly one km offshore.

Figure 4 (con't)
Filled circles represent recent collections that have been examined cursorily. Lower panel: offshore oblique hauls from 150-meters depth or greater to the surface; Pacific and Indian Ocean samples obtained with a standard one meter CalCOFI net.

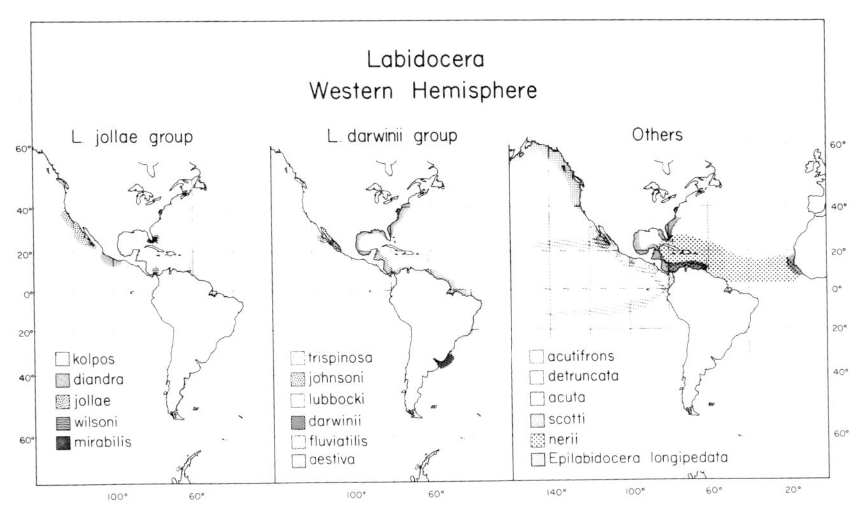

Figure 5. Distribution of **Labidocera** (Copepoda, Calanoida, family Pontellidae) in the Western Hemisphere. Coastal-zone species belonging to the **jollae** species group are on the left, those of the **darwinii** species group in the middle panel. **L. scotti** shown in the panel on the right belongs to the **darwinii** group. Distinctions among geographical populations that show morphological differences and are distinguished in the text as a or b forms are omitted from this figure. Distributions are based on published records expanded by unpublished data.

furca (Figs. 7e, g, i, k), and conspicuous differences in the development of the endopod of the fifth legs (Figs. 7f, h) contribute to the likelihood of qualitative functional differences in mating among species sharing the same sectors of the coastal zone. Finally the coupler of the spermatophore is markedly different, being essentially a ventral shield in both *jollae* and *kolpos* (Fig. 5j) and an encapsulating sleeve in *trispinosa* and *johnsoni* (Figs. 6e, 7k). Occupying a coastal sector that does not support an entrenched representative of the *darwinii* group, *diandra* of the *jollae* group has evolved a convergent encapsulating sleevelike coupler (4). This adaptation accompanying a group of other unique modifications of sexually influenced structures presumably interferes with hybridization between *diandra* and its two northern relatives, *jollae* and *kolpos*.

Special male asymmetrical adaptations appear on the right member of the second pair of legs of the *jollae* group (4, 8). However, in the *darwinii* group they appear on the right corner of the last thoracic segment (3). Differences in

TABLE 2.

Lengths of coastal-zone sectors occupied by species of the genus **Labidocera.** Estimates are approximate and do not include coastline irregularities.

	Degrees of latitude	km
West Coast Species		
jollae	15	1900
kilpos	7	650
diandra	9	1300
	$\bar{x}$ = 10.3°	$\bar{x}$ = 1280 km
trispinosa	15	1900
johnsoni a] **johnsoni** b	12	1750
lubbocki a	5	640
lubbocki b	25	3860
	$\bar{x}$ = 11.75°	$\bar{x}$ = 2080 km
East Coast Species		
scotti	30	8700
aestiva a	23	3200
aestiva b	11	2700
fluviatilis a	14	2900
fluviatilis b	45	6400
darwinii	15	2100
	$\bar{x}$ = 21.6°	$\bar{x}$ = 3460 km
mirabilis	?	?
wilsoni	?	?

the length of the left fifth leg of the male are of particular interest. Quantitative evidence in support of the widely held view that this appendage is used to position the coupler during copulation has been obtained from studies on the related family, *Centropagidae* (11). As shown below, species of *Labidocera* that co-occur tend to exhibit strong divergence in the length of the left fifth leg.

From available data I have made a preliminary comparison of morphological divergence among all known populations of American coastal zone *Labidocera.* With respect to each structure, the extent of divergence between species was ranked subjectively in terms of four indices: none, weak, moderate, and strong. In addition to general comparisons made between the two species groups, species within a group were compared relative to the presence or absence of geographical overlap. The results (Table 3) indicate an overwhelming concentration of moderate to strong divergence in sexually modified structures between species groups and, in addition, among geographically overlapping members of the same species group. In nonsexual structures, noticeable but weak divergence exists between species groups, but virtually none is apparent

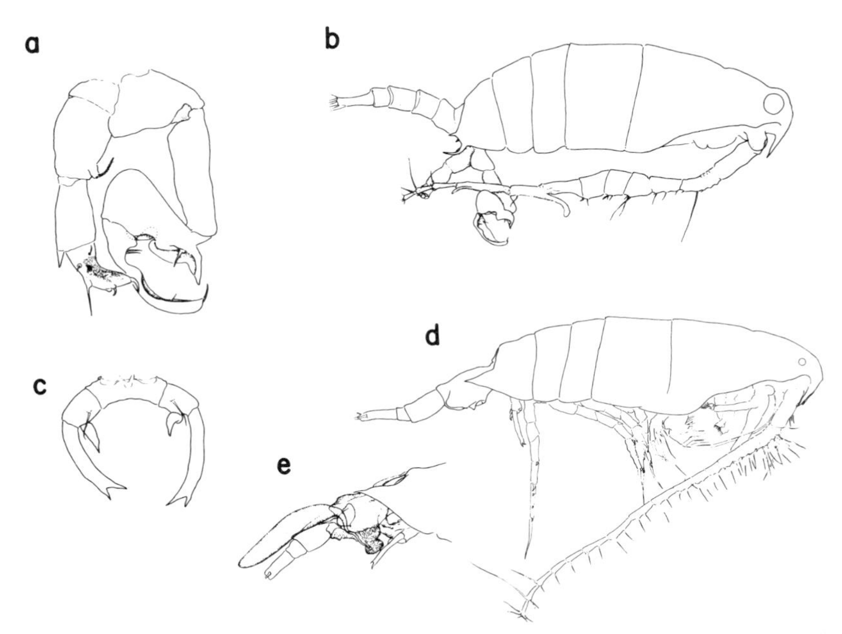

Figure 6. The sexually mature stages of **Labidocera trispinosa** Esterly, a typical representative of American coastal-zone species of the genus **Labidocera.** Male: a: fifth pair of legs, posterior; b: right lateral view of body, feeding and swimming appendages omitted, setae of caudal rami reduced. Female: c: fifth pair of legs, posterior; d: right lateral view of body, most setae of appendages omitted or reduced; e: right lateral view of abdomen bearing attached spermatophore.

within a species group irrespective of geographical relationship. It must be noted, however, that cephalic appendages were not subjected to fine-scale mensural study, as has been carried out on other calanoids (e.g., 12).

Comparative examination of cephalic appendages associated with swimming while feeding, namely the second maxillae and the maxillipeds, revealed a few minor differences of detail. Each homologous appendage is virtually identical in proportion among species within a species group and is very similar between groups. The cephalic appendages carry a large number of setae arranged in distinctive clusters (Table 4). *L. aestiva* and *trispinosa,* representing east and west coast subgroups of the *darwinii* species group, show no differences. *L. jollae,* representing the *jollae* group, deviates from the other two species only by the addition or subtraction of a single seta on any of the sets of homologous appendages.

Position of these appendages and their size in proportion to body size are also similar among the two species groups (Fig. 8). The second maxillae (Figs. 8e, h)

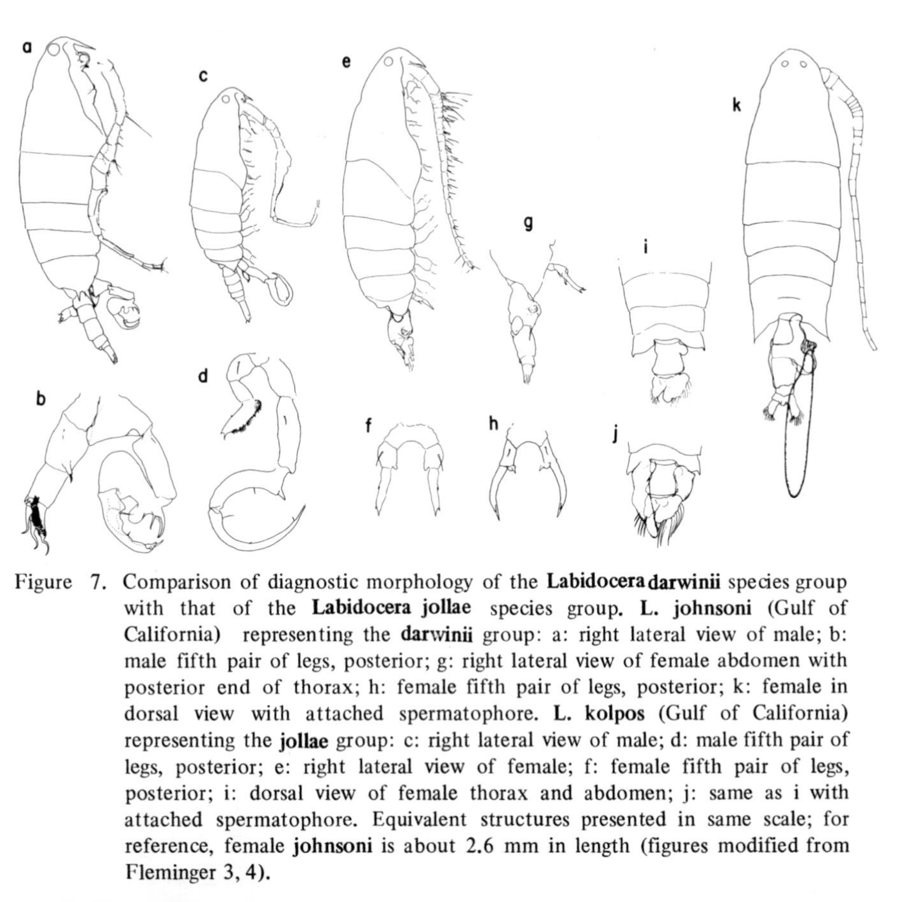

Figure 7. Comparison of diagnostic morphology of the **Labidocera darwinii** species group with that of the **Labidocera jollae** species group. **L. johnsoni** (Gulf of California) representing the **darwinii** group: a: right lateral view of male; b: male fifth pair of legs, posterior; g: right lateral view of female abdomen with posterior end of thorax; h: female fifth pair of legs, posterior; k: female in dorsal view with attached spermatophore. **L. kolpos** (Gulf of California) representing the **jollae** group: c: right lateral view of male; d: male fifth pair of legs, posterior; e: right lateral view of female; f: female fifth pair of legs, posterior; i: dorsal view of female thorax and abdomen; j: same as i with attached spermatophore. Equivalent structures presented in same scale; for reference, female **johnsoni** is about 2.6 mm in length (figures modified from Fleminger 3, 4).

TABLE 3.

Ranked morphological divergence in characters examined in American coastal-zone species of **Labidocera.** The four degrees of divergence are ranked subjectively on the basis of deviation from the basic pattern characteristic of the genus. Further discussion in text.

(O - none; 1 - weak; 2 - moderate; 3 - strong)

Ranked morphological divergence[1]

Structure	Between spp. of the **darwinii** and **jollae** groups	Between spp. within a species group	
		Geographically overlapping with other spp. group members	Geographically isolated from other spp. group members
Body segmentation	2	1	0
ThV corner[2]	3	3	1
♀ genital segment	3	3	1
Furcal rami	3	2-3	1
First antenna	3	1-2	1
♀ fifth legs	3	2	1
♂ fifth legs	3	2-3	1
Spermatophore	3	3	1
Second antenna	1	0	0
Mandible	2	1	1
First maxilla	1	0	0
Second maxilla	1	0	0
Maxilliped	1	0	0
Total length	0	0-2	0

1. 0 = none; 1 = weak; 2 = moderate; 3 = strong.

2. Right posterolateral corner of the thoracic segment bearing the fifth pair of swimming legs.

appear to be and are considered prehensile (1). They may scoop food particles as in *Acartia* and *Centropages,* though probably of a different size range. The most notable differences between the two species groups are a reduction in the dorsalmost three teeth (Figs. 8c, f) and the number of ctenoid setules on the distomedial seta of the second maxilla (Figs. 8e, h). These differences may relate to the generally omnivorous habits of the *darwinii* group and the more carnivorous habits of the *jollae* group (A. Barnett, personal communication and unpublished observations of the author).

TABLE 4.

Setation of cephalic appendages in species representing eastern (**aestiva**) and western (**trispinosa**) representatives of the **darwinii** species group and in **jollae** representing the **jollae** species group. Large numbers are the sum of the setae arranged in clusters along the length of the appendage and indicated in parentheses; two sets of parentheses indicate clustering along each ramus of biramous appendages.

	Second Antenna	Mandible	First Maxilla	Second Maxilla	Maxilliped
aestiva	33	21	55	25	18
	(1-2-2-9-7) (1-2-4-5)	(4-4-7) (3-3)	(15-3-4-3-2-2-5-11-1-9)	(6-3-3-3-3-7)	(2-3-3-3-2-1-1-3)
trispinosa	33	21	55	25	18
	(1-2-2-9-7) (1-2-4-5)	(4-4-7) (3-3)	(15-3-4-3-2-2-5-11-1-9)	(6-3-3-3-3-7)	(2-3-3-3-2-1-1-3)
jollae	34	21	54	26	17
	(1-2-2-9-7) (1-3-4-5)	(4-4-7) (3-3)	(15-3-4-3-2-5-10-1-9)	(7-3-3-3-3-7)	(2-3-3-3-2-1-1-2)

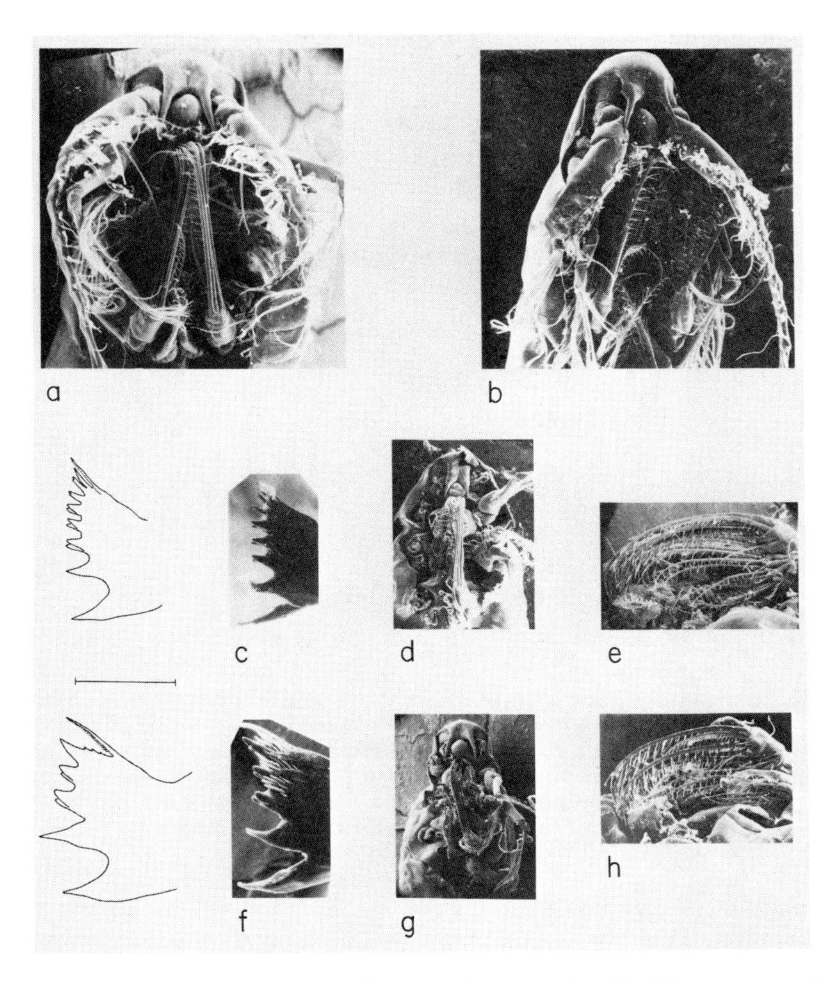

Figure 8. Feeding appendages in American coastal-zone species of **Labidocera;** comparison of similarities and differences between the **jollae** and **darwinii** species groups; a: **fluviatilis** b, cephalon, ventral, 103X; b: **lubbocki** a, cephalon ventral, 92X; c: **jollae**, mandible, 570X; d: **kolpos,** cephalon, ventral, 98X; e: **jollae,** second maxilla, 180X; f: **johnsoni,** mandible, 1060X; g: **trispinosa,** cephalon, ventral, 95X; h: **johnsoni,** second maxilla, 170X. Original magnifications given but scale varies among the individual figures. Ruled line pertaining to mandible represents 0.1 mm.

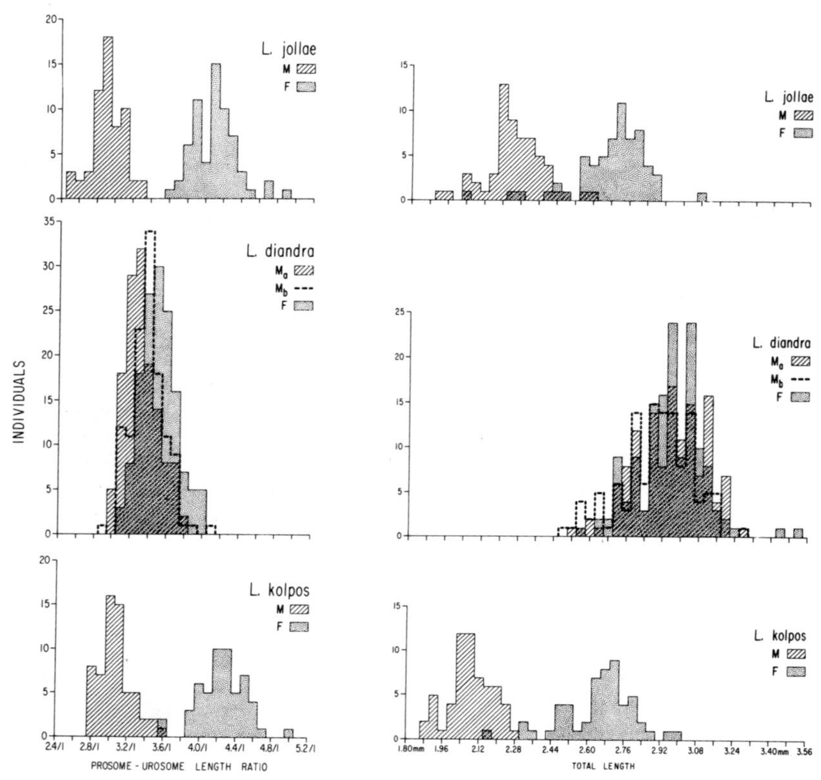

Figure 9. Frequency distribution of prosome (= cephalon plus thoracic segments) to urosome (= abdomen) length ratio (left) and overall length of body (right) of adult males and females of the **jollae** species group found in the western American coastal zone, from Fleminger (4). See text for discussion of patterns in size, body proportions, and geographical relationships and their bearing on prezygotic mating barriers. Abbreviations: M = male, Ma and Mb = male forms in **diandra**; F = female.

with the short overlapping geographical ranges found on the western coast, genital segments in the western *(trispinosa)* subgroup show large asymmetrical swellings near the genital pore (Figs. 12c, e, g) and a major proportional change in the length of the genital segment of *johnsoni* (Fig. 13), the species situated geographically between the more northern *trispinosa* and the tropical *lubbocki*. On the east coast, however, the genital segment tends to be essentially symmetrical and the genital pore midventral. External specializations appear only in the temperate species and include patches of sensory hairs and spinules along with large scattered pores on or near the genital segment (Figs. 12i, j, n, o). In the tropical species *fluviatilis,* the genital segment is devoid of prominent

It has frequently been suggested that body size may play a significant role in partitioning the food resource of potentially competing planktonic species. Among the more geographically crowded species of west coast *Labidocera* no appreciable difference in total length, irrespective of geographical relationship, is apparent (Table 5). Male mensural differences among species of the *jollae* group are of interest, however, if hybridization barriers include overall size and body proportions. Differences in the size of males and in the proportion of the prosome length (= cephalic and thoracic segments) to the urosome (= abdomen) length (Fig. 9) correlate with the geographical relationships of the three species. *L. diandra* occurs at the southern limits of *jollae,* which is found only in Pacific coastal waters, and of *kolpos,* known only from the eastern coast of the Gulf of California; the latter two species have no common boundary at present (Figs. 5, 10). In *kolpos* and *jollae,* males are much smaller relative to the female than is usual among species of *Labidocera* (Table 5) and differences in the prosome-urosome length ratio are also pronounced. In *diandra* males, however, total length and the prosome-urosome ratio are displaced to the extent that they show the same distributions as are shown by the females (Fig. 9). The *jollae* group shows other related patterns of geographical overlap and morphological divergence: in the asymmetry of the genital segment of females, including deviation of the genital pore from a mid-ventral position, and in the male and female sexually modified appendages as well (Fig. 10). The eastern coastal species *mirabilis* and *wilsoni* overlap and also show opposite asymmetry in the genital segment of females, in addition to strong divergence in the configuration of the clasping appendages of the males (8). The striking difference in the length of the left fifth leg in males of *wilsoni* and *mirabilis* is particularly interesting (Fig. 11A).

We find a similar example of divergence in the length of the left fifth leg relative to geographical overlap among the eastern species (the *aestiva* subgroup) of the *darwinii* group (Fig. 11B). Only two, *darwinii* and *fluviatilis* b, of the five populations in the *aestiva* subgroup, show extensive geographical overlap. From the limited data available the range of *L. darwinii* (roughly 38°S to 22°S) appears to be totally contained within the distribution of *fluviatilis* b, although this overlap is likely to represent a seasonally variable range extension by the latter. *L. fluviatilis* b is otherwise essentially tropical in distribution, whereas *darwinii* appears to be a warm-temperate form. The length of the exopod of the left fifth leg of *darwinii* is much shorter relative to body size in comparison to the proportions found in the other populations of this group. Proportional differences (i.e., apparent displacement) in the length of the left fifth leg of these co-occurring species are appreciable, being roughly 2 to 1 between *wilsoni* and *mirabilis* and 1.5 to 1 between *fluviatilis* b and *darwinii.*

The final examples selected for discussion involve the female genital segment and the spermatophore-coupler arrangement in the *darwinii* group. In accord

TABLE 5.

Overall length in mm (omitting setae on the caudal furca) in sexually mature stages of **darwinii** and **jollae** species group populations occurring on the western American coastal zone. Measurements made with the aid of an ocular micrometer at 80X magnification (3, 4 and unpublished data).

Species	Females				Males			
	$\bar{x}$	s	range	No.	$\bar{x}$	s	range	No.
jollae	2.69	±.152	2.07-3.07	(63)	2.26	±.121	1.93-2.61	(60)
diandra	2.95	±.140	2.57-3.49	(144)	2.95	±.148	2.53-3.25	(135)
kilpos	2.66	±.152	2.18-2.97	(54)	2.10	±.089	1.91-2.30	(61)
trispinosa	2.73	±.190	2.20-3.12	(188)	2.44	±.152	2.04-2.76	(165)
johnsoni	2.68	±.244	2.22-3.25	(278)	2.57	±.162	2.12-3.02	(179)
lubbocki a	2.69	±.190	2.36-3.17	(87)	2.50	±.179	2.29-3.02	(71)
lubbocki b	2.61	±.168	2.18-2.83	(71)	2.40	±.175	2.10-2.72	(60)

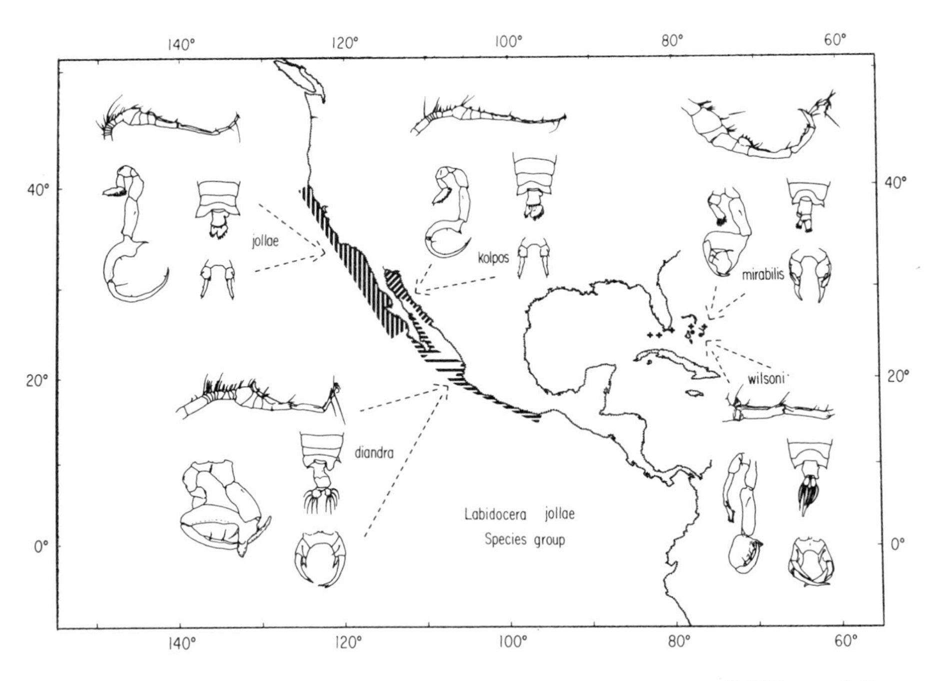

Figure 10. Geographical distribution and diagnostic morphology of the **Labidocera jollae** species group. shown for each species are posterior views of the fifth pair of legs of each sex, a dorsal view of the female abdomen and the male right first antenna. Equivalent structures have been drawn to the same scale. Distributions are based on published records expanded by unpublished data. Further discussion in the text.

sensory hairs and integumental pores; *darwinii* co-occurring with *fluviatilis* b has a prominent knob on the right side of the genital segment in addition to the integumental organs (Figs. 12k, m).

The correlation of strong morphological divergence with extensive geographical overlap is also seen in the spermatophore coupler of the *darwinii* group. In the *aestiva* (eastern) subgroup (Fig. 14) the coupler is a relatively simple shield-like structure covering the ventral survace of the genital segment. In four of the populations the spermatophore sac lies parallel to the left side of the female abdomen. In *darwinii,* however, which shares its range with *fluviatilis* b, the coupler is enlarged and somewhat asymmetrical and the sac lies to the right of the abdomen.

Among the geographically crowded *trispinosa* (western) subgroup the coupler encapsulates the abdomen, the extent varying with the species (Fig. 14). One eastern species derived from the *trispinosa* group, namely *scotti,* co-occurs with

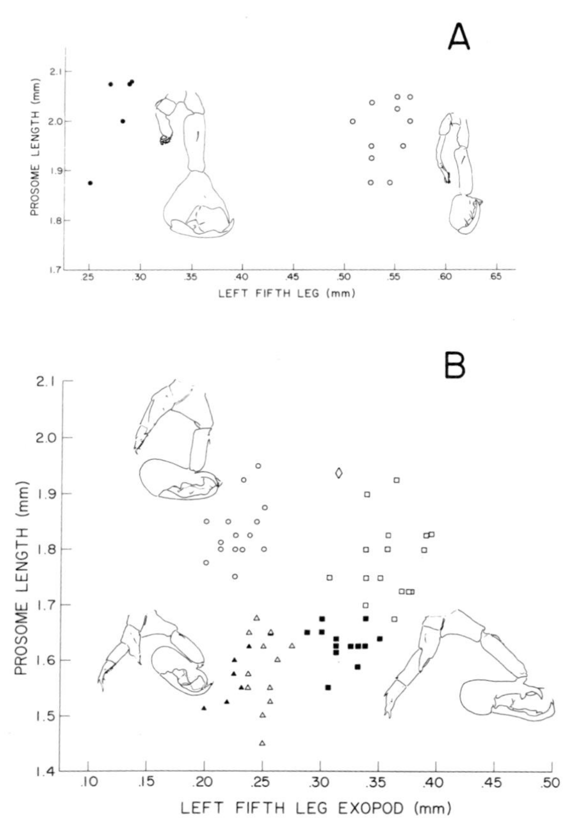

Figure 11. Scatter-diagram of male left fifth leg-length plotted against prosome length (= cephalon plus thoracic segments). A. **L. mirabilis,** filled circles, fifth pair of legs on left; **wilsoni,** open circles, fifth pair of legs on right. B. Scatter diagram of length of exopod of male left fifth leg plotted against prosome length. **L. aestiva** a, filled triangles; **aestiva** b, open triangles; unusually large **aestiva** taken in northern Caribbean, open diamond; male fifth pair of legs of **aestiva** to left of triangles; **fluviatilis** a, filled squares; **fluviatilis** b, open squares; male fifth pair of legs of **fluviatilis** to right of squares; **darwinii**, open circles, male fifth pair of legs to left of circles. All fifth pairs of legs drawn in posterior view and shown in approximately the same scale.

three populations of the *aestiva* subgroup (*aestiva* a and b, and *fluviatilis* a) and with members of the *mirabilis* subgroup as well. In *scotti,* the coupler achieves its most expansive development completely enveloping the entire abdomen (Fig. 14) but leaving a slot adjacent to the caudal furca where presumably the male holds the female during copulation. Another fundamental characteristic limited

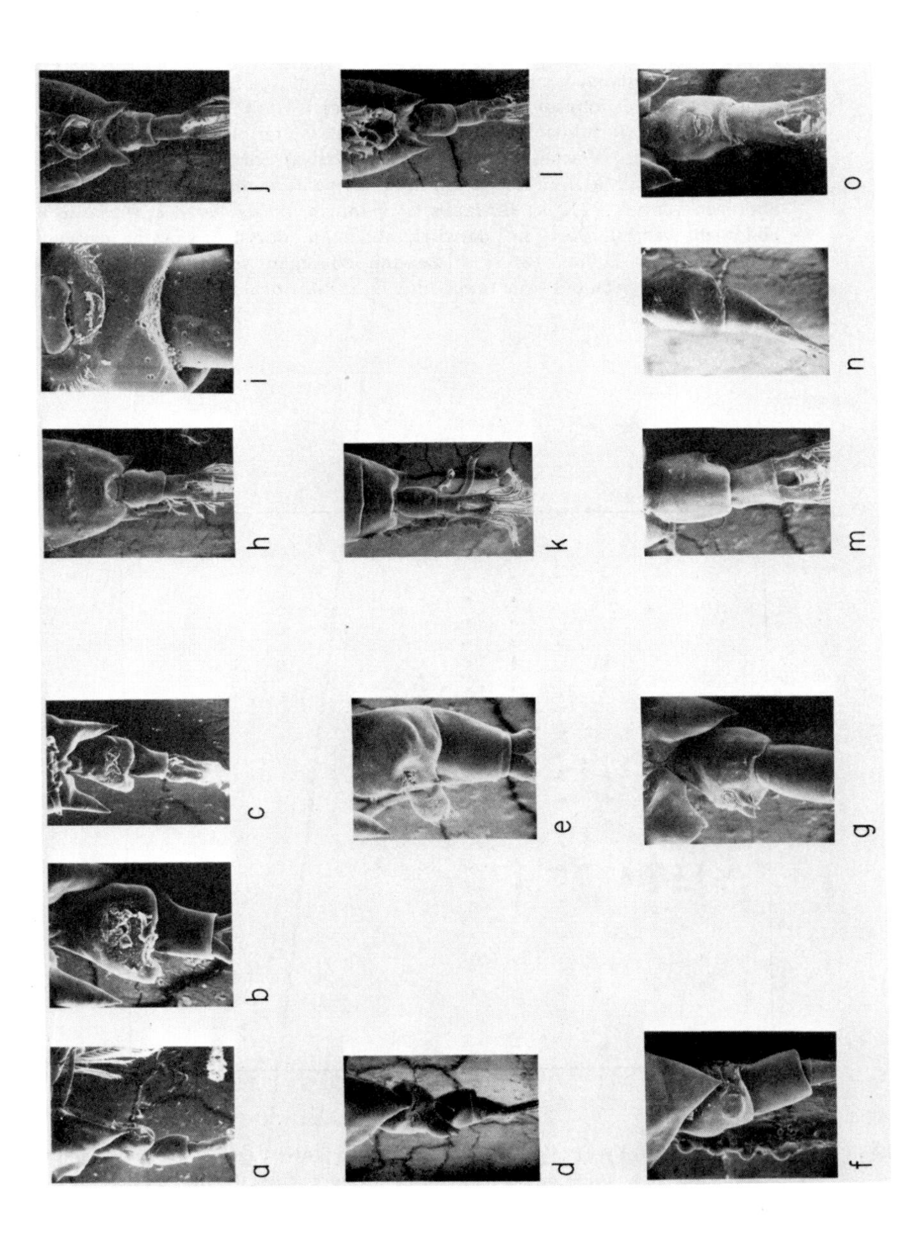

Figure 12.

Figure 12. Comparison of morphological divergence in the female genital segment among species of the eastern and western subgroups of the **Labidocera darwinii** species group. Western **(trispinosa)** subgroup: a: **trispinosa,** abdomen, right lateral, 109X; b: **trispinosa,** genital segment, ventral, 189X; c: **trispinosa,** abdomen, ventral, 107X; d: **johnsoni,** abdomen, right lateral, 109X; e: **johnsoni** abdomen, ventral, 180X; f: **lubbocki** b, abdomen, right lateral, 174X; g: **lubbocki** b, genital segment, ventral, 196X. Eastern **(aestiva)** subgroup: h: **aestiva** a, abdomen, dorsal, 90X; i: **aestiva** a, genital segment, ventral, 455X; j: **aestiva** a, abdomen ventral, 92X; k: **fluviatilis** b, abdomen, dorsal, 90X; l: **fluviatilis** b, abdomen, ventral, 92X; m: **darwinii,** abdomen, dorsal, 183X; n: **darwinii,** abdomen, right lateral, 185X; o: **darwinii,** abdomen, ventral, 174X. Original magnifications given but scale varies among the individual figures.

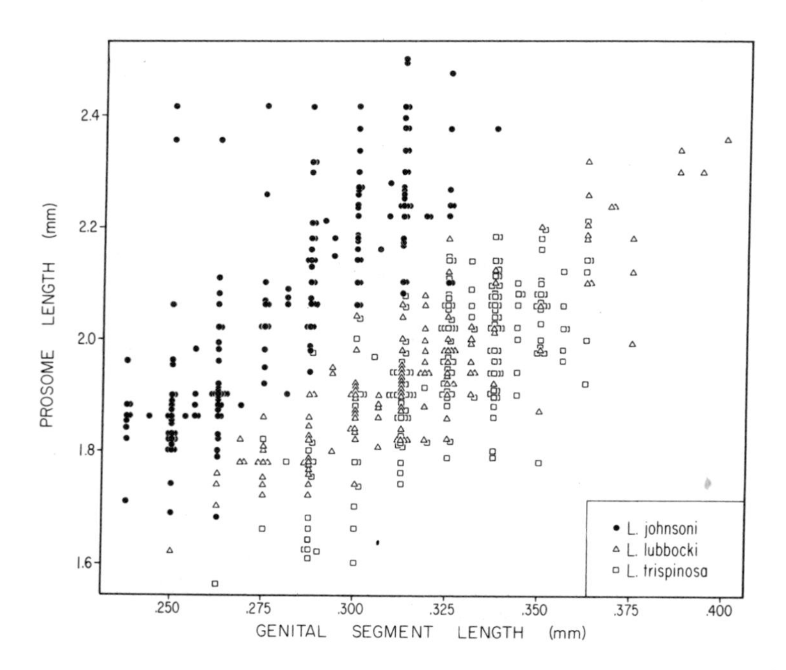

Figure 13. Scatter diagram of female genital segment length plotted against prosome length (= cephalon plus thoracic segments) for western coastal-zone species of **L. johnsoni, lubbocki,** and **trispinosa.** Measurements made with the aid of an ocular micrometer at 80X magnification from specimens selected at random.

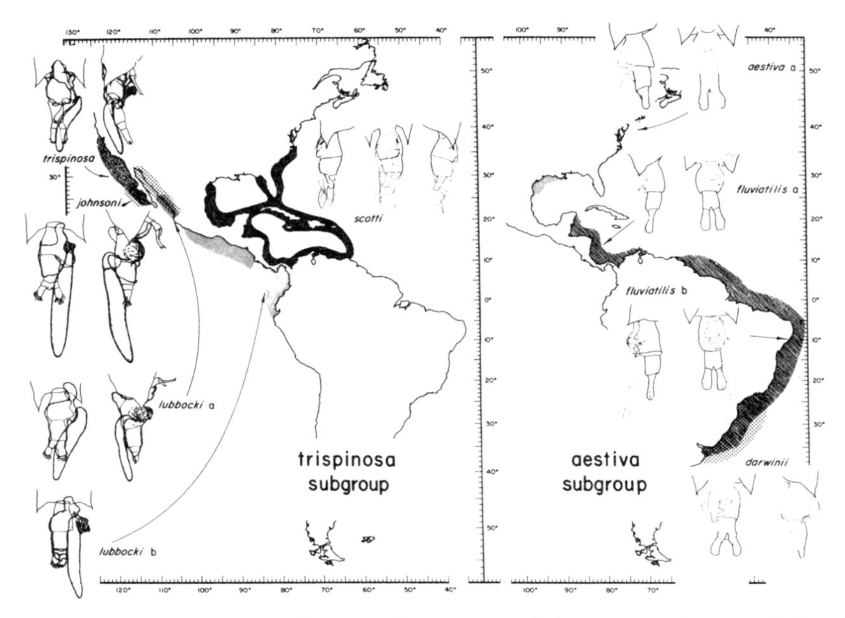

Figure 14. Known geographical range and appearance of the spermatophore attached to the female abdomen in populations of the **darwinii** species group. Eastern coastal-zone species of the **aestiva** subgroup are on the right panel, western coastal-zone species of the **trispinosa** group, including **scotti**, a derivative species found in the eastern coastal zone, on the left panel. Dorsal and right lateral views of the attached spermatophore and female abdomen shown of **trispinosa, johnsoni,** and **lubbocki** a, dorsal view shown of **lubbocki** b; left lateral, ventral and right lateral views shown of **scotti.** Left lateral and ventral views shown of **aestiva** a, **fluviatilis** a and b, ventral and right lateral views shown of **darwinii.** All exposed portions of the spermatophore covering the female integument are shaded. Figures are based on camera lucida drawings at roughly the same scale. Distributions represent a review of published records augmented by unpublished data; morphology is based primarily on unpublished data. Further discussion in text.

to the *trispinosa* subgroup is the apparent persistence of the spermatophore on mated females until eggs are to be shed (G. Theilacker, personal communication). Although the reason for retention of the spermatophore is a matter of speculation (cf. 11), if species of the *trispinosa* subgroup require a new spermatophore for each subsequent clutch of eggs, as is the case in *Acartia* (16), retention of the previous spermatophore until egg-laying would conserve spermatophores for unfertilized females. This would in effect conserve sperm, a function that may be inferred to be the object of divergence in all of the sexually modified morphological structures discussed above. As a group these

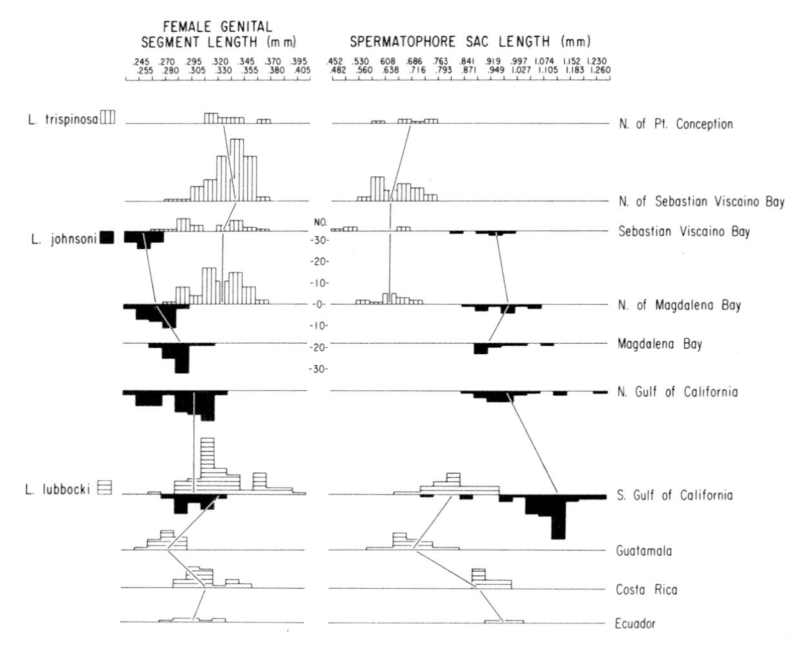

Figure 15. Frequency distribution of female genital segment length (left) and spermatophore sac length (right) taken from randomly selected specimens collected at specified localities in coastal zone waters. Lines connect median values of each geographical sample. Measurements made with the aid of an ocular micrometer at 80X magnification.

morphological specializations may be thought to serve in the formation of prezygotic reproductive barriers to sustain the homogeneity of gene pools. That is, presumably these species have proven to be better adapted to their particular habitat and geographical range than have hybrid offspring of pairings that have occurred in the past.

The numerous correlations between co-occurrence and morphological divergence support the idea that these divergences are a reinforcement of barriers against hybridization, i.e., essentially a form of character displacement. The question remains, however, as to the extent to which these patterns are fortuitous. The case against spurious correlations would be considerably strengthened by evidence of character displacement among overlapping populations that weakened or disappeared in regions outside the areas of geographical contact. Preliminary indications of such evidence are available from unpublished studies on the *trispinosa* subgroup. Trends in the frequency distribution of the length of the genital segment and in the length of the spermatophore sac suggest that these features are more dissimilar in regions where successive species of the subgroup co-occur (Fig. 15). From a broader view these data indicate that interspecific interactions leading to significant changes in the gene pools of zooplankton are amenable to study and that coastal-zone populations of copepods provide a promising subject for such investigations.

Considering the patterns observable in *Labidocera,* one must infer that interactions among the coastal-zone species have produced distinctive modifications of sexually modified structures but few or none in feeding structures. The results agree substantially with patterns noted in a number of oceanic calanoid taxa including *Calausocalanus* (9), several species of *Calanus* (12), *Pontellina* (7), and *Eucalanus* (5).

To conclude, what then is the pertinence of the biogeography and evolution I have been discussing relative to the contemporary zooplankton community? As a biogeographer I am strongly biased by historical perspective and contacts among co-occurring populations that affect their respective gene pools. This dimension of time has determined modern patterns of range and morphology among related populations. It provides an essential background for viewing special adaptations, contemporary relationships, and the significance of short-term perturbations within the zooplankton. For example, the lack of evidence that body size and feeding appendage morphology may partition food resources among congeneric species of calanoid copepods should induce more inquiry in the area of behavioral differences in feeding and the extent of predation control on copepod populations.

ACKNOWLEDGEMENTS

This study was aided by the generous contributions of many colleagues, past and present. John Isaacs must be singled out for special acknowledgment: his encouragement and infinite patience have made it all possible. I am grateful to Dr. T. Björnberg, University of Saõ Paulo, and the U. S. National Marine Fisheries Service for providing valuable collections from various sectors of the coastal zone. The study was funded by NSF Grants GB-12412 and GB-32076 and by the Marine Life Research Program of Scripps Institution of Oceanography.

REFERENCES

1. Anraku, M. and Omori, M.
1963 Preliminary survey of the relationship between the feeding habit and the structure of the mouthparts of marine copepods. **Limnol.**

Oceanogr., 8(1): 116-126.

2. Diamond, J. M.
1973 Distributional ecology of New Guinea birds. **Science, 179**: 759-769.

3. Fleminger, A.
1964 **Labidocera johnsoni**, species nov., Crustacea: Copepoda: Calanoida: Pontellidae. **Pilot register of zoology**, card. no. 3-A.

4. Fleminger, A.
1967 Taxonomy, distribution and polymorphism in the **Labidocera jollae** group with remarks on evolution within the group (Copepoda: Calanoida). **Proc. U. S. Nat. Mus.**, 120 (3567): 1-61.

5. Fleminger, A.
1973 Pattern, number, variability and taxonomic significance of integumental organs (sensilla and glandular pores) in the genus **Eucalanus** (Copepoda, Calonoida). **Fish. Bull.**, 71(4): 965-1010.

6. Fleminger, A. and Hulsemann, K.
1973 Relationship of Indian Ocean epiplanktonic calanoids to the world oceans. In **Ecological studies**, Vol. 3, p. 339-347. (ed. Zeitzschel, B.) Springer, Berlin, Heidelberg and New York.

7. Fleminger, A. and Hulsemann, K.
In press Systematics and distribution of the four sibling species comprising the genus **Pontellina** Dana (Copepoda: Calanoida). **Fish. Bull.**

8. —Fleminger, A. and Engchow Tan
1966 The **Labidocera mirabilis** species group (Copepoda: Calanoida), with description of a new Bahamian species. **Crustaceana**, 11(3): 291-301.

9. Frost, B. and Fleminger, A.
1968 A revision of the genus **Clausocalanus** (Copepoda: Calanoida) with remarks on distributional patterns in diagnostic characters. **Bull. Scripps Inst. Oceanogr.**, 12: 1-235.

10. Inman, D. L. and Brush, B. M.
1968 The coastal challenge. **Science**, 181: 20-32.

11. Lee, C. M.
1972 Structure and function of the spermatophore and its coupling device in the Centropagidae (Copepoda: Calanoida). **Bull. Mar. Ecol.**, 8(1): 1-20.

12. Mullin, M. M.
1969 Distribution, morphometry and seasonal biology of the planktonic copepods **Calanus tenuicornis** and **C. lighti** in the Pacific Ocean. **Pacific Sci.**, 23(4): 438-446.

13. Muromtsev, A. M.
1958 **The principal hydrological features of the Pacific Ocean.** Israel Progr. Sci. Trans. cat. no. 753, Jerusalem. (1963). 417 p.

14. Muromtsev, A. M.
1963 The principal hydrological features of the Atlantic Ocean. Gosudarstvennyi Okeanogr. Inst. Gidrometeorol. Izdatel'stvo, Leningrad. pp. 253-301. (In Russian.)

15. Pérez Farfante, I.
1969 Western Atlantic shrimps of the genus **Penaeuss.** U. S. Fish Wildl.

Serv., Fish. Bull., 67: 461-591.
16. Wilson, D. F. and Parrish, K. K.
1971 Remating in a planktonic marine calanoid copepod. Mar. Biol., 9: 202-204.
17. Wooster, W. S. and Reid, J. L., Jr.
1963 Eastern boundary currents. In **The Sea** Vol. 2: 253-280. (gen. ed. Hill, M. N.) John Wiley and Sons, New York and London.

NEKTONIC FOOD WEBS IN ESTUARIES[1]

Donald P. de Sylva[2]

ABSTRACT

Estuarine nekton is predominately fishes which comprise both surface-swimming species and species which are benthic but may feed in the overlying water column. Biotic and abiotic factors affect the distribution of estuarine nekton and its food; few studies of nektonic food webs have included concomitant environmental studies. Typical and atypical nektonic food webs are compared from England, the New York bight, South Florida, Guyana, Argentina, and South Viet Nam. Studies of nektonic food webs in estuaries should include analyses of biotic and abiotic factors, thorough diel and seasonal sampling, imaginative use and development of appropriate sampling gears, good preservation of samples, and laboratory and field studies which include ecological and behavioral studies of estuarine nekton.

INTRODUCTION

Nekton is defined in Webster's New International Dictionary (Second Edition) as "actively swimming pelagic organisms"; the definitions of estuaries have been repeatedly reviewed (30). Nekton excludes passively drifting aquatic animals and plants and presumably through definition, organisms which are more or less permanently associated with bottom communities. Nektonic organisms are readily characterized in the epipelagic or mesopelagic environments of the open sea, for it is largely through the food webs that these fishes are contiguous with

1. Contribution no. 1694, University of Miami, Rosenstiel School of Marine and Atmospheric Science.

2. Rosenstiel School of Marine and Atmospheric Science, University of Miami, Rickenbacker Causeway, Miami, Florida 33149.

littoral habitats or with abyssal bottom communities. Yet few nektonic species in estuaries can be considered truly independent of the estuarine bottom. Because of strong biotic and abiotic interrelationships characteristic of estuaries, based on present collections, it appears that most estuarine nektonic species live near or just off the bottom but periodically feed and swim in the water column just above the bottom.

Estuarine nekton comprises primarily fishes, but squids, scallops, and natant crustaceans, including crabs, lobsters, and shrimp have been considered as being nektonic in estuaries for short periods (41). Possibly seals and porpoises are also part of the estuarine nekton. In the sense that we consider them fast-swimming mid-water or surface species, truly nektonic estuarine fishes of North America include bluefish (*Pomatomus*), jacks (*Caranx*), ladyfish (*Elops*), menhaden (*Brevoortia*), shad (*Alosa*), mullett (*Mugil*), salmon (*Salmo* and *Oncorhynchus*), snooks (*Centropomus*), striped bass (*Morone saxatilis*), and tarpon (*Megalops atlantica*). None of these is restricted to an estuarine existence throughout its life history, and most are either anadromous or catadromous, utilizing the estuary largely for feeding either preceding or following migrations. A second group of estuarine nekton should be considered engybenthic (living close to the bottom), as these species are related to the bottom through their mode of life but also frequently feed in the overlying water column. These include the croaker (*Micropogon*), black drum (*Pogonias*), red drum (*Sciaenops*), white perch (*Morone chrysops*), seatrout (*Cynoscion*), silversides (*Menidia*), smelt (*Osmeridae*), some species of snapper (*Lutjanus*), spot (*Leiostomus*), sturgeons (*Acipenser*), toadfish (*Opsanus*), and flatfish (*Pseudopleuronectes, Paralichthys, Scophthalmus*).

ENVIRONMENTAL VARIABLES WHICH INFLUENCE ESTUARINE NEKTON

Nekton is closely dependent upon physical, chemical, biological, and geological characteristics of the estuary (41). Biotic and biochemical factors attract a nektonic organism to or repel it from the estuary, so that it may complete part of its life history there. Because environmental variables have usually been neglected in analyses of food habits of nektonic organisms and especially nektonic food webs, a brief review of major environmental factors is presented here.

Salinity

Salinity is probably the most important characteristic of the estuary. Wide fluctuations in salinity may occur annually, seasonally, or tidally, and in a given estuary salinity can vary extensively with depth or from one side to the other.

Nekton may be less affected by wide salinity variations than are the sessile organisms, but fish catches are nevertheless markedly influenced by seasonal changes in salinity (45). Especially in the tropics, such as in the Orinoco-Amazon watershed (34), West Africa (32), India (62), and South Viet Nam (19), heavy precipitation during the rainy or monsoon season causes marked salinity fluctuations which affect either the fishes or their food. For example, foods of estuarine fishes in India move from the muddy bottom to the surface during periods of high runoff (54), thus affecting availability of fishes to otter trawls.

Temperature

Temperature changes in estuaries, most noticeable in temperate and polar latitudes, affect population density. In tropical regions, annual temperature changes are less marked and changes in the commercial fish catch attributable to temperature variations are correspondingly smaller. Seasonal changes in upwelling of cold water into estuaries may deter migrations into or from estuaries, or may affect distribution or survival of food of estuarine nekton. Electric power plants located on estuaries add large quantities of heated water to the environment. While the beneficial effects of thermal additions, referred to by proponents as "thermal enrichment", may be specious, detrimental aspects may similarly be speculative (9, 16).

Oxygen

That anoxic waters occur in certain estuaries is well known (41). High biological activity below the halocline typical of most two-layer estuaries results in oxygen depletion and little opportunity for oxygen replenishment into deeper waters. Such conditions could cause oxygen to be a factor limiting the distribution of estuarine nekton or its food.

Bottom sediments

Although theoretically truly nektonic estuarine fishes are not affected by bottom sediments, those estuarine engybenthic-nektonic predators that feed on bottom organisms are probably correlated with the particle size and chemistry of bottom sediments, because epifauna and infauna on which the predators feed may be related to the physico-chemical nature of these sediments. Changes in feeding rates of oxeye (*Megalops cyprinoides*) in an Indian estuary occurred because a sand bar that formed seasonally at its mouth changed the estuarine hydrography and, hence, the composition of the food web (54).

Solar radiation and time of day

In India the percentage of predatory nektonic fishes with full stomachs is higher during periods of intense solar radiation (62), perhaps because prey organisms, normally distributed throughout the water column, avoid light by migrating toward the bottom where they become concentrated and thus more readily available to nektonic predators. This prey concentration can also actually concentrate predators near the bottom, making them more available to bottom trawls. Similarly, time of day is important not only to the behavior of estuarine nekton but also to gear used to catch them: when conditions are bright and the water is very clear, organisms may detect and thus avoid oncoming or even stationary nets.

Water transparency

The transparency of water is invariably neglected in studying estuarine nekton. Truly nektonic carnivores are sight-feeders and must see their prey. Turbid water detracts from their feeding ability and thus may deter nektonic predators in an estuary. Clear water can also make it easier for nekton to avoid nets and traps which may account for the fact that few collections of truly nektonic fishes and the few squids are taken in daytime in clear estuaries.

Moon phase

Anglers have long believed that moon phases affect the kind and number of fish they catch, but this clue seems to have eluded the scientist. Data in the scientific literature on the relationship between commercial fish catches and moon phases are noticeable by their virtual absence. Presumably, the connection between moon phase and diel migration of plankton, well-documented in pelagic environments (64), occurs in estuaries, especially where the water is clear and where lunar-tidal cycles are pronounced.

Tide

Little information is available on the relation between tidal stage or tidal speed and catch of estuarine nekton. Since tidal stage is usually strongly correlated with salinity and movement of pollutants and may cause stimulation of feeding rate in certain benthic organisms, swarming in estuarine plankton, or migration of nektonic prey, one might find that necktonic predators are influenced by the ebb and flow of estuaries. Tidal strength can also have a marked effect on the capture of nektonic organisms. For example, in the mangrove estuaries of South Viet Nam, current velocities of 1.7 meters per second were experienced during trawling operations (19), and it was frequently impossible to determine if the trawl was actually fishing on the bottom. When the speed of the towing vessel equalled the current speed, the net theoretically was not moving appreciably over the bottom. Under such circumstances, large nektonic organisms are not likely to be caught.

Nutrients and pollution

Little consideration has been given to the chemical nature of the waters in which estuarine nektonic organisms live or to the manner in which nutrients, including pollutants, might affect such organisms. Phosphates, nitrates, silicates, organic acids, trace elements, chelating compounds, and vitamins all may affect distribution and composition of food webs at lower trophic levels, which in turn could attract nektonic predators. Sewage and industrial wastes should be expected similarly to modify species composition and at least the lower food webs in estuarine systems. In the Delaware River estuary, benthic food webs of fishes have changed in the past 30 years; this change has been attributed to industrial pollution (18).

Plankton, micronekton, benthos, and competing predators

The distribution and availability of estuarine food organisms is certainly most important to nektonic predators, yet adequate biological inventories of estuaries are not often available, nor are adequate historical data on hand. Especially lacking are systematic collections with concomitant, adequate environmental data. Variations in higher trophic levels can occur through predation of lower levels as well. For example, striped bass compete with other secondary carnivores for food and thus indirectly influence the abundance of other species (27).

In summary, the above abiotic and biotic factors are responsible for determining occurrence, distribution, and abundance of nektonic predators in the estuary, their food, and the interrelationships among other organisms at various trophic levels. The need for, and lack of, data encompassing abiotic and biotic factors become whenever evident studies on food habits and food webs of estuarine nekton are undertaken.

STUDIES OF FOOD HABITS AND FOOD WEBS OF ESTUARINE NEKTON

While many studies have been made on the food of nekton, few have been specifically designed to analyze estuarine nekton, and most are really based on fishes collected in bottom trawls. True nektonic predators are usually not collected because they swim in the mid- or surface waters which are seldom fished by otter trawls. Also, most food studies have been done as an afterthought and not as part of a carefully planned experiment. One reason for this is that funds and personnel to carry out appropriate studies using modern methods are seldom available.

The research on nektonic food habits by geographic location and the scope of the studies are presented in Table 1. Although the analysis of the literature on food studies is incomplete, of the 76 references examined, 36 variously treat nektonic food webs in the estuary. Few of these studies permit comparison with other areas; none contains adequate environmental information. Most references cited have merely listed foods found in the stomachs of various estuarine nekton. However, some of these have gone into considerable detail; they usually raise more questions than they answer (51).

Generalized food webs (Table 2) are basically fueled by either a phytoplankton source of energy or a detrital source. Secondary trophic levels are benthos, either infaunal or epifaunal, or perhaps zooplankton or micronekton or both, and include primary carnivores, omnivores, or benthic herbivores (50). Middle carnivores include planktivores, benthophagous fishes and invertebrates, pelagic fishes, and invertebrates such as squid.

Recent advances in nektonic food studies have stressed detritus as a food source for tropical estuarine fishes (46, 51). Much detritus is produced by decaying mangrove and, perhaps *Thalassia* leaves, which are used by detritus feeders and thereafter by middle and higher carnivores. In Lake Pontchartrain, Louisiana, a subtropical-temperate estuarine ecosystem, detritus is the main energy source (11).

Some estuarine nekton depends solely upon plankton (Fig. 1). In a typical British estuary, the primary carnivores are mackerel (*Scomber*), herring (*Clupea*), and young gobies (*Gobius*). Although the secondary carnivores in this food web are not illustrated, adult gobies are eaten by flatfish (22). Undoubtedly flatfish, herring, and mackerel are eaten by other secondary carnivores such as dogfish and blue sharks.

A more complex food web occurs in the Delaware River estuary (Fig. 2), where numerous benthic and planktonic invertebrates are eaten by primary carnivores, consisting of silversides and anchovies, which in turn are eaten by secondary carnivores, such as bluefish, barracuda, and striped bass. Several species of sharks in this estuary undoubtedly represent tertiary carnivores, though shark stomachs were not specifically studied (18).

A synoptic complex food web (Fig. 3) of the striped bass (41) represents food webs from numerous localities along the Atlantic coast (27, 61). However, the complexity of this web is misleading, because many organisms depicted here are not found in the same habitat. Nevertheless, the web indicates the wide variety of organisms that can be eaten by the very adaptable striped bass.

A relatively simple food web occurs in the Knysna estuary (13) in South

TABLE 1

Literature on foods of estuarine nekton by geographical location; asterisks indicate those studies that include an analysis of the food web.

Author	Locality	Subject
Mulkana (45)	Rhode Island	nekton*
Townes (81)	Hudson River	nekton
de Sylva et al. (18)	Delaware R. estuary	nekton
de Sylva & Davis (17)	Delaware R. estuary and mid-Atlantic Bight	white marlin*
Hildebrand & Schroeder (26)	Chesapeake Bay	nekton
Hollis (27)	Chesapeake Bay	striped bass*
Dovel (20)	eastern U. S.	croaker*
June & Chamberlin (28)	eastern U. S.	menhaden*
Reintjes (67)	eastern U. S.	menhaden
Welsh & Breder (83)	eastern U. S.	nekton
Roelofs (69)	North Carolina	sciaenid*
Dawson (12)	South Carolina	spot
Rickards (68)	Georgia	tarpon
Odum (49)	Georgia	mullet*
Odum (51)	Georgia & Florida	mullet*
Harrington & Harrington (24)	Florida	tarpon*
Croker (10)	Florida	gray snapper
Odum (50)	Florida	nekton*
Odum & Heald (52)	Florida	nekton & benthos*
Marshall (39)	Florida	snook
Carr & Adams (6)	Florida	nekton*
de Sylva (14)	Florida	barracuda*
Springer & Woodburn (76)	Florida	nekton
Moody (44)	Florida	spotted trout
Naqvi (46)	Florida	blue crabs
Darnell (11)	Louisiana	nekton*
Pearson (55)	Texas	sciaenid
Gunter (23)	Texas	nekton
Miles (43)	Texas	nekton
Reid (65)	Texas	nekton
Reid et al. (66)	Texas	nekton
Simpson & Gunter (72)	Texas	nekton
Stevens (77)	California	striped bass
Stevens (78)	California	shad
Sasaki (70)	California	king salmon

TABLE 1 (Cont'd)

Authors	Locality	Subject
Radtke (60)	California	white sturgeon
Le Brasseur et al. (31)	British Columbia	chum & pink salmon
Manzer (37)	British Columbia	salmon*
Austin (2)	Puerto Rico	nekton*
Austin & Austin (3)	Puerto Rico	nekton*
Lowe (34)	Guyana	nekton*
Menezes (42)	Brazil	tarpon
Furtado (21)	Brazil	mullet
Ciechomski (8)	Argentina	anchovy
López & Castello (33)	Argentina	anchovy*
Olivier et al. (53)	Argentina	nekton*
Tobor (80)	Nigeria	catfishes*
Longhurst (32)	Sierra Leone	nekton*
Cadenat (5)	Senegal	nekton
Day (13)	South Africa	nekton*
Casabianca & Kiener (7)	Corsica	gobiids*
Pignalberi (56)	Italy	nekton*
Porumb (57)	Black Sea	bluefish*
Mayskiy (40)	Sea of Azov	nekton*
Yablonskaya (84)	Sea of Azov	nekton & benthos*
Zheltenkova (85)	Sea of Azov	benthos
Skazkina & Kostyuchenko (74)	Sea of Azov	gobies
Sirotenko (73)	Don R. estuary	shad
Shorygin (71)	Caspian Sea	nekton & benthos*
Spătaru (75)	Danube R. estuary	nekton*
Bhatt et al. (4)	India	sciaenid & polynemid
Noble (48)	India	sardines
Kaywade (29)	India	sardines
Rao (62)	India	sardines*
Rao (63)	India	sciaenid
Pandian (54)	India	oxeye*
Natarajan & Pàtnaik (47)	India	**Ambassis**
Qasim (59)	India	benthos & nekton*
Prasadem (58)	India	mullet
Lursinap (35)	Gulf of Thailand	**Rastrelliger**

TABLE 1. (Cont'd)

Authors	Locality	Subject
de Sylva & Michel (19)	South Viet Nam	nekton*
Marquez (38)	Philippines	goby
Macnae (36)	Indo-Pacific	mudskippers
Ah-kow-Tham (1)	Singapore	nekton*
Thomson (79)	western Australia	nekton
Walsh (82)	Hawaii	nekton*

TABLE 2

Classification of nektonic food webs in estuaries.

Phytoplankton-generated food webs:

phytoplankton → zooplankton → planktivorous pelagic & benthopelagic fish (shad)

phytoplankton → zooplankton → plantivorous fish → large fish predators (striped bass)

phytoplankton → phytoplanktonic fishes (menhaden): summer

phytoplankton → zooplankton → menhaden: winter

phytoplankton → zooplankton → large carnivores (manta)

phytoplankton (dinoflagellates) → mullet (49): alteration of usual feeding habits

Detritus-generated food webs:

detritus → benthos (epifauna) → benthophagous fishes

detritus → benthos (infauna) → benthophagous fishes

detritus → bentos → benthophagous fishes → large fish predators (sharks)

detritus → small benthos → larger invertebrates & small benthic fishes → large fishes

detritus → large detritivorous fishes (mullet): "telescoping" of food chain (51)

detritus → benthos → large predators

detritus → micronekton → intermediate predators (snappers, croaker)

detritus → zooplankton → small fishes and invertebrates

detritus → zooplankton → small fishes and invertebrates → larger fishes

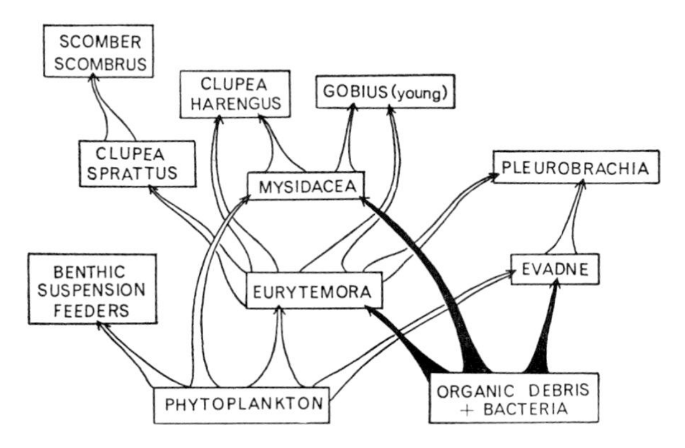

Figure 1. Simplified food web of the zooplankton in a British estuary (from Green [22]).

Africa (Fig. 4). Primary carnivorous fishes include the halfbeak (*Hyporhamphus*), a zooplankton feeder, a croaker (*Johnius*), and two porgies (*Lithognathus* and *Rhabdosargus*). The only secondary carnivore is a jack (*Hypacanthus*).

Two food webs occur in the littoral waters off the mouth of the La Plata estuary in Argentina (53). The first schema is a pelagic-demersal food web (Fig. 5) which is relatively simple, where the secondary carnivores are butterfish (*Stromateus*), anchovies (*Engraulis*), herring (*Clupea*), and a jack (*Parona*). Tertiary carnivores are the flathead (*Percophys*) and butterfish. The butterfish may feed directly on copepods and decapod larvae as a secondary carnivore, or it may be a tertiary carnivore by feeding on amphipods which in turn eat copepods and decapod larvae. The second schema represents a benthic-demersal food web (Fig. 6). It is more complex, with a wide variety of detritus-feeding infauna and epifauna being eaten by 12 taxa of primary carnivores representing 10 discrete feeding groups. These in turn are eaten by 6 groups of secondary carnivores which include 7 taxa, two of which (skates and dogfish) may also be primary carnivores.

Another example of a complex food web (34) occurs in the estuaries of Guyana (Fig. 7). The food relationships in this area represent a montage of 130 km of estuarine and littoral zones, and a wide variety of habitats, fishes, and invertebrates. This complexity also reflects the marked variation caused by fresh-water runoff during the rainy season.

A specialized food web is seen during the estuarine part of the life history of a barracuda (*Sphyraena barracuda*) in southern Florida (Fig. 8). During its first two years, this secondary carnivore spends most of its time within the estuary, feeding upon small forage fishes. During its third year it enters the coral-reef habitat, concurrently changing its food habits to include coral- and

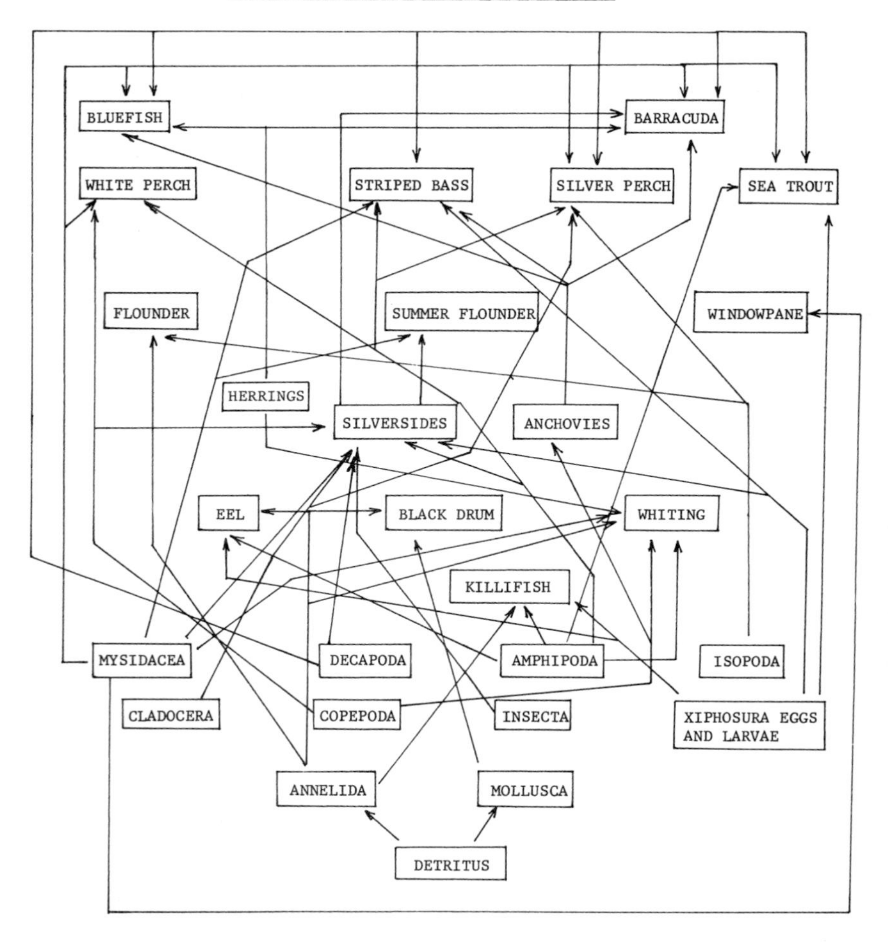

Figure 2. Principal pathways of energy flow in the Delaware River estuary (based on de Sylva et al. [18]).

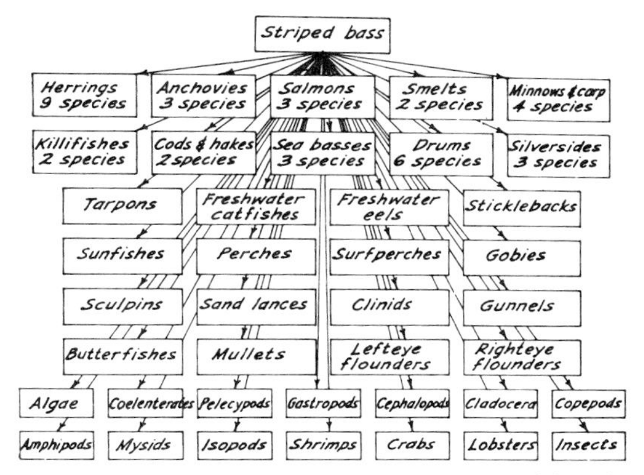

Figure 3. Simplified diagram of the food relationships of striped bass (**morone saxatilis**) in Chesapeake Bay (from McHugh [41]).

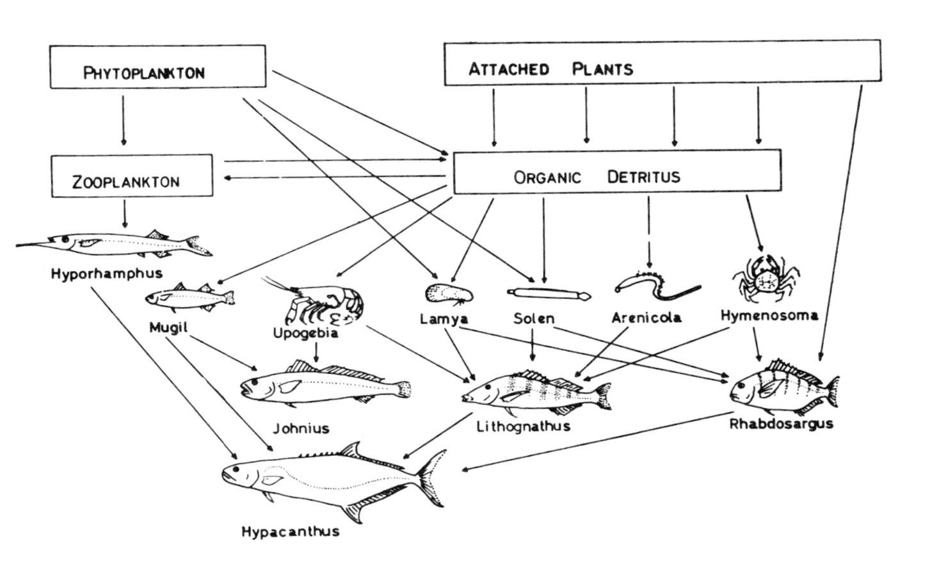

Figure 4. Trophic relations within Knysna estuary, Africa (from Day [13]).

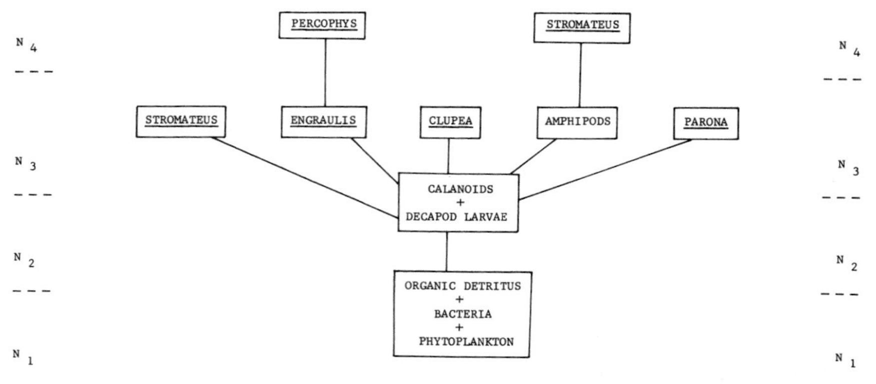

Figure 5. Pelagic-demersal trophic web in the littoral of Argentina (from Olivier et al. [53]).

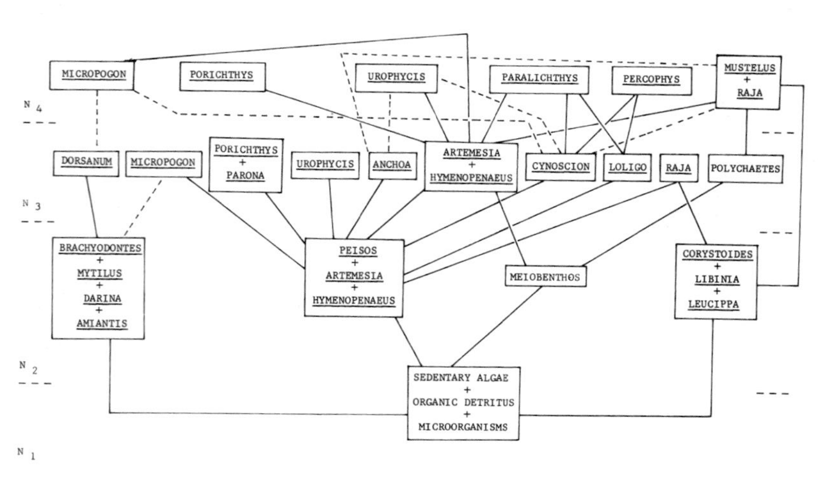

Figure 6. Benthic-demersal trophic web in the littoral of Argentina (from Olivier et al. [53]).

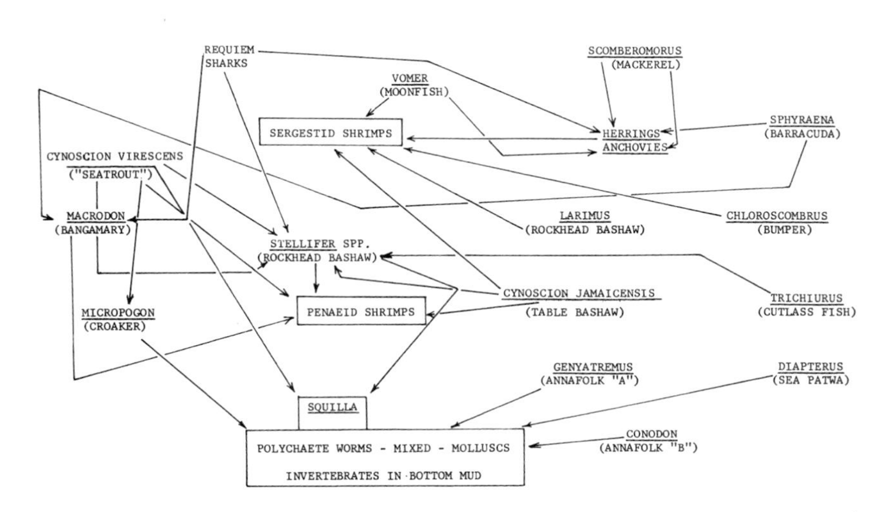

Figure 7. The main food relationships of the commonest species caught in trawls in Guyana (after Lowe [34]).

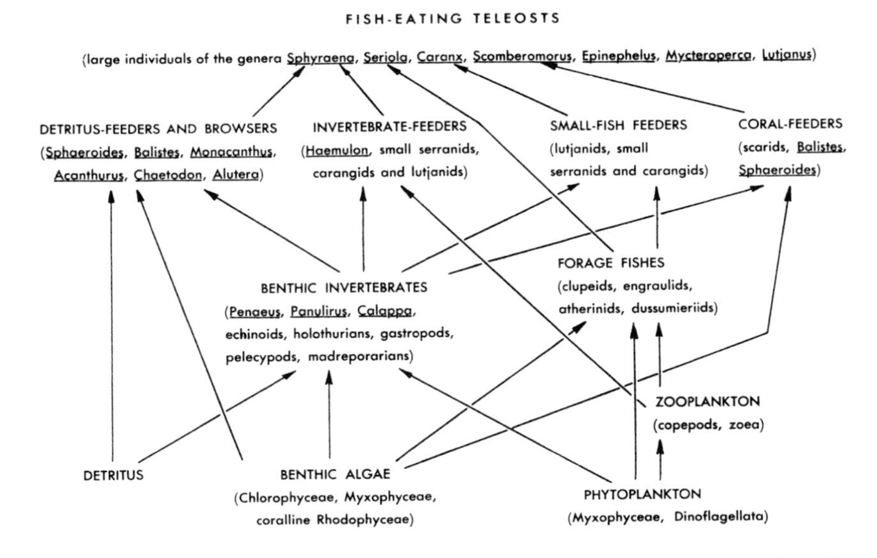

Figure 8. Generalized relations among organisms of the coral-reef habitat of southern Florida (from de Sylva [14]).

detritus-feeders (14). These foods are suspected of being implicated in the algae-based ciguatera cycle, which causes poisoning in man. Not until barracuda change their food habits does the possibility arise that they may be poisonous when eaten by man.

In the Indo-Pacific region, studies of food habits of estuarine fishes are scarce, though some have been made in Indian waters; however, few trophic relationships have been analyzed (54, 59, 62) (Table 1). Studies in the estuaries of South Viet Nam (19) have evaluated the effects of military defoliation from various herbicides on the estuarine ecology and fisheries dependent upon mangroves (50). In a nondefoliated coastal region near Vung Tau (Fig. 9), for

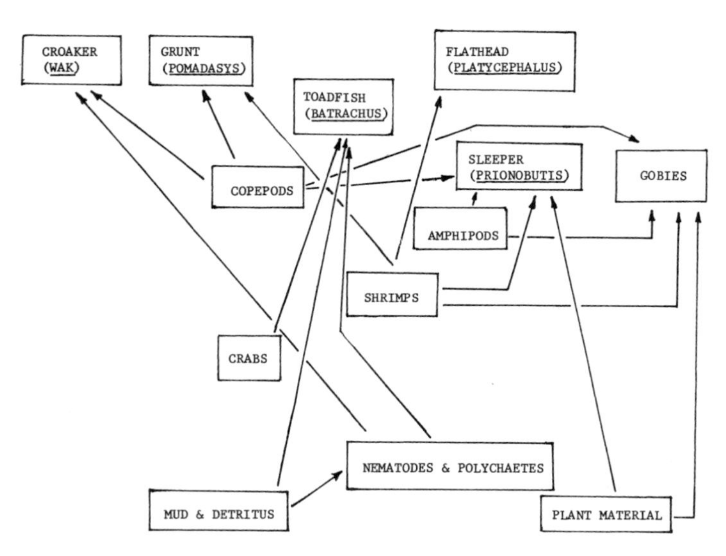

Figure 9. Food web in a nondefoliated estuary, Vung Tau, South Viet Nam in the wet season.

example, the food web during the wet season is reasonably complex, with several pathways and trophic levels apparent, which include secondary carnivores. The pattern does not vary greatly from season to season (Fig. 10) among the primary and secondary carnivores, except that during the wet season the primary source of energy is detritus, while during the dry season it is believed to be phytoplankton.

In the defoliated mangrove forests of the Rung Sat Zone, the food web during the wet season (Fig. 11) is simple. It was expected that the devastation of

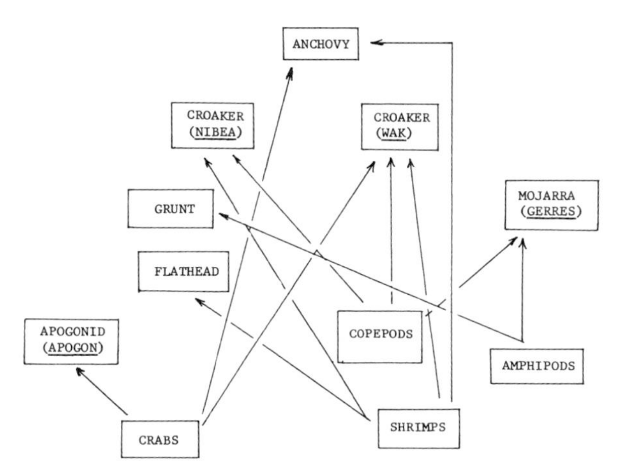

Figure 10. Food web in a nondefoliated estuary, Vung Tau, South Viet Nam in the dry season.

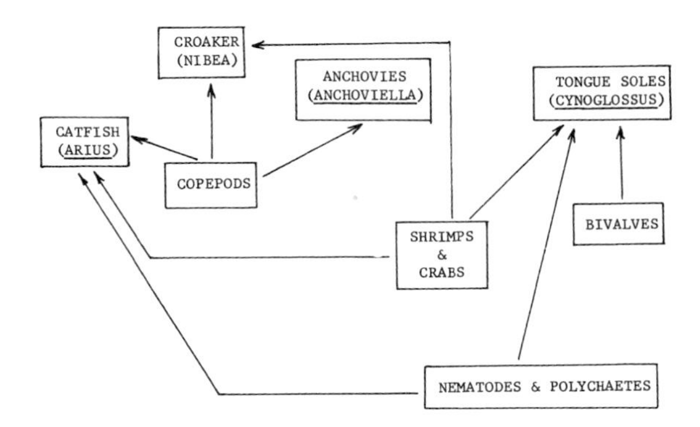

Figure 11. Food web in a defoliated estuary, Rung Sat Zone, South Viet Nam in the wet season.

mangroves would result in large amounts of detritus accumulating at the bottom of the estuaries, but because the strong tidal currents carry a lot of the wood, stems, and leaves of dead mangroves away from these waters before they can decompose to organic detritus, very little is available as a food source in the estuary. Undoubtedly, however, this detritus is carried downstream where it may be used as food in the littoral zone. During both the wet and dry seasons, the high turbidity of the water probably reduces photosynthesis, although diatoms were abundant and may represent the main source of energy. During the dry season in the Rung Sat Zone (Fig. 12), a more complex and diverse food web is

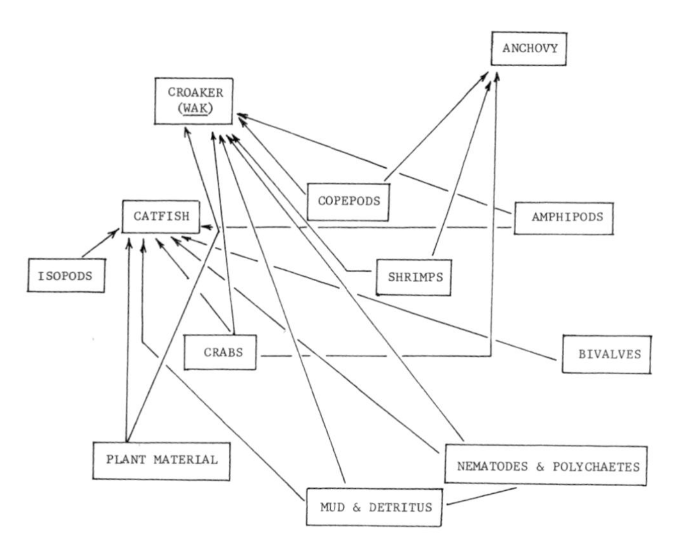

Figure 12. Food web in a defoliated estuary, Rung Sat Zone, South Viet Nam in the dry season.

apparent, and it may be that detritus becomes the major energy source, although the trophic relationships are still less complex than those found in the nondefoliated area of Vung Tau. During the dry season, when there is less runoff and salinities are higher, a greater variety of organisms can invade the defoliated estuarine region and serve as food for nektonic predators. Yet secondary carnivores occur during neither season, a fact that probably reflects the stressed ecosystem. Unfortunately, since studies of the Rung Sat Zone prior to to the wars in Indochina are not available, the ecosystem cannot be characterized. It is extremely difficult to evaluate the effects of defoliation of the mangroves, both because of the complete lack of background information and because of the complex effects of the wars (Fig. 13).

ONTOGENETIC AND OTHER VARIATIONS IN NEKTONIC FOOD WEBS

A subject long realized but poorly documented is the marked change in food habits as fish grow. Changes of food habits of fishes from Crystal River, Florida, were correlated with ontogenetic growth, and the dietary groups analyzed as follows (6):

I. Planktivore throughout
II. Transition from detritivore to planktivore
III. Transition from planktivore to herbivore
IV. Transition from planktivore to herbivore to carnivore
V. Transition from planktivore to carnivore
VI. Transition from planktivore to cleaner to carnivore
VII. Carnivore throughout
VIII. Transition from carnivore to omnivore
IX. Transition from detritivore to omnivore
X. Detritivore throughout

Many writers have discussed changes in food with size (6, 11, 45). Distinct changes occur with ontogenetic changes that in some cases can be correlated with changes in morphology, habitat, feeding time, and behavior (11). Food habits of fishes have been observed to change seasonally or even yearly (32). This phenomenon is more apparent in high latitudes, where extremes of water temperature are encountered (41). Yet fluctuations in food habits occur in tropical estuaries too, and result largely from salinity variations which cause changes in the estuarine productivity, the community structure, and the number of food organisms available to nekton. Conversely, where environmental fluctuations are minimal, changes in feeding intensity do not vary greatly (1). The food of estuarine nekton may change over its distributional range. For example, distinct variations in the food habits of salmon occur in different parts of Chatham Sound, British Columbia (37).

Studies of food habits of fishes seldom consider what effects the spawning season might have. Striped bass cease feeding during the spawning season (27). My observations of the food habits of fishes caught in trawls suggest that females with well-developed ovaries may not be able to feed because the large ovarian bulk greatly compresses the stomach, possibly making passage of food into the stomach physically difficult. As the gonads of certain fish species in Indian waters ripen (54), the fish become sluggish and easily fall prey to the oxeye (*Megalops cyprinoides*). These fish would, of course, result in an overestimation

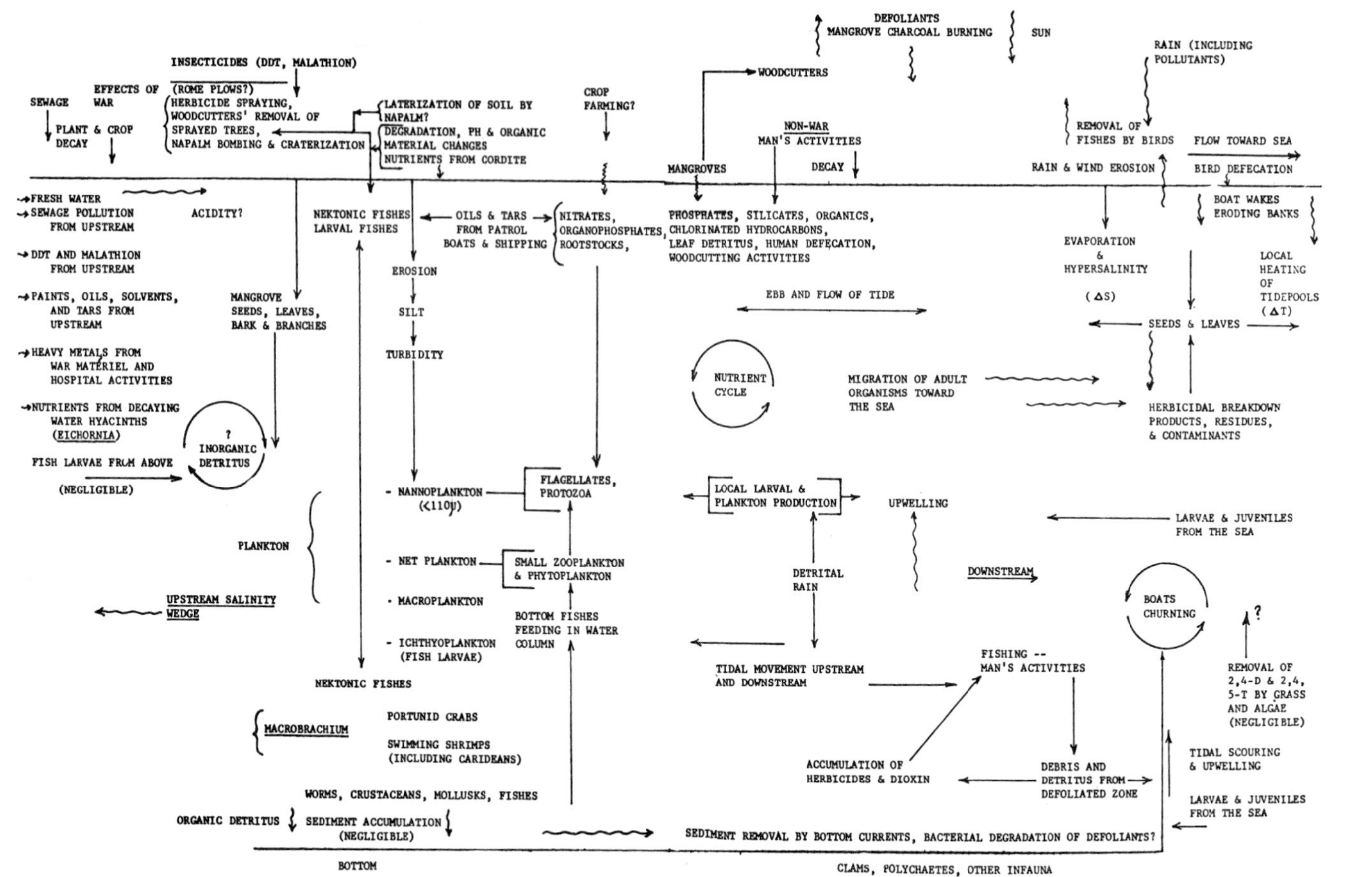

Figure 13. Hypothtical food web in the defoliated Rung Sat Zone, South Viet Nam,

of their importance in the predator's diet.

A biotic factor that is difficult to evaluate is the influence of predators on the movements of prey. During trawling operations off Guyana, whenever packs of carcharhinid sharks appeared, the prey fish scattered and catches became very patchy (34). Under such circumstances the prey fish might cease feeding, and conclusions drawn from such samples might be erroneous.

PROBLEMS OF SAMPLING ESTUARINE NEKTON

The chief problem in evaluating food webs of estuarine nekton is the lack of adequate sampling. Most sampling has been carried out using bottom trawls, though gill nets, traps, and trammel nets have occasionally been used. However, few studies of estuarine nekton report on swift-swimming fishes, either because they do not occur in estuaries or because they evade capture. The elusive tunas, mackerels, bonitos, bonefish, needlefish, halfbeaks, snooks, tarpons, many species of sharks and jacks, groupers, bluefish, cobia, sauries, and barracudas may well be important components of the estuarine nekton but it is difficult to capture them, especially the adults, by using conventional methods. Certainly, porpoises and seals must be considered as potentially important estuarine nekton, yet few have been captured for stomach analyses.

Another factor that makes food studies difficult to evaluate is the regurgitation of food by fishes when they are captured in nets. Fishes captured in trawls frequently disgorge their food (62), and there is differential regurgitation, the smaller fishes and fish-eating fishes tending to have empty stomachs more frequently (32). Fish taken in gill nets more often have empty stomachs than do those taken in trawls (77). The length of time a fishing gear is used is probably a factor as well. Gill nets or trawls fished for long periods undoubtedly yield more specimens with empty stomachs than those fished for short periods simply because there was more time for regurgitation to occur. In trawl catches, it would be expected that the rate of regurgitation would be proportional not only to the length of time of the trawl tow but also to the weight of the catch, because the greater the weight the greater the pressure exerted on the fishes' stomachs. Ideally, then, all fishing gear should be used for short periods if the greatest food retention in stomach is to be ensured. Short tows or fishing periods and rapid collection of stomach contents would also prevent digestion of food present in the stomach.

The use of trawls or nets in which specimens become greatly crowded can result in a greater availability of prey to predators which are also entrapped. Thus, biased estimates of food habits will be obtained, both quantitatively and, perhaps, even qualitatively. The use of short fishing periods would help to obviate this problem.

FUTURE STUDIES OF NEKTONIC FOOD WEBS IN ESTUARIES

To determine the dynamics of nektonic food webs in estuaries, carefully planned programs are needed using experimental designs which will relate trophic levels to one another and which will permit the evaluation of biotic and abiotic environmental factors, ontogenetic changes in food habits, and differences in location, time of day, and season. Studies should be carried out over 24-hour periods of continuous sampling to determine diel fluctuations in occurrence and kinds of estuarine nekton and variations in feeding intensity. At the least, attempts should be made to sample during both the day and night.

Consideration should be given to using a wide variety of collecting gear in addition to the traditional bottom trawls and gill nets, such as mid-water and surface trawls, drop nets, hoop nets, purse seines, large haul seines, stake nets, fyke nets, trammel nets, cast nets, setlines and longlines, spears, angling, rotenone, electricity, explosives (where permissible), attractant lights, and suction pumps. Quantification of nekton caught and quantitative evaluation of stomach contents would be a step toward sampling the *total* ecosystem. If present methods of collecting are shown to be adequate to delineate the estuarine nekton, then effort could be focused on developing good quantitative sampling methods for nekton and its food.

Large nets are necessary to capture large nekton. Where water is sufficiently deep, such as in the larger estuaries, especially in channels of the Mississippi, Columbia, Amazon, Hudson, and Saigon rivers, the Isaacs-Kidd mid-water trawl might be effective (25).

Stomachs from estuarine nekton should be properly preserved. Most food studies are based on fishes from trawls fished for commercial catches, which may have fished several hours. At best, the stomachs are immediately opened and the contents frozen or preserved in formalin. In the worst instances, as with stomachs from fishes caught by anglers, digestion continues long after the fish have been caught, and eight or ten hours may elapse before stomachs are examined. Sometimes angler-caught fish are chilled for a day or two before their stomachs can be examined; such material is almost valueless.

Laboratory and field studies are required to disclose the behavior of estuarine nekton (14). Because most estuaries are muddy, it is difficult or impossible for scuba divers or underwater television to make *in situ* observations; are difficult or impossible; nevertheless, these techniques should be used where water clarity is adequate. Ecological and behavioral studies in the field might lead to the development of better sampling methods for estuarine nekton. In the laboratory, facilities that simulate field conditions would enable behavioral observations and experiments to be made. Radioactive tracers could be used to quantify the rate of flux among trophic levels, and pollutants such as heavy metals and pesticides could be experimentally introduced and followed through the food web.

ACKNOWLEDGMENTS

Support for research in the United States was made possible through National Science Foundation grant no. G-8774; the Environmental Protection Agency, Contract FWQA 18050 DIU; and the Dingell-Johnson Fund, project no. F-13-R-1-2-3, job no. 10. The Maytag Chair of Ichthyology, University of Miami, is also gratefully acknowledged.

This research was also partially supported by the National Academy of Sciences, contract no. BA-23-73-47. I wish to thank various members of the NAS Committee on the Effects of Herbicides in Viet Nam for their advice and assistance in the course of our field work in South Viet Nam, specifically Dr. Anton Lang, Chairman; Dr. Fred Tschirley; Dr. Howard T. Odum; Professor Pham Hoang Hô; Dr. Joseph M. Carrier; and Dr. Philip Ross.

For various forms of support during the field and laboratory studies in Viet Nam, I wish to thank my colleagues and co-workers at the University of Miami, including Roger T. Hanlon, Eric Hildebrand, Raymond F. Hixon, James K. Low, John Markham, Susan McLean, Dr. Harding B. Michel, Shirley Pomponi, James F. Quinn, Dr. Porn Dispen Sari, Lewis N. Scotton, Robert C. Trumbull, Robert C. Work, and Bernard Yokel. Dr. Frederick A. Kalber assisted with much of the field and laboratory work in Delaware.

REFERENCES

1. Ah-kow-Tham
1950 Food and feeding relationships of the fishes of the Singapore Straits. **Fish. Publ. Col. Off. London,** 1(1): 1-35.

2. Austin, H. M.
1971 A survey of the ichthyofauna of the mangroves of western Puerto Rico during December, 1967-August, 1968. **Carib. Jour. Sci.,** 11(1-2) 27-39.

3. Austin, H. M. and Austin, S.
1971 The feeding habits of some juvenile marine fishes from the mangroves in western Puerto Rico. **Carib. Jour. Sci.,** 11(3-4): 171-178.

4. Bhatt, Y. M., Kutty, M. N., Subba Rao, K. V., and Punwari, D. M.
1967 'Ghol-Dara' fishery off Bedi Port in the Gulf of Kutch. **Indian Jour. Fish.** (1964), 11(A): 135-156.

5. Cadenat, J.
1954 Note d'ichtyologie ouest-africaine. VII. Biologie. Régime alimentaire. **Bull. Inst. Franc. Afrique Noire** (Sér. A), 16(2): 564-583.

6. Carr, W. E. S. and Adams, C. A.
1973 Food habits of juvenile marine fishes occypying seagrass beds in

the estuarine zone near Crystal River, Florida. **Trans. Am. Fish. Soc.,** 102(3): 511-540.

7. Casabianca, M. L. and Kiener, A.
1969 Gobiidés des étangs Corses: systématique, écologie, régime alimentaire et position dans les chaines trophiques. **Vie Milieu,** 20(A): 611-633.

8. Ciechomski, J. D. de
1967 Investigations of food and feeding habits of larvae and juveniles of the Argentine anchovy **Engraulis anchoita. Contr. Inst. Biol. Mar. Argentina,** 47: 72-81.

9. Coutant, Charles C.
1969 Thermal pollution – biological effects, a review of the literature of 1968. Battelle Mem. Inst. (Richland, Washington), BNWL-SA-2376. 43 p. (Unpublished manuscript.)

10. Croker, R. A.
1962 Growth and food of the gray snapper, **Lutjanus griseus,** in Everglades National Park. **Trans. Am. Fish. Soc.,** 91(4): 379-383.

11. Darnell, R. M.
1959 Food habits of fishes and larger invertebrates of Lake Pontchartrain, Louisiana, an estuarine community. **Publ. Inst. Mar. Sci. Texas,** 5: 353-416.

12. Dawson, C. E.
1958 A study of the biology and life history of the spot, **Leiostomus xanthurus** Lacépede, with special reference to South Carolina. **Contr. Bears Bluff Lab.,** 28: 1-48.

13. Day, J. H.
1967 The biology of the Knysna estuary, South Africa. In **Estuaries,** p. 397-407. (ed. Lauff, G. K.) **Am. Assoc. Adv. Sci. Publ.** 83. Washington, D. C.

14. de Sylva, D. P.
1963 Systematics and life history of the great barracuda, **Sphyraena barracuda** (Walbaum). **Stud. Trop. Oceanogr.** (Miami), 1: 1-179.

15. de Sylvia, D. P.
1969 Trends in marine sport fishery research. **Trans. Am. Fish. Soc.,** 98(1): 151-169.

16. de Sylvia, D. P.
1969 Theoretical considerations of the effects of heated effluents on marine fishes. In **Biological aspects of thermal pollution** p. 229-293. (eds. Parker, F. L. and Krenkel, P. A.) Vanderbilt University Press, Nashville, Tennessee.

17. de Sylva, D. P. and Davis, W. P.
1963 White marlin, **Tetrapturus albidus** Poey, in the middle Atlantic bight, with observations on the hydrography of the fishing grounds. **Copeia,** 1: 81-99.

18. de Sylva, D. P., Kalber, F. A., and Shuster, C. N.
1962 Fishes and ecological conditions in the shore zone of the Delaware River estuary, with notes on other species collected in deeper water. **Delaware Mar. Lab., Inf. Ser.,** 5: 164 pp.

19. de Sylva, D. P. and Michel, H. B.
1973 Effects of mangrove defoliation on the estuarine ecology and fisheries of South Viet Nam. Prel. Rep. Nat. Acad. Sci., Comm.

on the Effects of Herbicides in Viet Nam. 36 p. (Unpublished manuscript.)

20. Dovel, W. L.
1968 Predation by striped bass as a possible influence on population size of the Atlantic croaker. **Trans. Am. Fish. Soc.,** 97: 313-319.

21. Furtado, E.
1969 Alguns dados sôbre a alimentação de jovens do gênero **Mugil** Linnaeus no estado do Ceará. **Arq. Est. Biol. Mar. Univ. Ceard,** 8: 173-176.

22. Green, J.
1968 **The biology of estuarine animals.** Univ. Washington Press, Seattle. 401 p.

23. Gunter, G.
1945 Studies on marine fishes of Texas. **Publ. Inst. Mar. Sci. Texas,** 1(1): 1-190.

24. Harrington, R. W., Jr., and Harrington, E. S.
1960 Food of larval and young tarpon, **Megalops atlantica. Copeia,** 4: 311-319.

25. Harrisson, C. M H.
1967 On methods for sampling mesopelagic fishes. **Symp. Zool. Soc. Lond.,** 19: 71-126.

26. Hildebrand, S. F. and Schroeder, W. C.
1928 Fishes of Chesapeake Bay. **Bull. Bur. Fish.,** 43(1): 1-366.

27. Hollis, E. H.
1952 Variations in the feeding habits of the striped bass, **Roccus saxatilis** (Walbaum), in Cheapeake Bay. **Bull. Bingham Oceanogr. Inst.,** 14(1): 111-131.

28. June, F. C. and Chamberlin, J. L.
1959 The role of the estuary in the life history and biology of Atlantic menhaden. **Proc. Gulf Carib. Fish. Inst.,** 11th Ann. Sess., p. 41-45.

29. Kagwade, P. V.
1967 The food and feeding habits of the Indian oil sardine **Sardinella longiceps** Valenciennes. **Indian Jour. Fish.** (1964), 11(A): 345-370.

30. Lauff, G. K. (ed.)
1967 **Estuaries.** Am. Assoc. Adv. Sci. Publ. 83. Washington, D. C. 757 p.

31. Le Brasseur, R. J., Barraclough, W. E., Kennedy, O. D., and Parsons, J. R.
1969 Production studies in the Strait of Georgia, Part 3. Observations on the food of larval and juvenile fish in the Fraser River plume, Feb. to May, 1967. **Jour. Exp. Mar. Biol. Ecol.,** 3: 51-61.

32. Longhurst, A. R.
1957 The food of the commercial demersal fish of a West African estuary. **Jour. Anim. Ecol.,** 26(2): 369-387.

33. López, R. and Castello, J. H. P.
1967 Corvinas del rio de la Plata. **Mus. Argentina Cienc. Nat. 'Bernardino Rivadavia' Buenos Aires** (n.s.), 67: 1-14. Also, **Boln. Serv. Hidro. Naval,** 5: 14-27.

34. Lowe (McConnell), R. H.

1962 The fishes of the British Guiana continental shelf, Atlantic coas of South America, with notes on their natural history. **Jour Linn. Soc. London (Zool.),** 44(301): 669-700.

35. Lursinap, A.
1966 Analysis of stomach contents of **Rastrelliger neglectus** in Gulf o Thailand. **Tech. Pap. Indo-Pacif. Fish Coun.,** 11: 34-43.

36. Macnae, W. A.
1968 A general account of the fauna and flora of mangrove swamp and forests in the Indo-West-Pacific region. **Adv. Mar. Biol.,** 6 73-370.

37. Manzer, J. I.
1969 Stomach contents of juvenile Pacific salmon in Chatham Sounc and adjacent waters. **Jour. Fish. Res. Bd. Canada,** 26: 2219-2223.

38. Marquez, J. R. S.
1968 Age and size at sexual maturity of white goby (**Glossogobiu giurus**), a common species of fish of Laguna de Bay, with notes or its food habits. **Philippine Jour. Fish.** (1960). 8: 71-99.

39. Marshall, A. R.
1958 A survey of the snook fishery of Florida, with studies of the biology of the principal species, **Centropomus undecimalis** (Bloch). **Florida St. Bd. Conserv., Tech. Ser.,** 22: 1-37.

40. Mayskiy, V. N.
1939 [The effect of predators on the fish population of the Sea of Azov.] **Zool. Zhur.,** 18(2). (In Russian.)

41. McHugh, J. L.
1967 Estuarine nekton. In **Estuaries,** p. 581-620. (ed. Lauff, G. K.) Am. Assoc. Adv. Sci. Publ. 83. Washington, D. C.

42. Menezes, M. Ferreirade
1968 Sôbre a alimentação do camirupim, **Tarpon atlanticus** (Valenciennes), no estado do Ceará. **Arq. Est. Biol. Mar. Univ. Ceará,** 8: 145-149.

43. Miles, D. W.
1949 A study of the food habits of the fishes of the Aransas Bay area. **Ann. Rept. Mar. Lab., Texas Game, Fish, Oyst. Comm.,** 1948-1949: 129-169.

44. Moody, W. D.
1950 A study of the natural history of the spotted trout, **Cynoscion nebulosus,** in the Cedar Key, Florida, area. **Quart. Florida Acad. Sci.,** 12(3): 147-171.

45. Mulkana, M. S.
1966 The growth and feeding habits of fishes of two Rhode Island estuaries. **Gulf Res. Repts.,** 2: 97-167.

46. Naqvi, S. M. Z.
1968 Effects of predation on infaunal invertebrates of Alligator Harbor, Florida. **Gulf Res. Repts.,** 2: 313-321.

47. Natarajan, A. V., and Patnaik, S.
1968 Occurrence of mullet eggs in the gut contents of **Ambassis gymnocephalus** (Lacép.). **Jour. Mar. Biol. Assoc. India** (1967), 9: 192-194.

48. Noble, A.

1965 The food and feeding habits of the Indian oil sardine **Sardinella longiceps** Valenciennes at Karwar. **Indian Jour. Fish.,** 12(1A): 77-86.

49. Odum, W. E.
1968 Mullet grazing on a dinoflagellate bloom. **Chesapeake Sci.,** 9: 202-204.

50. Odum, W. E.
1970 **Pathways of energy flow in a South Florida estuary.** Doctoral dissertation, Univ. Miami, Florida 162 p.

51. Odum, W. E.
1970 Utilization of the direct grazing and plant detritus food chains by the striped mullet **Mugil cephalus.** In **Marine food chains,** p. 222-240. (ed. Steele, J. H.) Univ. California Press, Berkeley.

52. Odum, W. E. and Heald, E. J.
1972 Trophic analyses of an estuarine mangrove community. **Bull. Mar. Sci.,** 22(3): 671-738.

53. Olivier, S. R., Bastida, R., and Torti, M. R.
1968 Sobre el ecosistema de los aguas litorales de la Mar del Plata. **Serv. Hidrogr. Naval, Inst. Biol. Mar., Mar del Plata, Argentina,** 58: 1-46.

54. Pandian, T. J.
1969 Feeding habits of the fish **Megalops cyprinoides** Broussonet, in the Cooum backwaters, Madras. **Jour. Bombay Nat. Hist. Soc.,** 65: 569-580.

55. Pearson, J. C.
1929 Natural history and conservation of the redfish and other commercial sciaenids on the Texas coast. **Bull. U. S. Bur. Fish.** (1928), 44: 129-214.

56. Pignalberi, C.
1967 Importanza dei diversi gruppi di organismi nell' alimentazione estiva delle specie ittiche litorale do Lago Maggiore. **Mem. Ist. Italiano Idrobiol.,** 21: 89-103.

57. Porumb, I. I.
1968 Le rôle des jeunes **Pomatomus saltator** L. dans le chaîne trophique de la mer Noire. **Rapp. Comm. Int. Mer Méditerranée,** 19(2): 303-305.

58. Prasadem, R. D.
1970 Preliminary observations on food and feeding habits of the grey mullet **Mugil macrolepis** (Smith) Agues from Pulicat Lake, Madras. **Jour. Zool. Soc. India,** 22(1-2): 63-68.

59. Qasim, S. Z.
1970 Some problems related to the food chain in a tropical estuary. In **Marine food chains,** p. 45-51. (ed. Steele, J. H.) Univ. California Press, Berkeley.

60. Radtke, L. D.
1966 Distribution of smelt, juvenile sturgeon, and starry flounder in the Sacramento-San Joaquin delta with observations on food of sturgeon. **Fish. Bull. California,** 136: 115-129.

61. Raney, E. C.
1952 The life history of the striped bass, **Roccus saxatilis** (Walbaum).

Bull. Bingham Oceanogr. Coll., 14(1): 5-97.

62. Rao, K. S.
1967 Food and feeding habits of fishes from trawl catches in the Bay of Bengal with observations on diurnal variation in the nature of the feed. **Indian Jour. Fish.** (1964), 11(A): 277-314.

63. Rao, K. S.
1968 Some aspects of the biology of "ghol," **Pseudosciaena diacanthus** (Lacépède). **Indian Jour. Fish.** (1963), 10: 413-459.

64. Raymont, J. E. G.
1963 **Plankton and productivity in the oceans.** Pergamon Press, Oxford. 660 pp.

65. Reid, G. K., Jr.
1955 A summer study of the biology and ecology of East Bay, Texas. Pt. II. The fish fauna of East Bay, the Gulf Beach, and summary. **Texas Jour. Sci.**, 7(4): 430-453.

66. Reid, G. K., Jr., Inglis, A., and Hoese, H. D.
1956 Summer foods of some fish species in East Bay, Texas. **Southwestern Nat.**, 1(3): 100-104.

67. Reintjes, J. W.
1969 Synopsis of biological data on the Atlantic menhaden, **Brevoortia tyrannus. F. A. O. Fish Synopsis,** 42: 1-30.

68. Rickards, W. L.
1968 Ecology and growth of juvenile tarpon, **Megalops atlanticus,** in a Georgia salt marsh. **Bull. Mar. Sci.,** 18: 220-239.

69. Roelofs, E. W.
1954 Food studies of young sciaenid fishes, **Micropogon** and **Leiostomus,** from North Carolina. **Copeia,** 2: 156-200.

70. Sasaki, S.
1966 Distribution and food habits of king salmon, **Oncorhynchus tshawytscha,** and steelhead rainbow trout, **Salmo gairdnerii,** in the Sacramento-San Joaquin delta. **Fish. Bull. California,** 136: 108-114.

71. Shorygin, A. A.
1952 [Feeding and food relationships of fishes of the Caspian Sea.] Pishchepromizdat, Moscow. (In Russian)

72. Simpson, D. G. and Gunter, G.
1956 Notes on habitats, systematic characters and life histories of Texas salt water Cyprinodontes. **Tulane Stud. Zool.,** 4(4): 115-134.

73. Sirotenko, M. O.
1969 Food habits of Azov-Don shad. **Probl. Ichthyol.,** 9(4): 564-573.

74. Skazkina, Ye. P. and Kostyuchenko, V. A.
1968 Food consumption of the Azov goby (**Neogobius melanostomus** Pallas). **Probl. Ichthyol.,** 8(2): 238-245.

75. Spǎtaru, P.
1968 Relations trophiques chez les poissons du complexe d'étangs Crapina-Jijila (zone inondable du Danube). **Anal. Univ. Bucuresti (Biol.),** 17: 77-88.

76. Springer, V. G. and Woodburn, K. D.
1960 An ecological study of the fishes of the Tampa Bay area. **Florida**

Bd. Conserv., Prof. Paper Ser., 1: 1-104.

77. Stevens, D. E.
1966 Food habits of the striped bass, **Roccus saxatilis,** in the Sacramento-San Joaquin delta. **Fish. Bull. California,** 136: 68-96.

78. Stevens, D. E.
1966 Distribution and food habits of the American shad, **Alosa sapidissima** in the Sacramento-San Joaquin delta. **Fish. Bull. California,** 136: 97-107.

79. Thomson, J. N.
1957 The food of western Australian estuarine fish. **Fish. Bull. W. Australia,** 7: 1-13.

80. Tobor, J. G.
1969 Species of the Nigerian ariid catfishes, their taxonomy, distribution and preliminary observations of the biology of one of them. **Bull. Inst. Fond. Afrique Noire,** 31(A): 643-658.

81. Townes, H. K., Jr.
1937 Studies on the food organisms of fish. **Biol. Surv. New York St. Conserv. Dept.** (1936), 11: 217-230.

82. Walsh, G. E.
1967 An ecological study of an Hawaiian mangrove swamp. In **Estuaries,** p. 420-431. (ed. Lauff, G. K.) Am. Assoc. Adv. Sci. Publ. 83. Washington, D. C.

83. Welsh, W. W. and Breder, C. M., Jr.
1923 Contributions to the life histories of the Sciaenidae of the eastern United States coast. **Bull. U. S. Bur. Fish.** (1924), 39: 141-201.

84. Yablonskaya, Ye. A.
1955 [Possible changes in the food resources available to fishes in the Sea of Azov after regulation of river discharge.] **Vses. nauch. -issled. Inst. Ryb. Okeanogr.,** 31(1). (In Russian.)

85. Zheltenkova, M. V.
1955 [Nutrition and utilization of the available food resources by benthic fishes of the Sea of Azov.] **Vses. nauch.-issled. Inst. Ryb. Okeanogr.,** 31. (In Russian.)

CONSUMPTION AND UTILIZATION OF FOOD BY VARIOUS POSTLARVAL AND JUVENILE FISHES OF NORTH CAROLINA ESTUARIES

D. S. Peters and M. A. Kjelson[1]

ABSTRACT

Fish productivity may be limited by a variety of factors including diet composition, ingestion rate, and physical or chemical characteristics of the environment. We measured ingestion rate and diet composition in menhaden (*Brevoortia tyrannus*), spot (*Leiostomus xanthurus*), and pinfish (*Lagodon rhomboides*) and growth rates and food conversion efficiency in flounder (*Paralichthys lethostigma*). The summer diets of juvenile menhaden, spot, and pinfish averaged 40%-50% ash. The food ingested differed taxonomically between species, but the nitrogen and caloric content of the organic matter ingested was similar. Much of the nitrogen ingested by menhaden was probably in the form of microbes.

The daily rations of postlarval and juvenile pinfish, spot, and menhaden were estimated from data describing diel periodicity of gut contents and gastrointestinal evacuation rate. Evacuation rate constants were determined in the laboratory and used to estimate average evacuation rates under natural feeding conditions. High variability in gut content of postlarval fish caused imprecise estimates of their daily rations.

Feeding and growth rates of juvenile southern flounder were highest at high temperatures and low salinities. The salinity-producing maximum conversion efficiency increased as temperature decreased. Optimum feeding rate for

1. National Marine Fisheries Service, Atlantic Estuarine Fisheries Center, Beaufort, North Carolina 28516.

maximum efficiency was approximately 70%-90% *ad libitum* feeding and was relatively unaffected by temperature. Migration of summer flounder, southern flounder, and hogchokers may be toward temperature-salinity conditions under which maximum growth rates occur.

INTRODUCTION

Food consumption and utilization by fish are studied to increase our understanding of growth-related processes and to describe the impact of fish feeding on the ecology of the aquatic community. Although some fresh-water and marine species have been studied intensively, estuarine species have not. Little information is available regarding the physical, chemical, or biological factors that affect the growth of estuarine fish.

Production-related studies described in this paper are confined to determining diet composition, calculating food consumption rate, and describing environmental effects on the use of food for growth. Diet and feeding rates are described for pinfish, spot, and Atlantic menhaden – species that ingest large amounts of detritus. Since we cannot duplicate the diet of these detritivores in the laboratory, flounders were used in our studies relating environmental effects to growth.

FOOD CONSUMPTION

Since the quality of food ingested by fish populations influences the rates of food consumption and growth, knowledge of the type and nutritional quality of food consumed under natural conditions is important to our understanding of growth-related processes. Major problems include identifying food items in the gut and estimating their relative composition by weight or volume. Moderate-sized items can be identified to genus and species, provided adequate collections of local flora and fauna are available, but small, partially digested items may be impossible to identify. Organic detritus, a complex food resource consumed by many estuarine fish (12, 37), also is difficult to identify and measure. Detritus identification includes differentiating zooplankton, phytoplankton, and detritus (32), determining detrital origin (39), and differentiating between organic and inorganic material (61). The living microbial component, often coupled with nonliving detritus, is also difficult to identify and measure (13). Adenosine triphosphate (ATP) measurement (Ferguson and Murdoch, this volume) indicates the total microbial biomass, but we still have the problem of identifying and measuring the various microbes (bacteria, microalgae, protozoans, etc.) that may have different nutritional value for aquatic consumers.

Precise taxonomic identification is probably not necessary because nutrition-oriented descriptions of food consumption are satisfied by broad taxonomic categories. Animal, plant, and detritus categories were chosen under the assumption that materials within these categories have similar nutritional values that differ from the nutritional values in the other categories. Of particular nutritional interest are the organic matter, nitrogen, and caloric contents of ingested foods.

To evaluate the nature of food consumed by fish in the Newport River estuary, we have begun examining stomach contents of juvenile pinfish (*Lagodon rhomboides*), spot (*Leiostomus xanthurus*), and menhaden (*Brevoortia tyrannus*), three of the most abundant species present (56). Although samples for each species were collected on different days at a variety of locations and included fish of slightly different sizes, there was little variation within species in the percent of animal, plant, and detrital material in the guts (Table 1). Although spot and pinfisy may eat different animal species, they ingested nearly the same fraction of animal, plant, and detrital particles. The main food of menhaden was small detritus (< 10 μm) of unidentified origin, some of which may have been recently fragmented microalgae or protozoans.

To obtain some estimate of nutritional value, we analyzed stomach contents for sand, total ash, nitrogen, and caloric content (Table 2). Although pinfish and spot ingested considerable sand, and menhaden did not, there was little difference in the average composition of their food. The similarity in ash, nitrogen, and caloric contents was surprising in view of the much lower animal intake by menhaden. The nitrogen content of menhaden food was higher than

TABLE 1

Composition of food from the esophagus of juvenile Atlantic menhaden and from the stomachs of pinfish and spot. Each determination was from n samples of 10 fish.

Species (range total length mm)	n	Mean per cent by volume[1]			
		Amimal	Phytoplankton	Unidentified detritus	Vascular plant detritus
Menhaden (43-115)	8	0.4 ± 0.4	29 ± 6	70 ± 6	0.6 ± .05
Pinfish (47-87)	4	56 ± 1	8 ± 7	26 ± 5	10 ± 3
Spot (54-108)	5	59 ± 8	2.5 ± 1.3	30 ± 7	8 ± 1

1. ± one standard error

TABLE 2

Physico-chemical composition of food present in stomachs of Atlantic menhaden, pinfish and spot collected in the Newport River estuary, North Carolina. Determinations were made from n samples of 50-100 fish stomachs per sample.

Species (range total length mm)	n	Sand % dry weight	Total ash % dry weight	Nitrogen % ash-free dry weight	Energy cal mg^{-1} ash-free dry weight
Menhaden (85-115)	2	0	51.0 ± .5[1]	9.91 ± .42[1]	5.29 ± .06[1]
Pinfish (50-60)	5	24.7 ± 4.5[1]	49.5 ± 4.6	10.6 ± 1.0	5.31 ± .15
Spot (50-75)	6	20.8 ± 8.0	41.6 ± 5.3	10.1 ± .58	5.51 ± .33

1. ± one standard error

expected from investigations of phytoplankton (21, 51, 58) and detritus composition (6, 15, 34). The high nitrogen concentration found in menhaden stomachs may arise from ingestion of microbes; estuarine bacteria are as rich in nitrogen as estuarine animals (R. L. Ferguson, personal communication). The caloric value of food ingested by spot and pinfish was highest when annelids, which have a high caloric content (55), were abundant in the stomach. No other relationships were evident between taxonomic categories and nitrogen or caloric values.

Nitrogen content is an indication of the amount of protein ingested, even though a small amount of it may come from gastric secretion and some may be present in nonprotein forms. Protein content is frequently estimated as 6.25 times nitrogen content for mammals (31), but this factor may range from 5.7 to 8.0 for other animals (25). If we use a conservative estimate of 5 x %N = percent protein ingested, then all our fish consumed at least 50% protein (Table 2). Such a protein level does not appear to limit fish growth (9).

INGESTION RATE

Food-consumption rates are needed for a description of natural fish-production processes. These rates have been estimated in four ways, but none has been used extensively with estuarine fishes:

(a) Oxygen consumption is a frequently used indicator of energy or food requirements. Winberg (59) indicated that energy of diet = 1.25 (energy of growth + energy of metabolism). Unfortunately, it is difficult to determine natural growth because many estuarine fish migrate at size-dependent rates (14, 22), which biases population estimates of growth. Since the natural respiration rate is dependent on the feeding rate, the feeding regime chosen for respiration measurements will affect the estimate of energy or food ingested. Minimal estimates have been obtained without feeding, but these may vary considerably between investigators owing to variations in technique (24).

(b) Natural feeding rates may be estimated from laboratory feeding experiments (8) in which the natural growth rate is approximated. This method may be time-consuming, particularly if natural growth is slow.

(c) Natural food consumption also may be estimated by measuring assimilation efficiency, elimination rate, and body concentration of various materials, such as nitrogen (19), ^{137}Cs (29), and pesticides (S. Warlen and D. Wolfe, personal communication).

(d) Ingestion rate may be determined from stomach or gastrointestinal evacuation rate (1), because the average ingestion rate must equal the rate at which material leaves, whether by assimilation or expulsion. Evacuation is measured as the decrease in contents of the stomach or gastrointestinal tract and must be obtained in a food-free environment, since decreases in natural gut content may be less than true evacuation if the fish are consuming and evacuating food at the same time.

We estimated the daily rations of three species of postlarval and juvenile estuarine fishes by the evacuation method. We measured food evacuation rates in the laboratory and performed regression analyses on the data yelding rate constants which were used to calculate instantaneous evacuation rates. We determined the quantities of food present in the guts of selected species at 4-hour intervals throughout 24-hour cycles, and calculated instantaneous rates for each of the 4-hour periods in these cycles; the rates varied according to the amount of food present in the gut. Summing the evacuation during each of these periods produced an estimate of total evacuation of daily ration.

Periodicity of Feeding

We observed diel periodicity of gastrointestinal content in postlarval and juvenile pinfish, spot, and Atlantic menhaden. Juvenile spot had approximately the same amount in their guts at all times (Fig. 1), a fact that agrees with previous limited diurnal observations (11). Juvenile pinfish had considerable food in their guts during the day but very little after dark. Caldwell (7) found that pinfish stopped feeding at dusk. Juvenile menhaden maintained approximately the same food content from morning until after dark. Postlarval menhaden, pinfish, and spot, which are carnivorous sight-feeders, were found to have the greatest amounts of food (copepods) in their digestive tracts near

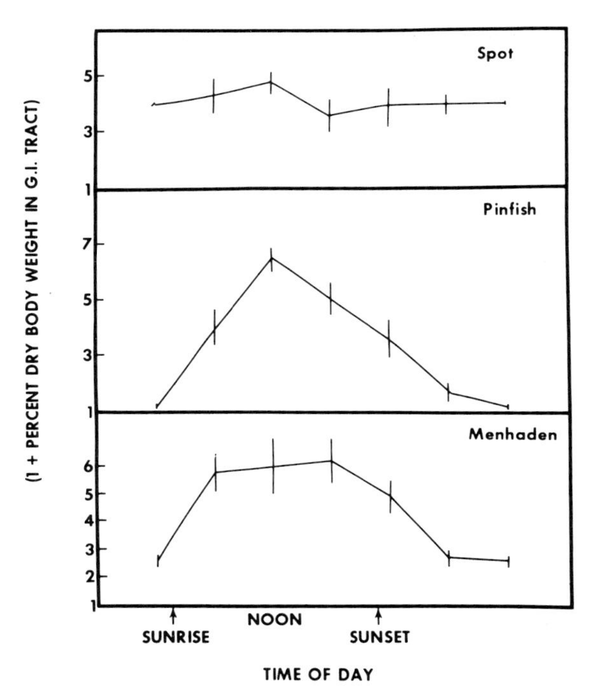

Figure 1. Diel periodicity of gastrointestinal content at 29 °C. Geometric means with one standard error are shown. Data were collected for each species in August 1972, June 1973, and August 1973.

midday (Fig. 2). Others have observed the same diel pattern in the gut contents of larval fish (2, 26, 49).

We replicated periodicity measurements of gastrointestinal contents of 20 mm spot in 1972 and 1973. The maximum observed in nature, 36 copepods per larva (Fig. 3), is near the maximum gut capacity (44 copepods) estimated from laboratory feeding experiments (unpublished data). Periodicity was similar in both years, but the absolute amounts of food present were much lower in 1972, suggesting that less food was available. Thayer et al. (54) have also concluded that zooplankton abundance may limit larval fish production.

Evacuation Rate

The amount of food evacuated during any time interval is directly dependent on the amount present in the gut. When no additional feeding is permitted, the amount of food remaining in the stomach at any time can be predicted from the following equation modified from Brett and Higgs (4) and Tyler (57):

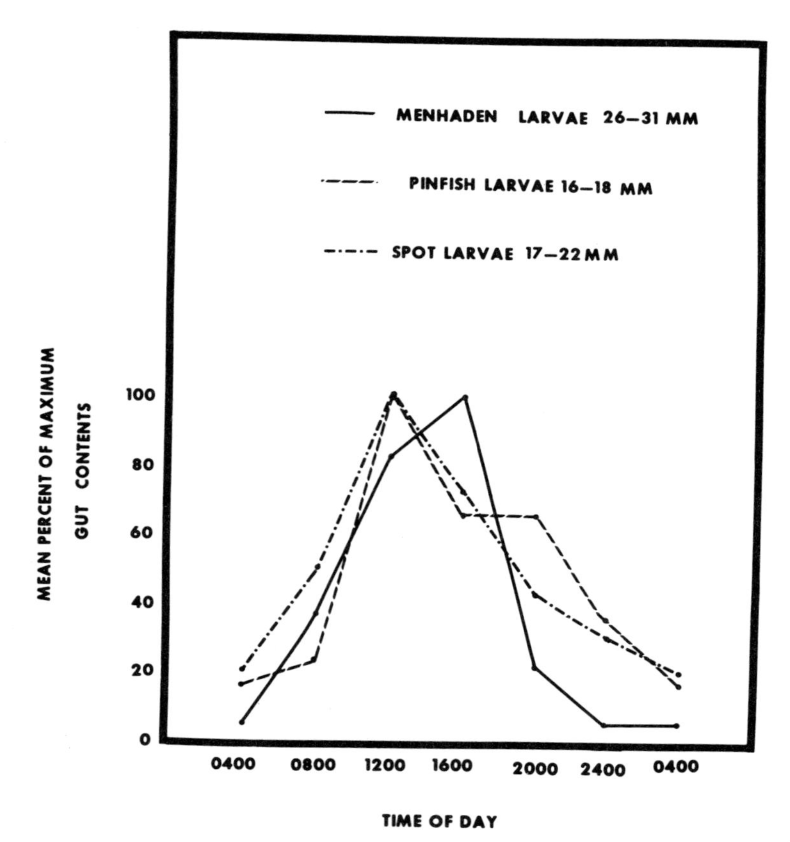

Figure 2. Diel cycle of gastrointestinal content in postlarval Atlantic menhaden, pinfish, and spot.

$$\log C = \text{Log } A + Bt \tag{1}$$

where C = content of gastrointestinal tract + 1, A = amount ingested + 1, B = evacuation rate constant, and t = time. By adding 1 to the amount ingested and to gut contents we were able to include in our calculations empty gastrointestinal tracts. From the above equation, with log base 10

$$C = e^{2.303\,(\log A + Bt)} \tag{2}$$

and the instantaneous evacuation rate

$$\frac{dC}{dt} = 2.303\ Be^{2.303\,(\log A + Bt)}$$

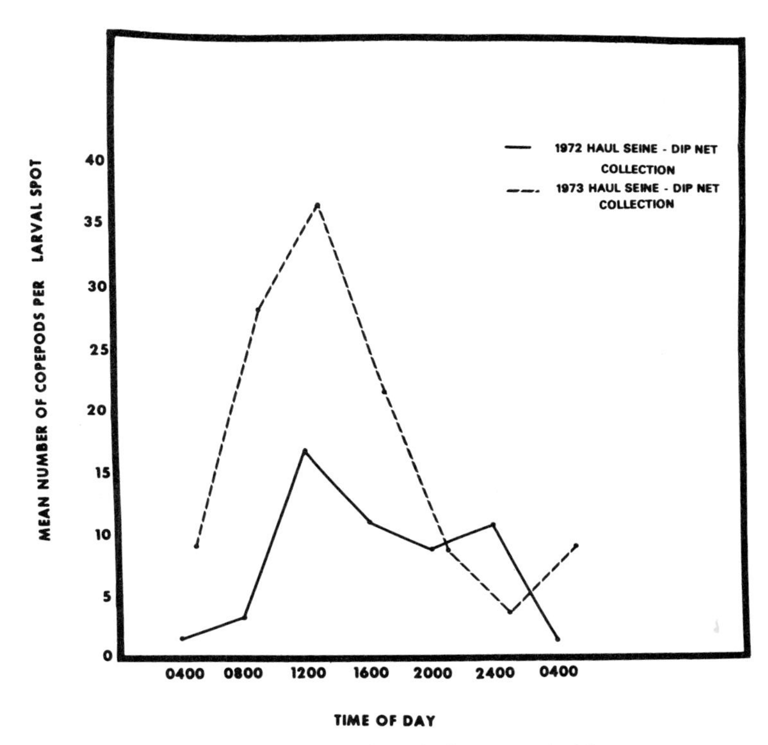

Figure 3. Variation in diel cycle of gastrointestinal contents in 20 mm spot.

or

$$\frac{dC}{dt} = 2.303\ BC \tag{3}$$

Gastric and gastrointestinal evacuation of single meals were summarized for several estuarine fish by linear regression of log-transformed data (47 and Peters, unpublished). We assumed that additional feedings would not change the evacuation rate constant, though the amount evacuated would increase. We did not test for the effect of additional feeding, but others have found extra meals had little (57) or no (18) effect on the rate constant. Evacuation by juvenile fishes does not always begin as soon as food is ingested (47). Species-dependent delays of up to 12 hours were apparent at low temperatures, but they decreased as temperature increased and were not evident at 30 °C. Inclusion of the lag term in the regression model (Fig. 4) increased the degree of fit at low temperatures (47).

Evacuation rate changes with temperature (4, 33). We determined rate

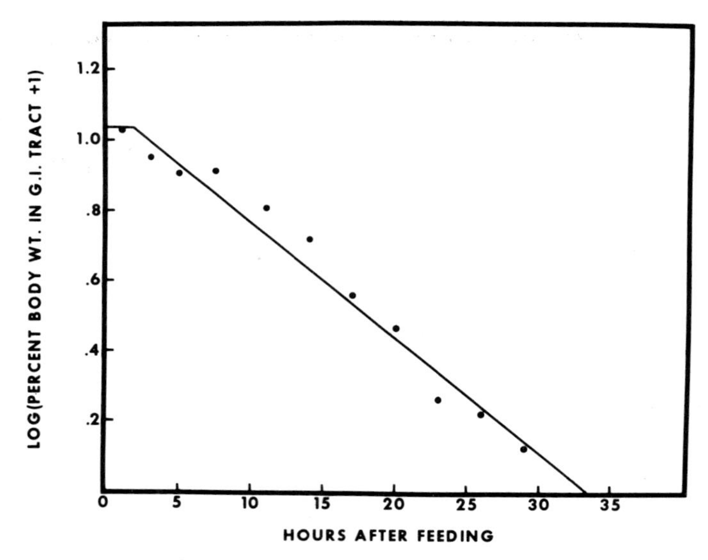

Figure 4. Gastrointestinal evacuation of commercial food by pinfish at 24 °C. Each point is the geometric mean of 5 observations. Log_{10} (% body wt. in G.I. tract + 1) = 1.037 - .033(X-2), where X = hours since feeding, lag = 2 hrs., $R^2 = 0.97$.

constants for juvenile fish at various acclimation temperatures and summarized them using a quadratic regression model (Fig. 5). With these estimates of the rate constants at any temperature and with measurements of the periodicity of gastrointestinal contents, we were able to calculate daily rations.

Daily Rations

To calculate daily rations, instantaneous evacuation rates were calculated from equation 3. Gastrointestinal periodicity of juveniles was measured at 29 °C and the evacuation rate constants (B at 29 °C) were calculated from equations in Figure 5. Postlarval evacuation rates were determined for naturally ingested food at 16 °C, a normal spring temperature. Gut-content measurements used to determine the instantaneous rates for juveniles are shown in Figure 1 and for postlarval spot in Figure 3. Similar data for postlarval menhaden and pinfish have been found by Kjelson (unpublished data). For each 4-hour period we calculated the average evacuation rate, which was the geometric mean of the rate at the beginning and end of the period. Since the periods lasted 4 hours, the estimate of food evacuated was four times the mean hourly evacuation rate. The total food evacuated per day, which was an estimate of the daily ration, was derived by summing the 4-hour evacuation estimates.

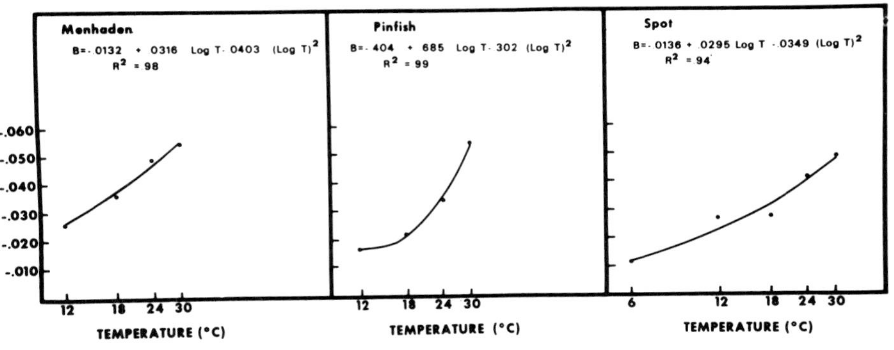

Figure 5. Food evacuation rates in Atlantic menhaden, pinfish, and spot as a function of temperature.

Our estimates of daily ration for juveniles were 9.5% of dry body wt per day for pinfish, 13.5% for menhaden, and 10.1% for spot. The postlarval estimates were less, 3.5% for pinfish, 4.9% for menhaden, and 4.3% and 9.0% for spot. The higher feeding rate in juveniles can be attributed to the sand and other inorganic materials they ingest. Juvenile daily ration expressed as percent organic matter would be similar to the larval estimates.

In light of the high variability in the ration estimates for larval spot, we must consider our estimates tentative. The large variation is caused by the variability of natural but contents. Since zooplankton has patchy distribution and postlarvae are dependent on local plankton abundance, one would expect high variability in postlarval feeding rates.

A variety of factors such as food quality, temperature, and handling may have caused inaccuracies in our estimates of daily ration. We found no difference in evacuation rates for commercial food and natural foods (47). However, quality may affect evacuation rates under some conditions.

Small fluctuations from acclimation temperatures may drastically affect evacuation rate. Using *Menidia* and the serial slaughter method (60) we determined that a 2 °C increase above a 24 °C acclimation reduced evacuation time by nearly 30% (47). Using a radioisotopic technique (46), we found a similar temperature effect in sea bass (*Centropristis striata*). Five fish fed and maintained at 25 °C evacuated their food in 43 hours (SE = 3.4), while others maintained at 25 °C before feeding and 27 °C after feeding evacuated in 31.8 hours (SE = 1.6). Although small changes in temperature obviously affect evacuation rate, the average effect of daily temperature changes on evacuation in nature is not known.

Since handling may affect determinations of gut content (26), we adjusted our measurement of natural gut content in postlarval menhaden by a factor of 2.5, a figure suggested by our laboratory observations (Table 3). Inaccuracy of

TABLE 3

The effects of handling on the retention of **Artemia nauplii** in digestive tracts of larval Atlantic menhaden, pinfish and spot. Rough handling is approximately equivalent to field capture by dip net.

Species (range total) length mm)	Mean Number retained[1]			
	Experiment 1		Experiment 2	
	Gentle	Rough	Gentle	Rough
Menhaden (28-32)	71 ± 15	29 ± 10	145 ± 10	76 ± 11
Pinfish (16-20)	37 ± 4	34 ± 5	35 ± 9	43 ± 6
Spot (19-23)	51 ± 5	47 ± 5	89 ± 7	92 ± 10

1. ± one standard error

this factor would obviously affect our estimate of daily ration. Since rough handling did not cause loss of food by other postlarval or any juvenile fish, we assume our capture technique had no significant effect.

Although in the evacuation model we considered only the amount of food in the gut, well-digested material might be evacuated more rapidly than an equal amount of newly ingested food. If the evacuation model is appropriate, we should be able to calculate feeding periodicity from the periodicity of gut contents. Food content at the beginning and end of each 4-hour period differs by an amount equal to feeding minus evacuation. To estimate food consumption of juvenile fishes during 4-hour periods, we added the 4-hour food evacuation estimates to the changes in content between the beginning and end of the periods. Obviously, negative feeding rates calculated for pinfish and menhaden immediately after dark (Fig. 6) are impossible. The discrepancy may have been due to inaccuracy of the evacuation constants, random error in measurement of natural gut contents, the effect of light on evacuation (60), or faster evacuation of well-digested food (4). If the degree of food digestion is important, perhaps a new model that includes a term for earlier slow digestion and later more rapid evacuation is needed. Use of such a model would result in different estimates of feeding periodicity (Fig. 6) but might have little effect on estimates of daily ration.

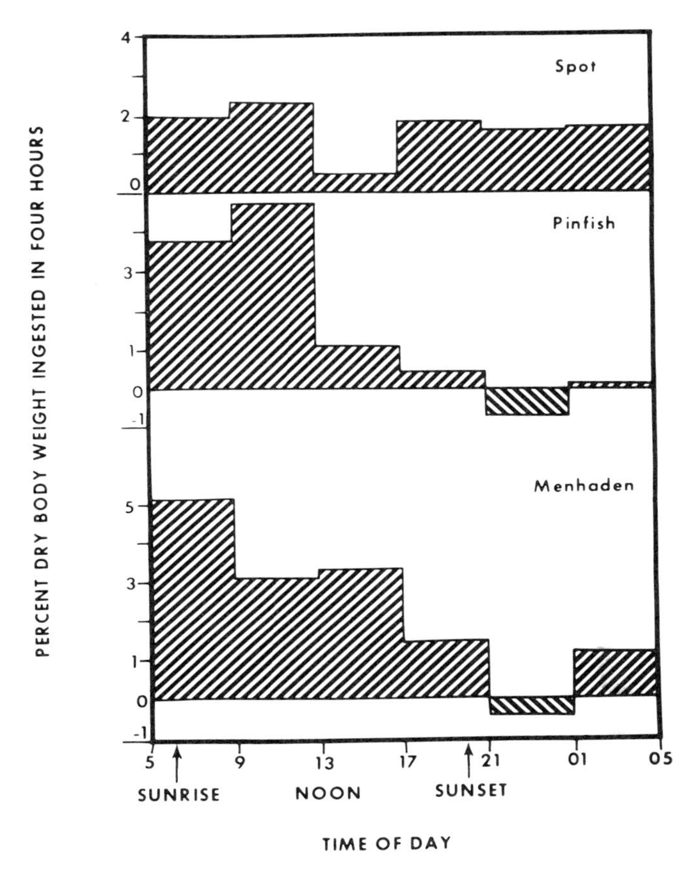

Figure 6. Diel cycle of food consumption (% dry body wt per 4 hr. period) by Atlantic spot, pinfish, and menhaden.

In spite of possible errors, our estimates of daily ration seem reasonable. Keast and Welsh (27) obtained minimum estimates of food-consumption rates which varied from 1.5% of body weight per day in *Fundulus* to 4.0% in *Ambloplites*. From data by Hoss (24) we calculated energy utilization rates for 25 g pinfish at 29 °C. In one day unfed fish respired 2.1% of their total energy content. Pinfish fed nearly to satiation once a day respire about 1.5 times as much as unfed fish (W. F. Hettler, personal communication) or about 3.2% of their body energy. Our estimate for pinfish ration was 9.5% of body weight per day, but much of this was ash; caloric intake was 4.5% of body energy. Assuming 3.2% of total energy goes for respiration, the remainder (28% of the ration) is either not assimilated or is laid down as new tissue. Depending on the assimilation efficiency, about 10%-20% of the energy ingested would be deposited as growth.

This estimate of conversion efficiency is consistent with reports on other rapidly growing young fish (5, 40). Little information is available concerning larval fish rations, but Rosenthal and Hemple (48) indicated that 20 mm herring larvae probably consume the equivalent of 100 *Artemia nauplii* per day. Since newly-hatched *Artemia* are one-half the size of copepods ingested by our larval fish (31), the herring larvae ingested the equivalent of 50 copepods, a number that agrees with our observations on spot, pinfish, and menhaden (Table 4).

Measurements of metabolic expenditures (Table 4) indicated several of our larval ration estimates were below maintenance levels, which may be indicative of natural food shortages. The menhaden estimate is highly dependent on the ill-defined factor used to adjust for handling effects. More accurate measurement of this conversion factor would probably provide better correlation with metabolic costs.

Our ration estimates do not indicate whether excess food energy is available to juvenile fish. The stomach of spot in nature never reaches the fullness obtained in laboratory feedings. In nature pinfish reached the high gut-content levels observed in the laboratory but did not maintain them throughout the day. Thus, caloric intake in spot and pinfish is probably limited by food availability. W. E. Odum (36) concluded that mullet, which are primarily detritivores, are not limited by food availability. The same may be true of menhaden.

TABLE 4

Daily rations calculated from feeding studies and routine O_2 consumption measurements at 16 °C for larval Atlantic mehaden, pinfish and spot in the Newport River estuary, North Carolina.

Species (range total length mm)	Number copepods/ fish/day	Percent of body weight	Calories/ fish/day	Calories/ fish/day estimated from O_2 consumption[1]
Menhaden (27-32)	53	4.9	1.2	3.0
Pinfish (16-20)	38	3.5	0.6	1.2
Spot (1972) (17-23)	47	4.3	0.8	1.2
Spot (1973) (17-23)	99	9.0	1.6	1.2

1. From Hettler and Hoss, unpublished data.

UTILIZATION OF FOOD FOR GROWTH

Given sufficient information on food consumption and natural ingestion rates, we can raise fish in the laboratory and obtain approximately natural growth rates under a variety of environmental conditions. The relative growth values of such tested environments will differ among species, but if known, could be useful in estuarine management. If growth in the laboratory is to reflect growth under natural conditions, the food must approximate natural food quality. The diet of a carnivore can be approximated in the laboratory, but estuarine detritivores and omnivores have complex diets that still are not adequately described. For this reason and for the convenience of handling discrete organismic food particles, our studies relating growth to environmental quality have been confined to carnivores.

Environmental factors may have different effects on processes such as assimilation, excretion, and respiration, but these effects need not be measured in assessing the relative growth value of a particular environment. To evaluate productivity only two measurements are needed, the amount of food consumed and total growth.

Environmental Effects on Feeding Rate

Size-specific feeding rates depend on the species and on physicochemical conditions and decrease as the fish grow. Within a restricted size range, maximum feeding rate is highly dependent on temperature. Food consumption increases with increasing temperatures until temperatures near the upper lethal limit are reached; it then declines (5).

A dominant temperature influence regulates feeding rate of juvenile (100 mg) southern flounder (*Paralichthys lethostigma*). We offered unlimited food to three replicate groups of three fish using previously described methods and experimental design (44). The feeding rates measured are summarized in Figure 7 and by the equation:

$$Y = -38.9 + 4.58T - .041T^2 + .089S + .008S^2 - .025TS \qquad (4)$$

where Y = % body weight ingested, T = temperature (^{o}C), and S = salinity ($^{o}/oo$). Analysis of variance indicated that the linear temperature term could account for 87% of the total variation in feeding rate. The optimum temperature for maximum feeding probably occurs above 30 ^{o}C, the highest temperature tested. In addition, the optimum temperature and the maximum feeding rate at the optimum temperature probably decrease as the fish grow (40, 41). The overriding temperature effect on feeding rate in southern flounder has also been shown in two other estuarine flatfish, summer flounder and hogchoker (44, 45),

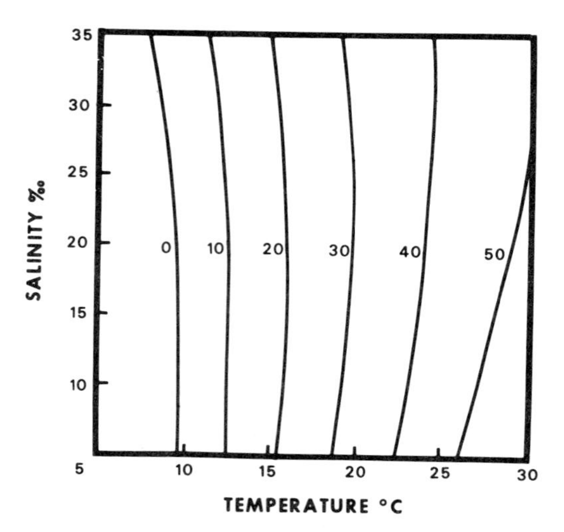

Figure 7. Estimation of per cent body weight ingested per day by **Paralichthys lethostigma** at **ad libitum** feeding based on the fitted response to three replications of nine observed combinations of temperature and salinity.

and in the Atlantic croaker (42).

Salinity may influence feeding rate (30) but in some species its importance is apparently minor. For example, salinity and other factors could account for only 5% of the variation in hogchoker feeding rate (45). In flounder a slight salinity difference is predicted at high temperatures, southern flounder consuming more at low salinity (Fig. 7) and summer flounder more at high salinity (44). Preliminary measurements of food evacuation rate by menhaden have not shown a significant salinity effect (Peters, unpublished data). Thus, unless stomach fullness changes with salinity, menhaden feeding rates would probably show little salinity effect.

The *ad libitum* feeding rate determined in the laboratory predicts the amount of food a fish would consume in nature if sufficient food were available. Unfortunately, effects of temperature, salinity and other factors on food availability in nature remain uncertain. Biomass data exist for some food types, but availability of food to the fish also depends on behavioral and other characteristics of both fish and food.

Growth Rate

Efforts to determine what effects environmental quality have on growth rate are complicated by the dependence of growth on feeding rate and ultimately on food availability. Experiments showing environmental effects on growth should include various feeding levels which encompass the likely feeding rate under altered conditions. Although it is not possible to predict growth rates, the way in which growth is affected can be shown.

The effects that temperature, salinity, and feeding level have on growth rate of southern flounder are shown in Figure 8. Temperatures ranged from 10-30 ^{o}C, salinities from 5 to 35$^{o}/oo$, and feeding rates from 30 to 90% of the *ad libitum* feeding rate (Fig. 7). Six fish were used in each of 20 treatments. Methods, desired experimental design, and approximate deviations from the design have been published (44). The regression relating caloric growth rate to the experimental variables was:

$$Y = .26758 + .01190T - .00027T^2 + .00578S + .00001S^2$$

$$-.00019TS + .00175F - .00002F^2 + .00015TF - .00004SF \qquad (5)$$

where Y = percent increase per day, T = temperature (^{o}C), S = salinity ($^{o}/oo$), and F = % *ad libitum* feeding rate. The high rates shown here decrease as the fish grows and the size-specific feeding rate declines. Such a decline in size-specific growth rate has been shown in young salmon (50).

Temperature and feeding rate were important in regulating growth. A regression containing only T and T^2 terms could account for 68% of the total variation. Inclusion of the additional regression terms, S, S^2, and TS increased R^2 to only 0.71. The optimum temperature is evidently above 30 ^{o}C except at feeding levels below those shown (Fig. 8). Growth increased with rations, as is indicated by the response surface diagram. The importance of food availability in predicting growth rate is shown by R^2, which increases from 0.71 without feeding terms to 0.95 for the equation shown, which includes the feeding terms. Fastest growth occurred at high temperatures, high feeding rates, and low salinities.

Conversion Efficiency

Both temperature and salinity influence conversion efficiency of southern flounder. The response surface diagram describing efficiency (Fig. 8) was calculated from the equation:

$$Y = -1.84973 + .08128T - .00091T^2 + .02690S - .00019S^2$$

$$- .00087TS + .02934F - .00014F^2 - .00032TF - .00001SF \qquad (6)$$

where Y = percent of assimilated food converted to growth, and the other terms are as defined earlier. Efficiencies reached 40% under favorable conditions and agree with determinations in other euryhaline fishes (30, 40, 44, 45). Although an optimum temperature is expected for maximum efficiency (5, 30, 41) in

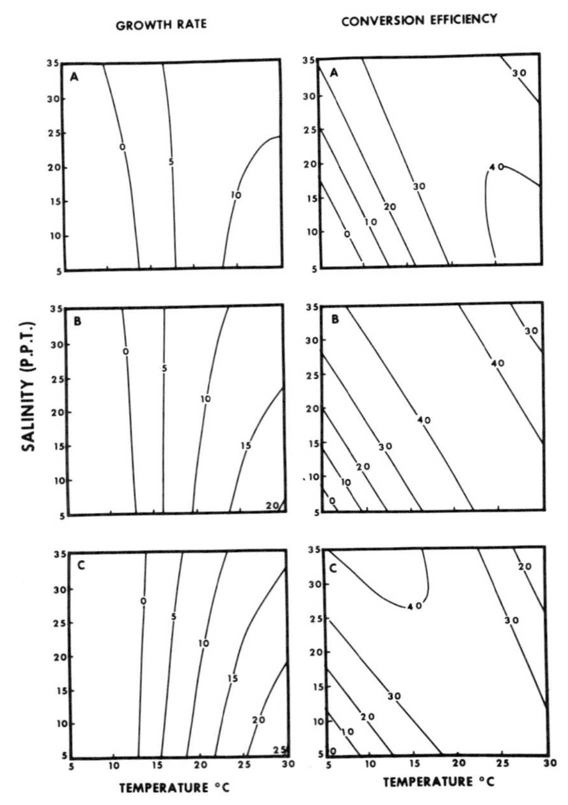

Figure 8. Estimation of caloric growth rate (per cent increase per day) and conversion efficiency in **Paralichthys lethostigma** based on the fitted responses observed under 15 different combinations of temperature and salinity and feeding level. A, B, and C equal 60%, 80% and 100% **ad libitum** feeding, respectively.

southern flounder, the optimum depends on salinity. As salinity decreases, temperature for maximum efficiency increases. This temperature-salinity interaction also occurs in summer flounder and hogchoker (44, 45).

Feeding rate is also important in determining conversion efficiency (5, 28). Adding feeding rate terms to an equation containing temperature and salinity terms increased R^2 from 0.45 to 0.80. Maximum efficiency occurred between 70% and 90% of maximum feeding rate (Fig. 9). Temperature showed little effect on the rate required for maximum efficiency. However, if we expressed food as per cent of body weight, the temperature effect would appear differently. Decreasing temperature decreases food consumption. Since maximum efficiency remains at nearly the same percentage of maximum consumption, the per cent of body weight which produces maximum efficiency

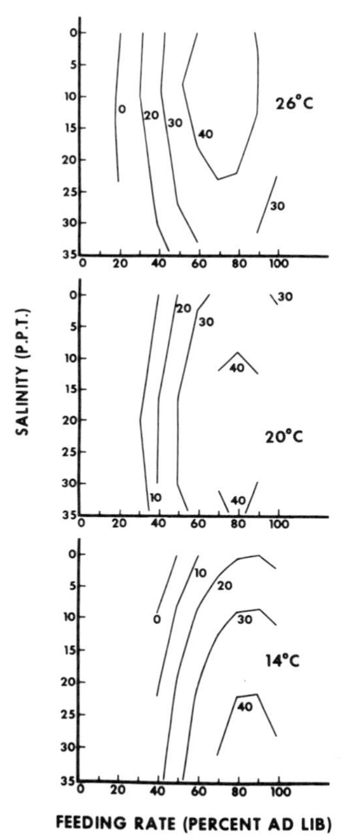

Figure 9. Estimation of caloric conversion efficiency in **Paralichthys lethostigma** based on the fitted response observed under 15 different combinations of temperature, salinity, and feeding levels.

should decline at nearly the same rate as appetite. Thus, feeding rate for maximum efficiency declines drastically with temperature or is relatively unaltered by it, depending on whether the feeding units are per cent of body weight or per cent *ad libitum* feeding rate.

BIOENERGETIC APPROACH TO FISH MIGRATION OR DISTRIBUTION

If growth in the natural environment is affected by the same factors we have observed in the laboratory, we should expect animals to possess bioenergetic adaptations to help them cope with these influences. For example, Brett (3) postulated that sockeye salmon maximize their growth by migrating daily between cold and warm water. From the maximum power hypothesis (35), which suggests that ideal conditions are those yielding maximum growth, we might expect flounder to occur most frequently in areas where growth is expected to be most rapid.

Both species of paralichthid flounders studied occur most frequently in areas which our data indicated would produce maximum growth, provided sufficient food was available. Both species migrate seasonally through a salinity gradient. Southern flounder move to low-salinity estuarine water in summer and back to high-salinity water in the winter (20, 53; W. Turner and G. Johnson, unpublished data). Summer flounder also move from the ocean to the estuary in the summer, but they remain in the high-salinity portion of the estuary (10, 38, 53; W. Turner and G. Johnson, unpublished data). Because of temperature effects, the inshore-offshore migration should increase growth the shallow inshore areas are warmer than offshore water in the summer and colder in the winter. Since increased temperature increases feeding and growth rates, inshore waters would be preferable in the summer and offshore areas in the winter.

Salinity effects are probably more important in flounder distribution during summer than during winter. The summer distribution patterns of the two species places them in different salinities which would maximize growth of each (43). The beneficial effects of low salinity on growth of southern flounder at high temperatures has been retested and verified (52). Deubler (16) and Deubler and White (17) indicated that cold water produced slightly better growth of both species at high salinities. Our observations generally indicate the same slightly beneficial effect of high salinity at low temperatures. Thus, the flounder which are offshore in the winter, probably because of temperature effects, may also benefit slightly from the higher salinity.

Southern and summer flounder exhibit the same temperature-salinity interaction affecting conversion efficiency. The southern flounder migrates in a pattern compatible with this interaction, but summer flounder does not. Movements of southern flounder maintain it in salinities that our data indicated would maximize conversion efficiency. Summer flounder do not move to low salinity and high temperature, which would increase their efficiency. Hickman (23) showed that seasonal changes in southern flounder osmoregulatory processes are related to the seasonal movement. However, there is no proof that changing osmoregulation costs are responsible for the conversion efficiency pattern.

Feeding rate information from the field and growth experiments on larger fish will be needed to establish a relationship between growth and distribution. The movement of postlarval flounder into estuaries in the winter appears disadvantageous, since colder estuarine water should reduce growth. Hogchokers also move into lower-salinity water for the winter, which reduces their food conversion efficiency (45). These otherwise unexplained movements may be in response to food supply, which overrides the temperature-salinity effects on conversion efficiency or growth. Since we measured growth only of small fish, we cannot be sure older fish would respond to temperature and salinity in the same way.

ACKNOWLEDGMENT

This research was supported jointly by the National Marine Fisheries Service and the U. S. Atomic Energy Commission under agreement no. AT(49-7)-5.

REFERENCES

1. Bajkov, A. D.
   1935 How to estimate the daily food consumption of fish under natural conditions. **Trans. Am. Fish. Soc.,** 65: 288-289.
2. Blaxter, J. H. S. and Holliday, F. G. T.
   1963 The behaviour and physiology of the herring and other clupeoids. In **Advances in marine biology,** vol. 1, p. 262-372. (ed. Russell, F. S.) Academic Press, New York.
3. Brett, J. R.
   1971 Energy responses of salmon to temperature. A study of some thermal relations in the physiology and freshwater ecology of sockeye salmon, **Oncorhynchus nerka. Am. Zool.,** 11(1): 99-113.
4. Brett, J. R. and Higgs, D. A.
   1970 Effect of temperature on the rate of gastric digestion in fingerling sockeye salmon, **Oncorhynchus nerka. Jour. Fish. Res. Bd. Canada,** 27: 1767-1779.
5. Brett, J. R., Shelbourn, J. E., and Shoop, C. T.
   1969 Growth rate and body composition of fingerling sockeye salmon, **Oncorhynchus nerka,** in relation to temperature and ration size. **Jour. Fish. Res. Bd. Canada,** 26: 2363-2394.
6. Burkholder, P. R.
   1956 Studies on the nutritional value of **Spartina** grass growing in the marsh area of coastal Georgia. **Bull. Torrey Bot. Club,** 83(5): 327-334.
7. Caldwell, D. K.
   1957 The biology and systematics of the pinfish **Lagodon rhomboides** (Linnaeus). **Bull. Florida State Mus.,** 2: 77-174.
8. Carline, R. F. and Hall, J. D.
   1973 Evaluation of a method for estimating food consumption rates of fish. **Jour. Fish. Res. Bd. Canada,** 30(5): 623-629.
9. Cowey, C. B. and Sargent, J. R.
   1972 Fish nutrition. In **Advances in marine biology,** p. 383-492. (ed.

Russell, F. S. and Yonge, M.) Academic Press, New York.

10. Dahlberg, M. D.
1972 An ecological study of Georgia coastal fishes. **U. S. Natl. Ma Fish. Serv., Fish. Bull.,** 70(20): 323-353.

11. Darnell, R. M.
1958 Food habits of fishes and larger invertebrates of Lak Pontchartrain, Louisiana, an estuarine community. **Publ. Ins Mar. Sci.** (Texas)), 5:353-416.

12. Darnell, R. M.
1961 Trophic spectrum of an estuarine community, based on studie of Lake Pontchartrain, Louisiana. **Ecology,** 42(3) 553-568.

13. Darnell, R. M.
1967 Organic detritus in relation to the ecosystem. In **Estuaries,** p 376-382. (ed. Lauff, G. H.) Am. Assoc. Adv. Sci. Publ. 83 Washington, D. C.

14. Darnell, R. M. and Wissing, T. F.
In press Nitrogen turnover and food relationships of the pinfish, **Lagodo rhomboides,** in a North Carolina estuary. In **Proceedings of th conference on physiological ecology of estuarine organisms** (Belle W. Baruch Coastal Res. Inst.) Univ. South Carolina Press Columbia.

15. de la Cruz, A. A. and Gabriel, B. C.
1973 Caloric, elemental, and nutritive value changes in decomposin **Juncus roemerianus** leaves.. **Assoc. Southeastern Biol. Bull.** 20(2): 47. (Abstract.)

16. Deubler, E. E.
1960 Salinity as a factor in the control of growth and survival of postlarvae of the southern flounder, **Paralichthys lethostigma. Bull. Mar. Sci. Gulf Carib.,** 10: 338-345.

17. Deubler, E. E. and White, J. C.
1962 Influence of salinity on growth of postlarvae of the summer flounder, **Paralichthys dentatus. Copeia,** 1962: 468-469.

18. Elliot, J. M.
1972 Rates of gastric evacuation in brown trout, **Salmo trutta** L. **Freshwater Biol.,** 2: 1-18.

19. Gerking, S. D.
1962 Production and food utilization in a population of bluegill sunfish. **Ecol. Monogr.,** 32: 31-78.

20. Gunter, G.
1945 Studies on marine fishes of Texas. **Publ. Inst. Mar. Sci.,** (Texas), 1(1): 1-90.

21. Haug, A., Myklestad, S., and Sakshaug, E.
1973 Studies on the phytoplankton ecology of the Trondheimsfjord. I. The chemical composition of phytoplankton populations. **Jour. Exp. Mar. Biol. Ecol.,** 11: 15-26.

22. Herke, W. H.
1971 Use of natural and semi-impounded Louisiana tidal marshes as nurseries for fishes and crustaceans. Doctoral dissertation, Louisiana State Univ., Baton Rouge. 241 p.

23. Hickman, C. P.
1968 Glomerular filtration and urine flow in the euryhaline southern flounder, **Paralichthys lethostigma,** in seawater. **Canadian Jour. Zool.,** 46: 427-437.

24. Hoss, D. E.
In press Energy requirements of a population of pinfish, **Lagodon rhomboides** (Linnaeus). **Ecology.**

25. Jacquot, R.
1961 Organic constituents of fish and other aquatic animal foods. In **Fish as food: production, biochemistry and microbiology.** Vol. 1: p. 145-210, (ed. Borstram, G.), Academic Press, New York.

26. June, F. C. and Carlson, F. T.
1971 Food of young Atlantic menhaden, **Brevoortia tyrannus,** in relation to metamorphosis. **U. S. Natl. Mar. Fish. Serv., Fish. Bull.,** 68: 493-512.

27. Keast, A. and Welsh, L.
1968 Daily feeding periodicities, food uptake rates, and dietary changes with hour of day in some lake fishes. **Jour. Fish. Res. Bd. Canada,** 25(6): 1133-1144.

28. Kerr, S. R.
1971 Analysis of laboratory experiments on growth efficiency of fishes. **Jour. Fish. Res. Bd. Canada,** 28: 801-808.

29. Kevern, N. R.
1966 Feeding rate of carps estimated by a radioisotopic method. **Trans. Am. Fish. Soc.,** 95: 363-371.

30. Kinne, O.
1960 Growth, food intake, and food conversion in a euryplastic fish exposed to different temperatures and salinities. **Physiol. Zool.,** 33(4): 288-317.

31. Kleiber, M.
1961 **The fire of life - an introduction to animal energetics.** John Wiley, New York. 454 p.

32. Krey, J.
1967 Detritus in the ocean and adjacent sea. In **Estuaries,** p. 389-394. (ed. Lauff, G. H.) Am. Assoc. Adv. Sci. Publ. 83. Washington, D. C.

33. Molnar, G., Tamassy, E., and Tolg, I.
1967 The gastric digestion of living predatory fish. In **The Biological basis of freshwater fish production,** p. 135-149. (ed. Gerking, S. D.). Blackwell Scientific Publications, Oxford.

34. Odum, E. P. and de la Cruz, A.
1967 Particulate organic detritus in a Georgia salt marsh ecosystem. In **Estuaries,** p. 383-388. (ed. Lauff, G. H.) Am. Assoc. Adv. Sci. Publ. 83. Washington, D. C.

35. Odum, H. T. and Pinkerton, R. G.
1955 Time's speed regulator: the optimum efficiency for maximum power output in physical and biological systems. **Am. Sci.,** 43: 331-343.

36. Odum, W. E.
1970 Utilization of the direct grazing and plant detritus food chains by the striped mullet, **Mugil cephalus.** In **Marine food chains,** p.

222-240. (ed. Steele, J. H.) Univ. California Press, Berkeley and Los Angeles.

37. Odum, W. E., Zieman, J. C., and Heald, E. J.
1973 The importance of vascular plant detritus to estuaries. In **Proceedings of the coastal marsh and estuary management symposium,** p. 91-114. (ed. Chabreck, R. H.) Louisiana State Univ. Div. Continuing Education, Baton Rouge.

38. Pacheco, A. L. and Grant, G. C.
1965 Studies of the early life history of Atlantic menhaden in estuarine nurseries. Part I. Seasonal occurrence of juvenile menhaden and other small fishes in a tributary creek of Indian River, Delaware, 1957-58. **U. S. Fish. Wildl. Serv., Spec. Sci. Rept. Fish.** 504. 32 p.

39. Parsons, T. R.
1963 Suspended organic matter in sea water. **Prog. Oceanogr.,** 1: 205-239.

40. Pandian, T. J.
1967 Intake, digestion, absorption and conversion of food in the fishes **Megalops cyprinoides** and **Ophiocephalus striatus. Mar. Biol.,** 1: 16-32.

41. Pandian, T. J.
1970 Intake and conversion of food in the fish **Limanda limanda** exposed to different temperatures. **Mar. Biol.,** 5:1-17.

42. Pendleton, E. C.
1973 Effects of temperature on growth, feeding efficiency, and thermal tolerance of juvenile croakers (**Micropogon undulatus**). M. S. thesis, North Carolina State Univ., Raleigh.

43. Peters, D. S.
1971 Growth and energy utilization of juvenile flounder, **Paralichthys dentatus** and **Paralichthys lethostigma,** as affected by temperature, salinity and food availability. Doctoral dissertation, North Carolina State Univ., Raleigh. 69 p.

44. Peters, D. S. and Angelovic, J. W.
1971 Effect of temperature, salinity and food availability on growth and energy utilization of juvenile summer flounder, **Paralichthys dentatus.** In **Proceedings of the third national symposium on radioecology,** p. 545-554. (ed. Nelson, D. J.) (U. S. Atomic Energy Commission.) Clearinghouse for Federal Scientific and Technical Information, U. S. Dept. Commerce, Springfield, Virginia.

45. Peters, D. S. and Boyd, M. T.
1972 The effect of temperature, salinity and availability of food on the feeding and growth of the hogchoker, **Trinectes maculatus** (Bloch and Schneider). **Jour. Exp. Mar. Biol. Ecol.,** 7: 201-207.

46. Peters, D. S. and Hoss, D. E.
1974 A radioisotopic method of measuring food evacuation time in fish. **Trans. Am. Fish. Soc.** 103: 626-629.

47. Peters, D. S., Kjelson, M. A., and Boyd, M. T.
1974 The effect of temperature on digestion rate in the pinfish, **Lagodon rhomboides;** spot, **Leiostomus xanthurus;** and silverside,

Menidia menidia. Proc. 26th Ann. Conf. Southeastern Assoc. Game Fish Comm. p. 637-643.

48. Rosenthal, H. and Hempel, G.
 1970 Experimental studies in feeding and food requirements of herring larvae (**Clupea harengus** L.). In **Marine food chains,** p. 344-364. (ed. Steele, J. H.) Univ. California Press, Berkeley and Los Angeles.

49. Schumann, G. O.
 1965 Some aspects of behavior in clupeid larvae. **Calif. Coop. Oceanic Fish. Invest. Rept.,** 10: 71-78.

50. Shelbourn, J. E., Brett, J. R., and Shirahata, S.
 1973 Effect of temperature and feeding regime on the specific growth rate of sockeye salmon fry **Oncorhynchus nerka,** with a consideration of size effect. **Jour. Fish. Res. Bd. Canada,** 30: 1191-1194.

51. Sick, L. V.
 1970 The nutritional effect of five species of marine algae on the growth, development and survival of the brine shrimp, **Artemia salina.** Doctoral dissertation, North Carolina State Univ., Raleigh. 43 p.

52. Stickney, R. R. and White, D. B.
 In press Effects of salinity on the growth of **Paralichthys lethostigma** postlarvae reared under aquacultural conditions. **Proc. 27th Ann. Conf. Southeastern Assoc. Game Fish Comm.** 1973.

53. Tagatz, M. E. and Dudley, D. L.
 1961 Seasonal occurrence of marine fishes in four shore habitats near Beaufort, N. C., 1957-60. **U. S. Fish Wildl. Serv., Spec. Sci. Rept. Fish.** 390. 19 p.

54. Thayer, G. W., Kjelson, M. A., Hoss, D. E., Hettler, W. F., Jr., and LaCroix, M. W.
 In press Influence of postlarval fishes on the distribution of zooplankton in the Newport River estuary. **Chesapeake Sci.**

55. Thayer, G. W., Schaaf, W. E., Angelovic, J. W., and LaCroix, M. W.
 1972 Caloric measurements of some estuarine organisms. **U. S. Natl. Mar. Fish. Serv., Fish. Bull.,** 71(1): 289-296.

56. Turner, W. R. and Johnson, G. N.
 1973 Distribution of relative abundance of fishes in Newport River, North Carolina. **U. S. Natl. Mar. Fish. Serv., Spec. Sci. Rept. Fish.** 666. 23 p.

57. Tyler, A. V.
 1970 Rate of gastric emptying in young cod. **Jour. Fish. Res. Bd. Canada,** 27: 1177-1189.

58. Vinogradov, A. P.
 1953 **The elementary chemical composition of marine organisms.** Sears Found. Mar. Res. Mem. 2. (trans. from Russian.)

59. Winberg, G. G.
 1956 Rate of metabolism and food requirements of fishes. **Fish. Res. Bd. Canada, Transl. Ser.** 194. 202 p. (trans. from Russian.)

60. Windell, J. T.
 1967 Rate of digestion in fishes. In **The biological basis of fresh water fish production,** p. 151-173. (ed. Gerking, S. D.) Blackwell

Scientific Publications, Oxford.

61. Wood, E. J. F.
1955 Fluorescence microscopy in marine microbiology. **Jour. Cons. Int. Explor. Mer,** 21: 6-7.

SOME ASPECTS OF FISH PRODUCTION AND CROPPING IN ESTUARINE SYSTEMS

Saul B. Saila[1]

ABSTRACT

The objectives of the study are to review some aspects of estuarine dynamics and productivity, to briefly describe requirements of estuarine fishes with a view to enhancing production, and to consider in a quantitive way some aspects of estuarine fish production and cropping.

The review of estuarine dynamics demonstrateds the high desirability of combining a knowledge of physical processes in estuaries with detailed knowledge of the behavior of organisms in order to make reasonable predictions about the consequences of environmental modifications. In many instances the requirements of fishes with respect to physiology and ecology are still not well enough known for predictably enhancing estuarine fish production. However, more intensive rearing of fish in estuaries seems promising if this effort is preceded by adequate knowledge of the organisms, the environment, and the possible consequences of overenrichment.

Application of some simple models of production and cropping are believed to be adequate for the purposes of computing fish production in estuaries and for maximizing the harvestable crop. Descriptions of these models are provided.

INTRODUCTION

The objectives of this report are to provide some background material and to describe some recent advances in estuarine studies, particularly as related to the

1. Graduate School of Oceanography, University of Rhode Island, Kingston, Rhode Island 02881.

nekton community. Specifically, the following material is considered: (a) a brief review of some aspects of estuarine dynamics and productivity from a fisheries point of view, (b) estuarine fishes and their requirements with a view to enhancing fish production in estuaries, and (c) some aspects of estuarine fish production and harvesting strategies.

The total biological productivity of estuaries can be very high and can support dense populations of animals, usually detritus-feeding benthic invertebrates. Some of these invertebrates, such as oysters and clams, are of high economic value and support active commercial fisheries. However, this report is concerned with nekton, defined as actively swimming pelagic organisms, which, as pointed out by McHugh (37), are dominated in numbers and biomass by fishes. In addition to the specific objectives of this report, an attempt is made to supplement portions of chapter IX of a previous symposium, Publication 83 of the American Association for the Advancement of Science entitled *Estuaries,* as well as portions of the American Fisheries Society's Special Publication No. 3 entitled *A symposium on Estuarine fisheries.* These studies include economically important invertebrate and vertebrate species.

ESTUARINE TYPES AND DYNAMICS

A number of definitions of an estuary have been proposed. Day (12), Ketchum (32) and McHugh (37) have considered any region as an estuary if it contains dilute sea water. This definition is meaningful from a biological point of view, because the term *estuarine* refers to marine organisms capable of living under a wide range of diluted salinities. However, the definition contains only one effective boundary, the coastline, and thus has limitations for such other applications as physical oceanography. When one is interested in estuarine nekton a somewhat more rigorous definition is often required. For example, a problem such as the quantitative estimation of the longitudinal dispersion of fish eggs or larvae requires another definition of an estuary. For such a problem a definition proposed by Pritchard (48) is suggested: "An estuary is a semi-enclosed body of water which has a free connection with the open sea and within which sea water is measurably diluted with fresh water derived from land drainage". McHugh (36) adopts the view that both the biologist's and oceanographer's definitions are useful when considering estuarine nekton. However, as Hickling (27) pointed out, these definitions are more applicable to north temperate conditions than to tropical or subtropical conditions.

By Pritchard's definition an estuary is a coastal body of water and excludes such bodies as the Mediterranean and Baltic seas. The conditions necessary for an estuary are the following: (a) There must be some mixing of sea water and fresh water, which have essentially different concentrations of dissolved salts. It is recognized that this mixing is relatively small in a typical salt-wedge estuary.

(b) Circulation must be strongly influenced by the presence and the shape of boundaries. (c) The area must be a coastal feature.

Estuaries may be classified in several ways. On a geological basis these are drowned river valleys, fjords, bar-built estuaries, and estuaries produced by tectonic processes.

The above classification is sometimes replaced by others developed by physical oceanographers. Stommel (57) has suggested that the causes of water motion, i.e., the wind, tide, and river flow, in estuaries be used as a classifying principle. On the basis of Stommel's suggestion Pritchard (47) developed a classification of estuaries based on the advection-diffusion equation, the so-called salt-balance equation. This classification provides for four kinds of estuaries based on circulation: type A or salt wedge estuary, which is highly stratified, with a wedge of saline water lying beneath outflowing low-salinity water; type B or partially mixed estuary; type C or vertically homogenous estuary but with a lateral gradient caused by the rotation of the earth; and type D, a sectionally homogenous estuary which retains only a longitudinal gradient from the sea to fresh water.

Actually many estuaries lie somewhere in between the types listed above, because they vary continously in many characteristics. Four physical factors affect the type of estuary: river flow, tidal current, estuarine width, and estuarine depth. A decrease in river flow or in estuarine depth will tend to move the estuary from type A to type D. Similarly, a decrease in tidal current or width will move the estuary from type D toward type A. Since two of the physical parameters are flows and the other two are lengths, Pritchard sketched the sequence of estuarine types as a function of two dimensionless parameters: L, which is defined as the depth of the estuary divided by its width; and V, which is defined as the strength of the tidal current divided by river flow. A schema for this presentation is:

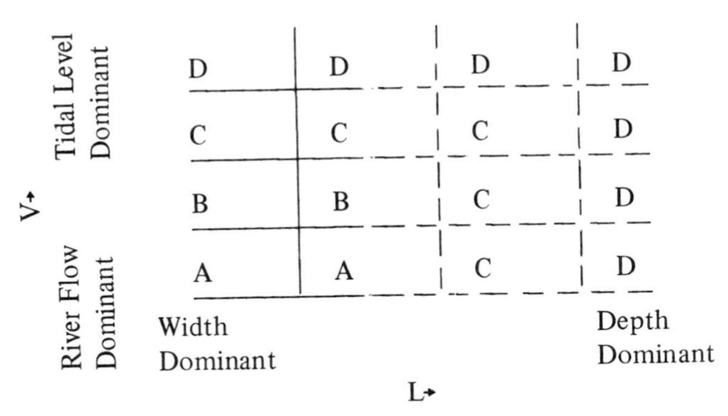

From this schema it is evident that the type A estuary is highly stratified and mixing is minimal; this type of estuary is found in regions with a small tidal range and a large river flow. When the tidal range is increased and the tidal flow involves volumes in excess of inflow, the estuary is less strongly stratified. In the northern hemisphere the rotation of the earth causes the water to move to the right side of an estuary as it flows to the sea, and inflowing saline water then comes nearer the surface on the left-hand side. This induces a somewhat tilted boundary typified by Pritchard's type B estuary. In type C estuaries there is usually no vertical stratification, but there are salinity gradients. Type C estuaries are relatively wide and have fairly high tidal velocities. If the estuary is sufficiently narrow the lateral gradient is destroyed by lateral mixing; type D estuaries thus lack both vertical and lateral salinity gradients and retain only the longitudinal gradient. It seems important to biologists concerned with estuarine flushing problems related to pollution and the dispersion of fish eggs and larvae to recognize that estuarine systems, in increasing order of complexity are: (a) the completely mixed estuary; (b) the segmented estuary completely mixed within each segment; (c) the estuary homogenous in cross-section but continuously varying in longitudinal direction, and (d) an estuary homogenous along a line at a given depth across the estuary but varying vertically and along the long axis.

Tidal currents in estuaries fall into three types according to Caldwell (5), whose classification relates to the nature of the entrance. When the entrance is large enough not to impede the progress of the oceanic tidal wave, the current strength and timing in relation of tides will depend on the length of the tidal wave. The length of the tidal wave in an estuary can be roughly estimated from the following expression given by Caldwell:

$$L = 48.1 \sqrt{d} \tag{1}$$

where L is the length in miles of a 12-hour, 25-minute tide from high tide to high tide, and d is the mean depth of the estuary measured in feet.

In Caldwell's first category the length of the tidal influence in an estuary exceeds a quarter of tidal wave-length and there is no restriction at the mouth; thus the incoming tide will flow unimpeded into the estuary with the fastest currents expected at high tide and slack water at mid-tide. In the second category the length of the area of tidal influence is less than one-quarter of the tidal wave-length. The inflowing water will pile up against the head of the area influenced by tides and the strength of the currents will be reduced before high tide. In this category the fastest current occurs at about mid-tide and slack water will occur at high tide. The third category of estuaries has a narrow mouth which impedes tidal inflow and the tidal range is smaller than in the adjacent open sea. Tidal currents in this case are fastest at high tide, because they depend on the hydraulic gradient through the mouth of the estuary, which is usually greatest at high tide. Information on tidal currents is considered a requisite of effective planning and execution of nekton sampling in estuaries.

The strength of estuarine currents also varies with the spring-neap tidal cycle. Spring tides occur when the relative positions of the sun and moon are most nearly in line, and neap tides result when the gravitational pulls of the sun and moon are nearly at right angles to each other. Nets used for fishing in estuaries may behave very differently and fishing effectiveness may vary considerably depending on currents, which are much higher during spring tides than neap tides.

The over-all productivity of estuaries and production of fish in estuaries are related to these water-mixing characteristics. Owing to density differences between the oceanic water entering near the bottom and river water leaving near the surface and the reduced velocities where the two kinds of water meet, suspended matter may become concentrated in certain portions of an estuary, usually near the mid-point. This concentration of suspended matter results in a zone of maximum turbidity which is of considerable biological significance. Dissolved as well as suspended materials may be involved in this maximum turbidity zone. Thus estuaries may form nutrient traps where much higher concentrations of nutrients are held in circulation than are in entering ocean or river waters. The phytoplankton crop is often not proportional to the increased nutrient loading because of the high turbidity. However, the total organic material consisting of primarily suspended detritus is extremely high and is directly related to the dense populations of detritus-feeding invertebrates. Among the estuarine nekton which are believed to use detritus directly are penaeid shrimp and the young of such fishes as menhaden, mullet, and pampano. The hydrologic features of estuaries which bring about accumulations of nutrients may also be responsible for some of the fertilizing effects of sewage discharged near the heads of estuaries. Mansueti (34), in a frequently cited and controversial paper, suggested that the influences of civilization, primarily through fertilization of the water by domestic sewage, may be indirectly responsible for the unusual increase in striped bass production in Chesapeake Bay. This general trend in increased catches of striped bass continued through 1969, the last year for which catch data are available.

Studies of the nutrient enrichment of Chesapeake Bay by Carpenter et al. (6) seems to increase justification for Mansueti's statement. These studies clearly showed enrichment of the upper bay by nitrogen and phosphorous which originated primarily from the Susquehanna River. Measurements of radio carbon productivity during the summer showed that about 0.2 μg of organic carbon were produced per liter per day in the upper layers of northern Chesapeake Bay. This rate of carbon fixation corresponds to an uptake of about 4 μg atoms of nitrogen per liter per day and 0.5 μg atoms of phosphorus per liter per day. The concentrations of these nutrients found by Carpenter et al. (6) were:

Inorganic nitrogen	8 μ g at liter^{-1}
Inorganic phosphates	0.1 μ g at liter^{-1}
Dissolved organic phosphate	0.8 μ g at liter^{-1}

The observed rate of carbon fixation would result in complete utilization of available nutrients in about 2 days. Since they do not disappear there is some indication of rapid renewal of phosphorous and nitrogen in the upper layer, and a removal of chlorophyll, probably through grazing by a large zooplankton population. This large zooplankton population may be responsible for the increased numbers of strong year-classes, because they serve as food during the initial early life-history stages of the striped bass. However, as indicated by Hedgpeth (26), additional uncontrolled enrichment by domestic sewage might be disastrous. It is believed that nutrient levels in the upper Chesapeake Bay are near the upper acceptable limits. Although phytoplankton production is high, the present grazing rate is believed to be high enough to prevent a significant build-up.

FATE OF ESTUARINE LARVAL NEKTON

The distinguishing feature of estuaries is the substantial influx of ocean water into a basin by means of solar and lunar tides. This influx introduces ocean salts, which mix with and are diluted by incoming fresh water from streams. The incursion of tides and the associated reversal of current are the primary mixing phenomena in most estuaries. The foregoing brief discussion of estuarine types referred to the relative magnitudes of the tidal current and the factors it depends upon.

The early life histories of several important fish species and bivalve mollusks consist of larvae found in estuaries. The fate of these larvae is significantly affected by the nature and magnitude of the advective and diffusive processes which result in larval dispersion. Cronin and Mansueti (8) have given some qualitative information on the mixing processes and life histories of estuarine fishes. However, quantitative data appear to be lacking.

Striped bass and winter flounder are two good examples of fish species of which the egg and larval stages are often found in estuaries. Important decisions concerning the construction of power plants in estuarine areas are dependent upon the predictions of potential entrainment losses of eggs and larvae. For example, the final environmental statement related to the operation of Indian Point Nuclear Power Plant No. 2 at Indian Point, New York, contains conflicting evidence concerning the magnitude and duration of entrainment losses of striped bass and their effects on the fishery. Other important decisions concerning man's utilization of estuaries for diverse purposes are equally dependent upon adequate knowledge of estuarine dynamics, population dynamics of organisms, and dispersal of adult, juvenile and larval fishes. Relatively little has been done to date to compare the importance of the various processes connected with estuarine larval dispersion. The following represents one effort in this direction.

The importance of advective dispersion depends upon the magnitude of water transport resulting from tidal forces, wind stress, and pressure gradients. A summary of some tidal dispersal coefficients computed for several estuaries by various techniques yields values ranging from 1-20 mi^2 (2.6-5.2 km^2) per day (59). The tidal dispersion coefficient decreases in general towards the head of an estuary.

Diffusion of planktonic estuarine organisms may result from turbulence (random bulk movements) and the movements of individuals. Diffusion is used herein in the context described by Pielou (44) who stated that it results from the undirected movements of animals that seem to be wholly random, analogous to Brownian motion. The radom self-propelled movement of individuals is termed nonturbulent diffusion. The following material has been taken with the permission of the authors from an unpublished manuscript by Sissenwine, Hess and Bearse of the University of Rhode Island, Graduate School of Oceanography. This study used the two-dimensional diffusion equation which can be derived from a random walk model:

$$\frac{\partial \phi}{\partial t}(x, y, t) = D \frac{\partial^2 \phi}{\partial x^2}(x, y, t) + D \frac{\partial^2 \phi}{\partial y^2} \quad (2)$$

where ϕ is the probability function for the position of a particle and D is the diffusion coefficient. Skellam (55) has determined that D is one-fourth the mean square displacement per unit time, which results in the following:

$$D = \frac{1}{4n} \Sigma r_i^2 / t_i \quad (3)$$

where r_i is the displacement during the interval t_i. The above equations demonstrate that the nonturbulent diffusion coefficient of a population may be estimated by determining the net movement of an organism from the population in an interval of time. Sissenwine et al. estimated the diffusion coefficients of winter flounder *Pseudopleuronectes americanus* and northern lobster *Homarus americanus* larvae from small samples of individuals under laboratory conditions. The results of these studies are admittedly preliminary, but nevertheless conservative. A value of 3.6 cm^2 s^{-1} with 95 percent confidence limits of 2.4-5.1 cm^2 s^{-1} was determined for winter flounder larvae which were 20 to 23 days old. The D value for lobster larvae 2-14 days old was found to be 1.9 cm^2 s^{-1} with 95 percent confidence limits of 0.64-1.74 cm^2 s^{-1}.

An empirical function expressing the turbulent diffusion coefficent of particles suspended in a fluid has been developed by Elder (15) from laboratory studies, and Csanady (9) has provided a detailed explanation of turbulent diffusion processes in the natural environment. Elder's results indicated that turbulent diffusion coefficients in nature should be typically greater than 1×10^4 cm^2 s^{-1}. Clearly nonturbulent diffusion coefficients estimated from winter flounder and lobster larvae are insignificant in contrast to the expected values for turbulent diffusion. That is, the estimated range of 2-4 cm^2 s^{-1} is nearly four orders of magnitude less than the turbulent diffusion at about 1×10^4 cm^2 s^{-1}. However, even the turbulent diffusion coefficient is small compared to the range for the coefficients of tidal dispersion, which have values of 3×10^5 to 6×10^6 cm^2 s^{-1} (1-20 mi^2 day^{-1}).

From the evidence it is suggested that models relating to the two-dimensional distribution of fish larvae in estuaries should include the coefficient of tidal dispersion and perhaps also turbulent diffusion, but may safely omit a term for the coefficient of nonturbulent diffusion.

A recent and extremely interesting empirical study by Graham (19) demonstrated a system of larval herring movements that retained them within the estuary despite its seaward residual flow. This carefully planned field study suggested that larval herring retained their position in the upper portion of the estuary by occupying landward net tidal flow near the bottom, moving upward upon reaching the limit of their landward penetration, then seaward in the net flow near the surface, and finally descending again into the landward flow. The results of this study suggested some form of vertical movement by herring larvae to maintain a preferred location in the estuary. A similar cyclic movement of the hogchoker *Trinectes maculatus* was reported in Chesapeake Bay by Dovel (14).

In summary, additional empirical studies of the type described by Graham are required for other estuarine species and additional information on the values of diffusivity coefficients for more species are needed before realistic models describing the fate of larval estuarine fishes can be developed. Furthermore, realistic models should involve a consideration of estuaries in three dimensions instead of two, as has been done to date. A further understanding of larval fish dispersion in estuaries will require more sophisticated models in combination with carefully planned and executed field studies.

ESTUARINE FISHES

Typically, estuarine fishes are composed of a mixture of endemic species (i.e., those restricted to the estuarine zone throughout their life histories), species which come in from the sea, and a relatively few species with osmoregulatory capabilities of penetrating to or from the fresh-water environment. Estuarine fishes, considered on a global basis, belong to a wide range of orders and families. Hildebrand and Schroeder (28), Dovel (14), Gunter (25), and Fierstine et al. (16) provide lists of species found in estuarine waters of the United States. Stroud (58) attempted to identify estuarine-dependent marine fishes common to coastal waters of the United States. He pointed out the need for more biological information, but suggested that about two-thirds of the annual catch of the Atlantic coast is made up of species believed to be estuarine-dependent. The bibliography of Srivastava (56) provides available information on the estuarine fishes of the Gulf of St. Lawrence, Canada. Pillay carefully listed the estuarine fishes of the Indian Ocean coastal zone (45) and of West African estuaries (46). Green (20) refers to some of the European species of estuarine fishes.

Available information on estuarine fishes tends to support the statement that faunal diversity is greater in tropical than in temperate latitudes. This phenomenon has already been demonstrated with marine snails along the east coast of the United States by Fischer (17). Table 1, which lists the number of fish species found in some estuaries along the east coast of the United States, shows that the diversity of estuarine fishes tends to increase as one proceeds from north to south. Presumably the better developed estuarine fish fauna in

TABLE 1

The number of species of fishes found in some estuarine areas along the Atlantic coast of the United States.

Location	Authority	Number of Species
Massachusetts (Quincy Bay)	Jerome, Chesmore and Anderson (30)	27
Massachusetts (Wellfleet Harbor)	Curley et al. (10)	35
Massachusetts (Pleasant Bay)	Fiske et al. (18)	36
New York (Great South Bay)	Briggs and O'Connor (3)	40
Delaware	deSylva et al. (13)	66
Georgia	Dahlberg and Odum (11)	70
Florida (Everglades Natl. Park)	Roessler (50)	103

southern latitudes reflects a more stable environment allowing for an essentially unbroken history of development and niche occupancy. McHugh (37) also indicated that the effect of low salinity on respiration and osmoregulation appeared to be less adverse at high temperatures than at low, tending to improve chances for a richer fauna in southern estuaries.

The important elements of the biology of fishes related to the estuarine environment are osmoregulatory ability, feeding habits, and breeding biology. Parry (43) has provided a comprehensive review of the osmoregulatory abilities of fishes. His breakdown of response to salinity changes includes: (a) species which survive only in conditions of isosmotic constancy, (b) species which tolerate some degree of environmental change by changes in the ionic concentration of their body fluids, (c) species or particular life-history stages of a species which may be semi-permeable allowing free water movement in or out in response to environmental changes, (d) species which reduce the permeability of their surfaces, and (e) species able to compensate actively for movements of water or solutes caused by external changes. Mechanisms available for the purposes of osmoregulation include excretion, swallowing the medium with active absorption of ions or water from the gut, and active absorption or excretion of specific ions by the whole epithelium or at specific sites. Green (20) stated that the outer reaches of estuaries do not present any major osmotic problems to many marine teleosts until the salinity falls to about one-third that of sea water.

McHugh (36), Clark (7), and Dovel (14) enumerated and listed the fish species considered to be estuarine-dependent at various life-history stages with partucular emphasis on the east coast of the United States. The list of fishes includes some 80 species that use estuaries for feeding or breeding. Massmann (35) pointed out important research needed for various estuarine-dependent sport fishes. Dovel (14) carefully summarized the known information concerning the breeding biology of important estuarine fishes found in the middle Atlantic region. He divided these fishes into fresh-water spawners, estuarine spawners, and marine species. As pointed out by McHugh (37) there seems to be no counterpart on the Pacific coast of the mass inshore and estuarine movement of larvae and juveniles of offshore spawning fishes into brackish nursery grounds. This phenomenon is a striking feature of the Atlantic coast and Gulf of Mexico estuaries. However, in Europe and Asia there are certain important species which seem also to use estuaries for early life-history stages. Important species with these kinds of life histories include many of the mullet species (family Mugilidae), the sea herring *Clupea harengus,* and the Bombay duck *Harpodon neherens.* Other estuarine organisms are listed by Hickling (27).

Hickling (27) has provided a recent and comprehensive review of estuarine fish-farming from a global point of view. His study includes descriptions of the imporatnt species cultured, sources of cultured organisms, food of cultivated organisms, and regional management plans for such diverse areas as Taiwan, Japan, Java, the Philippines, Italy, and India. Milne (38) has provided another overview, but with somewhat greater emphasis on European methodology. Vannuci (60) has reviewed the production potential of tropical lagoons and suggested that production could be enhanced by applying agricultural technology.

Although Walford and Wicklund (61) have generally condemned introduction and transplantation of fish species, several new fish species have been introduced into areas where food sources apparently were not efficiently exploited. Transplantation of such anadromous species as shad and striped bass to the west coast of the United States and salmon and trout from Britain to New Zealand are examples. The Aral Sea has apparently been colonized by Baltic herring, and a commercial fishery has been started as a result of transplantation (53).

The ability to tolerate rapid changes in the external environment must be coupled with the ability to regulate the internal environment if a fish is to become a permanent and successful estuarine dweller. Various flatfish species, such as the European flounder *Platichthys flesus* and the hogchoker *Trinectes maculatus,* are examples of truly estuarine fishes, because they are able to adapt very quickly to environmental changes. For these reasons some of the flatfishes may be desirable candidate species for enhancing fish production in estuaries, particularly in temperate regions. Saila (51) provided some quantitative estimates of the standing crop of winter flounder in some Rhode Island salt ponds. The contribution of these shallow estuarine areas was found to be very significant to the offshore fishery.

Detritus feeders are important benthic organisms in temperate estuaries. These include certain bivalve mollusks, crustaceans, and polychaetes. On both the European and the American continents there are flatfish species that feed on these organisms. Many of these flatfishes are of considerable interest to man as food and for sport. Some of these flatfish species are plaice, *Pleuronectes platessa;* European flounder, *Platichthys flesus;* dab, *Limanda limanda;* and brill, *Rhombus laevies,* from the European continent; and from the North American continent, winter flounder, *Pseudopleuronectes americanus;* summer flounder, *Paralichthys dentatus;* and hogchoker, *Trinectes maculatus.*

Some of the most successful experiments in raising marine fishes under controlled conditions have been carried out with plaice and flounder in Europe. A primary objective of Shelbourne's (54) artificial propagation studies of marine fish was to develop plaice-rearing to a mass-production technology. According to his analysis, one of the important ways in which hatchery output could be used is to release juvenile fish into partly-enclosed coastal waters (estuaries) in an attempt to establish new nursery grounds, preferably coupled with measures to increase the basic fertility of the area.

For plaice and perhaps for other estuarine-feeding flatfishes, the abundance of food acceptable to young fishes probably limits their production (biomass increase) on the nursery grounds. As already pointed out, enrichment of estuaries by domestic sewage has probably enhanced production of certain species. Both enclosed ponds and semienclosed bodies of water can be enriched by agricultural fertilizers. Intensive studies on two Scottish sea lochs by Gross (21, 22, 23, 24) demonstrated, not only an increase in the abundance of plankton and benthos, but also significant increases in the growth rate of transplanted and resident fish stocks. The concept of deliberately and predictably enriching estuarine nursery areas, specifically important flatfish growing areas, deserves much more careful study. A modeling study by Saila et al. (52) to determine the theoretical biomass of winter flounder necessary to increase significantly the fisheries in Rhode Island demonstrated that the apparent relationship between growth and mortality, on the basis of available data, was such that stocking juvenile (150-day-old) flounder would not be economically feasible in the study area without a consideration of other factors. However, this study did not involve estimating the effects of enhancing production by enrichment of nursery areas with nutrients or by reducing predation. Further studies are needed to better assess the feasibility of stocking estuaries to enhance production.

In summary, the potential for enhancing fish production in estuaries has not yet been fully realized. Flatfish species seem especially appropriate in temperate regions for experiments with controlled artificial enrichment, carefully designed transplantation experiments, predator control, and controlled breeding. Successful transplantation depends on a good understanding of the osmoregulatory abilities of the candidates species at various life-history stages. However, these techniques for enhancing production must be tempered with a thorough understanding of estuarine circulation and the ecological consequences of enhanced production, which includes reduced stability and overenrichment.

FISH PRODUCTION AND CROPPING IN ESTUARIES

In the previous sections some mention has been made of the biological productivity, or fertility, of estuaries. Productivity was considered to be the resultant of energy transfer and availability in the system, nutrient availability and uptake, species diversity, and ecosystem stability. It is possible to artifically enhance the fertility of estuaries and, at least in one instance, increased fish production was associated with enrichment of an estuary with sewage. The dangers of overenrichment have not been adequately documented. However, W. E. Odum (41) pointed out that intensive culture systems apparently are more easily disturbed by pollutants than natural ecosystems, and he warned of the possible adverse effects of intensive culture in estuaries.

If fish yields in estuaries are to be enhanced by any means, some understanding of and methods for quantifying the relations between the harvested crop (yield) and the production of the organisms is important. These relations should be based on direct estimates of growth and mortality, and there are methods for estimating these fairly precisely.

In estuaries, as in any other environment, the production of living matter is basically limited by inorganic nutrients, assuming conditions are otherwise suitable. In a simplified form a typical food web found in estuaries containing an exploited fish stock is:

inorganic nutrients → phytoplankton → benthic invertebrates → fish → man

Since fishes are near the right side in this schema, the indirect relationship between nutrients and fishes is apparent. Fish production and the harvest of fishes by man are limited by any factor which limits the production at lower levels on the food web. Thus, the relation between artificial enrichment of an aquatic environment and fish production is not as direct as is the relationship between fertilizers and terrestrial crop production.

The above simple schema can be elaborated in many ways, one of which involves model construction. Models involving energy transfer and an energy-circuit language developed by H. T. Odum (39) and applied by him (40) and Jannson (29) are being increasingly used to describe estuarine ecosystems. In addition, models of fish production developed by Paloheimo and Dickie (42) use ecological and physiological parameters related to the food supply of the organisms of interest.

This section briefly considers some quantitative relations only between fish and man in estuarine systems. These relationships assume that the cropping of fishes or other economically important organisms from estuaries will become increasingly intensive and somewhat analogous to fish farming or fish culture, as currently practiced. That is, the yield, or crop, will be derived from the stock during a relatively short interval of time and may form a considerable portion of total production under some circumstances. This is in contrast to most natural fisheries where fishing, or cropping, extends over a relatively long period of time and the yield is a smaller fraction of total production.

Studies by Ricker (49) and Allen (1, 2) provide much of the basis for the material that follows. The results of these studies seem appropriate for developing harvesting schemes for estuarine fishes under conditions of intensive culture.

For the purposes at hand, annual production is defined as the actual amount of new living matter produced during the year, either by growth of old individuals or by production and subsequent growth of new individuals, whether these individuals have survived to the end of the year or not. By this definition annual production is the algebraic sum of the amount dying or otherwise leaving the estuarine area and the increase in standing crop. It therefore follows that annual production bears no fixed relation to the standing crop at a given time or to the increase in the standing crop during the year.

Although the relationship between yields of fishes and fertility is not direct, it seems instructive to examine briefly the yields of fishes from brackish-water lagoons and estuaries on a global basis. Table 2, modified from data presented by

TABLE 2

Annual yields of fish from lagoons and estuaries without artifical feeding (modified from Hickling [27]). All author citations except those indicated by an asterisk are found in Hickling's original work.

Location	Author	Fish crop, kg hectare^{-1}
Étang de Biguglia, Morocco	Belloc, 1938	130
Russian limans	Ilin, 1954	About 50; but Pokrovski liman (60 acres), 33-750; Shabolatsky (3000 acres), 2-54; Sasik (19500 hectare), 0.8-11.6
Hawaii	Cobb, 1901	198
India	Pillay, 1954	112-168
	Pakrasi et al., 1964	858-1 244
Italy	de Angelis, 1960	90-170
	Beadle, 1946	60-100
	d'Ancona, 1954	150
Singapore	le Mare, 1950	**Tilapia,** 1670
	le Mare, 1949	Prawns 505, fish 250, total 755 lb acre^{-1} (847 kg ha^{-1})
Philippines	Rebanal, 1961	470
	Frey, 1947	500-1000
	Tang, 1967	500
Java	Schuster, 1952	147-627
Taiwan	Lin, 1968	940 in 1947 to 1863 in 1966, maximum 2500
Chesapeake Bay, U. S.	Mansueti, 1961* (34)	90
Sea of Azov	Zenkevitch, 1963	80
Southern North Sea	Korringa, 1967	52
Shallow bays (Texas)	Jones et al., 1963*(31)	121

Hickling (27), illustrates that the annual yields of fishes produced without supplementary feeding are reasonably constant over a wide geographic range and for a wide range of species. It also shows that some large natural systems (last four lines) have produced annual yields of fishes comparable to some culture ponds and lagoons without intensive management. The data of Table 2 reinforce the statement often found in the literature that estuaries are highly productive habitats which far exceed the open sea in yields to man per unit area.

Estimation of annual production involves estimates of the average size of the organisms and survival at a number of points in time. Where limited data are available (e.g., information on the numbers and size of organisms at the beginning and end of a period), some assumptions regarding the nature of growth and mortality must be made. The simplest assumption is that both growth and mortality curves are exponential in nature. These assumptions were made by Ricker (49) and Allen (1) to demonstrate that production (P) during the period t = 0 to t = T is given by:

$$P = \int_0^T N_o e^{-Mt} \, . W_o e^{Gt} \, . G . dt = \frac{G N_o W_o}{G\text{-}M} \left[e^{T(G-M)} - 1 \right] \tag{4}$$

where N_o is the original number of individuals, W_o is the initial weight of an individual at time t = 0, M is the instantaneous natural mortality rate, and G is the instantaneous rate of growth. If the yield is considered to be the standing crop at the end of some time period, then the yield (C) can be expressed as:

$$C = N_o W_o e^{T(G-M)} \tag{5}$$

A ratio of production (P) to crop (C) can be developed as follows:

$$\frac{P}{C} = \frac{G}{G\text{-}M} \left[1 - e^{T(M-G)} \right] \tag{6}$$

This result may be applied to estuarine fish populations to demonstrate the relationships between annual production and yield. The above relations hold only for a relatively short period of time owing to the assumption of exponential growth. When considered for an extended period of time, growth of fishes tends to follow a sigmoid curve.

Brody (4) has described an equation of the following form:

$$W_t = A - B_o e^{-kt} \tag{7}$$

where W_t is the weight of an organism or population weight at time t, A is the asympototic weight, B_o is the initial weight at time t = 0, and k is the relative growth rate.

Allen (2) successfully fitted this type of curve to a New Zealand trout population. He also derived an expression for production (P) during time T for the Brody growth function as:

$$P = \int_0^T N_o e^{-Mt} B_o e^{-kt} \;.k.dt = \frac{N_o B_o k}{M + k} \; [1 - e^{-(M+k)T}] \quad (8)$$

If we refer to crop as the stock at the end of some period, then crop (C) can be expressed as:

$$C = N.W. = N_o e^{-Mt} \; (A - B_o e^{-kt}) \quad (9)$$

A differentiated form of equation 9 has been solved for t to give a value for the maximum biomass as follows (Saila, unpublished data):

$$t^* = - \frac{1}{k} \log_e \left[\frac{MA}{(M+K) \;\; B_o} \right] \quad (10)$$

This expression can be used directly if the parameters of the decaying exponential growth function (equation 7) can be determined from empirical data. The value of t* (the time at which the maximum biomass occurs) can easily be transformed to size or age based on growth curves.

Kutty and Qasim (33) have already described methods for estimating the optimum age of exploitation for a fishery based on the Beverton and Holt yield equation incorporating the von Bertalanffy growth function when growth is either isometric or allometric. The most general expression derived by these authors is:

$$e^{k(t^* - t_o)} = \frac{nk+M}{M} \quad (11)$$

where k and t_o are paramenters of the von Bertalanffy function, M is the instantaneous natural mortality coefficient, n is an exponent in the length-weight relationship and t* is the age at which the biomass achieves its maximum weight. Equation 11 is solved for t* to determine the optimum age of exploitation. The value of t* obtained may then be substituted into the following expression to obtain the biomass in weight at time t*:

$$P_{wt} = Re^{-M(t^* - t_p)} W_{oo} 1 - e^{-k(t^* - t_o)^n} \quad (12)$$

where P_{wt} is the biomass in weight at time t*, t_p refers to the age recruits enter the fishing ground, W_{oo} is the third growth parameter, and R is the number of recruits entering the area.

In summary, some equations of increasing usefulness have been presented for the analysis of fish production and yield. The assumptions have been that the goal of management is the maximum harvestable crop and that a large proportion of the crop can be removed in a short period of time when the biomass of the stock reaches a maximum. A scheme such as described above seems suitable for intensive management of fish or other living resources in estuaries.

ACKNOWLEDGMENTS

Part of this work is the result of research sponsored by NOAA, Office of Sea Grant, U. S. Department of Commerce, under grant no. 04-3-158-3.

REFERENCES

1. Allen, K. R.
1949 Some aspects of the production and cropping of fresh waters. Rept. Sixth Sci. Congr. 1947. **Trans. Roy. Soc. New Zealand,** 77(5): 222-228.

2. Allen, K. R.
1950 The computation of production in fish populations. **New Zealand Sci. Rev.** (Sept.-Oct., 1950), p. 89.

3. Briggs, P. T. and O'Connor, J. S.
1971 Comparison of shore-zone fishes over naturally vegetated and sand-filled bottoms in Great South Bay, New York. **New York Fish Game Jour.,** 18(1): 15-41.

4. Brody, Samuel
1945 **Bioenergetics and growth.** Hafner Publishing Co., New York. 1023 p.

5. Caldwell, J. M.
1955 Tidal currents at inlets in the United States. **Proc. Amer. Soc. Civil Engr.,** 81(716): 1-12.

6. Carpenter, J. H., Pritchard, D. W., and Whalen, R. C.
1969 Observations of eutrophication and nutrient cycles in some coastal plain estuaries. In **Eutrophication: causes, consequences, corrections,** p. 210-221. Natl. Acad. Sciences, Washington, D. C.

7. Clark, J.
1967 Fish & man. Conflict in the Atlantic estuaries. **Spec. Publ. Amer. Littoral Soc.,** 5: 1-78.

8. Cronin, L. E. and Mansueti, A. J.
1971 The biology of the estuary. In **A symposium on the biological significance of estuaries,** 7 p. (eds. Douglas, P. A. and Stroud, R. H.) Sport Fishing Institute, Washington, D. C.

9. Csanady, G. T.

1973 **Turbulent diffusion in the environment.** Reidel Publishing Co., Boston, Mass. 248 p.

10. Curley, J. R., Lawton, R. P., Whittaker, D. K., and Hickey, J. M.
1972 A study of marine resources of Wellfleet Harbor. **Massachusetts Div. Mar. Fish., Monogr. Ser.,** 12: 1-37.

11. Dahlberg, M. D. and Odum, E. P.
1970 Annual cycles of species occurrence, abundance, and density in Georgia estuarine fish populations. **Amer. Midl. Nat.,** 83(2): 382-392.

12. Day, J. H.
1951 The ecology of South African estuaries. 1. A review of estuarine conditions in general. **Trans. Roy. Soc. South Africa,** 33: 53-91.

13. DeSylva, D. P., Kalber, F. A., and Shuster, C. N.
1962 Fishes and ecological conditions in the shore zone of the Delaware River estuary with notes on other species collected in deeper water. **Univ. Delaware Mar. Lab., Info. Ser., Publ.** 5: 1-164.

14. Dovel, W. E.
1971 Fish eggs and larvae of the upper Chesapeake Bay. **Spec. Rept. Nat. Res. Inst., Univ. Maryland,** 4: 1-71.

15. Elder, T. W.
1959 The disposition of a marked fluid in a turbulent shear flow. **Jour. Fluid Mech.,** 5: 544-560.

16. Fierstine, H. L., Kline, K. F., and Garman, G. R.
1973 Fishes collected in Morro Bay, California between January, 1968 and December, 1970. **California Fish Game,** 59(1): 73-88.

17. Fischer, A. G.
1960 Latitudinal variations in organic diversity. **Evolution,** 14: 64-81.

18. Fiske, J. D., Watson, C. E., and Coates, P. G.
1967 A study of marine resources of Pleasant Bay. **Massachusetts Div. Mar. Fish., Monogr. Ser.,** 5: 1-56.

19. Graham, J. J.
1972 Retention of larval herring within the Sheepscot estuary of Maine. **NOAA Fish. Bull.,** 70(2): 299-305.

20. Green, J.
1968 **The biology of estuarine animals.** Univ. Washington Press, Seattle, Washington. 401 p.

21. Gross, F.
1947 An experiment in marine fish cultivation.. 1. Introduction. **Proc. Roy. Soc. Edinburgh, B,** 63: 1-2.

22. Gross, F.
1947 An experiment in marine fish cultivation. V. Fish growth on a fertilized sea-loch (Loch Craiglin). **Proc. Roy. Soc. Edinburgh, B,** 63: 56-95.

23. Gross, F.
1950 A fish cultivation experiment in an arm of a sea-loch. 1. Introduction. **Proc. Roy. Soc. Edinburgh, B,** 64: 1-4.

24. Gross, F.
1950 A fish cultivation experiment in an arm of a sea-loch. V. Fish growth in Kyle Scotnish. **Proc. Roy. Soc. Edinburgh, B,** 64:

109-135.

25. Gunter, G.
1967 Some relationships of estuaries to the fisheries of the Gulf of Mexico. In **Estuaries,** p. 621-638. (ed. Lauff, G. H.) Am. Assoc. Adv. Sci. Publ. 83. Washington, D. C.

26. Hedgpeth, J. W.
1966 Aspects of the estuarine ecosystems. In **A symposium on estuarine fisheries,** p. 3-11. (ed. Smith, R. F., Swartz, A. H., and Massmann, W. G.) **Amer. Fish. Soc. Spec. Publ.** 3.

27. Hickling, C. F.
1970 Estuarine fish farminng. **Adv. Mar. Biol.,** 8: 119-213.

28. Hildebrand, S. F. and Schroeder, W. C.
1928 Fishes of Chesapeake Bay. **Bull. U.S. Bur. Fish.,** 43(1): 1-366.

29. Jansson, B. O.
1972 Ecoysystem approach to the Baltic problem. **Bull. Ecol. Res. Comm. (Swedish Nat. Res. Coun.),** 16: 1-82.

30. Jerome, W. C., Chesmore, A. P., and Anderson, C. O., Jr.
1966 A study of marine resources of Quincy Bay. **Massachusetts Div. Mar. Fish., Monogr. Ser.,** 2: 1-62.

31. Jones, R. S., Ogletree, W. B., Thompson, J. H., and Henniken, W.
1963 Helicopter-borne purse net for population sampling of shallow marine bays. **Publ. Inst. Mar. Sci. Texas,** 9: 1-6.

32. Ketchum, B. H.
1951 The exchange of fresh and salt water in estuaries. **Jour. Mar. Res.,** 10: 18-38.

33. Kutty, M. K. and Qasim, S. Z.
1968 The estimation of optimum age of exploitation and potential yield in fish populations. **Jour. Cons. perm. int. Explor. Mer.,** 32(2): 249-255.

34. Mansueti, R. J.
1961 Effects of civilization on striped bass and other estuarine biota in Chesapeake Bay and tributaries. **Proc. Gulf Carib. Fish. Inst., 14th Ann. Sess.,** p. 110-136.

35. Massmann, W. H.
1964 The fishes – a neglected aspect of estuarine research. **Trans. 29th North American Wildl. Nat. Res. Conf.,** p. 337-352.

36. McHugh, J. L.
1966 Management of estuarine fisheries. In **A symposium on estuarine fisheries,** p. 133-154. (eds. Smith, R. F., Swartz, A. H., and Massmann, W. H.) **Amer. Fish. Soc. Spec. Publ.** 3.

37. McHugh, J. L.
1967 Estuarine nekton. In **Estuaries,** p. 581-620. (ed. Lauff, G. H.) Am. Assoc. Adv. Sci. Publ. 83. Washington, D.C.

38. Milne, P. H.
1972 **Fish and shellfish farming in coastal waters.** Fishing News (Books) Ltd., London. 208 p.

39. Odum, H. T.
1967 Biological circuits and the marine systems of Texas. In **Pollution and marine ecology,** p. 98-157. (eds. Olson, T. A. and Burgess, F. J.) Interscience, New York.

40. Odum, H. T.
1968 Work circuits and system stress. In **Primary productivity and mineral cycling in natural ecosystems,** p. 81-138. Univ. of Maine Press, Orono.

41. Odum, W. E.
1973 The potential of pollutants to adversly affect aquaculture. **Proc. Gulf Carib. Fish. Inst., 25th Ann. Sess.,** p. 163-171.

42. Paloheimo, J. E. and Dickie, L. M.
1970 Production and food supply. In **Marine food chains,** p. 499-527. (ed. Steele, J. H.) Univ. California Press, Berkeley and Los Angeles.

43. Parry, G.
1966 Osmotic adaptation in fishes. **Biol. Rev.,** 41: 392-444.

44. Pielou, E. C.
1969 **An introduction to mathematical ecology.** John Wiley and Sons, New York.

45. Pillay, T. V. R.
1967 Estuarine fisheries of West Africa. In **Estuaries,** p. 639-646. (ed. Lauff, G. H.) Am. Assoc. Adv. Sci. Publ. 83. Washington, D.C.

46. Pillay, T. V. R.
1967 Estuarine fisheries of the Indian Ocean coastal zone. In **Estuaries,** p. 647-657. (ed. Lauff, G. H.) Am. Assoc. Adv. Sci. Publ. 83. Washington, D.C.

47. Pritchard, D. W.
1954 A study of salt balance in a coastal plain estuary. **Jour. Mar. Res.,** 13: 133-144.

48. Pritchard, D. W.
1967 What is an estuary: physical viewpoint. In **Estuaries,** p. 3-5. (ed. Lauff, G. H.) Am. Assoc. Adv. Sci. Publ. 83. Washington, D.C.

49. Ricker, W. E.
1946 Production and utilization of fish populations. **Ecol. Monogr.,** 16: 373-391.

50. Roessler, M.
1970 Checklist of fishes in Buttonwood Canal, Everglades National Park, Florida, and observations on the seasonal occurrence and life histories of selected species. **Bull. Mar. Sci.,** 20(4): 860-893.

51. Saila, S. B.
1962 The contribution of estuaries to the offshore winter flounder fishery in Rhode Island. **Proc. Gulf Carib. Fish. Inst., 14th Ann. Sess.,** p. 95-109.

52. Saila, S. B., Horton, D. B., and Berry, R. J.
1965 Estimates of the theoretical biomass of juvenile winter flounder **Pseudopleuronectes americanus** (Walbaum) required for a fishery in Rhode Island. **Jour. Fish. Res. Bd. Canada,** 22(4): 945-954.

53. Salah el-Din el-Zarkar
1963 Acclimation of **Solea vulgaris** (Ginn) in Lake Quarun, Egypt. **Jour. Cons. perm. int. Explor. Mer.,** 28: 126-136.

54. Shelbourne, J. E.
1964 The artificial propagation of marine fish. **Adv. Mar. Biol.,** 2: 1-83.

55. Skellam, J. G.
1951 Random dispersal in theoretical populations. **Biometrika,** 38: 196-218.

56. Srivastava, V. M.
1971 Fish of the Gulf of St. Lawrence. An unabridged bibliography. **Fish. Res. Bd. Canada Tech. Rept.** 261. 141 p. (Unpublished manuscript.)

57. Stommel, H.
1951 Recent developments in the study of tidal estuaries. **Woods Hole Oceanogr. Inst. Tech. Rep.,** Ref. No. 51-33. 53 p. (Unpublished manuscript.)

58. Stroud, R. H.
1971 Introduction to symposium. In **A symposium on the biological significance of estuaries,** p. 3-8. (ed. Douglas, P. A. and Stroud, R. H.) Sport Fishing Institute, Washington, D.C.

59. Thomann, R. W.
1972 **Systems analysis and water quality management.** Environmental Sciences Services Division. 286 p. (Unpublished manuscript.)

60. Vannucci, M.
1969 What is known about production potential of coastal lagoons. In **Coastal lagoons, a symposium,** p. 457-478. (eds. Castanares, A. A. and Phleger, F. B.) (UNAM-UNESCO) Univ. Nacional Autonoma de Mexico, Mexico, D.F.

61. Walford, C. A. and Wicklund, R.
1970 Introduction of exotic marine, euryhaline and anadromous organisms – a preliminary review. Advisory Council on Marine Resources Research. 6/71/WP2. 101 p. (Unpublished manuscript.)

THE EFFECTS OF POWER PLANTS ON PRODUCTIVITY OF THE NEKTON[1]

by S. G. O'Connor[2] and A. J. McErlean[3]

ABSTRACT

Steam electric generating stations, especially those with once-through cooling systems, can cause alterations in the productivity of fish populations through the effects of thermal discharges, chemical discharges, and impingement of adults or entrainment of juveniles at intake structures. These effects cannot be evaluated individually but must be analyzed in terms of changes in productivity, as indicated by standing-crop estimates, and changes in food-web dynamics, as indicated by diversity indices.

Diversity, standing crop, and mean biomass estimates of the Patuxent River estuary from 1962 until 1967 bear the common feature of a sinusoidal curve. Our data suggest a general environmental degradation as characterized by a decreasing diversity of species and fluctuating estimates of standing crop. Our study includes both pre-operational and post-operational data from the Chalk Point Generating Station; however, the generalized degradation cannot be directly attributed to the power plant.

INTRODUCTION

The effect of steam electric generating stations, especially those that use a

1. Contribution no. 558, Natural Resources Institute, University of Maryland, Solomons, Maryland 20688.

2. Department of Environmental Research, Chesapeake Biological Laboratory, Prince Frederick, Maryland. Present address: QLM Laboratories, Inc., 2 Burd Street, Nyack, New York 10960.

3. Department of Environmental Research, Chesapeake Biological Laboratory, Prince Frederick, Maryland. Present address: Office of Technical Analysis, EPA, Waterside Mall, 401 M Street, S. W., Washington, D. C. 20460.

once-through cooling system, has become a major concern in recent years. Electrical-power generation has doubled every ten years since 1900, and the rate of increase is expected to be even greater during the next decade (2, 47). Associated with increases in the production of electrical power are increased demands for water for use as a heat-transfer medium in the condenser systems of power plants. Recent trends toward increasing the size of individual power stations and the construction of more nuclear power plants as compared to fossil-fueled generating stations may cause the environmental impact of increased power production to fall most heavily upon estuarine systems. Since 1955, generating units have steadily increased in size (26). Moreover, these large units have been centralized into stations composed of several units. Most steam electric stations presently in operation have a capacity of 1,000 megawatts (MW) or less. Stations under construction are rated at up to 2,000 MW and future installations may reach 4,000-8,000 MW capacities (44). Although the total industrial effect of a large station may not differ from the combined effects of several small stations, it does have an intensified impact on the biota of the cooling water source. As much as 65% of the electricity generated in 1965 may be produced by nuclear fuel in 1980 (26), and a large number of nuclear stations are presently under construction. The thermal efficiency of light water reactors is about 34% compared with that of about 40% for equivalent capacity fossil fuel units. Thus, the amount of heat discharged to the cooling water used by nuclear power plants is greater than the amount discharged by fossil fuel power plants. In order to meet thermal criteria, nuclear units must use larger volumes of cooling water.

Increased generating capacity by centralization of large units and increased construction of nuclear fueled generating stations act together to require large volumes of cooling water. Only estuaries or the oceans can provide these large volumes. Consequently, a larger percentage of steam electric generating stations are being sited near estuaries. In 1980, 32% of U. S. power plants will be located on estuaries, compared to 22% in 1950 (49). This estimate may be slightly high because it does not take into consideration the current emphasis on the construction of cooling towers. Most cooling towers are sited near a fresh-water source, in order to avoid scale deposits from dissolved minerals in estuarine waters.

POTENTIAL POLLUTION EFFECTS OF STEAM ELECTRIC STATIONS

The multitude of effects that steam electric stations with once-through cooling systems have upon their sources of cooling water can be divided into three general categories: thermal-discharge effects, chemical-discharge effects, and effects of intake structures. Each of these categories involves all trophic levels and compounds itself with each increasing level. Their impact on nekton, particularly fishes, is a combination of direct influences toward individual animals and indirect influences accrued through the dynamics of the food web. The overall impact of power plants upon productivity of the nekton must be understood in terms of the responses of individual organisms but evaluated and measured in terms of changes in standing crop and in diversity of the food web.

Of the three categories, thermal-discharge effects have received the greatest attention. Fishes respond, both behaviorally and physiologically, to alterations in their thermal environment. Behavioral responses can be recognized by alterations in the distribution patterns of resident species. Trembley's study (57) of the Martins Creek Plant site was the first to document the alteration of fish-distribution patterns created by the introduction of a power-generating station. After the plant started operating, all fish species normally inhabiting the plant location were present at one season or another, but they tended to congregate near the plant during cold months and were absent from the plant site during warm months. This winter attraction and summer avoidance behavior has been observed repeatedly and is attributed to the presence of a warm-water discharge (1, 2, 40, 42, 57, 58). However, the abundance of prey organisms, some dead or stunned from passage through the cooling condensers, also acts as an attraction for large fishes (5, 15). Sports fishermen and occasionally the power industries have capitalized on the winter concentration of fishes near power-plant effluents. Winter angling pressure near warm-water effluents is often higher than that of unaffected estuarine areas (16, 28, 45, 46).

Alterations in fish-distributional patterns are very difficult to evaluate in terms of alterations in overall productivity. Increases or decreases of animals in the vicinity of a power-plant effluent may not indicate changes in the total productivity of an estuary. At minimum, however, alterations in distributions of fishes indicate reallocation of food resources. Conceivably the reallocation of food resources, coupled with increased primary productivity in warm-water areas, could bring about increased productivity of fishes resident in estuaries with thermal effluents (23, 41). Offsetting this possibility is the potential for fish kills within thermal plumes. Recorded fish kills directly attributable to high-temperature thermal effluents are few in number (1, 10); however, fishes acclimated to the temperatures of a thermal plume could undergo a lethal cold shock in the event a plant ceased to operate.

Physiological responses to temperature have been the subject of innumerable laboratory studies. Temperature is recognized as a controlling factor in the life of any poikilothermic organism. It can be lethal at levels higher or lower than the tolerance limits of a given species; it can restrict growth by modifying metabolic rates; or it can eliminate reproduction by affecting the rates of biochemical reactions (9, 14). The thermal requirements for reproduction are more restrictive than those for growth, while the requirements for growth are more restrictive than those for survival. Physiological responses of fishes to cold shock, heat shock, and chronic exposure to extremes of temperature are well documented (see 25, 51 for bibliographies). This information is vital to the determination of criteria for adequate water quality; however, the effect of sublethal physiological shock upon productivity of fish populations is largely theoretical.

In terms of Brett's temperature tolerance polygon (9) (Fig. 1), the presence or absence of a species is determined in part by its thermal requirements for survival. However, productivity levels within a species are determined on the short term by requirements for reproduction. The maximum productivity level that can be attained by a species in a single age group or a single season is limited by the growth rate achieved by that age class or during that season. Classically, production is defined as relative growth rate (G) multiplied by biomass (B):

$$P(\text{mg m}^{-2}\text{ day}^{-1}) = G(\text{mg gm}^{-1}\text{ day}^{-1}) \times B(\text{gm m}^{-2}) \tag{1}$$

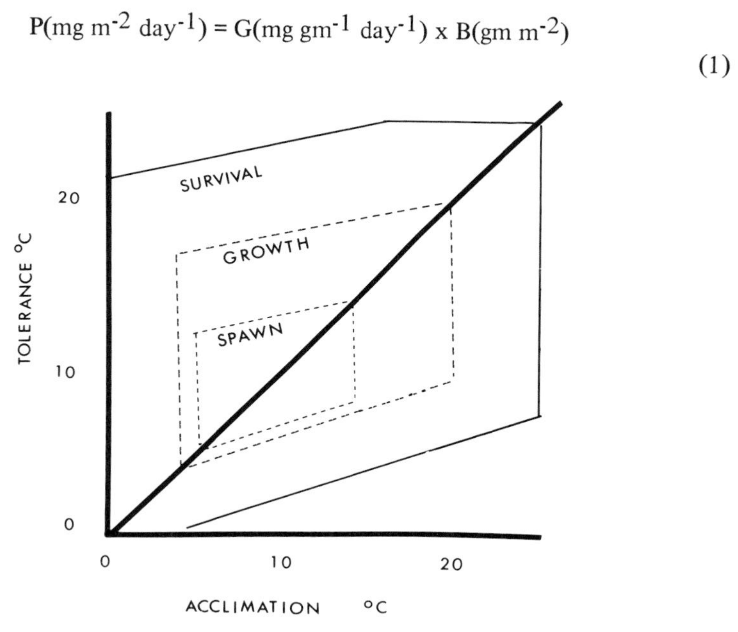

Figure 1. Temperature tolerance polygon for aquatic organisms, after Brett (9).

Therefore, optimal temperatures for growth would optimize productivity. In the long term, productivity of a species depends upon the total mass of animals formed, including growth of existing animals and production of new individuals. Maximal productivity is achieved only when ambient temperatures fall within the limits required for reproduction. Any temperature outside the range that allows reproduction inhibits production. Suboptimal, that is, reproduction-inhibiting, temperatures would reduce energy conversion efficiencies within the species, and thus reduce productivity.

Studies assessing the chemical-discharge effects upon fishes have been concerned primarily with chlorine, which is used widely as a biocide in generating-station cooling systems (see 6 for a review of power-plant chemicals). Hypochlorite or chloride gas is added intermittently to intake water to prevent the formation of organic deposits within condenser tubes. Chlorine and its compounds are toxic to aquatic organisms; therefore, during chlorination all organisms, including fish eggs and larvae, that pass through the plant are killed (39). The toxicity of the chlorine waste in estuaries depends on the concentration of residual chlorine, both the free form and chloramines. Toxic doses can be quite small; for example, 0.05 ppm are toxic to rainbow trout after 96 hours exposure (8). The direct effects of chemical discharges upon productivity cannot be estimated at this time.

The effects of intake structures alter productivity by entraining organisms small enough to pass through the condenser system of a power plant and impinging on organisms too large to pass through intake screens but not strong enough to avoid the intake current. Both entrainment and impingement can affect productivity directly by killing fish or fish eggs and larvae and indirectly by decreasing food supplies. Data on the effects of entrainment upon phytoplankton and zooplankton are difficult to evaluate. There is evidence of high mortality and decreases in primary production, but it is impossible to separate mortality caused directly by physical abrasion (38) from mortality caused by lethal temperatures within the plant (11) and by chlorination of cooling water (22, 61). In any case, the effect of localized decreases in primary producers and consumers is probably not a limiting factor to the production rate of the nekton. Plankters usually have a short generation period and a tendency to respond to decreases in species density by increasing reproductive rates. These characteristics probably allow localized losses in plankton productivity to be counterbalanced over the large estuarine areas used by fish populations.

Entrainment through a power plant can be lethal to both fish eggs and larvae (11, 32, 42). Moreover, there is some evidence that sublethal exposure to thermal, chemical, and mechanical shocks significantly affects the ability of an entrained larva to survive in the environment. One effect of sublethal exposure inside power plants is a reduced ability to avoid predation (18, 56). This effect may be compounded by increased concentration of predators in the region of discharge plumes. The cropping or predatory effect of power plants upon fish eggs and larvae has led to the development of predictive models capable of estimating the percent reduction in survival of striped bass larvae as a consequence of entrainment in plants along the Hudson River (19, 29). Two radically different estimates of mortality were predicted for the same power plant, Indian Point I. Lawler predicted a maximum reduction of 6% in the Hudson River population of striped bass (29). Goodyear predicted a reduction of 39.9% (19). Any reducation in the numbers of such a popular food and game fish is to be deplored; however, a 6% reduction would be extremely hard to quantify with field data, since the natural variation in the striped bass population may be greater than 6%. On the other hand, a decrease of 40% in the population of any species would be paralleled by losses in productivity and concurrent alterations in the food web.

Productivity may also be seriously affected by the impingement of adult or large juvenile fish. The number of fish impinged on intake screens can be extraordinarily high. Between November 6, 1969, and January 11, 1970, 1,310,345 fish of various species were killed on the intake screens at Indian Point I (62). This impingement kill could amount to 400 or more pounds of fish per day. This example is an extreme in reported fish impingement, but it illustrates the magnitude of the problem. Without question, fish impingement can bring about a decrease in the standing crop of an estuary. At present, the long-term effect of impingement upon productivity of the nekton cannot be estimated. Biological compensatory mechanisms in early stages as well as in adults may reduce or eliminate the effects of impingement. To be sure, the science of fisheries management depends upon the principle of cropping in order to produce maximally sustainable yields (52). However, it is difficult to interpret large impingement kills in terms of sustainable yields in total fish production, since the size of the standing crop in most estuaries is not known. It is also difficult to view without emotion a fish kill such as the one that occurred at Indian Point.

EVALUATING THE EFFECTS OF POWER PLANTS

The total impact of a power plant upon productivity in estuarine environments must be measured in terms of changes within the entire nektonic community. Productivity may be altered directly, as demonstrated by changes in total biomass, or indirectly, as demonstrated by changes in the dynamics of the food chain. Both effects must be evaluated; both are difficult to quantify. A description of the food chain for fish communities requires exhaustive sampling at all trophic levels, stomach-content analyses, and behavioral observations. A study of this nature requires years of gathering data, both before and after a plant starts operating, if alterations attributable to power plants are to be identified. A simple but informative approach is the use of species diversity measures to evaluate alterations in food webs.

The concept of species diversity is based on the relationship between the number of species present in a community (species richness) and the distribution of individual organisms among species (evenness of equitability). If an environment contains a large number of species, each represented by a few individuals, the diversity is high. Conversely, if an environment contains few species, each represented by many individuals, species richness and evenness are low. For some groups of organisms, including benthos and nekton, the diversity of a community reflects environmental conditions. Adverse or stressful conditions will reduce species diversity because some species will be unable to survive the stress and will be lost from the community, while other species will survive and increase in number because of decreased predation or decreased competition for food. Decreases in diversity related to sewage and industrial effluents have been documented for macroinvertebrates (24, 63, 65) and fish (59, 60). The relationship between pollution and reduced diversity indices has been suggested to be a biological indicator of pollution (3, 7, 48, 54, 55, 56, 66).

McErlean and his associates (35, 36) used diversity indices to estimate the effects of effluents from a steam electric generating station upon fish populations in the Patuxent River estuary. Localized effects were evaluated by three-dimensional graphs (Fig. 2). Each graph contained five west-east coordinates representing sampling locations by gear, and ten north-south coordinates representing sampling stations. Diversity was plotted vertically in the third dimension (Fig. 3). These graphs and statistical analyses (ANOVA) showed little evidence that the plant had a direct effect at any season.

Long-term effects were evaluated by plotting total collection means sequentially (Fig. 4). This graphic method does not eliminate seasonal fluctuations, but relies on the slope of the plotted trend. If there is no upward or downward trend the seasonal values oscillate around a horizontal baseline which represents the grand mean of all observations. When the collection means for each season ($N = 50$ for each point) are plotted by the trend-analysis technique, the Shannon-Weiner information theory index (34) demonstrates an alarming downward trend through the four-year study (Fig. 4). This downward trend is also apparent in trend analysis of number of species caught, species richness (33), and species evenness (50). Most probably this decrease in species diversity was caused by a general environmental degradation which reduced the structural complexity of the fish population in the Patuxent. Because the number of individuals and diversity of species in a community reflect the number of energy pathways within a food chain (3, 30), decreases in diversity indices indicate shortened food chains and simplified population structure. Simple systems are more likely to undergo crashes and booms, since the stabilizing effect of alternate energy chains is absent. Instead of diversified flow routes for energy, proportionately more energy is shunted into dominant species. Shunting energy into dominant species does not necessarily alter the total productivity of an estuary. However, when total productivity is dependent upon a few species, productivity levels are likely to undergo wide perturbations dependent upon year-class strength and other species parameters.

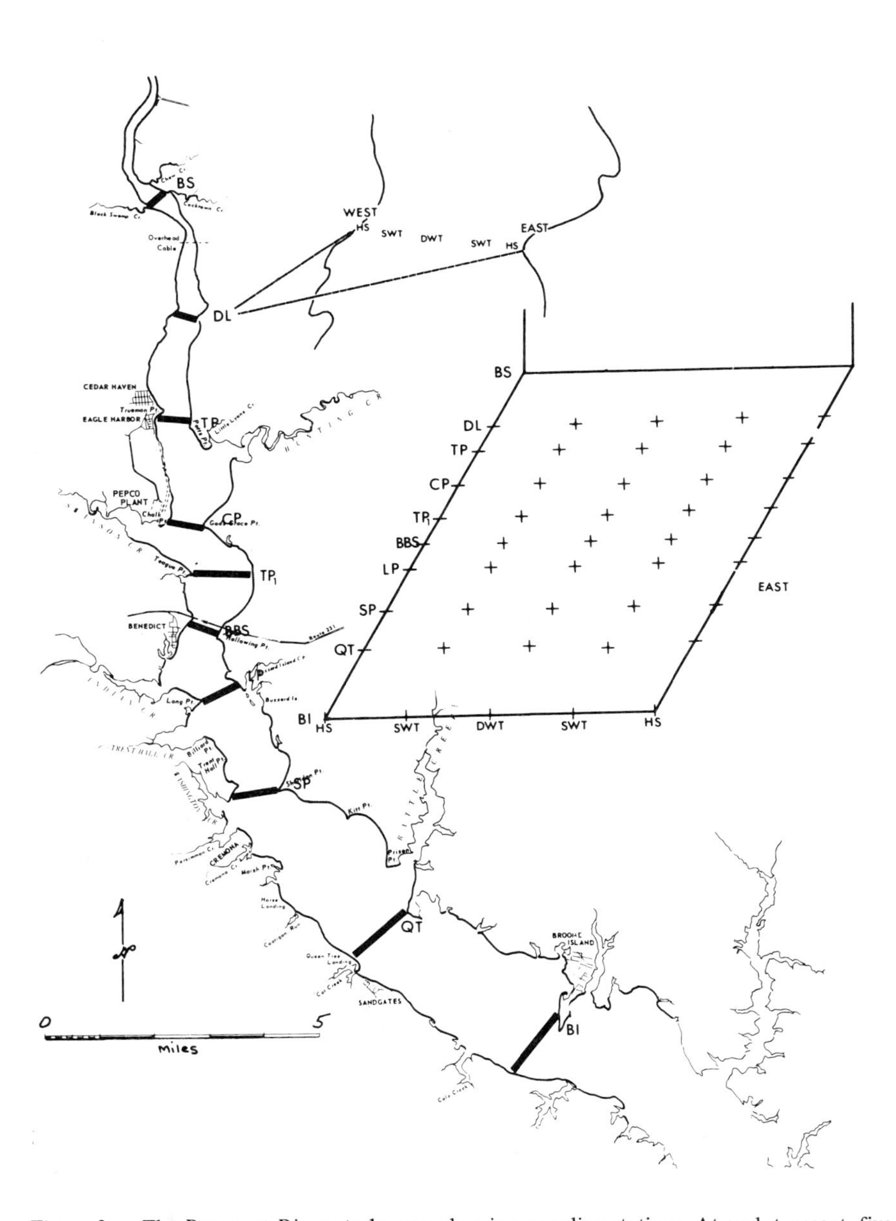

Figure 2. The Patuxent River study area showing sampling stations. At each transect, five samples were taken (see insert) for each collection. Also shown is the three-dimensional grid used to plot the various parameters (see Fig. 3). After McErlean, 1973.

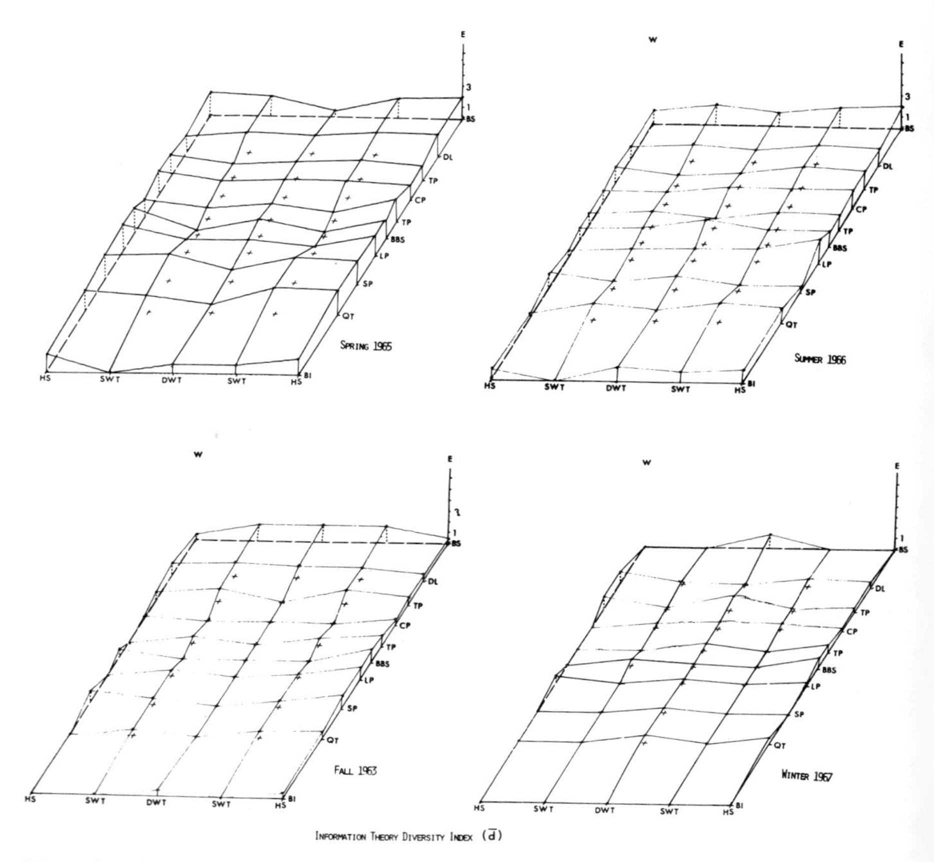

Figure 3. Representative graphs of information theory diversity index (d) for selected seasonal collections. (a) spring 1965, (b) summer 1966, (c) fall 1963, (d) winter 1967. After McErlean, 1973.

In order to evaluate the downward trend in species diversity that occurred in the Patuxent River estuary between 1962 and 1967, biomass estimates were calculated for the five most abundant species: *Morone americana* (white perch), *Trinectes maculatus* (hogchoker), *Fundulus heteroclitus* (mummichog), *Fundulus majalis* (striped killifish), and *Menidia menidia* (Atlantic silverside). These estimates were determined from length-weight regression equations calculated for each species (Table 1) (27). Ninety-one percent of the total catch per year was represented by the five most abundant species.

Biomass estimates are usually presented in terms of kilograms per hectare (kg ha^{-1}) or pounds per acre (lbs $acre^{-1}$) (Table 2). The mean value for all the sampling sites on the Patuxent ranged from a low of 7.3 lbs $acre^{-1}$ (8.0 kg ha^{-1})

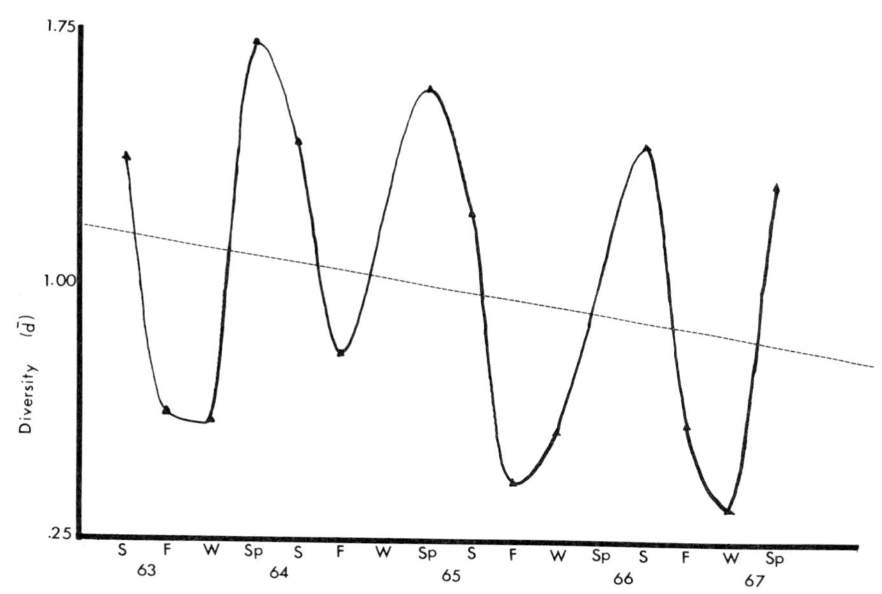

Figure 4. Trend analysis for information theory diversity (d). Each point is the mean for 50 observations. Curve fitting was done with a Hewlett-Packard calculating system (model 9100A) and utilized a polynomial fit for five equations in five unknowns. The trend line is a least-squares regression fit to the means. Notice seasonal cycle and downward trend with time. W, winter; Sp, spring; S, summer; F, fall. After McErlean, 1973.

TABLE 1

Length-weight regression equations for Patuxent River fishes

Species	Equation[1]		Correlation r
Morone americanas	Log W = 3.073	Log L – 4.713	.967
Trinectes maculatus	Log W = 0.0132	Log L + 0.034	.972
Fundulus majalis	Log W = 3.076	Log L – 5.022	.938
Fundulus heteroclitus	Log W = 3.066	Log L – 4.943	.968
Menidia menidia	Log W = 2.855	Log L – 4.915	.981

1. Total length in centimeters, weight in grams.

estimated in the fall of 1966 to a high of 190.6 lbs acre^{-1} (213.4 kg ha^{-1}) estimated in the fall of 1965. The mean estimate for pre-plant-operation samples was 34.8 lbs acre^{-1} (39.0 kg ha^{-1}) versus 44.9 lbs acre^{-1} (50.3 kg ha^{-1}) for

TABLE 2

Estimated biomass of Nekton

Type Place	Estimate lbs acre^{-1}	kg ha^{-1}	Source
Commercial landings, Gulf of Mexico	287	322	McHugh (37)
Commercial landings, Chesapeake Bay	193	216	McHugh (37)
Commercial landings, Japan Sea	25	28	Galtsoff (17)
Commerical landings, Banks of Nova Scotia	12	13.5	Graham and Edwards (20)
Commercial landings, Peru Current	330	370	Schaefer (53)
Haul seine, Rehoboth Bay	49-25	55-28	Derickson and Price (13)
Haul seine, Indian River Bay	68-33	76-37	Derickson and Price (13)

post-plant operation. These means are not statistically different.

Localized effects were evaluated by three-dimensional graphs of biomass corrected to the logarithms of a common linear collection effort. The use of a log transformation corrected to a common areal base reduces variation and requires a large absolute change in order to indicate effect. A whole log-unit change requires a tenfold increase or decrease in the raw estimate of biomass. No plant-related effect was evident in the three-dimensional graphs. For both pre- and post-operational years (1963 and 1966, respectively), spring and summer biomass estimates demonstrated a uniform distribution from shore to shore. Fall and winter graphs offer a contrast to the uniformity of spring and summer graphs (Figs. 5 and 6). In cooler weather, there was a tendency for concentration in mid-channel, deep-water locations, as well as an overall increase in biomass at upriver locations. Pre-plant-operation graphs (Fig. 5) indicate shelf and mid-channel concentration peaks; post-plant-operation graphs (Fig. 6) indicate more shore-to-shore uniformity in the vicinity of the plant. This can be interpreted as redistribution of fish with respect to the effluent.

Long-term analysis of mean biomass (Log N + 1 lbs acre^{-1}) is presented in Figure 7. The mean was calculated by weighing all sampling sites equally, regardless of the area represented by each site. That is, each shore-zone collection was weighed equally with each channel-zone collection even though each channel collection represented one hundred times the potential sampling

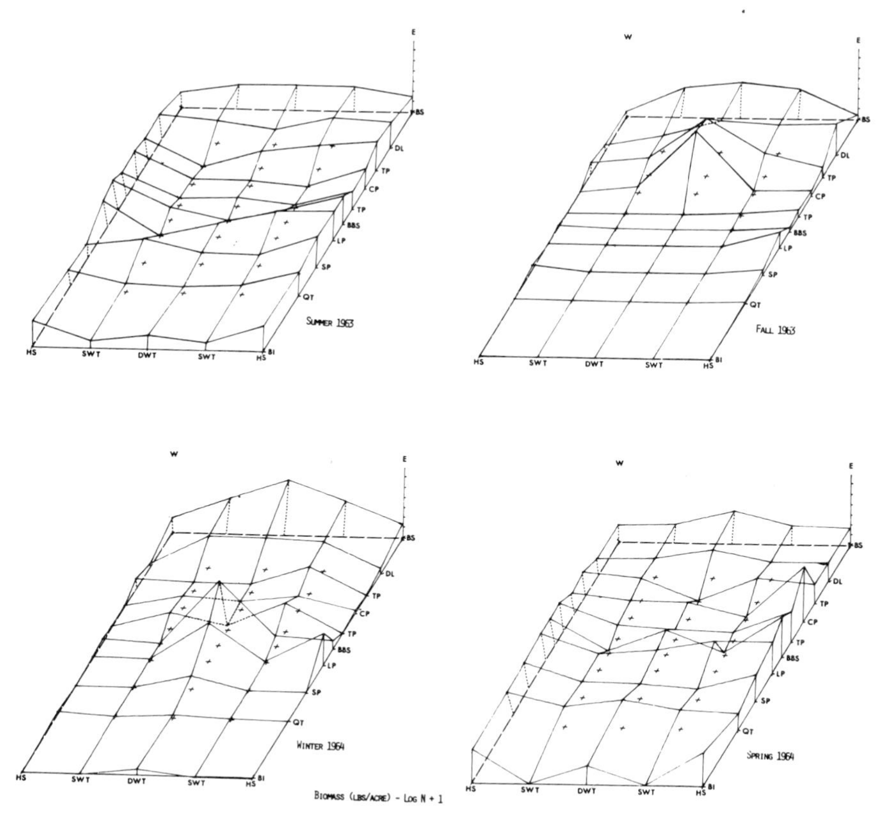

Figure 5. Representative graphs of biomass (Log N + 1) in lbs acre^{-1}. (a) summer 1963, (b) fall 1963, (c) winter 1964, (d) spring 1964. Plant pre-operative.

area of a shore-zone collection (Table 3). Moreover, the downriver stations represent a larger area than the upriver stations for all three gear types (Fig. 8). This uneven representation of area sampled is particularly meaningful in terms of the communities sampled by each gear type. By graphing mean values, shore zones populated by small fish species are weighted more heavily than channel zones populated by larger and commercially important species.

The trend analysis of mean biomass is similar to the trend analysis for diversity (compare Figs. 4 and 6) in that both parameters demonstrate annual seasonal cycles with sine-wave characteristics. Diversity and mean biomass are lower in fall and winter collections than in summer and spring collections. Unlike the trend analysis of diversity, which demonstrated a downward trend, there was no decrease in mean biomass of the Patuxent River through time. Viewed in the perspective of an environment undergoing a simplification of its food web, there was no increase in the production rate of the dominant fish

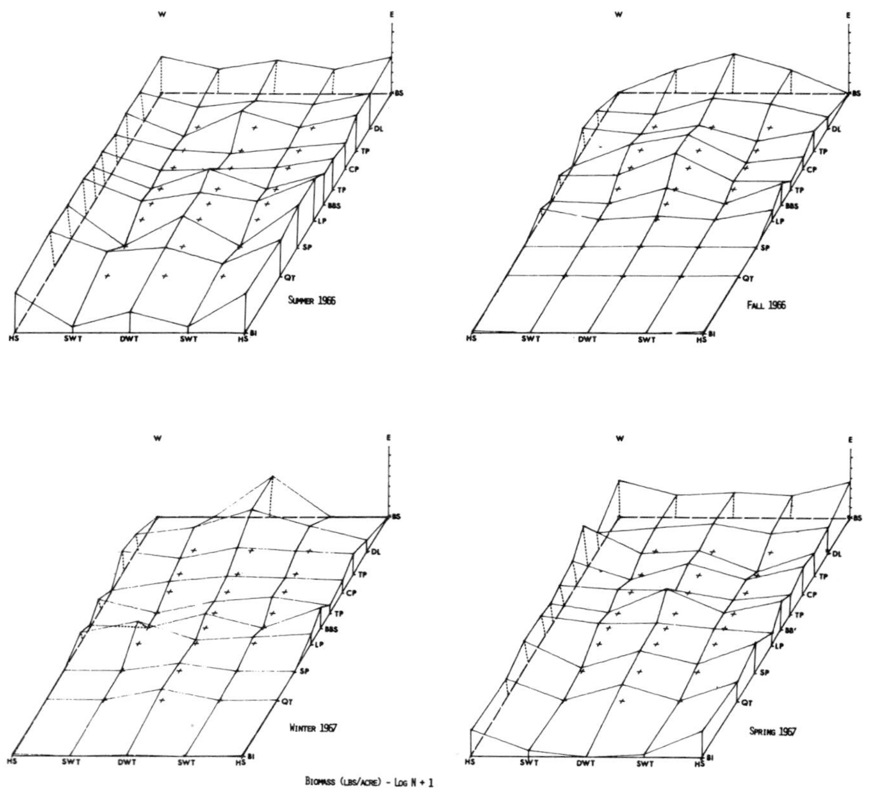

Figure 6. Representative graphs of biomass (Log N + 1) in lbs acre^{-1}. (a) summer 1966, (b) fall 1966, (c) winter 1967, (d) spring 1967. Plant post-operative.

species concurrent with the observed loss in diversity.

In order to put our biomass estimates into the perspective of the standing crop of the Patuxent River, the total river biomass was calculated from the weight collected at each sampling site using the formula:

$$B = \sum_{i=1}^{50} (Wi) \left(\frac{A_a}{A_s}\right) i \tag{2}$$

where B = total river biomass (lbs), Wi = weight of fish collected at sampling site i (lbs), A_a = total area available for sampling at site i (acres), and A_s = actual area sampled at site i (acres). Estimates were minimized by assuming that our collection gear was 100% efficient at all times. Obviously, this assumption is incorrect. The true standing crop for the total river area sampled (26,089 acres) may be as much as 40%-50% higher than our biomass estimates, depending upon the actual efficiency of our gear. In addition, our estimates do not include the

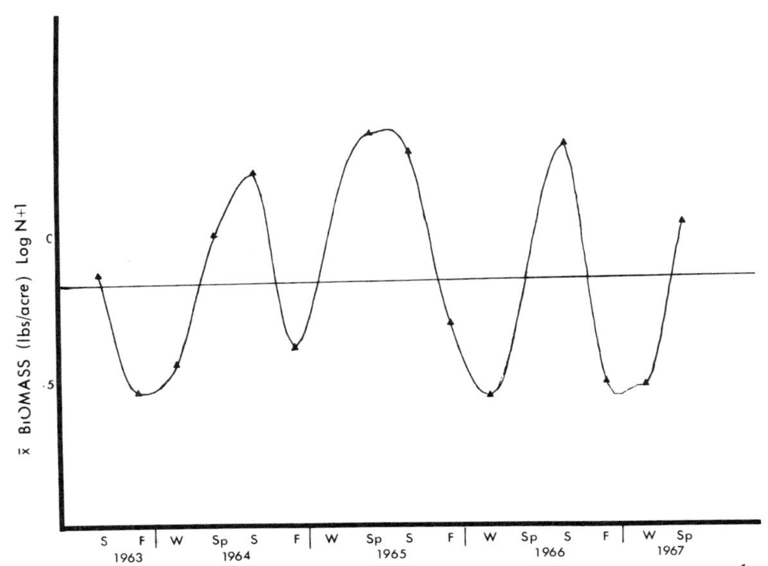

Figure 7. Trend analysis formean Patuxent River biomass (Log N + 1) in lbs acre^{-1}.

TABLE 3

Potential area sampled by each gear type

Zone	Gear	Potential area sampled (acres)
Shore	Haul seine (HS)	116
Shelf	Shallow-water trawl (SWT)	9,827
Channel	Deep-water trawl (DWT)	16,146
	Total river area	26,089

rare or migrating species which frequently visit the estuary. These seasonal migrants are more likely to be present during spring and summer months when the upper estuary is used as a spawning and nursery ground. Thus, the acutal standing crop during spring and summer collections may be as much as 60% higher than our biomass estimates.

Seasonal values for total river biomass were plotted consecutively in the manner of a trend analysis (Fig. 9). The picture of the Patuxent River presented by this graph is radically different from that presented by trend analysis of mean river biomass (Fig. 7). The total biomass fluctuated to an extreme high of

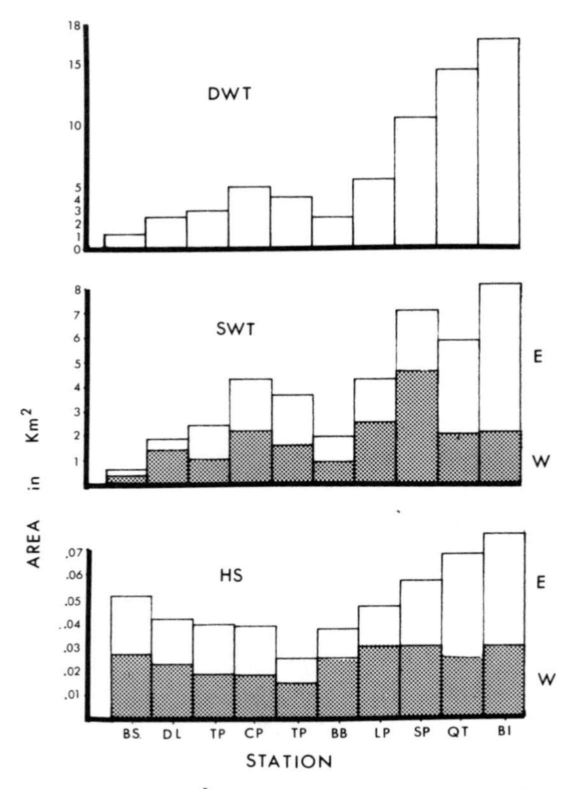

Figure 8. Histogram of area in Km^2 represented by each field sampling station. Note the 200-fold difference in the Y axis and the larger areas in downriver stations (BI & QT).

9,389,670 pounds in the fall of 1965, followed by a year of low standing-crop estimates. Such an extreme high could be interpreted as a boom followed by a crash. Although a suggestion of a seasonal cycle was apparent in 1963 and 1965, it is the inverse of the strong seasonal cycle apparent in the trend analysis of mean biomass. Total river biomass was high in fall and low in spring. This pattern might be expected, because growth becomes maximized in the fall (37).

How can the two graphs, one of total river biomass (Fig. 9) and one of mean collection biomass (Fig. 7), be integrated into a single clear picture of the Patuxent River? Figure 10 compares the total biomass estimates calculated for each gear type. The high values seen during fall and winter seasons in the trend analysis of total river biomass (Fig. 9) are proportional to large catches with deep-water trawl gear. In fact, the three-dimensional graphs of biomass per acre also indicate large deep-water trawl catches in localized areas (Fig. 5). White

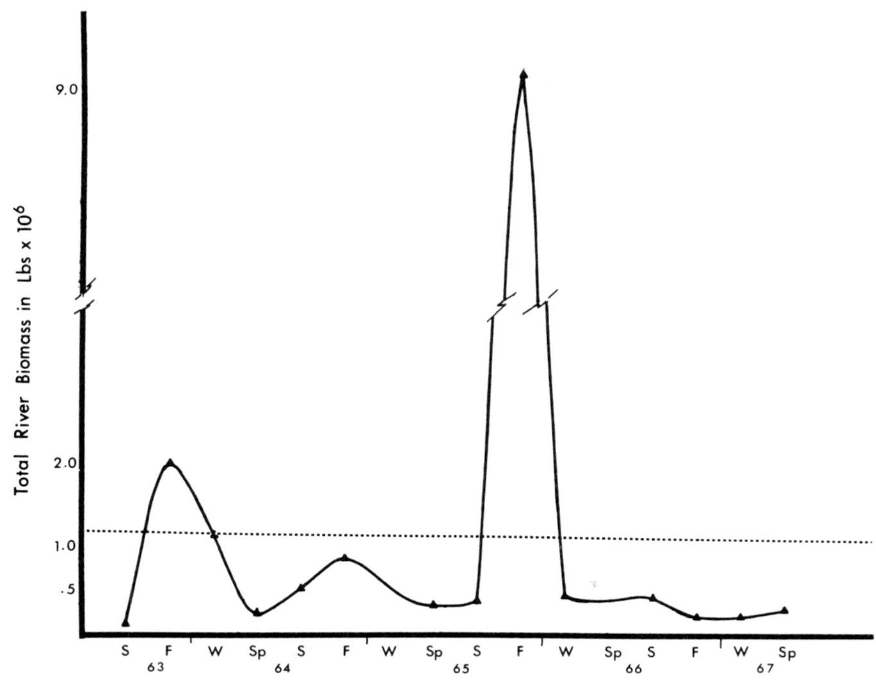

Figure 9. Trend analysis for total Patuxent River biomass in lbs x 10^6 for the area sampled (26,089 acres).

perch over-wintering in the deep channel dominated these catches. In view of these winter concentrations of fish, the low values calculated for mean cool-weather biomass (Fig. 7) are understandable. Mean values of the 50 sampling sites include zero values for many haul-seine sites. On the other hand, the total biomass estimates are weighted in favor of the deep-water trawl collections because this sampling gear was used for the greatest area of the river (Table 3). In summary, both graphs are important to our understanding of the estuary. The conservative graph of mean biomass (Fig. 7) represents the numerous small individuals which are abundant throughout the estuary during warm seasons. The more classic graph of total biomass or estimated standing crop (Fig. 9) represents the large individuals which concentrate in deep, warm holes during cool weather.

Unfortunately, the large fluctuations in total biomass coupled with the loss of seasonal cycles seen in Figure 9 underline the conclusion drawn from the decrease in fish-species diversity; the Patuxent River estuary has undergone a general environmental degradation. This degradation cannot be sepcifically attributed to the operation of the Chalk Point Power Plant, although the plant

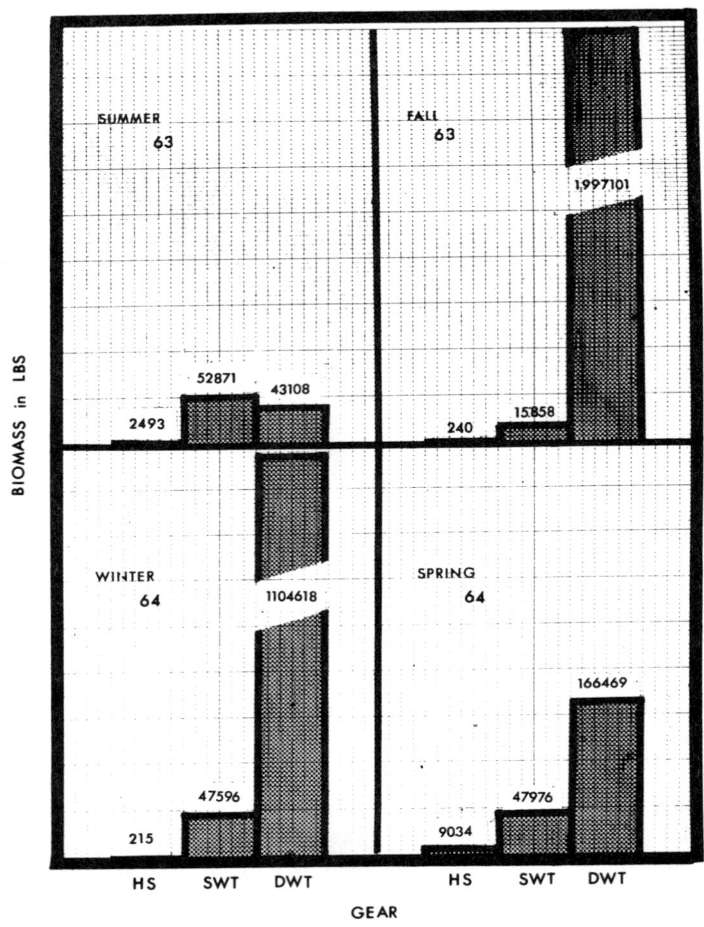

Figure 10. Histogram of biomass in lbs. collected by each gear type. HS = haul seine; SWT = shallow water trawl; DWT = deep water trawl. (a) summer 1963, (b) fall 1963, (c) winter 1964, (d) spring 1964.

undoubtedly contributed to the overall effect. The plant began intermittent operation in the fall of 1964.

FUTURE DIRECTIONS OF RESEARCH

Studies concerning the effects of power plants on estuarine fish populations have provided important insights into estuarine ecology. The data have been useful to managers and scientists alike. Hindsight suggests that many of the approaches would be changed were a similar study to be initiated today

Although a decade has elapsed since the Chalk Point studies commenced, little progress has been made in understanding the use of estuaries as heat sumps. Enough work has been done, however, to reorder the concerns of the 60s. Apart from catastrophic occurrences and from the situations in which ambient temperatures are naturally high, the thermal component has been de-emphasized. Heat *per se* may not be as disruptive as the physical and chemical factors that occur as a concomitant of water utilization. We need to redirect our efforts to solve these problems, the majority of which allow technological solutions.

Conceptually there is much to be gained from viewing power plants with once-through cooling as super nonspecific predators in an ecologic sense. Most materials in the water column, be they fish eggs or plankton, can be considered as grist to the intake screens or structures or to the condensers. Some ecologic change is a fact of industrialization. The extent of this change relative to the resiliency of the system is the practical measure upon which the effects of power plants should be judged.

A significant number of estuarine power plants are too efficient in their predation solely because they are badly placed within the gradient. Careful siting based on rigorous sampling and experienced judgment could have prevented such poor choices. Another group of power plants may or may not be inflicting significant ecological harm by entraining life history stages or screen impingement; the data available simply do not permit the significance of the damage to be calculated. For instance, condenser passage may add a few percentage points to the natural mortality coefficients. Our current level of understanding does not permit us to judge this increase relative to all causes of mortality.

There is a critical need to understand the stress or loading potential of natural systems in order to set intelligent guidelines for protection of the environment. A section of the new national legislation, Public Law 92-500, Sec. 316(a), deserves attention in this respect. To our knowledge, this is the first piece of legislation to promulgate an ecological standard, rather than a water-quality standard. This law does not mandate that oxygen be monitored at a specified level, it requires the protection and propagation of a " . . . balanced indigenous population of fish, shellfish and wild life". It also shifts the burden of proof to the user. While the application of 316(a) will assuredly be marked by some controversy, we are encouraged by this change in orientation. We also realize that this change will influence the spectrum of research activities and efforts.

We are optimistic that the scientific community will be able to respond to the conceptual challenges of 316(a) and to provide the direction required to solve these environmental problems, both in the design of new laboratory and field research programs and in the utilization of existing data.

ACKNOWLEDGEMENTS

Support for this research has come from the Maryland Department of Natura Resources, the Natural Resources Institute of the University of Maryland, the Cooling Water Studies Program of the State of Maryland, and the Marylanc Water Resources Research Center of the University of Maryland. Support fo manuscript preparation has come from QLM Laboratories, Inc.

REFERENCES

1. Alabaster, J. S.
1963 The effect of heated effluents on fish. **Intern. Jour. Air Wate Pollution,** 7: 541-563.

2. Allen, G. H., Boydstun, L. B., and Garcia, F. G.
1970 Reaction of marine fishes around warm-water discharge from ar atomic steam-generating plant. **Progr. Fish Culture,** Jan. 1970, p 9-16.

3. Armstrong, N. E., Gloyna, E. F., and Copeland, B. J.
1968 Ecological aspects of stream pollution. In **Advances in Wate Quality Improvement,** p. 83-95. (eds. Gloyna, E. F. anc Eckenfelder, W. W.) Univ. Texas Press, Austin.

4. Armstrong, N. E., Storrs, P. N., and Pearson, E. A.
1971 Development of a gross toxicity criterion in San Francisco Bay **5th Int. Water Pollution Res. Conf.** (San Francisco and Hawaii 1970.)

5. Barkely, S. W. and Perrin, C.
1972 The effects of the Lake Catherine Steam Electric Plant effluen on the distribution of fishes in the receiving embayment. **Proc Ann. Conf. Southeastern Assoc. Game Fish Comm.,** 25: 384-392

6. Basch, R. and Truchan, J.
1971 A caged fish study on the toxicity of intermittently chlorinate condenser cooling waters at the Consumers Power Company's J D. Weadock Power Plant. Bur. Water Manage., Mich. Wate Resour. Comm.

7. Bechtel, T. J. and Copeland, B. J.
1970 Fish species diversity as indicators of pollution in Galveston Bay Texas. **Contrib. Mar. Sci.,** 15: 103-132.

8. Becker, C. D. and Thatcher, T. O.
1973 Toxicity of Power Plant Chemicals to Aquatic Life. Battell Pacific Northwest Laboratories, Richland, Washington, 99352 WASH-1249.

9. Brett, J. R.
1960 Thermal requirements of fish – three decades of stud 1940-1970. In **Biological problems in water pollution,** p 110-117. (ed. Taft, R. A.) San. Eng. Center Tech. Rept. W60-3 (Unpublished manuscript.)

10. Clark, J. R.
1969 Thermal pollution and aquatic life. **Sci. Amer.,** 220: 19-25.

11. Countant, C. C.
1970 Biological aspects of thermal pollution. I. Entrainment and discharge canal effects. **Chem. Rubber Co. Critical Rev. Env. Contr.,** 1: 341-381.

12. Dahlberg, M. P. and Odum, E. P.
1970 Annual cycles of species occurrence, abundance, and diversity in Georgia estuarine fish populations. **Amer. Midl. Nat..,** 83: 382-392.

13. Derickson, W. K. and Price, K. S.
1973 The fishes of the shore zone of Rehoboth and Indian River Bays, Delaware. **Trans. Amer. Fish. Soc.,** 102: 552-562.

14. DeSylva, D. P.
1969 Theoretical considerations of the effects of heated effluents on marine fishes. In **Biological aspects of thermal pollution,** p. 229-293. (eds. Krenkel, P. A. and Parker, F. A.) Vanderbilt Univ. Press, Nashville, Tennessee. p. 14-17.

15. Drew, H. R. and Tilton, J. E.
1970 Thermal requirements to protect aquatic life in Texas reservoirs. **Jour. Water Pollution Contr. Fed.,** 42(4): 562-572.

16. Elser, H. J.
1965 Effect of warmed water discharge on angling in the Potomac River, Md. 1961-62. **Progr. Fish Cult.,** 27: 79-86.

17. Galtsoff, P. S.
1952 Food resources of the ocean. In **World population and future resources,** p. 108-118. (ed. Halt, P. K.) (Proc. Second Centenn. Acad. Conf.) American Book Co., New York. p. 108-118.

18. Goodyear, C. P.
1972 A simple technique for detecting effects of toxicants or other stresses on a predator prey interaction. **Trans. Amer. Fish. Soc.,** 101: 367-370.

19. Goodyear, C. P.
1973 Probable reduction in survival of young-of-the-year striped bass in the Hudson River as a consequence of the operation of Danskammer, Roseton, Indian Point Units 1 and 2, Lovett and Bowline steam electric generating stations. **U. S. Atomic Energy Commission, hearings on Cons. Edison Co. Indian Pt. nuclear generating unit 2, docket no. 50-247.** A-V 57-A-V 96. (Unpublished manuscript.) .

20. Graham, H. H. and Edwards, R. L.
1962 The world biomass of marine fishes. In **Fish in nutrition,** p. 3-8. (eds. Heen, E. and Kreuzer, R.) Fishing News Ltd., London.

21. Grimes, C. G.
1971 Thermal addition studies on the Crystal River Steam Electric Station. **Florida Dept. Nat. Res. Prof. Pap. Ser.,** 11: 1-53.

22. Hamilton, D. H., Jr., Flemer, D. A., Keefe, C. W., and Mihursky, J. A.
1970 Power plants: Effects of chlorination on estuarine primary production. **Science,** 169: 197-198.

23. Iles, R. B.
1963 Cultivating fish for food and sport in power station water. **New Scientist,** 119: 227-229.

24. Johnson, R. G.
1970 Variation in diversity within benthic marine communities. **Amer. Nat.,** 104: 285-300.

25. Kennedy, V. S. and Mihursky, J. A.
1967 Bibliography on the effects of temperature in the aquatic environment. Univ. Maryland, Nat. Resources Inst., Contr. No. 326. 89 p. (Unpublished manuscript.)

26. Krenkel, P. A. and Parker, F. L.
1969 Engineering aspects, sources, and magnitude of thermal pollution. In **Biological aspects of thermal pollution,** p. 10-72. (eds. Krenkel, P. A. and Parker F. A.) Vanderbilt Univ. Press, Nashville, Tennessee.

27. Lagler, K. F.
1956 **Freshwater fishery biology.** Wm. C. Brown Co., Dubuque, Iowa. p. 164-165.

28. Landry, A. M. and Strawn, K.
1973 Annual cycle of sportfishing activity at a warm water discharge into Galveston Bay, Texas. **Trans. Amer. Fish. Soc.,** 3: 573-577.

29. Lawler, J. P.
1972 Effects of entrainment and impingement at Indian Pt. on the population of the Hudson River striped basss. **U. S. Atomic Energy Commission,** hearings on Cons. Edison Co. **Indian Pt. nuclear generating unit 2, docket no.** 50-247. p. 80. (Unpublished manuscript.)

30. MacArthur, R. H.
1955 Fluctuations of animal populations and a measure of community stability. **Ecology,** 36: 533-536.

31. Mansueti, R. J.
1961 Movements, reproduction, and mortality of the white perch, **Roccus americanus** in the Patuxent estuary, Maryland. **Chesapeake Sci.,** 2: 9-36.

32. Marcy, B. C., Jr.
1971 Survival of young fish in the discharge canal of a nuclear power plant. **Jour. Fish. Res. Bd. Canada,** 28: 1057-1060.

33. Margalef, R.
1951 Diversidad de especies en comunidades naturales. **Pro. Ins. Biol. Apl.,** 9: 5-27.

34. Margalef, R.
1956 Informacion y diversidad especies en las comunidades de organismos. **Invest. Pesquera,** 3: 99-106.

35. McErlean, A. J., O'Connor, S. G., Mihursky, J. A., and Gibson, C. I.
1973 Abundance diversity and seasonal patterns of estuarine fish populations. **Estuarine Coastal Mar. Sci.,** 1: 14-36.

36. McEarlean, A. J., Wall, G. F., Gibson, C., Mihursky, J. A., and O'Connor, S. G.
1969 Tables listing fishes collected from the Patuxent Estuary by three collecting methods for the period 1962 to 1967. **Natural Resources Institute Ref.** No. 69-38. Univ. Maryland. (Unpublished manuscript.)

37. McHugh, J. L.
1967 Estuarine Nekton. In **Estuaries,** p. 581-620. (ed. Lauff, G. H.)

Am. Assoc. Adv. Sci., Publ. 83, Washington, D. C.

38. McNaught, D. C.
1972 The potential effects of condenser passage on the entrained zooplankton at Zion station. Prepared statement at Lake Michigan Enforcement Conf., Sept. 21, 1972. Chicago, Ill. U. S. Atomic Energy Commission, Dir. of Licensing in matter of Common Wealth Edison's Zion generating station, docket no. 5295-5304.

39. Merkens, J. C.
1958 Studies on the toxicity of chlorine and chloramines to the Rainbow trout. **Water Waste Treat. Jour.,** 7: 150-161.

40. Merriman, D. and staff
1965-1972 [Semiannual progress reports to Connecticut Yankee Atomic Power Company, Haddam, Conn.] (Unpublished manuscript.)

41. Mihursky, J. A.
1967 On possible constructive uses of thermal additions to estuaries. **Bio-science,** 17: 698-702.

42. Mihursky, J. A.
1969 Patuxent thermal studies - summary and recommendations. **Univ. Maryland Nat. Res. Inst. Spec. Publ.** 1: 20.

43. Mihursky, J. A. and Cronin, L. E.
1973 Balancing needs of fisheries and energy production. Presentation at **38th North American Wildlife Conference,** 18-21, March, 1973. **Natural Resources Institute,** Univ. Maryland. Contrib. no. 549. (In press.)

44. Mihursky, J. A., McErlean, A. J. and Kennedy, V. S.
1970 Thermal pollution, aquaculture and pathobiology in aquatic systems. **Jour. Wildl. Diseases,** 6: 347-355.

45. Moore, C. J. and Frisbie, C.
1972 A winter sport fishing survey in a warm water discharge of steam electric station on the Patuxent River, Md. **Chesapeake Sci.,** 13: 110-115.

46. Moore, C. J., Stevens, G. A., McErlean, A. J., and Zion, H. H.
1973 A sport fishing survey in the vicinity of a steam electric station on the Patuxent estuary, Maryland. **Chesapeake Sci.,** 14: 160-170.

47. Parker, F. L. and Krenkel, P. A.
1969 **Thermal pollution, status of the art.** Rpt. 3 Dept. of Environmental and Water Resources Engineering, Vanderbilt Univ., Nashville, Tennessee. p. 2-80. (Unpublished manuscript.)

48. Pearson, E. A., Storrs, P. N., and Silleck, R. E.
1970 A comprehensive study of San Francisco Bay. VIII. Summary, conclusions and recommendations. Univ. California Berkeley, Sanitary Eng. Res. Lab. Rep. No. 76-5. 85 p. (Unpublished manuscript.)

49. Picton, W. L.
1960 Water use in the United States 1900-1980. Bus. Def. Serv. Adm. U. S. Dept. Comm., GPO, Washington, D. C. p. 6.

50. Pielou, E. C.
1966 The measurement of diversity in different types of biological collections. **Jour. Theoretical Biol.,** 13: 131-144.

51. Raney, W. E.
1958 Handbook of computations for biological statistics of fish populations. **Bull. Fish. Res. Bd.** Canada, 119: 1-300.

52. Ricker, W. E.
1958 Handbook of computations for biological statistics of fish populations. **Bull. Fish. Res. Bd. Canada,** 119: 1-300.

53. Schaefer, M. B.
1965 The potential harvest of the sea. **Trans. Amer. Fish Soc.,** 94: 123-128.

54. Storrs, P. N., Pearson, E. A., Ludwif, H. F., Walsh, R. and Stann, E. J.
1968 Estuarine water quality and biologic population indices. **Proc. 4th Int. Conf. Water Poll. Res.** Prague.

55. Swartz, R. C.
1972 Biological criteria of environmental change in the Chesapeake Bay. **Chesapeake Sci.** (Suppl.), 13: 517-541.

56. Sylvester, J. R.
1972 Effect of thermal stress on predator avoidance in sockeye salmon. **Jour. Fish. Res. Bd. Canada,** 29: 601-603.

57. Trembley, F. J.
1960 Research project on effects of condenser discharge water on aquatic life. p. 87 (Unpublished manuscript.) Prog. Rept. 156-59. Lehigh Univ. (Bethlehem, Pa.)

58. Trembley, F. J.
1965 Effects of cooling water from steam electric power plants on steam biota. In **Biological problems in water pollution, 3rd Seminar,** p. 334-345. (ed. Tarswell, C. M.) PHS Publ. No. 999-WP-25.

59. Tsai, Chu-fa
1968 Effects of chlorinated sewage effluents on fishes in Upper Patuxent River, Maryland. **Chesapeake Sci.,** 9: 83-93.

60. Tsai, Chu-fa
1973 Water quality and fish life below sewage outfalls. **Trans. Amer. Fish. Soc.,** 102: 281-293.

61. U. S. Atomic Energy Commission
1972 Draft detailed statement on the environmental considerations related to the proposed issuance of an operating license to the Consumers Power Company, Inc., for the Palisades Nuclear Generating Plant, Docket No. 50-255. (Unpublished manuscript.)

62. U. S. Atomic Energy Commission
1972 Final environmental statement, Directorate of Licensing in the matter of Consolidated Edison's Indian Point Generating Unit, No. 2. Docket No. 50-247. (Unpublished manuscript.)

63. Wilhm, J. L.
1967 Comparison of some diversity indices applied to populations of benthic microinvertebrates in a stream receiving organic wastes. **Jour. Water Pollut. Control Fed.,** 39(10) Part 1: 1673-1683.

64. Wilhm, J. L.
1968 Use of biomass units in Shannon's formula. **Ecology,** 49: 153-156.

65. Wilhm, J. L. and Dorris, T. C.
1966 Species diversity of benthic macroinvertebrates in a stream receiving domestic and oil refinery effluents. **Amer. Wildl. Nat.,** 76: 427-449.

66. Wilhm, J. L. and Dorris, T. C.
1968 Biological parameters for water quality criteria. **Bioscience,** 18: 477-481.

STRUCTURAL AND FUNCTIONAL ASPECTS OF A RECENTLY ESTABLISHED *ZOSTERA MARINA* COMMUNITY

Gordon W. Thayer, S. Marshall Adams

and

Michael W. LaCroix[1]

ABSTRACT

Although the value of eel-grass productivity to an ecosystem has been recognized for over 50 years, little quantitative information is available on any major portion of the eel-grass community in North America, save for on the grass itself. The epifaunal and infaunal invertebrates and the fishes inhabiting a grass bed in the Newport River estuary are dominated by only a few species. The density and biomass of these groups are considerably greater than in the adjacent unvegetated portions of the estuary. Fishes using the grass bed appeared to exact some control over the density of the epifaunal community.

The macrofauna in the bed consume an amount of energy equivalent to 55% of the net production of eel-grass, phytoplankton, and benthic algae in the bed. There is sufficient available energy to support the estimated bacteria-microfauna-meiofauna compartment. The data further suggest that there is an excess of plant production in the bed, a portion of which is increasing the organic content of the sediments. The remainder is probably exported to the adjoining estuary. This export may be highly significant to the trophic function of the shallow estuarine system near Beaufort, since eel-grass is estimated to supply as much as 64% of the total production of phytoplankton, cord grass and eel-grass in this system.

1. National Marine Fisheries Service, Atlantic Estuarine Fisheries Center, Beaufort, North Carolina 28516.

INTRODUCTION

Eel-grass, *Zostera marina* L., occurs in temperate marine coastal waters throughout the northern hemisphere and is very important to overall productivity of the coastal zone. It is one of several flowering sea grasses that form structurally distinct estuarine communities or ecosystems composed of characteristic floral and faunal assemblages. In North America, eel-grass communities are to be found from the Bering Straits to Baja, California and from Greenland to Cape Fear, North Carolina.

Eel-grass beds are highly productive. Phillips (22) reported values ranging from 10 to 1200 grams dry weight m^{-2} yr^{-1}. Recent studies in the Beaufort, N. C., area (near the southern edge its range on the Atlantic coast) indicated that *Zostera* produces an average 350 g C m^{-2} yr^{-1} and that associated plants (*Halodule* and *Ectocarpus*) together contribute an additional 300 g C m^{-2} yr^{-1} (7). Thus, on an areal basis, macroscopic plants in eel-grass beds are more productive than the world averages for cultivated corn (412 g m^{-2}), rice (497 g m^{-2}), or the U. S. average for hay fields (420 g m^{-2}) (17).

This production is supplemented by the production of benthic microalgae and epiphytes in the beds, by phytoplankton in the overlying water, and by such shore-based vegetation as salt-marsh plants. More than 50 years ago it was suggested that vascular plants were responsible for a large proportion of the net primary production in shallow coastal waters (21). In the shallow estuarine system near Beaufort, eel-grass has been estimated to supply as much as 64% of the total production of phytoplankton, smooth cord grass (*Spartina alterniflora*) and eel-grass (35).

Few organisms, however, use the living plant as a food source (6); rather, detritus plus associated microbes active in the decay of the grass provide the basic source of nutrition for the majority of invertebrates present (1, 14, 18). This detrital material also supports faunal assemblages in other estuarine habitats (18). However, despite the extensive literature on eel-grass and other sea grasses (22, 39) and on the ecological consequences of eel-grass destruction, documented during and since the sudden and drastic decline of eel-grass stocks during the 1930s, few quantitative data are available on any major portion of the community or on the structure and function of eel-grass communities in North America. It is the purpose of this paper to describe the ongoing research at the Atlantic Estuarine Fisheries Center on the structure and function of eel-grass beds in the vicinity of Beaufort.

One representation of the eel-grass community is the simplified model shown in Figure 1. Energy and material budgets for the compartments included in this diagram are being determined and compared with those being developed for the adjacent unvegetated portions of the Newport River estuary. These data will provide quantitative information on the relative importance of vegetated and

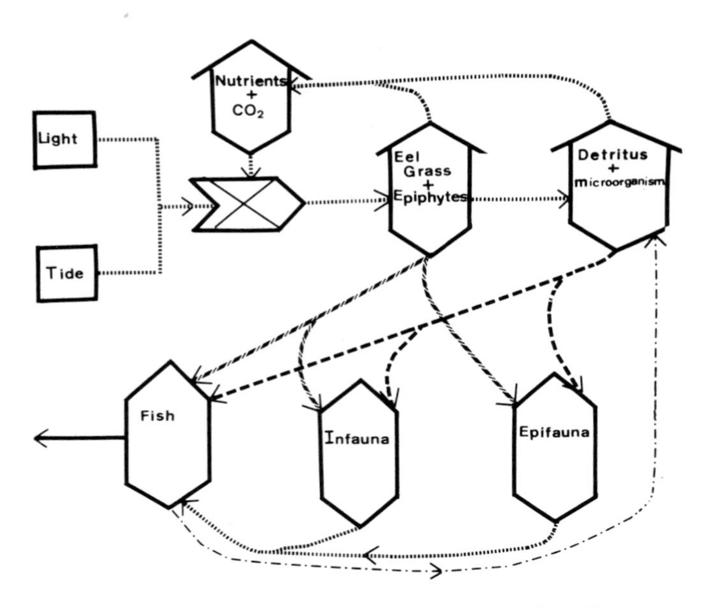

Figure 1. Simplified diagram of an eel-grass system showing the major compartments and flows. Squares are energy-producing forces, tanks represent energy storage compartments, and hexagons represent consumer units.

unvegetated estuarine areas and will provide a basis for evaluating the real benefits of eel-grass to the estuarine ecosystem. Our efforts to date have dealt primarily with the fish, infauna, and epifauna compartments, but work underway or planned includes sedimentation rates, nutrient cycles, the role of epiphytes, detrital cycles, microbial heterotroph activity, and export of materials from grass beds. In addition, the Center is engaged in developing a model of the flux of selected trace metals in eel-grass beds.

STUDY AREA

Figure 2 shows the Newport River estuary, Carteret County, North Carolina, with Morehead City on the west and Beaufort on the east. The insert in the figure is of Phillips Island and shows the location of an eel-grass bed in a semiprotected embayment on the southeast side of the island. The island also has an 8×10^3 m^2 mixed heronry of approximately 1800 birds of eight species of herons, ibises, and egrets.

The embayment, covering approximately 4×10^4 m^2, has a mean depth of 1.0 m at high tide and 0.3 m at low tide. Salinity averages 26 $^o/oo$ (range 19-32 $^o/oo$) and water temperature averages 20 ^oC (range 3.6-34.0 ^oC). Protection afforded by the island and submerged sand bars at the northeast and southeast

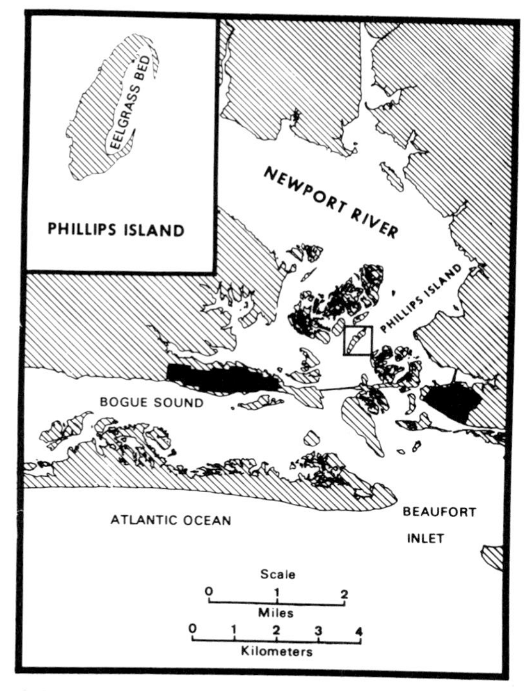

Figure 2. Diagram of the Beaufort-Morehead City, N. C., area, showing the location of the Newport River estuary and Phillips Island.

ends of the embayment minimize current velocities and scouring action within the embayment.

Prior to 1968 the eel-grass bed was not a permanent feature of this embayment, and at the beginning of our study (1969-1970) eel-grass covered approximately 30% of the embayment during periods of maximum density. Since then the bed has increased in size and, in 1973, it covered approximately 55% of the embayment.

METHODS

Sedimentation rate

In October 1972, 132 plastic-screen rectangles (10 x 10 cm; mesh openings 1.2 mm square) were anchored to the sediment surface with galvanized iron wire pins. Four screens were placed at each of 33 stations throughout the embayment. One-half of the screens were removed in April 1973, and the sediment which had accumulated on the screens was removed and freeze-dried.

Particle size (4) and organic carbon (16) were measured on each sample. Daily and annual sedimentation was computed in grams dry wt m^{-2}, and these data were converted to volumes based on particle size distribution and dry weight to volume factors of 493 mg cm^3 for mud and 1700 mg cm^3 for sand (38).

Eel-grass

Standing stocks of *Zostera* and its associated macroscopic algae were sampled from September 1969 to September 1970 and from September 1971 through October 1972. Six samples were collected monthly by clipping material enclosed in randomly tossed 0.25 m^2 quadrats. All material (eel-grass, algae, and detritus) was processed for wet, dry, and ash-free dry weight measurements and for caloric analyses (33).

Invertebrates

Epifauna were collected monthly from August 1969 through September 1970 using an epibenthic sled fitted with a mesh bag (1.2 mm diameter openings) and a cutting blade. Marsh (13) has reviewed many of the techniques used in sampling aquatic fauna. At least three randomly chosen one-meter-square areas were sampled each month. Infauna were sampled monthly from November 1970 through January 1972 using a post-hole digger which removed an area of approximately 0.014 m^2 to a depth of 18 cm. Each month 110 samples, or approximately 1.5 m^2, were taken. Organisms were separated from the sediment using a 1.2 mm mesh sieve, and identified, and the wet, freeze-dried, and ash-free dry weights and caloric content were measured.

Replicate respiration measurements of each invertebrate species were made on random subsamples of field populations using a differential respirometer at environmental temperatures. We used Millipore-filtered estuarine water (HA 0.45 m μ filters) containing streptomycine-SO_4 to prevent bacterial oxygen consumption (8). In each case two filtered water blanks were used.

Assimilation of energy by the invertebrate compartments was estimated from an assumed respiration to assimilation ratio of 0.75 for each species since the communities were composed primarily of herbivores and detritivores (30). We estimated the respiration of the bacteria-microfauna-meiofauna compartment from Pamatmat (20), who showed this compartment to respire a mean of approximately 11 ml O_2 m^{-2} hr^{-1} during the spring and summer. The mean standing crop biomass of this compartment was assumed to be equivalent to 0.2% of the carbon content of the sediment. The estimated biomass (4.7 g C m^{-2}) is at the upper end of the range of microbial biomass (0-5 g C m^{-2}) in the anaerobic sediments of the estuary; the proportion of the organic content of sediments represented by this compartment is similar to that observed in anaerobic sediments of the Newport River estuary (Ferguson and Murdoch, this symposium).

Fishes

The fish community frequenting the grass bed was sampled monthly from September 1971 through October 1972 using a portable drop-net (9, 11). Species composition, consumption, periodicity of feeding and production were analyzed. Consumption was computed using the equation

$$C = 1.25 (P + R)$$

where C is consumption in calories, P is production and R is respiration. The value 1.25 was used, since assimilation efficiency was assumed to be 80% (37). Production was computed using the model

$$P = GB$$

where P is production, G is the instantaneous coefficient of growth, and B is the mean biomass between sampling periods (25). Respiration rates were measured using a flowing-water respirometer (10) and a variety of ambient water temperatures and fish sizes. Caloric equivalents were computed using the oxycaloric equivalent 4.83 cal ml^{-1} O_2.

RESULTS

Sedimentation rates

The rates of sedimentation and percent carbon of the deposited material were variable, with deposition rates ranging from 1-57 mm yr^{-1} and carbon content of the deposited material ranging from 0.7-5.6% (Table 1). These values represent the ranges observed during the short term of the study, 170 days, and may not be applicable to long-term processes in the embayment. The range in sedimentation rate, however, is not unlike the range of sedimentation observed

TABLE 1.

Mean sedimentation rate and percent organic carbon of the material deposited in an embayment of Phillips Island. Ranges are shown in parentheses.

Location	mm Yr^{-1}	% carbon
Open estuarine side	6(1-15)	0.9(0.8-1.4)
Grassed portion	14(12-22)	1.8(0.7-2.9)
Island side	30(17-57)	2.3(0.9-5.6)

for marsh areas near Beaufort dominated by *Spartina alterniflora* (0.01-34.2 mm yr^{-1}) (36); this computation assumes that the material deposited on the marsh has a weight-to-volume ratio of an equal mixture of mud and sand (1100 mg cm^{-3}).

The data presented in Table 1 show the influence of eel-grass on the material deposited and the rate at which it is deposited. Both parameters were least near the estuary, intermediate in the vegetated zone, and highest near the island side of the embayment. The data are supported by our observations over the past several years that the island side is filling in, the vegetated area is rising relative to sea level, and the grass is migrating toward the estuary.

Eel-grass

Zostera biomass varied seasonally according to water temperature (Fig. 3) in a pattern similar to that noted in other estuarine areas (5, 7, 14). Maximum biomass during both collection periods occurred from June-Ausust, followed by a steady decline to minimum values during November-February. The biomass of

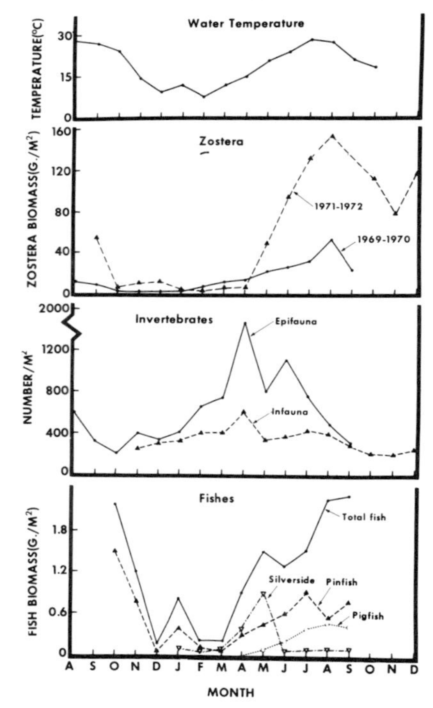

Figure 3. Seasonal distribution of water temperature, **Zostera** dry-weight biomass, mean epifaunal and infaunal abundances and total fish and major fish species biomass.

associated macroalgae followed a similar pattern, being greatest (32 g m^{-2}) during May-June and least during winter (0.1 g m^{-2}). Although 11 species of algae were collected, *Ectocarpus confervoides* and *Agardhiella tenera* dominated during winter-spring and *Hypnea* spp. dominated during summer-fall.

During 1969-1970, eel-grass biomass (1.0-50 g dry weight m^{-2}) was considerably smaller than that during 1971-1972 (1.6-153 g m^{-2}). Biomass values of more established beds, such as nearby Bogue Sound (175-545 g m^{-2}) (7) and the York River estuary (8-212 g m^{-2}) (14), are generally higher than our values. The rapid increase of eel-grass in the Phillips Island bed indicates that the embayment is evolving toward a permanent and more dense bed. Samples taken in this bed during the summer of 1973 (in excess of 200 g m^{-2}) confirmed this.

There was a seasonal variation in the caloric content of the grass, with maximum values of 5.73 $\pm$ 0.23 cal mg^{-1} ash-free dry weight during the period of flowering (April-May) and minimum values of 3.54 $\pm$ 0.31 (January) to 4.82 $\pm$ 0.38 (June). The overall mean for the year, 4.48 cal mg^{-1}, was similar to the 4.38 cal mg^{-1} reported for eel-grass in Alaska (15).

INVERTEBRATES

Epifauna

A total of 25,000 individuals of 45 species of epifaunal invertebrates was collected, 12 species representing 90% and 17 species representing 95% of all organisms collected (Table 2). Marsh (13, 14), in the only other available quantitative study of eel-grass bed epifauna on the east coast of the U. S., also showed that only a few species (22 of 100 species) comprised 95% of the epifauna in the York River. *Bittium varium,* which represented 48% of the epifauna at Phillips Island, also represented 36% of the total epifauna at one location in the York River (13). Shannon-Weaver diversity indices averaged 1.10 and ranged from 0.6-1.4 for the epifauna at Phillips Island compared to indices of 1.92-3.90 in the more mature beds of the York River. Our low species numbers and diversity indices relative to those in the York River suggest a young and evolving grass-bed community.

This epifaunal population is structurally distinct from the invertebrate populations of the adjacent Newport River estuary. The great majority of the species collected in the grass bed represent only a small fraction of the total biomass and abundances of the species present in the unvegetated portion of the estuary (23, 32).

The dominant taxa, based on total numbers collected, were gastropods (72%), amphipods (12%), pelecypods (7%), and decapods (6%). Analysis of feeding types, based on the literature, also revealed a dominance by one group. Deposit feeders represented 77% of the numbers; suspension feeders, 18%; and carnivore-scavengers, the remainder.

When the community was evaluated on an ash-free dry weight basis (AFDW), however, the dominance by one species or group was less striking. Ninety percent of the total AFDW collected (32.4 g) was contributed by 18 species. *Nassarius vibex* (18%), *Callinectes sapidus* (8.8%), *Mogula manhattensis* (8.0%) and *Bittium varium* (7.8%) were the most significant species. The dominant taxa, in terms of biomass, were gastropods (38%), decapods (30%), and pelecypods (11%) (Table 3). No feeding-type dominance was observed on a weight basis, each of the three major feeding types representing 30-36%.

There was a seasonal distribution in mean numerical abundance of epifauna, with maximum numbers during spring and early summer (Fig. 3). During fall through spring there was a weak but significant positive linear correlation between numbers and grass density ($r^2 = 0.64$; $F = 4.07$, significant at the 0.90 level). During the summer, however, invertebrate abundance generally declined, even though plant biomass increased. This decline coincided with an increase in fish biomass (Fig. 3), and during May through September there was a significant negative correlation between these parameters ($r^2 = -0.99$; $F = 122.21$, significant at the 0.95 level); thus, predation might have had some control over the epifaunal abundances we observed. Bird predation also might have been significant, since summer is the peak period of abundance on the surrounding island.

Ash-free dry weight biomass ranged from a mean of 0.21 g m^{-2} during September and October to 2.52 g m^{-2} during June (average 0.93 g m^{-2}) (Table 3). As evidenced by this low biomass and the species composition (Table 2), the epifauna was dominated by small species with an average organism having an AFDW of 1.56 mg. The energy available to predators at any one time was minimum during the fall (0.7 kcal m^{-2}) and maximum in spring (10.7 kcal m^{-2}), and averaged 4.2 kcal m^{-2} (Table 3).

The epifaunal population is changing, however, with bay scallops, *Argopecten irradians,* being the dominant organism in terms of biomass during 1973. Grass density during the summer increased about 5-fold and organismal density about 7-fold.

Infauna

The infaunal population had a low numerical abundance with a total of 6,930 individuals collected. It was composed of 40 species and was dominated by 5 species (*Nereis pelagica, Tellina versicolor, Solemya velum, Abra aequalis* and *Clymenella torquata*). Taxonomically, pelecypods represented 58% and

TABLE 2.

Epifaunal ranking by abundance.

Species	Rank	Number	Percent of fauna	Cumulative percent
Bittium varium	1	12092	48.11	48.11
Retusa canaliculata	2	2230	8.87	56.98
Mitrella lunata	3	1811	7.20	64.18
Carinogammarus mucronatus	4	1593	6.34	70.52
Hippolyte pleuracantha	5	905	3.60	74.12
Amphithoe longimana	6	779	3.10	77.22
Pyramidella fusca	7	770	3.06	80.28
Nassarius vibex	8	658	2.62	82.90
Caprella geometrica	9	628	2.50	85.40
Macoma tenta	10	494	1.96	87.36
Abra aequalis	11	390	1.55	88.91
Nereis pelagica	12	363	1.44	90.36
Chione cancellata	13	354	1.41	91.76
Paleomonetes vulgaris	14	243	0.97	92.73
Anadara transversa	15	211	0.84	93.57
Penaeus setiferus	16	167	0.66	94.23
Urosalpinx cinerea	17	163	0.65	94.88
Callinectes sapidus	18	134	0.53	95.41
Solen viridis	18	134	0.53	95.94
Anachis avara	20	133	0.53	96.47
Crangon septemspinosa	21	123	0.49	96.96
Cyathura carinata	22	118	0.47	97.43
Unidentified gastropod	23	113	0.45	97.88
Unidentified polychaete	24	88	0.35	98.23
Argopecten irradians	25	64	0.25	98.48
Crepidula convexa	26	58	0.23	98.71
Coryphella sp.	27	55	0.22	98.93
Crepidula fornicata	28	45	0.18	99.11
Panopeus herbstii	29	44	0.17	99.28
Oxyurostylis smithi	30	34	0.13	99.41
Balanus balanoides	31	32	0.13	99.54
Unidentified isopod	32	20	0.08	99.62
Ophioderma brevispina	33	19	0.07	99.69
Pagurus longicarpus	33	19	0.07	99.76
Mercenaria mercenaria	35	14	0.05	99.81

TABLE 2 (Continued).

Semele sp.	36	11	0.04	99.85
Molgula manhattensis	37	9	0.03	99.88
Polinices duplicata	38	5	0.02	99.90
Anomia simplex	39	4	0.01	99.91
Asterias forbesi	40	3	0.01	99.92
Nassarius obsoletus	41	1		
Limulus polyphemus	41	1		
Dosinia discus	41	1		
Callianassa stimpsoni	41	1		
Architectonica sp.	41	1		100.00
Total	45	25,135		

polychaetes 41% of the individuals, with nemertines, crustaceans and echinoderms together representing only 1% (Table 3). Most of the infaunal species were much less abundant in the remainder of the estuary. Species diversity ranged from 1.2-1.5 and is low compared to infauna of grass beds near Chesapeake Bay (19) and of unvegetated estuarine areas (3).

The infaunal invertebrates displayed a slight seasonal trend in numerical abundance with maximum numbers during spring (Fig. 3). Polychaetes dominated during fall-winter and pelecypods dominated during the remainder of the year. There also was a seasonal trend in AFDW biomass with maximum values during November-March (range 6.86-9.96 g m^{-2}) and lower values during the remainder of the year (range 2.97-6.51 g m^{-2}); the average biomass during the year was 6.6 g m^{-2} (Table 3). The energy equivalent of this biomass followed a similar trend and averaged 30.5 kcal m^{-2}; the specific energy (cal per mg) generally was minimum during fall through early winter and maximum during spring.

The infauna was dominated by deposit feeders, which comprised 53% and 44% of the abundance and biomass values, respectively; suspension feeders represented 42% and 35% respectively. The preponderance of deposit and suspension feeders, partially dependent upon detritus and its associated microbes in and over these fine sediments, is similar to the observations for fine sediments of the adjoining Newport River estuary (32).

Two features of the invertebrate populations of this eel-grass bed stand out. There was a low density of organisms and of biomass (Table 3). Total abundance averaged 923 individuals m^{-2} and total AFDW biomass averaged 7.5 gm^{-2}. The abundance values fell well below those for other areas. For example, Marsh (13) observed a mean epifaunal density of approximately 9600 m^{-2} and Orth (19) observed 15000 m^{-2} for the infauna of grass beds in the vicinity of Chesapeake

TABLE 3.

Mean values for structural and functional parameters of the invertebrate community and of the major invertebrate taxa collected from an eel-grass bed in the Newport River estuary. Data for infauna are for August 1969-September 1970; for epifauna, November 1970-January 1972.

	Structural Aspects			Functional Aspects		
	Number	Biomass	Energy	Assimilation	Respiration	Production
	($N\ m^{-2}$)	(g dry wt m^{-2})	(kcal m^{-2})		(cal m^{-2} day^{-1})	
Epifaunal community[1]	595	0.93	4.19	180.8	135.6	45.2
Amphipoda	56	0.05	0.21	12.3	9.2	3.1
Decapoda	37	0.27	0.98	59.9	44.9	15.0
Gastropoda	428	0.38	1.65	55.7	41.8	13.9
Pelecypoda	44	0.12	0.51	21.2	15.9	5.3
Infaunal community	328	6.61	30.5	641.6	481.2	160.4
Echinodermata	0.5	0.14	0.2	3.2	2.4	0.8
Decapoda	1.2	0.14	0.8	26.8	20.1	6.7
Nemertea	1.8	0.73	3.8	55.6	41.7	13.9
Pelecypoda	191	2.24	8.8	228.0	171.0	57.0
Polychaeta	134	3.36	17.1	328.0	246.0	82.0

[1] Includes **Balanus** sp., **Ophioderma brevispinam** and **Nereis pelagica** which occasionally were collected in the epifauna.

Bay. Comparative biomass estimates for invertebrate populations in eel-grass beds are not available.

Our values, however, are considerably larger than those of the adjacent unvegetated portions of the Newport River estuary. A recently completed spring survey of the estuary, encompassing intertidal, shallow subtidal and deeper areas, disclosed an invertebrate density of 170 individuals m^{-2} with a mean AFDW biomass of 5 g m^{-2} (32). Thus, not only is the species composition of the eel-grass bed different from that of the adjacent unvegetated estuary, but the density and biomass of invertebrates are greater in the grass bed.

Functional Aspects of the Invertebrate Populations

The functional aspects of benthic populations were examined in terms of metabolic and trophic relations. We have mentioned some of the trophic relations previously. Benthic metabolic studies generally have shown that the macrofaunal compartment is responsible for only a small fraction of total benthic metabolism, the largest fraction being contributed by the bacterial-microfauna-meiofauna compartment.

Our metabolic measurements indicated that the macroinvertebrate populations of this grass bed respired about 620 cal m^{-2} day^{-1} (130 ml O_2 m^{-2} day^{-1}); the epifauna contributed 136 cal m^{-2} day^{-1}, and the infauna the remainder (Table 3). The microbial compartment was estimated to respire 1280 cal m^{-2} day^{-1}, or 67% of the total respiration of the invertebrate and microbial compartments. The estimated proportion of the total respiration is within that noted for other benthic systems (20, 28, 29).

Individual and macroinvertebrate population metabolism tended to be least during winter and greatest during May-July. Although decapods comprised only 6% of the mean epifaunal density, 29% of the biomass, and 23% of the mean standing crop of energy, they contributed 33% of the epifaunal respiration. The dominant gastropods (71% of the numbers, 41% of the tissue biomass, and 39% of the mean standing crop of energy) contributed 31% of the epifaunal respiration. Infaunal metabolism was dominated by polychaetes and pelecypods, 51% and 35% of the total, respectively, and the crustaceans, nemertines, and echinoderms (1% of the numbers and 15% of the biomass and energy values) contributed 13.5% of infaunal respiration. Thus, dominance in terms of density, biomass or energy does not necessarily signify a similar functional dominance.

Our estimates of assimilation indicated that the total macroinvertebrate community assimilated an average 820 cal m^{-2} day^{-1}, the epifauna contributing 181 cal m^{-2} day^{-1} and the infauna the remainder (Table 3). Production estimates, based on the difference between estimated assimilation and measured respiration, indicated a production of approximately 200 cal m^{-2} day^{-1}. Consumption estimates, based on an assimilation-to-ingestion ratio of 0.40, averaged 450 and 1600 cal m^{-2} day^{-1} for the epifauna and infauna, respectively, values which on a yearly basis are equivalent to 11% and 38%, respectively, of the annual net production of eel-grass, phytoplankton, and benthic algae in this system (1550 kcal m^{-2}). The net production of eel-grass (950 kcal m^{-2} yr^{-1}) is based on the work of Dillon (7), that for phytoplankton(470 kcal m^{-2} yr^{-1})from Williams (34) and Thayer (31), and that for benthic algae (125 kcal m^{-2} yr^{-1}) on unpublished data from R. B. Williams.

Our estimates of energy available to these populations does not include the contribution of benthic microalgae and epiphytes to the productivity of the system. The latter value is probably large relative to production by included plant compartments, since in many instances total epiphyte biomass may equal that of the grass blades (39).

FISHES

Temporal Distribution

During our survey 33 species of fish were collected from the eel-grass bed. Pinfish (*Lagodon rhomboides*), pigfish (*Orthopristes chrysopterus*), Atlantic silverside (*Menidia menidia*), and anchovies (*Anchoa mitchilli* and *Anchoa hepsetus*) dominated the ichthyofauna and accounted for about 80% of the total annual biomass of fishes.

There was a seasonal distribution in both numerical abundance and biomass of fishes in this grass bed (Fig. 3). Pinfish was the dominant species, accounting for approximately 50% of the total standing-crop biomass. During most of the year the distribution of total biomass was similar to that of the pinfish population. The fall peak was the result of an influx of adult pinfish, whereas biomass during summer was composed of a combination of juvenile pinfish, pigfish, and silver perch (*Bairdiella chrysura*), which entered the grass bed as larvae in early spring through early summer and grew at an exponential rate throughout this period (Table 4). The temporal distribution in biomass was related not only to influx and growth but was significantly correlated to temperature patterns ($r^2 = 0.65$; $F = 5.79$, significant at the 0.95 level) and also to the density of eel-grass ($r^2 = 0.68$; $F = 15.33$, significant at the 0.95 level) (Fig. 3). The latter presumably in part was the result of increased food availability and increased cover afforded by the grass. The diversity of fish, based on weights, ranged from a Shannon-Weaver index of 0.34 in December to 1.87 in July (average 1.16).

The biomass of fishes ranged from 0.08-2.18 g dry wt m^{-2} and averaged 1.33 g m^{-2} over the year. Kjelson et al. (11) estimated a mean biomass for *Spartina* marsh-pond areas in the Newport River estuary (13% of the total estuary) of 5.9 g wet wt m^{-2} (approximately 1.2 g dry wt m^{-2}) and 1.5 g wet wt m^{-2} (approximately 0.3 g dry wt m^{-2}) for the estuary as a whole. Thus, the mean standing-crop biomass of fishes in the eel-grass bed is similar to that ot marsh-pond habitats in the estuary and considerably larger than the biomass for the estuary as a whole.

Production

The production values for the major fish species and their size classes are summarized in Table 4. Production increased exponentially during spring and summer to a maximum during late summer. There was little fish production during the winter months, and adult silversides showed negative production during this period. The negative production values for silversides, juvenile filefish, anchovies, and pipefish suggest either that the fish were losing weight or that there was a differential replacement of large fish by smaller fish during the nonproductive months as a result of immigration and migration processes.

The relatively high production value, 20 g wet wt m^{-2} (4.3 g dry wt m^{-2}), is greater by an average of 25% than production values for fish populations reported from a variety of habitats (12). The high production of fish in this eel-grass bed apparently resulted from the large number of young-of-the-year fish which grew at an exponential rate, the availability of adequate and diverse food sources, and the relative stability of the eel-grass habitat in terms of protection from predators and diversity of niches (22, 24).

Of the total annual production for all species, juvenile pinfish contributed 30% and adult pinfish 22%. Juvenile pigfish contributed 18%. Thus, these two species accounted for approximately 70% of the annual production of fishes in the bed.

Respiration

The total amount of energy used in metabolism by the fishes during the year was 72 kcal m^{-2} (Table 5). Anchovies were responsible for 30% of the total, yet accounted for only 9% of the average fish biomass. Pinfish also contributed 30% of the metabolism, but this species was responsible for 49% of the population biomass. Thus, as was noted for the invertebrate populations, structural dominance and functional dominance are not necessarily synonymous.

Consumption

The energy consumed by the fish population using this bed was 103 kcal m^{-2} yr^{-1} (Table 5). Thus, the total fish population consumed an amount of energy equivalent to approximately 7% of the annual net production of eel-grass, phytoplankton, and benthic algae in the bed (1550 kcal m^{-2}), a value which is somewhat less than that estimated to be consumed by the epifauna in the bed.

TABLE 4.

Production in g dry wt m^{-2} of all species of fish in the Phillips Island eel-grass bed each month.

Month	1	2	3	4	5	6	7	8	9	10	11	12	13	14	15	16	Monthly totals
Sept. 71	—	—	—	—	0.068	—	—	—	—	—	0.280	—	—	—	—	—	0.348
Oct. 71	—	—	—	—	—	0.032	-0.011	—	—	—	0.261	0.041	—	—	—	—	0.323
Nov. 71	—	—	—	—	—	—	—	—	—	—	—	—	—	—	—	—	—
Jan. 72	0.042	—	—	—	—	—	0.009	—	—	—	—	—	—	—	—	—	0.043
Feb. 72	0.005	—	—	—	—	—	-0.008	—	—	0.005	—	—	—	—	—	—	0.002
March 72	0.0009	—	—	—	—	—	-0.008	—	—	—	—	—	—	—	—	—	-0.007
April 72	0.013	—	—	0.014	0.005	0.007	-0.023	—	—	—	0.156	0.013	—	—	—	—	0.185
May 72	0.127	0.063	0.026	-0.029	-0.011	-0.012	0.041	—	—	—	0.191	—	—	—	—	—	0.396
June 72	0.423	0.23	—	0.014	0.007	0.013	0.008	—	0.007	—	—	—	0.040	0.003	0.003	—	0.748
July 72	0.440	0.367	—	0.067	-0.229	0.005	—	—	0.073	—	—	—	0.064	0.007	—	0.074	0.868
Aug. 72	0.240	0.124	—	—	0.776	0.009	—	—	0.100	—	—	—	0.100	—	—	—	1.350
Sept. 72	—	—	—	—	—	—	—	—		—	—	—	—	—	—	—	—
Species	—																
totals	1.291	0.784		0.066	0.616	0.054	0.001		0.180	0.005	0.888	0.054	0.204	0.010	0.003	0.074	Yearly prod.=4.3

Species Code

1. Juvenile pinfish
2. Juvenile pigfish
3. Adult silver perch
4. Juvenile filefish
5. Anchoa
6. Pipefish
7. Adult silverside
8. Adult spot
9. Gag - grouper
10. Larval spot
11. Adult pinfish
12. Flounder
13. Juvenile silver perch
14. Postlarval filefish
15. Juvenile silverside
16. Subadult pigfish

TABLE 5.

Yearly mean values for the structural and functional parameters measured for the fish species collected from an eel-grass bed in the Newport River Estuary, N. C.

	Structural Aspects		Functional Aspects			
	Biomass	Energy	Production	Respiration	Consumption	P/C
	g dry wt $m^{-2}yr^{-1}$	kcal $m^{-2}yr^{-1}$	kcal $m^{-2}yr^{-1}$			
Juvenile pinfish	0.215	0.968	5.81	11.94	22.18	0.26
Adult pinfish (< 145 mm)	0.335	1.508	4.00	7.00	13.75	0.29
Adult pinfish (> 145 mm)	0.100	0.450	2.59	2.45	6.29	0.41
Anchovy	0.120	0.540	3.93	21.69	32.04	0.12
Juvenile pigfish	0.069	0.310	3.52	4.67	10.25	0.34
Adult pigfish (< 145 mm)	0.053	0.240	0.33	2.06	2.97	0.11
Adult pigfish (> 145 mm)	0.038	0.170	—	2.42	—	—
Silver perch	0.056	0.253	1.17	3.24	5.51	0.21
Silverside	0.080	0.360	—	6.32	—	—
Filefish	0.016	0.072	0.47	2.44	3.63	0.13
Pipefish	0.023	0.102	0.30	0.56	1.07	0.28
Grouper	0.026	0.117	0.81	0.77	1.97	0.41
Spot	0.101	0.460	0.18	1.95	2.66	0.07
Flounder	0.064	0.289	0.24	0.67	1.13	0.21
Goby	0.011	0.050	—	0.11	—	—
Others	0.026	0.120	—	4.15	—	—
Total:	1.333	6.009	23.35	72.44	103.45	—
Mean	—	—	—	—	—	0.24

The amount of energy consumed by a species is probably the best indicator of the functional importance of that species in the community, since consumption reflects both production capacities and metabolic requirements of a species. Therefore, pinfish were not only dominant in the structure of the fish population (49% of the biomass) but also were functionally dominant in terms of total consumption (Table 5). The anchovy population, although 9% of the fish biomass, was second in importance in the function (31% of annual consumption compared to 41% for pinfish) of the fish population.

Ecological Efficiency

The ecological efficiency, or the ratio of total production to total food consumed, assuming that all of production is taken up by predators (12), was calculated for each major fish species (Table 5). The mean ecological efficiency for this population, 24%, is high compared to values of 5%-15% for other fish populations (2, 12) and aquatic environments in general (26), and in part was the result of the large proportion of juveniles in the population; Mann (12) also has noted highest ecological efficiencies in populations dominated by juvenile fishes. Our comparatively high efficiencies suggest that this eel-grass bed is an efficient system that provides resident fish with superior shelter, food, and protection (22, 24). These fishes therefore would spend proportionally less of their assimilated energy coping with environmental extremes, searching for food, and escaping from predators, and hence may use a great proportion of consumed energy for growth and production.

SYNTHESIS

We have synthesized the data presented in Tables 3 and 5 into an energy-flow diagram based on the mean yearly values for standing crops, production, respiration, and consumption (Fig. 4). The budget is incomplete because we do not know what contribution epiphytes and benthic microautotrophs make to the productivity of the system, nor do we know what proportion of the energy sources is used by the various taxonomic groups and species. In addition, average literature values for respiration/assimilation and assimilation/consumption have been used for the invertebrates. We have not included all of the known flows to consumer units in the diagram; e.g., the flow from phytoplankton to the consumers is omitted.

Slobodkin (26, 27) suggested that the steady-state ratio of yield to energy consumed (ecological efficiency) in aquatic environments should have a value between 5% and 20%. If we assume that all the production is used by predators, as Mann (12) has done for fish populations, since we have no measure of the proportion to yield and to other types of mortality, our gross-growth efficiency

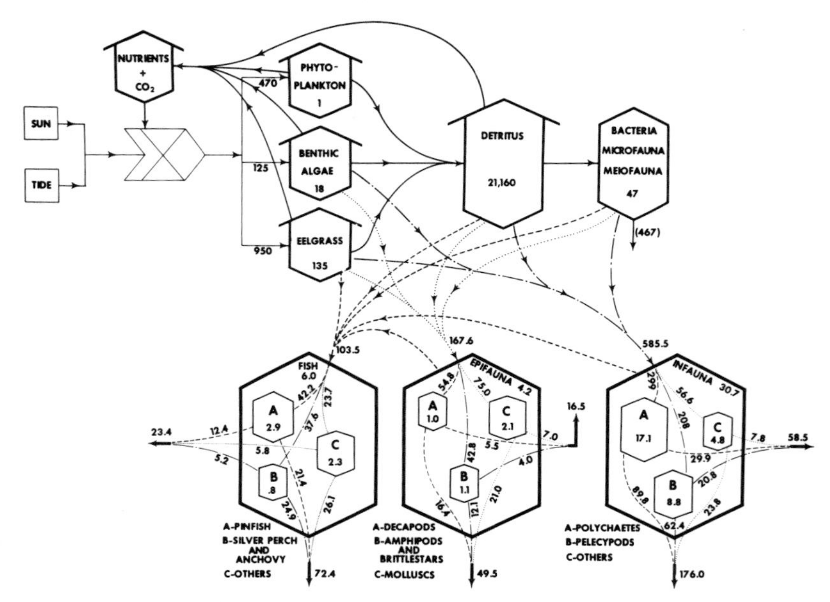

Figure 4. Diagram of the annual flow of energy in an eel-grass bed in the Newport River estuary. Flows (indicated by arrows) are in kcal m^{-2} yr^{-1}. Smaller hexagons in their consumer units represent the major species or taxonomic groups. Flows into producer compartments are net production, and flows into each consumer compartment represent consumption and are formed of flows from the major food sources for each compartment. Outflows at the bottom of the consumer compartments represent metabolic energy losses, and the remaining outflows are production values; in each case the flows from the major species or groups are summed to form outflows.

(production/consumption) can be used as a measure of ecological efficiency. Our gross-growth efficiency for the macrofauna of this eel-grass bed is 12%, a value which is intermediate in Slobodkin's range; the net efficiency for the macrofauna is 26%. The gross-growth efficiency for the macrofauna and for fishes alone (24%) suggests that the eel-grass system is a fairly efficient system.

Our estimates also suggest that the macrofauna consume an amount of energy equivalent to approximately 55% of the net production of eel-grass, phytoplankton, and benthic algae in the system; the proportion of plant production consumed by the macrofauna is probably high because of the lack of information on the contributions made by epiphytes and benthic microautotrophs. Thus, there appears to be an abundance of available energy in the grass bed, and the invertebrate populations should be able to expand beyond the biomass and densities we observed. Failure of these herbivores and detritivores to expand to the limits of their estimated food reserves suggests that estimates of energy available may be too high because of the refractory nature of the material or because the herbivores are predator limited. Fishes appear to exact some control over the abundance of the epifaunal community, and this may be true for the infaunal population.

The rate of deposition of material in the bed (Table 1) and the presence of a large standing crop of detritus (21,160 kcal m^{-2}, including suspended detritus, 18 kcal m^{-2}, detritus greater than 1 mm within the sediment to a depth of 18 cm, 242 kcal m^{-2}, and available organic carbon less than 1 mm to a depth of 18 cm, 20,900 kcal m^{-2}) (Fig. 4), suggest sufficient available energy for the microbial compartment; there was also a net increase in the carbon content of the sediments of 0.4% between October 1972 and April 1973. Therefore, there probably in an excess of plant production in the system which is probably exported to the adjoining estuary, thus providing detrital materials to that system. Although eel-grass covers only a small part of the Newport River estuary (approximately 1%), it does cover approximately 17% of the estuarine area near Beaufort (35) where it contributes an estimated 64% of the total production of phytoplankton, cord-grass, and eel-grass. Thus, export of detrital material from grass beds may be highly significant to the trophic function of the estuarine system near Beaufort.

Phillips (22) has discussed needed research on eel-grass and other sea-grasses and Odum, Zieman, and Heald (18) have discussed the needs for research on the role of detritus in estuarine systems, but the need for research on structural and functional aspects of eel-grass systems and the role of these systems in the trophic function of estuaries has been largely ignored. We have described some of these aspects and, in part, the role of these beds in supporting the trophic functions of estuaries near Beaufort.

The species diversity of eel-grass communities coupled with spatial and temporal variations in biomass and abundance make it difficult to describe this community. Much has been written about gross changes at many levels of the community since the decline of eel-grass stocks during the 1930s, but little quantitative information is available on most of the compartments of this system (except for the grass itself), on the success of the community, and on the role the community plays in supporting the trophic structure of the estuaries of which they are a part. As a result, any subtle changes that result from human activities may pass unnoticed or be ascribed to natural variation. Neither the relative effects of natural processes nor the influence of man's activities on these processes can yet be quantified, but they should be considered carefully in developing priorities for man's ultimate use of the coastal zone.

ACKNOWLEDGEMENTS

This research was supported jointly by the National Marine Fisheries Service and the U. S. Atomic Energy Commission under agreement no. AT(49-7)-5.

REFERENCES

1. Adams, S. M. and Angelovic, J. W.
   1970 Assimilation of detritus and its associated bacteria by three species of estuarine animals. **Chesapeake Sci.,** 11: 249-254.
2. Backiel, T.
   1971 Production and food consumption of predatory fish in the Vistula River. **Jour. Fish. Biol.,** 3: 369-405.
3. Boesch, D. F.
   1972 Species diversity of marine macrobenthos in the Virginia area. **Chesapeake Sci.,** 13: 206-211.
4. Bouyoucos, G. J.
   1936 Directions for making mechanical analyses of soil by the hydrometer method. **Soil Sci.,** 42: 225-229.
5. Conover, J. T.
   1958 Seasonal growth of benthic marine plants as related to environmental factors in an estuary. **Publ. Inst. Mar. Sci.** (Texas), 5: 97-147.
6. Day, J. J.
   1967 The biology of Knysna Estuary, South Africa. In **Estuaries,** p. 397-407. (ed. Lauff, G. H.). Am. Assoc. Adv. Sci. Publ. 83, Washington, D. C.
7. Dillion, C. R.
   1971 A comparative study of the primary productivity of estuarine phytoplankton and macrobenthic plants. Doctoral dissertation, Univ. North Carolina, Chapel Hill.
8. Hargrave, B. T.
   1969 Epibenthic algal production and community respiration in the sediments of Marion Lake. **Jour. Fish. Res. Board Canada,** 26: 2003-2026.
9. Hellier, T. R., Jr.
   1958 The drop-net quadrat, a new population sampling device. **Publ. Inst. Mar. Sci.** (Texas), 5: 165-168.
10. Hettler, W. F., Jr.
    1972 Respiration of estuarine fish and crustaceans. In **Annual report to the Atomic Energy Commission,** p. 360-367. NMFS Atlantic Estuarine Fisheries Center, Beaufort, North Carolina. (Unpublished manuscript.)
11. Kjelson, M. A., Johnson, G. N., Garner, L., and Watson, J. E.
    1973 On the success of portable drop nets and surface and bottom trawls on estimating absolute fish densities. In **Annual report to the Atomic Energy Commission,** p. 140-168. NMFS Atlantic Estuarine Fisheries Center, Beaufort, North Carolina. (Unpublished manuscript.)
12. Mann, K. H.
    1965 Energy transformations by a population of fish in the River Thames. **Jour. Anim. Ecol.,** 34: 253-275.
13. Marsh, G. A.
    1970 A seasonal study of **Zostera** epibiota in the York River, Virginia. Doctoral dissertation, College of William and Mary, Williamsburg, Virginia.

14. Marsh, G. A.
1973 The **Zostera** epifaunal community in the York River, Virginia. **Chesapeake Sci.,** 14: 87-97.

15. McRoy, C. P.
1970 Standing stocks and other features of eelgrass **(Zostera marina)** populations on the coast of Alaska. **Jour. Fish. Res. Board Canada,** 27: 1811-1821.

16. Morgans, J. F. C.
1956 Notes on the analysis of shallow-water soft substrate. **Jour. Anim. Ecol.,** 25: 367-387.

17. Odum, E. P.
1959 **Fundamentals of ecology.** W. B. Sanders, Philadelphia. 546 p.

18. Odum, W. E., Zieman, J. C., and Heald, E. J.
The importance of vascular plant detritus to estuaries. In **Coastal marsh and estuary symposium,** Louisiana State Univ., 17 July 1972.

19. Orth, R. J.
1971 Benthic infauna of eel-grass, **Zostera marina,** beds. **M. S. thesis,** Univ. Virginia, Charlottesville.

20. Pamatmat, M. M.
1968 Ecology and metabolism of a benthic community on an intertidal sandflat. **Int. Rev. ges. Hydrobiol.,** 53: 211-298.

21. Peterson, C. J. G.
1918 The sea bottom and its production of fish food. A survey of the work done in connection with valuation of the Danish waters from 1883-1917. **Rept. Danish Biol. Sta.,** 25: 1-82.

22. Phillips, R. C.
1969 Temperate grass flats. In **Coastal ecological systems of the United States,** Vol. 2, p. 737-773. (eds. Odum, H. T., Copeland, B. J., and McMahan, E. A.) Rep. Fed. Water Poll. Cont. Adm. Contract RFP 68-128. (Unpublished manuscript.)

23. Price, T. J., Williams, R. B., Thayer, G. W., and Roberts, T. G.
1972 Species composition and biomass of benthos and particle size distribution of sediments in the open water areas of the Newport River estuary. In **Annual report to the Atomic Energy Commission,** p. 166-185. NMFS Atlantic Estuarine Fisheries Center, Beaufort, North Carolina. (Unpublished manuscript.)

24. Reid, G. K., Jr.
1954 An ecological study of the Gulf of Mexico fishes in the vicinity of Cedar Key, Florida. **Bull. Mar. Sci. Gulf Carib.,** 4: 1-94.

25. Ricker, W. E.
1946 Production and utilization of fish populations. **Ecol. Monogr.,** 16: 347-391.

26. Slobodkin, L. B.
1960 Ecological energy relationships at the population level. **Amer. Nat.,** 94: 213-236.

27. Slobodkin, L. B.
1962 Energy in animal ecology. In **Advances in ecological research,** vol. 1, p. 69-99. (ed. Cragg, J. B.) Academic Press, New York.

28. Smith, K. L., Jr.
1971 Structural and functional aspects of a sublittoral community. Doctoral dissertation, Univ. Georgia, Athens.

29. Smith, K. L., Jr., Burns, K. A., and Teal, J. M.
1972 **In situ** respiration of benthic communities in Castle Harbor, Bermuda. **Mar. Biol.,** 12: 196-199.

30. Teal, J. M.
1962 Energy flow in the salt marsh ecosystem of Georgia. **Ecology,** 43: 614-624.

31. Thayer, G. W.
1971 Phytoplankton production and the distribution of nutrients in a shallow unstratified estuarine system near Beaufort, N. C. **Chesapeake Sci.,** 12: 240-253.

32. Thayer, G. W., Price, T. J., LaCroix, M. W., and Montgomery, G. B.
1973 Structural and functional aspects of invertebrate communities in the Newport River estuary. In **Annual report to the Atomic Energy Commission,** p. 98-135. NMFS Atlantic Estuarine Fisheries Center, Beaufort, North Carolina. (Unpublished manuscript.)

33. Thayer, G. W., Schaaf, W. E., Angelovic, J. W., and LaCroix, M. W.
1973 Caloric measurements of some estuarine organisms. **Fish. Bull.,** 71: 289-296.

34. Williams, R. B.
1966 Annual phytoplanktonic production in a system of shallow temperate estuaries, p. 699-716. In **Some Contemporary Studies in Marine Science** (ed. Barnes, H.). George Allen and Unwin Ltd., London.

35. Williams, R. B.
1973 Nutrient levels and phytoplankton productivity in the estuary. In **Proceedings of the coastal marsh and estuary symposium,** p. 59-89. (ed. Chabreck, R. H.) Louisiana State Univ. Div. Continuing Education, Baton Rouge.

36. Williams, R. B. and Murdoch, M. B.
1972 The rate of sedimentation in salt marshes near Beaufort, N. C. In **Annual report to the Atomic Energy Commission,** p. 278-292. NMFS Atlantic Estuarine Fisheries Center, Beaufort, North Carolina. (Unpublished manuscript.)

37. Winberg, G. G.
1956 Rate of metabolism and food requirements of fishes. (Trans. from Russian). **Fish. Res. Board Canada, Transl. Ser.** 194, 202 p. (Trans. **Nauchnye Trudy Belorusskova Gosudarstrennova Univ.** V. I. Lenina, Minsk, 253 p.)

38. Wolfe, D. A., Cross, F. A., and Jennings, C. D.
1973 The flux of Mn, Fe, and Zn in an estuarine ecosystem. In **Radioactive contamination of the marine environment,** p. 159-175. International Atomic Energy Agency, Vienna.

39. Wood, E. J. F., Odum, W. E., and Zieman, J. C.
1969 Influence of sea grasses on the productivity of coastal lagoon. In **Coastal lagoons, a symposium,** p. 495-502. (eds. Ayala-Castanares, A. and Phleger, F. B.) (UNAM-UNESCO) Univ. Nacional Autonoma Mexico, Mexico, D. F.

QUANTITATIVE AND DYNAMIC ASPECTS OF THE ECOLOGY OF TURTLE GRASS, *THALASSIA TESTUDINUM*

Joseph C. Zieman[1]

ABSTRACT

In recent years the vast beds of sea grasses bordering the temperate and tropical coastlines have been recognized as a valuable resource. Techniques were developed to measure the production and seasonal dynamics of *Thalassia testudinum,* the dominant marine grass of the Caribbean. Conventional productivity measures are subject to error from gas storage within the leaves. Production of leaf material varied from 0.3 to 10.0 g dry wt m^{-2} day^{-1} in the south Florida area, with mean values of 2.3-5.0 g m^{-2} day^{-1}. Leaf growth rates averaged 2-5 mm day^{-1} with maximum values exceeding 10 mm day^{-1}. The rhizomes of *Thalassia* were found 5-25 cm in the sediment, and roots penetrate to 4-5 m. Leaves constituted 15-22% of the total plant biomass, and leaf standing crops were found from 30-650 g dry wt m^{-2}, with average values of 126 and 280 in inshore and off shore waters, respectively. Leaf densities averaged 3460-4300 blades per m^2. *Thalassia* was found to have an optimum temperature near 30°C and an optimum salinity near 30°/oo. Standing crop varied by about 50% throughout the year, with leaf decreases during flower and fruit production and again in the fall. *Thalassia* produced about 6.8 crops of leaves per year. Few were directly grazed. The leaves decayed rapidly, losing 65% of their original weight in seven weeks. Drying and agitation greatly accelerated decomposition. Turtle grass contributes to the maintenance of the high production of estuaries in many other ways also.

1. Department of Environmental Sciences, University of Virginia, Charlottesville, Virginia 22903.

INTRODUCTION

In recent years, as marine scientists have paid increasing attention to the living resources of the shallow coastal seas, the vast beds of sea grasses (*Spermatophyta*) bordering the temperate and tropical coastlines have been recognized as important resources. The significance of sea grasses was documented early in the classic studies of Peterson (31), but little work on the ecological importance of these highly productive and easily studied areas followed until recently.

There are several species of sea grasses in the Caribbean area (32, 37), but the dominant is one turtle grass, *Thalassia testudinum,* König. This sea grass, a member of the family Hydrocharitoceae, was first described and named by König in 1806 from staminate flowers. Rydberg (35) also described it and gave an excellent account of its early history. Orpurt and Boral (29) reported on the anatomy and germination of the seed, and Tomlinson and Vargo (47) and Tomlinson (44, 45, 46) produced an extensive series of studies of the anatomy and morphology of turtle grass.

Strawn (41) studied the zonation of sea grasses with respect to tide levels, and Phillips (32) gave a comprehensive account of the distribution and ecology of sea grasses in the vicinity of St. Petersburg, Florida. Moore (23) discussed the distribution of *Thalassia* in the United States, and den Hartog (12) reviewed the biology and ecology of *Thalassia testudinum* in his exhaustive monograph on the sea grasses of the world. In recent years, it has become apparent that *Thalassia* contributes significantly and in numerous ways to the maintenance of coastal ecosystems of the tropics and near-tropics (50).

Much attention has been drawn to the importance of sea-grass beds as nurseries for the rearing and sheltering of small fishes and invertebrates (2, 14, 15, 28, 42). While there are records of some animals feeding directly on the green *Thalassia* leaves (33, 34, 40, 50), since the demise of the green sea turtle there has been little direct grazing pressure on *Thalassia* beds except in certain localized situations, such as areas with high concentrations of sea urchins (40). Thus most *Thalassia* leaves remain untouched until they die and are cast off from the plant. They then decay and become detritus (6, 22, 52), and it is through the detritus food web that 90% or more of the energy of *Thalassia* makes its way to higher trophic levels. The leaves have other functions, such as the well-known abilities of a *Thalassia* bed to slow water speed, increase sedimentation, and stabilize the sediment surface for both inorganic sediments (9, 36, 54) and organic matter (50), and as a substrate for epiphytic algae (17). The studies in this paper are concerned with contributing to the growing body of knowledge of *Thalassia* and sea grasses in general. They were carried out primarily in the south Florida area from 1965 until the present time (Fig. 1), although observations were made throughout the Caribbean area, especially in

Figure 1. Study areas in south Florida.

the Bahamas and at Glovers Reef, off British Honduras.

The long-term aims at the beginning of this continuing study were four: (1) the environmental conditions responsible for the distribution of *Thalassia* beds, (2) growth and production rates and dynamics of the grass beds, (3) the rate of conversion of the sloughed or torn-off blades into detritus, and (4) the value of the clean blades and detritus as food. This paper will attempt to relate some of the progress that has been made, as well as some of the problems encountered.

Thalassia testudinum has a large standing crop of leaves, but most of the biomass of the plant is in the sediments and quite difficult to sample adequately owing to the depth of penetration of the root system (18, 54). Leaves usually constitute 15%-22% of the total plant biomass, although this varies between 10% and 45% (1, 3, 18). In a study in central Biscayne Bay, Florida, Jones (18) found a relatively constant ratio of 1:2:2:2 for leaves: short shoots:rhizomes:roots.

The density of *Thalassia* leaves as well as biomass and production are related to a variety of environmental paramenters. Much of the current data are observational, rather than experimental, and the interrelation of factors is poorly understood. *Thalassia* is a tropical marine plant requiring temperatures ranging from 20°-30°C (12, 18, 23, 32, 37, 53) and favoring salinities between $24^{o}/oo$ and 35 $^{o}/oo$ (18, 19, 32, 37, 53). Turtle grass is an active producer and requires high levels of light to photosynthesize. Various reports put the maximum depths recorded at 15-20 m, but it is generally conceded that the maximum depth for the development of large beds is around 10 m (8, 12, 32). While most authors feel this is due to light limitation (18, 32), there is a suggestion that pressure plays an important part in determining the lower limits (7, 8, 12). Nutrient sources have only recently begun to receive attention (30) and the source and quantity of the required nutrients is still unclear. McRoy and Barsdate (20) have shown that *Zostera* is capable of pumping nutrients from the sediments to the surface waters, from where they are exported, and it is quite likely that *Thalassia* beds can do the same thing.

Thalassia beds may also serve as nutrient sources in other ways. In most of the Caribbean, vast schools of snappers (Lutjanidae) and grunts (Pomadasyidae) live during the day in the shelter of coral reefs, but move onto the adjacent grass beds at night to feed (38, 39). As they return to the reef in the morning, they transport nutrients from the grass beds to the reef environment through excretion and defecation. *Thalassia* leaves have been found in drifting rafts far at sea and at abyssal depths 500-1000 km from the nearest *Thalassia* beds (21).

The most troublesome aspect of the quantitative ecology of *Thalassia* is determination of its true productivity (48). Most studies concerned with the productivity of *Thalassia* have used one of the oxygen-evolution techniques. Oxygen evolution methods, either light and dark bottle or benthic chambers (3, 18), and the downstream 0_2 method (25, 26, 27) have an undetermined but potentially significant error because gases are stored within the leaves of *Thalassia*. This storage within aquatic macrophytes has been shown to cause considerable error when production is measured with exchange methods owing to the internal recycling of the gases within the interstitial chambers of the plants (11). *Thalassia* leaves are essentially rectangular in cross-section during the early morning, but as production increases during the day they have been observed to swell to as much as 200%-250% of their early morning volume owing to the internal production of gases (primarily 0_2) at a much greater rate than can be exchanged. When the leaves reach a certain volume, bubbles burst out from the edges of the leaves in little streams, since *Thalassia* leaves do not possess stomata (7, 12, 32). This bubbling can be quite dramatic. By early afternoon on a shallow, calm *Thalassia* bed with little tide flow, the bed can readily be heard to hiss from the rapid bubbling. The bubble evolution can be so rapid a swimmer snorkling in these beds has the feeling of diving in a newly opened bottle of beer.

To avoid the difficulties and uncertainties encountered in previous productivity determinations (48), the technique described in this paper was used. It measures only the production of new-leaf material, which although less than net production, is the energy which is important in providing the food source for higher trophic levels, primarily through the detritus food web.

METHODS

Growth and Production

Population densities were measured by placing a 20 x 20 cm square frame in a

Thalassia bed, by means of scuba, without disturbing the community. The quadrat was visited every 2-3 weeks, and growth and production were measured according to the following technique:

After an area within a *Thalassia* bed was selected, a square wire frame, usually 10 cm on a side, was pushed into the sediment to mark the study area (Fig. 2).

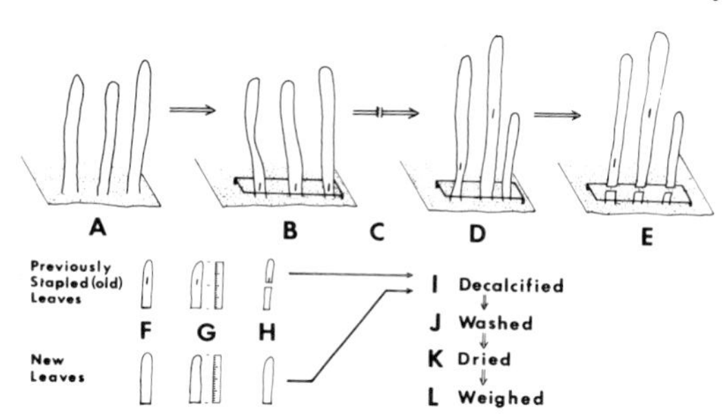

Figure 2. Technique for marking **Thalassia**: An area within a bed is selected (A), marked with a quadrat, and the blades are stapled at the level of the quadrat (B). Two - three weeks later (C-D) the leaves are harvested at the quadrat level (E), and separated into old and new (without staples) leaves (F). These are measured (G) and the new growth is separated (H) from the old material. The new growth is decalcified (I), washed (J), dried at 105°C (K), and weighed (L).

All of the grass blades within the quadrat were marked individually with a staple from a small stapler, which was slightly modified to prevent the leaves from being crushed, and the number of marked blades was recorded. The staple was placed so that its lower end was level with the frame, because the sediment level might vary over several weeks.

After a period of 2-4 weeks, the blades within the quadrat were at the level of the wire frame, placed in a plastic bag, and taken to the laboratory to be processed. There, they were divided into two groups, the first group containing the new leaves, that is, the leaves that had appeared since the time of marking and consequently had no staple. These were counted, their length and width were measured, and the presence or absence of round tip, the amount of epiphytic growth, and the general health of the leaves were recorded. The second group contained the blades that had been stapled. In this group the distance from the base of the leaf to the staple was measured, in addition to taking the same data as for new leaves. In the marked leaves most, but not all, of the growth occurred between the base of the leaf and the staple, whereas for new leaves, growth was the same as the total length.

After measurements were taken, the old leaves were cut below the staple, the new growth being thereby removed. This new growth was placed on a sieve, washed, and dipped in a bath of 5% phosphoric acid to remove the carbonates and epiphytes adhering to or encursted on the leaves. This important step is frequently overlooked. *Thalassia* often has large amounts of carbonate sediments adhering to the leaves, which give artificially high standing crop values. Also many epiphytes do not adhere directly to the *Thalassia* leaf, but to the carbonate skeleton of the calcareous red alga *Melobesia*. As the slight acidification dissolves the carbonates, most of the epiphytes are removed also. The leaf material was then thoroughly washed under a stream of running water, oven-dried at 105°C, and weighed. The same procedure was then repeated with the new leaves and with the excised tips of the old leaves. The numbers of blades present when the quadrat was both marked and collected were recorded. The net change of the *Thalassia* community for the period was the difference between these two numbers.

In order to determine the main region of growth on *Thalassia* blades, staples were placed at the base of the blade and successively at 3-cm intervals up its length, with the aid of a plastic template. When the blades were harvested, the distance from the base of each blade to the first staple was measured, as previously described; then the distance between each of the successive staples was measured. Any deviation from 3 cm indicated elongation or shrinkage of the leaf.

As the plant grows, leaves are successively put out from a meristematic region which is enclosed among the innermost leaves (32, 46, 47). Leaf replacement was measured by stapling together at the base all of the blades from an individual branch. This marked all the existing leaves in time, and any growth of these leaves in the stapled group would be shown by changes in the relative position of its staple. By harvesting the entire branch it was possible to maintain the leaves in their correct positions. Since new leaves originate from inside existing ones, and since younger leaves have higher growth rates, the relative age of individual leaves could be determined both from their positions and their growth rates.

Decomposition

Comparisons of the rates of which plant materials decay are often difficult owing to the variety of techniques used and such other complications as grazing of the samples. Wet weights are highly variable and give an indeterminate initial point, but drying the material to obtain an initial dry weight causes considerable internal damage and chemical change. With *Thalassia*, most of the grass decays on the estuary floor and does not dry out prior to decomposition. Some percentage, as yet undetermined, washes up on the beach and decays there; however, this area constitutes an entirely distinct regime for decay from the sediment surface.

Despite its disadvantages, drying material prior to weighing does give an accurate starting point. A careful wet weight and subsequent calculations of

The length-frequency distribution of *Thalassia* leaves is not normally distributed about a mean, but is a linear distribution from the minimum to the maximum. In areas and times of high turnover, a new leaf is produced by a shoot as frequently as every 10 days, one new leaf per 14-16 days per shoot being the average. During colder months, output is much slower.

About 80%-90% of the growth of the leaf occurs in the basal portion, but there continues to be some elongation throughout the length of the leaf. Figure 5 shows the increase in length to be greatest at the base and to decrease regularly towards the tip of the leaf. The growth rate of *Thalassia* leaves can be quite rapid. Leaves have been observed to grow 1 cm per day for 15-20 days, but rates of growth of 2.5-4 mm per day are the most common. Average growth rate increases with blade width, so that the rate of a 10-mm-wide blade (about 5 mm per day) averages 2.5 times greater than the rate of a 5 mm leaf (about 2 mm per day). The weight per unit length of a *Thalassia* blade (Figure 6) is relatively

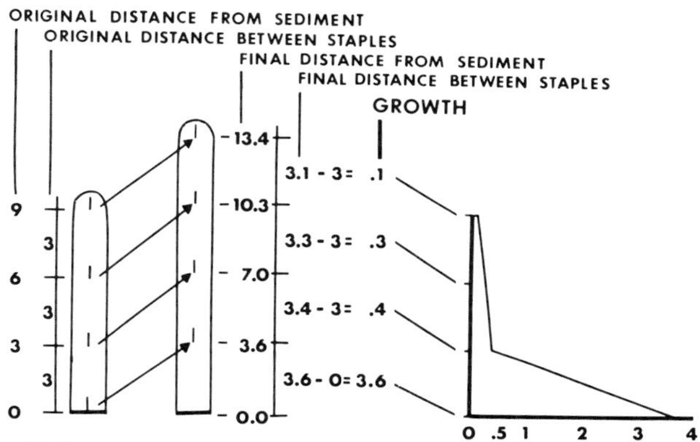

Figure 5. Most blade growth is at the leaf base, but some is due to elongation throughout the leaf.

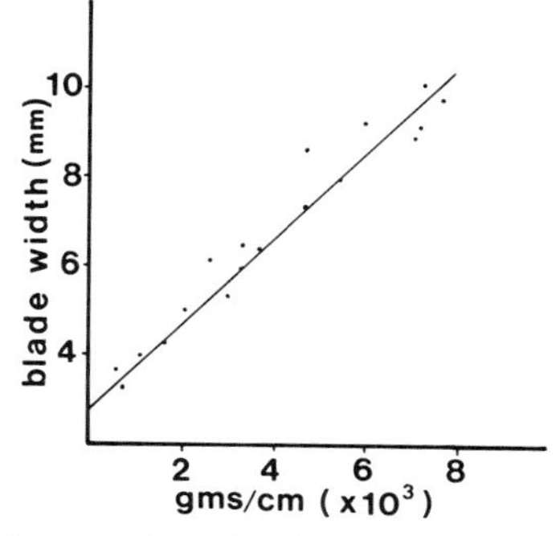

Figure 6. The weight of a 1 cm section of various widths of **Thalassia** blades.

constant, owing to the rectangular cross-section of the leaf.

Standing Crop

Thalassia standing crops can be expressed in terms of biomass (g m^{-2}) and in terms of leaf population densities (leaves m^{-2}). As communities mature, leaf density increases, as does the average width and biomass of the leaves. Biomass and production are indicators of the amount of material made available for higher trophic levels; leaf density also relates to other important functions of the turtle-grass community. The ability of the community to reduce water flow, promote sedimentation, and stabilize the sediment surface is dependent largely on leaf density.

Figure 7, modified from Zieman (54) shows the increase in the elevation of the sediment surface caused by increasing density of turtle-grass leaves. Here a small bed raised the sediment level 3 cm as a result of its ability to buffer currents. This figure also shows the relationship of blade density and length to sediment depth, indicating that adequate sediment is necessary to develop a proper grass bed. The requirement for a deep sediment seems to be related to the need for a large root system for nutrient intake, which could be expected of a plant with a high productivity.

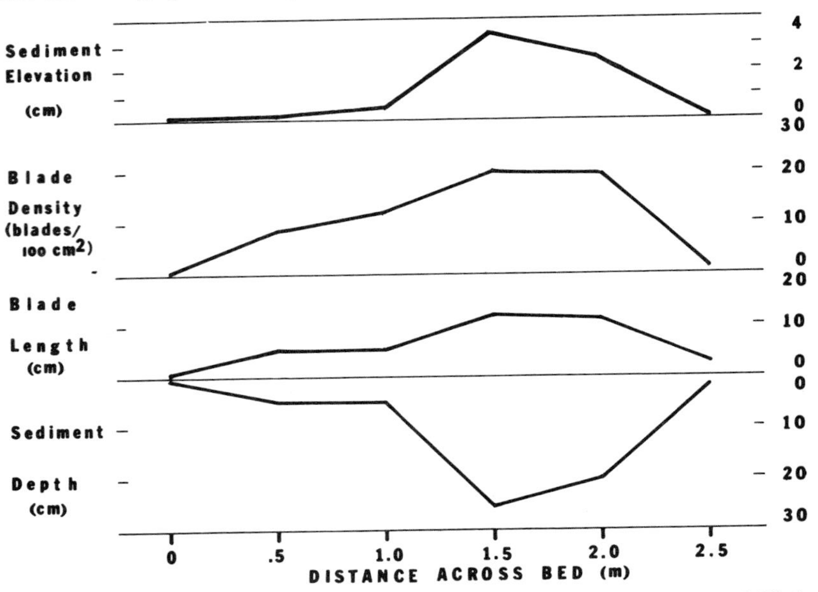

Figure 7. Relationships between depth of sediment, length and density of **Thalassia** blades, and sediment elevation caused by sediment stabilization in a bed off Key Largo, Florida.

An increasing density of leaves increases the surface area of the community and offers a greater substrate for epiphytes and epifauna, as well as providing a denser habitat and greater protection for the organisms that use the grass beds as a nursery and for shelter.

Leaf densities of *Thalassia* vary widely; the maximum density encountered has been 8800 blades m^{-2}. Inshore stations in western Biscayne Bay varied from 1200-5800 blades m^{-2}, the average being 3460 blades m^{-2}. The offshore stations were generally denser, with extremes of 1800-7000 blades m^{-2} and an average of 4300 blades m^{-2}. Turtle-grass beds tend to be relatively homogeneous in density over broad areas, generally tapering off in density at the margins, although the transition may be abrupt where blowouts or scours occur (36, 50). Areas of denser, more robust growth are, however, often seen within grass beds. Many of these dense patches have been explored and found to be either over a deeper depression in the bedrock, allowing deeper sediment development, or in the vicinity of a buried log or other organic debris, which was presumably acting as a nutrient source.

Reported standing crop values are listed in Table 1 but are difficult to interpret owing to the variety of techniques used by researchers.

TABLE 1

Standing crops and productivities of **Thalassia testudinum**

Blade Biomass (g dry wt m^{-2})		Total Plant (g dry wt m^{-2})		Productivity		Location and Reference
range	mean	range	mean	range	mean	
10 - 740	340	– –	850[1]	– –	4.7[2]	Cuba (3)
138 - 857	450	539 - 7388	2807	– –	–	Puerto Rico (5)
700 - 1800	830	2200 - 4300	2830	.9 - 2.5[2]	1.9[2]	Florida (18)
– –	–	– –	–	1.8 - 6.0[3,4]	3.3[3,4]	Barbados (30)
30 - 230	126	– –	–	.3 - 6.8[3]	2.3[3]	Florida, inshore (52,53)
80 - 650	280	– –	–	2.2 - 10.0[3]	5.0[3]	Florida, offshore (52,53)

1. Estimate from author's data
2. Grams C m^{-2} day^{-1}
3. Grams dry weight m^{-2} day^{-1}
4. Leaf production only

Values of standing crop biomass of leaves in southern Florida vary greatly. In shallow inshore areas, where fluctuations of such environmental paramenters as salinity and temperature cause stress, values ranged from 30-230 g dry wt m^{-2} with an average of 126 g m^{-2}. In the offshore beds near Soldier Key and Key Largo the values ranged from 80-650 g dry wt m^{-2} with an average of 280 g m^{-2}; these areas are less subject to land run-off and the wide salinity and temperature variations of a shallow, coastal lagoon.

Burkholder et al. (5) found standing crop values of 539-7388 g dry wt m^{-2} for whole plants with an average of 20807 g m^{-2}. The standing crop of leaves varied from 138-857 g dry wt m^{-2} with an average near 450 g m^{-2}. Jones (18) working in Bear Cut, Biscayne Bay, Florida, found leaf standing crops of 700-1800 g dry wt m^{-2} with an average of 830 g m^{-2} and a root and rhizome biomass of 1500-2500 g dry wt m^{-2}. Buesa (3) reported leaf densities in the coastal shelf of western Cuba to range between 10-740 g dry wt m^{-2}, with an average around 340 g m^{-2}. These figures all agree remarkably well. It is felt that most measurements of *Thalassia* leaf standing crop are overestimates owing to the fact that the leaves were apparently not cleaned of carbonates and epiphytes before they were dried.

Productivity

One aspect of the ecology of *Thalassia* upon which all authors agree, despite the wide variety of methods used, is the high organic productivity. Table 1 shows some representative productivity values derived from the literature. Most directly comparable with the results presented here are those of Patriquin (30), as they are figures for leaf production only. All other figures represent productivity of *Thalassia* communities; owing to the methods utilized, these figures include both epiflora and usually benthic flora productivity, and they are subject to gas-storage error. Much of the production of *Thalassia* goes into the robust rhizome system and is not readily used as a food source. It is felt that what values of net leaf production best reflect is the contribution of *Thalassia* to the trophic structure of the community.

The concept of *turnover* is central to the study of production ecology. Turnover rate is the fraction of the system that is produced per unit time; it can conveniently be expressed as a per-cent-per-day change. Its inverse, turnover time, is the number of days required to replace the standing crop (2% change per day = 50 days turnover time = 7.3 crops per year). The *Thalassia* blade population usually showed a change of 1-3% per day. The mean rate was 1.9% per day which gives an average turnover time of 54 days and indicates the production of some 6.8 crops of leaves per year, on the average. In other studies, only Jones (18) and Buesa (3) measured both total net production and total standing crop so that the turnover of the whole plant could be estimated. Jones' (18) average figures of about 2500 g dry wt m^{-2} (1 kg C m^{-2}) total plant and 700 g C m^{-2} yr^{-1} productivity for *Thalassia* would give a total plant turnover of 0.7 crops per year and a turnover time of 17 months. Buesa (3) measured average standing crops of *Thalassia* leaves at about 340 g dry wt m^{-2}. He found that leaves were about 40.7% of the entire plant, indicating a standing crop of 850 dry wt m^{-2} or about 340 C m^{-2}, although this value is twice the proportion normally ascribed to leaves. Using his productivity value of 4.7 C m^{-2} day^{-1} would give a total plant turnover of 72 days, or 5 complete crops per year, an exceptionally high figure for an entire plant. However, an overestimate could be caused by underestimating root biomass, by overestimating productivity, by including epiphytic algal production, or by all three.

Temperature and Salinity

Nearly 200 values of *Thalassia* productivity from a variety of locations in Florida measured by the technique described here were plotted against temperature and salinity, and the results are shown in Figures 8 and 9. Figure 8 shows that *Thalassia*, a tropical plant, has a definite temperature optimum at 30 °C and that production drops off to near zero at temperatures below 20 °C and above 35 °C. This agrees with the observations of Phillips (32) and Sculthorpe (37) that *Thalassia* is restricted to areas where the temperature rarely falls below 20 °C and loses leaves if temperatures drop below 20 °C or exceed 30 °C. Jones (18) also found that productivity was maximal near 30 °C and dropped off between 30 and 33 °C.

Thalassia showed a salinity optimum in these studies (Fig. 9) of 30 °/oo, with decreased productivity above and below this. *Thalassia* can tolerate salinities as low as 3.5-5 °/oo (37) and as high as 60 °/oo (19) for short periods of time, but

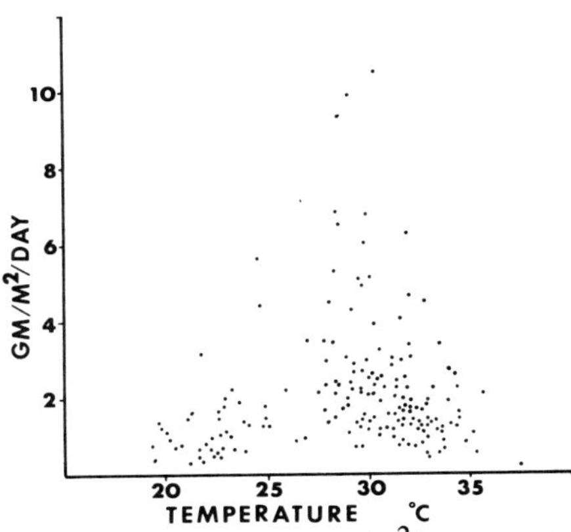

Figure 8. Scattergram of productivity (gm dry wt/m^2/day) as a function of temperature. Note the optimum near 30°C and the decreases above and below this value.

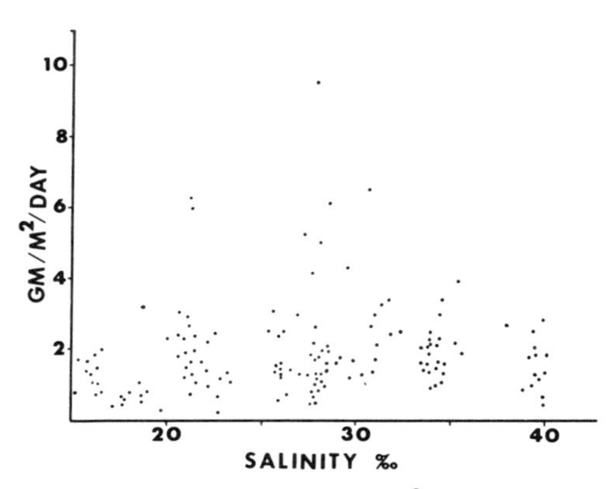

Figure 9. Scattergram of productivity (gm dry wt/m^2/day) as a function of salinity.

usually undergoes defoliation.

Undoubtedly temperature and salinity (and such other factors as light, turbidity, nutrients, and depth) are synergistically coupled as shown by Hammer (10) for salinity and CO_2, but the data for detailed analyses of these factors are lacking.

In addition to the salinity and temperature stresses associated with shallow subtidal flats, turtle grass may at times be subject to CO_2 limitation (10). During times of slow water movement over shallow ($<$0.5 m) grass flats, the pH of the water has been observed to reach a value of 9.4, an effect also observed in grass beds by Wood (50). Theoretically HCO_3^- ions have been removed from the water, and bubble evolution soon ceased, although the leaves remained somewhat swollen. Thus turtle grass can, under certain conditions, overcome the CO_2 buffering capacity of sea water, which may be a large source of much error associated with productivity measurements made in any enclosed chambers.

Seasonality

Although *Thalassia* undergoes seasonal fluctuation, it always retains a significant winter population, especially when compared with the highly seasonal, temperate-zone correlate, *Zostera,* which virtually disappears in winter. Figure 10 shows the seasonal fluctuation of *Thalassia* over a period of 18 months, from March 1969 to September 1970, at a station in the southwestern portion of Biscayne Bay at a depth of about 2m. The data from this station are representative of the trends found in stations throughout the south Florida area. The community showed general seasonal trends with maximum densities, both in numbers of leaves present (leaves per m^2) and biomass (g per m^2), in May through July of each year, and minimum in the fall months, from late August to

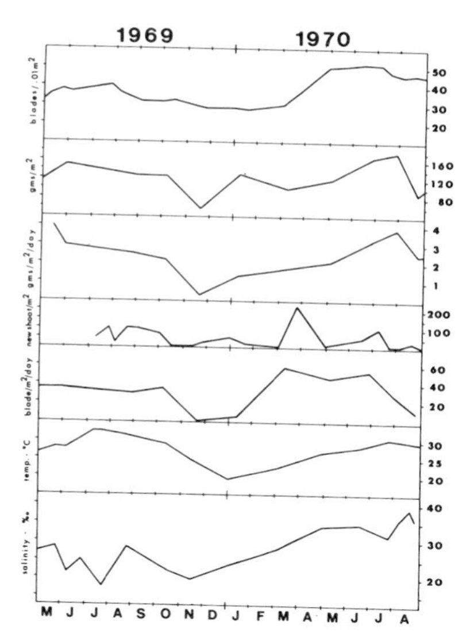

Figure 10. Seasonality of various biological features of the **Thalassia** beds at a station in southern Biscayne Bay, for an 18-month period. Temperature and salinity at the station are also shown.

November. Jones (18) also reported maximal standing crops in May-July and minimum amounts during September - January and found the maximum to be about twice the minimum amount. The fall decline in plant abundance, which has been noted by Phillips (32) and Jones (18), among others, is more pronounced in terms of biomass than leaf numbers, indicating that it is the older, larger leaves that are lost. This is supported by examination of the dead and floating material cast off from the sea-grass beds at this time. The net production of the blades declines simultaneously. During 1969 this decline was associated with declining temperature and salinity, but in 1970 it occurred with hypersaline conditions and normal summer maximum temperatures. The years 1969 and 1970 were quite in contrast. In 1969, rainfall in the study area was 224 cm, 35% greater than normal, and salinities were considerably below normal in the south Florida estuaries, while during the same months in 1970, conditions were hypersaline. It has been reported that *Thalassia* is sensitive to high (18, 32) and low (23, 32) temperatures and to high (32) and low (10, 23, 32, 43) salinities. The data here agree, showing that *Thalassia* populations and production declined when temperature and salinity departed significantly from the previously described optima.

At this station flowers of *Thalassia* were found during May of 1969, and fruits were present in June and July, which was typical for *Thalassia* beds in Biscayne Bay (16, 18), but no reproductive structures were seen in 1970. A common feature of the curves of new-leaf output and net production of leaves is a depression during the time of flower and fruit production when the plants divert energies to sexual reproduction. This was seen at all stations where reproduction was observed and at many where reproduction was not observed: lack of observation of reproductive structures is not proof the plant was not reproducing. This depression is not seen in Figure 10 because the marker bouy was lost at that time.

The rate of production of flowers and fruit is small and observation is difficult. During July-September 1969, it was observed that several germinating seeds had settled to the bay bottom and begun to put out leaves and roots. These areas were marked and the progress of the seedlings noted; however, none survived for more than 3 months. In these studies, no *Thalassia* seeds surviving past several months were observed, and no plants that had obviously grown from recent seedlings were seen. It is believed that seedling success is a relatively rare event, and that most *Thalassia* growth is due to vegetative reproduction.

Production of new shoots or upright branches was an indication of active rhizome growth and showed a bimodal distribution, with peak activities in early spring and midsummer.

Although seasonal changes in *Thalassia* are related to temperature and salinity, such factors as day length, are important in the life cycle of aquatic hydrophytes, although their effect has been studied for relatively few plants (37).

Decomposition

The previously discussed studies of growth and production of the grass beds give some indications of the potential of this community as a producer. Observations made during these studies, as well as data from the literature, indicate that the grass blades of the *Thalassia* beds are not grazed directly to any extent (33, 50, 52). Experiments conducted using wire enclosures over grass beds indicated little grazing, and this is supported by direct observation. As *Thalassia* leaves grow, become senescent, and die, they often remain attached to the shoot by tough fibers. Careful inspection reveals that few leaves (usually less than 1%) are torn or cut off before becoming senescent. It is common, however, to find several 3-5 mm crescent-shaped cuts removed from the blades by a small scarid, *Sparisoma radians,* a common inhabitant of the *Thalassia* beds (34, 35). Except under isolated local conditions, the leaf material usually dies, decays, and then is consumed as detritus; thus it is important to determine the rate at which the senescent and sloughed-off leaves are converted into useable detritus.

In all experiments the weight loss of decaying *Thalassia* leaves was greatest in the early days of decomposition, at which time the microscopically examined leaves of *Thalassia* appeared much the same as fresh leaves. When examined according to the method of Wood and Oppenheimer (51), nearly all of the chloroplasts fluoresced.

A *Thalassia* blade resembles a flattened cylinder with the epidermal cells on the outer walls (32, 46). If decay was allowed to proceed in an undamaged leaf, it progressed slowly down the blade from the torn ends where much breakdown of cell structure was visible and cell contents disappeared. Adjacent to this area considerable microbial activity by bacteria, small flagellates, ciliates, and several fungal mycelia were observed. As the blades decayed further in an undisturbed condition for the next several weeks, the zones of necrosis and decay moved down the undamaged blades, leaving empty spaces in the areas behind them.

Although decay progressed according to the above pattern in most leaves, a few leaves decayed throughout their entire length. Upon examination, these proved to be torn or broken to some degree, whereas in the leaves whose cuticles were essentially intact, decay proceeded at a slower rate, indicating that an intact surface prevents microbes from entering the leaf.

The results of the decay experiments are shown in Figure 11 and are compared with other decomposition rates taken from the literature for communities which seem to function in the same manner as turtle-grass communities (4, 13, 24, 52), that is, areas of high production with little grazing which function primarily through the detritus food web. The *Thalassia* experiments were carried out until about 65% of the initial weight was lost (63.2% of response = 1 time constant, TC). The *Thalassia* blades in the tanks decayed as shown with 63.2% weight loss at 7 weeks, a rather rapid decay rate. In this experiment the three mesh sizes used for litter bags (0.1 mm, 1 mm, 5 mm) had no effect on loss rate, since under calm conditions loss was not due to fragmentation. Dried *Thalassia* leaves lost weight extremely rapidly (1 TC = 1 week), which was to be expected owing to the chemical changes accompanying drying.

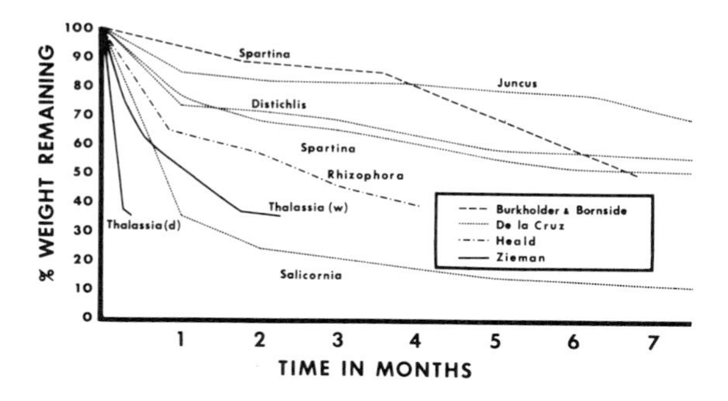

Figure 11. Comparison of the decay rates of **Thalassia** and other estuarine plants. (d) indicates material dried prior to decomposition: (w) indicates material prepared according to the method outlined here.

Litter bags that were totally submerged on a beach had nearly the same decay pattern as those in the tanks, but the rate was somewhat faster (1 TC = 6 weeks). The bags in the bay were continually subjected to wave motion, which eventually tore the leaf surfaces and allowed the entrance of decay organisms, although the small mesh size used here (1.1 mm) excluded many grazers. Still, the condition of the leaves, cells, and cuticles, and the way in which decay progressed inward from tears and lesions, resembled the situation in the tanks.

The samples from the mid- and high-water marks had very rapid decay rates (TC = 2 weeks). These leaves were alternately subjected to wetting and drying, and the desiccation speeded up weight loss by killing the cells and causing internal chemical changes. When the grass was wet during high tide, many cellular constituents simply leached out. The drying also shrank and cracked the cuticle of the blades, thus facilitating the entrance of microorganisms into the leaf. The cell walls and fibrous strands remained and gave structure to the leaves, but the cell contents were gone. Despite the small mesh size of the litter bags, a few amphipods were found in the collected samples.

The decay rates for the submerged *Thalassia* leaves most closely reflected the natural conditions of *Thalassia* decay. Although great windrows of leaves are often present on beaches after storms (43), most of the dead material decays on the estuary floor and is not subject to desiccation or violent agitation, although amphipods and similar grazers are present (6).

Interestingly, the mixture of *Thalassia* leaves and epiphytic *Ulva* decayed rapidly (TC = 3 weeks), largely owing to the rapid breakdown of the algal portion of the sample. Inspection indicated that the *Thalassia* was in an advanced state of decay also, possibly induced by the high numbers of decomposers associated with the less refractile algae.

CONCLUSIONS

Sea grasses exert a controlling influence on shallow estuarine areas in a variety of ways. Although it is different to make exact comparisons of *Thalassia* productivity measured by a wide variety of techniques, all the data indicate that *Thalassia* is an exceedingly active producer, but that there is little direct grazing pressure on the green leaves. Compared with other systems which function similarly, such as salt marshes, the turnover in terms of both new leaf production and decomposition is quite rapid. High production, however, is but one of the many functions of turtle-grass in the tropical and subtropical estuarine system.

ACKNOWLEDGEMENTS

Portions of this work are from a master's thesis and a doctoral dissertation which were directed by the late E. J. Ferguson Wood, to whom a great debt will forever be owed.

REFERENCES

1. Bauersfeld, P., Kifer, R. R., Durrant, N. W., and Sykes, J. E.
   1969 Nutrient content of turtle grass (**Thalassia testudinum**). **Proc. Intl. Seaweed Symp.,** 6: 637-645.
2. Breuer, J. P.
   1962 An ecological survey of the lower Laguna Madre of Texas, 1953-1959. **Publ. Inst. Mar. Sci. Texas,** 8: 153-183.
3. Buesa, R. J.
   1972 Producción primaria de las praderas de **Thalassia testudinum** de la plataforma noroccidental de Cuba.a. **I.N.P. Cont. Inv. Pesqueras Reunión Bal. Trab.,** 3: 101-143.
4. Burkholder, P. R. and Bornside, G. H.
   1957 Decomposition of marsh grass by aerobic marine bacteria. **Bull. Torrey Bot. Club,** 84(5): 366-383.
5. Burkholder, P. R., Burkholder, L. M., and Rivero, J. A.
   1959 Some chemical constituents of turtle grass, **Thalassia testudinum. Bull. Torrey Bot. Club,** 86(2): 88-93.
6. Fenchel, T.
   1970 Studies on the decomposition of organic detritus derived from the turtle grass **Thalassia testudinum. Limnol. Oceanogr.,** 15(1): 14-20.
7. Gessner, F.
   1968 Die Zellwand mariner Phanerogamen. **Mar. Biol.,** 1: 191-200.
8. Gessner, F. and Hammer, L.
   1961 Investigaciones sobre el clima de la luz en las regiones marinas de la costa venezolana. **Bol. Inst. Oceanogr.,** 1(1): 263-272.
9. Ginsburg, R. N. and Lowenstam, H. A.
   1958 The influence of marine bottom communities on the depositional environment of sediments. **Jour. Geol.,** 66(3): 310-18.
10. Hammer, L.
    1968 Salzgehalt and Photosynthese bei marinem pflanzen. **Mar. Biol.,** 1: 185-190.
11. Hartman, R. T. and Brown, D. L.
    1967 Changes in internal atmosphere of submersed vascular hydrophytes in relation to photosynthesis. **Ecology,** 48(2): 252-258.
12. Hartog, C den
    1970 **The sea-grasses of the world.** North-Holland Publishing Co., Amsterdam. 275 p.
13. Heald, E. J.
    1971 The production of organic detritus in a South Florida estuary.

Univ. Miami Sea Grant Tech. Bull., 6: 110. (Unpublished manuscript.)

14. Hoese, H. D.
1960 Juvenile penaeid shrimp in the shallow Gulf of Mexico. **Ecology,** 41(3): 592-593.

15. Hoese, H. D. and Jones, R. S.
1963 Seasonality of larger animals in a Texas turtle grass community. **Publ. Inst. Mar. Sci. Texas,** 9: 347-357.

16. Hopper, B. E. and Meyers, S. P.
1967 Foliicolous marine nematodes on turtle grass, **Thalassia testuidnum,** Konig, in Biscayne Bay, Florida. **Bull. Mar. Sci.,** 17(2): 471-517.

17. Humm, H. J.
1964 Epiphytes of the sea grass, **Thalassia testudinum,** in Florida. **Bull. Mar. Sci. Gulf Carib.,** 14(2): 306-341.

18. Jones, J. A.
1968 Primary productivity by the tropical marine turtle grass, **Thalassia testudinum,** Konig, and its epiphytes. Doctoral dissertation, Univ. Miami, Miami, Florida. 196 p.

19. McMillan, G. and Moseley, F. N.
1967 Salinity tolerances of five marine spermatophytes of Redfish Bay, Texas. **Ecology,** 48(3): 503-506.

20. McRoy, C. P. and Barsdate, R. J.
1970 Phosphate absorption in eelgrass. **Limnol. Oceanogr.,** 15(1): 14-20.

21. Menzies, R. J., Zaneveld, J. S., and Platt, R. M.
1967 Transported turtle grass as a source of organic enrichment of abyssal sediments off North Carolina. **Deep-Sea Res.,** 14: 111-112.

22. Meyers, S. P. and Hopper, B. E.
1967 Studies on marine fungal-nematode associations and plant degradation. **Helgolander wiss. Meeresunters.,** 15: 270-281.

23. Moore, D. R.
1963 Distribution of the sea grass, **Thalassia,** in the United States. **Bull. Mar. Sci. Gulf Carib.,** 13(2): 329-342.

24. Odum, E. P. and de la Cruz, A. A.
1967 Particulate organic detritus in a Georgia salt marsh-estuarine ecosystem. In **Estuaries,** p. 383-388. (ed. Lauff, G. H.) Am. Assoc. Adv. Sci. Publ. 83. Washington, D. C.

25. Odum, H. T.
1963 Productivity measurements in Texas turtle grass and the effects of dredging an intracoastal channel. **Publ. Inst. Mar. Sci. Texas,** 9: 48-58.

26. Odum, H. T., Burkholder, P. R., and Rivero, J.
1959 Measurement of productivity of turtle grass flats, reefs, and the Bahia Fosforescente of southern Puerto Rico. **Publ. Inst. Mar. Sci. Texas,** 6: 159-170.

27. Odum, H. T. and Hoskin, G. M.
1958 Comparative studies on the metabolism of marine waters. **Publ. Inst. Mar. Sci. Texas,** 5: 16-46.

28. O'Gower, A. K. and Wacasey, J. W.
1967 Animal communities associated with **Thlassia, Diplanthers,** and sand beds in Biscayne Bay. I. Analysis of communities in relation to water movements. **Bull. Mar. Sci.,** 17(1) 175-210.

29. Orpurt, P. A. and Boral, L. L.
1964 The flowers, fruits and seeds of **Thalassia testudinum** König. **Bull. Mar. Sci. Gulf Carib.,** 14(2): 296-302.

30. Patriquin, D. G.
1972 The origin of nitrogen and phosphorus for growth of the marine angiosperm **Thalassia testudinum. Mar. Biol.,** 15: 35-46.

31. Petersen, C. G. J.
1918 The sea bottom and its production of food. **Danish Biol. Sta. Rept.,** 25: 1-62.

32. Phillips, R. C.
1960 Observations on the ecology and distribution of the Florida sea grasses. **Prof. Pap. Florida Bd. Conserv.,** 2: 1-72.

33. Randall, J. E.
1965 Grazing effect on sea grasses by herbivorous reef fishes in the West Indies. **Ecology,** 46(3): 255-260.

34. Randall, J. E.
1967 Food habits of reef fishes of the West Indies. **Stud. Trop. Oceanogr.** (Miami), 5: 665-847.

35. Rydberg, P. A.
1909 The flowers and fruit of the turtle-grass. **Jour. New York Bot. Gardens,** 10: 261-265.

36. Scoffin, T. P.
1970 The trapping and binding of subtidal carbonate sediments by marine vegetation in Bimini Lagoon, Bahamas. **Jour. Sed. Petrol.,** 40(1): 249-273.

37. Sculthorpe, C. D.
1967 **The biology of aquatic vascular plants.** Arnold Publ., London, 618 p.

38. Starck, W. A. II
1971 The biology of the gray snapper **(Lutjanus griseus,** Linnaeus) in the Florida Keys. **Stud. Trop. Oceanogr.** (Miami), 10: 11-150.

39. Starck, W. A. II and Davis, W. P.
1966 Night habits of fishes of Alligator Reef, Florida. **Ichtyologica.,** 38(4): 313-356.

40. Stevenson, R. A. and Ufret, S. L.
1966 Iron, manganese, and nickel in skeletons and food of the sea urchins, **Tripneustes esculentus** and **Echinometra lacunter. Limnol. Oceanogr.,** 11(1): 11-17.

41. Strawn, K.
1961 Factors influencing the zonation of submerged monocotyledons at Cedar Key, Florida. **Jour. Wildl. Manag.,** 25(2): 178-189.

42. Tabb, D. C., Dubrow, D. L., and Manning, R. B.
1962 The ecology of northern Florida Bay and adjacent estuaries. **Florida Bd. Conserv., Tech. Ser.,** 39: 1-81.

43. Thomas, L. P., Moore, D. R., and Work, R. C.
1961 Effects of Hurricane Donna on the turtle grass beds of Biscayne

Bay, Florida. **Bull. Mar. Sci. Gulf Carib.**, 11(2): 191-197.

44. Tomlinson, P. B.
1969 On the morphology and anatomy of turtle grass, **Thalassia testudinum** (Hydrocharitaceae). II. Anatomy and development of the root in relation to function. **Bull. Mar. Sci.**, 19(1): 57-71.

45. Tomlinson, P. B.
1969 On the morphology and anatomy of turtle grass, **Thalassia testudinum** (Hydrocharitaceae). III. Floral morphology and anatomy. **Bull. Mar. Sci.**, 19(2): 286-305.

46. Tomlinson, P. B.
1972 On the morphology and anatomy of turtle grass, **Thalassia testudinum** (Hydrocharitaceae). IV. Leaf anatomy and development. **Bull. Mar. Sci.**, 22(1): 75-93.

47. Tomlinson, P. B. and Vargo, G. A.
1966 On the morphology and anatomy of turtle grass, **Thalassia testudinum** (Hydrocharitaceae). I. vegetative morphology. **Bull. Mar. Sci.**, 16(4): 748-761.

48. Wetzel, R. G.
1965 Techniques and problems of primary productivity measurement in higher aquatic plants and periphyton. **Mem. Ist. Italiano Idrobiol.**, 18(Suppl.): 249-267.

49. Wood, E. J. F.
1959 Some aspects of the ecology of Lake Macquarie, N. S. W., with regard to an alleged depletion of fish. VI. Plant communities and their significance. **Australian Jour. Mar. Freshw. Res.**, 10(3): 322-40.

50. Wood, E. J. F., Odum, W. E., and Zieman, J. C.
1969 Influence of sea grasses on the productivity of coastal lagoons. **In Coastal lagoons, a symposium,** UNAM-UNESCO. p. 495-502. Univ. Nacional Autonoma Mexico, Mexico, D. F.

51. Wood, E. J. F. and Oppenheimer, C. A.
1962 Note on fluorescence microscopy in marine microbiology. **Zeit. allgem. Mikrobiol.**, 2(2): 164-165.

52. Zieman, J. C.
1968 A study of the growth and decomposition of the sea-grass **Thalassia testudinum.** M. S. thesis, Univ. Miami, Miami, Florida. 50 p.

53. Zieman, J. C.
1970 The effects of a thermal effluent stress of the sea-grasses and macro-algae in the vicinity of Turkey Point, Biscayne Bay, Florida. Doctoral dissertation, Univ. Miami, Miami, Florida. 129 p.

54. Zieman, J. C.
1972 Origin of circular beds of **Thalassia** (Spermatophyta: Hydrocharitaceae) in South Biscayne Bay, Florida, and their relationship to mangrove hammocks. **Bull. Mar. Sci.**, 22(3): 559-574.

THE ROLE OF RESUSPENDED BOTTOM MUD IN NUTRIENT CYCLES OF SHALLOW EMBAYMENTS

Donald C. Rhoads[1], Kenneth Tenore[2], and Mason Browne[2]

ABSTRACT

The bottom of Buzzards Bay, Massachusetts, below a depth of about 13 meters is dominated by silt-clay sediment. Water flowing near the bottom is turbid because of resuspension of the bioturbated mud surface by tidal-current scour. The turbid-water layer extends about 3 meters above the bottom and may extend higher during spring tides. Maximum turbidity develops during mid-ebb tide. A seasonal turbidity cycle also exists. Minimum values of suspended carbon were measured in April and May when bioturbating activities of the benthos were at a minimum.

We have followed the concentration of particulate organic carbon (POC), particulate organic nitrogen (PON), and chlorophyll *a* in both surface and bottom water at one station over an 11-month period. Seasonal changes in these variables in surface and bottom waters showed two maxima related to peak phytoplankton production: one in the winter (December-January) and another in the autumn (September-October). Throughout most of the year, concentration of these consumer nutrients is higher in bottom water than at the surface.

Although primary production takes place in surface water, detritus from this production sinks and is concentrated on the muddy sea floor, stimulating bacterial growth. The significance of this nutrient-rich detrital resource for benthic productivity and aquaculture is discussed.

1. Department of Geology and Geophysics, Yale University, New Haven, Connecticut 06520.

2. Department of Biology, Woods Hole Oceanographic Institution, Woods Hole, Massachusetts 02543.

INTRODUCTION

The concentration of suspended solids in many estuaries and coastal embayments is higher in bottom water than at the surface, especially where the water column passes over muds that have undergone intensive bioturbation. The mechanism for resuspension of muds is related to turbulent mixing by tidal flow. Turbidity structure may change in space and time according to state of the tide, surface productivity, run-off, density structure of the water column, and the metablolic activity and trophic composition of the benthos. A detailed review of biologic and hydrographic mechanisms leading to development of near-bottom turbidity maxima has been given by Rhoads (10, 11).

The high concentration of suspended particles in bottom water of Buzzards Bay, Massachusetts, was first described by Young (21), who placed sediment traps in a vertical array above a muddy bottom. Traps located near the bottom were found to contain more sediment than traps near the surface. Benthic invertebrates were also found to be collected in bottom traps. Inspection of the entrapped sediment showed that much of it was in the form of fecal pellets of benthic deposit feeders. Subsequent studies have elucidated the relationship of this turbidity layer to benthic reworking by deposit feeders (13) and the relationship of turbidity cycles to tidal flow (10).

The potential significance of suspended particles near the bottom as a food source for commercially important suspension-feeding mollusks was initially investigated by placing metal racks on the bottom in about 20 meters of water within the turbidity layer. Trays of the hard clam, *Mercenaria mercenaria* were placed in Buzzards Bay and racks of the American oyster *Crassostrea virginica* in Long Island Sound (10, 12). Growth rates of the introduced mollusks were found to be equal to, or greater than, growth rates of control populations in shallow water near the shore. The purpose of the study reported here was to measure consumer nutrient concentrations (particulate carbon and nitrogen and chlorophyll *a*) of resuspended bottom muds over a seasonal and tidal cycle in Buzzards Bay. The term *nutrient* as used in this paper refers to particulate consumer food. Comparisons are made between Buzzards Bay and other embayments in vertical and temporal distribution of detrital consumer nutrients. The food value of resuspended mud in Buzzards Bay is compared to optimal concentrations of particulate carbon for three commercially important bivalves.

STUDY AREA

All of the data reported for Buzzards Bay came from Station H-7 located near a bell buoy in 15.5 meters of water at the western entrance to Woods Hole passage (Fig. 1). We chose this station for intensive study because it is characteristic of the silt-clay facies of Buzzards Bay in sediment type,

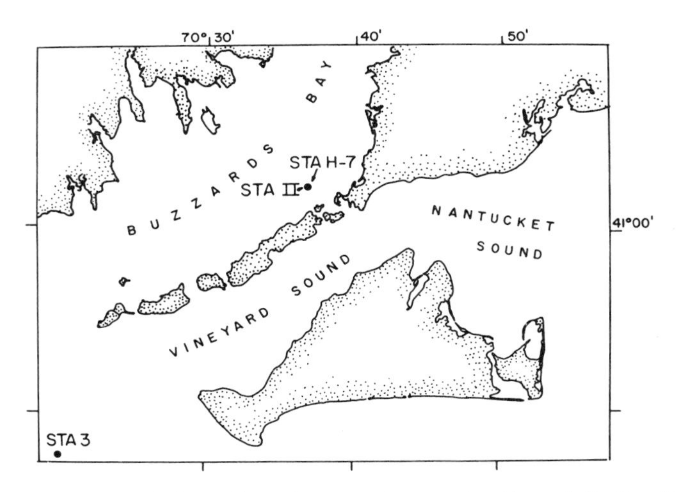

Figure 1. Map of Buzzards Bay, Massachusetts, showing location of Station H-7 (this study), Station II of Peck (8), and Station 3 established by Lillick (4).

hydrography, and benthic biology. The mud bottom at H-7 is reworked by deposit feeders and resuspended by tidal current scour.

Sampling in Long Island Sound was done at two sites, Station NWC located in central Long Island Sound off New Haven, Connecticut, in 17 meters of water, and Station DS, located nearby in 23 meters of water (Fig. 2). Both stations are in a silt-clay facies.

METHODS

Sampling of the water column for suspended solids and chemical analyses was done once each month starting on December 22, 1972, and ending on September 11, 1973. All measurements were made three hours after slack high water during peak ebb flow. Measurements of turbidity and nutrient concentrations related to stage of the tidal cycle were made during spring tide in Buzzards Bay on August 17, 1973, over a 7-hour period. Comparative measurements were made in Long Island Sound on August 14, 1973, over a 3-hour period during spring tide.

Turbidity was measured by lowering a 10 cm path-length transmissometer to the bottom (Hydro Products, San Diego, California). Continuous readings of transmissibility were made on an x-y recorder. Threshold concentrations for this instrument are about 10 mg l^{-1} of suspended bottom particles; suspensions of lower concentration are recorded as 100% transmissibility. Continuous salinity profiles were measured with a Hydro Products salinometer, and thermal profiles with a Hydro Products thermistor with deck read-out.

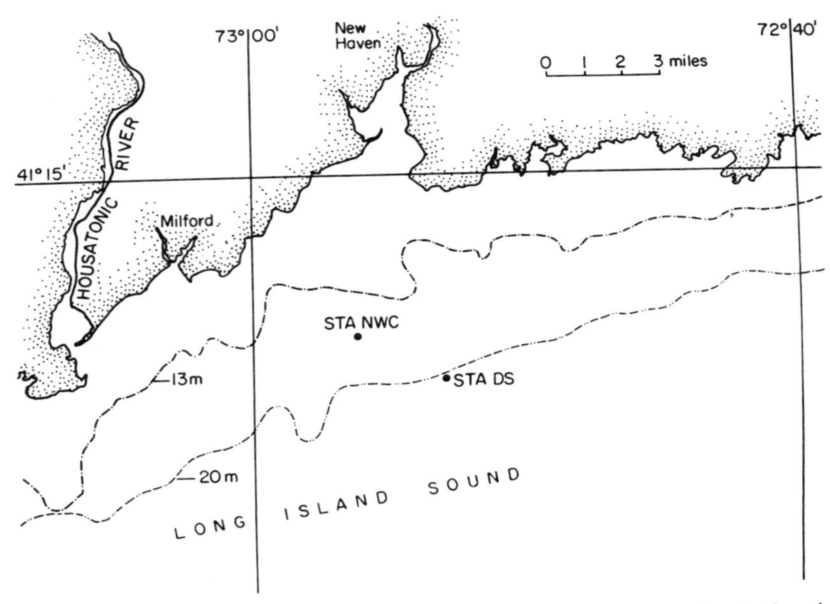

Figure 2. Map of central Long Island Sound off the Connecticut coast showing locations of Stations NWC and DS.

Sea water was collected in 5-liter Niskin sampling bottles from one meter below the surface and one meter above the bottom. A fraction of the collected water was filtered through Gelman glass fiber filters and frozen for later analysis. Particulate carbon and nitrogen were determined with a Perkins Elmer Model 240 Elemental Analyzer and chlorophyll *a* concentrations by fluorescence (5, 19).

RESULTS

Seasonal variations

Figure 3 shows the seasonal changes in chlorophyll *a*, particulate organic carbon (POC), and particulate organic nitrogen (PON) in surface and bottom waters at station H-7. During the winter and early spring, values of each of these variables were higher in bottom water than in surface water. There was a general decrease in both bottom and surface values from January to minimum annual concentrations in March and April. Concentrations increased during early to midsummer and declined again in July and August. This was followed by another increase in late summer and early fall. Concentrations of both particulate carbon and nitrogen remained significantly higher in bottom water

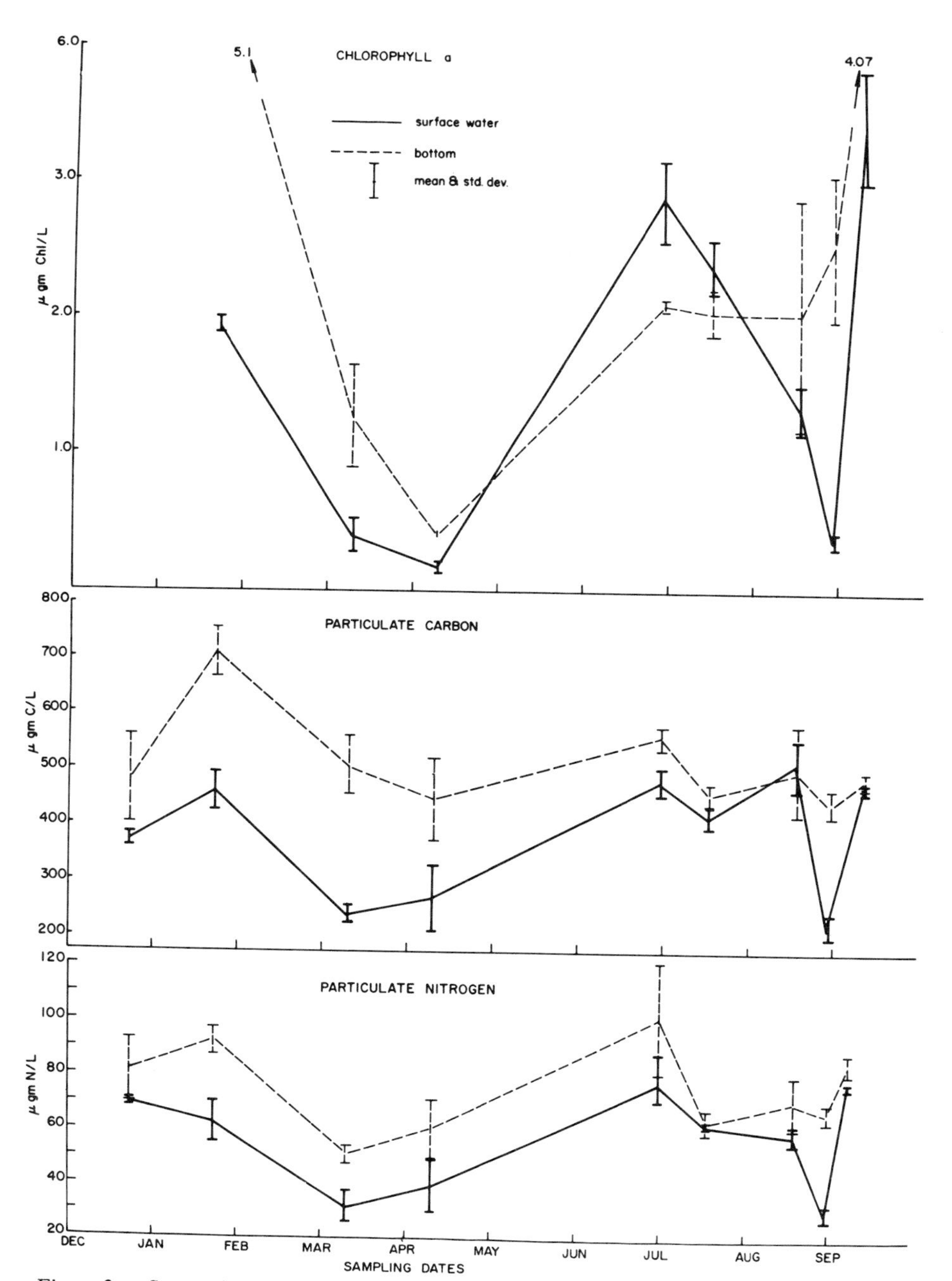

Figure 3. Seasonal changes in chlorophyll **a**, particulate carbon, and particulate nitrogen in surface and bottom water at Station H-7, Buzzards Bay, Massachusetts.

than in surface water except during the months of July, August, and September when bottom and surface values converged. Surface-water chlorophyll was higher than bottom-water chlorophyll only during early summer.

Seasonal changes in carbon-to-chlorophyll ratio and carbon-to-nitrogen ratio are given in Table 1. All of the data plotted in Figure 3 and tabulated in Table 1 represent samples taken about three hours after high water. Maximum bottom turbidity is developed during peak ebb tide (12).

TABLE 1

Seasonal changes in ratios of particulate carbon to chlorophyll **a** and particulate carbon to particulate nitrogen at Station H-7, Buzzards Bay, Massachusetts.

Sampling date	Sampling depth[1]	POC:chl a	POC:PON
Dec. 22, 1972	Surface	---	5.4:1
	Bottom	---	6.0:1
Jan. 22, 1973	Surface	241:1	7.4:1
	Bottom	138:1	7.6:1
March 9	Surface	603:1	7.7:1
	Bottom	400:1	9.8:1
April 9	Surface	1588:1	6.8:1
	Bottom	1076:1	7.4:1
June 29	Surface	151:1	6.0:1
	Bottom	546:1	5.5:1
July 16	Surface	176:1	6.7:1
	Bottom	250:1	6.4:1
August 31	Surface	617:1	7.3:1
	Bottom	166:1	6.2:1
Sept. 11	Surface	240:1	5.8:1
	Bottom	138:1	6.1:1

1. Surface samples from one meter below surface; bottom samples from one meter above bottom.

Tidal cycle variations

In order to evaluate short-term variations in chlorophyll, POC and PON, station H-7 was sampled over a tidal cycle during spring-tide conditions on August 17, 1973. In Figure 4 vertical profiles of turbidity are plotted above the tidal curve. To facilitate comparison of these profiles over time, we have

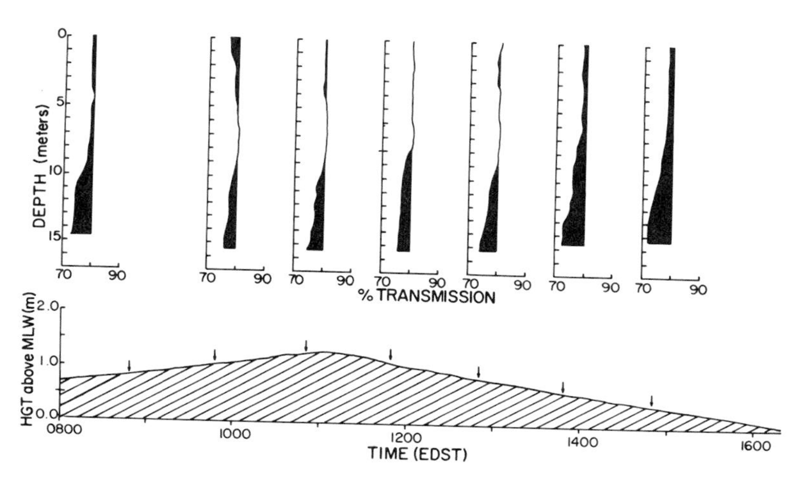

Figure 4. Vertical profiles of water turbidity over a tidal cycle on August 17, 1973, at Station H-7, Buzzards Bay, Massachusetts. Transmissivity values ≤ 80% are shaded to facilitate comparison. Time of measurement of turbidity indicated by the relative positions of the profiles above the tidal curve.

darkened that part of the turbidity curve falling below 80% transmittance. The shaded area of the curve is confined to the lower half of the water column during high water. Midwater had the highest transmissibility. Surface productivity presumably accounted for reduced surface transmissibility, especially at time 1000. Approximately three hours after high water, tidal currents suspend and mix the surface of the loosely compacted mud into the water column, some of it reaching the sea surface. This was reflected in an increase in both bottom and surface concentrations of POC (Fig. 5). Chlorophyll values were considerably higher than surface water during ebb flow. The range of PON was more conservative, but nitrogen values generally followed the tidal curve with surface values uniformly lower than bottom values. Peak nitrogen concentrations were found in bottom water during mid-ebb flow. The carbon-to-chlorophyll ratio of surface to bottom water is given for the August 17th tidal cycle in Table 2. The POC:PON value of bottom water over the tidal cycle is plotted in Figure 6.

DISCUSSION

Peck was the first to attempt a systematic study of primary productivity in Buzzards Bay (8). He found that diatoms dominated the planktonic flora and that the plankton was more abundant and evenly distributed in the water column at the head of the bay than at the mouth. One of his stations (II) is located in the general area of the Station H-7 occupied in this study (Fig. 1). He

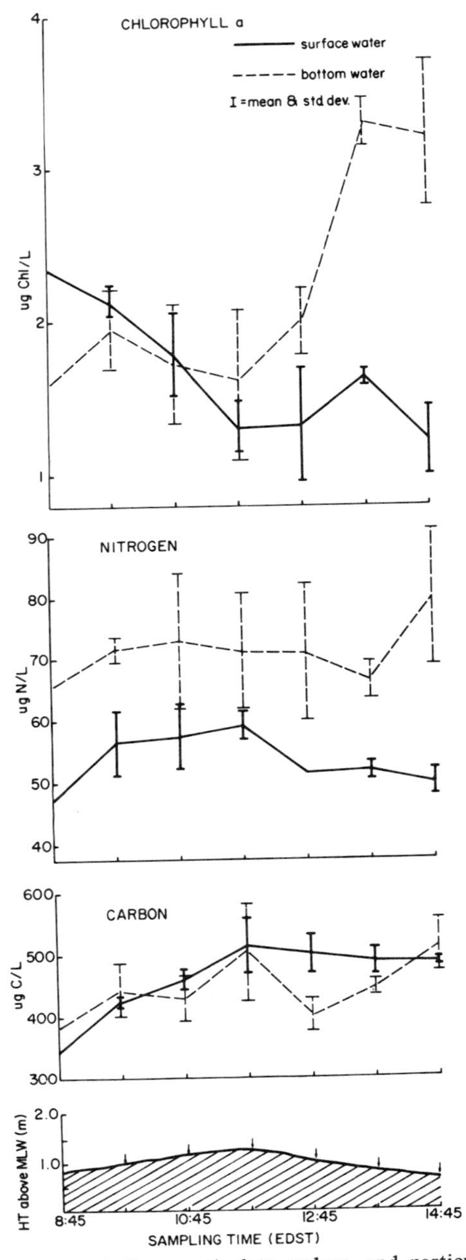

Figure 5. Changes in chlorophyll **a**, particulate carbon, and particulate nitrogen over a tidal cycle at Station H-7, Buzzards Bay, Massachusetts, on August 17, 1973.

TABLE 2

Changes in carbon to chlorophyll **a** ratio in surface and bottom water over a tidal cycle at Station H-7, Buzzards Bay, Massachusetts, on August 17, 1973.

Time	Stage of tide	Sampling depth[1]	POC:chl **a**
8:45	mid-flood	surface	145:1
		bottom	240:1
9:45	flood	surface	199:1
		bottom	227:1
10:45	slack flood	surface	256:1
		bottom	249:1
11:45	ebb	surface	384:1
		bottom	282:1
12:45	ebb	surface	374:1
		bottom	196:1
13:45	mid-ebb	surface	298:1
		bottom	137:1
14:45	ebb	surface	399:1
		bottom	159:1

1. Surface samples from one meter below surface; bottom samples from one meter above bottom.

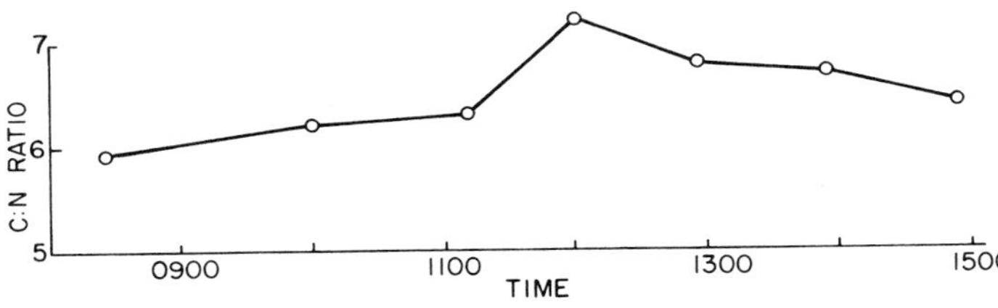

Figure 6. Particulate carbon to particulate nitrogen ratio of bottom water over a tidal cycle at Station H-7 in Buzzards Bay, Massachusetts, on August 17, 1973. The increase in carbon after 1100 marks ebb flow when bottom mud is strongly mixed into the water column.

reported higher numbers of diatoms in bottom water than at the surface at this station during September, 1894.

Fish measured the relative abundance of phytoplankton over an annual cycle in Great Harbor, Woods Hole, during the period 1922-23 and found two maxima in production: a fall bloom in September-October and a winter bloom in

December-January (2). He also noted that the surface of the bottom was covered with large numbers of diatoms and suggested that these represented resting stages. Fish's productivity cycle compares favorably with that represented by our chlorophyll data (Fig. 3).

Few comparative data exist for consumer nutrient cycles in the Bay. Lillick (4) occupied a sampling site near our Station H-7 for two months and measured inorganic plant nutrients as part of a more extensive productivity study in Vineyard Sound (Fig. 1). He indicated that observations at the Buzzards Bay station were essentially the same as those at his Station 3 (Fig. 1) and so abandoned further study in Buzzards Bay. Station 3 was located SW of Gay Head, Martha's Vineyard, in 90 meters of water. The substratum in this area consists of coarse-grained sediments. Figure 7 shows data taken at a depth of 15 meters at Station 3 over a 14-month period. Samples were taken at the surface and near the bottom. The 15-meter sampling depth is the best for showing seasonal variations in productivity. Lillick's data are consistent with those of Fish (2), showing a fall and winter bloom. Diatoms again dominated the flora. Throughout the summer, inorganic nitrate in the bottom water was found to be higher than at the surface. Nitrates built up in the water column in the winter, reaching peak concentrations in early January. The number of cells at the bottom of the water column were occasionally found to be higher than at the surface.

From data presented here, and those of previous workers in Buzzards Bay and adjacent waters, we interpret the annual maxima of chlorophyll to reflect a winter phytoplankton bloom in December-January and a fall flowering in September-October. A minor bloom is also present during the early summer (April-June). The higher concentration of chlorophyll in bottom water in the winter and fall probably represents unconsumed diatoms that have settled to the bottom where they are periodiclly mixed back into the water column by tidal flow. Young recorded an annual range of 300-1500 cells per cm^2 (pennate diatoms) in the silt-clay region of Buzzards Bay (21). Most of this silt-clay facies lies below a depth of 13 meters. Only 1% of surface illumination is recorded at about 10 meters, placing most of the settled diatoms near, or below, the compensation depth.

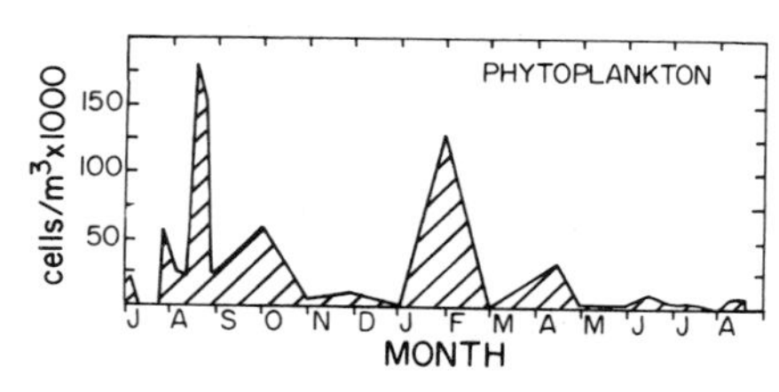

Figure 7. Seasonal changes in primary production at Lillick's Station 3, Vineyard Sound, Massachusetts. Data from 15 meters below the surface (4).

The high concentration of POC in bottom water was related to the resuspension of bottom sediment. One could identify some of this detritus as coming from the seafloor, because it consisted of fecal pellets of benthic consumers (21). Photomicrographs of this detritus have been published by Rhoads (10). The detritus had a bacterial population associated with it contributing to the high values of PON recorded for the turbid-water zone. Seasonal changes in the carbon-to-nitrogen ratio showed that the bottom water had more PON associated with it than had the surface water during late June, July, and August, when bacterial reduction of the detritus was high. This same period was associated with peak rates of benthic invertebrate growth and metabolism.

The plankton cycle of Buzzards Bay is shared with other semi-confined bodies of coastal water such as Narragansett Bay, yet the vertical structure of the water column studied by Smyda at two stations at the mouth of Narragansett Bay did not show the inverted nutrient stratification present in Buzzards Bay (14). Smyda did indicate, however, that there was a huge excess of phytoplankton production over that consumed. This excess probably ends up on the bottom.

The vertical distribution of phytoplankton has been determined for central Long Island Sound by Conover (1). Much of the central Sound, like Buzzards Bay, has fine-grained muds reworked by deposit-feeders. The winter phytoplankton bloom occurred about two months later in Long Island Sound than in Buzzards Bay (February-March). Chlorophyll *a* was more abundant in bottom water than at the surface only after termination of periods of active phytoplankton growth and throughout the fall months. Chlorophyll *a* values of both bottom and surface water in Long Island Sound were higher than those encountered in Buzzards Bay (Fig. 8). Some parts of central Long Island Sound off New Haven, Connecticut (Sta. DS), were similar to Buzzards Bay in vertical turbidity structure (Fig. 9). During periods of density stratification this turbidity rose in the water column to the bottom of the halocline where it was prevented from reaching the surface. Generalizations about the Sound's turbidity structure must be made with care, however, as Station NWC, located NW of Station DS (Fig. 2), showed very weak bottom suspension on account of lower tidal current velocities. The bottom at Station NWC is highly bioturbated, uncompacted, and capable of being resuspended, yet tidal velocities at this site were insufficient to strongly mix this sediment up into the water column. Figure 10 shows the turbidity structure at Station NWC over part of a tidal cycle during spring tide conditions. Surface production was higher at this station than bottom water. Clearly, nutrient stratification in central Long Island Sound is complex and may vary temporally and spatially depending upon spatial gradients and temporal fluctuations in bioturbation and hydrography.

Inverted nutrient stratification is not limited to Buzzards Bay and parts of Long Island Sound. This condition has been described for the York River

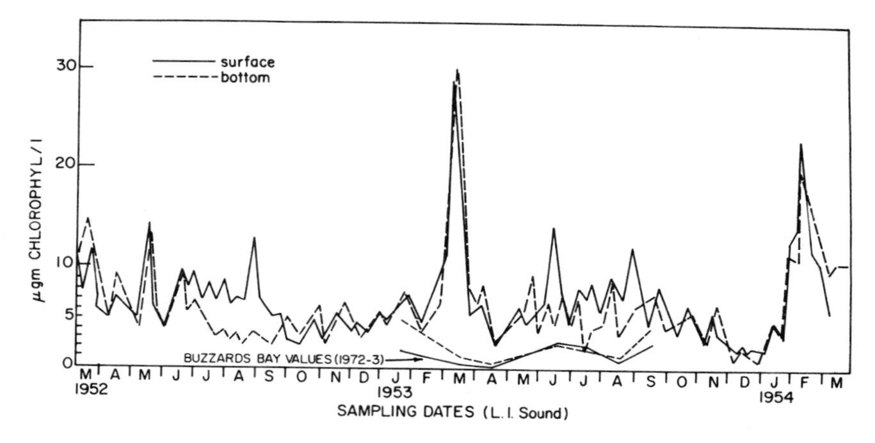

Figure 8. Seasonal changes in chlorophyll **a** in surface and bottom water in central Long Island Sound. Data from Conover (1). Data from Buzzards Bay, Massachusetts (this study), is superimposed for comparison.

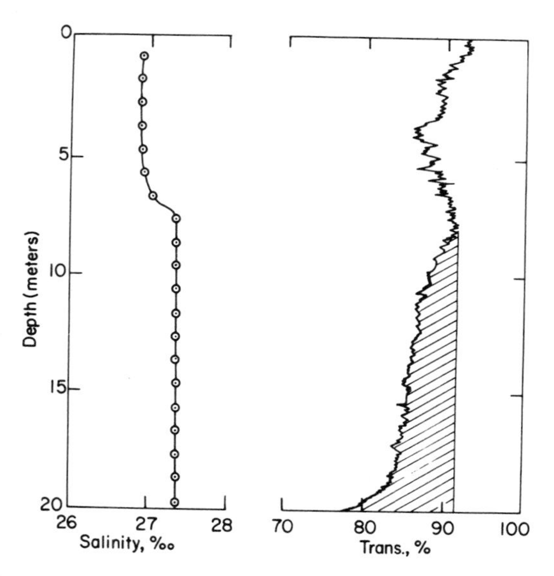

Figure 9. Vertical profiles of salinity and transmissivity at Station DS, Long Island Sound. Resuspended mud is confined below the halocline.

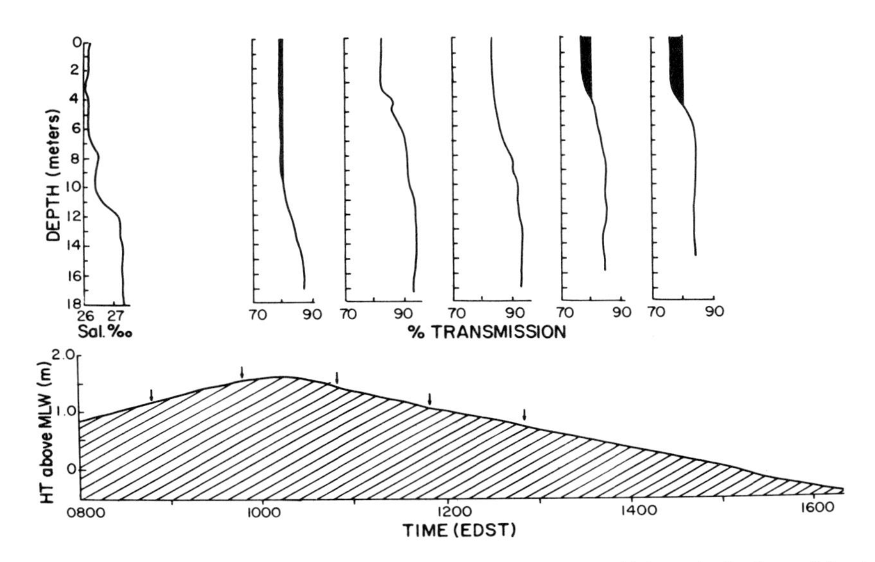

Figure 10. Vertical profiles of salinity and turbidity over a tidal cycle in Long Island Sound, Station NWC, August 14, 1973. Surface water was higher in POC, PON, and chlorophyll than bottom water due to weak tidal mixing.

estuary, Virginia (7, 20), and for the tropical Cochin Backwater in India (9). The implications of this phenomenon for benthic productivity would seem to be sufficiently important to consumer productivity to justify speculation on the differential effect of surface versus benthic nutrient stratification. Where surface nutrients and productivity are located well above the bottom (surface stratified), nutrients reach the bottom only after plankton die and settle to the seafloor as detritus. One might expect benthos living in this system to be responsive to the periodic influx of food following major plankton blooms. The response could be manifested by fluctuations in benthic biomass and species diversity related to instability in the input of food resources. For instance, the dominant spawning of the benthos coincides with the burst of productivity in estuaries (6, 16, 18) as the planktonic larvae take advantage of the increase in a suspended food source. In addition, this material might eventually settle on the bottom at about the same time the larvae settle to a benthic existence and be a further food source for the organisms (15). Under conditions like those described here for Buzzards Bay, the bottom serves as a nutrient reservoir and may dampen the effects of surface productivity cycles. The dampening effect may be manifested by large standing crops and a stable population structure. This mode of coupling between benthic and planktonic productivity may be an important one and may explain, in part, some synchronous aspects of the population dynamics of both planktonic and benthonic species.

The recycling and release of nutrients to the water column from the sediment depend largely on the ease with which the muddy seafloor can be resuspended. Seasonal cycles in bioturbation and tidal current turbulence control this factor. Primary productivity, in turn, feeds back to the benthic system by the sedimentation of plankters to the bottom. Bacteria use this detritus as a food source. Both living and dead diatoms, as well as benthic bacteria, produce detrital food for benthic consumers. This hypothesis supports the ideas of Levinton regarding the relative stability of suspension- and deposit-feeding populations (3). Suspension-feeding populations may be more unstable than deposit-feeding populations because they feed upon a resource which fluctuates in abundance and quality according to plankton production cycles. Deposit feeders, on the other hand, feed on the detrital reservoir, which is more predictable. We must add, however, that the quality of deposited detritus appears also to fluctuate as measured by the POC:PON ratio (Table 1).

Embayments, like Buzzards Bay, which show an inverted nutrient column represent systems that may provide an opportunity to maximize consumer productivity for molluscan aquaculture. For aquaculture, suspension feeders must be inserted into nutrient-rich bottom water. Natural populations of suspension feeders have been relatively unsuccessful in colonizing soft unstable muds because of the problem of feeding in highly turbid suspension (trophic group amensalism) (13). This problem may be overcome, however, by suspending these mollusks a meter or so above the bottom on racks located within the turbidity layer. Initial studies in Buzzards Bay have shown that trays of the quahog *Mercenaria mercenaria* have high growth rates compared to populations grown nearshore (12). Similar results were obtained near Station DS in Long Island Sound using the oyster *Crassostrea virginica* (10).

One advantage of a bottom raft culture is the favorable nutritive conditions of the suspended detritus near the bottom. Tenore and Dunstan measured the feeding and biodeposition rates of mussels, oysters, and clams in experimental trays receiving controlled additions of phytoplankton at different concentrations of particulate carbon (17). These trays were continuously flushed with temperature-controlled sea water. They found that, whereas the feeding rate of the bivalves increased linearly with increasing food concentrations, the biodeposition of feces and pseudofeces increased logarithmically (Fig. 11). At low food concentrations, greater amounts of the filtered food were egested as pseudofeces and not ingested by the bivalves. Optimum food concentrations for these three bivalves fell within a range of 300 to 500 μg POC liter^{-1} of sea water. The period May-October in Buzzards Bay corresponds closely to the maximum period of growth for these bivalves. During this time, average carbon concentration of bottom water samples was 435, with a range of 297 to 562 μg POC liter^{-1} (Fig. 11). Shellfish placed on bottom racks would be presented food at optimum concentration levels. These support racks would at the same time

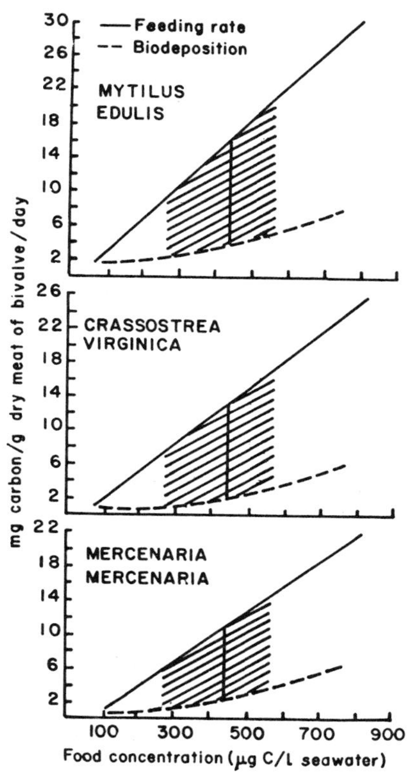

Figure 11. Feeding and biodeposition rates of three bivalve species. From Tenore and Dunstan (17). Values of POC in bottom water at Station H-7 in Buzzards Bay are plotted on these graphs. The vertical line is the mean annual POC value at Station H-7 and the cross-hatched area the range of values. The values of POC at Station H-7 fall within the range of most efficient feeding for these bivalves.

hold the bivalves above the sediment-water interface to avoid siltation and burial.

A major conclusion of our study relates to the importance of measuring levels of resuspended detritus in estimating nutritive material available to suspension-feeding benthos. Any determination of the standing crop of food available to the herbivore trophic level must include not only phytoplankton carbon but also detrital carbon. The periodic resuspension of detritus from the bottom results in the possibility of further use by suspension feeders of recycled organic matter. The process of resuspension may further allow greater heterotrophy by aerobic bacteria as this detritus is resuspended into an aerated water column. This probably results in greater mineralization of the detritus than if the sediment remained on the sea floor. The presence of bacterial coatings on the detritus probably enhances its food value to consumers. Such carbon cycling may be a major importance in increasing secondary production.

ACKNOWLEDGMENTS

We wish to thank Mr. R. Colburn, skipper of the R. V. *Asterias,* for his assistance in collecting samples in Buzzards Bay. Mr. J. Baker, director of the Connecticut State Shellfish Laboratory, Milford, Connecticut, provided us with ship time to obtain comparative data in Long Island Sound. Dr. R. Gordon, Yale University, provided us with turbidity-depth recording equipment. Mr. M. Reid operated this equipment at sea for us. This research was supported by N.S.F. grant nos. GI 32140 and GA 39911.

REFERENCES

1. Conover, S. A. M.
   1956 Oceanography of Long Island Sound, 1952-1953, IV. Phytoplankton. **Bull. Bingham Oceanogr. Coll.,** 15: 62-112.
2. Fish, C. J.
   1925 Seasonal distribution of the plankton of the Woods Hole region. **Fish. Bull. U. S. Fish Wildl. Serv.** 41: 91-179.
3. Levinton, J.
   1972 Stability and trophic structure in deposit-feeding and suspension-feeding communities. **Am. Nat.,** 106: 472-486.
4. Lillick, L. C.
   1937 Seasonal studies of the phytoplankton off Woods Hole, Massachusetts. **Biol. Bull.,** 73: 488-503.
5. Lorenzen, C. J.
   1966 A method for the continuous measurement of **in vivo** chlorophyll concentration. **Deep-Sea Res.,** 13: 223-227.
6. Muus, B. J.
   1967 The fauna of Danish estuaries and lagoons. **Med. Danmarks fish. Havunders.,** 5: 1-316.
7. Patten, B. C., Young, D. K., and Roberts, M. H., Jr.
   1966 Vertical distribution and sinking characteristics of seston in the lower York River, Virginia. **Chesapeake Sci.,** 7: 20-29.
8. Peck, J. I.
   1896 The source of marine food. **Fish. Bull. U. S. Fish Wildl. Serv.,** 15: 351-368.
9. Qasim, S. Z. and Sankaranarayanan, V. N.
   1972 Organic detritus of a tropical estuary. **Mar. Biol.,** 15: 193-199.
10. Rhoads, D. C.
    1973 The influence of deposit-feeding benthos on water turbidity and nutrient recycling. **Am. Jour. Sci.,** 273: 1-22.
11. Rhoads, D. C.
    In press Organism-sediment relations on the muddy seafloor. In **Oceanography and marine biology** (1974) (ed. Barnes, H.)
12. Rhoads, D. C. and Pannella, G.

1970 The use of molluscan shell growth patterns in ecology and paleoecology. **Lethaia,** 3: 143-161.

13. Rhoads, D. C. and Young, D. K.
1970 The influence of deposit-feeding organisms on sediment stability and community trophic structure. **Jour. Mar. Res.,** 28: 150-178.

14. Smyda, T. J.
1957 Phytoplankton studies in lower Narragansett Bay. **Limnol. Oceanogr.,** 2: 342-359.

15. Stephens, K., Sheldon, R. W. and Parsons, T. R.
1967 Seasonal variations in the availability of food for benthos in a coastal environment. **Ecology,** 48: 852-855.

16. Tenore, K.
1972 Macrobenthos of the Pamlico River estuary, North Carolina. **Ecol. Monogr.,** 42: 51-69.

17. Tenore, K. and Dunstan, W. M.
1973 Comparison of feeding and biodeposition of three bivalves at different food levels. **Mar. Biol.,** 21: 190-195.

18. Thorson, B.
1957 Benthic communities. In **Treatise on marine ecology and paleoecology,** Vol. I, p. 461-534. (ed. Hedgpeth, J.) Geol. Soc. Am. Mem. 67. Waverly Press, Baltimore, Maryland.

19. Yentsch, C. S. and Menzel, D. W.
1963 A method for the determination of phytoplankton chlorophyll and phaeophytin by flourescence. **Deep-Sea Res.,** 10: 221-231.

20. Young, D. K.
1968 Chemistry of Southern Chesapeake Bay sediments. **Chesapeake Sci.,** 9: 254-260.

21. Young, D. K.
1971 Effects of infauna on the sediment and seston of a subtidal environment. **Vie Milieu,** Suppl. 22: 557-571.

PART III.

THE ESTUARINE SYSTEM: ESTUARINE MODELING

Convened By:
Richard Williams
Biological Oceanography Section
National Science Foundation
Washington, D.C. 20050

A PRELIMINARY ECOSYSTEM MODEL OF COASTAL GEORGIA *SPARTINA* MARSH

R. G. Wiegert[1]

R. R. Christian[2]

J. L. Gallagher[3]

J. R. Hall[3]

R. D. H. Jones[4]

R. L. Wetzel[3]

ABSTRACT

A model of a coastal Georgia salt marsh was constructed using biologically realistic equations which represented the transfer of carbon among 14 compartments, 7 biotic and 7 abiotic. A simulation run with this model for 5 years showed a larger transfer of carbon from air to materials such as dissolved and particulate organic matter in the open water and sediments. Five-year simulation runs made after varying, successively, each of the 81 model

1. Department of Zoology, University of Georgia, Athens, Georgia, 30602.

2. Department of Microbiology, University of Georgia, Athens, Georgia 30602.

3. University of Georgia Marine Institute, Sapelo Island, Georgia, 31327.

4.Department of Statistics, University of Georgia, Athens, Georgia 30602.

parameters by +100% or -50% showed that only 21 were "sensitive" enough to change the final system steady state organic accumulation by $\pm$ 0.5%. These "sensitive" parameters involved the components *Spartina* shoots, *Spartina* roots, *algae,* particulate organic sediments, sediment saprophages (both aerobic and anaerobic), and open-water heterotrophs. In addition to pinpointing sensitive parameters the model suggests scarcity of space (as opposed to scarcity of material resource) as the major limiting factor for most components. Both of these conclusions need careful scrutiny as field research and the development of more complex models proceed.

INTRODUCTION

A realistic simulation model of an ecosystem can aid in organizing a research program by 1) exposing important existing gaps in the physical and biological data, and 2) pinpointing "sensitive" areas, i.e. those parameters that act as amplifiers or governors on the system, producing large changes in fluxes or state variables in response to small changes in their own values. Such identification of sensitive parameters permits a concentrated effort to decrease the error associated with their measurement.

To be useful for these objectives, a model need be neither complex and detailed, nor thoroughly validated. Its structure, however, must be realistic. By this we mean that the equations describing the interactions between the state variables or compartments must incorporate *at least* the major types of biological feedback control that may be operative over the entire range of values that the given state variables could reasonably be expected to assume in the real system or be forced to assume (via initial conditions or perturbations) in the model. If the equations do not satisfy this criterion, conclusions about the effects of perturbations of parameter values have little worth. We want to emphasize that this argument is independent of the magnitude of the perturbations contemplated for the model. For example, slight alterations in the equilibrium density of a prey population can often produce significant changes in the predator density and vice versa. But these mutual changes are observed in a model *only* when the flow of matter-energy from prey to predator is a function of *both* the prey population and the predator population. The logical conclusion from this argument is that equations that are realistic in the sense defined above cannot be linear when the flux pathway is either biotic-biotic or abiotic-biotic. Such pathways must always be represented by a recipient-determined but donor-recipient controlled type of equation (18), where the primary determination of the amount of matter-energy flux is provided by the density and feeding rate of the recipient biotic population, subject to controls imposed by scarcity of the donor (resource limitation) or by density-dependent interactions within the recipient population (space limitation) or by both.

We have constructed the preliminary model discussed in this paper in a manner sufficiently realistic to answer three specific questions: 1) On the basis of present knowledge of the biology of the coastal Georgia *Spartina* marsh, is this ecosystem a potential source of supply of organic matter to the offshore waters? 2) Which *biological* groups are most responsible for the production and decomposition of carbon in this ecosystem? and 3) Which of the many biological parameters are potentially important but as yet poorly known and how might research effort be allocated efficiently for improving our predictive accuracy regarding the seasonal dynamics of the ecosystem as well as its response to perturbations?

GENERAL DESCRIPTION OF THE MODEL

Our preliminary model of the flux of carbon in a coastal Georgia marsh ecosystem comprises 14 state variables or compartments, 7 living (biotic) and 7 non-living (abiotic) components (Table 1). The units of flow used in the simulation runs are gC x m^{-2} x yr^{-1} and the units of standing stock are gC x m^{-2}, based on a column extending from 1 meter below the surface of the bottom mud to a height of 50 km. No attempt was made in this first version to incorporate explicitly the effects of spatial heterogeneity or tidal effects, although they are implicit in many of the parameter values. Temporal heterogeneity in biomass distribution, i.e. diel and seasonal movements of detritus and movements of animals were not modelled, but many of the model parameters are varied on a seasonal, 4-times-a-year basis, changing on days 0, 91, 181, and 271 (Table 2).

The compartments of the model are listed in Table 1 and, together with the labelled flow pathways, are shown diagrammatically in Figure 1. The latter are listed sequentially (by donor, then recipient) in Table 2. In Figure 1 and Table 2 a D (donor) label indicates a donor-determined, donor-controlled pathway, i.e. one where the flux of carbon is the product of a specific rate of transfer (gC x gC^{-1} x time) multiplied by the standing crop of the donor (gC x m^{-2}). These flows are characteristic of transfers between compartments where the recipient provides no feedback to the donor. Such relationships are common in biotic-abiotic and abiotic-abiotic fluxes, at least within the normal range of variation encountered in or forced upon the abiotic component.

In this paper we first describe the general biological relationships between donor and recipient and indicate how these are represented in the equational structure of the model (a copy of the Fortran program with comments is available from the senior author). This is followed by a brief description of the compartments (variables) of the model and of selected parameter values. Then we discuss the various compartments and flow pathways plus the sensitivity of

TABLE 1.

Biotic and abiotic compartments of the salt-marsh ecosystem model. State variables are designated by the symbol. For definitions of α_{jj} and γ_{jj}, see text.

Symbol	Name or Designation	gc x m-2 Std. Stock Day 0 (Jan. 1)	α_{jj} (gc x m^{-2})	γ_{jj} (gc x m^{-2})	Sources
X1	CO_2 - air	875.0	—	—	12
X2	Algae	8.8	14.0	15.0	12
X3	**Spartina** shoots	48.2	145.0	176.0	
X4	Heterotrophs-air	0.1	0.25	1.25	11, 14
X5	**Spartina**-std. dead	118.0	—	—	
X6	DOM-water	4.0	—	—	
X7	Heterotrophs-water	5.4	7	10.0	1, 4, 6
X8	Anaerobic heterotrophs-sediment	10.0	180.0	200.0	
X9	POM - water	2.0	—	—	
X10	CO_2 - water	30.0	—	—	
X11	DOM - water	4.0	—	—	
X12	**Spartina** roots	58.5	—	—	3
X13	POM - benthic	17.5×10^3	—	—	
X14	Aerobic heterotrophs-sediment	5.4	7	10.0	10, 14, 16, 19

the parameters.

FLUX EQUATIONS

The relationships between the 14 compartments of the salt-marsh ecosystem were initially assigned to either the donor-determined or the recipient-determined category. The equations describing these relationships follow the models proposed by Wiegert (17, 18).

A donor-determined relationship is illustrated by

$$F_{ij} = \tau_{ij} X_i \qquad (1)$$

where: F_{ij} = the total flux of carbon from compartment *i* to *j*

(gC x m^{-2} x day^{-1}).

Parameter values for the fluxes (Fij) of the salt-marsh ecosystem model. A single value designates a constant. Time dependent variable values are given clockwise from upper left as winter (day 0), spring (day 91), summer (day 181) and fall (day 271). R type fluxes are recipient-determined, D type are donor-determined. Exception to the standard control terms are indicated by an asterisk. For definitions, see text.

Fluxes (F_{ij})	Flux Type	τ_{ij}	α_{ij}	γ_{ij}	μ_{ij}	ρ_{ij}	η_{ij}	ε_{ij}	ν_{ij}	Sources
1,3	R	.22 | .31 .17 | .12	40	20						
1,10	R*	.7		30						
2,6	D						.01			7
2,7	R	.8	11.0	1.0				.5	.3	
2,10	D					.007 | .021 .125 | .014				13
3,1	D					.094 | .121 .071 | .053				15
3,4	R	.25	1.0	0				.25		11,14
3,5	D				.015 | .017 .016 | .024					
3,6	D						.002 | .003 .002 | .001			
3,12	D*	.094 | .136 .075 | .054								
4,1	D					.1				14
4,6	D						.01			
4,9	D				.23 | .01 .01 | .23					11,14
5,6	D	.001 | .002 .002 | .002								
5,7	R	.8	8.0	4.0				.25	.1	
5,9	D	.008 | .024 .018 | .022								
6,7	R	.8	6.0	4.0					.3	
7,6	D						.01			5
7,9	D				.01					

TABLE 2 (continued)

Fluxes (F_{ij})	Flux Type	τ_{ij}	α_{ij}	γ_{ij}	μ_{ij}	ρ_{ij}	η_{ij}	ε_{ij}	ν_{ij}	Sources
7,10	D					.10				
8,10	D					.10				
8,11	D						.80			
8,13	D				.01					
9,7	R	.35	1.5	1.0				.25	.30	
9,13	D	.01								
10,1	D*	.7		30.0						
10,2	R	.082 | .257 / 1.5 | .171	7.8	.1						8,9
11,6	D	.008								
11,8	R	.8	10.1	.1					.8	
11,14	R	.8	10.3	.3					.5	
12,3	D*	.002 | .006 / 0 | 0								
12,10	D					.058 | .088 / .053 | .041				15
12,11	D						.002 | .003 / .002 | .001			
12,13	D				.018 | .025 / .020 | .013					
13,8	R	.179	.001	2200				.8	.2	
13,14	R	.35	2400.0	2200				.25	.5	
14,10	D					.1				
14,11	D						.01			
14,13	D				.01					

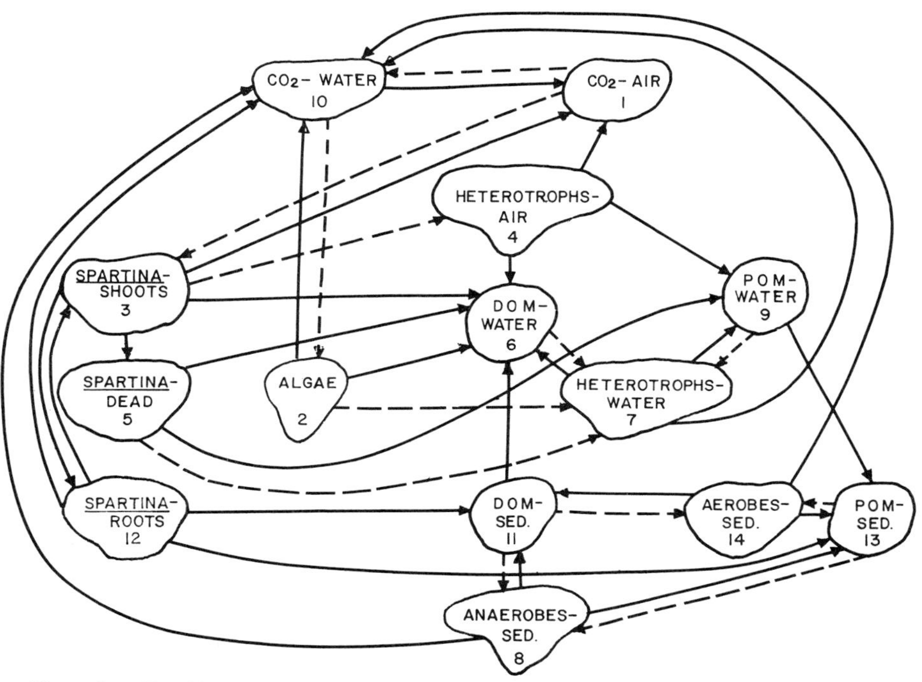

Figure 1. Trophic web of the salt-marsh model. Solid lines indicate pathways that are donor-determined and donor-controlled. Dashed lines indicate pathways that are recipient-determined, but donor-recipient controlled (see text and Table 2).

τ_{ij} = the maximum specific transfer rate of carbon

(gC x gC^{-1} x day^{-1}).

χ_i = the standing stock of carbon in compartment *i*

(gC x m^{-2}).

A recipient-determined flux is illustrated by

$$F_{ij} = \tau_{ij}\chi_j \left[1 - f_{ij} - \left\{1 - (\rho_{ij} + \eta_{ik}) \, / \, \tau_{ij}(1-\varepsilon_{ij})\right\} f_{jj}\right]_+ \qquad (2)$$

given:

$$f_{ij} = \left[1 - \frac{1}{\alpha_{ij}\sigma_{ij}}(\chi_i - \sigma_{ij})_+\right]_+ \qquad (3)$$

$$f_{jj} = [1 - \frac{1}{\alpha_{jj} - \alpha_{jj}}(\sigma_{jj} - X_j)]_+ \quad (4)$$

$$\text{and } (\theta)_+ = \begin{cases} \theta \text{ if } \theta > 0 \\ 0 \text{ if } \theta \leq 0 \end{cases}$$

where: τ_{ij} = the maximum specific (gC x gC^{-1} x day^{-1}) ingestion rate of recipient *j*.

ϱ_j = specific loss rate of gaseous carbon compounds (gC x gC^{-1} x days^{-1}).

ε_{ij} = egestion loss (a proportion of F_{ij}).

$\mathfrak{y}$ = specific excretion rate of carbon compounds (gC x gC^{-1} x days^{-1}).

σ_{ij} = the minimum density of donor *i* that is available for ingestion by *j*.

α_{ij} = the density of donor *i* below which resource limitation of ingestion by the recipient begins, i.e. whenever $X_i \geq \alpha_{ij}$, $f_{ij} = 0$.

σ_{jj} = the maximum density of recipient *j* that can be maintained in the "space" available.

α_{jj} = the density of the recipient *j*, above which space limitation of ingestion by the recipient begins, i.e. whenever $X_j \leq \alpha_{jj}$, $f_{jj} = 0$.

The two feedback control terms, f_{ij} and f_{jj}, respectively, represent the limitation imposed by scarcity of resources and that imposed by scarcity of any factor whose magnitude is a function of physical space. Either or both may act to control the density of a biotic component at or around the equilibrium point. In the absence of information to the contrary, the biological assumptions implicit in the mathematical forms of equations 3 and 4 are assumed to hold. These are: 1) that there are definite density thresholds below which a species shows no effects of crowding or a resource becomes unavailable; and 2) that densities exceeding these thresholds produce directly proportional effects on the realized ingestion rate. An exception to the second assumption has been described for the space control of photosynthesis in thermal blue-green algae (2). In the present model we found assumption 1) to be inadequate for representing feedback control of anaerobes in the sediments (X8).

When a biotic component both lives on and feeds from an abiotic organic resource, in this case sediment particles, available space and available material resources are to a large extent correlated, and cannot be represented adequately by separate feedback controls. An organic particle, as opposed to a dissolved substance, represents a "package" of concentrated resource to the saprophage on its surface. Its "value" in promoting growth is a function of the composition and is independent of the concentration of particles themselves. Thus, the realized growth must depend on the degree of crowding on the particles themselves, i.e. realized growth must be a function of the two standing stocks, X_i and X_j and the unavailable material σ_{ij}. An appropriate modification of the resource feedback control equation seemed to be:

$$f_{ij} = [1 - \alpha_{ij} \frac{X_i - \delta_{ij}}{X_j} +] + \tag{5}$$

where: α_{ij} = the ratio of the standing stock of the recipient to the available standing stock of the donor at the point where feedback limitation of growth first begins.

A feedback control of ingestion, such as equation 5, requires a decrease in X_j as X_i decreases, but permits some ingestion at all densities of X_i greater than δ_{ij}, i.e. a saprophage can, given enough time, completely utilize the available portion of the resource.

State Variables and Flux Parameters

The 14 state variables or compartments listed in Table 1 comprise the major functional and/or trophic groupings used in the salt-marsh ecosystem model. The choice of these 14 variables should be so constrained that: 1) each compartment represents a potentially important entity, one whose dynamics should be considered in the assignment of research priority; 2) each of the major potential input-output sources of carbon to the system is made a separate state variable; and 3) the total number of state variables is restricted to less than 15, in order to keep to a minimum the time and expense of the preliminary development of the model. Despite some compromises, we are satisfied that criteria 1) and 2) were met by the 14 compartments we have identified.

The standing stock of carbon in the air (X1) was based on the total in a column from the marsh surface to a height of 50 km. Continual mixing was assumed, since the natural CO_2 concentration varies little and CO_2 depletion is usually only a transient and local phenomenon in the immediate vicinity of plant surfaces. *Spartina alterniflora* L. shoots, *Spartina* - std. dead, and *Spartina* roots (X3, X5, X12) are the functional components of the dominant primary producer in the marsh. The algae (X2) represent both benthic forms (those on the sediment surface) and phytoplankton. The heterotrophs in air (X4) comprise mainly grasshoppers and homoptera; predators of this group were not modeled. Heterotrophs in water (X7) include a number of quite different groups of organisms from fish to bacteria, and thus are among the most difficult for which to estimate or measure parameter values. The anaerobic heterotrophs in the sediment (X8) are mainly bacteria. The aerobic heterotrophs in the sediment (X14) comprise those macroscopic saprophages that live in or on the sediments plus the aerobic bacteria operating in the aerated zone around *Spartina* roots and at the surface of the sediments.

Not only was it necessary to determine the initial conditions for the state variables, but values had to be assigned to some 81 biological parameters of the model equations. Insofar as possible these were measured values published in the open literature. In many cases we had to resort to unpublished data from current Sapelo Island studies. Sometimes information was so scarce that the parameter value represents little more than an educated guess. The literature sources are listed, along with the initial conditions and parameter values, in Tables 1 and 2. Space does not permit a discussion of all of these; we have confined our discussion to those which turned out, in the sensitivity analyses, to be important (*see* Table 3).

The maximum specific gross photosynthesis rate of *Spartina* (τ 1, 3) represents the mean of rates under optimum conditions of light and nitrogen availability, based on extensive unpublished data from Sapelo Island.

The specific rate of transfer of carbon from the shoots to the roots of *Spartina* (τ 3, 12) was estimated.

The maximum specific rate of algal gross photosynthesis (τ 10, 2) was estimated from the literature (8, 9) and represents a mean of rates for both benthic and planktonic forms.

The specific rate of utilization of POM in sediments by anaerobes (τ13, 8) was estimated from Sapelo Island data. A specific growth rate of 0.069 gC x gC^{-1} x $days^{-1}$, representing one cell division every10 days, was incremented by the "respiration" rate (0.1) and the estimated mortality rate (0.01).

The density of algae at the point where space limitation of growth begins (α2, 2) was estimated on the basis of no demonstrable effect of crowding until 90% of the maximum density was reached (12).

The density of *Spartina* shoots at the point where space limitation of growth begins (α 3, 3) was assumed equal to three times the winter (initial) standing stock. This assumption was justified from examination of the annual growth curves from Sapelo Island.

The α value associated with the resource limitation of anaerobes in the sediment was discussed earlier. Refractory material in the sediments of the Sapelo marsh was estimated to be 2200 gC x m^{-2} (α 13,8). Then α 13, 8 was calculated, using the mean annual standing stock of sediment and 15 gC x m^{-2} of bacteria as the density at which limitation begins (*see* equation 5).

The four α_{jj} values for algae, *Spartina*, heterotrophs in water, and aerobic heterotrophs in sediments were derived as follows: 2,2 from (12); α 3,3 from Sapelo Island data; α 7,7 was based on (1, 4, 6); and α 14,14 was based on (10, 14, 16, 19). The value of α 13,8 was discussed above.

The respiration, mortality, and excretion rates listed in Table 3 were all specific rates with units gC x gC^{-1} x $days^{-1}$. Wherever possible they were based on published data (*see* Table 2).

TABLE 3.

Results of a simple sensitivity analysis of the preliminary salt-marsh ecosystem model. Each simulation run was terminated at five years. The first two rows show the final steady-state values (gC x m^{-2}) of the closed and open systems. The following rows (percentages) show the final standing stocks of carbon in each compartment after five years as a percentage of the final standing stocks of the open system run. The perturbed parameter of each run and the percentage by which it was changed from the value of Tables 1 and 2 are given in columns 1 and 2 of the bottom portion of Table 3. A blank indicates no change.

	Parameter Change	X2	X3	X4	X5	X6	X7	X8	X9	X10	X11	X12	X13	X14	Total Carbon (X1 omitted)
	gC x m^{-2}														
Closed system	---	15	16	.1	20	789	10	21	103	36	.5	21	17338	10	18379
Open system	---	15	50	.1	72	1128	10	25	226	39	.5	63	20569	10	22205
	Percentages														
τ 1,3	200		241		120	23		13	40	17	20	165	12		14
”	-10	-26	-100	-91	-96	-85	-1	-38	-93	-19	-37	-100	-36	0	-40
τ 3,12	10	-1	-100		-94	-56	-1	-27	-92	-13		-100	-26		-18
”	-50		214		103	19		-5	32	1	20	22	-4		-2
τ 10,2	-50	-36			-2	-26	-1	-2	-7	1			-1		-3
τ 13,8	200							60		6	-20		-13	1	-12
”	-50					2		-88		-9	240		15	-3	14
α 2,2	-50	-27				-20			-2	1			-1		-2
α 3,3	20		17		18	8		5	14	2		16	4		5
”	-50		-38		-41	-19		-11	-31	-5		-38	-10		-11

TABLE 3 (continued)

	Parameter Change	X2	X3	X4	X5	X6	X7	X8	X9	X10	X11	X12	X13	X14	Total Carbon (X1 omitted)
	Percentages														
α 13,8	200							69		6	-20		-16		-14
”	-50							-46		-4	40		9		8
γ 2,2	200	78				84				-1					4
γ 3,3	200		23		27	24		14	30	3		23	12		13
”	-20		-4		-5	-6		-3	-6	-1		-4	-3		-3
γ 7,7	200	-3			-13	-90	98	-2	-5				-2		-6
γ 13,8	200							-9			20		2		2
γ 14,14	200							-8		3	40		-7	93	-6
ρ 3,1	10	-1	-95		-92	-42		-22	-85	-12		-94	-22		-24
ρ 7,10	200	-2			-8	-59		-2	-5	1			-2		-6
ρ 8,10	200					3		-96		-9	480		16	-6	14
ρ 12,10	200					-1		-11	-1	-1		-49	-10		-10
ρ 14,10	200							-6		2			-6	-7	-5
ε 2,6	200	-1				95				-1					5
μ 3,5	50		-75		-32	-24		-14	-18	-8		-66	-14		-14
μ 8,13	500							-22		-2	40		5		5
μ 12,13	200							15		-1		-23	23		12

SIMULATION

From the initial conditions given in Table 1, the changes in each compartment were simulated for five years, using an integration interval of 1/3 day. First, the marsh was regarded as a closed system whose total carbon content held constant. Next, an open system was simulated in which the carbon content of the air was constantly reset to the initial condition. This prevented CO_2 content of the air from ever becoming limiting to any photosynthetic compartment. Finally, in a simple sensitivity analysis, the open-system marsh was simulated repeatedly, each time a change being made in one of the parameters of the model. In general, a parameter was first doubled in value, then halved, unless such changes produced extinction, in which case more moderate perturbations were tried. In some cases only an increase was tried. In a few instances the small initial value of the parameter made a fivefold increase advisable. No simulation of tidal exchange was needed to evaluate the model responses with respect to the questions posed initially, and carbon was allowed to accumulate in place in the abiotic compartments wherever such accumulations had no effect on the dynamics of the model. All results are shown in Figures 2 and 3 and Table 3.

From a perusal of Figure 2 and Table 3, we conclude that enclosing a section of atmosphere above a salt marsh would cause: 1) a recurring annual shortage of carbon in the air; 2) severe limitation of *Spartina* growth and a low annual mean for this component; 3) a continual increase in the DOM in the water; 4) after five years an annual mean POM concentration in the water 50 times greater than the initial conditions; and 5) less than a 1% change in the POM in the sediment after five years. Thus we see the marsh as a potential sink for atmospheric CO_2, which is fixed and exported in the form of DOM and POM via the tidal exchange.

However, because of rapid turbulence exchange in the atmosphere, *Spartina* in nature would never be faced with the CO_2 limitations imposed by the closed-system model. Thus the amounts of material estimated to be fixed and stored as DOM and POM in the previous simulation are low. Simulating the five-year fixation-storage cycle of the marsh with atmospheric CO_2 held constant produced a somewhat different picture (*see* Figure 3 and Table 3). By the end of the fifth year, either a steady state or a constant annual pattern of fluctuation was established in all compartments, except for DOM in open water (X6) and POM in sediments (X13). Net carbon flux for the marsh as a whole was from air to organic matter during the winter, spring, and summer quarters, and from organic matter to air only during the fall, when the standing crop and gross photosynthesis of *Spartina* were low.

By the end of the fifth year, the winter standing stock of dead *Spartina* (X5) had decreased to 60% of the initial winter value. Dissolved organic matter in the open water had increased by 1121 $gC \times m^{-2}$ and was still increasing. Particulate

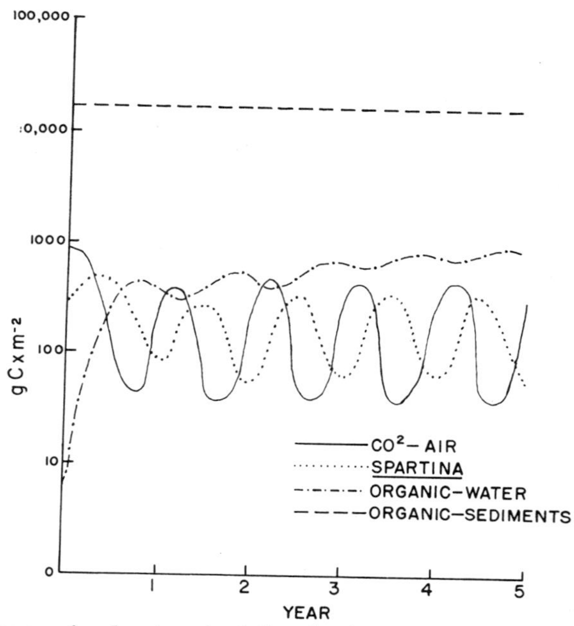

Figure 2. Five-year closed-system simulation showing seasonal and annual changes in CO_2 in air (X1), **Spartina** standing stock (X3 + X5 + X12), total orgainic detritus in water (X6 + X9), and total organic detritus in sediments (X11 + X13).

organic matter in the water had stabilized at twice the value of the closed-system simulation; a mean annual value of more than 200 gC x m^{-2} representing approximately a hundredfold increase over the initial value. Because of the settling rate (τ 9, 13), this resulted in an increase in the POM in the sediments (X13) of 3069gC x m^{-2}. Overall, the system increased its carbon content by 4411gC x m^{-2} or 882gC x m^{-2} per year. If tidal exchange had been simulated, this would represent the amount of net export by the salt marsh to the offshore waters.

A crude impression of the validity of this "typical" five-year simulation was obtained by comparing the annual growth pattern of the biotic compartments with standing crop data from the field. The latter are not extensive for most groups and the purpose of this report was not to validate the model. But some general comments are justified to help interpret the sensitivity measurements discussed in the following section: 1) the simulated values for algae (X2) are too high, perhaps by a factor of 2; 2) *Spartina* (X3, X5, X12) reaches simulated peak standing stocks at a time and in a manner generally comparable with the field populations; 3) the grazers (X4) are arbitrarily forced to a low overwintering limit in fall and winter. Since it has no predators in the present model, the group

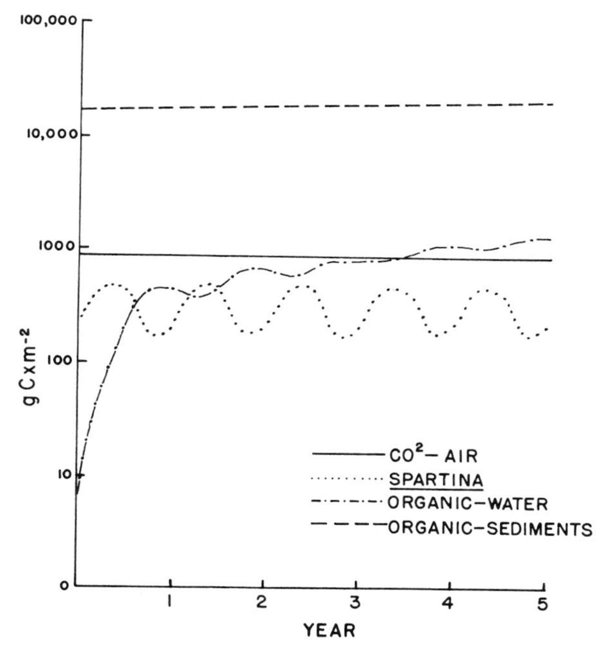

Figure 3. Five-year open-system simulation showing seasonal and annual changes in CO_2 in air (X1), **Spartina** standing stock (X3 + X5 + X12), total organic detritus in water (X6 + X9), and total organic detritus in sediments (X11 + X13).

expands in spring and summer to the maximum limit preset in accordance with the field data. The aerobic heterotrophs of open water (X7) and of sediment (X14) are heterogeneous groups for which few data on seasonal stocks were available. Finally, the anaerobic bacteria of the sediments (X8) exhibited almost no seasonal changes in the model simulation and, although field data are scarce, this component may reach densities that are from three to four times as great as the peak density in the simulation runs.

Using the open-system simulation as the "type" or standard run, we next varied each of the parameters in turn and compared the effect on final standing stocks to determine whether the system as a whole was sensitive to change in a given parameter and, if so, which state variables were being affected. Of the 81 parameters in the model, perturbations of 21 produced a change in final standing stock (at the fifth year) of more than 0.5%. These we considered important or "sensitive" parameters. They are listed in Table 3, together with the way in which they were varied and the changes in final standing stock that the change caused. Some parameters, for example those associated with *Spartina* shoots (X3), affected many other compartments of the model as well as the total standing stock. Others affect only a few compartments.

Many parameters were surprisingly sensitive. The maximum gross photosynthesis rate of *Spartina* (τ 1, 3) turned out to be so close to the sum of all loss rates that a decrease of only 10% in its value caused *Spartina* to become extinct, whereas doubling its value produced a 14% increase in the final standing stock of *Spartina.*

Variations in only 4 α densities had any pronounced effect on the marsh's production of organic mater; all were concerned with *Spartina,* algae, or anaerobic saprophages in the sediments. Similarly, the ϑ densities which most affected the buildup of organic matter were those that set density maxima for the two primary producers, the algal predator (X7) and the anaerobic saprophages in the sediment. The first three were limited primarily by lack of space and thus the sensitive parameter was ϑ_{jj}. The latter, limited by resources in the manner discussed earlier, was controlled by ϑ ij. In addition, the ϑ density setting the limit on aerobic heterotrophs in the sediment caused a 6% decrease when it was doubled. Increasing X14 increased X11 and decreased X8 slightly. Despite this reduction in X8, the 93% increase in X14 increased the overall utilization of X13. This instance illustrates an important point, namely that even in such a small model as the present one, which consists of 14 compartments, the results of perturbations cannot be predicted intuitively. Thus the simulation provides useful insight obtainable in no other way.

Five respiration terms were sensitive parameters; those representing losses from *Spartina* (shoots and roots), algal predators, as well as anaerobic and aerobic saprophages in the sediments.

Only one excretion term was of sufficient magnitude that multiplying by 2 affected significantly the organic production of the marsh, η 2, 6, the excretion of DOM by algae. However, most excretion rates were set very low simply because we know little about them.

Mortality parameters varied greatly in their sensitivity. Of those listed in Table 3 the first, mortality of *Spartina* shoots (μ 3, 5), produced a -14% change in final standing stock when increased only 50%. Mortality of *Spartina* roots (μ 12, 13) caused a 5% increase upon being incremented by 100% and the last (μ 8, 13) had to be multiplied fivefold to achieve a 12% final state system increase.

Several of these sensitivities were expected. *Spartina* shoots and roots, for example, constitute a large proportion of the living biomass of the marsh and are heavily interconnected with the other components (the connectivity of this model is high, 20.4% of the 196 possible pathways exist). Algae, although low in mean standing stock, have a very rapid turnover and a high excretion rate, and thus have a considerable effect on DOM in the water. Because the anaerobic saprophages subsist in part on the huge store of POM in the sediment, they are important. However, little is known about the parameters governing the levels and turnover of this group. Thus the predictions from the model will have considerable impact on the direction of some of the research projects at Sapelo

Island.

Apart from their practical input into the continuing research program, however, the results of this crude sensitivity analysis illustrate a few important general aspects of population growth and regulation. During at least part of the yearly cycle, all of the biotic components of the marsh model were limited primarily by scarcity of a space-related factor, i.e. the f_{jj} feedback controls were greater than zero. Under such conditions, when the maximum value is only slightly greater than the realized population-ingestion rate, changes either way in τ usually have a marked effect, and increasing the loss rates (respiration, mortality, etc.) decreases the equilibrium density. However, where the maximum τ is much greater than the realized rate of ingestion, changes in τ may or may not have a marked effect, depending on the values of α_{jj} and $\eth_{jj}$, i.e. on how close the equilibrium density was to the maximum $\eth_{jj}$ density. Thus varying the maximum gross photosynthesis rate of *Spartina* (1, 3) produced great changes in equilibrium density, but perturbations of the maximum ingestion rate of heterotrophs in water (X7) caused little change in the equilibrium density of X7 because it was initially within 1% of the maximum and the realized rate of ingestion was much lower than the maxima.

When the theoretical maximum τ is very large compared to the realized rate of ingestion, increasing the loss rates initially has no effect on the equilibrium density. Thus doubling the loss of algae (X2) had little demonstrable effect on the equilibrium density of algae or the productivity of the system, but increasing the respiration of *Spartina* shoots by only 10% caused a 24% decrease in the final system standing stock. Indeed, when several life history stages are modelled as separate compartments, and the space limitation occurs mainly in only one of them, increasing the mortality of this limited stage may actually result in an increased density of one or more of the other stages (18).

This preliminary salt-marsh model showed little sensitivity to changes in parameters affecting the abiotic variables, a reflection of the (usually) minor role of resource scarcity as a feedback control. Thus none of the α_{ij} or $\eth_{ij}$ densities associated with resource feedback control (f_{ij}) were "sensitive" parameters. This undoubtedly is in part an artifact of the model, and it will repay us to carefully scrutinize the biological structure of the model to see: 1) where nutrient limitation should be playing a role; and 2) what parameters are thus possibly in error or what additional structure is needed. Information feedback from nitrogen availability to photosynthesis rate would be a good example of the latter type of addition.

ACKNOWLEDGMENTS

We would like to acknowledge the assistance of the other members of the Sapelo coherent area research team. Those who played a major role in the organization and/or criticism of earlier drafts are: M. Elrod, D. Frankenberg, L. Maurer, L. R. Pomeroy, R. Reimold, S. Rice, W. Sottile, and W. Wiebe.

The work on computer simulation was carried out by R. G. Wiegert when he attended a workshop held in the summer of 1973 at Imperial College Field Station, Silwood Park, Ascot, Berkshire, England. Computer time and some travel support were provided by a grant from the Office of Resource and Environment in the Ford Foundation to T.R.E. Southwood, Imperial College.

The development of the model was supported by N.S.F. grant no. GA 35793X1. to L. R. Pomeroy. Computer time and some travel support were provided by Imperial College, University of London.

REFERENCES

1. Bahr, L.
1973 The role of the oyster community in the salt marsh. Doctoral dissertation, University of Georgia.

2. Fraleigh, P. C. and Wiegert, R. G.
A model explaining successional change in standing crop of thermal blue-green algae. **Ecology** (Submitted)

3. Gallagher, J. L.
Sampling macro-organic matter profiles in salt marsh plant root zones. **Soil Sci. Soc. Amer. Proc.** (In press.)

4. Jacobs, J.
1968 Animal behavior and water movement as co-determinants of plankton distribution in a tidal system. **Sarsia,** 34: 355-370.

5. Johannes, R. and Satomi, M.
1967 Measuring organic matter retained by aquatic invertebrates. **Jour. Fish. Res. Bd. Canada,** 24(11): 2467-2471.

6. Kuenzler, E.
1961 Structure and energy flow of a mussel population in a Georgia salt marsh. **Limnol. Oceanogr.,** 6: 191-204.

7. Nelewajko, C. and Lean, R. S.
1972 Growth and excretion in planktonic algae and bacteria. **J. Physiol.,** 8: 361-366.

8. Pomeroy, L. R.
1959 Algal productivity in salt marshes of Georgia. **Limnol. Oceanogr.,** 4: 386-397.

9. Ragotskie, R. A.
1959 Plankton productivity in estuarine waters of Georgia. **Mar. Sci.,** 6: 146-158.

10. Shanholtzer, S. F.
1973 Energy flow, foood habits and population dynamics of **Uca Pugnax** in a salt marsh system. Doctoral dissertation, University of Georgia.

11. Smalley, A. E.
1960 Energy flow of a salt marsh grasshopper population. **Ecology,** 41: 672-677.

12. Steeman-Nielsen, E.
1957 The chlorophyll content and the light utilization in communities of plankton algae and terrestrial higher plants. **Physiol. Plantarum,** 10: 1009-1021.

13. Steeman-Nielsen, E. and Hansen, V. K.
1959 Measurement with the carbon 14 technique of rates of respiration in natural population of phytoplankton. **Deep-Sea Res.,** 5: 222-233.

14. Teal, J. M.
1962 Energy flow in the salt marsh ecosystem of Georgia. **Ecology,** 43: 614-624.

15. Teal, J. M. and Kanwisher, J.
1961 Gas exchange in a Georgia salt marsh. **Limnol. Oceanogr.,** 6: 388-399.

16. Teal, J. M. and Wieser, W.
1966 The distribution and ecology of nematodes in a Georgia salt marsh. **Limnol. Oceanogr.,** 11(2): 217-222.

17. Wiegert, R. G.
1973 A general ecological model and its use in simulating algal-fly energetics in a thermal spring community. In **Insects: studies in population management,** Vol. 1, (eds. Geier, P. W., Clark, L. R., Anderson, D. J., and Nix, H. A.) Occasional Papers, Ecol. Soc. of Austr., Canberra.

18. Wiegert, R. G.
1974 Simulation modelling of the algae-fly components of a thermal ecosystem: effects of spatial heterogeneity, time delays and model condensation. In **Systems analysis in Ecology,** Vol. 3., (ed. Patten, B. C.) Academic Press, New York. (In press.)

19. Wolf, P., Shanholtzer, S. and Reimold, R.
Population estimates for **Uca pugnax** on the Duplin estuary marsh. **Crustaceana** (In press.)

THE *A POSTERIORI* ASPECTS OF ESTUARINE MODELING[1]

Robert E. Ulanowicz
David A. Flemer
Donald R. Heinle
Curtis D. Mobley[2]

ABSTRACT

This exercise is the application of an analytical method for systematically modeling ecosystems data to observations made on a naturally eutrophic, mesohaline planktonic microcosm. The theory and experimental design are briefly outlined and the particular steps in the acutal modeling process follow. Then there is a discussion as to how the whole endeavor can be refined to culminate in models with predictive capabilities.

INTRODUCTION

The type of marine ecosystem most useful to man on a per-unit basis is the mesohaline, or estuarine system. For this reason, there has been quite a strong demand from management and commercial sources for models that can predict the time behavior of these baseline systems, and the evolution of perturbations upon them. Predictive models however are elusive, and any new philosophical approach which would hasten the development of models well founded upon observation would be most welcome.

"Good data are the precursors of good models" (12), and one would expect that any models with good precision and prediction would be of an *a posteriori*

1. Contribution No. 532, Natural Resources Institute, University of Maryland, Solomons, Maryland 20688.

2. Chesapeake Biological Laboratory, Natural Resources Institute, University of Maryland, Solomons, Maryland 20688.

nature in the sense of Watt (16). Generally, such models are accepted to be ones in which data are regressed to determine the parameters of a given model. But *a posteriori* models can be more precisely defined. To be specific, it is useful to consider Dale (3) and Ross' (13) chronology of the problem-solving process: (A) the lexical phase, or the delimination of the entities or parts; (B) the parsing phase, or the choice of relationships between entities of interest; (C) modeling, or the specification of the mechanism by which these relationships take place; and (D) the analysis or validation of the model. Regression usually occurs at step (D), after the first three phases have been initially executed on *a priori* grounds. In a truly *a posteriori* approach, regression should take place before steps (B) and (C) and should strongly influence the subsequent steps.

Bellman (1), in treating regression as the most significant step in the modeling process, referred to a procedure with such emphasis as solving the "inverse problem" in ecological modeling. It remained, however, for Mobley (9, 10) to couple regression with statistical hypothesis testing and construct an algorithm whereby regression is actually antecedent to parsing and modeling.

From the foregoing, it might seem as if the development of *a posteriori* models has languished because of the lack of an adequate theoretical framework. In reality, however, a dearth of good data on multi-species population dynamics has probably been the rate-limiting step. One can readily cite the two-species competition data of Gause (4) and Park (11), but it is not easy to find equally adequate information on the time dynamics of all the compartments involved in the biomass cycle of an ecosystem. That this vacuum exists is not surprising, given the complexities of most real systems, the manifold of exchanges of any particular open system with its neighbors, and the unpredictability of the intrinsic variables that drive the system.

In planning the data collection, the authors tried to surmount the latter two difficulties by enclosing a system and subjecting it to nearly constant salinity, temperature, and incident light. In an effort to assemble populations large enough to reduce sampling noise, the system was limited to planktonic species.

The two sections immediately following outline the theory and experimental methods in order to provide background for the example analysis, in which the modeling process is demonstrated.

THEORY

The inverse problem, as viewed by the mathematician, is as follows: Given observations on several populations, N_i at l separate times, how does one construct a model that will fit the data as closely as possible, and yet include only those interactions which the data indicate are significant?

Since the object of this approach is to make a few *a priori* assumptions as possible about the significant interactions, it will be assumed that all possible interactions of a given class are initially pertinent. For the sake of mathematical tractability, the class of models under consideration will be of the form

$$\frac{d\,Ni}{dt} = \sum_{p=1}^{k} b_p f_p \tag{1}$$

where the b_p are the parameters which multiply the arbitrary functions f_p. The f_p model the various phenomena which influence the change in the population of species i and may be any single-valued function of any population, time, or extrinsic variable, e.g.

$$f_p = f_p(N_1, N_2, \ldots, N_n, t, \text{etc.}) \tag{2}$$

In addition, any phenomenon may be represented in equation 1 by more than one f_p, where the different f_p are alternative mathematical representations of the same effect.

This general model must first be regressed to fit the data. Subsequently, each f_p must be examined for its contribution to the fit, and those terms with minimal contribution are then systematically dropped from the model. This systematic exclusion is accomplished with the help of statistical hypothesis testing, and effects both parsing and modeling (since alternative mathematical expressions may be evaluated against one another). The full mathematical exposition of the regresseion and hypothesis testing is to be found in Mobley (9, 10); only a brief outline of these methods is given below.

The method of regression used is specially adapted to the overall analysis and is fundamentally different from standard linear regression. To demonstrate what is involved, it is useful to write (1) in full.

$$\begin{aligned}
\left.\frac{d\,Ni}{d\,t}\right|_{t_1} &= b_1 f_1(t_1) + b_2 f_2(t_1) + \ldots + b_k f_k(t_1) + e_1 \\
\left.\frac{d\,Ni}{d\,t}\right|_{t_2} &= b_1 f_1(t_2) + b_2 f_2(t_2) + \ldots + b_k f_k(t_2) = e_2 \\
&\vdots \\
\left.\frac{d\,Ni}{d\,t}\right|_{tl} &= b_1 f_1(tl) + b_2 f_2(t_2) + \ldots + b_k f_k(tl) + el
\end{aligned} \tag{3}$$

The left-hand sides in equation (3) are the observed rates of change as calculated from the series of observed populations by any one of a number of standard techniques. The first k terms on the right-hand side yield the rate of change at t_i as predicted by the hypothesized interaction f_j. Finally, the e_i are the differences between the observed and predicted rates of change; i.e., the "error" to be minimized.

Equations (3) can consisely be written in matrixvector form as:

$$\overline{Y} = \overline{\overline{X}}\,\overline{B} + \overline{E} \tag{4}$$

where:

$$Y_j = \frac{dN_i}{dt}(\text{at } t = t_j) \tag{4a}$$

$$X_{ij} = f_j\,(t_i) \tag{4b}$$

$$B_j = b_j \tag{4c}$$

$$E_j = e_j \tag{4d}$$

In general Y is a vector in n-dimensional Euclidean space, whereas the column vectors of matrix $\overline{X}$ span a subspace of dimension m, V_m ($m \leq n$). The special case n = 3, m = 2 is illustrated graphically in Figure 1. All possible vectors $\overline{\overline{X}}\overline{B}$ will lie in subspace V_m (the plane defined by $\overline{X}_1$, $\overline{X}_2$ in Figure 1). It is readily seen from Figure 1 that the magnitude of the error vector $\overline{E}$ will be minimal when $\overline{E} = \overline{E}^* \perp V_m$. Or equivalently,

$$\overline{\overline{X}}^t\overline{E} = \overline{0} \tag{5}$$

applying (2.6) to (2.5) one arrives at the familiar solution

$$\overline{B}^* = [\overline{\overline{X}}^t\overline{\overline{X}}]^{-1}\quad[\overline{\overline{X}}^t\overline{Y}] \tag{6}$$

Standard linear regression theory is an implementation of equation 6.

The execution of (6) is difficult when the dimension of $\overline{X}$ becomes large and it is impossible when the columns of $\overline{\overline{X}}^t\overline{\overline{X}}$ are linearly dependent. To circumvent these difficulties, Mobley works instead with an algorithm that minimizes the magniture of $\overline{E}$ by, equivalently, maximizing the value of $\cos^2\theta$. This method begins with the assumption $\overline{B} = \overline{0}$ and determines which element of $\overline{B}$ can be changed by what amount δ such that the resulting increment in $\cos^2\theta$ is the maximum obtainable by changing only one element of $\overline{B}$. The element of $\overline{B}$ thus located is incremented by the appropriate value δ, and the method is reiterated with the revised B until the calculated increment in $\cos^2\theta$ is negligible. The vector $\overline{B}$ must converge to $\overline{B}^*$ since $\cos^2\theta$ is bounded from above by unity.

The above iterative procedure is useful in more than just obviating the numerical difficulties that accompany equation 6. In practice it often happens that the required fit is achieved without ever changing certain elements of $\overline{B}$

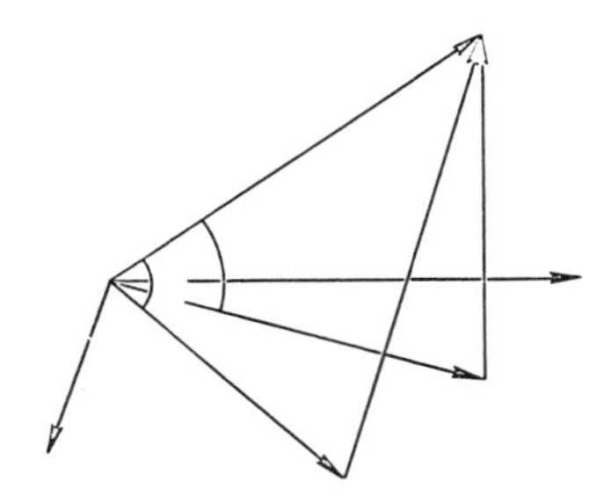

Figure 1. Illustration of $\bar{Y} = \bar{\bar{X}}\bar{B} + \bar{E}$.

from their initial values of zero. The corresponding f_p may be dropped from further consideration, thus effecting a preliminary contribution to the parsing.

Having fit the general model to the data, it is necessary to ascertain which of the many f_p may be neglected without significantly affecting the ability of the resultant system to mimic the observations. For the sake of brevity, Ω will denote the set of assumptions that the data can be represented by the general model, and that the error vector of the fit, E*, has as elements normally distributed, statistically independent, random variables with means of zero and identical, constant, albeit unknown, variances. A measure of the inability of the full model to perfectly represent the data will be designated by

$$\mathfrak{z}_{\Omega} = \bar{E}^t\bar{E}. \tag{7}$$

H will denote the hypothesis under test; namely, that any given combination of q elements of B may be set equal to zero (i.e., H denotes the hypothesis that the interactions modeled by the corresponding f_p are insignificant). Finally, the error-sum-of-squares of the resultant reduced model, $W (W = \Omega \cap H)$, will be labeled $\mathfrak{z}_w$.

One sees intuitively that H should be rejected if $(\mathfrak{z}_w - \mathfrak{z}_{\Omega})$ becomes large. To be more quantitative, it is possible to employ the likelihood-ratio statistic

$$F = \frac{(\mathfrak{z}_w - \mathfrak{z}_{\Omega})/q}{\mathfrak{z}_{\Omega}/(n-k)} \tag{8}$$

where, under W, F is a central F-variable with q and (n-k) degrees of freedom. For any chosen level of significance, say .05, one can determine a value, $F_{.05}$, such that if $F > F_{.05}$, the hypothesis H may be rejected with 95% confidence.

On the other hand, if $F < F_{.05}$, H is accepted, which implies that the reduced model can provide as adequate a mathematical description of the observed population dynamics as can the general model. It is possible, however, that this accepted hypothesis is in reality false. It is useful then to compute the power of the test; that is, the probability that if H is false, it would have been correctly rejected. If H is supposed false, then Ω (but not W) is assumed to hold and the F in (8) becomes a non-central F-variable, F'. The power of the test, β, is given by

$$\beta = \text{Prob}\,(F' > F_{.05}) \tag{9}$$

One observes that as the confidence level is raised to insure that a valid reduced model is not falsely rejected, it becomes more probable that a spurious reduced model will be accepted as a good representation of the data (i.e., the higher the confidential level, the lower the power). Selection of a significance level commensurate with the noise of the data is one of the few magnitudes the modeler must estimate.

By systematic iteration of the apparatus outlined above, the full model can be pruned until it consists only of terms which contribute significantly to the model's ability to mimic the data. An example of the application of this procedure to planktonic data is presented in Section 4, and the biological significance of both the modeling process and the ensuing model is discussed in Section 5.

METHODS

The microcosms under study were contained in 757-liter polyethlyene cylinders sheltered in the Chesapeake Biological Laboratory pier house. Twelve hours of daylight were simulated by a 500 W General Electric quartz-iodide, wide-flood lamp placed 1.5 m above each cylinder. Lighting was attenuated to a satisfactory level at the surface of 0.18 cal m^{-2} min^{-1} with layers of window screening. Twilighting was provided each cylinder by a single 20 W cool-white fluorescent light which preceded and followed the intense lighting by ½ hour. Temperature varied diurnally by about $2^{o}C$ around a mean of $22^{o}C$. The tanks were stirred four times a day by large polyethylene propellers driven at 24 rpm for intervals of one hour.

To fill the tanks initially, Bay water from a depth of 0.5 m was pumped through a 28 μ plankton net into a reservoir and thence simultaneously fed to the three experimental chambers. Natural populations of rotifers, protozans, and algae were thereby introduced. The final salinity of the systems was 9 ^{o}/oo. The copepod *Eurytemora affinis* (Poppe) was added to the system from mass cultures grown by the techniques of Heinle (5). When predators were desired, the selective planktivore, *Menidia menidia* (Linnaeus) larvae, were added.

The systems were allowed to evolve in the batch mode for a period of 15 days. They were sampled at the same time each day by dipping water from the surface with a bucket after the stirrers had run for 10 to 20 minutes. Each sample was analyzed for the following: total seston, particulate carbon, particulate nitrogen, particulate carbohydrate, total phosphorous, dissolved organic phosphorous, dissolved inorganic phosphorous, ammonia, nitrates, nitrites, chlorophyll *a*, primary productivity (C^{14} uptake), algal taxonomy and cell count, herbivore biomass, salinity, temperature, and dissolved oxygen. A number of the above variables were dominated by sampling and analytical noise. Discernible trends appeared, however, on the abiotic level in the dissolved nitrogen ($NO_3 + NO_2$), on the producer level as active chlorophyll *a*, and on the second trophic level in the total herbivore biomass. For brevity, only the analyses of these quantities germane to the model derived in Section 4 are referenced below.

Part of the water samples collected were filtered through GF/C glass fiber filters. The filtrate was analyzed for nitrite and nitrate by the procedure of Strickland and Parsons (14). Chlorophyll *a* was extracted from the filter pads with a 90% acetone solution and the resulting concentration was measured on a Turner fluorometer after Yentsch and Menzel (18) and Holm-Hansen (7). Calibration determination of active chlorophyll *a* followed the procedure of Lorenzen (8) and Yentsch (17).

Herbivore biomass was estimated by measuring necessary dimension of all the herbivores found in the material retained from a one-liter sample by 63 μ pore size net. Previously established length-weight relationships were used for the copepods (6). Dimensions of representative samples of rotifers and protozoans were measured, their volumes thereby estimated, and dry weights arrived at by assuming a water content of 80%. With few exceptions, copepods dominated the herbivore biomass.

EXAMPLE ANALYSIS

Data taken during two weeks of observation of a microcosm are displayed in Table 1. The compartments to be modeled are those previously identified—nitrate plus nitrite (species 2), active chlorophyll *a* (species 3), and herbivore biomass (species 1). A predator was also present, but its biomass remained unchanged during the course of the experiment and is deleted from further analysis. Data were taken for 13 days, although each component was not measured on each day. Since the modeling algorithm requires the populations to be measured simultaneously, linear interpolation between adjacent measurements is used to provide missing values.

To minimize the *a priori* assumptions imposed upon the mathematical form of the model, a generalized first-order, quadratic, ordinary differential equation is employed as the "full" model for each species:

TABLE 1.

The populations of the three ecosystem components modeled. Component 1 is total herbivore biomass in μg dry wt./liter, Component 2 is dissolved nitrogen in μg at N/liter, and Component 3 is active chlorophyll **a** in μg/liter. An asterisk denotes values obtained by interpolation.

TIME IN DAYS	COMPONENT 1	COMPONENT 2	COMPONENT 3
0	0	8.	5.4
1	3.	8.	4.7
2	7.	7.	7.8
3	5.	4.	8.0
4	10.	5.5*	9.5
5	40.	7.	8.5
6	55.*	6.	10.1
7	70.	7.*	11.8*
8	73	8.	13.5
9	56.	7.*	13.0
10	67.*	6.	10.1
12	89.	4.*	8.6*
13	69.*	3.	7.8

$$\frac{d\ Ni}{d\ t} = A_i + \sum_{j=1}^{3} B_{ij}N_j + \sum_{j=1}^{3} C_{ij}N_iN_j + \sum_{\substack{j=1\\ j\neq i}}^{3} D_{ij}N_j^2 \tag{10}$$

Referring to (1), the correspondences $b_1 = A_i$, $f_1 = 1$, $b_2 = B_{i1}$, $f_2 = N_i$, etc. are obvious. Equation (10) may be viewed as generalized Lotka-Volterra interactions or may be regarded as the result of expanding the f_p as Taylor series and retaining only the linear and lowest order non-linear terms.

It is important to remember that when using equation 10, it is not the populations that are being modeled, but rather their derivatives. These derivatives are not directly measurable and must be estimated by some suitable scheme. For the present example, the derivative at a given time t_i is estimated by averaging the slopes of the straight-line segments connecting the population at t_i to the two adjacent populations at t_{i-1} and t_{i+1}.

Once the three final single-component models have been obtained, the equations must be simultaneously integrated to yield the predicted populations. Only a small change in a population derivative can sometimes lead to a large change in the population, especially after long integration times. Thus it often occurs that a restricted model which by acceptance of the associated hypothesis is equivalent to the full model, yields upon integration a population significantly different from the population obtained from integration of the full model. Since one goal of our analysis is a model with predictive power, there is imposed the additional requirement that a derivative model, when integrated, closely mimic the data.

To proceed with the actual modeling, equation 10 is applied to each component in turn. The iterative algorithm for solving the "inverse problem" yields the sets of least-squares parameters for the three general models as tabulated in Table 2. The predicted populations obtained by simultaneous integration of these general models are exhibited in Figure 2. To gain some idea as to which terms of the general model best represent the dominant ecosystem

TABLE 2

The values of the least-squares parameters for the general models of equation (10).

COMPONENT 1		COMPONENT 2		COMPONENT 3	
A_1	$= -4.518 \cdot 10^{1}$	A_2	$= 9.967 \cdot 10^{-4}$	A_3	$= 2.484$
B_{11}	$= -2.859 \cdot 10^{-1}$	B_{21}	$= -3.068 \cdot 10^{-2}$	B_{31}	$= -6.048 \cdot 10^{-2}$
B_{12}	$= 2.636 \cdot 10^{1}$	B_{22}	$= 8.386 \cdot 10^{-3}$	B_{32}	$= -1.553 \cdot 10^{-1}$
B_{13}	$= 0$	B_{23}	$= 1.264 \cdot 10^{-1}$	B_{33}	$= -4.891 \cdot 10^{-2}$
C_{11}	$= -1.811 \cdot 10^{-3}$	C_{21}	$= 7.897 \cdot 10^{-3}$	C_{31}	$= 3.910 \cdot 10^{-3}$
C_{12}	$= 1.756 \cdot 10^{-1}$	C_{22}	$= -9.478 \cdot 10^{-3}$	C_{32}	$= 2.158 \cdot 10^{-1}$
C_{13}	$= -4.630 \cdot 10^{-2}$	C_{23}	$= -9.988 \cdot 10^{-3}$	C_{33}	$= -9.443 \cdot 10^{-2}$
D_{12}	$= -2.502$	D_{21}	$= -1.342 \cdot 10^{-4}$	D_{31}	$= 2.053 \cdot 10^{-4}$
D_{13}	$= -2.036 \cdot 10^{-1}$	D_{23}	$= -9.999 \cdot 10^{-3}$	D_{32}	$= -1.088 \cdot 10^{-1}$

interactions, it is useful to test the series of hypotheses resulting when the model parameters are deleted, one at a time, from the general model. For brevity, only one component, say, the herbivore biomass, is discussed. The series of hypotheses are

$$H_1: A_1 = 0$$
$$H_2: B_{11} = 0$$
$$\vdots$$
$$H_9: D_{13} = 0$$

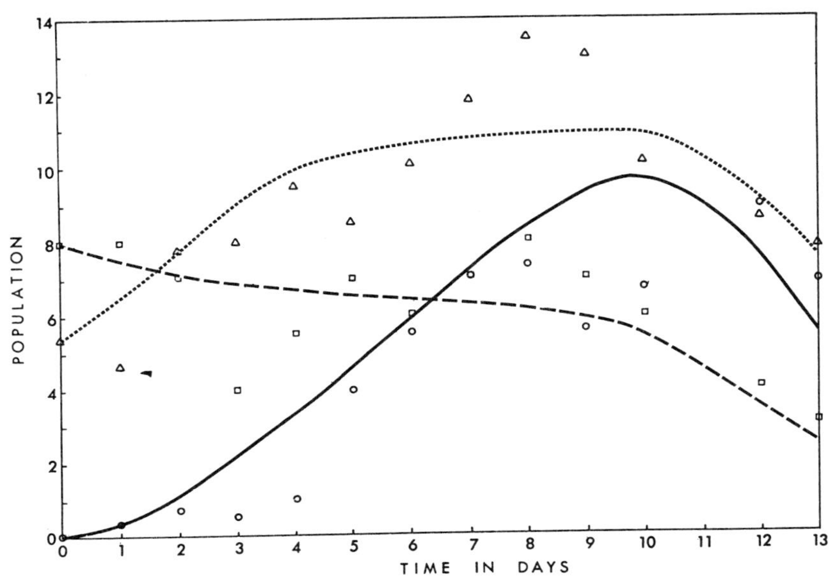

Figure 2. The observed populations and the populations predicted by the general models of equation (10). o and —— are the observed and predicted total herbivore biomasses, respectively, in μg dry wt./liter times 10^{-1}; □ and – – –are the observed and predicted dissolved nitrogen concentration in μg at N/liter; and Δ and - - - - are the observed and predicted active chlorophyll *a* concentrations in μg/liter.

The results of the tests of $H_1, \ldots, H_9$ when made at the .25 significance level are shown in Table 3.

Table 3 shows that H_1 is rejected (with a .75 probability that the rejection is correct), H_2 is accepted (with a .71 probability that if H_2 were really false it would have been correctly rejected), and so on. These results immediately suggest testing the hypothesis

$$H_{10}: B_{11} = B_{13} = C_{11} = C_{13} = D_{13} = 0$$

And indeed, H_{10} is accepted with a power of .59. A further series of tests can determine whether any more parameters can be deleted from the general model:

$$H_{11}: A_1 = B_{11} = B_{13} = C_{11} = C_{13} = D_{13} = 0$$

$$\vdots$$

$$H_{14}: B_{11} = B_{13} = C_{11} = C_{13} = D_{12} = D_{13} = 0$$

TABLE 3

Results of the first hypothesis tests on Component 1. In the case of acceptance, the power of the test is given. The significance level was .25.

HYPOTHESIS	REJECTION OR POWER
H_1: $A_1 = 0$	rejected
H_2: $B_{11} = 0$	.71
H_3: $B_{12} = 0$	rejected
H_4: $B_{13} = 0$	.65
H_5: $C_{11} = 0$	.73
H_6: $C_{12} = 0$	rejected
H_7: $C_{13} = 0$	.77
H_8: $D_{12} = 0$	rejected
H_9: $D_{13} = 0$	.63

H_{11}, H_{12}, and H_{14} are all rejected. However, H_{13}, which attempts to delete C_{12} in addition to the parameters already deleted by acceptance of H_{10}, is accepted with a power of .63. This may seem a surprising result since H_6: $C_{12} = 0$ was rejected. It must be remembered that the degrees of freedom for the two F-tests are different; thus, the possibility of such a result. It is found, however, that upon integration, the populations predicted by the model resulting from H_{13} deviate considerably (about 50%) from the observed populations, and for this (subjective) reason the model is excluded from further consideration. Thus, the final model for component has the structure

$$\frac{d\ Ni}{d\ t} = A_1 + B_{12}N_2 + C_{12}N_1N_2 + D_{12}N_2{}^2 \qquad (11)$$

It is to be noted that five of the nine terms of the general model have been found to play no significant role in the modeling. Each of the remaining terms does contribute significantly to the integrated model's ability to mimic the observed data.

The analysis proceeds similarly for the other two ecosystem components, the results being

TABLE 4.

The values of the least-squares parameters for the final component models of equations (11) and (12).

COMPONENT 1		COMPONENT 2		COMPONENT 3	
A_1	$= -1.171 \cdot 10^{2}$	B_{23}	$= 1.489 \cdot 10^{-2}$	A_3	$= 4.623$
B_{12}	$= 4.404 \cdot 10^{1}$	C_{21}	$= 5.286 \cdot 10^{-3}$	B_{31}	$= -1.158 \cdot 10^{-2}$
C_{12}	$= -9.998 \cdot 10^{-3}$	C_{23}	$= 1.498 \cdot 10^{-2}$	B_{32}	$= -1.378 \cdot 10^{-2}$
D_{12}	$= 3.647$	D_{21}	$= -3.418 \cdot 10^{-4}$	C_{32}	$= 2.223 \cdot 10^{-1}$
				C_{33}	$= -9.034 \cdot 10^{-2}$

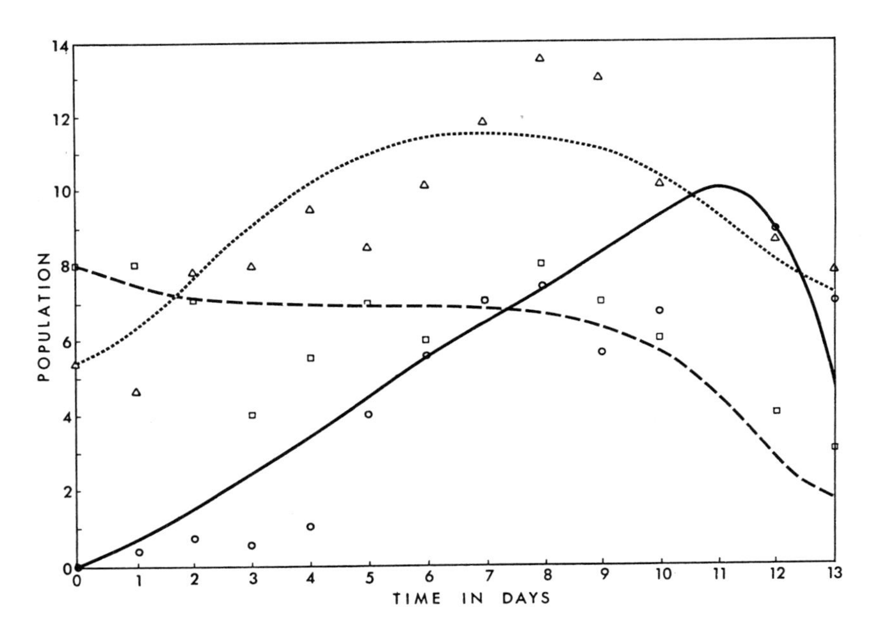

Figure 3. The observed populations and the populations predicted by the final models of equations (11) and (12). o and —— are the observed and predicted total herbivore biomasses, respectively, in μg dry wt./liter times 10^{-1}; □ and - - — are the observed and predicted dissolved nitrogen concentrations in μg at N/liter; and Δ and - - - - - - are the observed and predicted active chlorophyll *a* concentrations in μg/liter.

$$\frac{d\,N_2}{d\,t} = B_{23}N_3 + C_{21}N_2N_1 + C_{23}N_2N_3 + D_{21}N_1{}^2$$

$$\frac{d\,N_3}{d\,t} = A_3 + B_{31}N_1 + B_{32}N_2 + C_{32}N_3N_2 + C_{33}N_3{}^2 \qquad (12)$$

The sets of least-squares parameters for equations (11) and (12) are displayed in Table 4, and the populations obtained from equations 11 and 12 simultaneously integrated with these parameter values are shown in Figure 3.

DISCUSSION AND FUTURE WORK

Having derived a suitable mathematical representation of observed population interactions, it is tempting to proceed with a bioligical interpretation of each term in the model. Such analysis must proceed with caution engendered by the following observations. When the matrix $\bar{\bar{X}}^t\bar{\bar{X}}$ is non-singular (as it is in this instance) the vector $\bar{B}^*$ of least-squares parameters is unique. The approach to this vector is, however, (a) non-unique, and (b) at times, mathematically very stiff. As an example of the latter difficulty, iteration of the regression algorithm may proceed until the increment of $\cos^2\theta$ is, say, 10^{-6}. Continuation of the iteration until the $\cos^2\theta$ is incremented by only 10^{-12} may result in substantial changes in certain elements of B. Generally speaking, the noisier the data, the greater such difficulties become.

Looking at equation 11, it can be said with confidence that A_1 is necessary to mimic the data and, as such, has physical and/or biological significance (in this case the loss of copepod biomass to the planktivores and sediments). To say, however, that such loss is 117.5 μg dry wt./ l /day, as indicated by the last regression, is stretching the point, in view of the noise in the given data.

Equation (11) reveals the somewhat surprising independence of the copepods to the phytoplankton standing crop. Of course, this is not to say that the herbivores can do without their food source. It reveals, however, that under the conditions to which the microcosm was subjected, the pertinent observable kinetics occur between copepods and nitrogen. Evidently, phytoplankters are in a condition of minimum response to nutrients and grazing pressure. Caperon (2) suggests that such minimal responses are characteristic of eutrophic ecosystems, and the thread of evidence presented here indicates the possibility that grazers may be of prime importance in assessing the stability of natural eutrophic mesohaline waters to additional nutrient loading.

When strong interactions are present (viz., the N_1 - N_2 coupling above), it is not surprising that all terms including one or the other of the populations will be kept as significant in the final reduced model. This is only an indication that the actual relation is not adequately described by any single term. If the reduced model is viewed as a Taylor approximation of the actual functionalities, then clues to the actual function are at least provided.

Thus, at this time, only semi-quantitative information is yielded by the modeling exercise. However, the pathway toward a model with quantitative predictability is clear and will direct the thrust of future work at this Laboratory. An evident necessity is a reduction in the noise of the data, or at least programmed redundancy to offset such noise. This will be accomplished by enlarging the microcosm and sampling it more frequently. An extension of the period of measurement and an increased effort at replication will also add to the confidence level of the final model.

Work also is being initiated on the construction of an algorithm to assess the confidence interval of each individual parameter calculated from the data. Also, a sensitivity analysis after Ulanowicz (15) is planned to help elucidate the temporal sequence of interations as they occur in the derived model.

In conclusion, the authors feel that this work constitutes a beginning toward the systematic derivation of empirical models capable of quantitative application to management problems. In addition, the information gathered during the process outlined above should engender a number of hypotheses regarding the mechanism of important interactions (*a priori* models).

ACKNOWLEDGMENTS

The work described herein was supported by the National Science Foundation's Committee on Research Applied to National Needs (RANN), Grant Number GI-29906, to the Chesapeake Research Consortium (CRC). Computer time was provided by the United States Army Corps of Engineers' Contract DACW31-70-C-0077 to the University of Maryland Computer Science Center. The authors wish to thank Ms. Shelley Sulkin and Mr. Rogers Huff for their excellent support in obtaining the data, and Ms. Margaret Roper for typing the manuscript.

REFERENCES

1. Bellman, R., Kagiwada, H., and Kalaba, R.
   1966 Inverse problems in ecology. **Jour. Theoretical Biol.,** 11: 164.
2. Caperon, J., Cattell, S. A., and Krasnick, G.
   1961 Phytoplankton kinetics in a subtropical estuary: eutrophication. **Limnol. Oceanogr.,** 16: 599.
3. Dale, M. B.
   1970 Systems analysis and ecology. **Ecology,** 51(1): 1.
4. Gause, G. F.
   1934 **The Struggle for Existence.** Williams and Wilkins, Baltimore.
5. Heinle, D. R.
   1969 Culture of calanoid copepods in synthetic seawater. **Jour. Fish.**

Res. Bd. Canada, 26: 150.

6. Heinle, D. R.
   1969 Effects of temperature on the population dynamics of estuarine copepods. Doctoral dissertation, University of Maryland, College Park, Maryland.
7. Holm-Hansen, O., Lorenzen, C. J., Holmes, R. W., and Strickland, J. D. H.
   1965 Fluorometric determination of chlorophyll. **Jour. Conseil, Perm. Intern. Explor. Mer.,** 30: 3.
8. Lorenzen, C. J.
   1967 Determination of chlorophyll and pheo-pigments: spectro-photometric equations. **Limnol. Oceanogr.,** 12: 343.
9. Mobley, C. D.
   1972 Statistical hypothesis testing as applied to the inverse problem in ecosystems analysis. N.R.I. reference no. 72-84, Natural Resources Institute, Solomons, Maryland.
10. Mobley, C. D.
    1973 A Systematic Approach to Ecosystems Analysis. (In preparation.)
11. Park, T.
    1948 Experimental studies of interspecies competition. **Ecol. Monogr.,** 18: 265.
12. Patten, B. C.
    1968 Mathematical models of plankton production. **Int. Revue ges. Hydrobiol.,** 53: 357.
13. Ross, D. T.
    1967 The AED approach to generalized computer-aided design. **Proc. Ass. Comput. Mach. National Meeting,** 367.
14. Strickland, J. D. H., and Parsons, T. R.
    1965 A manual of sea-water analysis. **Fish. Res. Bd. Canada Bull. No. 125.** Ottawa.
15. Ulanowicz, R. E.
    1971 Transport through hemoglobin solutions: a sensitivity analysis. **Jour. Theoretical Biol.,** 32: 185.
16. Watt, K. E. F.
    1962 Use of mathematics in population ecology. **Ann. Rev. Ent.,** 7: 243.
17. Yentsch, C. S.
    1965 In **Primary Productivity in Aquatic Environments.** (ed. Goldman, C. R.) Mem. 1st Ital. Idrobiol., 18. Univ. California Press, Berkeley.
18. Yentsch, C. S., and Menzel, D. W.
    1963 A method for the determination of phytoplankton chlorophyll and phaeophytin by fluorescence. **Deep-Sea Res.,** 10: 221.

UTILITY OF SYSTEMS MODELS: A CONSIDERATION OF SOME POSSIBLE FEEDBACK LOOPS OF THE PERUVIAN UPWELLING ECOSYSTEM[1]

John J. Walsh[2]

ABSTRACT

A non-linear mathematical model of the upwelling ecosystem off the coast of Peru is used to examine two questions of significance to biological oceanography: the regeneration of nitrogen and its availability to phytoplankton, and the oxidation of detrital carbon. Predictions of the model agree well with observational data not used in its construction. The model suggests that nitrate uptake by phytoplankton is suppressed not only by ammonia but also by urea and possibly also by amino acids. Oxygen utilization predicted by the model plus a few reasonable assumptions agree with the observed gradient and suggest that most of the remineralization of fecal pellets occurs in the upper 30 m of the water column.

INTRODUCTION

Systems analysis and simulation models have become increasingly important analytical tools to complement empirical and experimental methods in attempts to understand the dynamic behavior of aquatic ecosystems. A possible misunderstanding may be involved in the presently fashionable concept

1. Contribution no. 771 from the Department of Oceanography, University of Washington, Seattle, Washington 98195.

2. Department of Oceanography, University of Washington, Seattle, Washington 98195.

"model," however, *i.e.,* if a system is sufficiently well understood, and thus amenable to mathematical description, the system's future behavior should be predictable. Such prescient knowledge of a model depends almost entirely on the questions asked, the available data, and the scientist's intuition upon which the models are based. The levels of prediction and resolution, the criteria of validation, and the use of sensitivity analyses to determine the most important processes of the model (not necessarily of the real world) for management decisions are all important factors. These factors involve, however, the application mode of systems analysis in the possible ultimate pay-off from building complex biological simulation models as part of environmental research efforts. Instead, I will address some proximate pay-off, inherent in the research mode of a systems approach to marine sciences (46).

Building models of total ecosystems, in contrast to individual populations or trophic levels, more sharply focuses questions about how individual components of the ecosystem are linked to each other, for the structure of coupling between state variables is extremely important in determining the total behavior of a system. Paradigms and data that are advanced to explain the behavior of an individual component are generally lacking in resolution with respect to both other living variables and the spatial and temporal synopticity of any one variable. A systems model provides a conceptual hypothesis within which different sets of data may be explored, it exposes which critical observations have not been made, and it suggests laboratory and environmental experiments to be performed. As an example, I will consider some questions about two couplings or feedback loops, the regeneration of nitrogen and the oxidation of detrital carbon, which arose from construction of a systems model of the Peruvian upwelling ecosystem.

THE CARBON/NITROGEN MODEL

Various aspects of the model have been described in detail (48, 49), and only its logic is presented here to provide a perspective for discussion of the carbon and nitrogen feedback loops. Ecosystems appear to be non-linear, time dependent, spatially heterogeneous, and synergistic phenomena. It seems reasonable that if simulation models are to be used to analyze such a system, they should then contain these properties. Upwelling areas involve sharp transition between oligotrophic and eutrophic ecosystems with large gradients in the nutrient, phytoplankton, and herbivore concentrations. Within such a large range of values, the specific rates (hr^{-1}) of nutrient uptake and of invertebrate and vertebrate ingestion appear to follow nonlinear rectangular hyperbolae (8, 16, 32) of the respective rates versus nutrient or prey concentration. A Michaelis-Menten expression of the uptake of nutrients and of light regulation, as well as a similar formulation of herbivore ingestion with a modified term to include a grazing threshold, are thus assumed in the model. The time step or interval of calculation is one hour, and long-term (months) predation on the herbivores is assumed to be a constant linear loss over the time regime (10-20 days) of the model.

The phytoplankton uptake of nutrient and incorporation of carbon during photosynthesis are assumed to vary sinusoidally during the daylight hours. The phytoplankton are not allowed to assimilate these elements during the night, and nocturnal grazing occurs instead. The anchoveta and zooplankton are both assumed to be herbivorous during autumn off Peru, while the anchoveta are not assumed to be phytophagous in winter, and in both seasons, diel vertical migration of the metazoans is represented by sine and cosine functions at various depths of the water column. Different steady state, seasonal water circulation and eddy diffusion patterns are assumed in relation to the observed wind and density fields during autumn and winter.

It is not likely that every m^3 is the same in coastal regimes, and the x (offshore) and z (vertical) dimensions are explicitly represented in the advective terms of the model, while the alongshore effects in the y dimension are parameterized with an eddy diffusion term. The circulation pattern consists of a 20 m offshore Ekman layer, a uniform return flow from 20 to 50 m, and an offshore decay of the vertical velocity within 20 km of the coast. There are 5 layers in the vertical (0 to 50 m), 10 layers downstream (100 km), and the small-scale horizontal patchiness of phytoplankton and herbivores is parameterized within the 10 km grid mesh by the grazing threshold assumption.

Within the submodel of phytoplankton growth as a multiple function of light, nitrogen, silica, and phosphorus, an hourly expression of Liebig's law of the minimum represents synergistic effects:

the resource which most limits the growth of phytoplankton at each hour is selected by first comparing the ratio of dissolved elements of the model to the requirements of nutritionally unlimited diatoms; if nitrogen is chosen, the sum of ammonia, urea, and amino acid concentrations is allowed to depress nitrate uptake with more of the reduced nitrogen preferentially utilized by the algae; the lowest of the elemental uptake rates is then compared with the light uptake rate as a function of depth and phytoplankton biomass; finally the lowest of these last two rates is then used as an estimate of phytoplankton growth at that hour of computation in the model.

At the herbivore trophic level, excretion is an important form of rapid nitrogen recycling and it is assumed to be a density-dependent function of grazing without a basal metabolism correction. In contrast, silica is not assumed to be returned to the system through herbivore excretion. Fecal pellet production is also a function of grazing, with a formation rate of 20% of ingestion, *i.e.*, an 80%

assimilation efficiency is assumed for the herbivores.

The above complex model, involving relationships of wind, water circulation and diffusion, light, nitrate, a combination of ammonia, urea and amino acids, silicate, phosphate, carbon, phytoplankton, zooplankton, and fish, was built with data from an autumn 1969 cruise of the *Thomas G. Thompson* (3) off Punta San Juan, Peru. The results of an autumn 1966 cruise of the *Anton Bruun* (1) and a winter 1969 cruise of the *Thomas Washington* (4) have been used to test the model. Only the results of the autumn model are considered in the discussion, however, and more detailed analysis is presented by Walsh (49).

RESULTS

The downstream distributions of the predicted state variables agree with the validation data taken in the same season three years before; the model's solutions for nitrate, silicate, phytoplankton carbon and nitrogen values in the upper 10 m layer are shown in comparison with the *Bruun* data (Figs. 1-4). The model's atomic C:N ratio of phytoplankton was a mean of 4.8, which also agrees fairly well with a mean of 4.0 for all samples on the *Bruun* cruise, of 5.6 on the *Thompson* cruise, and of 6.1 from an *Unanue* (2) cruise in 1966 to an area just south of the other two cruises. The match of independent data sets with the model's C:N ratio and spatial distribution of state variables suggests that this mathematical description might be a reasonable approximation of the biological dynamics of the euphotic zone of the Peruvian ecosystem for this season. I will now explore what consequences of feedback structure are implied by such a fit of observation and theory.

Nitrogen recycling

The total nitrogen uptake submodel involves a V_{max} assumption of 0.08 hr^{-1} for recycled nitrogen uptake and of 0.08 hr^{-1} for nitrate uptake in the absence of ammonia or other reduced nitrogen compounds within the Michaelis-Menten

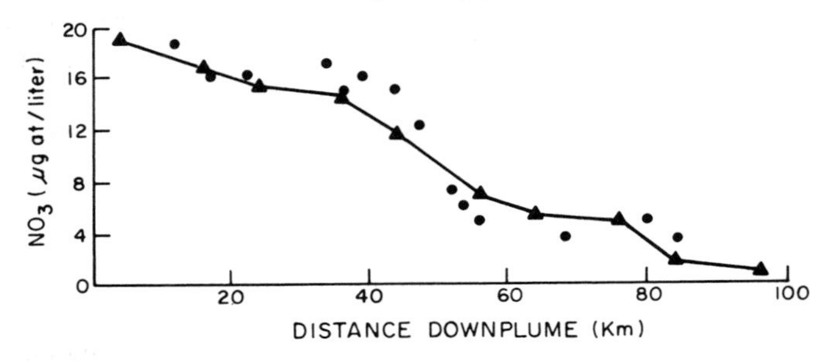

Figure 1. Simulated (▲) and observed (•) dissolved nitrate of the 0-10 m layer.

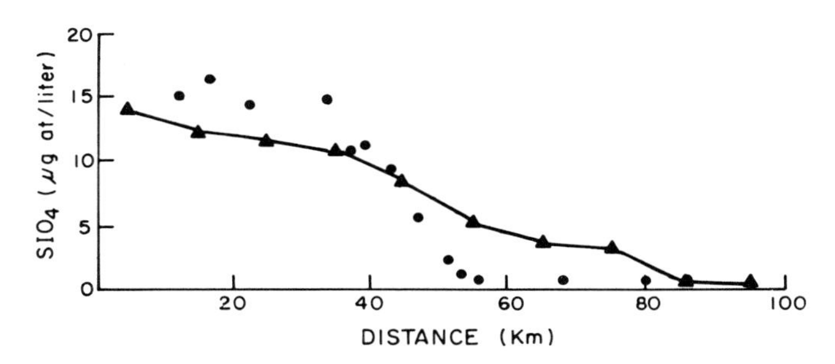

Figure 2. Simulated (▲) and observed (•) dissolved silicate of the 0-10 m layer.

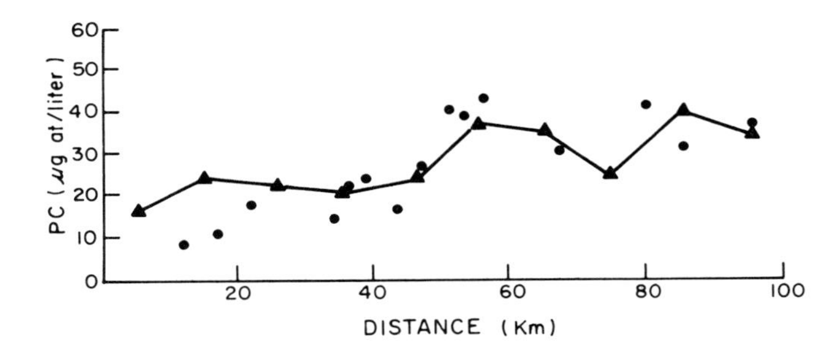

Figure 3. Simulated phytoplankton (▲) and observed particulate (•) carbon of the 0-10 m layer.

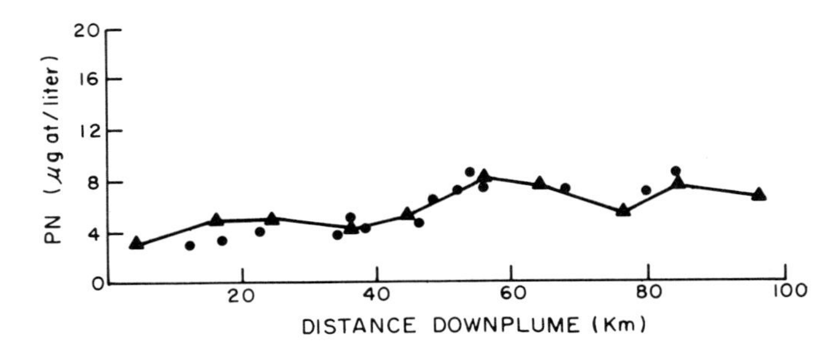

Figure 4. Simulated phytoplankton (▲) and observed particulate (•) nitrogen of the 0-10 m layer.

expression:

$$\text{Nutrient uptake} = \frac{(V_{max})\ (\text{Nutrient})\ (\text{Phytoplankton})}{(K_t + \text{Nutrient})} \quad (1)$$

where V_{max} is the maximum specific uptake rate (hr^{-1}) and K_t is the half saturation constant or substrate concentration (μ gat/liter) at which half the maximal rate occurs. If ammonia, urea, or amino acid are present, the V_{max} of NO_3 is depressed in the model as a linear function of these concentrations. This assumption of a 12 hour, daylight nitrogen V_{max} of 0.08 hr^{-1} or a doubling rate of 1.4 divisions/day (assuming exponential growth) is based on chemostat studies of NH_3 uptake by *Skeletonema costatum* in the laboratory (Conway, personal communication) and on shipboard experiments in the Californian current of phytoplankton grown separately on nitrate, ammonia, and urea (9). The rate of growth of phytoplankton during autumn 1969 in the Peruvian current system, estimated from the rate of ^{14}C uptake, was 1.2 divisions/day (5).

A maximum nitrate uptake of 0.04 hr^{-1} was determined, however, with ^{15}N incubation experiments on natural populations collected from nearshore stations during the *Thompson* cruise. This rate decreased to 0.02 hr^{-1} at offshore stations where higher *in situ* ammonia concentrations may occur as a result of density-dependent excretion fluxes. Previous models (47) that involved phytoplankton growth based on just the measured ^{15}N nitrate uptake rate failed to grow enough phytoplankton to match observed values off Peru. It is possible in the rich Peruvian upwelling area, where reduced nitrogen compounds are rapidly excreted by herbivores, that even the inshore phytoplankton populations were suppressed by ammonia when the ^{15}N nitrate experiments were performed on the *Thompson.* The few ammonia ^{15}N incubation studies that were carried out on this and other cruises (21, 23), however, suggest that the sum of nitrate and ammonia uptake fluxes would still not account for a total nitrogen uptake flux that could yield a ^{14}C:^{15}N uptake ratio of 5:1.

If the model's C:N ratio, the observed particulate carbon, particulate nitrogen, and ^{14}C:^{15}N uptake data are reasonable, then presumably other forms of reduced nitrogen besides NH_3 might be taken up in the Peruvian system. The historical nutrient data were re-examined to assess the feasibility of a larger, positive nitrogen feedback loop. Not all the dissolved nitrogen compounds that might be responsible for such a hypothetical NO_3 suppression have been identified, but ammonia, urea, and amino acids (*i.e.,* creatine, etc.) may be among the most important for phytoplankton uptake (9, 42). These nitrogen sources accounted for 82% to 84% of the total nitrogen excretion of both the Peruvian and Californian anchovy (24), and one might expect to find these forms of nitrogen in the water column off Punta San Juan, where the Peruvian anchovy may be the most important herbivore in the autumn ecosystem (49).

Different methods were used on the *Bruun* (35) and the *Thompson* (31) cruises to determine ammonia concentrations, and in a comparative study of these two techniques in the equatorial Pacific euphotic zone (20), the *Bruun* Richards and Kletsch method (35) appeared to overestimate NH_3 by about three times the *Thompson* Prochazkova method (31). A similar relationship may occur in the Peruvian euphotic zone, for the mean (1.84 μ gat/liter) of the *Bruun* determinations of NH_3 was 2.75 times the mean (0.67 μ gat/liter) of the *Thompson* determinations. (A third method was used on the *Thomas Washington* cruise (4) to this area, but possible bacterial contamination of the filters casts serious doubt on the validity of these determinations and the data are not considered in this paper.) The Prochazkova method is presumably NH_3 specific. Most of the other dissolved organic nitrogen compounds which may be measured with the Richards and Kletsch method could be amino acids, for little urea is supposedly converted to ammonia in this technique. The range of free amino acids thus far measured in sea water is 0.2 to 2.0 μ gat/liter and phytoplankton have been demonstrated to use some of these as a nitrogen source (42). Thus, although the Richards and Kletsch method does not measure all amino acids, it is possible that the 1.17 μ gat/liter difference between averages of the *Bruun* and *Thompson* observations could be attributable to amino acids.

McCarthy (22) and Remsen (34) both measured urea off Punta San Juan at the same time, the former on the *Thompson,* the latter on the *Gosnold.* McCarthy's measurements were made with a urea-specific urease method that yielded a mean of 0.28 μ gat N/liter for the euphotic zone. Remsen's method is not urea specific, but is supposedly negative for a number of amino acids and ammonia, and yielded a mean of 3.46 μ gat N/liter for surface waters. McCarthy's mean of 0.28 μ gat urea N/liter is approximately 1/3 of the NH_3 specific mean of 0.67 μ gat/liter, which is about the same ratio that was observed in the above anchoveta excretion experiments. Remsen's samples were frozen, and then analyzed, while McCarthy's samples were analyzed fresh, and there may be interference in Remsen's method from material such as citrulline, allantoin, thiourea, and unknown dissolved organic nitrogen compounds. McCarthy's estimate may be the closer approximation of just the urea concentration off Punta San Juan in autumn.

A minimum concentration of ammonia, urea, and amino acids within the autumn euphotic zone off Punta San Juan might then be 0.67 μ gat N/liter (NH_3) + 0.28 μ gat N/liter (urea) + 1.17 μ gat N/liter (amino acids) or 2.12 μ gat N/liter of recycled nitrogen. This is a small fraction of the dissolved organic nitrogen (DON) in the water column. The total amount of DON can be estimated from the *Thompson* dissolved organic carbon (DOC) average of 2.35 mg/liter (22 samples) for the euphotic zone. Using a 5 to 1 C:N weight ratio, a mean of 0.47 mg DON/liter is obtained or 34 μ gat DON/liter, of which at least

1.45 μgat N/liter may be the herbivore excretory products, urea and amino acids. A mean of 31.8 μgat DON/liter was actually observed in surface water at six of the *Unanue* stations and agrees with the above computed estimate in suggesting a large, diverse source of reduced nitrogen is present off Peru which may lead to suppression of nitrate uptake by phytoplankton.

Concentrations of DON at the *Unanue* stations declined rapidly with depth until 3.0 μgat/liter were observed at 50 m, however, neither the source nor the utilization rate of this high amount of DON is clear. I am not aware of any measurements that suggest DON (other than urea and trimethylamine oxide) is a major component of herbivore excretory products (7, 50). Regeneration of DON from decomposing phytoplankton, kept in the dark, occurs at a rate of 0.001 μgat $liter^{-1}$ hr^{-1} (14), in contrast to 0.014 μgat $liter^{-1}$ hr^{-1} from shipboard cultures of live phytoplankton (9). The role of most of the DON in relation to phytoplankton uptake is not clear, however, for no correlation of DON concentrations was found with phosphorus and nitrogen metabolism in the above shipboard experiments (9). We are presently conducting experiments to see how important these reduced components of the nitrogen pool are in alleviating what may be general nitrogen limitation of productivity in coastal waters (13, 38).

Carbon recycling

The role of excretory products as a positive nitrogen feedback loop, in possibly inducing higher production of the Peruvian ecosystem through energetically cheaper nitrogen sources, raised questions about the role of oxygen stress, generated by detrital remineralization, as a negative carbon feedback loop. The nutrient uptake and photosynthesis assumptions, coupled with additional estimates of herbivore biomass, growth and excretion rates, led to the mean grazing flux in the model which was 80% of the mean measured daily phytoplankton productivity. If it is further assumed that this herbivore ingestion flux of phytoplankton and the subsequent fecal pellet production are the major sources of detrital material (river run-off is negligible and bubble formation is ignored), a herbivore assimilation efficiency of 80% generates a predicted fecal pellet input of 0.8 μgat N $liter^{-1}$ day^{-1} or 4 μgat C $liter^{-1}$ day^{-1} over the upper 30 m off Punta San Juan in autumn. The impact that such an organic load might have on the Peruvian ecosystem is evaluated below in a budget of the hypothetical oxygen demand and vertical distribution of the model's detritus, of the additional biological respiratory demands, and of the possible supply of oxygen through vertical diffusive fluxes.

Determination of the oxygen demand of the fecal pellets involves additional assumptions, for, although a number of C:N analyses have been made on fecal pellets of macro-invertebrate omnivores and on those of one teleost, *Mugil cephalus* (11, 12), few studies have involved a known food source fed to a herbivore and subsequent elemental analysis of these fecal pellets. The elemental composition, and thus the oxygen utilization rate, may vary from one type of herbivore fecal pellet to the next. For example, Johannes and Satomi (18) fed *Nitzschia closterium* to *Palaemonetes pugio* and observed an atomic C:N:P ratio of 30:6:1 for the fecal pellets, but they did not report the cellular composition of this diatom. A phytoplankton N:P ratio greater than 6:1 would be expected in this experiment, however, unless the algae were under severe nutrient limitation. Butler, Corner, and Marshall (6, 7) have recently performed experiments involving elemental analyses of both prey and fecal pellets which show the reverse result: a systematic 1½ times increase from the N:P ratio of phytoplankton food source (~ 16:1) over that of the zooplankton fecal pellets (~ 25:1), with a zooplankton excretion ratio of 11:1. The high ratio of fecal pellets in this second study was attributed to increased efficiency of phosphorus retention by the zooplankton. Zooplankton fecal pellets should be a minor export of the Peruvian system, however, compared to the high biomass and nutrient turnover of the anchoveta during autumn.

In agreement with inverse size relationships of N and P metabolism (17, 51), the specific nitrogen excretion rate of zooplankton in the model was 2 to 3 times that of the anchoveta, and the maximum specific grazing rate of the zooplankton was 3 to 4 times that of the anchoveta. It also appears that, per unit time, zooplankton retain at least twice as much phosphorus as do fish, for an N:P excretion ratio of 5:1 was found for the Peruvian anchoveta (50). The N:P ratio of the anchoveta fecal pellets is then assumed to be ~ 3 times less than that of the above zooplankton studies (6, 7), or 8:1 with the assumption that anchoveta do not retain as much phosphorus per unit time as the zooplankton because of the relatively slower metabolic processes of the nekton.

The oxygen demand of autumn fecal pellets can then be computed with a phosphorus flux of 0.1 μgat P liter^{-1} 24 hr^{-1} and an O_2:P atomic utilization ratio (33) of -276:1 to yield a flux of 13.8 mgat O_2 m^{-3} 12 hr^{-1}, evenly distributed over the upper 60 m of the water column. This calculation is also based on the model's assumption of a diel distribution of herbivores in the upper 30 m during night and in the lower 30 m during the day, or a fecal pellet flux of 0.05 μgat P liter^{-1} 12 hr^{-1} within each depth interval. The fecal pellets are further assumed to remain within this upper 60 m layer to be oxidized, for the alongshore, subsurface counter currents at 25 to 50 m depths off Peru are about 16 cm/sec (41) or ~ 100 times that of the mean sinking rate of 17.5 x 10^{-2} cm/sec (10, 28, 40) of zooplankton fecal pellets. This strong shear zone, just below the 20 m, oxygen laden Ekman layer, may thus form a partial barrier to the descent of detrital material. The observed vertical distribution of oxygen, in fact, supports these assumptions with a mean of 4.55 ml/liter over 0 to 20 m, then a drop to 0.55 ml/liter over 20 to 60 m, and a low level of ~ 0.20 ml/liter from 60 m to the bottom at the 15 *Bruun* stations. The much larger number of *Thompson* Stations exhibit the same vertical pattern of oxygen distribution.

If the fecal pellets are remineralized mainly in the upper 60 m, however, the oxygen utilized for consumption of the suspended detritus as well as that for respiration of organisms between 20 and 60 m in the model must at least balance the oxygen supplied through vertical diffusion in order to avoid a shallow subsurface anoxic layer. An estimate of the daily vertical flow of oxygen, evaluated at 25 m, can be made with a finite difference expression for the vertical diffusion of oxygen down into the 20 to 60 m layer. The vertical diffusion flux can be estimated by

$$\frac{\partial (K_z \frac{0_2}{z})}{\partial z} \cong \frac{K_{z,\,k-\frac{1}{2}} \frac{(0_{2,\,k-1} - 0_{2,\,k})}{\Delta z}}{\Delta z} - \frac{K_{z,\,k+\frac{1}{2}} \frac{(0_{2,\,k} - 0_{2,\,k+1})}{\Delta z}}{\Delta z} \qquad (2)$$

where $K_{z,k-\frac{1}{2}}$ is the vertical eddy coefficient at 30 m interface, $K_{z,\,k+\frac{1}{2}}$ is the vertical eddy coefficient at 20 m interface, Δz is 10 m (*i.e.*, the distance across the 20-m interface between 15 m and 25 m and across the 30-m interface between 25 m and 35 m), and $0_{2,\,k-1,\,k,}$ and $_{k+1}$ are the observed mean oxygen concentrations at 35, 25, and 15 m depths, respectively.

Consideration of the vertical eddy coefficient, K_z, in relation to the physical processes of the model (49) led to an assumed value of 5 cm^2/sec in the upper 20 m and a much smaller value of 1 x 10^{-3} cm^2/sec below 20 m in the return flow. The first term of the above diffusion flux can then be deleted, for the value of $K_{z,\,k-\frac{1}{2}}$ has been assumed to be about 1/1000 that of $K_{z,\,k+\frac{1}{2}}$. The second term, upon evaluation, then becomes:

$$\frac{\partial (K_z \frac{\partial 0_2}{\partial z})}{\partial z} \cong \frac{K_{z,\,k+\frac{1}{2}} \frac{(0_{2,\,k+1} - 0_{2,\,k})}{\Delta z}}{\Delta z} = \frac{5\ cm^2/sec\,(4.55 - 0.55 ml/l)}{(10m)^2} \qquad (3)$$

or

$$\frac{\partial (K_z \frac{\partial 0_2}{\partial z})}{\partial z} \cong 153.6 \text{ mgat } 0_2\ m^{-3}\ day^{-1} \text{ at 25 m depth.} \qquad (4)$$

Such an oxygen input of ~ 155 mgat 0_2 m^{-3} day^{-1} across the 20 m interface might be sufficient to meet hypothetical subsurface demands, for a vertically homogeneous distribution of detritus and organisms over 20 to 60 m leads to a 14 mgat 0_2 m^{-3} day^{-1} demand for the fecal pellets and an additional 11 mgat 0_2 m^{-3} day^{-1} respiration demand for the living organisms [total plankton = 0.352 mgat 0_2 m^{-3} day^{-1}, (29); anchoveta = .873 mgat 0_2 m^{-3} hr^{-1} for a 12 hour period (50)], or a total demand of ~ 25 mgat 0_2 m^{-3} day^{-1} at each of the 25, 35, 45, and 55 m grid points.

The sharp oxygen gradient below 20 m and deleterious effects of low oxygen tension on consumer metabolism (19, 44) suggest, however, that both the fecal pellets and the herbivores might instead be concentrated within a smaller vertical layer at 20 to 30 m. Such a vertically heterogeneous case would then imply that ~ 100 mgat O_2 m^{-3} day^{-1} might be utilized at 25 m, but in either the homogeneous or patchy situation, the predicted oxygen utilization of the upper layers would not lead to an oxygen deficit. This oxygen budget suggests that all of the vertical demands could thus be met in agreement with the observed oxygen gradient if in fact: (a) the N:P ratio of anchoveta fecal pellets is not lower than 8:1; (b) that K_z is ~ 5 cm^2/sec in the upper layer; (c) that either the respiratory rates of the anchoveta and zooplankton do not increase under conditions of lower oxygen tension, or that the herbivores spend most of their time above and within 20 to 30 m; and (d) that most of the detrital material is remineralized in this upper layer as oppposed to coprophagy, or a continual build-up of detritus within the water column, or a rain of fecal pellets to the bottom.

With respect to the last implication of vertical patchiness of detritus, a high correlation of bacteria and yeast populations has been observed (45) with subsurface phytoplankton numbers at the density interface of Antarctic Intermediate Water, north of the Antarctic convergence. Similarly the high PN:Chl *a* ratios of 5:1 and 10:1 observed at 30 m depth during the *Thompson* cruise, in contrast to a mean of 1.67:1 for the euphotic layer, suggest that detrital nitrogen may occur at the lower depths of the aphotic zone. In support of such a vertical patchiness hypothesis, Menzel's data (25) suggest that there was little variation of particulate organic carbon below 75 m, regardless of surface concentrations at the *Bruun* stations. His data indicate that the 75 m to 150 m integrated values are approximately 10% of those of the upper layer, implying that little detrital material remains in the water column below 75 m. As discussed previously, there was also a sharp vertical decline at the *Unanue* stations in the measured dissolved organic nitrogen, from 34 μ gat DON/liter at the surface to 3 μ gat DON/liter at 50 m.

There is undoubtedly some leakage of detrital material from the surface system to the Peruvian bottom, but the rate of input to the benthos is not clear. If, in contrast to the above discussion, all of the anchoveta fecal pellets were to sink through a still water column at a rate of 150 m/day (17.5 x 10^{-2} cm/sec), the sea bottom above 150 m depth (within 10 km to 20 km of the coast) would eventually be littered with fecal pellets (26, 39). The bottom should then exhibit anoxic conditions, for the integrated oxygen demand of the fecal pellets extruded over the upper 60 m of the water column during 24 hours would be 828 mgat O_2 m^{-2} day^{-1}, all of which would arrive on the nearshore bottom each day, with little evidence of such downward transport in the water column. In comparison with these hypothetical oxygen demands of the bottom, however, the mean *in situ* utilization rate obtained at these depths during the *Thompson* cruise off Peru (30) was 14.3 O_2 mgat m^{-2} day^{-1}.

If the low observed rates of oxygen utilization on the Peruvian sea bottom are correct, they may result from benthos metabolism rather than detrital remineralization. The mean benthic biomass of 7 samples, also taken on the *Thompson* cruise between depths of 85 to 300 m (36), was 454 μg C/m^2. If a C:N ratio of 5:1 and an O_2:N utilization ratio (15) for Long Island Sound benthos, under a minimum of 40% oxygen saturation, are assumed, an estimate of 132 mgat O_2 m^{-2} day^{-1}, or 10 times the measured value, is obtained for just the benthos respiration without consideration of oxygen demands of fecal pellets on the bottom.

The low oxygen tension near the Peruvian sediments (3% saturation) undoubtedly suppresses respiration of the benthos and the fecal material may pile up, rather than be remineralized, because the sediments have a high 4% organic content (Rowe, personal communication); but if the majority of the fecal pellets were not remineralized in the surface layer and if this arrival rate to the bottom was as fast as hypothesized above, anoxic conditions would presumably prevail. An oxygen budget of the upper waters does suggest that an appropriate combination of the model's flux of fecal pellets and estimates of herbivore respiration and vertical diffusion of oxygen could possibly account for the observed vertical distributions of both oxygen and particulate material off Peru. One could continue to speculate about future anoxic situations and associated fish kills, but the present number of assumptions of the model is sufficient to consider tests of the above hypotheses in a research mode, using the results of the model for experimental design of future cruises, rather than prematurely switching to the application mode of systems analysis.

CONCLUSION

After a second round of simulation models of the Peruvian upwelling ecosystem, we are now in a better position to describe how we think the system works and what we do or do not know about it. It may come as no surprise that not much is known about the catabolic processes of a pelagic ecosystem, *i.e.*, the nitrogen and carbon recycling, and one's suspicions may be confirmed with the number of ifs and buts involved in such a systems analysis with ensuing speculation. Without a quantitative hypothesis to rigorously isolate the lack of data, however, only qualitative assessments of the behavior of an ecosystem can be made. We can now perhaps go beyond the description of carbon-nitrogen coupling posed in this upwelling model to more specific hypotheses about the spatial and temporal structure of feedback loops within the Peruvian ecosystem, which may then be tested in future field studies. Certainly, more sophisticated experiments are now being designed on the role of dissolved organic nitrogen in phytoplankton nutrition and the role of detritus in creation of anoxic conditions. If none of the new hypotheses bear any relation to presumed reality (*i.e.,* what we can measure), then the original assumptions of the model must, of course, be re-examined with this additional information on the reliability of the input data and our present theories.

Estuarine ecologists may have sufficient data on all facets of their coastal system to allow them to proceed past the research mode to the application mode of systems analysis. Then it will be possible to ask more intricate questions, such as species shifts at various trophic levels as a result of either eutrophiction, pesticide input, switches in nutrient limitation, or oxygen stress. If data are not available in a form to answer these questions, however, a number of models of estuarine systems have been recently built (43) and a wealth of field studies have been conducted over the last 30 to 40 years in what are relatively well defined and accessible systems (27). It would, then, perhaps, be time to integrate these models and field studies to re-examine the available data in a comparative systems approach to estuarine ecosystems in order to see if our present understanding can lead to their rational utilization in relation to demands of urban societies. If this is not yet an attainable goal, it is possible that the state of the art of aquatic systems analysis is such that future estuarine studies might benefit from concurrent systems model efforts in experimental design of additional research.

ACKNOWLEDGMENTS

I would like to thank Mr. James Anderson and Drs. John Caperon, Carl Lorenzen, Francis Richards, Gilbert Rowe, and Richard Williams for helpful critique of this manuscript. This research was supported by grant GB 35880 X as part of our International Biological Program (IBP) Upwelling Biome program and grant GX 33502 as part of our International Decade of Ocean Exploration (IDOE) Coastal Upwelling Ecosystems Analysis (CUEA) program.

REFERENCES

1. Anonymous
1966 Cruise Report, Research Vessel **Anton Bruun,** Cruise 15. Special Report 5, Marine Laboratory, Texas A&M University, Galveston, Texas.

2. Anonymous
1967 Data Record and Comments, Cruise FGG. 66/1. Research on the Marine Food Chain; Progress Report January-February 1966, Part III, IMR 67-9, University of California, San Diego, California. 152 p.

3. Anonymous

1970a Biological production in upwelling ecosystems. Data Report, Part 1: Hydrography and productivity. Special Report 42, Department of Oceanography, University of Washington, Seattle, Washington. 97 p.

4. Anonymous
   1970b Data Record, Cruise Piquero, Leg 7, Section 1, Physical chemical and production measurements off the coast of Peru, 28 May-22 June 1969 aboard the R/V **Thomas Washington.** Research on the Marine Food Chain; Progress Report, July 1969-June 1970, Part III, IMR 7-5, University of California, San Diego, California.

5. Barber, R. T., Dugdale, R. C., MacIsaac, J. J., and Smith R. L.
   1971 Variations in phytoplankton growth associated with the source and conditioning of upwelling water. **Inv. Pesq.,** 35(1): 171-194.

6. Butler, E. I., Corner, E. D. S., and Marshall, S. M.
   1969 On the nutrition and metabolism of zooplankton. VI. Feeding efficiency of **Calanus** in terms of nitrogen and phosphorus. **Jour. Marine Biol. Assn. U. K.** 49: 977-1003.

7. Butler, E. I., Corner, E. D. S., and Marshall, S. M.
   1970 On the nutrition and metabolism of zooplankton. VII. Seasonal survey of nitrogen and phosphorous excretion by **Calanus** in the Clyde Sea area. **Jour. Marine Biol. Assn. U. K.** 50: 525-560.

8. Dugdale, R. C.
   1967 Nutrient limitation in the sea: dynamics, identification, and significance. **Limnol. Oceanogr.,** 12(4): 685-695.

9. Eppley, R. W., Carlucci, A. F., Holm-Hansen, O., Kiefer, D., McCarthy J. J., Venrick, E., and Williams, P. M.
   1971 Phytoplankton growth and composition in shipboard cultures supplied with nitrate, ammonium, or urea as the nitrogen source. **Limnol. Oceanogr.,** 16(5): 741-751.

10. Fowler, S. W., and Small, L. F.
    1972 Sinking rates of euphausid fecal pellets. **Limnol. Oceanogr.,** 17(2): 293-296.

11. Frankenberg, D., Coles, S. L., and Johannes, R. E.
    1967 The potential trophic significance of **Callinassa major** fecal pellets. **Limnol. Oceanogr.,** 12(1): 113-120.

12. Frankenberg, D., and Smith, K. L.
    1967 Coprophagy in marine animals. **Limnol. Oceanogr.,** 12(3): 443-450.

13. Goldman, J. C., Tenore, K. R., and Stanley, H. I.
    1973 Inorganic nitrogen removal from wastewater: effect on phytoplankton growth in coastal marine waters. **Science,** 180: 955-956.

14. Grill, E. V., and Richards, F. A.
    1964 Nutrient regeneration from phytoplankton decomposing in seawater. **Jour. Mar. Res.,** 22(1): 51-69.

15. Harris, E.
    1959 The nitrogen cycle in Long Island Sound. **Bull. Bingham. Oceanogrr. Coll.,** 17(1): 31-65.

16. Holling, C. S.
    1966 The functional response of invertebrate predators to prey

density. **Mem. Entomol. Soc. Can.,** 48: 1-85.

17. Johannes, R. E.
1964 Phosphorous excretion and body size in marine animals: microzooplankton and nutrient regeneration. **Science,** 146: 923-924.

18. Johannes, R. E. and Satomi, M.
1966 Composition and nutritive value of fecal pellets of a marine crustacean. **Limnol. Oceanogr.,** 11(2): 191-197.

19. Longhurst, A. R.
1967 Vertical distribution of zooplankton in relation to the eastern Pacific oxygen minimum. **Deep-Sea. Res.,** 14(1): 51-64.

20. MacIsaac, J. J.
1967 Ammonia determination by two methods in the northeast equatorial Pacific Ocean. **Limnol. Oceanogr.,** 12(3): 552-554.

21. MacIsaac, J. J. and Dugdale, R. C.
1972 Interactions of light and inorganic nitrogen in controlling nitrogen uptake in the sea. **Deep-Sea. Res.,** 19(3): 209-232.

22. McCarthy, J. J.
1970 A urease method for urea in seawater. **Limnol. Oceanogr.,** 15(2): 309-312.

23. McCarthy, J. J.
1972 The uptake of urea by natural populations of marine phytoplankton. **Limnol. Oceanogr.,** 17(5): 738-748.

24. McCarthy, J. J. and Whitledge, T. E.
1972 Nitrogen excretion by anchovy (**Engraulis mordax** and **E. ringens**) and jack mackerel (**Trachurus symmetricus**). **Fish. Bull.,** 70(2): 395-401.

25. Menzel, D. W.
1967 Particulate organic carbon in the deep sea. **Deep-Sea Res.,** 14: 229-238.

26. Moore, H. B.
1931 The muds of the Clyde Sea Area III. Chemical and physical conditions; rate and nature of sedimentation; and fauna. **Jour. Marine Biol. Assn. U.K.** 17(2): 325-358.

27. Odum, H. T., Copeland, B. J., and McMahon, E. A.
1969 Coastal ecological systems of the United States. Unpublished report of Contract RFP68-128 to the Federal Water Pollution Control Administration, Marine Science Institute, University of Texas, Port Aransas, Texas, 1-1187 p.

28. Osterberg, C., Carey, A. G., and Curl, H.
1953 Acceleration of sinking rates of radionuclides in the ocean. **Nature,** 200: 1276-1277.

29. Packard, T. T.
1969 The estimation of the oxygen utilization rate in seawater from the activity of the respiratory electron transport system in plankton. Doctoral dissertation, University of Washington, Seattle, Washington, 115 p.

30. Pamatmat, M.
1971 Oxygen consumption by the seabed IV. Shipboard and laboratory experiments. **Limnol. Oceanogr.,** 16(3): 536-550.

31. Prochazkova, L.
1964 Spectrophotometric determination of ammonia as rubazoic acid with bispyrazolone reagent. **Anal. Chem.,** 36: 865-871.

32. Rashevsky, N.
1959 Some remarks on the mathematical theory of nutrition of fishes. **Bull. Math. Biophysics.,** 21: 161-183.

33. Redfield, A. C., Ketchum, B. H., and Richards, F. A.
1963 The influence of organisms on the composition of sea-water. In **The Sea,** Vol. II, p. 26-77. (ed. Hill, M. N.) Interscience Publishers, New York.

34. Remsen, C. C.
1971 The distribution of urea in coastal and oceanic waters. **Limnol. Oceanogr.,** 16(5): 732-740.

35. Richards, F. A. and Kletsch, R. A.
1964 The spectrophotometric determination of ammonia and labile amino compounds in fresh and seawater by oxidation to nitrate, In **Recent researches in the fields of hydrosphere, atmosphere, and nuclear geochemistry,** p. 65-81. (eds. Yake, Y. M. and Koyama, T.) Maruyen Go., Tokyo.

36. Rowe, G. T.
1971 Benthic biomass in the PISCO, Peru upwelling. **Inv. Pesq.,** 35(1): 127-136.

37. Ryther, J. H., Menzel, D. W., Hulburt, E. M., Lorenzen, C. J., and Corwin, N.
1971 The production and utilization of organic matter in the Peru coastal current. **Inv. Pesq.,** 35(1): 43-60.

38. Ryther, J. H. and Dunstan, W. M.
1971 Nitrogen, phosphorus, and eutrophication in the coastal marine environment. **Science,** 171: 1008-1013.

39. Schrader, H. J.
1971 Fecal pellets: role in sedimentation of pelagic diatoms. **Science,** 174: 55-57.

40. Smayda, T. J.
1969 Some measurements of the sinking rate of fecal pellets. **Limnol. Oceanogr.,** 14(4): 621-625.

41. Smith, R. L., Enfield, D. B., Hopkins, T. S., and Pillsbury, R. D.
1971 The circulation in an upwelling ecosystem: The Pisco Cruise. **Inv. Pesq.,** 35(1): 9-24.

42. Stephens, G. C. and North B. B.
1971 Extrusion of carbon accompanying uptake of amino acids by marine phytoplankton. **Limnol. Oceanogr.,** 16(5): 752-757.

43. Tracor, Inc.
1971 Estuarine modelling: an assessment. Water Pollution Control Research Series. U. S. Govt. Printing Office, Washington, D. C., Stock No. 5501-0129, 1-497 p.

44. Vinogradov, M. E. and Voronina, N. M.
1961 Influence of the oxygen deficit on the distribution of plankton in the Arabian Sea. **Okeanologiya,** 1: 670-678.

45. Walsh, J. J.
1971 Relative importance of habitat variables in predicting the distribution of phytoplankton at the ecotone of the Antarctic

Upwelling Ecosystem. **Ecol. Monogr.,** 41: 291-309.

46. Walsh, J. J.
1972 Implications of a systems approach to oceanography. **Science,** 176: 969-975.

47. Walsh, J. J. and Dugdale, R. C.
1971 A simulation model of the nitrogen flow in the Peruvian upwelling system. **Inv. Pesq.,** 35(1): 309-330.

48. Walsh, J. J. and Howe, S. O.
Protein from the sea: A comparison of the simulated nitrogen and carbon productivity of the Peruvian upwelling ecosystem. In **Symposium of Systems Ecology, the modelling and analysis of ecosystems,** (ed. Patten, B. C.) Academic Press, New York. (In press.)

49. Walsh, J. J.
A spatial simulation model of the Peruvian upwelling ecosystem. **Deep-Sea Res.** (In press.)

50. Whitledge, T. E. and Packard, T. T.
1971 Nutrient excretion by anchovies and zooplankton in Pacific upwelling regions. **Inv. Pesq.,** 35(1): 243-250.

51. Zeuthen, E.
1947 Body size and metabolic rate in the animal kingdom with special regard to the marine micro-fauna. **Compt.-rend. Lab., Carlsberg, Ser. chim.,** 26(2): 17-161.

RELATIONSHIP BETWEEN MORPHOMETRY AND BIOLOGICAL FUNCTIONING IN THREE COASTAL INLETS OF NOVA SCOTIA

K. H. Mann[1]

ABSTRACT

Comparison is made of some system properties in three inlets of Nova Scotia, Canada. St. Margaret's Bay has mainly rocky shores and is fairly open to the ocean. Bedford Basin has rocky shores but is very sheltered and exchanges with the ocean through a narrow channel. Petpeswick Inlet is a long, narrow estuary with extensive areas of shallow water which, at low tide, are drained to expose mud flats and sand flats. These morphometric factors influence the type and amount of primary production and the nature of the exchange with coastal waters. In St. Margaret's Bay about 75% of primary production is by seaweeds, especially *Laminaria*. In Bedford Basin phytoplankton production is the most important, while in Petpeswick Inlet production by marsh grass *Spartina alterniflora* and eelgrass *Zostera marina* predominates. St. Margaret's Bay has a slowly moving exchange with the coastal waters, driven by river run-off. Much of the export is at higher trophic levels. Bedford Basin exports more than half of its daily phytoplankton production by means of the twice-daily tidal flushing through the narrow outlet. Petpeswick Inlet appears to utilize its primary production within the inlet, but to export large amounts of dissolved nitrogen compounds.

It appears to be impossible to model these inlets as ecosystems until the exchange terms are more precisely known. Theoretical reasons are given why large, detailed, dynamic simulation models of these inlets would have

1. Department of Biology, Dalhousie University, Halifax, Nova Scotia, Canada.

unacceptably wide confidence limits on their outputs and therefore would be useless for prediction.

INTRODUCTION

The coast of Nova Scotia is heavily indented with estuaries and embayments, many of which appear to be eminently suitable for ecosystem analysis and subsequent model-building. Their physical size (largest dimension 10-20 km long) and taxonomic complexity tempt a small but well-equipped group of marine ecologists to regard them as profitable areas for the study of ecosystem properties.

Three of these inlets, St. Margaret's Bay, Bedford Basin, and Petpeswick Inlet have been subjected to fairly intensive study by the staff of the Bedford Institute of Oceanography and by the staff and students of Dalhousie University. Yet, no detailed dynamic simulation model has been produced for any one of them. It is the consensus of those involved that the construction of such a model, in the present state of the art, would not be profitable.

In this paper I intend to justify that point of view by two main arguments. The first is a practical argument: we are still discovering hitherto unsuspected interactions between the biological and physical components of these systems, and any attempt to model the systems before these interactions have been identified would lead to erroneous models. The second is a theoretical argument: systems models constructed entirely from the properties of the constituent species cannot predict well the properties of the total system.

OBSERVATIONS ON THREE NOVA SCOTIAN INLETS

These observations are concerned with the interactions between the physical and the biological properties of three contrasting inlets. St. Margaret's Bay (Fig. 1) is about 16 km from the mouth to the head of the bay and about 5 km wide at the mouth. The maximum depth is about 70 m, but there is a sill at a depth of 40 m. Hence, there is a fairly free, but not unlimited, exchange with the open coastal waters. Bedford Basin in only about one tenth the size of St. Margaret's Bay, but its depth is about the same, and it is separated from the sea by a channel, 10 km long, relatively narrow, forming a sill at a depth of only 20 m. Hence, there is a much more limited exchange with the open sea. Petpeswick Inlet is long, narrow, and very shallow, so that at every tide large areas of sand and mud flats are exposed. Except in the inner end of the inlet, where there is a basin 26 m deep, the exchange of water between the inlet and the open coast is very free indeed.

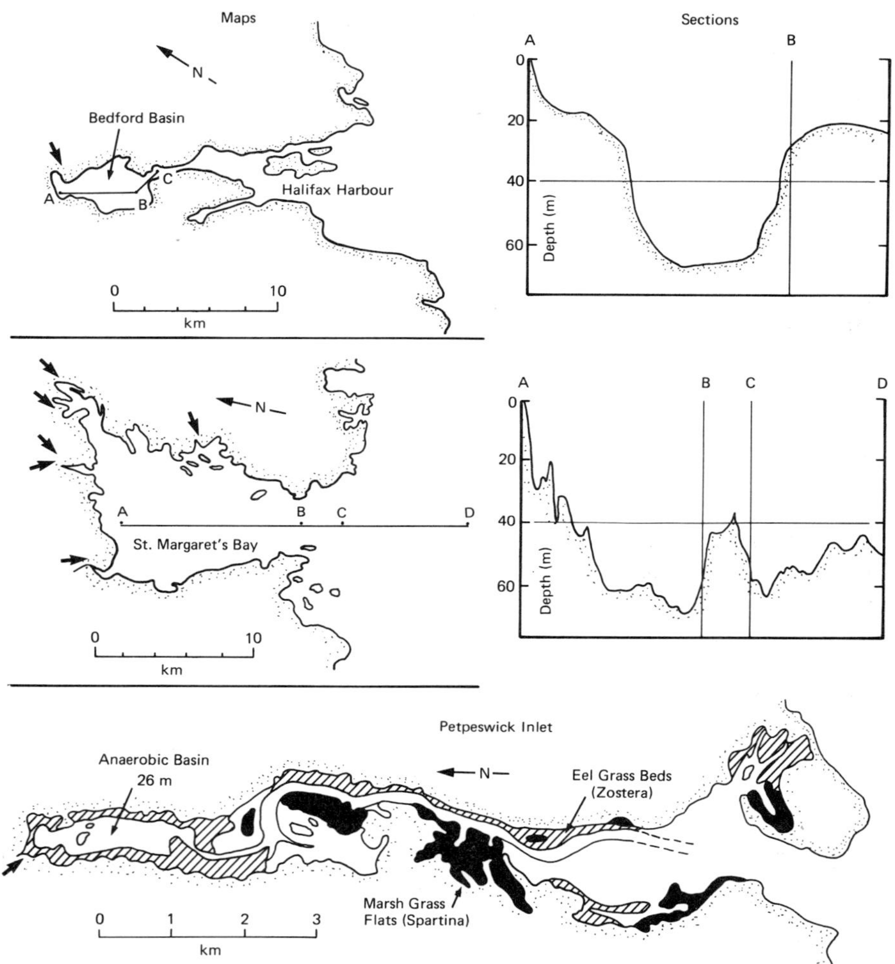

Figure 1. Three inlets on the east coast of Nova Scotia, Canada, and vertical profiles of two of them. Arrows indicate inputs of fresh water.

Primary Production

These three inlets differ markedly in the type of primary production that occurs. St. Margaret's Bay has a moderately rich phytoplankton, fixing about 200 g C m^{-2} yr^{-1} (15), and in this respect its waters are similar to those off the coast. In addition, however, in the intertidal and subtidal zones, it has luxurious growths of seaweed to a depth of about 25 m. The seaweed zone is dominated by *Laminaria,* which is 1968-69 and 1969-70 was shown to fix about 1750 g C

m^{-2} yr^{-1} (7). Although the seaweed was confined to the shallow margins, it was estimated that it produced in St. Margaret's Bay three times as much organic matter as did the phytoplankton. Hence, in St. Margaret's Bay, which is characterized by steep, rocky shores and fairly strong wave action, the primary production is dominated by seaweeds.

In Bedford Basin the shores are also steep and rocky, but wave action along the shores is much more limited, since the inlet is relatively enclosed and is sheltered by surrounding hills, Also, Bedford Basin receives considėrable input of treated and untreated sewage effluent. As a result, nutrient and chlorophyll concentrations are 2-3 times higher than in St. Margaret's Bay, and light penetration is much reduced. Seaweed beds are poorly developed and primary production is dominated by phytoplankton. In spite of the higher phytoplankton biomass, production appears to be limited by self-shading and the annual figure is only about 10% higher than in St. Margaret's Bay (17).

Petpeswick Inlet's intertidal flats are colonized by extensive areas of *Spartina alterniflora* and *Zostera marina*, the latter extending a considerable distance into the subtidal zone. *Spartina* fixes annually 200-300 g C m^{-2} and *Zostera* 100-200 g C m^{-2}. This production is supplemented by phytoplankton production and by algal production on the surfaces of the macrophytes and the mud, but the extent of this has not been determined.

Exchange at the mouth of the inlets

An important component of any model of a coastal inlet is exchange with the open sea. Analysis of such exchange requires not only an understanding of the physical oceanography of the situation, but a marrying of physical and biological observations on a one-to-one basis, to an extent that has not often been achieved.

Our best information on this exchange is available for Bedford Basin, where the transport of chlorophyll and zooplankton were studied intensively over a 25-hour period (16). The entrance to the basin (C in Fig. 1) was divided into 6 sectors: 3 depth intervals on the northeast side, and 3 on the southwest side. A current meter was placed in each sector, and the biomass of chlorophyll and zooplankton in each sector was determined once per hour for 25 hours (2 tidal cycles). In this way, the export of chlorophyll from the basin could be calculated for each sector and each hour. The total figure was compared with estimates of primary production within the basin, and it was found that the amount exported was 58% of the day's production. The remainder was accounted for by zooplankton grazing, and the budget was balanced to within 7%.

Movement of water through the entrance to the basin may be categorized into two types: (a) the mean flow, which is fresh water from the Sackville River that, together with its entrained salt water, tends to move steadily outwards at the surface, with a compensating inflow at a lower level; (b) the fluctuating flow, which is water that moves inwards on the rising tide and outwards on the falling tide, twice a day in each direction. It was found that the major part (65%) of the transport of chlorophyll-bearing organisms occurred via fluctuations in the flow. What seemed to be happening was that water coming in on the rising tide, having a relatively low chlorophyll concentration, was mixing with the water of high chlorophyll concentration in Bedford Basin. The resulting mixture, which went out on the falling tide, had a higher chlorophyll concentration than that which came in (Fig. 2). Hence, it may be said that a large proportion of the phytoplankton production of the basin was exported primarily by the tidal flushing, and that this was a direct result of the configuration of the basin. When similar calculations were made for zooplankton, it was found that almost all the export, (over 99%) was via fluctuations in the flow.

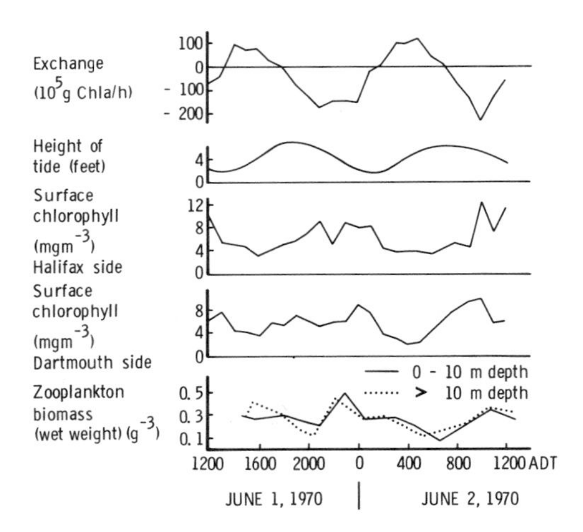

Figure 2. Data for mouth of Bedford Basin, Nova Scotia. Chlorophyll exchange (positive into basin), surface chlorophyll concentrations on either side of the inlet, and zooplankton biomass, with tidal height shown for comparison. After Platt and Conover (16).

The entrance to Bedford Basin is only 400 m wide and 20 m deep, yet the interpretation of events there over a 25-hour period required several weeks of intensive planning, the concerted effort of 15 persons, and many weeks of analysis. St. Margaret's Bay is about 5 km wide at the mouth, so there was no question of mounting a monitoring effect of comparable spatial density. Instead, we must resort to general statements derived from a study of the physical oceanography of the bay as a whole. Fresh-water inflow at a number of points at the head of the bay (Fig. 1) causes a slow outflow at the surface and upwelling from deeper waters of the bay (20). In addition, wind-induced currents with a strong offshore component cause rather rapid exchange of large segments of water (18). The net result of these processes is that the bay is flushed every 30-40 days on average, or about 10 times per year.

The estimated nitrogen requirements of the primary producers in the euphotic zone are about 60 g N m^{-2} yr^{-1}. Of this, it is thought that 7 g comes from the rivers, 12 g from zooplankton excretion, and 14 g from decomposition of organic matter in the euphotic zone, leaving 26 g to be supplied by upwelling (20).

The composition of the zooplankton in St. Margaret's Bay is distinctly different from that of the nearby open coastal waters. This is not surprising in view of the high level of primary production, especially by seaweeds, and a residence time for water in the bay of 30-40 days. There is no evidence of a mechanism that would cause an export of primary production, as in Bedford Basin. Instead, it appears that there is a slow, steady export of organic matter, much of it in the form of zooplankton and young fish. The rate of export varies a little acccording to the fresh-water run-off from the land, and is occasionally accelerated for a short period by strong wind-induced currents.

At the mouth of Petpeswick Inlet exchange of particulate and dissolved material was monitored throughout a tidal cycle on five separate occasions in the summer of 1971. Since a considerable proportion of the water in the inlet is exchanged on every tide, it was expected that there would be a heavy export of living and dead particulate organic matter. In fact, it was found that very little particulate matter was exported, and on some occasions there was a net import.

On the other hand, it was found that there was a consistent and heavy export of dissolved nitrogen compounds (21). The inlet contains large areas of marsh-grass and eelgrass beds. It appeared that the organic matter produced in these areas was being trapped and utilized by the fauna and microflora on the inlet, so that only the mineralized remains were available for export. It has been shown (14) that some sea-grass beds are sites of heavy nitrogen fixation, and preliminary indications (21) are that comparable amounts of nitrogen fixation are occurring in the salt marshes of Petpeswick Inlet.

From the foregoing evidence, it is clear that any attempt to model or predict the accumulation of nutrients, primary production, or secondary production in one of these inlets, without rather detailed knowledge of the exchange terms, is doomed to failure. Further, it is clear that the details of the kind and amount of primary production and the kind and amount of material being exported from the system depend to a great extent on the morphometry of each inlet and its relationship to the three driving forces of exchange: river run-off, tides, and winds.

K. H. MANN

THE NEED FOR AN HOLISTIC APPROACH TO MODELING

Although Odum (12) defines systems ecology as the formal approach to holism, a great deal of the literature on ecosystem models is basically reductionist in its approach. What are the features of an ecosystem that distinguish it as a *system?* According to Weiss (22) a system is an organization characterized by three basic properties: (a) it is hierarchical in its structure; (b) the variance of properties of the total system is less than the sum of the variances of the component subsystems; (c) there are internal flexibilities which enable the system to accept changes in its components.

The hierarchical nature of living systems is clearly understood in relation to organisms and their subsystems. The division of an organism into organs, tissues, cells, and molecules as different levels of organization is understood and accepted. It is also understood that each level in the hierarchy is valid as a field of study, in partial isolation from other levels. One can make and test hypotheses about RNA in the liver cells of a rat without knowing very much about the total behavior patterns of the animal. Conversely, one can make and test hypotheses about the behavior of the animal without invoking details of molecular biology. However, it is not useful to extrapolate too far between levels. It would not be very good strategy to attempt to predict the behavior of a rat by studying the properties of all its component cells, and using the cells as functional components of a dynamic simulation model.

There is an analogous series of levels in the hierarchy of an ecosystem. The basic unit is an organism; organisms are grouped into populations and populations into communities. However, a common method of constructing dynamic simulation models of ecosystems is to use species populations as the functional units. The behavior of the populations is simulated partly by using parameters that are characteristics of individuals. An attempt to model the behavior of an ecosystem from the properties of individual organisms is roughly analogous to an attempt to model the behavior of an organism from the properties of its cells.

A recent assessment of eleven environmental modeling efforts (9) documents the tendency to reductionism: (a) The predominate modeling strategy is to decompose the problem into a set of sybsystems, and further reduce each subsystem into its components until each component is recognized as a segment of some discipline that can be understood by an individual; (b) Models are sometimes considered to be validated when all components of the model are scientifically understood. Thus, the thrust to decompose the system until the components are small enough to analyze and resolve is very strong; (c) Once all segments are understood by their respective disciplines, the pieces of the model can be integrated upward and the total model used for analysis. The models referred to in this assessment embraced disciplines in both the natural and social sciences, but models of natural ecosystems are commonly constructed on the same principles.

In a symposium entitled "Beyond reductionism: new perspectives in the life sciences", (4) a distinguished group of biologists, most of them working at the cellular or organismal level, showed how the process of dissecting a biological system into its components results in a loss of information about that system. The lost information is not restored by simple juxtaposition of the components after analysis. They showed, for example, how the success of the reductionist approach to molecular structure of genes and cells has, in effect, diverted attention from much-needed studies on the system properties of cells and tissues.

The ecological modeling process described above seems to be exactly analogous, and it is probable that the technique can never lead to good predictive models of ecosystems because it fails to take into account the emergent properties of the whole systems. If one wishes to predict the behavior of ecosystems one must study the properties of whole systems, not the properties of some organisms at a much lower level in the hierarchy.

Another aspect of dynamic simulation modeling that seems to have received insufficient attention, is the confidence limits of the output. Dynamic simulation has proved extremely useful in physical sciences, where the parameters of the models are physical coefficients often known with great precision. However, the relationships between organisms and their environment or between one organism and another do not have the same kind of invariant qualities. The variance of biological parameters is not mainly a consequence of inaccurate measurement, but is an important property of species populations which enables them to respond to stress. Much of the variance is genetically determined, the remainder is a phenotypic response of organisms to environmental conditions encountered during growth or development. The role of this variability in the functioning of ecosystems is as yet far from clear, and deserves intensive study, but in our work on the inlets of Nova Scotia we see seaweed populations adapting their morphology to the degree of wave exposure and light intensity (1) and sessile invertebrates adapting their larval behavior to ensure optimum survival in a mosaic of environmental conditions (2). Phytoplankton cells adjust their nutrient uptake parameters to prevailing conditions (5), and the production parameters of a fish-population change in response to fishing pressure (6).

Until we are able to make accurate assessments of the degree of variability in biological paramenters and the way in which the mean values are likely to change in response to changing environmental conditions, it is unlikely that detailed ecosystem models will have any predictive power. An example of the variability in one of the relationships we have investigated (11) is the production model for benthic invertebrates of the sublittoral seaweed zone. Production was calculated from the relationship demonstrated by McNeill and Lawton (10) between population respiration and population production. Population density was calculated from 165 quadrat samples. The relationship between body size, temperatures, and respiration was determined at 2-month intervals for a year, in the laboratory. The confidence limits, taken as two standard errors of the mean, were accumulated through successive steps in the calculation. The upper confidence limit on the estimate of the daily respiration rate of an individual was 17% above the mean. For the estimate of the respiration of an individual for a year it was 21% above the mean. For the respiration of the whole population for a year it was 60% above the mean. The final step, introducing the production/respiration relationship, brought the upper limit of confidence in the production estimate to 446% above the mean.

The second characteristic of systems, mentioned earlier, is that the variance of some properties of the total system is less than the sum of variances of the component subsystems. In other words, systems have self-regulating properties which enable them to exhibit an overall stability greater than the stability of the component parts. Evidence for this is seen in the phytoplankton production of St. Margaret's Bay and Bedford Basin. As already mentioned, Bedford Basin has much higher levels of nutrients and supports much denser phytoplankton populations. The annual sequence of development of species populations is quite different in the two basins, yet their annual production differed by no more than 10% (17), and was probably determined by the fact that the annual light input to the two systems was similar. Another example is in the yield of trout from a range of Canadian lakes. Kerr and Martin (3) showed that yield per unit area as a function of basic productivity was relatively constant, in spite of the fact that the sizes of the fish and even the number of trophic levels in their food chains differed markedly from lake to lake.

This last example illustrates the third property of systems: that they are able to accept changes in their functional components and still perform their characteristic functions. The changes may range from small genetic shifts in the properties of a species population, to replacement of one species by another. Such changes occur not only in an evolutionary time scale, but in relatively short periods when the physical environment is changing. This renders doubly difficult the problem of predicting future events in an ecosystem, if the predictions are made by a model that is built on the properties of particular species.

My conclusion from this is that we have a long way to go before we can make good quantitative predictions about ecosystems. We can only hope to do this if we take an holistic system view, studying and measuring the properties of systems as wholes. As yet, we probably do not know which are the best properties to measure and predict, but it is clear that total annual primary productivity is more easily predicted than the population dynamics of any constituent species, and that import/export terms are better handled in terms of total particulate or total dissolved organic matter than on a species basis. The tropho-dynamic approach has served well for several decades, but now the literature contains so many examples of animals that take their food from two, three, or more trophic levels that the concept of trophic levels as functional units may well be nearing the end of its usefulness.

What shall we put in its place? There is no clear answer at present, but recent work in ecosystem modeling (8, 13, 19) suggests that particle size, particle distribution, photosynthetic pigment indices, carbon-to-nitrogen ratios, and, perhaps, amino-acid spectra are more important attributes of functional components than latin names and lists of gut contents.

ACKNOWLEDGMENTS

This paper relies heavily on the work of colleagues in the Bedford Institute of Oceanography during 1967 and 1972. I wish to thank them for their friendly collaboration and stimulating discussions: particularly, Drs. L. M. Dickie, R. J. Conover, T. Platt, and W. H. Sutcliffe Jr.

REFERENCES

1. Chapman, A. R. O.
   Genetic basis of morphological differentiation in some **Laminaria** populations. **Mar. Biol.** (In press)
2. Doyle, R. W.
   1973 Choosing between darkness and light: the ecological genetics of the photic behavior of planktonic larvae. **Mar. Biol.** (In press.)
3. Kerr, S. R. and Martin, N. V.
   1970 Tropho-dynamics of lake trout production systems. In **Marine food chains,** p. 365-376 (ed. Steele, J. H.). Oliver and Boyd, Edinburgh.
4. Koestler, A. and Smythies, J. R. (eds.)
   1969 **Beyond reductionism: new perspectives in the life sciences.** Hutchinson, London.
5. MacIsaac, J. J. and Dugdale, R. C.
   1969 The kinetics of nitrate and ammonia uptake by natural populations of marine phytoplankton. **Deep-Sea Res.,** 16: 45-57.
6. MacKinnon, J. C.
   1972 Production dynamics of a marine flatfish population. Doctoral thesis, Dalhousie University, Halifax, Canada.
7. Mann, K. H.
   1972 The ecological energetics of the seaweed zone in a marine bay on the east coast of Canada. II. Productivity of the seaweeds. **Mar. Biol.,** 14: 199-209.
8. Mann, K. H.
   1972 The analysis of aquatic ecosystems. In **Essays in Hydrobiology,** p. 1-14 (eds. Clark, R. B. and Wootton, R. J.) University of Exeter Press, Exeter, England.

9. Mar, B. W. and Newell, W. T.
1973 Assessment of selected RANN environmental modeling efforts. Report to National Science Foundation, Washington, D. C.

10. McNeill, S. and Lawton, J. H.
1970 Annual production and respiration in animal populations. **Nature,** (London), 225: 472-474.

11. Miller, R. J., Mann, K. H., and Scarratt, D. J.
1971 Production potential of a seaweed-lobster community in Eastern Canada. **Jour. Fish. Res. Bd. Canada,** 28: 1733-1738.

12. Odum, E. P.
1971 **Fundamentals of ecology.** 3rd ed. W. B. Saunders Co., Philadelphia.

13. Parsons, T. R. and LeBrasseur, R. J.
1970 The availability of food to different trophic levels in the marine food chains. In **Marine food chains,** p. 325-343 (ed. Steele, J. H.). Oliver and Boyd, Edinburgh.

14. Patriquin, D. G. and Knowles, R.
1972 Nitrogen fixation in the rhizosphere of marine angiosperms. **Mar. Biol.,** 16: 49-58.

15. Platt, T.
1971 The annual production of phytoplankton in St. Margaret's Bay, Nova Scotia. **Jour. Cons. Int. Explor. Mer.,** 33(3): 324-334.

16. Platt, T. and Conover, R. J.
1971 Variability and its effect on the 24h chlorophyll budget of a small marine basin. **Mar. Biol.,** 10: 52-65.

17. Platt, T. and Irwin, B.
1971 Phytoplankton production and nutrients in Bedford Basin, 1969-70. **Fish. Res. Bd. Canada Tech. Rept.** No. 247.

18. Platt, T., Prakash, A. and Irwin, B.
1972 Phytoplankton, nutrients and flushing of inlets on the coast of Nova Scotia. **Naturaliste Can.,** 99: 253-261.

19. Smith, F. E.
1972 Spatial heterogeneity, stability and diversity in ecosystems. **Trans. Conn. Acad. Arts. Sci.,** 44: 307-335.

20. Sutcliffe, W. H. Jr.
1972 Some relations of land drainage, nutrients, particulate material and fish catch in two eastern Canadian bays. **Jour. Fish. Res. Bd. Canada,** 29: 357-362.

21. Sutcliffe, W. H. Jr.
(Unpublished data.)

22. Weiss, P.
1969 The living system: determination stratified. In **Beyond Reductionism,** p. 3-55 (eds. Koestler, A. and Smythies, J. R.) Hutchinson, London.

THE ESTUARINE ECOSYSTEM(S) AT BEAUFORT, NORTH CAROLINA[1]

by

Douglas A. Wolfe[2]

ABSTRACT

Metallic elements cycle in estuarine ecosystems among the sediments, the biota, and the dissolved and suspended components of the water. Environmental variables, such as temperature, salinity, and pH, affect the distribution and flux of the metals. Prediction of the distribution of contaminant additions of radioactive or stable isotopes requires a thorough understanding of the normal flux of the element through and within the system. Using the Newport River estuary, Beaufort, North Carolina, as a typical tidally dominated, drowned-river estuary, I estimated the sizes of various biological and physical reservoirs of manganese, iron, and zinc, as well as the annual fluxes of these elements resulting from run-off, tidal flushing, fishery exploitation, and biological emigration. Biological processes were most significant in the case of zinc, for which as much as 10% of the annual input from the watershed may be removed by emigration and commercial harvest. Recycling of zinc within the trophic structure of the ecosystem, however, amounted to about 16 times the annual river input, and the estuary appears to conserve zinc efficiently. The major complexities of progressing from this static "annual budget" concept to a dynamic model of the flow of trace metals through the estuarine ecosystem arise because the physicochemical speciation of the metal itself must be superimposed upon food-chain transfers, i.e., whether it occurs in a soluble or particulate form, enters stable organic complexes, or is strongly adsorbed onto sediments. The rates and extents to which these forms are altered by the transitional estuarine

1. Research supported jointly by NMFS and USAEC agreement no. AT(49-7)-5.

2. National Marine Fisheries Service, Atlantic Estuarine Fisheries Center, Beaufort, North Carolina 28516.

environment must also be delineated to model successfully the fate of a contaminant metal introduced as a particular chemical species into the estuarine system.

INTRODUCTION

This paper represents mainly a progress report on our continuing efforts at the Atlantic Estuarine Fisheries Center to understand the flow of materials and energy in the shallow estuaries and embayments near Beaufort, North Carolina. If we accept Gordon's (24) definition of a *model* as "the body of information about a system gathered for the purpose of studying the system", then most of us attending this conference are in the business of modeling, merely by virtue of having collected bits of the already large body of information available on the structure and function of estuarine systems. If, on the other hand, we approach estuarine research with the view to predicting the changes that an ecosystem may undergo in response both to normal climatic fluctuations and to man-induced environmental modification, then we must (for reasons of practicality) develop quantitative dynamic models – the out-puts of which can be verified or disproven. One desired end product of our researches at Beaufort is a predictive mathematical model of the distribution and flux of metalic elements in the local estuarine ecosystem. Toward this end, we have developed a coordinated, long-term, interdisciplinary ecological research program which is now in its fourth year of formal data collection, but which had its conceptual origins within the Bureau of Commercial Fisheries' Radiobiological Laboratory several years earlier.

A major basis for our research approach is the belief that the estuary is in fact an ecosystem, not just an agglomeration of individual organisms or species residing in the aquatic or sedimentary milieu. If this is so, then the flows of energy and materials in this ecosystem should follow certain generalized operational principles irrespective of the precise species composition, as has been suggested by various recent studies of ecosystem structure (40, 41, 58, 87). Accordingly, we believe it is feasible for a small organization such as ours to undertake a comprehensive study of an estuarine ecosystem by attacking the required reserch in small pieces over an extended period of time and integrating our research results with data from investigators in other relevant areas. Despite the various inadequacies of the modeling and systems analysis approach (34) and our lack of any great experience in applying this approach to ecosystems, it still appears to be the most promising approach for testing our present knowledge of these ecosystems, for identifying gaps in our knowledge, and for summarizing knowledge of ecosystem operation into a form useful for prediction.

In the following sections, I will (a) review briefly the progress and current status of our ecosystem research as it applies to the distribution and cycling of metallic elements, (b) discuss other research findings as they apply to our understanding of elemental cycling, and (c) suggest preliminary approaches and their associated problems for quantitative modeling of various aspects of elemental cycling in estuaries.

THE ESTUARINE ECOSYSTEM(S) AT BEAUFORT, NORTH CAROLINA

Our area of study encompasses nearly 500 km^2 of shallow sounds, embayments, "river", and marshes extending behind the Outer Banks of North Carolina approximately 60 km to the northeast and 35 km to the west of Beaufort, North Carolina. Within that broad area, we have, for logistic reasons, focused our ecological sampling primarily on the Newport River estuary, a 31 km^2 embayment which receives fresh-water run-off from a watershed of about 340 km^2 consisting principally of pine-cypress pocosins, low-lying pine forest, and agricultural lands (Fig. 1). Periodic maintenance dredging is conducted by the U. S. Army Corps of Engineers in the Intracoastal Waterway which connects Bogue Sound through the Newport River estuary to Adams Creek Canal. The Port Terminal turning basin at Morehead City and the channels connecting it to the open ocean through Beaufort Inlet are maintained at a depth of 10-11 m, but most other channels in the area are much shallower, and the mean depth in the Newport River estuary is less than 1 m at low tide. Tidal amplitude is about 0.8 m near the ocean inlets and somewhat less at the heads of the estuaries. For further qualitative descriptions of the area, *see* (11, 69, 85).

To assess the pathways by which metallic elements are transferred between components of the estuarine ecosystem, it is necessary to determine the trophic structure and energy flow of the system. Consequently, we have undertaken a wide variety of ecological and experimental studies to describe the standing crops and productivity of the major biota in the system.

The primary production of the estuarine system results mainly from phytoplankton (60, 69, 71), *Spartina alterniflora* (72, 73), and other macrophytes *Zostera marina,* (15, 64, 71, 86), and benthic algae (22). In the estuarine system near Beaufort, annual net primary productivity averages on the order of 190 g C/m^2 and is probably divided approximately as follows: macrophytes (*Zostera, Halodule,* and *Ectocarpus*) 64%, phytoplankton 28%, and *S. alterniflora* 8% (71). Productivity of benthic microalgae, 6.3 g $C/m^2 \cdot yr$ (22), has not been included in these estimates. Production in high marshes, e.g. *Juncus roemerianus* (72, 75) probably contributes very little to overall estuarine productivity because these areas are flooded irregularly, usually only during storms or spring tides. Phytoplankton productivity is probably limited at certain times by either nitrogen or phosphorus in this system, but N:P ratios suggest that nitrogen is more frequently the limiting nutrient (60, 61).

The standing crop of zooplankton is quite variable and is small relative to

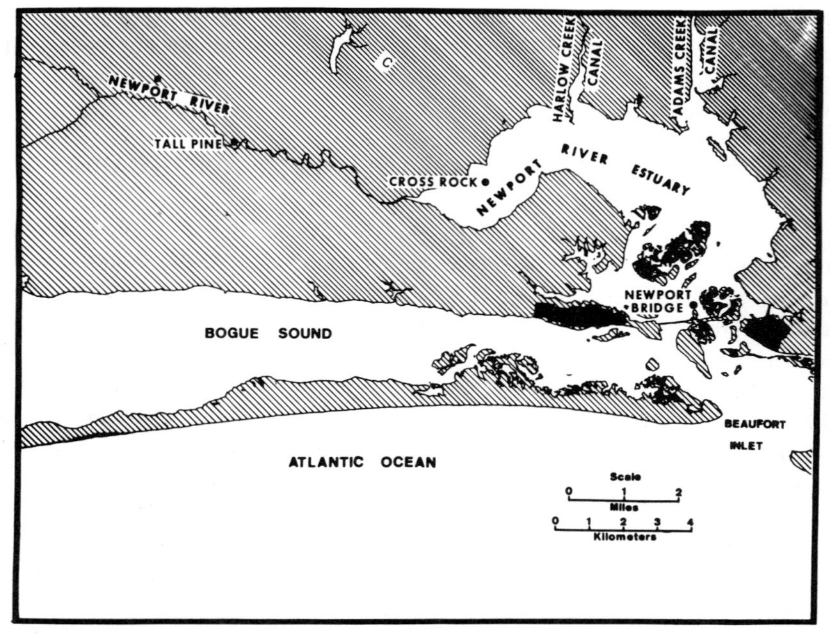

Figure 1. The Newport River Estuary, showing major waterways and stations where samples of sediments and water were taken for elemental analysis (12). Cape Lookout, North Carolina, is located 13 km to the east. For the purposes of the estimates in this paper, the Newport River estuary was arbitrarily delimited on the ocean side at the Morehead City-Beaufort causeway, located immediately south of the Newport Bridge station. Thus, the area connecting Bogue Sound, Newport River, and Beaufort Inlet was excluded from consideration.

theoretical populations which could be supported by the phytoplankton production (77). Thayer et al. (63), however, present evidence that the zooplankton population is sometimes predator-limited, since the fluctuations of zooplankton abundance are related to the seasonal influx of larval fish. Estimated numbers and feeding rates of larvae (31) are sufficient to produce the observed spring decrease in zooplankton numbers. Grazing by ctenophores (38) might also reduce the standing crop of zooplankton.

Less is known concerning the abundance of fish, measurements of larval fish abundance and species composition (63) are reasonably consistent with larval numbers obtained for the Newport River pinfish population (26) and with estimates based on the abundance of juvenile menhaden in the lower channel of the Newport River (85). The distribution and relative abundances of juvenile and adult fishes in the Newport River have been estimated by a variety of fishing gears (65), and preliminary research on the quantitative abundance of fishes and of motile macroinvertebrates, e.g. crabs and shrimp, is in progress (30). Data on the local commerical fisheries shellfish catches, by species and by area of capture, are available from the NMFS Statistics Office, located in Beaufort, North Carolina.

Macrobenthic infauna have been surveyed in one very small intertidal area (76), and are currently the subject of an extensive quantitative sampling program by Price, Williams, and Thayer (unpublished data) encompassing the lower intertidal zone, the extensive shallow portions of the estuary throughout the entire salinity regime, as well as the deep channels. In conjunction with this study, Ferguson (22) is conducting a sampling program for microbial plus meiofaunal biomass distribution, using ATP as a measure of total biomass in sediments and water throughout the estuary. Concurrent chlorophyll analyses provide the basis for microalgae estimates, and the difference provides the heterotrophic biomass.

The eelgrass community represents a specialized assemblage within the overall estuarine system and has received separate attention. Thayer et al. (64) have described the pathways and annual energy flows from eelgrass to detritus to consumers and, ultimately, to the resident fish populations of the community. Eelgrass communities are known to be heavily preyed upon by fishes, shorebirds, and large wading birds, and a study of this aspect of ecosystem energetics is now underway.

Measurements of caloric contents are available for a variety of estuarine animals (62) and are proceeding for other species throughout their life cycles. Rates of food consumption, assimilation efficiency, and growth efficiency, or respiration, must be known, however, if the inputs of energy or metallic elements into any trophic component are to be estimated, and so far, data are quite incomplete, especially for herbivores or detritivores (1). Respiration was measured for most of the major animal species of the eelgrass community, and energy flow between trophic compartments was then calculated using an estimate of mean respiration/growth for all invertebrate species in conjunction with measured consumption and production for the fish (64). By far the most complete data on efficiency of energy utilization are available for fish (25, 28, 31, 45, 46, 48, 49) whose respiration and feeding rate can be described as functions of temperature and food availability for a number of species.

After we have determined the pathways and the rates of energy flow, we must then determine the concentrations of the metallic elements in the system compartments. We have focused our attention primarily on the biologically essential transition elements Mn, Fe, and Zn, and have analytical data on a variety of organisms, including *S. alterniflora* (74), selected mollusks (79, 80) and Cross (unpublished data), selected polychaetes (11), and various size classes for a number of estuarine and marine fishes (9, 13, 14). Trace-element distributions have been followed in the sediments and water at selected stations in the Newport River estuary (11), and the influence of salinity and other environmental factors on the bioaccumulation of trace metals has been noted in both field (81, 83) and laboratory (19, 82) studies using radionuclides. Earlier efforts to understand the flow of metallic elements through the estuarine system were based on the observed behavior of radioisotopes in experimental ponds (16, 17). We are now extending these studies to a general consideration of the entire ecosystem, and are attempting to consider all the important reservoirs of elements and all the significant rate-influencing factors involved (84).

In modeling the amounts of a metallic element present in any biological compartment at any time, we must determine biological turnover rates for the element – a process analogous to respiration in energy-flow models. In numerous laboratory experiments, we have employed radioisotopes to determine biological turnover of metallic elements (2, 3, 27), sometimes as a function of environmental parameters (19, 82). The exchange rate of zinc between estuarine sediments and water has also been estimated using ^{65}Zn as a tracer (18). Further insight on the availability and turnover of elements has been gained from the distribution and retention of fallout isotopes by estuarine organisms (79, 83). We are currently assessing other approaches to determining consumption and assimilation of metals by organisms (14). We have already conducted prototype studies to evaluate the effect of physico-chemical state on biological availability of zinc in estuarine systems (12) and to test the feasibility of using radioisotope accumulation data as the basis for turnover measurements of stable elements (12, 78) and Willis (unpublished data).

INFLUXES AND EFFLUXES OF METALS IN THE NEWPORT RIVER ESTUARY

From our estimates of the standing crops of major organisms in the estuarine system and the elemental concentrations in these organisms, we determined (85) the reservoir sizes for the trace elements Mn, Fe, and Zn in the Newport River estuary. Coupling these data with compositional data on sediments, fresh-water, and estuarine water and with estimates of sedimentation rates, run-off, tidal flushing, seasonal biological emigration, and commercial harvest provided us with crude estimates of some of the more significant fluxes of the elements into and out of the total ecosystem (85). The results of these calculations are presented in the form of flow diagrams in Figure 2.

The derivation of values shown in Figure 2 for standing crop and fluxes of metallic elements for the compartments' water, sediments, *S. alterniflora*, and biota were explained previously (85), except that the standing crops of mollusks (oysters, clams, and scallops) are now thought to have been overestimated. These standing crops were calculated from the commercial yield, but only after a probably excessive adjustment for the unknown non-commerical shellfish yield

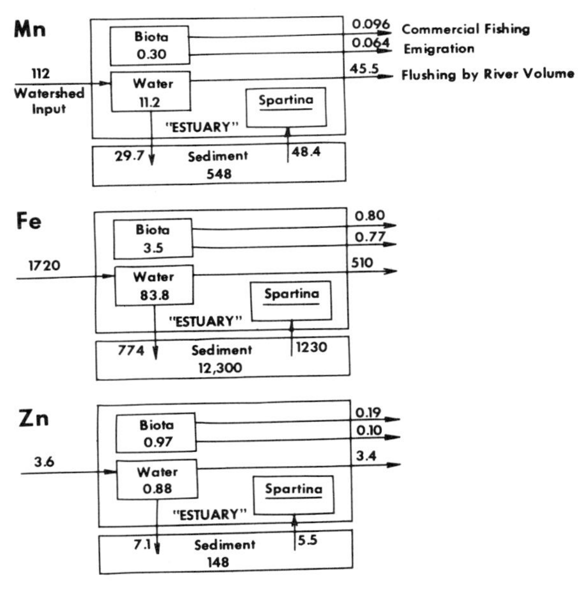

Figure 2. Annual inputs and exports of Mn, Fe, and Zn for the Newport River estuary, modified from Wolfe et al. (85). The "estuary" is considered to have an area of 31 km^2 and a half-tide depth of 1.3 m. Units in boxes are mg element/m^2 and on arrows are mg element/m^2. yr. The sediment compartment represents the acid-extractable element in the top 2 cm only. See text for further explanation.

from the estuary (85). Accordingly, standing crop of clams and oysters were reduced to 30% and that for scallops was reduced to one-third of the values used initially. This revision modified the estimates for Mn and Fe only slightly, but greatly reduced the value for Zn in the standing crop of biota, were much of the zinc occurs in just one species, oysters. The species composition of the biota otherwise remains unchanged from that presented earlier. The lower estimate for annual menhaden catch (i.e., 50,000 kg) from the Newport River estuary was used in computing the flux of metallic elements from the biota to man via the commerical catch, as shown in Figure 2.

RECYCLING WITHIN THE SYSTEM

The estuarine fauna probably obtain their supply of Mn, Fe, and Zn primarily through their food (43). The lower levels of the trophic system are depicted in Figure 3, with provisions for: (a) primary production by phytoplankton and *S. alterniflora,* (b) conversion of part of this production to detritus by herterotrophic microorganisms, (c) direct consumption of a portion of the phytoplankton production by zooplankton and other herbivorous macrobiota,

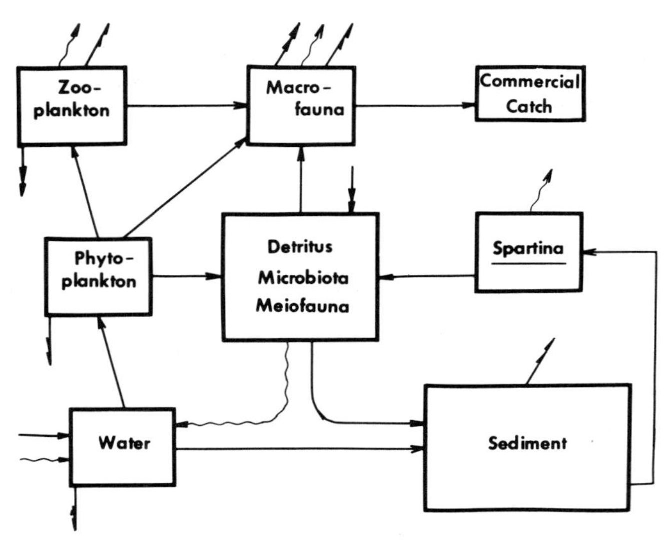

Figure 3. Principal compartments and pathways involved in the transfer of metallic elements in salt-marsh estuaries. Arrows not directly connecting compartments are defined as follows:
1. ➤ = Flushing (river volume only) plus emigration
2. ⁓ = Excretion of soluble metals to water
3. ➶ = Deposition of unassimilated materials to detritus.

and (d) consumption of detritus either directly or indirectly after microbial assimilation. Selection of these compartments and pathways is discussed subsequently.

Primary production has been studied extensively in the Beaufort area. Williams (72) calculates that phytoplankton account for 28%, *S. alterniflora* 8%, and *Z. marina* 64% of the total of these groups in the sounds and estuaries of the area. Effective production of *Z. marina* in the Newport estuary is very difficult to estimate, but is probably much less than Williams' estimate. Thayer et al. (64) estimate that *Z. marina* covers less than 1% of the Newport River estuary, whereas Williams bases his calculations on an average of 17% coverage for the total estuarine area. Since 45% of the grass productivity may not be utilized by the eelgrass macrofaunal community, much of this production is likely exported beyond the limits of the eelgrass beds. Tidal exchange probably imports *Z. marina* detritus into the Newport River estuary, therefore, from the adjacent sounds (64). For estimating metal fluxes, however, I used the minimum value of 1% coverage which, at the mean annual production rates reported (64), would make *Z. marina* account for only 1% of the sum of primary production of phytoplankton plus *S. alterniflora* plus *Z. marina* in the estuary (Table 1). *Z. marina* was therefore omitted from Figure 3.

TABLE 1.

Net annual primary production in Newport River estuary

Source	Effective area	Productivity on effective area	Productivity on total area	Reference
	km^2	---g C/m^2. yr---		
Phytoplankton	31	110	110	(69)
Spartina	6.2	470	94	(74) and Williams personal communication)
Zostera	0.3	95	0.92	(64)

Judging from experimental work with radiophosphorus, both *S. alterniflora* and *Z. marina* obtain their mineral nutrients from subsurface anaerobic sediments, and similar processes appear to work in the case of Zn (36, 37, 51, 52). The subsurface sediment compartment can probably be represented as a nearly infinite compartment of the three trace metals, with exchange mediated by the shallow-surface sediments, except during intermittent periods of direct contact with the over-lying water, such as might occur in subsurface burrows or at times of unusually high scouring or resuspension of surface sediments.

We may assume that phytoplankton, benthic microalgae, macroalgae, *Z. marina* and *S. alterniflora* productivity together constitute the bulk of detritus formation in the Newport River estuary. The intimate association of detritus, sediments, and their attendant microflora and microfauna has thus far precluded separation of these compartments and meaningful individual chemical analysis, but this may not be necessary for our present purposes. The compartment labeled sediments in Figure 3 will be used to include all these indistinguishable reservoirs to a depth of 2 cm.

Williams (unpublished data) has synthesized available data on primary productivity, trophic structure, and fisheries catch into a balanced steady-state annual budget of energy flow in this ecosystem. In his model, a mean ecological efficiency or net growth efficiency of about 18% was required at each trophic interaction in order to sustain the fisheries yield from the system based on production of phytoplankton, *Z. marina,* and *S. alterniflora.* This high result led Williams to believe that primary production from additional sources must be significant in supporting this yield, and we are currently considering approaches for studying benthic algal productivity and detrital inputs from the watershed.

The "biota" or "macrofauna" compartment, as depicted in Figures 2 and 3, includes 16 identified major species, plus 3 "other categories" - i.e., other filter feeders, deposit feeders, and other fish (85). Since these species are in different trophic levels, the flow diagram in Figure 3 is not directly comparable to trophic-level modeling schemes. We have considered constructing a model with 40 interconnecting elemental compartments individually identified (84, 85). Since this conceptual model exceeds our presently available data and capability, we are examining the fluxes outlined in Figure 3 as a more feasible first approach.

SOURCES OF METALS FOR THE BIOTA

The major pathways for metals incorporation by the biota are from the sediments to *S. alterniflora* and from the water to phytoplankton (Fig. 3). The magnitude of the *S. alterniflora* source has already been estimated (74) in Figure 2. In order to estimate metal fluxes through the phytoplankton compartment, values for mean standing crop and annual production in the Newport River estuary (Table 1) were coupled with compositional values for plankton collected in the Gulf of Panama with a no. 20 plankton net, i.e., 0.017 μg Mn/mg C, 3.33 μg Fe/mg C, and 0.50 μg Zn/mg C (35). These values were assumed typical for phytoplankton – an assumption which is compatible with our estimates of total element suspended and dissolved in the water column (11, 85) for manganese and iron, but not for zinc. By this method, the standing crop of phytoplankton plus seston was estimated to contain more zinc (0.95 mg/m^2) than was analyzed in whole unfiltered water (0.88 mg/m^2), whereas for Mn and Fe the phytoplankton estimates represented only small percentages of the total element found in the water column. Furthermore, the productivity of phytoplankton accounted for 1.6%, 21%, and 1500%, respectively, of the values shown in Figure 2 for the Mn, Fe, and Zn inputs to the Newport River estuary from the watershed. Most of the Zn present in the water column is probably associated with plankton or other suspended particulate matter, especially during periods of high productivity. The large annual zinc requirement of the phytoplankton is not met by run-off and therefore zinc must be returned to the water column from excretion and turnover of Zn by the zooplankton and macrofauna, from biological or chemical turnover within the detritus or sediment compartments, or from tidal exchange of sea water. This flux of metals into primary producers also is large relative to the biological fluxes of metals out of the estuary (Fig. 2), suggesting significant losses along the trophic system. We have focused our further evaluation and discussion on zinc alone, and examined possible avenues for this loss from the biota back to water or detritus compartments within the system.

Assimilative Losses

Metallic elements ingested with food materials are assimilated with efficiencies different from that of carbon. Some elements (for example, cerium in fishes) may not be assimilated at all, but passed quickly through the gastrointestinal tract (27, 47). Assimilation efficiencies are not available from the literature for many elements in most organisms. Zinc probably has been studied most intensively, and Table 2 shows representative values for zinc assimilation for various marine organisms feeding on different food types. Zinc assimilation by fish appears to depend on species and on the food type consumed, but is in the range of 5%-36%. Carbon assimilation, however, is generally closer to 90% (46).

Once assimilated, loss of a metallic element may occur through the kidneys into the urine, by hepato-pancreatic secretions back into the gastrointestinal tract, by direct transfer across body surfaces such as gills or integument as, for example, in mucous secretions, or deposition in integumentary layers subsequently lost through molting. Urinary or hepato-pancreatic excretions may be indistinguishable from losses of unassimilated material in the feces. Reichle (53) emphasizes that instantaneous dietary inputs of radiotracers are essential for accurate distinction of assimilated and unassimilated isotope. Even with experimentally feasible short-term intakes of radioisotopes, however, the presence of multiple compartments with rapid turnovers can cause errors in interpretation of the experimental data (8).

Biological Turnover of Assimilated Metals

The retention of assimilated metals usually can be simulated with a multiple exponential function consisting of components with distinguishable turnover rates. If all compartments within an organism are equally labeled, the retention of a radioisotopic tracer can be used to describe the turnover rates and compartment sizes involved. Again, however, if multiple compartments with similar turnover rates are involved, interpretation may be difficult - partly as a result of the intrinsic error in experimental measurements coupled with the statistical resolution of multiple-rate functions from a single loss curve. Similarly, incomplete labeling of all compartments leads to inaccurate estimates of relative compartment size, because long-lived compartments cannot become completely labeled from a short exposure to the radiotracer and their size is thus underestimated relative to those with faster turnovers. Incomplete labeling may result also from exposure only to soluble radiotracer as opposed to a food source. This phenomenon is demonstrated by comparing ^{65}Zn retention by *Mytilus galloprovincialis* after exposures of 1-21 days (29) with the ^{65}Zn retention found for *Mytilus californianus* after life-long exposure to ^{65}Zn from the Columbia River (88). With short expossures, ^{65}Zn retention consists of multiple exponential functions, with the apparent proportion of the longest-lived component increasing with exposures of 1, 7, and 21 days. For *M.*

TABLE 2.

Assimilation of ingested zinc by marine organisms

Species	Fraction assimilated	Conditions	Reference
Meganyctiphanes norvegica	0.1-0.2	Dietary intakes estimated as sum of measured efflux; defecation rate based on **Artemia** diet	(59)
Zooplankton	0.20	Calculated from relative concentration factors of zooplankton and phytoplankton	(33)
Fundulus heteroclitus	0.05	^{65}Zn retention 24 hr after consuming labeled plant "detritus"	Hoss (unpublished data)
Fundulus heteroclitus	0.30	^{65}Zn retention 24 hr after consuming labeled shrimp	Hoss (unpublished data)
Pleuronectes platessa	0.10	^{65}Zn retention 4 days after consuming labeled mussel or **Nereis**	(43)
Pleuronectes platessa	0.28	^{65}Zn retention 4 days after consuming labeled shrimp	(43)
Pleuronectes platessa	0.36	^{65}Zn retention 4 days after consuming labeled **Nereis**	(44)

californianus, however, only a single exponential function was observed with a biological half-life of 76 days (Table 3). In macroorganisms (Crustacea, Mollusca, and Osteichthyes), the biological half-life of the predominant long-lived compartment for zinc probably exceeds 100 days (Table 3) and may approach 2 years in organisms such as large Dover sole *Microstomus pacificus* (67) and the soft parts of oysters *Crassostrea gigas* (57).

Sediment-water Interactions

Zinc is transferred from the detritus-microbe compartment into each of 3 other compartments: (a) sediments, (b) water, and (c) (detritivorous) biota (Fig. 3). This scheme is analogous to that depicted by Pomeroy et al. (52) for phosphorus-cycling in salt-marsh estuaries. These fluxes are intended to represent (a) deposition of detrital material, and microbial or chemical remineralization, followed by release into interstitial waters and adsorption by sediment minerals; (b) microbial or chemical remineralization or excretion of soluble organic-metal ligands into the overlying water column, as well as resuspension of finely divided detrital material; and (c) consumption of detritus by detritivores. Quantitative information on the magnitudes of these fluxes is totally lacking for metallic elements, and indeed there is some uncertainty concerning even the direction of the net annual flux between certain of these compartments.

The net flux from water to sediments in Figure 2 represents an estimated mean sedimentation rate from the water column (85), but does not consider any net fluxes caused by exchange processes on the sediments. For example, Bradford (5) postulated that in Chesapeake Bay, a net release of Zn to the water column from the sediments occurred under certain salinity conditions, whereas Duke et al. (18) found a slow, continuous depletion of zinc in estuarine water maintained under laboratory conditions in contact with estuarine sediments where the instantaneous rate of Zn exchange between the sediments and water was estimated at 17 ± 4 $\mu g\ Zn/m^2 \cdot hr$. Thus, small changes in relative influx-efflux could produce large annual net fluxes in either direction. Under variable regimes of salinity, temperature, and microbial productivity, it is conceivable that the direction of net flux may periodically reverse in the exchange equilibrium between sediments and water.

Ustach (66) found that the turnover of previously incorporated ^{65}Zn from decomposing *S. alterniflora* proceeded with a biological half-life of about 2.6 days compared to 68 days for loss of organic material. Williams and Murdoch (74) found a net increase in Mn, Fe, and Zn in standing dead *S. alterniflora* compared to the living plants, an increase that could not be accounted for simply by loss of organic material. Since Ustach did not measure stable Zn in his samples of decomposing *S. alterniflora,* it is not known whether the high rate of

TABLE 3.

Biological half-lives for zinc turnover in marine organisms

Species	$T_{b½}$ (days)	Conditions	Reference
Mytilus edulis	48-60	$\leq$ 20 days uptake; animals unfed	(68)
Mytilus californianus	76 $\pm$ 3.5	life-long uptake	(88)
Crassostrea gigas	300	life-long uptake	(56)
	300	(late summer to early fall) 3 day uptake; shells excluded	(57)
	850	(late fall to summer)	(57)
Euphausia pacifica	140	15 day uptake	(23)
Anonyx sp.	90-140	6-10 day uptake; inverse temperature relation 3-12° C	(10)
Homarus vulgaris	60-270		(7)
Micropogon undulatus	138	single intraperitoneal injection	(3)
Pleuronectes platessa	295-313	267 day uptake	(43)
Microstomus pacificus	635	free-living population	(67)

turnover was associated with the net accumulation (74) or with a subsequent loss process. If, however, the rapid turnover of ^{65}Zn seen by Ustach represented Zn loss from *S. alterniflora,* then the detritus may not represent a significant source of Zn to detritivores, even through it represents an important component of the trophic system. The high living microbial biomass (both heterotrophic and phototrophic) associated with sediments (22) could provide a retentive capacity for metallic elements in detritus deposited on or buried under estuarine sediments. If estuarine meiofauna exhibit metal assimilation efficiencies similar to those seen for macroorganisms (Table 2), considerable amounts of these elements must be returned directly to the water or sediments as unassimilated material. Thus, large amounts of zinc and other metals from *S. alterniflora* may be remineralized and released to interstitial or overlying water instead of being consumed by the macrobiota.

ESTIMATION OF ZINC FLUX WITHIN NEWPORT RIVER ECOSYSTEM

On the basis of the preceding discussion, annual fluxes can be estimated for most of the flows depicted in Figure 3. For this purpose, several assumptions must be made about trophic structure and flow efficiency within the system. I made the following assumptions:

a. Each consumption step involves an assimilation efficiency of 0.2 with the unassimilated 0.8 being returned to the detrital compartment as feces (Table 2).

b. The zooplankton population regenerates itself an average 18 times per year, or a mean-production/standing-crop ratio of 0.05/day, as estimated from a mean-production/assimilation ratio of 0.4 (39), data on standing crop and respiration of zooplankton (77), a mean zooplankton standing crop of 21 mg dry weight/m^3 (63), and a zinc concentration of 144 ppm on dry weight (Cross, unpublished data).

c. Biological half-life of Zn in zooplankton is 7.6 days (turnover time of 11 days). Effective turnover time was estimated to be at least 3 times that observed (32) in very-short-term ^{65}Zn loss experiments with marine zooplankton.

d. The macrofauna compartment regenerates with a mean-production/standing-crop ratio of 2.0/year (55).

e. An average of 2 trophic interactions is effected on all zinc consumed by the macrofauna compartment. (Thus, net assimilation for the compartment is 0.2 x 0.2 = 0.04).

f. Biological half-life of Zn in macrofauna is 300 days (Table 3).

g. All detrital Zn, whether it originates directly from phytoplankton and *S. alterniflora* or from previously cycled fecal matter, is assimilated by the microbiota and meiofauna with a net efficiency of 40% prior to becoming available to consumers. This is comparable to 25% of the detrital zinc being directly available to consumers, and the remainder undergoing a 0.20 assimilative reduction by meiofauna before macrofaunal consumption, consistent with the conclusion that most detrital organic material is converted to microbial and meiofaunal biomass before consumption (21, 22).

Assumptions a-c established the Zn requirement for the zooplankton, and assumptions a and d-f established the Zn requirement for the macrofauna. The fluxes into and from the detritus compartment were then computed (assuming a steady-state for the phytoplankton compartment) by a set of simultaneous equations which considered also assumption g and the macrofaunal requirement.

This process resulted in the net annual fluxes shown between the compartments in Figure 4. Several features of this annual budget should be mentioned. First, the magnitude of fluxes produced by assumptions a-g are compatible with each other and with fluxes previously estimated independently (85) (Fig. 2). Next, the amount of zinc cycled annually through the trophic structure is approximately 16 times the amount that enters the estuary from the watershed. If most of the unassimilated Zn from the detrital compartment is

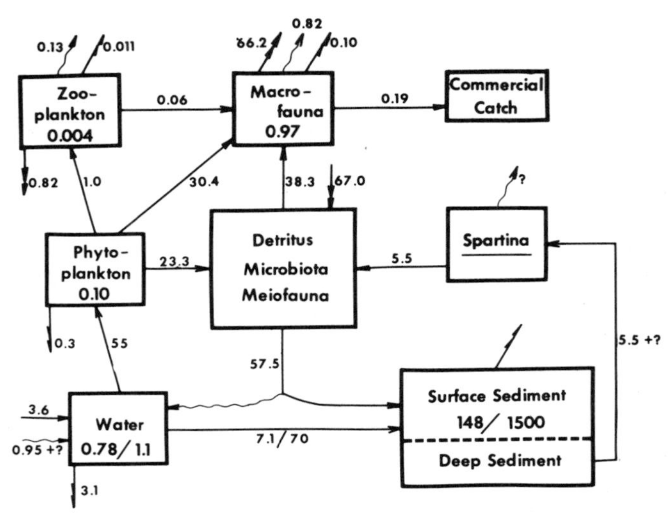

Figure 4. Annual budget of zinc in the Newport River estuary, based on assumption formulated in text. Values inside compartments are mg Zn/m^2 for the 31 km^2 estuary at half-tide depth of 1.3 m. Values on arrows represent annual fluxes between compartments, mg Zn/m^2. yr. Different arrows are defined in legend for Figure 3.

returned to the water column, the biological zinc cycle would represent almost a closed system, requiring only about 0.6 mg Zn/m^2 to replace the Zn lost annually from the system through flushing, emigration, and commerical exploitation. Zn influx to the zooplankton required by assumptions a-c exceeded the outputs caused by excretion, turnover, and flushing by 0.06 $mg/m^2 \cdot yr$ which was assumed to be available for consumption by the macrofauna compartment. An additional 1.45 mg Zn/m^2 consumed by macrofauna is not represented on the outputs from the compartment and would be available to support recreational fishing and other non-commercial yield from the system, such as consumption by birds. Assumption g can be varied over a wide range of net efficiencies (0.24 - 0.56) for transfer of detrital zinc to detritivore food within the detritus-microbiota-meiofauna compartment. As this net assimilation increases, the proportion of zinc influx to biota from detritus increases and that from phytoplankton decreases, with a corresponding diversion of phytoplankton-fixed zinc to detritus.

Microalgal productivity has not been assessed in the development of this scheme because of the lack of definitive information on both production rates (22) and elemental composition. Thus, it cannot be estimated accurately whether the detrital inputs of zinc from phytoplankton, *S. alterniflora*, and fecal deposition are sufficient to support this primary productivity as well as the heterotrophic productivity proceeding within the detritus-microbiota-meiofauna compartment. It is significant, however, that the largest estimated input of zinc to this compartment is associated with a very small input of assimilable carbon, and thus it is believable that the microorganismic requirements may be met by the sources shown in Figure 4. If, on the other hand, benthic microalgal productivity proves to be very large relative to phytoplankton productivity (and zinc concentrations are similar in benthic microalgae and phytoplankton), then zinc must be supplied from interstitial water within the detritus-sediment milieu or from sea water entering from the ocean. Either of these mechanisms would result in a larger flux of zinc from detritus to biota, and an increased diversion of phytoplankton-fixed zinc from biota to detritus.

Thus, zinc is conserved within the Newport River estuary ecosystem. Conservation of zinc within the ecosystem was discussed earlier (79) as a factor contributing to the long effective half-life of ^{65}Zn from fallout observed in oysters (*Crassostrea virginica*) from Beaufort. Specific activity of ^{65}Zn in oysters decreased between 1964 and 1966 with an effective half-life of 276 days. Since the physical half-life of ^{65}Zn is only 245 days, the oysters must have been accumulating ^{65}Zn during this period, even though atmospheric deposition of fallout ^{65}Zn was decreasing simultaneously. In order to maintain a high availability of ^{65}Zn to oysters, the ^{65}Zn must have been recycled effectively within the ecosystem without significant dilution from large pools of stable zinc. This observation, therefore, supports the premise of the "closed system" for zinc in the Newport River estuary, and suggests that neither of the very large reservoirs of potentially available zinc, i.e., the sediments and the ocean, is directly involved in the rapid biological cycle. This conclusion conflicts substantially with the observation mentioned previously that a rapid, instantaneous exchange of zinc occurs between sediment and water (18), and suggests that biologically-available Zn in estuarine water represents a different compartment from that involved in the sediment-water equilibrium. This possibility is discussed further in the following section.

THE PROBLEM OF CHEMICAL FORMS

Different physical or chemical forms of an element may behave in the environment as distinct compartments with different biological availabilities and with slow exchange or equilibration between forms. The presence of more than one elemental form is indicated for the compartments identified as water and sediments in Figure 4. For water, the smaller value is based on the dithizone-reactive zinc in unfiltered, untreated estuarine water, whereas the higher value is the zinc that is reactive after acid treatment. The difference between the two values presumably represents zinc that was strongly adsorbed onto small particulate matter or that was bound in stable soluble organic complexes. For sediments (and for the flux from water to sediment) the two values in Figure 4 represent 0.1 *N* HC1-extractable zinc and total zinc (determined after fusion of the sediment with Na_2CO_3), respectively (11). The acid-extractable zinc is probably more representative of a readily exchangeable element which would be biologically available, whereas the difference between the two values undoubtedly includes much zinc that was firmly trapped within the crystalline lattices of the clay minerals and unavailable to the remainder of the system.

There is a growing body of evidence that zinc occurs in more than one physico-chemical state in sea water, and that organinisms (particularly phytoplankton) accumulate some forms in preference to others (4, 12, 54, 89). Three forms have been distinguished: ionic, 10%-20% of total; particulate, 30%-50%; and complexed, 40%-50% (6, 50). While there is general agreement that exchange and equilibration are very slow between ionic and complexed forms (12, 50), no consensus has yet been achieved on the form or forms that are more available to the biota. Cross et al. (12) hypothesized a large pool of complexed zinc which was highly available to phytoplankton and which prevented equilibration of ^{65}Zn (initially added in the ionic form) with the cells even after 1 year. Bernhard and Zattera (4) found that added ionic ^{65}Zn was somewhat more available and added EDTA-complexed ^{65}Zn was somewhat less available to phytoplankton than stable Zn already present in the medium, leading Piro et al. (50) to postulate that organisms may selectively accumulate ionic zinc. Kuenzler (32) used filtration and ion-exchange techniques to demonstrate that ^{65}Zn eliminated by copepods occurred not only in a particulate form, but also as exchangeable and non-exchangeable soluble zinc. It is obvious that for predictive purposes we must improve our understanding of how the physico-chemical speciation of metallic elements in the environment affects the biological availability at all trophic levels.

APPROACHES TO MODELING THE SYSTEM

NEEDS FOR FURTHER RESEARCH

The annual budget for zinc in the Newport River estuary (Figure 4) identifies several compartments–water, sediments, detritus and attendant organisms, and macrofauna–each of which, in fact, consists of multiple components. To progress from this model to a dynamic mathematical description of metals flow in the ecosystem, certain of these subcompartments might have to be distinguished. Probably the most important distinction required is the separation of chemical forms of zinc according to relative biological availability. If only a fraction of the total zinc present at any time is available to biota and that fraction represents a form of zinc different from that which enters sediment-water processes, then transfer coefficients based on the total zinc will be useless in modeling unless the fractional distribution remains constant all the time. Since we are interested in predicting metal distribution and turnover after contaminative additions to the system, regardless of the form of the added metal, it seems unreasonable to place this constraint upon the model.

The non-living and living components of the detritus compartment can be estimated separately (22), but again, there are major analytical problems in distinguishing the elemental concentrations associated either with these components themselves or with the inorganic phase of the sediment. The subcompartments of the macrofauna can be clustered by feeding type or trophic level. Most of the data required for this distinction are already available, and the two major requirements for modeling elemental flux within the macrofauna are the determination of turnover rates for all relevant chemical forms of the element, and assimilation efficiencies for representative species of each trophic compartment. The flux of metal between biotic compartments usually occurs with consumption, but it will be necessary to distinguish larval fish from adults because of their different feeding habits. In this case, the elemental flux between larvae and adults would be a multiple function of consumption, growth, and survival. If assimilation efficiencies and turnover rates are sufficiently uniform among similar biota (e.g. primary and tertiary carnivores), then trophic levels can readily be combined in the model.

It is apparent that, as depicted in Figure 4, the flux of Zn represents relatively long-term processes and would be valid only for chronic environmental zinc concentrations. If we are to predict such changes in Zn distribution or flux as might occur with an acute environmental release of Zn or ^{65}Zn, we will probably need to consider rapid exchange processes for nearly all compartments, resulting in two-compartment animals (rapid exchange plus bioaccumulation), as discussed by Eberhardt (20). Before designing a model to predict the compartmental distribution and flux of zinc or any other metal in an estuarine ecosystem such as the Newport River, it will be necessary to specify the range of conditions under which it will be applied. O'Neill (42) found that for a simple 3-compartment system, a linear model provided more accurate predictions than did a non-linear form, so long as the system was near steady state. Thus, linear models may suffice for the system at steady-state, while a non-linear approach might be superior for conditions of acute release.

Finally, we are forewarned by O'Neill (42) that the predictive capability of any mathematical model is limited by the basic uncertainty of the scientific measurements on which the model is based. Thus, for very complex systems (such as zinc cycling in the Newport River estuarine ecosystem) it may prove impossible to predict distributions through time in particular biological compartments without considerable expansion of our understanding of the processes involved and major inprovements in our ability to measure the relevant ecological and chemical parameters. Meanwhile, the continuing exercise of formulating our "present" view of trace metals cycling in estuaries helps to identify gaps in our overall understanding of the system, as we attempt to produce dynamic models with predictive capability.

REFERENCES

1. Adams, S. M. and Angelovic, J. W.
1970 Assimilation of detritus and its associated bacteria by three species of estuarine animals. **Chesapeake Sci.,** 11(4): 249-254.
2. Baptist, J. P. and Lewis, C. W.
1969 Transfer of ^{65}Zn and ^{51}Cr through an estuarine food chain. In **Symposium on Radioecology,** p. 420-430. (eds. Nelson, D. J. and Evans, F. C.) USAEC CONF-670503. Oak Ridge, Tennessee.
3. Baptist, J. P., Hoss, D. E., and Lewis, C. W.
1970 Retention of ^{51}Cr, ^{59}Fe, ^{60}Co, ^{65}Zn, ^{85}Sr, ^{95}Nb, ^{114m}In, and ^{131}I by the Atlantic croaker (**Micropogon undulatus**). **Health Phys.,** 18: 141-148.
4. Bernhard, M. and Zattera, A.
1969 A comparison between the uptake of radioactive and stable zinc by a marine unicellular algae. In **Symposium on Radioecology,** p. 389-398. (eds. Nelson, D. J. and Evans, F. C.) USAEC CONF-670503. Oak Ridge, Tennessee.
5. Bradford, W. L.

1972 A study on the chemical behavior of zinc in Chesapeake Bay water using anodic stripping voltammetry. Doctoral dissertation. The Johns Hopkins Univ. 153 pp. Univ. Microfilm 72-24, 958.

6. Branica, M., Bernhard, M., and Piro, A.
Zinc in seawater. II. Determination of physico-chemical states of zinc in seawater. Cited by Piro (50). (In press.)

7. Bryan, G. W.
1966 The metabolism of Zn and ^{65}Zn in crabs, lobsters and freshwater crayfish. In **Radioecological Concentration Processes,** p. 1005-1016. (eds. Aberg, B. and Hungate, F. P.) Pergamon, New York.

8. Conover, R. J. and Francis, V.
1973 The use of radioactive isotopes to measure the transfer of materials in aquatic food chains. **Mar. Biol,** 18: 272-283.

9. Cross, F. A. and Brooks, J. H.
1973 Concentrations of manganese, iron and zinc in juveniles of five estuarine-dependent fishes. In **Proceedings of the Third National Symposium on Radioecology,** p. 769-775. (ed. Nelson, D. J.) USAEC CONF-710501-P2. NTIS, Springfield, Virginia.

10. Cross, F. A., Dean, J. M., and Osterberg, C. L.
1969 The effect of temperature, sediment and feeding on the behavior of four radionuclides in a marine benthic amphipod. In **Symposium on Radioecology,** p. 450-461. (eds. Nelson, D. J. and Evans, F. C.) USAEC CONF-670503. Oak Ridge, Tennessee.

11. Cross, F. A., Duke, T. W., and Willis, J. N.
1970 Biogeochemistry of trace elements in a coastal plain estuary: Distribution of manganese, iron, and zinc in sediments, water andpolychaetous worms. **Chesapeake Sci.,** 11(4): 221-234.

12. Cross, F. A., Willis, J. N., and Baptist, J. P.
1971 Distribution of radioactive and stable zinc in an experimental marine ecosystem. **Jour. Fish. Res. Bd. Canada,** 28(11): 1783-1788.

13. Cross, F. A., Hardy, L. H., Jones, N. Y., and Barber, R. T.
Relationship between total body weight and concentrations of Mn, Fe, Cu, Zn and Hg in white muscle of bluefish (**Pomatomus saltatrix**) and a bathyl demersal fish (**Antimora rostrata**). **Jour. Fish. Res. Bd. Canada.** (In press.)

14. Cross, F. A., Willis, J. N., Hardy, L. H., Jones, N. Y., and Lewis, J.
Role of juvenile fish in cycling of Mn, Fe, Cu and Zn in a coastal plain estuary. This symposium. (In press.)

15. Dillon, R. C.
1971 A comparative study of the primary productivity of estuarine phytoplankton and macrobenthic plants. Doctoral thesis. Univ. North Carolina, Chapel Hill.

16. Duke, T. W.
1967 Possible routes of zinc-65 from an experimental estuarine environment to man. **Jour. Water Pollut. Contr. Fed.,** 39(4): 536-542.

17. Duke, T. W., Willis, J. N., and Price, T. J.
1966 Cycling of trace elements in the estuarine environment. I.

Movement and distribution of zinc 65 and stable zinc in experimental ponds. **Chesapeake Sci.,** 7(1): 1-10.

18. Duke, T. W., Willis, J. N., and Wolfe, D. A.
1965 A technique for studying the exchange of trace elements between estuarine sediments and water. **Limnol. Oceanogr.,** 13(3): 541-545.

19. Duke, T., Willis, J., Price, T., and Fischler, K.
1969 Influence of environmental factors on the concentration of ^{65}Zn by an experimental community. In **Symposium on Radioecology,** p. 355-362. (eds. Nelson, D. J. and Evans, F. C.) USAEC CONF-670503. Oak Ridge, Tennessee.

20. Eberhardt, L. L.
1973 Modeling radionuclides and pesticides in food chains. In **Proceedings of the Third National Symposium on Radioecology,** p. 894-897. (ed. Nelson, D. J.) USAEC CONF-710501-P2. NTIS, Springfield, Virginia.

21. Fenchel, T. M.
1969 The ecology of marine microbenthos. IV. Structure and function of the benthic ecosystem, its chemical and physical factors and the microfauna communities with special reference to the ciliated protaozoa. **Ophelia,** 6:1-182.

22. Ferguson, R. L. and Murdoch, M. B.
Microbial biomass in the Newport River estuary. This symposium. (In press.)

23. Fowler, S. W., Small, L. F., and Dean, J. M.
1971 Experimental studies on elimination of zinc-65, cesium-137 and cerium 144 by euphausiids. **Mar. Biol.,** 8: 224-231.

24. Gordon, G.
1969 **System Simulation.** Prentice-Hall, Inc. Englewood Cliffs, New Jersey. 303 p.

25. Hoss, D. E.
1968 Rates of respiration of estuarine fish. **Proceedings of the 21st Annual Conference, Southeastern Association Game Fish Commissioners,** 1967, p. 416-423.

26. Hoss, D. E.
Energy requirements of a population of pinfish **Lagodon rhomboides** (Linnaeus). **Ecology.** (In press.)

27. Hoss, D. E. and Baptist, J. P.
1973 Accumulation of soluble and particulate radionuclides by estuarine fish. In **Proceedings of the Third National Symposium on Radioecology,** p. 776-782. (ed. Nelson, D. J.) USAEC CONF-710501-P2. NTIS, Springfield, Virginia.

28. Hoss, D. E., Coston, L. C., and Hettler, W. F., Jr.
1971 Effects of increased temperature and postlarval and juvenile estuarine fish. **Proceedings of the 25th Annual Conference, Southeastern Association Game Fish Commissioners,** 1971, p. 635-642.

29. Keckes, S., Ozretic, B., and Krajnovic, M.
1968 Loss of Zn^{65} in the mussel **Mytilus galloprovincialis. Malacologia,** 7(1): 1-6.

30. Kjelson, M. A. and Johnson, G. N.
Description and evaluation of a portable drop-net for sampling nekton populations. **Proceedings of the 27th Annual Conference, Southeastern Association Game Fish Commissioners**, 1973. (In press.)

31. Kjelson, M. A., Peters, D. S., Thayer, G. W., and Johnson, G. N.
The general feeding ecology of larval fishes in the Newport River estuary. U. S. Natl. Mar. Fish. Serv., **Fish. Bull.** (In press.)

32. Kuenzler, E. J.
1969 Elimination of iodine, cobalt, iron and zinc by marine zooplankton. In **Symposium on Radioecology**, p. 462-473. (eds. Nelson, D. J. and Evans, F. C.) USAEC CONF-670503. Oak Ridge, Tennessee.

33. Lowman, F. G., Rice, T. R., and Richards, F. A.
1971 Accumulation and redistribution of radionuclides by marine organisms. In **Radioactivity in the Marine Environment**, p. 161-199. National Academy of Sciences, Washington, D. C.

34. Mann, K. H.
1970 The dynamics of aquatic ecosystems. **Advan. Ecol. Res.**, 6: 1-81.

35. Martin, J. H.
1969 Distribution of C, H, N, P, Fe, Mn, Zn, Ca, Sr, and Sc in plankton samples collected off Panama and Columbia. **Bioscience**, 19(10): 898-901.

36. McRoy, C. P. and Barsdate, R. J.
1970 Phosphate absorption in eelgrass. **Limnol. Oceanogr.**, 15(1): 6-13.

37. McRoy, C. P., Barsdate, R. J., and Nebert, M.
1972 Phosphorus cycling in an eelgrass (**Zostera marina** L.) ecosystem. **Limnol. Oceanogr.**, 17(1): 58-67.

38. Miller, R. J. and Williams, R. B.
1972 Energy requirements and food supplies of ctenophores and jellyfish in the Patuxent River estuary. **Chesapeake Sci.**, 13(4): 328-331.

39. Mullin, M. M.
1969 Production of zooplankton in the ocean: The present status and the problems. **Oceanogr. Mar. Biol. Ann. Rev.**, 7: 293-314.

40. Odum, E. P.
1969 The starategy of ecosystem development. **Science**, 164: 262-270.

41. Odum, H. T., Copeland, B. J., and McMahan, E. A. (eds.)
1969 Coastal ecological systems of the United States. Report to the FWPCA. (Inst. Marine Sciences, Univ. North Carolina). 1878 p.

42. O'Neill, R. V.
1973 Error analysis of ecological models. In **Proceedings of the Third National Symposium on Radioecology**, p. 898-908. (ed. Nelson, D. J.) USAEC CONF-710501-P2. NTIS, Springfield, Virginia.

43. Pentreath, R. J.
1973 The roles of food and water in the accumulation of radionuclides by marine teleost and elasmobranch fish. In **Radioactive Contamination of the Marine Environment**, p. 421-436. International Atomic Energy Agency, Vienna.

44. Pentreath, R. J.
1973 The accumulation and retention of ^{65}Zn and ^{54}Mn by the plaice, **Pleuronectes platessa** L. **Jour. Exp. Mar. Biol. Ecol.,** 12(1): 1-18.

45. Peters, D. S. and Boyd, M. T.
1972 The effect of temperature, salinity, and availability of food on the feeding and growth of the hogchoker **Trinectes maculatus** (Bloch and Schneider). **Jour. Exp. Mar. Biol. Ecol.,** 7: 201-207.

46. Peters, D. S. and Angelovic, J. W.
1973 Effect of temperature, salinity, and food availability on growth and energy utilization of juvenile summer flounder **Paralichthys dentatus.** In **Proceedings of the Third National Symposium on Radioecology,** p. 545-554. (ed. Nelson, D. J.) USAEC CONF-710501-P2. NTIS, Springfield, Virginia.

47. Peters, D. S. and Hoss, D. E.
A radioisotopic method of measuring food evacuation time in fish. **Trans. Am. Fish. Soc.** (In press.)

48. Peters, D. S. and Kjelson, M. L.
Consumption and utilization of food by various postlarval and juvenile estuarine fishes. This Symposium. (In press.)

49. Peters, D. S., Kjelson, M. A., and Boyd, M. T.
The effect of temperature on food evacuation rate in pinfish (**Lagodon rhomboides**), spot (**Leiostomus xanthurus**) and silverside (**Menidia menidia**). **Proceedings of the 26th Annual Conference, Southeastern Association Game Fish Commissioners,** 1972. (In press.)

50. Piro, A., Bernhard, M., Branica, M., and Verzi, M.
1973 Incomplete exchange reaction between radioactive ionic zinc and stable natural zinc in sea water. In **Radioactive Contamination of the Marine Environment,** p. 29-45. International Atomic Energy Agency, Vienna

51. Pomeroy, L. R., Johannes, R. E., Odum, E. P., and Roffman, B.
1969 The phosphate and zinc cycles and productivity of a salt marsh. In **Symposium on Radioecology,** p. 412-419. (eds. Nelson, D. J., and Evans, F. C.) USAEC CONF-670503. Oak Ridge, Tennessee.

52. Pomeroy, L. R., Shenton, L. R., Jones, R. D. H., and Reimold, R. J.
1972 Nutrient flux in estuaries. **Limnol. Oceanogr. Spec. Symp.,** Vol. I: 274-291.

53. Reichle, D. E.
1969 Measurement of elemental assimilation by animals from radioisotope retention patterns. **Ecology,** 50(6): 1102-1104.

54. Rona, E., Hood, D. W., Muse, L., and Buglia, B.
1962 Activation analysis of manganese and zinc in sea water. **Limnol. Oceanogr.,** 7: 201-206.

55. Sanders, H. L.
1956 Oceanography of Long Island Sound, 1952-1954. X. The biology of marine bottom communities. **Bull. Bingham Oceanogr. Coll.,** 15: 345-414.

56. Seymour, A. H.

1966 Accumulation and loss of zinc-65 by oysters in a natural environment. In **Disposal of Radioactive Wastes into Seas, Oceans and Surface Waters,** p. 605-619. International Atomic Energy Agency, Vienna.

57. Seymour, A. H. and Nelson, V. A.
1973 Biological half-lives for zinc and mercury in the Pacific oyster **Crassostrea gigas.** In **Proceedings of the Third National Symposium on Radioecology,** p. 849-856. (ed. Nelson, D. J.) USAEC CONF-710501-P2. NTIS, Springfield, Virginia.

58. Slobodkin, L. B., Smith, F. E., and Hairston, N. G.
1967 Regulation in terrestrial ecosystems, and the implied balance of nature. **Am. Nat.,** 101: 109-124.

59. Small, L. F., Fowler, S. W., and Keckes, S.
1973 Flux of zinc through a macroplanktonic crustecean. In **Radioactive Contamination of the Marine Environment,** p. 437-452. International Atomic Energy Agency, Vienna.

60. Thayer, G. W.
1971 Phytoplankton production and the distribution of nutrients in a shallow unstratified estuarine system near Beaufort, N. C. **Chesapeake Sci.,** 12(4): 240-253.

61. Thayer, G. W.
Identity and regulation of nutrients limiting phytoplankton production in the shallow estuaries near Beaufort, N. C. **Oecologia.** (In press.)

62. Thayer, G. W., Schaaf, W. E., Angelovic, J. W., and Lacroix, M. W.
1973 Caloric measurements of some estuarine organisms. U. S. Natl. Mar. Fish. Serv., **Fish. Bull.,** 71(1): 289-296.

63. Thayer, G. W., Hoss, D. E., Kjelson, M. A., Hettler, W. F., Jr. and Lacroix, M. W.
Biomass of zooplankton in the Newport River Estuary and the influence of postlarval fishes. **Chesapeake Sci.** (In press.)

64. Thayer, G. W., Adams, S. M., and Lacroix, M. W.
Structural and functional aspects of a recently established **Zostera marina** community. This Symposium. (In press.)

65. Turner, W. R. and Johnson, G. N.
Distribution and relative abundance of fishes in Newport River, North Carolina. U. S. Natl. Mar. Fish. Serv., **Spec. Sci. Rep. Fish.** (In press.)

66. Ustach, J. F.
1969 The decomposition of **Spartina alterniflora.** M.S. thesis, North Carolina State Univ., Raleigh. 26 pp.

67. Vanderploeg, H. A.
1973 Rate of zinc uptake by dover sole in the northeast Pacific Ocean: Preliminary model and analysis. In **Proceedings of the Third National Symposium on Radioecology,** p. 840-848. (ed. Nelson, D. J.) USAEC CONF-710501-P2. NTIS, Springfield, Virginia.

68. Van Weers, A. W.
1973 Uptake and loss of ^{65}Zn and ^{60}Co by the mussel **Mytilus edulis** L. In **Radioactive Contamination of the Marine Environment,** p. 385-401. International Atomic Energy Agency, Vienna.

69. Williams, R. B.
1966 Annual phytoplanktonic production in a system of shallow temperate estuaries. In **Some Contemporary Studies in Marine Science,** p. 699-716. (ed. Barnes, H.) George Allen and Unwin Ltd., London.

70. Williams, R. B.
1972 Steady-state equilibriums in simple, nonlinear food webs. In **Systems Analysis and Simulation in Ecology,** Vol. 2, p. 213-240. (ed. Patten, B. C.) Academic Press, New York.

71. Williams, R. B.
Nutrient levels and phytoplankton productivity in the estuary. **Coastal Marsh and Estuary Symposium,** Louisiana State Univ., 17 July 1972. (In press.)

72. Williams, R. B.
Plant productivity in coastal marshes. **Delaware Academy of Science Symposium.** 16 November 1972. (In press.)

73. Williams, R. B. and Murdoch, M. B.
1966 Phytoplankton production and chlorophyll concentration in the Beaufort Channel, North Carolina. **Limnol. Oceanogr.,** 11(1): 73-82.

74. Williams, R. B. and Murdoch, M. B.
1969 The potential importance of **Spartina alterniflora** in conveying zinc, manganese, and iron into estuarine food chains. In **Symposium on Radioecology,** p. 431-439. (eds. Nelson, D. J. and Evans, F. C.) USAEC CONF-670503. Oak Ridge, Tennessee.

75. Williams, R. B. and Murdoch, M. B.
1972 Compartmental analysis of the production of **Juncus roemerianus** in a North Carolina salt marsh. **Chesapeake Sci.,** 13(2): 69-79.

76. Williams, R. B. and Thomas, L. K.
1967 The standing crop of benthic animals in a North Carolina estuarine area. **J. Elisha Mitchell Sci. Soc.,** 83(3): 135-139.

77. Williams, R. B. and Murdoch, M. B., and Thomas, L. K.
1968 Standing crop and importance of zooplankton in a system of shallow estuaries. **Chesapeake Sci.,** 9(1): 42-51.

78. Willis, J. N., III
1973 Specific activities of carrier-free radioisotope preparations. **Int. Jour. Appl. Radiat. Isot.,** 24(6): 354.

79. Wolfe, D. A.
1970 Levels of stable Zn and ^{65}Zn in **Crassostrea virginica** from North Carolina. **Jour. Fish. Res. Bd. Canada,** 27(1): 47-57.

80. Wolfe, D. A.
1970 Zinc enzymes in **Crassostrea virginica. Jour. Fish. Res. Bd. Canada,** 27(1): 59-69.

81. Wolfe, D. A.
1971 Fallout cesium-137 in clams (**Rangia cuneata**) from the Neuse River estuary, North Carolina. **Limnol. Oceanogr.,** 16(5): 797-805.

82. Wolfe, D. A. and Coburn, C. B., Jr.
1970 Influence of salinity and temperature on the accumulation of cesium-137 by an estuarine clam under laboratory conditions.

Health Phys., 18: 499-505.

83. Wolfe, D. A. and Jennings, C. D.
1973 Iron-55 and ruthenium-103 and -106 in the brackish-water clam **Rangia cuneata.** In **Proceedings of the Third National Symposium on Radioecology,** p. 783-790. (ed. Nelson, D. J.) USAEC CONF-710501-P2. NTIS, Springfield, Virginia.

84. Wolfe, D. A. and Rice, T. R.
1972 Cycling of elements in estuaries. U.S. Natl. Mar. Fish. Serv., **Fish. Bull.,** 70(3): 959-972.

85. Wolfe, D. A., Cross, F. A., and Jennings, C. D.
1973 The flux of Mn, Fe, and Zn in an estuarine ecosystem. In **Radioactive Contamination of the Marine Environment,** p. 159-175. International Atomic Energy Agency, Vienna.

86. Wolfe, D. A., Thayer, G. W., and Williams, R. B.
Ecological effects of man's activities on temperate estuarine eelgrass communities. **Coastal Zone Workshop Background Paper,** Woods Hole, Massachusetts, May 22-June 3, 1972. (In press.)

87. Woodwell, G. M.
1970 Effects of pollution on the structure and physiology of ecosystems. **Science,** 168: 429-433.

88. Young, D. R. and Folsom, T. R.
1967 Loss of Zn^{65} from the California sea-mussel **Mytilus californianus. Biol. Bull.,** (Woods Hole) 133: 438-447.

89. Zirino, A. and Healy, M. L.
1971 Voltametric measurement of zinc in the northeastern tropical Pacific Ocean. **Limnol. Oceanogr.,** 16(5): 773-778.

AN ECOLOGICAL SIMULATION MODEL OF NARRAGANSETT BAY – THE PLANKTON COMMUNITY

James N. Kremer[1]

and

Scott W. Nixon[1]

ABSTRACT

A deterministic numerical simulation model is being developed for the Narragansett Bay, Rhode Island, estuarine ecosystem. A hydrodynamic model provides for spatial circulation of the biological and chemical components between eight vertically and horizontally averaged sections of the bay. In addition to tidal circulation and fresh-water input, the local regimes of light and temperature are used as forcing functions. Interactions within and between the compartments are formulated according to current biological and ecological theory and data. Attempts have been made to include a degree of biological detail, while retaining generality. For example, phytoplankton mechanistic formulations include production integrated over depth and time-of-day and physiological adaptation to stochastic variations in light levels. Zooplankton are subdivided into adults, eggs, and juveniles. Feeding occurs on phytoplankton and detrital carbon, as well as on juveniles and eggs by the adults. Daily rations are obtained from the available food supply according to an hyperbolic satiation effect and unrespired assimilation results in egg production by adults. After a temperature-dependent time lag, the eggs hatch into the juvenile compartment where growth toward adulthood is modulated by temperature and available food. Other trophic levels such as ctenophores, meroplankton, fish, and the

1. Graduate School of Oceanography, University of Rhode Island, Kingston, Rhode Island 02881.

benthos are handled less deterministically and act as forces operating on the system. Nutrient regeneration occurs at all levels and provisions are made for the sources of external nutrient input, including sewage, rivers, and tides.

INTRODUCTION

For the past several years there has been a multidisciplinary effort at the University of Rhode Island to develop a mathematical simulation model of man and nature in Narragansett Bay. The emerging result is an eclectic model that contains an input-output treatment of economic activity and waste effluents (13), a two-dimensional hydrodynamic model (20), and our own mechanistic formulation of ecological dynamics. While the structure of the overall model and the coupling of the economic, hydrodynamic, and ecological subsections has been described (33), the formulation of expressions to describe the ecological system remains particularly challenging and elusive. Our decision to take a mechanistic approach to the ecosystem analysis has produced some of the most difficult, but also some of the most interesting, problems. As Patten (37) has observed, "Nothing seems more natural . . . after a half century or so of biological analysis devoted to understanding organisms by understanding their parts, than to try in synthesis to put the understood parts back together again." Of course, there is a broad spectrum of detail, and models that include a great deal of the fine structure of the system probably stand little chance of retrieving emergent properties of system behavior (37). While our model goes further in the elaboration of compartment details and mechanism of response than did those of many earlier efforts (9, 30, 39, 43), it still does not include individual species, internal nutrient pools (8, 17), elaborate grazing formulations (7), or other details of behavior and physiology. We included only as much functional detail as we considered necessary to put the model on a sound experimental and theoretical basis. Sensitivity analysis and verification will undoubtedly modify many of the present concepts and suggest additional interpretations. As anyone involved with this kind of work knows, ecological models change. In many ways, they are as dynamic in their conception as in their execution. The flow diagram and most of the equations of the Narragansett Bay model are tied to the natural history of the bay only through initial conditions, forcing functions, and the selection of values for some constants. There is available (33) a brief description of the ecology of the system being modeled that may put into perspective some of the abstraction that follows.

THE ECOLOGICAL MODEL

The intent of the model is to generate a numerical simulation that will include seasonal and spatial variation for a number of compartments in the bay system.

While we have found short-term diurnal simulation useful in some estuarine models (34), the basic time step of the bay model is one day. Not only is this choice compatible with the time base of most empirically determined rate constants, but it enables simulations on the order of weeks, months, or years. To provide for spatial variation around the bay, the system has been divided into eight large subsections. The choices of initial conditions for biomass, nutrient levels, extinction coefficients, and other variables, as well as for the forcing functions of temperature, sewage input, advective mixing, and some biological compartments, vary appropriately for each subsection. While the flow diagram for the ecological model includes phytoplankton, zooplankton adults, juveniles and eggs, a benthic bivalve, fish, fish larvae, carnivorous ctenophores, ammonia, nitrite, nitrate, phosphate, silicate, and detrital organic matter (Fig. 1), only the plankton and nutrients are mechanistically simulated. The seasonal variation in biomass for higher trophic levels is included as an empirical function of observed patterns. In the model this biomass is coupled with feeding and nutrient-regeneration formulations so that their impact on the simulated compartments can be calculated. Deterministic models for at least two of the

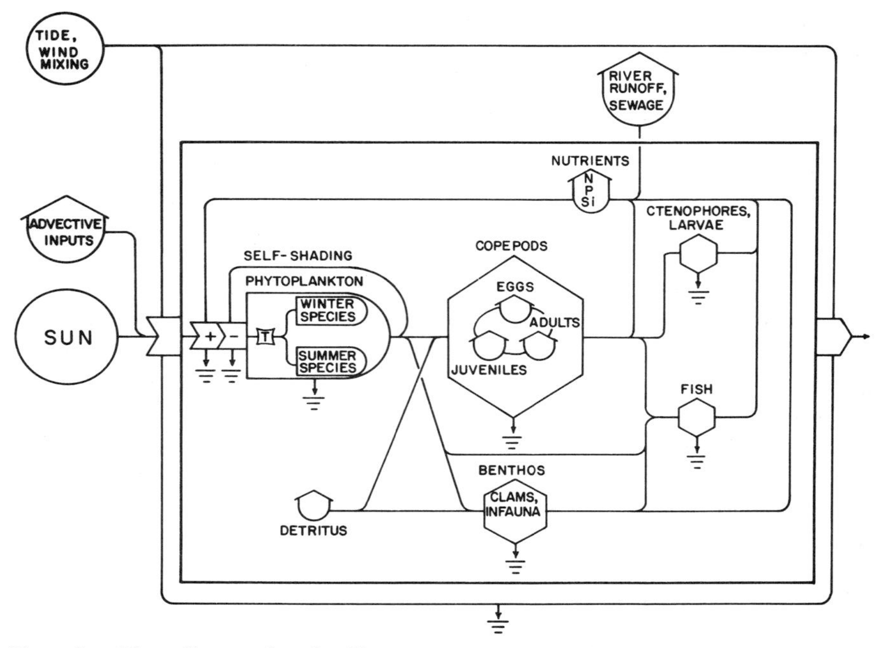

Figure 1. Flow diagram for the Narragansett Bay ecological model. Dynamics of the phytoplankton, zooplankton, and nutrients are mechanistically simulated, while the other biological compartments, and physical and chemical inputs are forces that operate on the system.

higher trophic levels, clams and ctenophores, are now being developed. All the biomass compartments and flows are expressed as carbon, and conservation of mass is observed in all processes. The balance and conservation of nutrients is also maintained throughout the model.

The general procedure in formulating equations for the biological processes has been to specify a maximum rate for each process under optimum conditions. The departure of environmental conditions in the model from this optimum is then determined and the maximum rates multiplied by a series of dimensionless limitation fractions. The result is a series of differential equations which are evaluated each day. The final ecological model is a combination of statistically supported formulations based on experimental data and current theory. An appealing consequence is that the form of the physiological and experimental data used to develop the model is completely different from that of the time-series field data used to test and verify it.

Since the phytoplankton and zooplankton compartments form the basis of this and of many other estuarine models, a detailed discussion of the formulation of these compartments may best indicate the kinds of problems involved in this work and suggest solutions. The importance of circulation in estuarine systems and the fact that it is often omitted from ecological models make it worthwhile to mention some of our problems in coupling hydrodynamic and ecological models. We will not attempt, however, in this paper to discuss the entire model. A discussion of compartments not considered in detail here has already been published (33).

TIDAL CIRCULATION AND MIXING

The hydrodynamic model of circulation in the bay is based on a grid system that contains 324 operational elements, each 925m on a side. With a field that large in a relatively shallow estuary such as Narragansett Bay, the model must run on a time step of about four minutes in order to remain stable and to achieve the desired accuracy in calculations (20). The result is a model with a high computer-time-to-real-time ratio that is not suitable for long-term simulations. For this and other reasons, it was not possible or desirable to couple the ecological model directly to the hydrodynamic equations. Yet it was important to get a good estimate of how the complex circulation in the bay effected the advective transport of suspended and dissolved materials among the eight large ecological subsystems that we had defined. A working solution to the problem was obtained by using a salinity model developed by K. Hess that was based on his detailed hydrodynamic model (20). Each of the small elements in one of the eight large subsections of the bay was assigned a salt concentration of unity, while all the small elements in the remaining seven subsections were initialized at a concentration of zero. Short-term runs of two tidal cycles were then used to estimate the transfer between each subsection and the other seven, as well as across the boundaries. Fortunately, the daily exchange rate between any two subsections appeared to be a virtually linear function of tide height over the range of amplitudes observed in the bay, so that simple regression equations could be determined. These regressions gave the daily exchange coefficient for each pair of subsections in response to the tidal forcing function. While this treatment lacks the hydrodynamic and mathematical elegance of a finite-element model, it is relatively simple and rapid, and appears to give realistic approximations.

PHYTOPLANKTON

Following the procedure of Di Toro et al. (9) and others, the maximum daily growth rate for the phytoplankton is determined as a function of temperature (Fig. 2). As the model is presently being used, this relationship is expressed with an equation given by Eppley (10):

$$G_{MAX} = e^{0.063T-0.16} \tag{1}$$

where G_{MAX} is the daily growth rate (base e), and T is the water temperature (C). The water temperature is calculated from a cosine function chosen to fit the observed seasonal pattern in the bay (after Busser (3)) and modified slightly for each subsection to account for a 3-5 C temperature gradient in the bay:

$$T = 11.0-10.5 \text{ COS } [2\pi \text{ (DAY}-30)/365] \tag{2}$$

where T is the daily water temperature (C), and DAY is the Julian Calendar date. By choosing the exponent in the G_{MAX} equation to represent empirical values for cold-water and warm-water plankton, or for diatom and flagellate species groups, it may be possible to generate an approximate seasonal succession that recognizes two or more types of group physiology. The level of detail chosen could be carried through the model by separating nutrient uptake constants, grazing parameters, and so forth.

Once the maximum growth rate is determined, the rate is depressed according to a complex formulation that includes available light and the relationship between light and growth (Fig. 2). The maximum clear-sky, total daily-incident radiation is calculated for this latitude using a cosine function fitted to the solstices in period and amplitude based on sun-altitude data from U. S. Navy hydrographic tables:

$$RADN_{MAX} = 677.5-371.5 \text{ COS } [2\pi \text{ (DAY}+10)/365] \tag{3}$$

where $RADN_{MAX}$ is the clear-sky radiation (ly/day). The maximum clear-sky radiation is then depressed according to a variable cloud factor based on the monthly means and standard deviations of cloud cover calculated from a 12-year U.S. Weather Bureau record for the Narragansett Bay area. A stochastic cloudiness factor produced daily by a Gaussian number-generator conforming to the monthly distributions is then combined with the clear-sky maximum according to Sverdrup et al. (46):

$$RADN = RADN_{MAX}(1.0-0.71C) \quad (4)$$

where C is the cloudiness factor from 0-1.0.

The result is a pattern of incident radiation, variable on a day-to-day basis within the model, which agrees closely with average seasonal trends in the bay, and which plays an important role in a provision for light acclimation by the phytoplankton that is discussed below. This formulation has the added advantage of allowing atypical years or seasons to be easily simulated by simply altering the cloud-cover statistics.

The analysis of the distribution of this daily radiation over the photoperiod and through the water column, and the resulting effect on the maximum growth rate becomes more complicated. Non-optimum illumination is a major factor in lowered productivity of the phytoplankton. A basic formulation presented by Steele (43) relating photosynthesis of phytoplankton to light intensity has been substantiated and modified by numerous workers (12, 23, 40, 47), and demonstrates the tendency for production to increase approximately in proportion to light at low intensities, approach a maximum rate at an optimum intensity, and decrease again at high-light levels. While many *in situ* measurements of the relationship between light and photosynthesis at all seasons in the bay indicate that surface inhibition may be sharper than that indicated by Steele, his formulation gives a close approximation to our field data. A weakness of many earlier uses of this type of formulation was the arbitrary choice of an optimum light intensity, I_{opt}, which has been experimentally demonstrated to be highly variable within and between species (21, 31, 41). Further, acclimation of populations to varying light intensities is well documented and is potentially an important consideration (44, 49). Steeman Nielsen et al. (44) report that such acclimation occurs within two or three days, the most rapid change taking place during the first day. Our experimental data suggest that fully acclimated natural bay populations frequently demonstrate maximum production at a depth of one meter. Based on this, we calculated the I_{opt} parameter as moving, weighted average of the previous three days mean light intensity at the one-meter depth. Let

$$I' = RADN \cdot e^{-k} \quad (5)$$

where k is the extinction coefficient (1/m), and I' is the incident light at 1 meter (ly/day). Then

$$I_{opt} = (0.7\,\overline{I}_1' + 0.2\,\overline{I}_2' + 0.1\,\overline{I}_3') \tag{6}$$

where $\overline{I}_j'$ is the average light level at 1 meter, j days earlier. Preliminary simulations indicate that the resulting I_{opt} is substantially variable during the year, in contrast to single values used in other models (9).

The extinction coefficient is affected by the phytoplankton present in the water column. The self-shading effect on k of an increase in phytoplankton has been expressed by Chen (4):

$$[k = k_o + 0.17P\,] \tag{7}$$

where k_o = the coefficient with no phytoplankton (1/m), and P is the phytoplankton concentration (mg dry wt/1). This equation agrees with Riley's more complex one (39) for the levels of phytoplankton in the bay.

The remainder of the light-productivity formulation concerns the integration of the photosynthesis-light equation over time-of-day and throughout the water column. Di Toro et al. (9) have developed an equation for the exact double integral of Steele's equation, taking into account the extinction coefficient of the water, k, the photoperiod length as a fraction of the day, f, and the depth of the water column (assuming complete mixing), d. Their formulation, however, assumed a square-wave light input during the photoperiod equal to the daily average. Our analysis of this assumption suggests that, over a wide range of photoperiods and radiation levels, their formulation overestimates production by about 15% when compared to a sinusodial halfwave daily light pattern with the same total energy input. With this correction, the expression for the limitation of the production of the phytoplankton due to non-optimum light, integrated over the water column throughout the day, is:

$$LTLIM = \frac{0.85\ e\ f}{kd}\left[e^{\left(-\frac{I_o}{I_{opt}}\cdot e^{-kd}\right)} - e^{\left(-\frac{I_o}{I_{opt}}\right)} \right] \tag{8}$$

The second factor operating to reduce productivity is nutrient limitation. In our model we consider phosphate, silicate, and two pools of nitrogen (ammonia, and nitrite plus nitrate), using the familiar Michaelis-Menton or Monod formulations from enzyme kinetics. This theory has been extended and used by a number of workers (11, 15, 29, 35) to represent the nutrient limitation on growth of phytoplankton. Based on the choice of the half-saturation constant for growth, K_s, for each nutrient, the growth rate possible with a given nutrient concentration is defined relative to the maximum rate with abundant nutrients:

$$NUTLIM = \frac{G}{G_{max}} = \frac{[n]}{k_{s_n} + [n]} \quad (9)$$

where NUTLIM is the fraction of the maximum growth possible with the nutrient concentration, [n].

It has been suggested that the limitation factors for additional nutrients be multiplied together to yield the combined effect (9, 24). This approach is subject to a number of problems. For example, since the relationship is asymptotic for each nutrient, the respective limitation terms will always be less than 1.0, even with relatively high levels of the nutrient. Thus, simply the attempt to include many nutrients in the model necessarily lowers the productive rate. There is also a potential for error in using Michaelis-Menton equations to represent uptake or growth kinetics of mixed species (48), though this is not a great problem in Narragansett Bay where one or two species strongly dominate the community. More complex nutrient-interaction schemes (8, 17) may be introduced later but, as now written, the model determines which nutrient most limits growth on the basis of K_s and [n]. Since this is an intensive property, the total uptake of the limiting nutrient by the plankton must be calculated and the growth further reduced, if the supply of the limiting nutrient cannot meet the projected demand.

The final expression for phytoplankton growth is a simple combination of light and nutrient limitations (equations 8 and 9) acting on the potential maximum rate (equation 1):

$$G_P = (G_{max})\,(LTLIM)\,(NUTLIM)$$

$$P = P\ e^{(G_p - g_{za + zj} - g_{benthos} - g_{fish})} + Q_p \quad (10)$$

where P is phytoplankton (mg C/1), G_P is the daily growth rate, g represents herbivorous grazing, and Q is advective exchange.

ZOOPLANKTON

In the model the adult zooplankton are assumed to feed upon the eggs and juveniles of their own compartment, as well as on the available detritus and phytoplankton (Heinle, personal communication, and 38) (Fig. 2). The feeding response of copepods to varied food concentrations has been widely studied. Early work emphasized a passive filtering mode and postulated a temperature-dependent volume swept clear regardless of food concentrations (1, 16). More recent studies (30, 36, 45) support a satiation pattern where the

PHYTOPLANKTON COMPARTMENT

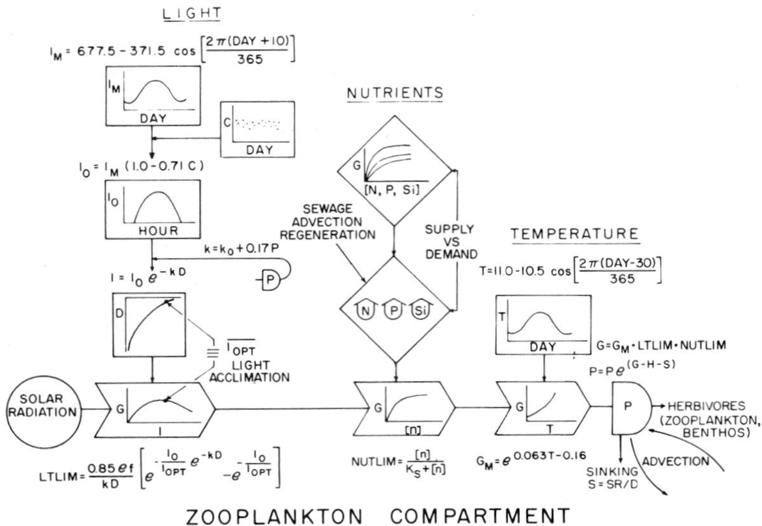

ZOOPLANKTON COMPARTMENT

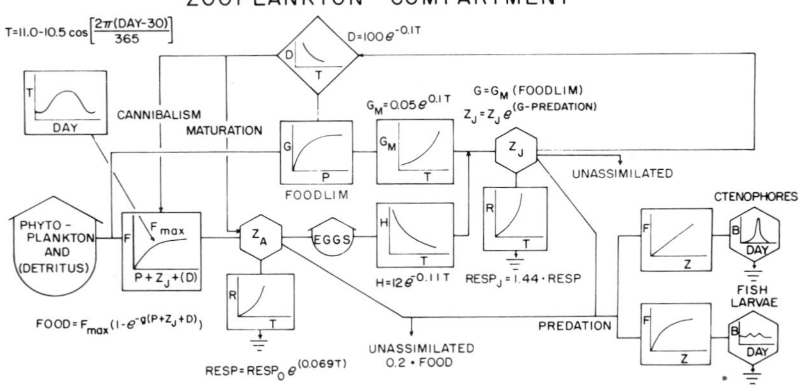

Figure 2. Mathematical and symbolic representation of the flows and controls operating on the phytoplankton and zooplankton compartments of the model. Details in text.

ration approaches a maximum at some critical food concentration, as described for fish by Ivlev (22) with the equation:

$$R = R_{max} (1 - e^{-kP}) \tag{11}$$

where R is the daily ration, less than the maximum, R_{max}, achieved at the food concentration P, and k is a constant controlling the rate of approach to R_{max}. While this form of the equation is sometimes modified by the addition of a threshold food concentration below which feeding does not occur (30, 36), or simplified to two intersecting straight lines (14), its basic mechanism is increasingly used in ecological models.

The reported values in the literature, however, make the application of this equation difficult. Estimates of k and R_{max} that we have made from reported feeding experiments that dealt with this problem are extremely variable, with k ranging from 0.4 (reference 50) to 25.0 (reference 45) 1/mg carbon and R_{max} from 0.013 (reference 45) to 4.30 (reference 42) mg phytoplankton carbon/mg zooplankton carbon/day. Our own experiments on natural populations in Narragansett Bay, while falling well within these ranges, have not resolved this critical choice of parameters, even when the ration is normalized with respect to R_{max}. Figure 3 graphically demonstrates these marked contrasts, though it is important to note that different experimental organisms and conditions undoubtedly contribute to the variability.

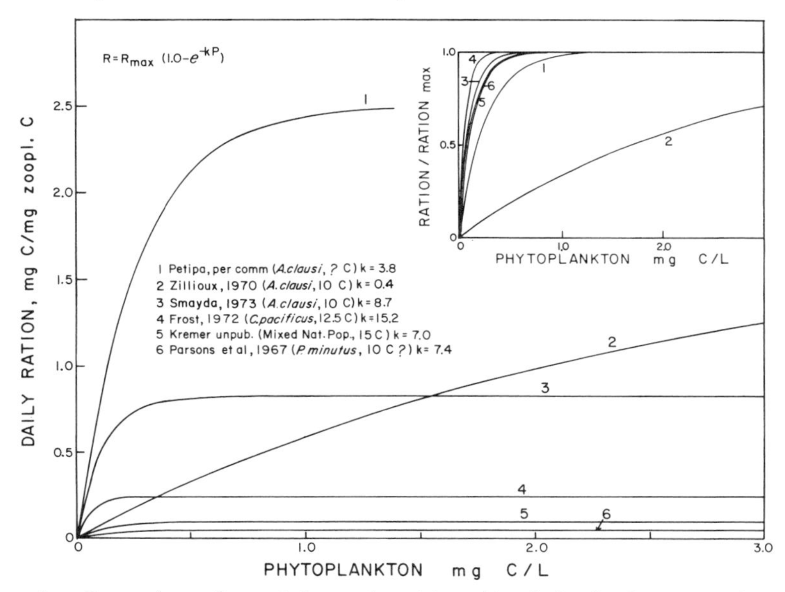

Figure 3. Comparison of reported experimental results relating food concentration to the daily ration of zooplankton. Original data were converted to standard units as accurately as possible and fitted to the feeding equations of Ivlev (22). The insert shows the ration normalized with respect to the maximum, allowing direct comparison of the exponent, k.

A critical assumption implied by using equation 22 is that the food concentration is not lowered by the grazing itself. While this is acceptable during bloom conditions, it is clearly a potential error during critical periods of low or initially increasing phytoplankton. The magnitude of this error and the possibility of an accurate estimate of the integrated effect are under investigation.

An additional complication is the effect of temperature on R_{max}. Although there have been reports of the realized ration varying with temperature for estuarine copepods (38, 42), no direct study of the effect of temperature on the potential R_{max} appears to be available. Species changes due to seasonal succession and individual temperature acclimation probably lower the Q_{10} from the conventional 2.0 for biological reactions, and the sensitivity of models to this parameter must be evaluated. Initially, however, a Q_{10} of 2 was applied to published R_{max} values. This choice is supported by some feeding data on *Acartia clausi* in Narragansett Bay at 4 and 10C (42). The feeding of the zooplankton adults, then, is represented in the model by equation 21, with the maximum ration defined as:

$$R_{max_T} = R_{max_O} \; e^{(0.069\,T)} \qquad (12)$$

A large body of work suggests that assimilation of the ingested ration is remarkably high, from 70%-100% under most natural conditions (6, 25, 28). In line with this, a constant assimilation of 0.80 is used in the model.

The adult respiration rate is described by a simple exponential function of temperature:

$$RESP_T = RESP_0 \; e^{(0.069\,T)} \qquad (13)$$

where $RESP_0$ is the zero degree respiration (1/day) with a Q_{10} of 2. A variety of data are available but again the condition and organisms used in the experiments make a choice of $RESP_0$ difficult. Some authors have reported data apparently supporting a linear rate increase over a 10-20C range (5, 16). However, these data may be approximated by the exponential (equation 13) with an intercept (resp. rate at 0 C) of 0.04-0.07 of the body weight per day. This range exceeds the early estimate by Riley (39) of $Resp_0 = 0.018\ day^{-1}$ and an estimate taking into account the allometric size relationship using the equation of Comita (in reference 6) of 0.006 per day. In general, a $Resp_0$ within the bounds of 1%-8% of the body weight per day can be supported.

Within the adult subsection, the temperature and the available food dictate the maximum and realized ration, respectively. The assimilated fraction is balanced against the respiration requirement resulting in a positive or negative increment. If negative, the adults are effectively starving, and the biomass is reduced accordingly. If the unrespired assimilation is positive, the adults are reproducing and the excess represents egg production.

In the zooplankton formulation to this point, the critical balance between the acquired ration and respiration clearly determines egg formation and ultimately the production of the compartment. The choices of parameters are not unique, and a balance can be struck by reciprocal variations within a reasonable range. In this way extreme values reported for one parameter may be eliminated in view of others. For example, early indications are that R_{max} estimates of greater than 100% of the body weight daily are inconsistent with reasonable choices of k and respiration rate. The high R_{max} estimates from the literature consistently result in an extreme zooplankton peak exceeding the average pattern in the bay (26) by up to two orders of magnitude. We investigated the sensitivity of R_{max} to k with a $RESP_0$ of 7% daily and a Q_{10} of 2 on R_{max} and respiration. The R_{max} at 0 C that results in the reported amplitude is essentially constant at 1%-2% above $RESP_0$ for all values of k above 3 1/mg C, although the precise value for each k is quite sensitive. The most satisfactory simulation so far is with this R_{max}, Q_{10}'s of 1.5 and a k of 25 1/mg C. Extensive interpretation would be premature, but the lower Q_{10} seems reasonable in view of seasonal succession and acclimation, and the high k may indicate that temperature responses are more important in regulating copepod growth rate than food limitation by the Ivlev formulation.

Any excess unrespired adult assimilation enters the pool of incubating eggs. The time required for eggs to hatch has been determined for a number of estuarine species (32) and appears to be an exponential function of temperature:

$$H = 12.0\ e^{-0.110T} \tag{14}$$

where H = hatching time (days)

After this time lag, the appropriate amount of carbon enters the juvenile section of the compartment, where growth begins. Throughout incubation and development, eggs and juveniles are subject to predation by carnivores as well as by adults. Maturation to adulthood involves a daily growth rate operating over a development time, both of which are temperature-dependent. Based on the available data for estuarine species (18, 19), the following equations were developed:

$$D = 100\ e^{-0.10\ T} \tag{15}$$

$$G_{max} = 0.05\ e^{0.10\ T} \tag{16}$$

where D = development time from hatching to maturity (days). G_{max} = maximum daily growth rate (1/day).

While the exponential pattern is strongly supported, the choice of the equation parameters is arbitrary within limits imposed by the data. The reciprocal exponents result in a constant net growth to maturation, regardless of temperature. Seasonal size variations may be reflected by altering this balance. Although the compartment grows as a homogeneous carbon pool in the model, choice of the parameters was facilitated by an extensive analysis of the pattern of an individual's growth from egg to adult. This analysis also revealed a simple estimate of the effect of constantly changing temperature on the predicted total development time. Little error was introduced when the projected development time was taken as the average of the initial estimate, and the estimate was based on the temperature at that later time.

As with the adults, a food limitation effect, based in this case only on the abundance of phytoplankton, may reduce the daily growth from the temperature-dependent maximum potential rate. The net effect of this decreased daily growth factor is simulated by slowing the development to adulthood in proportion to the degree of food limitation. Since this correction is not predictable in the context of the model, as was the effect of changing temperatures on the development time, a small error is unavoidably introduced in the accuracy of the development-time estimate.

Using allometric respiration estimates (6) throughout the development stages, an average rate of 1.44 times the respiration of the adults was calculated for the juvenile compartment. This respiration is combined with the adjusted growth rate and a correction for 80% assimilation to calculate the projected ration. Since this estimate does not consider zooplankton standing stock, finite difference approximations of the total grazing impact may exceed the available food supply. To avoid this, the ration requirements of the juveniles, as well as of the adults and carnivores, are expressed as a filtering rate for the total population with respect to the available food supply. This instantaneous rate is then applied over the daily interval to determine the final ration. The realized ration apportioned throughout the juvenile compartment represents the final growth before predation:

$$RTN_{estimate} = [\,(G_{ZJ_{est.}} + RESP)/0.8]\; ZJ \tag{17}$$

$$F = RTN_{estimate}/FOOD \tag{18}$$

$$RTN_{realized} = FOOD(1.0 - e^{F}) \tag{19}$$

$$G_{ZJ} = (0.8\; RTN_{realized})/(ZJ) - RESP \tag{20}$$

where F is the instantaneous daily grazing rate by the juveniles. Except for time-lag considerations, then, the equations for the zooplankton compartment are:

$$\text{ADULTS:}\quad ZA = ZA\, e^{(-g_{carn.} - [RESP])} + ZJ_{mature} + Q_{ZA} \qquad (21)$$

$$\text{EGGS:}\quad ZE = ZE\, e^{(-g_{carn.} - g_{ZA})} + [URA_{ZA}] - ZE_{hatch} + Q_{ZE} \qquad (22)$$

$$\text{JUVENILES:}\quad G_{ZJ} = (G_{max})\,(FOODLIM)$$

$$ZJ = ZJ\; e^{(G_{ZJ} - g_{carn.} - g_{ZA})} - ZJ_{mature} + Q_{ZJ} \qquad (23)$$

where ZJ_{mature} is the portion of juveniles maturing, ZE_{hatch} is the portion of eggs hatching, g is grazing attributed to adults or other carnivores, RESP is respiration of the adults (juvenile respiration is implicit in G_{MAX}), URA is unrespired adult assimilation, if any, Q is the advective influence on each compartment, and [] represent a contribution depending on the balance of assimilation and respiration of the adults.

HIGHER TROPHIC LEVELS

A variety of species make up the meroplankton of the bay and exert predation pressure on the smaller zooplankton. The most important members of the compartment are larval fish and the ctenophore, *Mnemiopsis ledyii.* The abundance of fish larvae shows a sequence of pulses throughout the year, while the ctenophore develops one or two large sharp peaks in late summer. Both populations are input to the model as step-functions that approximate field observations (Fig. 2). The ctenophores are represented as passive filterers based on laboratory data (P. Kremer, personal communication 27) but the fish larvae feed with a temperature-dependent maximum ration reduced by the availability of food in a manner similar to the copepods (2, 22).

The benthos in Narragansett Bay is dominated by a large population of hard clams which, during the warmer months, has a substantial impact on the phytoplankton. Temperature-dependent filtering rates and an energetic loss-gain estimate based on the available food concentration determines this flow within the model (33).

While the modeling is far from completed, initial indications suggest that stability will not be a problem. The equations respond within expected ranges and appropriate seasonal patterns with reasonable parameter choices. However, a great deal of analysis is required before final verification against our field data and, optimistically, the simulation of a variety of manipulative experiments.

ACKNOWLEDGMENTS

As with any large ecosystem study, many people have contributed their time and help to our modeling project. In the work reported here, we are particularly grateful to Frank White, George Brown, Bill Palm , Kurt Hess, and Fred Short for help with the physical circulation and with some aspects of mathematical analysis. Patricia Kremer contributed much to the development of the model and made her unpublished data on ctenophores available to us. Unpublished data on zooplankton growth and development was provided by Donald Heinle, Ed Zillioux, and T. S. Petipa. Candace Oviatt, Perry Jeffries, Harlan Lampe, Ted Smayda, Nelson Marshall, and others made numerous suggestions that were helpful in developing the model. Terry Smith assisted with data-processing and sensitivity-analysis. The figures were prepared by Mickey Leonard. The modeling project is supported by the Office of Sea Grant Programs, NOAA.

REFERENCES

1. Anraku, M.
1964 Influences of the Cape Cod Canal on the hydrography and on the copepods in Buzzards Bay and Cape Cod Bay, Massachusetts. II: Respiration and feeding. **Limnol. Oceanogr.,** 9: 195-206.

2. Blaxter, J. H. S.
1965 The feeding of Herring larvae and their ecology in relation to feeding. **California Cooperative Oceanic Fisheries Investigations Reports,** 10: 79-88.

3. Busser, J.
1967 Number language in natural history using examples of the phytoplankton and Narragansett Bay. Doctoral thesis, Graduate School of Oceanography, Univ. Rhode Island.

4. Chen, C. W.
1970 Concepts and utilities of ecologic modeling. **Jour. of Sanitary Engineering Division,** ASCE, Vol. 96, No. SA5.

5. Conover, R. J.
1956 Biology of **Acartia clausi** and **A. tonsa. Bull. Bingh. Ocean. Coll.,** 15: 156-233.

6. Conover, R. J.
1968 Zooplankton – life in a nutritionally dilute environment. **Am. Zool.,** 8: 107-118.

7. Cushing, D. H.
1968 Grazing by herbivorous copepods in the sea. **Jour. Cons. perm. int. Explor. Mar.,** 32: 75-82.

8. Droop, M. R.
1973 Nutrient limitation in osmotrophic protista. **Am. Zool.,** 13: 209-214.

9. Di Toro, D. M., O'Connor, D. J., and Thomann, R. V.
1971 A dynamic model of phytoplankton populations in the Sacramento-San Joaquin delta. **Adv. in Chem. Series,** 106: 131-180.

10. Eppley, R. W.
1972 Temperature and phytoplankton in the sea. **Fish. Bull.,** 70: 1063-1085.

11. Eppley, R. W., Carlucci, A. F., Holm-Hansen, O., Kiefer, D., McCarthy, J. J., Venrick, E., And Williams, P. M.
1971 Phytoplankton growth and composition in shipboard cultures supplied with nitrate, ammonium, or urea as the nitrogen source. **Limnol. Oceanogr.,** 16: 741-751.

12. Fee, E. J.
1969 A numerical model for the estimation of photosynthetic production, integrated over time and depth, in natural waters. **Limnol. Oceanogr.,** 14(6): 906-911.

13. Feld, S. and Rorholm, N.
Economic growth and generation of waterborne wastes-Narragansett Bay, Rhode Island. Sea Grant Publications Series, Univ. Rhode Island, Kingston, Rhode Island. (In press.)

14. Frost, B. W.
1972 Effects of size and concentration of food particles on the feeding behavior of the marine planktonic copepod **Calanus pacificus. Limnol. Oceanogr.,** 17: 805-815.

15. Fuhs, G., Demmerle, S., Canelli, E., and Chen, M.
1972 Characterization of P-limited plankton algae. **Nutrients and Eutrophication. Limnol. Oceanogr. Special Symposia,** Vol. 1, 113-132 p.

16. Gauld, D. T. and Raymont, J. E. G.
1953 The respiration of some planktonic copepods II. The effect of temperature. **Jour. Mar. Biol. Assoc. U. K.** 31: 447-460.

17. Grenney, W. J., Bella, D. A., and Curl, H. C., Jr.
1973 A theoretical approach to interspecific competition in phytoplankton communities. **Am. Naturalist,** 107: 405-425.

18. Greze, V. N. and Baldina, E. P.
1964 Population dynamics and annual production of **Acartia clausi** Gieshr. and **Centrophages kroyeri** Giesbr. in the neritic zone of the Black Sea. **Trudy Sevastopal' skoi Biologicheskio stantsii,** Akad. Nauk Ukrain. SSR 17: 299-261. (Fish Res. Bd. Canada Trans. 1967).

19. Heinle, D. R.
1969 The effects of temperature on the population dynamics of estuarine copepods. Doctoral thesis, Univ. Maryland.

20. Hess, K. and White, F.
A numerical tidal model of Narragansett Bay. (In press.)

21. Ignatiades, L. and Smayda, T. J.
1970 Autecological studies on the marine diatom **Rhizosolenia fragilissima** Bergon. I. The influence of light, temperature, and salinity. **Jour. Phycology,** 6: 332-339.

22. Ivlev, V. S.
1955 **Experimental ecology and nutrition of fishes.** Pishchepromizdat, Moscow. (Trans. Scott, D.) Yale Univ. Press, New Haven, 1961.

23. Jitts, H. R., McAllister, E. D., Stephens, K., and Strickland, J. D. H.
1964 The cell division rates of some marine phytoplankters as a function of light and temperature. **Jour. Fish. Res. Bd. Canada,** 21: 139-157.

24. Kelly, R. A.
1973 Conceptual ecological model of Delaware estuary. In **Systems analysis and simulation in ecology,** Vol. 3. (ed. Patten, B. C.) Academic Press, New York (In press.)

25. Marshall, S. M. and Orr, A. P.
1965 On the biology of **Calanus finmarchicus** VIII. Food uptake, assimilation and excretion in adult and stage V **Calanus. Jour. Mar. Biol. Assoc. U.K.,** 34: 495-529.

26. Martin, J. H.
1965 Phytoplankton-zooplankton relationships in Narragansett Bay. **Limnol. Oceanogr.,** 10: 185-191.

27. Miller, R. J.
1970 Distribution and energetics of an estuarine population of the ctenophores, **Mnemiopsis leidyi.** Doctoral thesis, North Carolina State Univ. Raleigh, North Carolina.

28. Mullin, M. M.
1969 Production of zooplankton in the ocean: the present status and problems. **Oceanogr. Mar. Biol. Ann. Rev.,** 7: 293-314.

29. MacIsaac, J. J. and Dugdale, R. C.
1969 The kinetics of nitrate and ammonia uptake by natural populations of marine phytoplankton. **Deep-Sea Res.,** 16: 45-57.

30. McAllister, C. D.
1970 Zooplankton rations, phytoplankton mortality and the estimation of marine production. In **Marine food chains,** p. 419-457. (ed. Steele, J. H.) Univ. California Press, Berkeley and Los Angeles.

31. McAllister, C. D., Shah, N., and Strickland, J. D. H.
1964 Marine phytoplankton photosynthesis as a function of light intensity: A comparison of methods. **Jour. Fish. Res. Bd. Canada,** 21: 159-181.

32. McLaren, I. A., Corkett, C. J., and Zillioux, E. J.
1969 Temperature adaptation of copepod eggs from the Arctic to the Tropics, Unpub. manuscript, Nat. Mar. Water Qual. Lab., EPA, Narragansett, Rhode Island.

33. Nixon, S. W. and Kremer, J. N.
Narragansett Bay – The development of a composite simulation model for a New England estuary. In **Models as Ecological Tools: Theory and Case Histories.** (eds. Hall, C. and Day, J.) William Kaufmann, San Francisco. (In press.)

34. Nixon, S. W., Oviatt, C. A., and Kremer, J. N.
1973 A simulation model of diurnal dissolved oxygen. In S. W. Nixon and C. A. Oviatt, **Ecology of a New England saltmarsh. Ecological Monogr.,** (In press.)

35. Paasche. E.
1973 Silicon and the ecology of marine plankton diatoms. I. **Thalassiosira pseudonama** grown in a chemostat with silicate as limiting nutrient. **Marine Biology,** 19: 117-126.

36. Parsons, T. R., LeBrasseur, R. J., and Fulton, J. D.
1967 Some observations on the dependence of zooplankton grazing on the cell size and concentration of phytoplankton blooms. **Jour. of Ocean. Soc. Jap.,** 23: 10-17.

37. Patten, B. C.
1971 **Systems Analysis and Simulation in Ecology,** Vol. 1. Academic Press, New York.

38. Petipa, T. S.
1966 Relationship between growth, energy metabolism, and ration in **A. clausi.** Physiol. of marine animals. p. 82-91. (Trans. Paranjape, M. A.) **Akademiya Nauk SSSR,** Oceanographical Commission.

39. Riley, G. A.
1963 Theory of food-chain relations in the ocean. In **The Sea,** Vol. 2, p. 438-463. (ed. Hill, M. N.) Interscience, New York.

40. Ryther, J. H.
1956 Photosynthesis in the ocean as a function of light intensity. **Limnol. Oceanogr.,** 1: 61-70.

41. Smayda, T. J.
1969 Experimental observations on the influence of temperature, light and salinity on cell division of the marine diatom, **Detronula confervacea** (Cleve) Gran. **Jour. Phycology,** 5: 150-157.

42. Smayda, T. J.
1973 The growth of **Skeletonema costatum** during a winter-spring bloom in Narragansett Bay, Rhode Island. **Norw. Jour. Bot.,** 20: 219-247.

43. Steele, J. H.
1965 Notes on some theoretical problems in production ecology. In **Primary production in aquatic environments,** p. 383-398. (ed. Goldman, C. R.) Univ. California Press, Berkeley.

44. Steeman Nielsen, E., Hansen, V. K., and Jorgensen, E. G.
1962 The adaptation to different light intensities in **Chlorella vulgaris** and the time dependence on transfer to a new intensity. **Physiologia Plantarum,** 15: 505-1517.

45. Sushchenya, L. M.
1970 Food rations, metabolism and growth of crustaceans. In **Marine food chains,** p. 127-141. (ed. Steele, J. H.) Univ. California Press, Berkeley, and Los Angeles.

46. Sverdrup, H. V., Hohnson, M. W., and Fleming, R. H.
1942 **The Oceans,** Prentic-Hall, Englewood Cliffs, New Jersey.

47. Vollenweider, R. A.
1965 Calculation models of photosynthesis-depth curves and some implications regarding day rate estimates in primary production measurements. In **Primary production in aquatic environments,** p. 427-457. (ed. Goldman, C. R.). Univ. California press, Berkeley.

48. Williams, P. J. Le B.
1973 The validity of the application of simple kinetic analysis to heterogeneous microbial populations. **Limnol. Oceanogr.,** 18: 159-164.

49. Yentsch, C. S. and Lee, R. W.
1966 A study of photosynthetic light reactions, and a new interpretation of sun and shade phytoplankton. **Jour Mar. Res.,** 24: 319-337.

50. Zillioux, E. J.
1970 Ingestion and assimilation in laboratory cultures of Acartia. Tech. Rep., Nat. Mar. Water Qual. Lab., EPA, Narragansett, Rhode Island. (Unpublished data.)

A TROPHIC LEVEL ECOSYSTEM MODEL ANALYSIS OF THE PLANKTON COMMUNITY IN A SHALLOW-WATER SUBTROPICAL ESTUARINE EMBAYMENT

by

John Caperon[1]

ABSTRACT

A non-linear, five-component (nutrient, plant, herbivore, carnivore, and detritus) model is developed for the plankton community in a shallow-water marine environment, Kaneohe Bay, Hawaii. Predicted steady-state standing stocks for each component are compared with observed values in the object ecosystem, and the two sets of values agree very well. This is a non-trivial requirement in model validation. A similar comparison with the predictions of a six-component system of the same form, but including both a primary and secondary carnivore level, showed much poorer agreement with observed values, indicating that the community is functioning primarily as a one-carnivore ecosystem.

The steady-state response of both the five-component model and the six-component model to different levels of the dissolved-nutrient input rate to the system are compared qualitatively with ecologically similar adjacent communities with different enrichment levels. Here again the comparison argues compellingly for a one-carnivore, five-component system. Attention is drawn to the fact that the nutrient-plant response to enrichment in the context of both a five-component system and a six-component system is distinctly different from what a phytoplankton ecologist who is habitually preoccupied with two-component plant-nutrient systems might expect.

1. University of Hawaii, Hawaii Institute of Marine Biology, Coconut Island, P. O. Box 1346, Kaneohe, Hawaii 96744.

INTRODUCTION

A trophic-level ecosystem model, much like that described by Smith (34) has been developed for the plankton community of a shallow-water, marine environment, Kaneohe Bay. We did this only because initially we wanted to understand, quantitatively, the distribution in time and space of the phytoplankton. This meant understanding the time-space distribution of the phytoplankton grazers, which in turn required a similar knowledge of the carnivores; it appears that no major component can be omitted, since knowledge of the dynamics of any trophic level requires knowledge of the rest of the ecosystem in nearly the same detail. Our earliest attempts at understanding phytoplankton in the bay were successful only when the data were examined in the context of the whole planktonic ecosystem (6). We know of no short cut.

The trophic-level approach was selected because it appeared to be the simplest model that met our requirement for a non-linear system. It was a natural choice because it is very difficult to make growth-rate observations of component phytoplankton populations and they are almost never made. One has to begin at the bottom of the food chain with trophic-level data and look for similar data at the herbivore and carnivore levels. It is generally as easy or easier to make the kind of field observations necessary to develop mass-balance equations at the trophic level than at the population level. With this approach we were dealing with a rather small set of tractable equations, and much of our data on the ecosystem are in a form directly compatible with a trophic-level model. Also, there was a compelling interest in finding out whether or not the kind of generalization required by this type of model would be valid – at least in a descriptively useful sense. If so, then the implications of differences in food-chain length are quite profound and must receive ecological attention (34). Only a rather approximate description of the ecosystem is required to find out whether or not it functions principally as a one- or a two-carnivore system: the responses of the two are so very different at every trophic level.

Kaneohe Bay is a subtropical embayment approximately 55 square km in area (12.7 x 4.3 km) with depths to 15 meters (Fig. 1). It is separated from the open ocean by the Mokapu Peninsula and by an extensive coral reef. The southeastern sector, as shown in Figure 1, has rather restricted communication with the rest of the bay and the open ocean, and receives municipal waste from two outfalls. The discharge rate from these two sources has more than doubled during the past decade, with the result that this end of the bay is highly eutrophic. The northern sector enjoys relatively good communication with the open ocean and is free of significant waste discharge. There is thus a strong persistent gradient from eutrophy in the southeast to oligotrophy in the northwest (6).

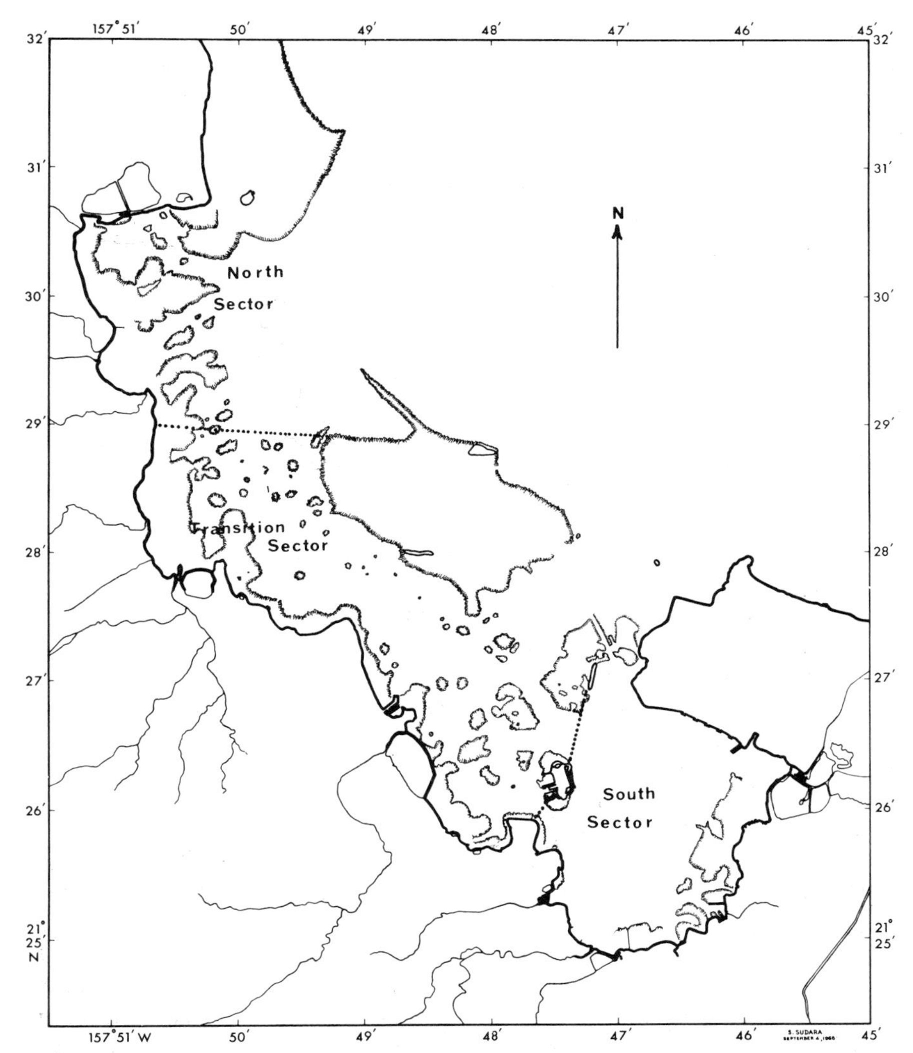

Figure 1. Kaneohe Bay. The three sectors used in the analysis are defined by the shoreline and the dotted partitions.

The watershed of the bay measures approximately 46.6 square km, and is drained by a series of small streams of highly variable discharge depending on rainfall with a strong seasonal component to the variability. These streams enter the bay at several points along its long axis. Density stratification of the bay water caused by fresh-water run-off occurs only after heavy rain, which is occasional, and, under normal trade-wind conditions, lasts only for a day or two.

The run-off is a second source of nutrient enrichment that can be distinguished from municipal waste in the bay water by its seasonal characteristics, by the spatial distribution of the inputs, and by the fact that the nitrogen in run-off is in the nitrate ion, as distinct from the ammonium ion characteristic of municipal waste.

The ecology of the zooplankton community in the bay is relatively well studied. Piyakarnchana (25), Peterson (24), and Clutter (8), have done general studies of the zooplankton. While these do not include direct observation of growth, feeding, or nutrient-regeneration rates they do provide a rather complete description of the species composition, distribution, and abundance. Rowe (32) has studied the feeding rate of the ctenophore *Pleurobrachia pileus* – an important carnivore in the bay – and Szyper, (37) studied the feeding rate of herbivorous zooplankton from the bay – principally *Paracalanus parvus* and *Oithona* sp.

General studies of phytoplankton primary productivity by Doty and Capurro (10), Caperon et al. (6), and Krasnick (16) have been completed, and this work is presently being extended by other workers. Murphy (22) has described the species composition, distribution, and relative abundance of the phytoplankton in the bay.

An extensive study of the physical oceanography is available in Bathen (3), which describes the general circulation within the bay and gives temporal and spatial descriptions of temperature, salinity, and phosphate. Over a period of eight months, Young, Burbank, and Morphew (39) measured the nutrients – ammonia, nitrate, nitrite, phosphate, dissolved organic phosphorus, dissolved organic nitrogen, and oxygen – at ten stations throughout the bay and in all major streams that empty into the bay. Steinhilper (35) studied the particulate organic carbon and nitrogen throughout the bay over a three-month period coincident with the primary productivity study of Caperon, et al. (6).

These studies brought our understanding of the plankton community to the point where a preliminary ecosystem model could be constructed and its validity checked against field observations. It appeared (6) that the bay plankton could be treated as three relatively homogenous communities: a highly eutrophic south sector and a relatively oligotrophic north sector separated by a transition sector in mid-bay (Fig. 1). The present model makes no finer analysis of the spatial variability treating each of these three sectors as well-mixed reactors. We focused our attention on the southern sector, using the transition and northern sectors only as qualitatively representative of what the southern sector was like during earlier stages of the eutrophication process.

Interest in the study has been stimulated by the decision to discontinue waste discharge into the bay in approximately two years. Thus we had the opportunity to study a relatively well-documented ecosystem that had been and would be subjected to large-scale perturbations, one that affords both spatial and temporal control environments relative to the eutrophication process. An opportunity to study a natural ecosystem under both steady-state and perturbed conditions is rate, and, if it did nothing else, the model focused our attention on the dynamics of the system with a precision and acuity not likely to be achieved otherwise.

MODEL DESCRIPTION

Ecosystems can be as simple as a single-species bacterial population whose growth rate is limited by supply of an essential nutrient, or as complex as a community of several thousand interacting populations. Simple two-component systems (population-substrate) have been extensively studied, both theoretically and experimentally, and two quite independent lines of investigation have converged upon the same general relationship between the limiting substrate concentration in the environment and the rate of substrate utilization by the dependent population.

Microbiologists using the Michaelis-Menten enzyme kinetics paradigm have found that the hyperbola

$$y = ax/(b+x) \tag{1}$$

adequately describes specific substrate utilization rate, y, as a function of substrate concentration, x, where a and b are constants. This equation has proven adequate for single-species populations of bacteria, yeast, unicellular algae, and protozoan ciliates utilizing such diverse kinds of substrate as light, dissolved organic and inorganic nutrient, and bacterial populations. The application of equation 1 to so many different kinds of substrate-population pairs, together with theoretical consideration of the basis of Michaelis-Menten kinetics in such instances, led to speculation that it might be applicable to higher organisms, i.e., to predator-prey and host-parasite situations (5). Most of this microbiological work is reviewed by Strickland (36).

In fact, this same expression had been developed independently both by Holling (12), who considered the feeding behavior of animals, and by Rashevsky (26) in theoretical considerations of Ivlev's (14 and 15) fish-feeding experiments. Royama (33) presents a review of much of the work that has been done on the application of this hyperbolic equation to predator-prey and host-parasite situations.

The extension from simple two-component models to complex multi-component systems involving a number of populations at each trophic level could be accomplished using equation 1 as a basis for forming the component-differential equations. These would be non-linear, first-order, differential equations, one equation for each component population or substrate considered. These are essentially mass-balance equations, one for each system component. A system large enough to represent a typical natural community could easily involve 100 or 200 such differential equations. Such a system is not very tractable mathematically or ecologically.

Systems involving a large number of component-differential equations are tractable if the component equations are linear. When the substrate concentration, x, is large relative to the constant, b, in equation 1, the associated differential equation is linear to a good approximation. Under this condition the substrate utilization rate, Y, is proportional to the concentration of the "predator" population, x_p, and $Y = ax_p$. When x approaches the size of b, equation 1 must be used and we have $Y = yx_p$. When x is much less than 6, then $Y = (a/b)xx_p$ with good approximation. The last two cases involve non-linear differential equations. Since we wish to study the ecosystem response to large-scale changes in the environment, a linear approximation is not likely to be adequate. We therefore examine other alternatives to simplify our system.

Ecologists, since Lindeman (18), have used the concepts of energy and material flow between trophic levels as simplifying system analysis concepts in considering ecosystems. Smith (34) used equation 1 to represent the relationship between substrate concentration and substrate utilization rate, where substrate is one trophic level and the user is the next higher trophic level. There are some experimental data to justify the use of equation 1 in this context. Natural mixed-species populations of marine phytoplankton using dissolved nitrate and ammonium ions have been shown to follow this relationship reasonably well (19). Whittaker (38) presents data that suggest the applicability of equation 1 to describe the primary productivity of the plants in various terrestrial communities as a function of mean annual precipitation. The application of equation 1 to higher trophic levels, e.g., plant-herbivore, herbivore-carnivore and primary-carnivore/secondary-carnivore rests principally with single-species/single-species observations and a few with single-species/mixed species substrate observations. Rigler (31), McMahon and Rigler (21), Reeve (28), Adams and Steele (1), and Parsons et al. (23) give examples of herbivorous zooplankton feeding on phytoplankton that can be described by equation 1. Reeve (29), Holling (13), LeBrasseur et al. (17), and Brocksen et al. (4) give data from higher trophic level situations that can be described by this equation. Although there are substantial data to suggest its use as a trophic-level/trophic-level relation, such use should be viewed as hypothetical at present. We accept this hypothesis and make use of equation 1 to describe material flux between trophic levels. This leads to a relatively simple ecosystem model.

A more general form of equation 1 is

$$y_i = a_i(x_{i-1} - p_i) / (b_i + x_{i-1} - p_i) \tag{2}$$

where i = 1, 2, 3, 4 represents the ith trophic level, y, x, a, and b are as defined previously, and p represents the substrate concentration at which the substrate utilization rate falls to zero. There is evidence that this rate falls to zero in some instances, at least, at finite substrate concentration, e.g., Parsons et al. (23) and Caperon and Meyer (7). We define a yield coefficient, q_i as the amount of substrate required to produce a unit of population at the ith trophic level. A block diagram for a one-carnivore ecosystem is given in Figure 2. This system is described by the following set of differential equations:

$$\dot{x}_1 = x_1 y_1(x_o) / q_1 - x_1 D - x_2 y_2(x_1) \quad (3)$$

$$\dot{x}_2 = x_2 y_2(x_1) / q_2 - x_2 D - x_3 y_3(x_2) \quad (4)$$

$$\dot{x}_3 = x_3 y_3(x_2) / q_3 - x_3 D \quad (5)$$

$$\dot{x}_4 = r_2 x_2 y_2(x_1) + r_3 x_3 y_3(x_2) - (k+D) x_4 \quad (6)$$

$$\dot{x}_o = R - D x_o + s_2 x_2 y_2(x_1) + s_3 x_3 y_3(x_2) - x_1 y_1(x_o) + k x_4 \quad (7)$$

In these equations $\dot{x}_i$ represents dx_i/dt, and y_i is written as $y_i(x_{i-1})$ to show explicitly the functional dependence. R represents the rate of addition of new dissolved mineral substrate per unit volume, and D represents a physical loss rate that applies to the populations at all trophic levels. x_o represents a pool of essential dissolved mineral nutrient, x_1 the phytoplankton, x_2 the herbivorous zooplankton, x_3 the carnivorous zooplankton, and x_4 represents the detritus pool (both particulate and dissolved organic). Note that the transfer rate, $x_i y_i(x_{i-1})$ depends upon the population concentration in the i and i-1st trophic levels. Extension of the system to include an additional trophic level is straightforward. Each equation is simply a mass-balance equation, and can be viewed as

population change = population size (specific growth rate - specific dilution rate) - consumption by the next trophic level

or

$$dx_i/dt = (x_i y_i / q_i) - x_i D - x_{i+1} y_{i+1}$$

since

specific growth rate = (substrate utilization rate)/q.

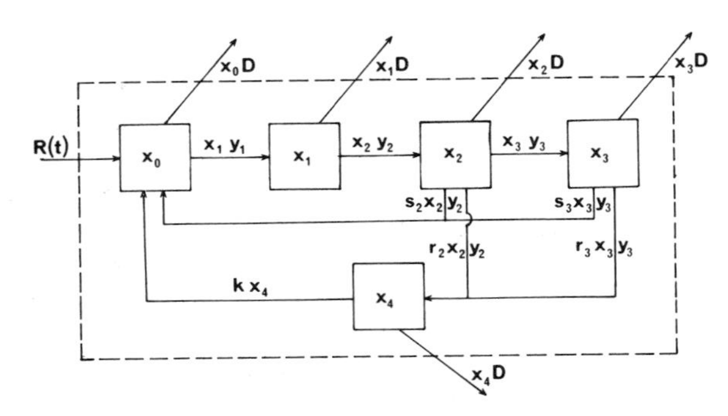

Figure 2. A block diagram of the one-carnivore model, with x_0, x_1, x_2, x_3 and x_4 representing dissolved nutrient, phytoplankton, herbivore, carnivore, and detritus standing crops, respectively.

Alternatively we can write

population change = substrate utilization rate - excretion loss - substrate utilization by the next trophic level - dilution loss

or

$$dx_i/dt = x_i y_i - [(q_i - 1)/q_i]\, x_i y_i - x_{i+1}\, y_{i+1} - x_i D$$

which is the same as the above equation.

The fractional loss of the uptake to defecation and excretion, $(q_i\text{-}1)/q_i$, is both to the detritus and to the nutrient pool. The constants s_i and r_i partition this loss to the nutrient and detritus pools, respectively. Thus $(s_i+r_i) = (1\text{-}1/q_i)$.

In a planktonic ecosystem D can be considered to represent advection and diffusion processes. Often R and D can be considered the same and represent simply the dilution rate. The model assumes that the ecosystem is so well mixed that the variables x_0, x_1, x_2, x_3, and x_4 represent the trophic level population size in concentration units: population per unit volume or population per unit area. When a hydrodynamic model becomes available we will be able simply to replace the Dx_i terms in the equations with the time-dependent advection and diffusion terms, and the model will be completely general with respect to time and space variability. For the present we focus on an ecosystem that can be assumed to be a well-mixed habitat. The southern sector of the bay meets this requirement quite well.

NUTRIENT LIMITATION

We chose nitrogen as a measure of standing stock: x_0, x_1, x_2, x_3, and x_4 have units of mg(N)/m^3 and all terms in equations 3 through 7 have units mg(N)/m^3/hr. This choice, together with the form of equation 3, implies that the concentration of available nitrogen controls the growth rate of the phytoplankton population. This control is modified by the availability of light which, in turn, is partially dependent upon the plant population size. We made the simple assumption that the light regime plays a role in setting the value of a_1, the maximum specific utilization rate. The identification of a limiting nutrient is quite difficult and our efforts have been limited to choosing between phosphorus and nitrogen because silicon is abundant in the bay.

The ratio of nitrogen to phosphorus, N/P, in waters coming into the bay compared to growth requirements for phytoplankton strongly suggests that nitrogen would become limiting before phosphorus would. We assume that available nitrogen consists of NO_3^-, NO_2^- and NH_4^+. Phosphorus is equated to orthophosphate. In the water of the open ocean adjacent to the bay, this N/P ratio ranges from 0.85 to 2.7. In the land run-off entering the bay it ranges from 0.93 to 9.01. The mean value for the secondary treated municipal waste discharged into the bay is 5.38. The N/P ratio for phytoplankton material is quite variable. Redfield et al. (27) give values of 5.6, 30.9, and 2.9 for normal, phosphorus-deficient and nitrogen-deficient *Chlorella* cells, respectively, and they use 15 as a "normal" ratio for marine plytoplankton. Ignoring regeneration rates for the present, it appears that the algae in the bay would deplete their nitrogen supply sooner than their phosphorus supply.

Laboratory cultures of *Cyclotella nana, Monochrysis lutheri,* and *Dunaliella tertiolecta* were grown in a standard sea-water enrichment medium as growth rate controls and in open-ocean sea water with 5% by volume municipal waste water added. Growth rates for the two sets of data were quite comparable but particulate N/P ratios, after growth in the waste-water-enriched culture had stopped, were 2.0, 3.2, and 3.1 for *Cyclotella, Monochrysis,* and *Dunaliella,* respectively; in the standard laboratory medium the values were 9.3, 7.5, and 31.2 for the same organisms. Thus phytoplankton dependent on sewage for nutrients are in the nitrogen-deficient category upon exhaustion of their supply of nitrogen. Both the chemistry of the waters entering the bay and the response of phytoplankton to these waters indicate a shortage of nitrogen relative to phosphorus.

MODEL PARAMETER VALUES USED IN THE SIMULATION

A list of the values used for each ecological and environmental parameter required by the model is given in Table 1. Most of the ecological parameters are taken from the literature and most seem quite appropriate to the Kaneohe Bay ecosystem. In only one case, the rate constant for nutrient regeneration from detritus, k, did we simply guess at a value. The model in its present form is not

TABLE 1

Model Parameters – Values Used and Sources

	x_1 Plants	x_2 Herbivores	x_3 Carnivores
$a_i(hr^{-1})$ Maximum Substrate Utilization Rate Constant	0.07	.015	.019
$b_i(mg/m^3)$ Half Saturation Constant	14	10	52
q_i Yield Coefficient	1	4	4
p_i Zero Intercept	0	0	0
s_i Partition of Excrement to Nutrient	—	0.3	0.3

Nutrient regeneration from detritus, rate $k(hr^{-1})$: small compared to D

Dilution Rate, $D(hr^{-1})=0.0007$

Nitrogen Input Rate, $R(mg/m^3/hr) = 0.03$ to 0.15

particularly sensitive to this value.

To determine the dilution factor, D, we assumed that the southern sector of the bay is well mixed, and that there is no net flux of nitrogen to or from the benthos. The dilution rate then is given by $D = (dN/dt)/N$, where N is the total amount of nitrogen in the southern sector and dN/dt is the input rate. Stream run-off contributed an average of 22.4×10^6 mg N/day, and municipal waste discharge contributed 180×10^6 mg N/day in 1969 (39). The average content of fixed nitrogen – particulate plus dissolved – at this time was 12×10^9 mg N (35). This gives a dilution rate, D, of 0.0007 hr^{-1} and a specific input rate, R, of 0.11 mg $N/m^3/hr$ for the southern sector, with a volume of $73.6 \times 10^6 m^3$.

None of the parameters used in the model have been as exhaustively examined as the set associated with phytoplankton, using nitrogen in the form of ammonium or nitrate ion. The a_1 value is an observed mean value for Kaneohe Bay. Under substrate limiting conditions q_1 is set equal to 1 (7), and this should be an adequate value for the purposes of this model. There is considerable difference between published values of b_i for laboratory populations (7, 11) and for field data (19). The value chosen is consistent with field observations.

The herbivore substrate utilization rate parameters, a_2, b_2, q_2, p_2, and r_2, given in Table 1 were taken from the observations of feeding and excretion by *Calanus helgolandicus* feeding on *Biddulphia sinensis*, (9). The single exception is the q_2 value which is based on data from this same source, except that it represents the case where *Calanus* was feeding on natural food rather than on *Biddulphia*. The a_2 value used is in close agreement with another estimate of this parameter by Parsons et al. (23) for *Pseudocalanus minutus* and *Oithona* sp. Since as much as 30% of this trophic level in Kaneohe Bay consists of the microcopepods *Paracalanus* sp. and *Oithona simplex*, we are encouraged to feel these values are applicable. They agree relatively well with more indirect observations of the feeding kinetics of the microcopepods in Kaneohe Bay (2, 37).

We relied entirely on LeBrasseur et al. (17) for our estimates of a_3, b_3, and p_3 for the carnivore trophic level. These results are for juvenile salmon feeding on zooplankton, and we supplied our own factors for conversion of fish length to wet weight and for conversion of the original wet weight data to organic nitrogen. The values so obtained are given in Table 1. We hoped for more directly relevant data from marine systems, but little is available. Rowe (37) gives a_3 values ranging from 0.020 to 0.040 hr^{-1} for *Pleurobrachia pileus* (an important component of the planktonic carnivores in Kaneohe Bay) feeding on *Artemia nauplii*. It is considered likely that this rate is a bit high on this unnatural food, and the value used does fall at the lower end of this range. We used the same value for q_3 as was used for q_2, although Reeve (30) determined a value of 3 for the carnivore *Sagitta hispida*. We also took $s_3 = s_2$ as our estimate of this value, since we were unaware of a comparable observation for a carnivorous zooplankter. A model with two carnivorous trophic levels was also examined. In this case, both the primary and secondary carnivores were assigned the same parameter set used for the carnivores in the one-carnivore model. All other parameters were the same.

In the present application of this model we confine our interest to steady-state solutions. The left-hand side of equations 3-7 can be set equal to zero to yield five algebraic equations in five unknowns, x_0, x_1, x_2, x_3, and x_4. Equation 5 can be solved directly for x_2, the steady-state herbivore population concentration in terms of the dilution rate, D, and the parameters describing the feeding kinetics of the carnivores, a_3, b_3, q_3, and p_3. Thus x_2 is independent of the forcing function R(t). The system is reduced to a system of four equations in four unknowns which can be reduced algebraically to a single fourth-order equation in one unknown. The roots of this equation must be found numerically, and the remaining x_i's are obtained by substitution. The steady-state values of x_i must satisfy the equation $R = D(x_0+x_1+x_2+x_3+x_4)$, obtained by summing equations 3-7 with $dx_i/dt = 0$. Solutions, so obtained, are presented in Figure 3 for a one-carnivore system for the series of values of R = .03, .04, . . . , .15 for the ecological parameters give in Table 1.

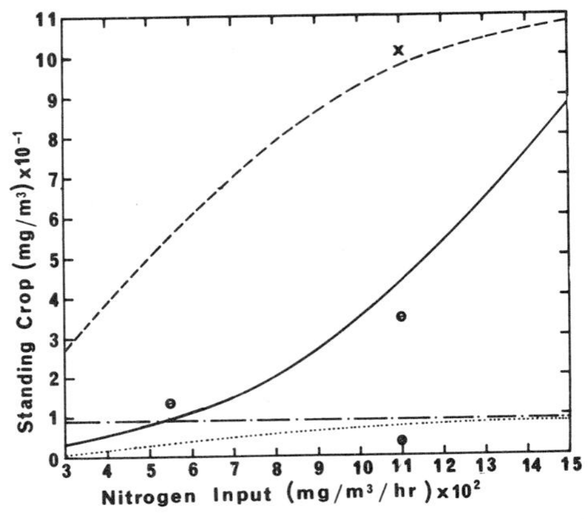

Figure 3. The standing crop of detritus (dashed line), phytoplankton (solid line), herbivore (dash-dot line), and carnivore (dotted line) as a function of the forcing function or input of nitrogenous nutrient to the ecosystem model described in Figure 1. The cross, circle, and filled circle show observed values of detritus, phytoplankton, herbivore and carnivore, respectively, (the latter two values are a single point on this scale).

RESULTS

A summary of the observed mean biomass at each trophic level in each of the three sections of the bay is presented in Table 2. The values are means of a one-year series of samples, with an approximately two-week sampling interval. Zooplankton and some of the nutrient and detritus data were collected in 1969, while the phytoplankton and other nutrient and detritus data are from a 1970 series. The herbivore and carnivore data represent enumeration of zooplankton samples taken with paired vertical hauls of a 333-μ mesh net and a 64-μ mesh net with 333-μ prescreening (8). The zooplankton data presented represent a single representative time series of samples for each sector. The mean nitrogen content per individual was determined by analyzing representative samples in the F and M model 185 CHN analyzer for most of the important species in the bay. These data were used to convert counts to nitrogen concentration for each species. Component populations were assigned to either the herbivore or carnivore trophic level following Peterson (24). Fortunately, only one omnivore was considered a numerically significant member of the zooplankton, so little difficulty was encountered in establishing with good approximation the standing

TABLE 2

Trophic Level Comparison of Sectors of Kaneohe Bay in Term of Concentration of Fixed Nitrogen (mg/m^3).

	South Sector	Middle Sector	North Sector
Dissolved Nutrient	3.8	3.8	4.1
Phytoplankton	34.5	15.3	7.6
Herbivorous Zooplankton	3.54	3.27	1.18
Carnivorous zooplankton	3.30	1.18	0.26
Detritus (dissolved plus particulate)	108	83	68

crop of each of these trophic levels by summing the component populations. We did not make any distinction between primary and secondary carnivores.

The plant standing crop values are based on *in vivo* fluorescence surveys of the whole bay (6, 16). It has been shown that chlorophyll *a* is linearly related to *in vivo* fluorescence within the bay. Chlorophyll *a* data are converted to nitrogen biomass on the basis of ratios developed for algal growth under nitrogen-limiting conditions in continuous culture (7).

Detritus values are based on surveys of the particulate carbon and nitrogen in the bay by Steinhilper (35). This work covered nine surveys over a three-month period and was contemporaneous with the work reported by Caperon et al. (6). The chlorophyll values for these same samples were converted to particulate nitrogen as described above. Total particulate nitrogen minus plant nitrogen was taken as particulate detritus. The dissolved organic nitrogen component of the detritus stock was taken from data of Young et al. (39).

There is some question as to which forms of nitrogen serve as a nutrient source under both nutrient-limiting and nutrient-saturating conditions. We included only ammonium ion in the nutrient pool. Possibly both nitrate and urea should be included, but there is evidence that the presence of ammonium inhibits utilization of the latter two species – see, for example, McCarthy (20).

The data presented in Table 2 are mean values and, to the degree that a steady state exists, can be considered to represent steady-state values for the bay. The principal source of nutrient to the system (municipal waste discharge) is constant on a day-to-day basis, and the low latitude of the area minimizes seasonal variability in the light regime. Only the sporadic and seasonal character of the land run-off contributes significant perturbations to the system, and these are brief and mostly limited to the winter months. Thus it would appear that comparison between observed data and steady-state solutions in the model would be instructive. Since accurate values of the dilution factor, D, and the forcing function, R, are available only for the southern sector this is the only area for which a direct comparison is possible. Table 3 compares the predictions for a primary-carnivore and a secondary-carnivore model with the data from the southern sector. The observed and predicted one-carnivore values are in remarkable agreement, in view of the uncertainties in some of the model parameters and the limited data available to us. This agreement in no way validates the model or the assumptions that have gone into it. It does tell us that we still have a viable and, indeed, a much more interesting hypothesis: that this kind of model may well be instructively and predictively useful.

TABLE 3

A comparison of the predictions of a one- and a two-carnivore food chain model with observed data.

	Observed	Predicted One-carnivore system	Predicted Two-carnivore system
Nutrient	3.8	0.7	9.0
Phytoplankton	34.8	42.4	5.0
Herbivore	3.5	8.9	26.8
Carnivore	3.3	7.5	3.1
Detritus	108	98	113.3

The comparison between the secondary-carnivore system and observed values is much less satisfactory. Since we have no observed data that discriminate between first- and second-order carnivores, the model output for these two trophic levels are summed to give the comparison value tabulated in Table 3. The comparison between observation and prediction is excellent for carnivores, good for detritus, not so good for nutrient, and quite unacceptable for plants and herbivores. Apparently the plankton community cannot be represented by this type of model with two carnivorous trophic levels. Even an extensive parameter study involving a_1, a_2, b_1, and b_2 failed to discover a set of values for these four parameters that gave a steady-state model solution that compared well with observations.

Clearly, the more critical test comes from comparison of model results with observations for different values of the forcing function R(t). Using historical records of waste-discharge volume, we can make reasonable estimates of R(t) over the past decade. Limited data are available for historical comparison. The 1960 value for the phytoplankton standing crop in the southern sector (10) was 13.4 mg(N)/m^3. This corresponds to an estimated value of R(1960) = 0.55

$mg/m^3/hr$ and a predicted value of the phytoplankton standing crop of 9.1 $mg(N)/m^3$ in the one-carnivore model and 5.8 $mg(N)/m^3$ in the two-carnivore model.

Since no other historical data were available for comparison, we used the spatial variability within the bay to indicate qualitatively what changes have taken place in the southern sector. The same set of planktonic species occurs throughout the bay, though areas differ in relative abundances (22, 24). Presumably the degree of eutrophication that exists in the transition and northern sectors represents stages that the southern sector has gone through during the last thirty years with increasing urbanization in the watershed and increasing amounts of municipal waste disposal in this sector. Figure 3 gives the results of the primary-carnivore model's response to increasing values of the forcing function R(t). It is assumed that the changes took place slowly so that the steady-state model is applicable over this time history in which R(t) took on different values. The nutrient concentration cannot be shown on the scale used in Figure 3. It decreases monotonically from a high value of 2.54 mg/m^3 at R(t) = 0.03 to a low of 0.43 mg/m^3 at R(t) = 0.15. We want to make a qualitative comparison of the effect of changing the forcing function as represented by Figure 3 which we think of as time variation in R, and the effect of spatial variation in R as represented by north-to-south differences in nutrient enrichment. Since in the second case we have a variable dilution factor, D, we cannot expect precise comparison between Figure 3 and the trend of changes shown in Table 2, but the sense or direction of the change, at least, should be the same.

The clear increase in phytoplankton in going from the northern sector to the southern sector is consistent with the one-carnivore prediction. The relative invariance of the herbivores in the three sectors is also consistent with the one-carnivore prediction. The marked increase in carnivores in going from north to south is also a characteristic of the one-carnivore predictions. The two-carnivore model shows none of these changes: with increasing eutrophication (i.e., north to south) it predicts instead a slight increase in carnivores. Although the observed nutrient values are higher than predicted by the one-carnivore model, their relative invariance is more consistent with the one-carnivore system (range 2.54 to 0.43 mg/m^3) than the two-carnivore (range 2.0 to 18.1) for the R(t) = 0.03 to 0.15 range of the forcing function. Thus, qualitative comparisons, using spatial variability to indicate the character of the temporal change that is concentrated in the south sector, argue compellingly for the one-carnivore model and strengthen our confidence in its usefulness.

In terms of comparative ecosystem studies and applied problems in resource management, the implications of these modeling efforts are quite profound. Does exploitation by man complement the feeding by an existing trophic level or does it add an additional trophic level? How will an introduced species affect the length of a food chain? What will be the impact of municipal waste discharge on a given ecosystem? Indeed, can we in any way predict the impact of such environmental changes without the kind of tool this model represents? It does not seem so. Even the very much more narrow objectives of the phytoplankton ecologists stated in the introduction are insoluble without his kind of effort: if nutrient is added to an ecosystem at an increased rate, will the standing crop of nutrient increase with a concomitant decrease in the plant standing crop, or will it be just the reverse? Either choice is surprising to the phytoplankton ecologist who has really dealt with only two component systems – substrate and plant population. We insist that our model must become more detailed and be much more rigorously tested before we feel confident that we have a tool that can provide quantitative answers to such questions, but we do feel justified in raising such questions and in insisting that they are critical to the study of ecology.

ACKNOWLEDGMENTS

Ms. Judith Meyer provided programming assistance in some early versions of the model and was a valuable sounding board for most of the ideas that went into this work. Mr. James P. Szyper worked up the carnivore data for all three sections of the bay and the herbivore data for the transition and northern sections. The assistance of both of these individuals is gratefully acknowledged.

REFERENCES

1. Adams, J. A. and Steele, J. H.
1966 Shipboard experiments on the feeding of **Calanus finmarchicus.** In **Some Contemporary Studies in Marine Sciences,** p. 19-35. (ed. Barnes, H.) George Allen and Unwin Ltd., London.

2. Bartholomew, E. F.
1973 The production of microcopepods in Kaneohe Bay, Oahu, Hawaii. M.S. thesis. Univ. Hawaii, Department of Oceanography.

3. Bathen, K. H.
1968 A descriptive study of the physical oceanography of Kaneohe Bay, Oahu, Hawaii. M.S. thesis. Univ. Hawaii, Department of Oceanography, 353 pp.

4. Brocksen, R. W., Davis, G. E., and Warren, C. E.
1970 Analysis of trophic processes on the basis of density-dependent functions. In **Marine food chains.** p. 468-498. (ed. Steele, J. H.) Univ. California Press, Berkeley and Los Angeles.

5. Caperon, J.
1967 Population growth in microorganisms limited by food supply. **Ecology,** 48: 715-722.

6. Caperon, J., Cattell, S. A., and Krasnick, G.
1971 Phytoplankton kinetics in a subtropical estuary: Eutrophication. **Limnol. Oceanogr.,** 16: 599-607.

7. Caperon, J. and Meyer, J.
1972 a. Nitrogen-limited growth of marine phytoplankton. Part I:

Changes in population characteristics with steady-state growth rate. **Deep-Sea Res.,** 19(9): 601-618.

8. Clutter, R. C.
1969 Plankton ecology. In **Estuarine pollution in the state of Hawaii. Part II: Kaneohe Bay Study.** Technical Report No. 31 of the Water Resources Research Center. Univ. Hawaii, Honolulu, Hawaii.

9. Corner, E. D. S.
1972 On the nutrition and metabolism of zooplankton. VIII. The grazing of **Biddulphia** cells by **Calanus helgolandicus. Jour. Mar. Biol. Ass. U. K.,** 52: 847-861.

10. Doty, M. S. and Capurro, L. R. A.
1961 Productivity measurements in the world ocean. Part I. **IGY Oceanogr. Rep. 4.** 298 pp.

11. Eppley, R. W., Rogers, J. N., and McCarthy, J. J.
1969 Half saturation constants for uptake of nitrate and ammonium by marine phytoplankton. **Limnol. Oceanogr.,** 14: 912-920.

12. Holling, C. S.
1959 Some characteristics of simple types of predation and parasitism. **Can. Ent.,** 91: 385-398.

13. Holling, C. S.
1965 The functional response of predators to prey density and its role in micicry and population regulation. **Ent. Soc. Can.,** 45: 60 p.

14. Ivlev, V. S.
1944 The time of hunting and distance traveled by the predator in relation to the population density of the prey. **Zoologicheskij Zhurnal,** 23: 139-144.

15. Ivlev, V. S.
1955 **Experimental ecology and nutrition of fishes.** Moscow Pishchepromizdat, Moscow. (Transl. Scott, D.) Yale Univ. Press, New Haven, 1961.

16. Krasnick, G. J.
1973 Temporal and spatial variations in phytoplankton productivity and related factors in the surface waters of Kaneohe Bay, Oahu, Hawaii. M.S. thesis. Univ. Hawaii, Department of Oceanography.

17. LeBrasseur, R. J., Barraclough, W. E., Kennedy, O. D., and Parsons, T. R.
1969 Production studies in the Strait of Georgia. Part III. Observations on the food of larval and juvenile fish in the Fraser River plume, February to May, 1967. **Jour. Exp. Mar. Biol. Ecol.,** 3: 51-61.

18. Lindeman, R. L.
1942 The trophic-dynamic aspect of ecology. **Ecology,** 23: 399-418.

19. MacIsaac, J. J. and Dugdale, R. C.
1968 The kinetics of nitrate and ammonia uptake by natural populations of marine phytoplankton. **Deep-Sea Res.,** 16: 45-57.

20. McCarthy, J. J. and Eppley, R. W.
1972 A comparison of chemical, isotopic and enzymatic methods for measuring nitrogen assimilation of marine phytoplankton. **Limnol. Oceanogr.,** 17: 371-382.

21. McMahon, J. W. and Rigler, F. H.
1963 Mechanisms regulating the feeding of **Daphnia magna** Straus.

Can. Jour. Zool., 41: 321-332.

22. Murphy, C.
1972 An annual cycle of phytoplankton populations in Kaneohe Bay, Oahu. M.S. thesis, Univ. Hawaii, Department of Botany. 109 pp.

23. Parsons, T. R. et al.
1969 Production in the Strait of Georgia. Part II. Secondary production under the Fraser River plume. February to May, 1967. **Jour. Exp. Mar. Biol. Ecol.,** 3: 39-50.

24. Peterson, W. T.
1969 Species diversity and community structure of the macrozooplankton of Kaneohe Bay, Oahu. M.S. thesis, Univ. Hawaii, Department of Oceanography.

25. Piyakarnchana, T.
1965 The plankton community of Kaneohe Bay, Oahu. Doctoral dissertation. Univ. Hawaii, Department of Zoology.

26. Rashevsky, N.
1959 Some remarks on the mathematical theory of **nutrition** of fishes. **Bull. Math. Biophysics,** 21: 161-183.

27. Redfield, A. C., Ketchum, B. H., and Richards, F. A.
1963 The influence of organisms on the composition of sea water. In **The Sea,** Vol. II, p. 26-77. (ed. Hill, M. N.) Interscience Publishers, New York.

28. Reeve, M. R.
1963 Growth efficiency in **Artemia** under laboratory conditons. **Biol. Bull.,** 125(1): 133-145.

29. Reeve, M. R.
1964 Feeding of zooplankton, with special reference to some experiments with **Sigitta. Nature,** 201(4915): 211-213.

30. Reeve, M. R.
1970 The biology of Chaetognatha I. Quantitative aspects of growth and egg production in **Sagitta hispida.** In **Marine food chains.** (ed. Steele, J. H.) Univ. California Press, Berkeley and Los Angeles.

31. Rigler, F. H.
1961 The relation between concentration of food and feeding rate of **Daphnia magna** Straus. **Can. Jour. Zool.,** 39: 857-868.

32. Rowe, M. D.
1971 Some aspects of the feeding behavior of the ctenophore **Pleurobrachia pileus.** M.S. thesis, Univ. Hawaii, Department of Oceanography.

33. Royama, T.
1971 A comparative study of models for predation and parasitism. **Researches on Population Ecology,** Supplement No. 1, Society of Population Ecology, c/o Entomological Lab., Kyoto Univ., Japan. 91 p.

34. Smith, F. E.
1969 Effects of enrichment in mathematical models. In **Eutrophication: causes, consequences, correctives.** National Acad. Sci., Washington, D.C.

35. Steinhilper, F. A.
1970 Particulate organic matter in Kaneohe Bay, Oahu, Hawaii. M.S. thesis, Univ. Hawaii, Department of Oceanography.

36. Strickland, J. D. H.
1971 Microbial activity in aquatic environments. In **The Twenty-First Symposium of the Society for General Microbiology: Microbes and biological production,** Cambridge Univ. Press.

37. Szyper, J. P.
1972 Zooplankton grazing in Kaneohe Bay, Hawaii. M.S. thesis. Univ. Hawaii, Department of Oceanography.

38. Whittaker, R. H.
1970 **Communities and Ecosystems.** The Macmillan Company, Collier-Macmillan, Ltd. London.

39. Young, R. H. F., Morphew, K. L., and Burbank, N. C., Jr.
1969 In **Estuarine pollution in the State of Hawaii, Part II. Kaneohe Bay study.** Technical Report No. 31 of the Water Resources Research Center, Univ. Hawaii, Honolulu, Hawaii.

EDUCING AND MODELING THE FUNCTIONAL RELATIONSHIPS WITHIN SUBLITTORAL SALT-MARSH AUFWUCHS COMMUNITIES – INSIDE ONE OF THE BLACK BOXES

John J. Lee, John H. Tietjen, Norman M. Saks[1],

George G. Ross[2], Howard Rubin[3] and William A. Muller[4]

ABSTRACT

Salt-marsh aufwuchs communities are extremely complex assemblages of > several hundred species of bacteria, fungi, algae, protozoa, and micrometazoa. At the present time, information on the population structure is too fragmentary to be used in modeling and predicting community structure. Studies of foraminifera, nematodes, and microalgae grown in gnotobiotic culture suggest it is unreasonable to assume that measurements made on one species will be good estimators for similar parameters of other species or of the assemblage as a whole. Nutritional experiments suggest many meiofauna are selective feeders

1. Department of Biology, City College of New York, Convent Avenue at 138th Street, New York, New York 10031.

2. Department of Computer Science, City College of New York, Convent Avenue at 138th Street, New York, New York 10031.

3. Departments of Biology and Computer Science, City College of New York, Convent Avenue at 138th Street, New York, New Yor 10031.

4. Department of Biology, City College of New York, Convent Avenue at 138th Street, New York, New York 10031. Presently affiliated with New York Institute of Technology, Wheatly Road, Westbury, New York.

and gain different nutritional values from the food they assimilate. If the phenomena of balanced diets and special growth factors are widespread among meiofauna, we may be forced to reconsider certain aspects of energy flow and modeling. The development of a model of the trophic dynamics and life cycle alternatives of just one species from the community has proved extremely challenging. We used our simulation of the trophic dynamics of *Allogromia laticollaris* to organize our present information and to serve as a generator of predictions to be further tested experimentally.

INTRODUCTION

Ecological modeling can be and has been viewed from many different perspectives. Modeling has enormous contemporary appeal, and many feel that it will lead to new insights or depths of understanding of ecological processes and eventually be of predictive value in ecosystem management. Because of the difficulties in securing "hard" data, ecosystems are statistically very "noisy" and so exceedingly complex that faithful modeling is uncertain. We cannot expect a model to have general qualities unless we are willing to sacrifice precision and realism (33). The modeler's first task, therefore, is to define his interest and reduce to manageable size and complexity those ecosystem components which he wishes to study. Ideally the conceptual organization of data for building a model focuses attention on a relationship(s) heretofore ignored and is structured to test an hypothesis(es) about how some natural system operates.

Many modelers have concentrated their attention on the thermodynamic aspects of whole ecosystems and the interrelationships of component subsystems. Each subsystem hierarchy is assumed to behave as the result of combined actions or interactions of a lower order. Populations of organisms are organized on the basis of their ascribed trophic characteristics and their collective dynamics described by energy input-output relationships (in caloric terms). While there is logic in grouping taxonomically disparate populations into functional groups (e.g., detritus feeders) to make the system mathematically more tractable, many problems remain in application. Some organisms may fit into more than one category; others may move from one category to another in different life-history phases. Regrettably two of the key qualities of living organisms are lost to many modelers: the uniqueness of a species (or the delicate biological distinctions between species); and the capacity of species to regulate, within bounds, their life processes, including the rates at which they accumulate and dissipate energy. Dynamic aspects of community structure (i.e., fluctuating abundance of component species and changing food-web relationships) and the collective attributes of communities beyond those of subordinate populations can also be lost. Thus, as abstraction and generality are approached, details collapse into lumped variables and parameters, and information about the system being modeled is invariably forfeited.

How then do we represent the little compartments in large ecosystem models? Is it reasonable to assume that parameters of transfer that have been measured for one species will be good estimators for parameters of other species and of the compartment as a whole? To answer these questions we chose to study the very large assemblages of micro- and meio-flora and fauna which inhabit salt-marsh sublittoral epibenthic communities and those of shallow embayments.

INSIDE ONE OF THE BOXES

Population Structure and Related Field Studies

According to one estimate (36), sublittoral benthic communities down to a depth of 100 m have as many as 10^6 meiofauna per m^2. Even higher densities have been reported in shallow estuarine habitats (47, 53). Within the meiobenthic assemblages, species diversity may be quite high (12, 13, 54). Within a 1 cm^3 may live, at any one instant, ~ 10^5 or 10^6 organisms belonging to several hundred different species. One wonders why there are so many different species of bacteria, algae, protozoa, and micrometazoa in these communities. Ecological reasoning argues that a large diversity of species should be present because the organisms are small regardless of their position in the food web (19).

Before we can begin to define and model functional relationships among these small organisms and between them and the rest of the system, we have to know something about the population structure of aufwuchs communities. Regrettably only "soft" and fragmentary data exist. For example, let us look at the population structure and distribution of one of the best-studied assemblages, the foraminifera. For the best part of 5 years (~ 6 man years) we studied and analyzed the distribution and population structure of foraminifera during three growing seasons in a relatively small (~ 750 m^2) but very productive sublittoral part of a Long Island salt marsh (29, 34). Our approach to the problem of environmental grain (see review, 59) was to study the distribution of foraminifera in a great many small samples (0.28 g dry weight) which approached the size of an operational habitat ($\lesssim$ 1.5 cm^3) for these animals. Even within the small study area natural resources were found to be unevenly distributed and subject to seasonal and climatic changes. The sedimentary substrate was sand covered with a fine layer of silt. Grain size was generally correlated with rivulet flow at ebb tide but the picture was complicated by the fact that wind and wave action in this relatively shallow embayment (~ 1.5 m at MHT) caused the grains to be re-sorted. Parts of the study site are exposed for ~ 30-50 min at ebb tide, depending upon the characteristics of the wind and the tide.

Macrophyte growth is always patchy at the field station during the growing season (May to September) and characterized by patterns of seasonal succession of macrophytes. Autumn storms generally scour the area and macrophyte growth is renewed in late spring of the follwing year. Some of the factors influencing the occurrence and distribution of particular species were discerned but even careful analysis of > 750 samples did not yield "hard" data that could be used for deterministic modeling. Foraminifera were non-randomly distributed in the samples; 40 of the 758 samples contained more than 60% of the total foraminifera harvested. The population structure of the foraminiferan assemblage was quite different in each of the years it was studied. Climatic conditions might have caused the differences, since one summer terminated a long drought, another was "exceptionally rainy", and, the third was "normal", but in the absence of replicate studies, these statements lack statistical validity.

Community history seemed to be important in determining community structure. Three species, *Allogromia laticollaris, Elphidium subarticum,* and *Rosalina leei,* were common in the drought year, fairly rare the following year, and not detected in the population two years later. *Ammonia beccarii* on the other hand, was twice as abundant (~ 40% of the population) as any other species in both the drought and rainy years, but was replaced in dominance by another species, *Elphidium incertum,* in the following "normal" year. Predictably, species diversity of foraminifera was lowest during the drought year. Large standing crops of foraminifera were found, but not uniformly, in the sublittoral epiphytic communities of some species of macrophytes but only rarely in those of others. Statistical treatment of the data on the most abundant species suggested some foraminiferan species (e.g. *Protelphidium tisburyensis*) were more likely to be associated with the epiphytic communities growing on particular macrophytes (e.g. *Enteromorpha*) whereas other species, (i.e. *Quinqueloculina* spp) were less apt to be so distributed. Some foraminiferan species (*Ammonia beccarii, Elphidium* spp) were broadly distributed on all macrophytes except *Fucus* or *Codium.* Blooms of some species of foraminifera were much more frequent on patches of decaying *Enteromorpha* than on young green patches. Some species (*Ammobaculities dilatatus* and *Quinqueloculina lata*) were found in the eipiphytic community but not below in the sediments; others (*Trochammina inflata* and *Ammotium salsum*) were much more abundant in the sediments than in the water column above. Still other species (*Elphidium incertum* and *Protelphidium tisburyensis*) were abundant in both. Changes in sediment grain size apparently affected the horizontal and vertical distribution of some of the psammolittoral species. As have others (3, 4, 8, 9, 20, 41, 44), we found seasonal succession of foraminiferan species. There was no detectable correlation between gross abundance of food (epiphytic community dry weight) and the size of foraminifera community. Probability and coincidence tables were constructed with the aim of testing whether various species of foraminifera were distributed at random to each other or whether some shared common distribution patterns. Positive and negative correlations and independent distributions were found.

Less is known about the distribution of other major groups of organisms in the community; some preliminary studies of the bacteria (22), nematodes (55) and diatoms (24, 25) have been completed. The diatom assemblage of the aufwuchs community is much more complex than commonly believed (48). More than 250 species or varieties of diatoms have been recognized collectively in aufwuchs communities from Towd Point, Southampton, Long Island; Jamaica Bay, New York; and Great Sippewissett Marsh, Falmouth, Massachusetts. Field and correlated laboratory studies (24, 25, 51) suggest that many individual diatom species and the assemblage as a whole take up and release significant quantities of organic materials. Field studies also indicate seasonal changes in the nutrients available and in the abilities of aufwuchs algae as a group to use them. In a study just completed (25) we were able to show that changes in the population structure of the diatom assemblage were reflected also by changes in qualitative and quantitative differences in the growth of individual species of diatoms on differential media tested. More aufwuchs algal species and greater total numbers of algae grew on agar plates enriched with crude extracts of *Enteromorpha* than on plates enriched with other potential nutrient sources. This suggests that some of the organic substrates used by individual species originate within the community, and are recycled over very short distances among the community members. The nutritional patterns of the dozen aufwuchs algae studied in detail are individually so varied as to suggest that some measure of stability in this assemblage is achieved through heterotrophic diversity.

Laboratory Studies

Perhaps we will never have assessments of the population structure within the box that are detailed enough to allow us to understand and model it. What about functions of the box as a whole; can they be modeled? To date the environmental responses of less than two dozen aufwuchs, foraminifera, nematodes, microcrustacea, and ciliates have been studied in the lboratory, and far fewer under gnotobiotic conditions (1, 2, 5-7, 15-18, 21, 26, 28, 30, 35, 37, 38, 40-43, 52, 56-58, 60). The information available is hardly comforting to the modeler. For illustrative purposes, we will restrict this part of our discussion to organisms which were isolated from one field station, Towd Point, Southampton, Long Island, New York. The effects of just two abiotic variables, salinity and temperature, tested singly on five representative species (Figs. 1, 2) illustrate the range and the difficulties encountered in drawing modeling generalizations about the responses of meiofauna to these abiotic variables. In the real world many abiotic factors are varying at the same time. Muller (37) has recently studied the effects on the generation time of three species of foraminifera by varying several abiotic factors simultaneously. As could be expected, the generation times of the foraminifera were lengthened under these conditions. Food was kept constant in these experiments. Although *Allogromia laticollaris,* for example, has wide tolerances to individual abiotic environmental variables (Figs. 1, 2) its reproductive capacity is severely restricted at or near the limits of several variables. At 10°C and 15°C the reproduction of *A. laticollaris* occurred only between 20°/oo-30°/oo salinity and was only 16% of the biotic potential measured at 25°C-30°C. At the same temperature, 10°C, reproduction occurred only at pH 8.0, although at 25°C reproduction takes place from pH 5.0-9.5.

Various types of feeding relationships have long occupied the attention of ecological modelers. Food quantity has, perhaps, received more attention than food quality because it is mathematically simpler to handle. The Lotka-Volterra assumption that the rate of interaction between two species is directly proportional to the product of their populations has much merit as a first-order approximation in many situations. In tracer-feeding experiments using foraminifera, for example, uptake of food was erratic below a threshold concentration (~ 10^3 cells/10 ml culture) and was directly proportional to concentration within a range of 10^4-10^6 organisms per experimental culture (26). Many other factors also affect meiofaunal feeding behavior. For example, small *Allogromia laticollaris* (150-200 μm in diameter) ate many more food organisms than did larger ones (~ 350-400 μm in diameter). They also grew much faster. Sometimes the difference was as great as 200%. Characteristics of the algae also may affect feeding rate. At one extreme *Ammonia beccarii,* a foraminiferan, ate almost 5 x more washed, log-phase *Chlorococcum* sp than washed cells from stationary phase cultures. Crowding of the herbivores (intraspecific competition) and interspecific competition can sharply reduce feeding rates. For example, individual animals in crowded cultures of *Allogromia laticollaris* (> 100 animals) ate > 70% less *Nitzschia acicularis* and *Phaeodactylum tricornutum* than did single animals (37). In the presence of *Rosalina leei, A. laticollaris* ate only 39% of the *Amphora* sp (strain 5), 23% of the *Amphora* sp (strain RF 8), 7% of the *Phaeodactylum tricornutum,* and 86% of the *Nannochloris* sp (strain 41) that it would eat when incubated alone (7). None of the meiofauna we have studied thus far are indiscriminate feeders (37, 38, 56, 57). We found that the microherbivores we studied vary greatly in their abilities to graze on and assimilate the organisms they ingest (Fig. 3). As can be seen in the illustrative table, when food concentration is above threshold levels it does not limit feeding. Only a small fraction of the algae present are eaten. Microscopic examination of the worm guts after feeding has shown that many algae are not digested during their passage through the digestive tract and pass

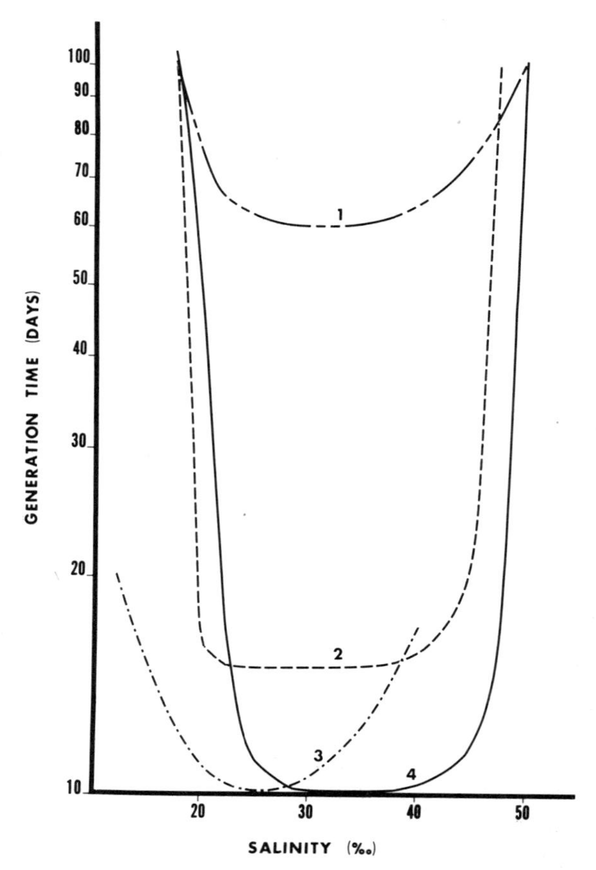

Figure 1. The effect of salinity on the reproduction of five salt-marsh meiofauna at 26°C.: **Rosalina leei, Spiroloculina hyalina, Monystera denticulata, Allogromia laticollaris.** (Data from 37, 56, 58).

from the system intact (58).

Growth experiments, particularly those employing gnotobiotic cultures, are really the only means to assay the nutritional quality of food organisms. Only one marine aufwuchs nematode has yet been grown in axenic culture, but several other species of marine nematodes and foraminifera have been grown in monoxenic or dixenic culture. Each of the species studied thus far has particular nutritional requirements satisfied by very few organisms in its diet. In the cases of *Rhabditis marina,* a marine nematode, and *Allogromia* sp (strain NF), a foraminiferan, continuous monoxenic culture required the presence of a marine pseudomonad (30, 55, 58). Although many species of foraminifera ingest particular species of algae to meet their bulk carbon budget, they all seem to

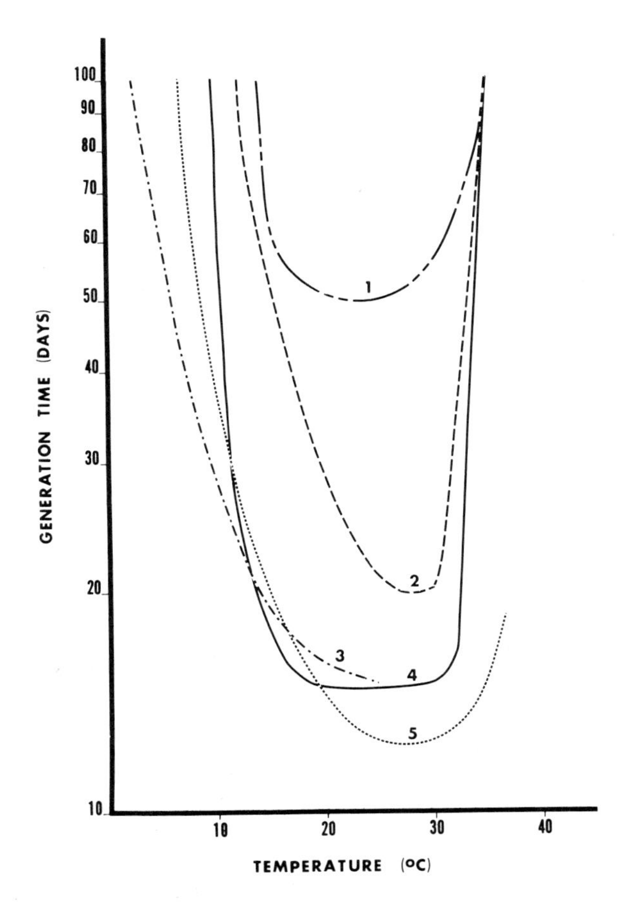

Figure 2. The effect of temperature on the reproduction of five salt-marsh meiofauna. Salinity at optimum for each species: **Rosalina leei, Spiroloculina hyalina, Monhystera denticulata, Allogromia laticollaris, Rhabditis marina.** (Data from 37, 56, 58).

have specific requirements essential for growth and reproduction found only in selected species of bacteria or in rate-limiting quantities in the algae which have thus far been examined (30, 38). Aside from absolute growth requirements, there is an increasing body of evidence to suggest that the organisms gain different nutritional values from the food they assimilate. Gnotobiotic clones of several species of foraminifera grow and reproduce much more rapidly in the presence of certain mixtures of algae than they do when fed other mixtures or only one algal species (23, 27, 38). If the phenomena of balanced diets and special growth factors are widespread among meiofauna, we may be forced to

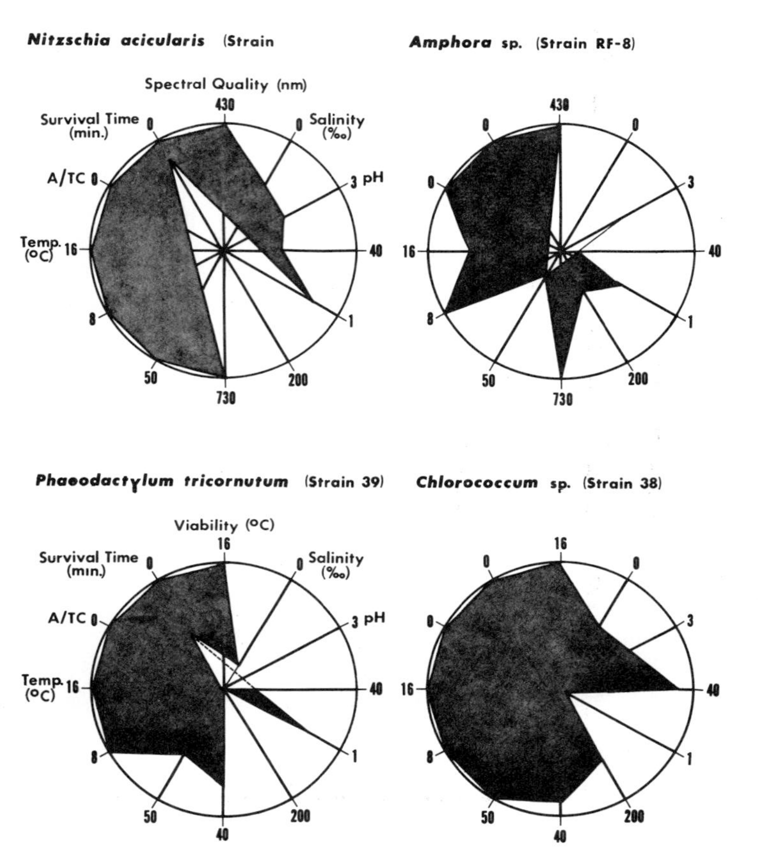

Figure 3. Graphic representation of several aspects of the niches of typical aufwuchs algae. (Data from 50, 51, and work in progress).

reconsider certain aspects of modeling. From a thermodynamic point of view, different food organisms will have to be considered not only in caloric terms but also in terms of the information content they hold for their consumer. Herbivores and predators would thus be cybernetic processors (49). Since many predators can feed on a number of different species, the task of evaluating the informational contribution from each prey would be formidable. Tied in with this aspect, there may be the phenomenon of prey switching (39), but at this time we have no information on this point for any meiofaunal organism.

Another important modeling consideration is the energetic filter function of feeding. Various investigators have used different methods to assess ecological growth efficiencies ($E_e = P/I$). We have measured E_e in three foraminiferan species (28) and have begun to do so in the nematodes we have in culture (Table

1). Ecological growth efficiency in the foraminifera is not a constant. It is highest in the first weeks of a culture, during the logarithmic growth phase (5%-21%), and gradually declines to levels of $< 1\%$ in later phases. In our opinion, data, coupled with other lines of evidence, suggest that foraminifera and perhaps other meiofauna are well adapted to play brief and very specialized roles in the rapidly changing food webs of the aufwuchs communities of which they are an important part. The high productivity of the aufwuchs community could be achieved through rapid cycling of very specialized species.

Though we recognize the importance of interspecific competition and predation on the aufwuchs community we have too little information to assess it for modeling purposes. Initial studies in our laboratory (25, 37, 51) indicate that there are so many variables involved that we cannot, at this time, predict the outcome of such interactions. Not only is the community structure channeled by the larger ecological and physical restraints of the system of which it is a part, but the community itself restricts the degrees of freedom of its component populations. Levandowsky's recent generalizations (32) may eventually be of great help in reducing competitive consideration to more manageable proportions.

Niche consideration and niche theory are really beyond the bounds of this symposium, nevertheless we contend that ecosystem processes can best be understood and modeled if the identities of individual organisms can be maintained. We have found Muller's method (37) a useful tool for summarizing and visualizing a large amount of niche-related data.

Experimental Manipulations of Communities in the Field

This past summer we made an initial attempt to experimentally manipulate aufwuchs community structure in the field under conditions as close to ambient as practical. Plastic culture flasks with 1-μm pore size nylon filters bound to their surface contained experimental communities and allowed free exchange between culture fluids and the surrounding sea water. In one experiment chunks of heavy metals were added to the vessels. In other experiments, aufwuchs communities were transplanted between field stations. Preliminary results were encouraging, since both qualitative and quantitative changes in the diatom population were observed. Species diversity indices were poor indicators of experimental results (Table 2). The most interesting results we found were the changes in community structure effected by the presence of various metallic ions. In different ways the population structure of each of the experimental diatom assemblages differed from the control population. Recurrent group analysis (57) of the diatom population of each of the experimental flasks has identified groups of diatom species sensitive to particular metallic ions or groups of metallic ions, those organisms which replace them, and those unaffected.

TABLE 1.

Consumption of algae by three species of nematodes.

ALGAE***	Rhabditis marina					Monhystera denticulata				
	No. algae/worm		grazing factor	assimi-lation	mgm/	No. algae/worm		grazing factor	assimi-lation	mgm/
	24 hr	48 hr	% *	% **	worm/day	24 hr	48 hr	% *	% **	worm/day
Rf-1	$2.4x10^1$	$2.7x10^1$	0.003	100	$3x10^{-8}$	$3.5x10^1$	$2.3x10^1$	0.006	64	$3x10^{-7}$
BL-44	$1.2x10^1$	$8.7x10^0$	0.002	70	$1x10^{-8}$	$2.3x10^1$	$7.5x10^0$	0.002	32	$9x10^{-6}$
BL-42	$9.8x10^2$	$6.6x10^2$	0.37	67	$3x10^{-6}$	$6.5x10^1$	$2.5x10^1$	0.004	38	$1.2x10^{-7}$
39	$3.5x10^2$	$3.4x10^2$	0.04	95	$2x10^{-5}$	$5.5x10^1$	$4.9x10^1$	0.005	88	$3x10^{-6}$
8	$2.7x10^2$	$1.5x10^2$	0.01	58	$2x10^{-5}$	$3.6x10^2$	$1.7x10^2$	0.036	47	$2x10^{-5}$
13	$1.8x10^3$	$1.5x10^3$	0.19	88	$1x10^{-4}$	$1.5x10^2$	$2.0x10^2$	0.037	100	$1.3x10^{-5}$
38	$4.9x10^2$	$3.4x10^2$	0.07	70	$3x10^{-7}$	$1.2x10^2$	$1.9x10^2$	0.27	100	$1x10^{-7}$
41	$9.2x10^2$	$6.4x10^2$	0.09	70	$4x10^{-6}$	$1.0x10^2$	$1.6x10^2$	0.02	100	$1x10^{-5}$
RF-8	$5.0x10^2$	$3.5x10^2$	0.12	71	$3x10^{-6}$	$4.4x10^1$	$2.4x10^2$	0.033	100	$2x10^{-6}$
Archaias	$2.2x10^2$	$7.1x10^1$	0.01	31	$6x10^{-8}$	$8.3x10^1$	$1.4x10^1$	0.002	60	$1x10^{-8}$
symbiont 94	$8.3x10^2$	$2.0x10^2$	0.25	24	$4x10^{-6}$	$1.8x10^2$	$2.6x10^1$	0.003	14	$4.6x10^{-7}$

* No. of algae ingested/total inocula fo food x 100

** % of algae digested of those ingested/worm

*** RF-1 **Nitzschia brevirostris**; BL-44 **Fragillaria** sp; BL-42 **Navicula diversistriata**; 39-**Phaeodactylum tricornutum**; 8-**Nitzschia acicularis**; 13-**Dunaliella salina**; 38-**Chlorococcum** sp; 41-**Nannochloris** sp; RF-8 **Amhora** sp; 94-**Chlamydomonas** sp.

TABLE 1 (continued)

Chromadorina germanica

ALGAE	No. algae/worm 24 hr	48 hr	grazing factor % *	assimilation % **	mgm/worm/day
RF-1	1.5×10^3	1.9×10^3	0.75	100	2×10^{-5}
BL-44	9.4×10^3	7.8×10^3	20.5	83	9×10^{-6}
BL-42	9.4×10^3	4.5×10^3	1.8	47	2×10^{-5}
39	8.7×10^3	3.1×10^3	0.9	36	2×10^{-4}
8	6.3×10^3	5.2×10^3	1.7	82	6×10^4
13	1.3×10^3	1.2×10^3	0.1	95	8×10^{-5}
38	2.7×10^3	2.2×10^3	0.6	79	2×10^{-6}
41	1.5×10^4	1.2×10^4	1.2	83	8×10^{-4}
RF-8	1.5×10^3	1.4×10^3	0.4	86	1×10^{-5}
Archaias symbiont	6.1×10^3	4.5×10^3	1.5	73	9×10^{-6}

TABLE 2.

Experimental manipulation of aufwuchs population structure in the field. Sample results.

		Total No. species	Organisms per ml	Species Diversity Index	% Similarity * Woods Hole Week 6	Jamaica Bay Week 6	Iron Week 6	Zinc Week 6
Inoculum	Jamaica Bay	55	1×10^4	0.413	85.6	81.3	83.2	82.8
	Southampton	66	1×10^4	0.271	86.1	88.0	89.1	88.6
	Woods Hole	80	1×10^4	0.462	69.0	71.3	63.3	64.1
	Mixed Community	68	1×10^4	0.377	76.0	70.7	65.4	68.0
After 6 Weeks Incubation at Woods Hole	Jamaica Bay	71	19×10^4	0.391	58.1	—	41.9	78.9
	Southampton	87	44×10^4	0.478	64.5	53.1	66.1	73.3
	Woods Hole	75	22×10^4	0.447	—	58.1	58.6	68.7
	Mixed Community	96	10×10^4	0.318	65.1	63.1	48.2	45.5
	Iron	74	9×10^4	0.376	58.6	41.4	—	59.4
	Copper	77	2×10^4	0.377	59.2	41.4	35.5	61.5
	Zinc	98	19×10^4	0.338	68.7	78.9	59.4	—
	Chromium	53	4×10^4	0.360	57.2	52.8	45.8	74.1
	Lead	67	6×10^4	0.328	61.3	49.0	36.1	59.2

$$\text{*Similarity} = \frac{\Sigma\,|PCT(I, K) - PCT(I, L)|}{PCT(I, K)^2}$$

is the percentage of total population accounted for by species I (in column K) compared to the species in a second group (column L). The dashed lines above indicate the reference group to which all the other populations were compared.

MODELING

Modeling has a place in our studies. We have used it as an essential tool to help determine what experiments would be relevant, thereby increasing the efficiency of the research effort. The relationships of even a single meiofaunal animal we have studied are too complex to be realistically modeled at this time. The model is an attempt to organize our present information concerning the trophic dynamics and life cycle of a single species and to serve as a generator of predictions that are to be further tested experimentally, i.e., we expect new areas of investigation to arise from the results of model simulations.

We are presently trying to simulate and understand some of the trophodynamics of one of the foraminifera, *Allogromia laticollaris,* which we have studied most extensively (23, 27, 37; Fig. 4, and equations). This little animal usually reproduces asexually by schizogomy but can, when fed various different diets, reproduce by binary fission, cytotomy, and budding. Sexual reproduction is rare. Some of these options are reflected in the flow diagram (Fig. 4). Data on the growth and reproduction of > 20,000 individual organisms (23, 27) have given us some basis for evaluation of the constants a_1, a_2, a_3, a_4, c_2, c_3, c_4, c_5. *Allogromia laticollaris* requires the presence of at least a minimum amount of food X 1, a bacterium, for growth and reproduction. In monoxenic cultures when only X 1 is present or when X 1 is the major food source, binary fission is favored (pathway c_2y_2). Otherwise, $(a_2\text{-}c_2)y_2$ is the major pathway. Smaller animals Y 1, Y 2 feed at much higher rates than do medium-sized (Y 3) or larger ones (Y 4 and Y 5). Constants for food organisms X 2-X 6 have been measured experimentally (23, 26, 27, 37). Reproduction is higher in the presence of certain mixtures of foods than in others. Threshold values for feeding, intraspecific competition (crowding), and differences in the ecological growth efficiency of different life-cycle states have also been simulated.

CONCLUSION

It is clear that we have discussed only a few aspects of the niches of some salt-marsh and estuarine protozoa. The negative interactions (predation, parasitism, and antagonism) have not been discussed (10, 11, 45, 46).

Many questions about the distribution of meioflora and meiofauna and their niches remain unanswered. Patchiness is an ecological strategy that diversifies the habitat and reduces competition. We have studied the dietary habits of meiofauna and the characteristics of the algal assemblage suggest that the two groups should be linked both temporally and spatially. Critical demonstrations are still needed on this point. How are the meiofaunal life cycles adapted to the dynamic changes in the community? Do species of meiofauna grow and

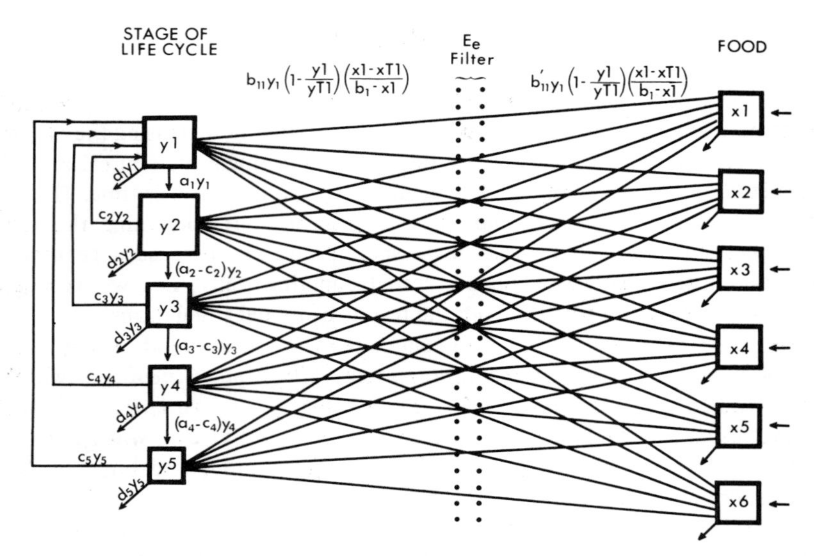

Figure 4. Schematic representation of the trophic dynamics and life cycle of **Allogromia laticollaris.**

reproduce in response to particular algal blooms? Animals that do so would maximize energy yield by decreasing search time per food organism eaten and, temporarily, would have very high ecological efficiencies, but they might pay for this later in energy of maintenance between blooms of food. What triggers reproduction, algal blooms, or anticipation of food for offspring? If a meiofaunal species has a long generation time, will its offspring be adapted to feed on the new organisms available in the community at the time of their release or hatching, or will they grow slowly and survive until the next bloom? Probably both types of feeding strategies have evolved among various meiofaunal groups. One step further removed is the role of substrate in developing particular algal assemblages. Some small-scale patchiness may depend on the type of substrate present.

Studies of intraspecific and interspecific competition have barely begun. Presumably intraspecific competition is an important aspect of life cycles leading to or promoting dispersion of offspring. *A priori* adaptive strategies evolve toward maximizing total net energy yield. The energy expense of foraging for food may be reduced when animals are not grouped too closely, but when the number of contacts between mates falls below a certain level, other mechanisms, resulting in the expenditure of energy, must be operative.

Interspecific competition depends upon many variables that have yet to be systematically studied by meiofaunal specialists. In the simplest case, two or more herbivorous species feed simultaneously on the same food species. For competition to take place there must be less food available than the two species require. In aufwuchs communities a number of species of herbivores or omnivores may attack the algal species that are temporarily in abundant supply. The extent to which the same species can also exploit other sources of food would seem to be of great adaptive value as a hedge against competitive pressure, since they could shift their alternate foods. Most herbivores in the aufwuchs community probably feed on several species of prey even though each has "preferred" food, so some overlap must be expected. Quantification, or even precise definition of competitive interactions, will not be easy, but should be attempted.

As our views of meiofaunal ecology and its modeling have gradually evolved into more quantitative analyses and more detailed approaches, the measuring and modeling techniques have become more sophisticated and correspondingly more time-consuming. After looking into the modeler's meiofaunal or aufwuchs black box, one can only conclude that many interesting but very incompletely understood processes are occurring inside. Some may wish to throw up their hands in disgust, and others will want to slam the box shut and never again think about the complexities and puzzles writhing around inside. We will do neither. It is with some trepidation and some hope that we look forward to progress in this field.

LEGEND

d_i = mortality and respiration $\frac{\text{mgC}}{\text{mgCday}}$

a_i = total transfer from box i; measured by age distribution variation with time experiments $\frac{\text{mgC}}{\text{mgCday}}$

c_i = transfer from box i to box 1 by reproduction; measured by age distribution variation with time experiments $\frac{\text{mgC}}{\text{mgCday}}$

g_i = natural growth constants for food $\frac{\text{mgC}}{\text{mgCday}}$

b'_{ij} = grazing constant (food j eaten by *Allogromia* i) $\frac{\text{mgC}}{\text{mgCday}}$

b_{ij} = assimilation constant (*Allogromia* i assimilates food j) $\frac{\text{mgC}}{\text{mgCday}}$

Note: $b_{ij} = b'_{ij} \cdot E(i, j)$

$E_{(i,j)}$ = ecological efficiency of *Allogromia* i on diet of food organism j

b_i = constant reflecting limiting ability of predator to take advantage of surplus food i $\frac{mgC}{mgCday}$

Note: We have assumed same b_i for all ages of *Allogromia*

e_i = natural crowding constant of food i in absence of predator $\frac{mgC\ ml}{(mgC)^2 day}$

Xi(t) = dynamic concentration of food i. i = 1,6

Yi(t) = dynamic concentration of *Allogromia* i. i = 1,5

XT_i = threshold concentration of food i for predation $\frac{mgC}{ml}$. Note independence of age of predator. i = 1,6.

YT_i = predator concentration at which it crowds itself and ceases to feed. $\frac{mgC}{ml}$ i = 1,5.

X 1 = *Pseudomonas* sp
X 2 = *Nitzschia acicularis* (strain 8)
X 3 = *Phaeodactylum tricornutum* (strain 39)
X 4 = *Cylindrotheca closterium* (strain 9)
X 5 = *Nannochloris* sp (strain 41)
X 6 = *Chlorococcum* sp (strain 40)

$$\frac{dy\,1}{dt} = a_1y_1 - d_1y_1 + c_2y_2 + c_3y_3 + c_4y_4 + c_5y_5 + [y_1(1-(\frac{y_1}{y_{T\,1}})]\ [b_{11}(\frac{x\,1\ -\ XT\,1}{b_1 + X\,1})$$

$$+ b_{12}(\frac{X\,2 - XT\,2}{b_2 + X\,2}) + b_{13}(\frac{X\,3 - XT\,3}{b_3 + X\,3}) + b_{14}(\frac{X\,4 - XT\,4}{b_4 + X\,4}) + b_{15}(\frac{X\,5 - XT\,5}{b_5 + X\,5}) + b_{16}(\frac{X\,6 - XT\,6}{b_6 + X\,6})]$$

$$\frac{dy\,2}{dt} = -a_2y_2 - d_2y_2 + a_1y_1 + [y_2(\frac{1 - y_2}{y_2T})]\ [b_{21}(\frac{X\,1 - XT\,1}{b_1 + X\,1}) + b_{22}(\frac{X\,2 - XT\,2}{b_2 + X\,2})$$

$$+ b_{23}(\frac{X\,3 - XT\,3}{b_3 + X\,3}) + b_{24}(\frac{X\,4 - XT\,4}{b_4 + X\,4}) + b_{25}(\frac{X\,5 - XT\,5}{b_5 + X\,5}) + b_{26}(\frac{X\,6 - XT\,6}{b_6 + X\,6})],$$

$$\frac{dy\,3}{dt} = -a_3y_3 - d_3y_3 + (a_2 - c_2)y_2 + [y_3(1 - \frac{y_3}{y_3T})]\ [b_{31}(\frac{X\,1 - XT\,1}{b_1 + X\,1}) + b_{32}(\frac{X\,2 - XT\,2}{b_2 + X\,2})$$

$$+ b_{33}(\frac{X\,3 - XT\,3}{b_3 + X\,3}) + b_{34}(\frac{X\,4 - XT\,4}{b_4 + X\,4}) + b_{35}(\frac{X\,5 - XT\,5}{b_5 + X\,5}) + b_{36}(\frac{X\,6 - XT\,6}{b_6 + X\,6})]$$

$$\frac{dy\,4}{dt} = -a_4y_4 - d_4y_4 + (a_3 - c_3)y_3 + [y_4(1 - \frac{y_4}{y_4T})]\ [b_{41}(\frac{X\,1 - XT\,1}{b_1 + X\,1}) + b_{42}(\frac{X\,2 - XT\,2}{b_2 + X\,2})$$

$$+ b_{43}(\frac{X\,3 - XT\,3}{b_3 + X\,3}) + b_{44}(\frac{X\,4 - XT\,4}{b_4 + X\,4}) + b_{45}(\frac{X\,5 - XT\,5}{b_5 + X\,5}) + b_{46}(\frac{X\,6 - XT\,6}{b_6 + X\,6})]$$

$$\frac{dy5}{dt} = -c_5y_5 - d_5y_5 + (a_yc_y)y_4 + [y_5(1-\frac{y_5}{y_5T})]\,[b_{51}(\frac{X1 - XT1}{b_1 + X1}) + b_{52}(\frac{X2 - XT2}{b_2 + X2})$$

$$+ b_{53}(\frac{X3 - XT3}{b_3 + X3}) + b_{54}(\frac{X4 - XT4}{b_4 + X4}) + b_{55}(\frac{X5 - XT5}{b_5 + X5}) + b_{56}(\frac{X6 - XT6}{b_6 + X6})]$$

$$\frac{dX1}{dt} = g_1X1 - [(\frac{X1 - XT1}{b_1 + X_1})]\,[b'_{11}y_1(1-\frac{y1}{yT1}) + b'_{12}y_2(1-\frac{y2}{y_2T}) + b'_{13}y_3(1-\frac{y3}{y_3T})$$

$$+ b'_{14}y_4(1-\frac{y4}{y_4T}) + b'_{15}y_5(1-\frac{y5}{y_5T})] - e1X1^2$$

$$\frac{dX2}{dt} = g_2X_2 - [(\frac{X2 - XT2}{b_2 + X2})]\,[b'_{21}y_1(1-\frac{y1}{yT1}) + b'_{22}y_2(1-\frac{y2}{yT2}) + b'_{23}y_3(1-\frac{y3}{y_3T})$$

$$+ b'_{24}y_4(1-\frac{y4}{y_4T}) + b'_{25}y_5(1-\frac{y5}{y_5T})] - e\,2X2^2$$

$$\frac{dX3}{dt} = g_3X_3 - [(\frac{X3 - XT3}{b_3 + X3})]\,[b'_{31}y_1(1-\frac{y1}{yT1}) + b'_{32}y_2(1-\frac{y2}{y_2T}) + b'_{33}y_3(1-\frac{y3}{y_3T})$$

$$+ b'_{34}y_4(1-\frac{y4}{y_4T}) + b'_{35}y_5(1-\frac{y5}{y_5T})] - e3X3^2$$

$$\frac{dX\,4}{dt} = g_4X_4 - [(\frac{X\,4 - XT\,4}{b_4 + X\,4})]\ [b_{41}'y_1\,(1 - \frac{y\,1}{yT\,1}) + b_{42}'y_2\,(1 - \frac{y\,2}{y_2T}) + b_{43}'y_3\,(1 - \frac{y\,3}{y_3T})$$

$$+ b_{44}'y_4\,(1 - \frac{y\,4}{y_4T}) + b_{45}'y_5\,(1 - \frac{y\,5}{y_5T})]\ -e4X4^2$$

$$\frac{dX\,5}{dt} = g_5X5 - [(\frac{X\,5 - XT\,5}{b_5 + X\,5})]\ [b_{51}'y_1\,(1 - \frac{y\,1}{yT\,1}) + b_{52}'y_2\,(1 - \frac{y\,2}{y_2T}) + b_{53}'y_3\,(1 - \frac{y\,3}{y_3T})$$

$$+ b_{54}'y_4\,(1 - \frac{y\,4}{y_4T}) + b_{55}'y_5\,(1 - \frac{y\,5}{y_5T})]\ -e5X5^2$$

$$\frac{dX\,6}{dt} = g_6X_6 - [(\frac{X\,6 - XT\,6}{b_6 + X\,6})]\ [b_{61}'y_1\,(1 - \frac{y\,1}{yT\,1}) + b_{62}'y_2\,(1 - \frac{y\,2}{y_2T}) + b_{63}'y_3\,(1 - \frac{y\,3}{y_3T})$$

$$+ b_{64}'y_4\,(1 - \frac{y\,4}{y_4T}) + b_{65}'y_5\,(1 - \frac{y\,5}{y_5T})]\ -e6X6^2$$

ACKNOWLEDGMENTS

We acknowledge the valuable technical assistance rendered by Marie McEnery, Judith Garrison, Eileen Kennedy, Johnny Hagen, and Carmine Mastropaolo. Various phases of the work were supported by NSF grant GB 19245 and US AEC contract COO (11-1)3254, US AEC document no. COO 3254-16.

REFERENCES

1. Arnold, Z.
1954 Culture methods in the study of living foraminifera. **Jour. Paleontol.,** 28: 78-101.
2. Arnold, Z.
1964 Biological observations on the foraminifer **Spiroloculina hyalina. Univ. Cal. Publ. Zool.,** 72: 1-93.
3. Boltovskoy, E.
1964 Seasonal occurrences of some living foraminifera in Puerto Deseado (Patagonia, Argentina). **Jour. Con. Perm. Int. Explor. Mer.,** 39: 136-145.
4. Boltovskoy, E. and Lena, H.
1969 Microdistribution des foraminifères benthoniques vivants. **Rev. Micropaleontol.,** 12: 177-185.
5. Bradshaw, J. S.
1955 Preliminary laboratory experiments on the ecology of foraminiferal populations. **Micropal.,** 1: 351-358.
6. Bradshaw, J. S.
1957 Laboratory studies of the rate of growth of the foraminifera **Streblus beccarii** (Linné) var. **tepida** (Cushman). **Jour. Paleontol.,** 31: 1138-1147.
7. Bradshaw, J. S.
1961 Laboratory experiments on the ecology of foraminifera. **Contrib. Cushman Found. Foram. Res.,** 12: 87-106.
8. Brooks, A. L.
1967 Standing crop, vertical distribution, and morphometrics of **Ammonia beccarii. Limnol. Oceanogr.,** 12: 667-684.
9. Buzas, M.
1969 Foraminiferal species densities and environmental variables in an estuary. **Limnol. Oceanogr.,** 14: 411-422.
10. Connell, J. H.
1970 A predator-prey system in the marine intertidal. I. **Balanus glandula** and several predatory species of **Thais. Ecol. Monogr.,** 40: 49-78.
11. Connell, J. H. and Orias, E.
1964 The ecological regulation of species diversity. **Am. Naturalist,** 48: 399-414.
12. Coull, B. C.
1970 Shallow water meiobenthos of the Bermuda platform. **Oecologia,** 4: 325-357.

13. Coull, B. C.
1972 Species diversity and faunal affinities of meiobenthic copepoda in the deep sea. **Mar. Biol.,** 14: 48-51.

14. Fager, E. W.
1957 Determination and analysis of recurrent groups. **Ecology,** 38: 586-595.

15. Gerlach, S. A. and Schrage, M.
1971 On the importance of marine meiofauna for benthic communities. **Oecologia,** 6: 176-190.

16. Gerlach, S. A. and Schrage, M.
1972 Life cycles at low temperatures in some free-living marine nematodes. **Veröff. Inst. Meeresforsch. Bremerh.,** 14: 5-11.

17. Hopper, B. E. and Myers, E. H.
1966a Aspects of the life cycle of marine nematodes. **Helogländer wiss. Meeresunters.,** 13: 444-449.

18. Hopper, B. E. and Myers, E. H.
1966b Observations on the bionomics of the marine nematode **Metoncholaimus** sp. **Nature,** 209: 899-900.

19. Hutchinson, G. E. and MacArthur, R. H.
1959 A theoretical ecological model of size distribution among species of animals. **Am. Naturalist,** 93: 117-125.

20. Jepps, M.
1942 Studies in **Polystomella lamarck. Jour. Mar. Biol. Assoc.,** 25: 607-666.

21. Johannes, R. E.
1964 Uptake and release of dissolved organic phosphorus by representatives of a coastal marine ecosystem. **Limnol. Oceanogr.,** 9: 224-234.

22. Kennedy, E.
1974 An experimental approach to the physiological ecology of bacteria epiphytes on **Enteromorpha intestinalis.** (In progress.)

23. Lee, J. J. and McEnery, M. E.
1970 Autogamy in **Allogromia latiocollaris** (Foraminifera). **Jour. Protozool.,** 17: 184-195.

24. Lee, J. J., McEnery, M. E., Kennedy, E. M., and Rubin, H.
1973 Educing the functional relationships among the diatom assemblages within sublittoral salt marsh epiphytic communities. **Bull. Ecol. Res. Comm.** (Stockholm), 17: 387-397.

25. Lee, J. J., McEnery, M. E., Kennedy, E. M., and Rubin, H.
1974 A nutritional analysis of a sublittoral epiphytic diatom assemblage from a Long Island salt marsh. (Ms submitted for publication.)

26. Lee, J. J., McEnery, M., Pierce, S., Freudenthal, H. D., and Muller, W. A.
1966 Traces experiments in feeding littoral foraminifera. **Jour. Protozool.,** 13: 659-670.

27. Lee, J. J., McEnery, M. E., and Rubin, H.
1969b Quantitative studies on the growth of **Allogromia laticollaris** (Foraminifera). **Jour. Protozool.,** 16: 377-395.

28. Lee, J. J. and Muller, W. A.
1973 Trophic dynamics and niches of salt marsh foraminifera. **Am.**

Zool., 13: 215-233.

29. Lee, J. J., Muller, W. A., Stone, R. J., McEnery, M. E., and Zucker, W.
1969a Standing crop of foraminifera in sublittoral epiphytic communities of a Long Island salt marsh. **Mar. Biol.**, 4: 44-61.

30. Lee, J. J. and Pierce, S.
1963 Growth and physiology of foraminifera in the laboratory: Part 4 - monoxenic culture of an allogromiid with notes on its morphology. **Jour. Protozool.**, 10: 404-411.

31. Lee, J. J., Tietjen, J. H., Stone, R. J., Muller, W. A., McEnery, M., Saks, N. M., Mastropaolo, C., and Kennedy, E.
1973 The effects of environmental stress on the community structure, productiviy, energy flow, and mineral cycling in salt-marsh epiphytic communities. In **Proceedings of the Third National Symposium on Radioecology**, Vol. 2, p. 816-828. (ed. Nelson, D. J.) U. S. Atomic Energy Comm., Oak Ridge, Tennessee.

32. Levandowsky, M.
1972 Ecological niches of sympatric phytoplankton species. **Am. Naturalist**, 100: 71-78.

33. Levins, R.
1966 The strategy of model building in population biology. **Am. Sci.**, 54: 421-431.

34. Matera, N. J. and Lee, J. J.
1972 Environmental factors affecting the standing crop of foraminifera in sublittoral and psammolittoral communities of a Long Island salt marsh. **Mar. Biol.**, 14: 89-103.

35. McEnery, M. and Lee, J. J.
1970 Tracer studies on calcium and strontium mineralization and mineral cycling in two species of foraminifera, **R. leei** and **Spiroloculina hyalina. Limnol. Oceanogr.**, 15: 173-182.

36. McIntyre, A. D.
1969 Ecology of marine meiobenthos. **Rev. Biol.**, 44: 245-290.

37. Muller, W. A.
1972 Graphic representation of niche and its application to salt marsh littoral foraminifera. Doctoral dissertation. City Univ. of New York. 40 pp.

38. Muller, W. A. and Lee, J. J.
1969 Apparent indispensability of bacteria in foraminiferan nutrition. **Jour. Protozool.**, 16: 471-478.

39. Murdoch, W. W.
1969 Switching in general predators: Experiments on predation specificity and stability of prey populations. **Ecol. Monogr.**, 39: 335-354.

40. Murray, J. W.
1963 Ecological experiments in forminifera. **Jour. Mar. Biol. Assoc. U.K.**, 43: 621-642.

41. Murray, J. W.
1967 An ecological study of the **Thecamoebina** of Christchurch Harbour, England. **Jour. Nat. Hist.**, 1: 377-387.

42. Myers, E. H.
1935 Culture methods for the marine foraminifera of the littoral zone.

Trans. Amer. Microscop. Soc., 54: 264-267.

43. Myers, E. H.
1936 The Life cycle of **Spirillina vivipara** with notes of morphogenesis, systematics, and distribution of the foraminifera. **Jour. Roy. Soc.**, 61: 120-146.

44. Myers, E. H.
1943 Life activities of foraminifera in relation to marine ecology. **Proc. Am. Phil. Soc.**, 86: 439-458.

45. Paine, R. T.
1966 Food web complexity and species diversity. **Am. Naturalist**, 100: 65-75.

46. Paine, R. T.
1969 The **Pisaster-Tegula** interaction: prey patches, predator food preferences, and intertidal community structure. **Ecology**, 50: 950-961.

47. Rees, C. B.
1940 A preliminary study of the ecology of a mud flat. **Jour. Mar. Biol. Assoc. U.K.**, 24: 185-199.

48. Round, F. E.
1971 Benthic marine diatoms. **Oceanogr. Mar. Biol. Ann. Rev.**, 9: 83-139.

49. Rubin, H.
1974 An experimental approach to the assessment of the informational aspects of energy flow in marine ciliates. (In progress.)

50. Saks, N. M., Lee, J. J., Muller, W. A., and Tietjen, J. H.
1974 The growth of salt-marsh microcosms subjected to thermal stress. In **Symposium on Thermal Ecology.** (In press.)

51. Saks, N. M., Stone, R. J., and Lee, J. J.
Auxotrophic and hererotrophic nutritional budget of salt-marsh epiphytic algae. (Ms in preparation.)

52. Sliter, W. V.
1970 **Bolivina doniezi** Cushman and Wickenden in clone culture. **Contrib. Cushman Found. Foram. Res.**, 21: 87-99.

53. Tietjen, J. H.
1969 The ecology of shallow water meiofauna in two New England estuaries. **Oecologia**, 2: 251-291.

54. Tietjen, J. H.
1971 Ecology and distribution of deep-sea meiobenthos off North Carolina. **Deep-Sea Res.**, 18: 941-957.

55. Tietjen, J. H.
1974 Axenic culture and nutritional requirements of the marine nematode. **Rhabditis marina.** (Unpublished data.)

56. Tietjen, J. H. and Lee, J. J.
1972 Life cycles and marine nematodes. Influence of temperature and salinity on the development of **Monhystera denticulata** Timm. **Oecologia**, 10: 167-176.

57. Tietjen, J. H. and Lee, J. J.
1973 Life history and feeding habits of the marine nematode. **Chromadorina macrolaimoides** Steiner. **Oecologica**, 12: 303-314.

58. Tietjen, J. H., Lee, J. J., Rullman, J., Greengart, A., and Trompeter, J.
1970 Gnotobiotic culture and physiological ecology of the marine nematode. **Rhabditis marina** Bastian. **Limnol. Oceanogr.,** 15: 535-543.

59. Vandermeer, J. H.
1972 Niche theory. **Ann. Rev. Ecol. System,** 3: 107-132.

60. Wieser, W. and Kanwisher, J.
1961 Ecological and physiological studies on marine nematodes from a salt marsh near Woods Hole, Massachusetts. **Limnol. Oceanogr.,** 6: 262-270.

Index

A

Adams, S. Marshall, 518-540
Andren, Anders W., 64-107
ATP, 229-250

B

Benthic organisms, 372-391
Bousfield, E. L., 325-351
Bradford, W. L., 188-214
Bricker, Owen P., 3-27
Broenkow, W. W., 153-187
Browne, Mason, 563-580

C

Caperon, John, 691-709
Carbon, 229-250, 583-601
Carpenter, J. H., 188-214
Chesapeake Bay, 3-27
Christian, R. R., 583-601
Conomos, T. J., 153-187
Copeland, B. J., 287-302
Copepods, 392-419
Correll, David L., 108-136
Cross, F. A., 45-63

D

Davies, J. M., 251-264
deSylva, Donald, 420-447
Detritus, 217-228, 265-286

E

Estuarine morphometry, 634-644
Estuary
  Georgia, 303-324, 998-1007
  Hawaiian, 691-709
  Long Island, 710-734
  Nova Scotia, 634-644
  Rhode Island, 672-690
  Scotland, 251-264
  South Carolina, 303-324
  St. Lawrence, 325-351

F

Faust, Maria A., 108-136
Ferguson, Randolph L., 229-250
Filteau, G., 325-351
Fish, 45-63, 420-447, 448-472, 473-493, 494-517
Flemer, David A., 602-616
Fleminger, Abraham, 392-419

Florida, 265-286, 352-371, 541-562
Flood webs, 265-286, 420-447

G

Gallagher, J. L., 217-228, 583-601
Gamble, J. C., 251-264
Gene flow, 372-391
Gentes, P., 325-351
Geochemistry, 64-107
Georgia estuary, 303-324
Georgia marshes, 217-228
Grant, V., 188-214

H

Haines, Evelyn B., 303-324
Hall, J. R., 583-601
Hardy, L. H., 45-63
Harrison, W. G., 287-302
Harriss, Robert C., 64-107
Hawaiian estuary, 691-709
Heald, Eric J., 265-286
Heavy metals, 28-44
Heinle, Donald R., 602-616
Hobbie, J. E., 287-302

J

Jones, N. Y., 45-63
Jones, R. D. H., 583-601

K

Kjelson, M. A., 448-472
Kremer, James N., 672-690

L

LaCroix, Michael W., 518-540
Lee, John J., 710-734
Lewis, J. M., 45-63
Lindberg, Steven E., 64-107
Linthurst, R. A., 217-228
Long Island estuary, 710-734
Long Island Sound, 28-44

M

Mangrove, 265-286
Mann, K. N., 634-644
Marshes, Georgia, 217-288
McCaffery, Richard J., 28-44
McErlean, A. J., 494-517
Mercury, 64-107, 137-152
Metals, heavy, 28-44, 45-63, 64-107, 137-154, 188-214
Microbes, 229-250
Mobley, Curtis D., 602-616
Models, 583-601, 602-616, 617-633, 672-690, 691-709, 710-734
Muller, William A., 710-734
Murdoch, Marianne B., 229-250

N

Narragansett Bay, 672-690
Natural selection, 372-391
Nekton, 420-447, 494-517
Nixon, Scott W., 672-690
North Carolina estuary, 45-63, 229-250, 287-302, 448-472, 518-

540, 645-671
Nova Scotia estuary, 634-644
Nutrients, 108-136, 153-187, 287-302, 303-324, 563-580

O

O'Connor, S. G., 494-517
Odum, William E., 265-286
O'Neill, M., 325-351
Organisms, benthic, 372-391

P

Peters, D. S., 448-472
Peterson, David, 153-187
Pfeiffer, W. J., 217-228
Phosphorus, 108-136
Phytoplankton, 251-264
Plankton, 672-690, 691-709
Population dynamics, 251-264, 325-351, 372-391, 392-419, 448-472, 473-492
Power plants, 494-517
Primary production, 217-228, 265-286, 287-302, 303-324, 518-540, 541-562
Production, secondary, 448-472, 473-492, 494-517

R

Reeve, Michael R., 352-371
Reimold, R. J., 217-228
Rhoads, Donald C., 563-580
Rhode Island estuary, 672-690
Ross, George C., 710-734
Rubin, Howard, 710-734

S

Saila, Saul B., 473-493
Saks, Norman M., 710-734
Salt marshes, 217-228, 583-601, 645-671, 710-734
Salt marsh, heavy metals, 137-152
San Francisco Bay, 153-187
Scheltema, Rudolf S., 372-391
Scotland estuary, 251-264
Scrivani, E. P., 153-187
Secondary production, 448-472, 473-492, 494-517
Sediments, 563-580
Sediment, water exchange, 3-27
Severn, David J., 108-136
Silica, 153-187
South Carolina estuary, 303-324
Spartina, 217-228, 583-601
Steele, J. H., 251-264
St. Lawrence estuary, 325-351
System modeling, 583-601, 602-616, 617-633, 634-644, 672-690, 691-709, 710-734

T

Tenore, Kenneth, 563-580
Thalassia, 541-562
Thayer, Gordon W., 518-540
Thomson, John, 28-44
Tietjen, John H., 710-734
Troup, Bruce N., 3-27
Turekian, Karl K., 28-44

U

Ulanowicz, Robert E., 602-616
Upwelling, 617-633

W

Walsh, John J., 617-633
Water chemistry, 108-136, 188-214, 251-264, 287-302, 303-324
Wetzel, R. L., 583-601
Wiegert, R. G., 583-601
Willis, J. N., 45-63
Windom, Herbert L., 137-152
Wolfe, Douglas A., 645-671

Z

Zieman, Joseph C., 541-562
Zooplankton, 251-264, 325-351, 352-371, 392-419
Zostera, 518-540

B
C 7
D 8
E 9
F 0
G 1
H 2
I 3
J 4